Applied GROUNDING & BONDING

© 2011 National Joint Apprenticeship and Training Committee for the Electrical Industry

All rights reserved. No part of this material shall be reproduced, stored in a retrieval system, or transmitted by any means whether electronic, mechanical, photocopying, recording, or other- wise without the express written permission of the NJATC.

1 2 3 4 5 6 7 8 9 – 11 – 9 8 7 6 5 4 3 2

Printed in the United States of America

Contents

CHAPTER 1 Introduction

Introduction 2	Path for Current through the Human Body 5
Grounding and Bonding for Safety 2	Grounding and Bonding Concepts Together 7
Earth in the Circuit 2	Performance *Code* Language 8
Grounding Concepts..3	Grounded Electrical Systems 9
Bonding Concepts 4	Ungrounded Electrical Systems..14
Minimizing Shock Hazards...4	

CHAPTER 2 Circuit Basics and Overcurrent Protection

Introduction 22	Overcurrent Protection Basics...28
Circuit Fundamentals 22	Amperes Operate Overcurrent Protective Devices..........30
Ohm's Law 22	Time and Current..31
Opposition to Current in Circuits....................................26	Safety by System Design 32
Current in Circuits (Normal and Fault Current) 27	Equipment Grounding Conductor Capacity 40

CHAPTER 3 Using the *National Electrical Code®*

Introduction 46	Section 250.2 Definitions 57
Building a Solid Foundation 46	Article 200—Identification and Use of Grounded Conductors...58
NEC Arrangement and Application.................................46	Article 250—Arrangement and Use................................59
Enforcement and Approvals 48	Part I—General...59
Requirements, Exceptions, Alternatives, and Information..48	Table 250.66—Sizing Grounding Electrode Conductors...60
Permissive *Code* Language 49	Table 250.122—Sizing Equipment Grounding Conductors...61
Explanatory Information 50	Special Occupancies, Equipment, and Conditions 62
Use of Defined Terms..50	
Code-Making Panel Responsibilities.................................51	
NEC Article 100—Definitions of Grounding and Bonding Terms..52	

CHAPTER 4 Grounding Electrodes and the Grounding Electrode System

Introduction 68	Mandatory Grounding Electrodes 72
Grounding Electrode Defined 68	Types of Grounding Electrodes 72
Purpose and Performance of Electrodes 69	Grounding Electrode Installation Requirements 77
Grounding Electrode System Requirements 71	Lightning Protection Fundamentals 84
Establishing a Grounding Electrode System...... 71	

CHAPTER 5 Requirements for Grounded Conductors at Services

Introduction 94	Functions (Purposes) of the Grounded Service Conductor 101
Grounded Utility Supply Systems 94	Grounded Neutral Conductor 102
First Line of Defense 95	Requirements for Service Equipment (Listing) 107
Grounding Scheme for Services 96	Grounded Conductor (Neutral) Disconnect Requirement for Services 109
Grounded Conductor Routing and Connections 98	
Main Bonding Jumpers in Service Equipment 99	Requirements for Services Supplied by Ungrounded Systems 110
Dual-Fed Service Equipment 100	Marking Equipment for Ungrounded Systems 111
Minimizing Impedance in Service Grounded Conductors 100	Grounding of Service Raceways and Enclosures 111

CHAPTER 6 Grounding Electrode Conductors

Introduction 118	Grounding Electrode Conductor Installation 124
The Path to Ground 118	Effectiveness (Integrity) of the Grounding Path 131
Purpose of Grounding Electrode Conductors 119	Grounding Electrode Conductor Connection Locations 132
Current in Grounding Electrode Conductors 120	Grounding Electrode Conductor Connections............... 134
Grounding Electrode Conductor Material 120	Magnetic Field Concerns 136
Sizing Grounding Electrode Conductors 121	
Grounding Electrode Conductors for DC Systems 123	

CHAPTER 7 Bonding Requirements

Introduction 144	Cleaning Coated Surfaces 149
Definitions of Bonding Terms 144	Bonding Jumper and Bonding Conductor Length 149
Bonding Performance Criteria 145	Equipment Bonding Jumpers (Function and Purpose) 150
Maintaining Continuity 147	
Bonding Connections (Wire-Type Conductors) 147	

CHAPTER 7 — Bonding Requirements—cont.

Sizing Requirements for the Supply Side and Load Side 150	Installation and Size of Equipment Bonding Jumpers (Load Side) 156
Service Bonding Rules (Line Side) 151	Expansion Fittings and Loose Joined Metal Raceways 157
Reducing Washers 153	Bonding Metal Piping Systems 158
Boxes with Concentric or Eccentric Knockouts 154	Bonding Structural Metal Building Frames 162
Sizing Supply-Side Bonding Jumpers (Wire Types) 154	Bonding Lightning Protection Systems 163
General Equipment Bonding Rules (Load Side) 155	

CHAPTER 8 — Equipment Grounding Conductors

Introduction 168	Equipment Grounding Conductor Installations 175
Equipment Grounding Conductor Defined 168	Equipment Grounding Conductor Connections 177
Purpose of Equipment Grounding Conductors 169	Equipment Grounding Conductor Identification 178
Equipment Grounding Conductor Material 170	Equipment Grounding Conductor Sizing 179
Types of Equipment Grounding Conductors 170	Current in Equipment Grounding Conductors 185

CHAPTER 9 — Grounding Electrical Equipment

Introduction 190	Receptacle Grounding Connections 198
Purpose of Grounding Equipment 190	Receptacle Replacements 203
General Grounding Rules for Equipment 190	Grounding Appliances Using the Grounded Conductor 205
Conductor Enclosure and Raceway Grounding Requirements 194	Auxiliary Grounding Electrode Requirements 206
Methods of Grounding Equipment (Part VII of Article 250) 195	Grounding Nonelectrical Equipment 206
	Equipment Grounded by Secure Metal Supports 207
Connections of Equipment Grounding Conductors 198	Use of the Grounded Conductor for Grounding 207

CHAPTER 10 — Isolated/Insulated Grounding Circuits and Receptacles

Introduction 214	Isolated Grounding Circuits 218
Electrical Noise in Grounding Circuits 214	Use of Auxiliary Grounding Electrodes 222
Purpose of Isolated Grounding Circuits and Receptacles 216	Grounding and Bonding in Information Technology Centers 223
Objectionable Currents in Grounding Paths 217	Signal Reference Structures (Grids) 224
Power Quality System Grounding Analysis 218	Surge Protection 226

CHAPTER 11 Grounding at Separate Buildings or Structures

Introduction 232	Supplied by Separately a Derived System........................238
Definitions..232	Building Disconnecting Means Requirements...............239
Supplying Power to Separate Buildings or Structures ...233	Metal Water Pipe Bonding and Other Bonding.............240
Purpose of Grounding and Bonding at Separate Buildings or Structures...233	Disconnecting Means Remote from Building or Structure 240
Grounding Electrode Required 234	Buildings or Structures Supplied by Separately Derived Systems...241
Grounding Electrode Conductor 235	
Feeder and Branch Circuit Requirements.......................236	Buildings or Structures Supplied by an Ungrounded System 242
Ungrounded Systems Supplying Services 237	Buildings or Structures Supplied by Generators 243

CHAPTER 12 Grounding Electrical Systems

Introduction 250	Mandatory System Grounding 253
Definitions..250	Optional System Grounding.............................257
System Grounding...251	System Grounding Prohibited 261
Methods of System Grounding.........................252	High-Impedance Grounded Neutral Systems 261
System Grounding Requirements 253	Ungrounded Systems (Concepts).....................262

CHAPTER 13 Grounding and Bonding for Separately Derived Systems

Introduction 268	Outdoor Source 279
Definitions..268	Ungrounded Systems 280
Determining a Separately Derived System.....................269	Generators and Transfer Equipment 281
Grounding Requirements..................................270	Small Wind Electrical Systems 286
Grounded Systems...270	Grounding DC Systems.....................................287
Grounding Electrodes..275	Ungrounded DC Systems..................................289
Bonding Water Piping and Building Steel......................278	

CHAPTER 14 Special Occupancies and Conditions

Introduction 296	Special Rules for Agricultural Installations.....................309
Special Rules for Hazardous Locations...........................296	Mobile and Manufactured Home Grounding and Bonding Rules 313
Special Rules for Health Care Facilities 300	

CHAPTER 15 Grounding for Special Equipment

Introduction 320	Information Technology Equipment and Sensitive
Purpose of Grounding Equipment............................ 320	Electronic Equipment 327
Electric Signs and Outline Lighting Systems 320	Grounding and Bonding Requirements for Swimming
Electric Cranes and Elevators 326	Pools and Similar Installations 329
	Grounding Requirements for Solar PV Systems 341

CHAPTER 16 Grounding and Bonding for Limited-Energy Systems

Introduction 350	Intersystem Grounding and Bonding....................... 354
Performance and Concepts 350	Common Grounding and Bonding Rules
Definitions 350	for Communications Systems 356
Grounding and Bonding Performance 351	Grounding and Bonding at Mobile Homes 360
Connecting to a Grounding Electrode 352	Radio and Television Equipment and Antennas 361
Grounding Electrode Conductor Installation 353	Overvoltages and Lightning Events 363

CHAPTER 17 Ground-Fault Circuit Interrupters and Equipment Ground-Fault Protection

Introduction 368	Equipment Ground-Fault Protection 374
Ground-Fault Circuit Interrupters 368	

CHAPTER 18 Grounding Rules for Medium- and High-Voltage Systems

Introduction 388	Portable or Mobile Equipment Grounding............... 392
Requirements for Grounding Systems..................... 388	Grounding Equipment.. 393
Grounding Methods for Systems Over 1 kV 388	Substation Grounding Requirements 394
Solidly Grounded Systems 389	Conductor Shielding and Stress Reduction 397
Impedance Grounding.. 391	Grounding through Surge Arresters 399

ANNEXES

Annex A Investigation and Testing of Footing-Type Grounding Electrodes for Electrical Installations..........404

Annex B Steel Conduit and EMT for Equipment Grounding..414

Annex C UL Guide Information for Electrical Equipment (2009 White Book) 425

Annex D Short-Circuit Current Calculations 443

Annex E NEC Table 8 (NFPA) 454

INDEX

Index 455

Acknowledgments

Technical Review:
Dewig, Jim, Evansville Electrical JATC
Dobrowsky, Paul, Innovative Technology Services
Minder, Tom, Alaska Electrical JATC
Ohde, Harold C., Chicago Electrical JATC

NJATC Staff Contributors:
Palmer Hickman
Director of Code and Safety Training
and Curriculum Development
Technical Editor

Content, Photograph, and Illustration Contributions:
Cogburn Brothers, Inc.
Colgan, Rob, NECA
Cooper Bussmann
Dollard, Jr., James T., IBEW Local 98
ERICO Corporation International
Fluke Corporation
Harger Lightning and Grounding
Insulated Cable Engineers Association
IEEE
Johnston, Michael, NECA
Maddox, Rick, Clark County, NV
McGovern, Bill, City of Plano, TX
Morse, Electric Inc.
National Electrical Contractors Association (NECA)
National Fire Protection Association
Pass and Seymour/Legrand
Schneider Electrical USA, Inc.
Siemens
Southwire Company
Thomas and Betts
The Conduit Committee of the Steel Tube Institute of North America
Underwriters Laboratories Inc.
Washington DC Joint Apprenticeship and Training
Young Electric Sign Company (YESCO)

Features

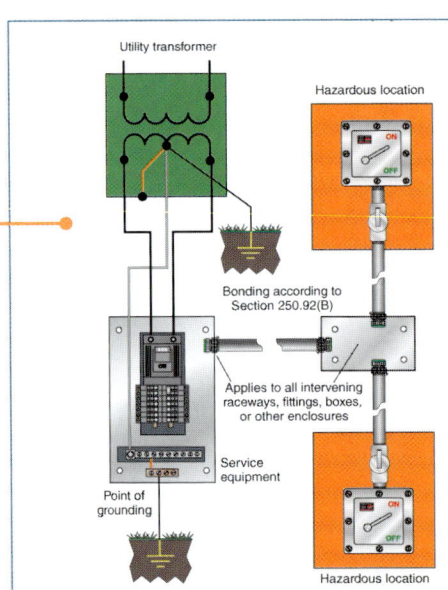

Definitions and important text are "torn" straight from the 2011 NEC.

Photos and graphics visually illustrate concepts from the chapter text.

Features — cont.

Pull quotes direct readers to and reiterate the most important points of the text.

Summary helps students process information.

Review Questions help students test the knowledge gained from the chapter.

Introduction

Safe electrical circuits and systems depend on *Code*-compliant electrical grounding and bonding. Grounding and bonding are integral functions that must perform effectively to ensure safety for persons and property as required by the National Electrical Code (*NEC*). Why is this subject so intimidating and mysterious to so many in the electrical field? There is an apparent need to eliminate the myths that cloud the subject of electrical grounding. Grounding and bonding does not have to be as complicated or difficult as many make it out to be. This training program breaks down this subject and simplifies it for instructors and students. Concepts are much easier to learn once one achieves an understanding of how and why circuits and systems work the way they do. This is a simple way to explain theory. One must have a good working knowledge about electrical circuits and current to understand how proper grounding and bonding result in circuits and systems that are essentially safe. It is equally important to develop a thorough understanding of the *NEC* and how to apply it to electrical installations. There are many words and terms used in the electrical field, some slang, some defined in the *NEC*. The best approach to training students is to use defined terms for a common understanding of the rules. This training material places significant emphasis on use and understanding of defined grounding and bonding terms. Using slang terms such as *ground wire* and *stingers* can result in misapplication of Code requirements. This textbook takes a fundamental approach to taking the mystery out of this topic. Keeping it simple is a far better approach in the classroom and when working on electrical equipment.

This material is built on the principles of explaining basic concepts and establishing a solid foundation in the subject of grounding and bonding. Electrical workers are trained in the classroom to understand theoretical operation of circuits and systems. Hands-on knowledge is gained in the field through the on-the-job segment of apprenticeship programs. The National Joint Apprenticeship and Training Committee (NJATC) Applied Electrical Grounding and Bonding program is designed for an optimal teaching and learning experience with the instructor, veteran and student in mind. Essential performance concepts are presented in a straight forward and concise fashion making a challenging subject easy to master. Effective use of visual support, including actual photos of products and from the field assist workers and students of the *Code* in developing their knowledge of grounding and bonding while enhancing their abilities to accurately apply the *NEC* grounding requirements. The complex issues related to grounding and bonding systems are made simple through easy to read text and detailed explanations and diagrams that ensure an interesting learning experience. It is the author's hope that the information in this training program be passed on to others entering this rewarding field to build upon an industry second to none in skills, attitude, and knowledge.

The objective of this training material is to assist students of the electrical industry in developing and mastering the subjects of electrical grounding and bonding.

Introduction—cont.

This textbook starts by introducing basic concepts and graduates to more complex applications of grounding and bonding circuits used in electrical wiring systems. The program is arranged in a fashion similar to that of a construction project. In other words, the starting point is in the ground (at the grounding electrodes), and the subjects are covered carefully and in detail through the entire electrical system—all the way to the final branch circuit outlet. It is helpful to build knowledge of a subject in an order similar to the order of the tasks and phases of a construction project. Before we get into how to build the grounding and bonding circuits and systems, we will review various fundamentals to reinforce current electrical knowledge and establish a good foundation for the more complex topics that follow.

About this Book

Applied Grounding and Bonding is an authoritative textbook on one of the most critical topics in our industry. It is authored by Michael J. Johnston, NECA's Executive Director of Standards and Safety. Mike is recognized as one of the leading experts in the field of grounding and bonding and presently serves the industry as the Chairman of the *NEC* Technical Correlating Committee. Prior to this appointment as TCC Chair, Mike most recently served as Chairman and a longtime member of *NEC* Code Panel 5 where Article 250, Grounding and Bonding, resides in the *NEC*.

The textbook is uniquely designed to follow the installation the way a contractor and an electrical worker would install the grounding and bonding system in actual practice. Topics covered include "traditional" topics such as service, feeder, and branch circuit grounding and bonding, as well as more specialized topics such as grounding and bonding in health care facilities, hazardous locations, and for lightning protection.

Author

Mr. Michael J. Johnston is the executive director of standards and safety for the National Electrical Contractors Association (NECA). Prior to working with NECA, Mike worked for the International Association of Electrical Inspectors as the director of education, codes, and standards. He also worked as an electrical inspector and electrical inspection field supervisor for the City of Phoenix, AZ. Mike achieved a BS in Business Management from the University of Phoenix. He served on *NEC* CMP-5 in the 2002, 2005, and 2008 cycles and as the chair of CMP-5 representing NECA for the 2011 *NEC* cycle. He currently serves as the chair of the *NEC* Technical Correlating Committee. Mike is a member of the IBEW and has experience as an electrical journeyman wireman, foreman, and project superintendent. Johnston is an active member of IAEI, the NFPA Electrical Section, Education Section, the UL Electrical Council, and National Safety Council.

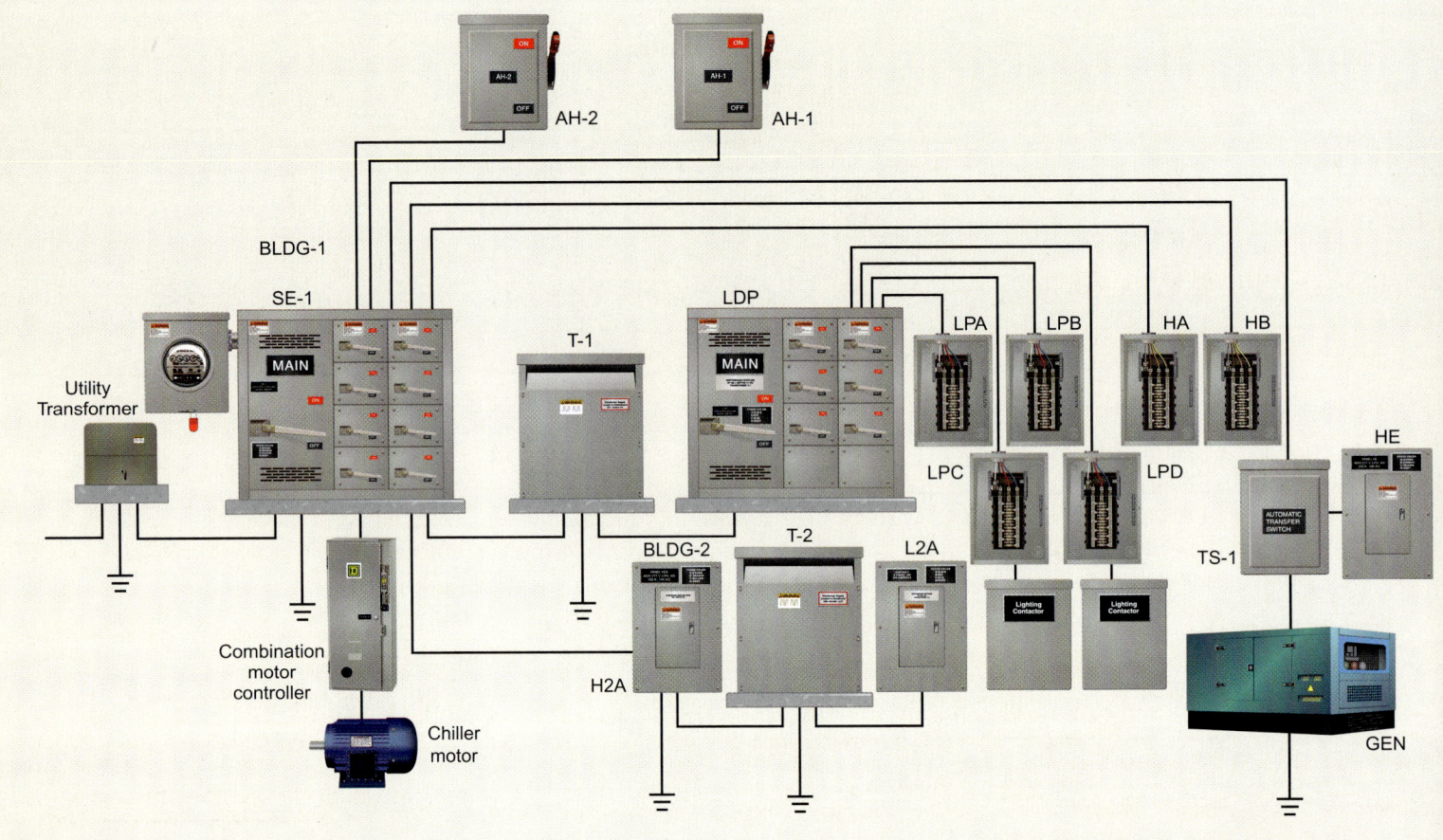

CHAPTER 1

Introduction

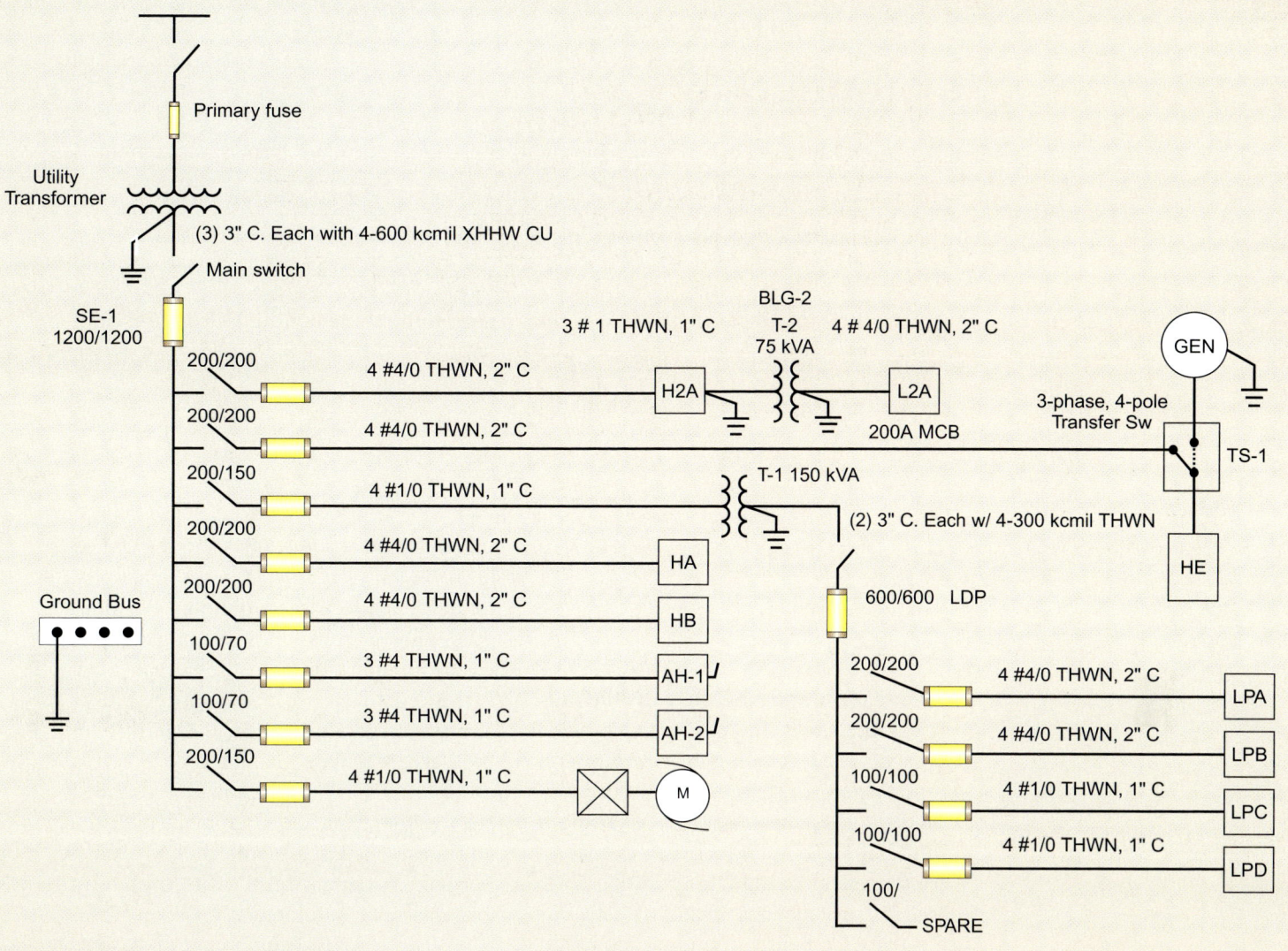

Objectives

- Recognize and understand key grounding and bonding terms
- Understand the role of the Earth in the electrical grounding system
- Understand the purpose of electrical grounding and bonding and fundamental grounding concepts of systems and equipment
- Understand fundamentals related to bonding and connecting conductive parts and equipment together to establish electrical continuity and conductivity
- Understand grounding and bonding concepts that perform together and simultaneously in electrical wiring systems
- Understand performance requirements for electrical grounding and bonding and application of *NEC* requirements that address what grounding and bonding are intended to accomplish

Outline

Grounding and Bonding for Safety

Earth in the Circuit

Grounding Concepts

Bonding Concepts

Minimizing Shock Hazards

Path for Current through the Human Body

Grounding and Bonding Concepts Together

Performance *Code* Language

Grounded Electrical Systems

Ungrounded Electrical Systems

Introduction

Basic grounding and bonding concepts and the related performance provisions contained in the *National Electrical Code® (NEC®)* are knowledge areas that an electrical worker must be familiar with. Concepts such as what is ground, what is grounding, what is electrical bonding, and understanding how and why these circuits work will help the electrical worker master knowledge in the area of electrical grounding and bonding. Grounding and bonding are essential for safe electrical systems. The concepts are fairly straightforward, yet many find them complex and challenging.

Grounding and Bonding for Safety

Section 90.1 of the *NEC* contains the purpose of the *Code*. The *NEC* rules provide a practical safeguarding of persons and property from hazards that might be associated with electricity use (*NEC* 90.1). Both grounding and bonding concepts are necessary for safe electrical wiring installations. Electrical installations include safety circuits that accomplish grounding and bonding together and simultaneously. One function complements and supports the other. When non–current-carrying conductive parts of equipment are bonded together and then connected to the ground, bonding and grounding are accomplished and enhanced electrical safety is the result.

Earth in the Circuit

From the beginning, the discovery of electricity brought many challenges and questions. Just as there were difficult debates and discussions about distribution of alternating current (AC) systems as compared with direct current (DC) systems, equally challenging decisions had to be made about electrical grounding. After much debate and experimentation, grounding electrical equipment and systems prevailed as the most popular and effective method of building safety into electrical systems, at least in the United States. Many arguments were made for and against electrical grounding. Without grounding, the possibility of completing an electrical circuit through the human body is reduced; this is because when equipment and systems are grounded, the Earth is in the electrical circuit, making shock hazards more likely. On the other hand, grounded electrical systems and equipment offer the advantages of circuit protection by causing fuses or circuit breakers to operate in ground-fault conditions.

When electrical equipment or systems are grounded, the Earth is in the circuit and part of the electrical system. The terms *ground* and *grounded (grounding)* are defined in Article 100 of the *NEC* as follows:

Ground. The earth.[1]

Grounded (Grounding). Connected (connecting) to ground or to a conductive body that extends the ground connection.[2]

One of the most elementary concepts to remember is that throughout the *Code*, *electrical grounding* refers to a connection to the Earth—to this planet. This is accomplished by using what is referred to in the *Code* as a grounding electrode. **See Figure 1-1.**

Therefore, with the understanding that the Earth is part of any circuit or system that is grounded, the quality of the Earth as a conductor must be discussed. The Earth is a large but very poor conductor. **See Figure 1-2.** The Earth should not be relied on to carry any current, be it normal current or fault current. The Earth offers significant opposition to current in circuits, including grounding and bonding circuits. This opposition to current is known as either *resistance* or *impedance*. In DC circuits, the primary opposition to current is *resistance,* whereas in an AC circuit, the opposition to current is referred to as *impedance*.

Grounding Concepts

To understand electrical grounding concepts, getting back to basics is a must. Grounding is a function of connecting to the Earth. Buildings and structures depend on a good connection to the Earth, a foundation or footing. **See Figure 1-3.**

The electrical systems are built on a similar foundation. The foundation of a grounded electrical system and grounded equipment is the grounding electrode—the direct connection to the Earth. When an object qualifies as ground, any connection to it renders a grounding function. Since the *Code* defines ground as the Earth, if equipment or a system is grounded, it has to be connected to this planet—to Earth.

For example, the frame of an automobile does not qualify as ground because there is no connection to Earth; this is contrary to what many people think. It is the negative side of the vehicle battery, not ground. The same holds true for aircraft. The frame of a jet airliner may be used as the reference for the 400-cycle power system onboard, but it is not ground, as some people might say. Ground is the Earth, by definition. When there is no connection to the ground (the Earth), there is no grounding.

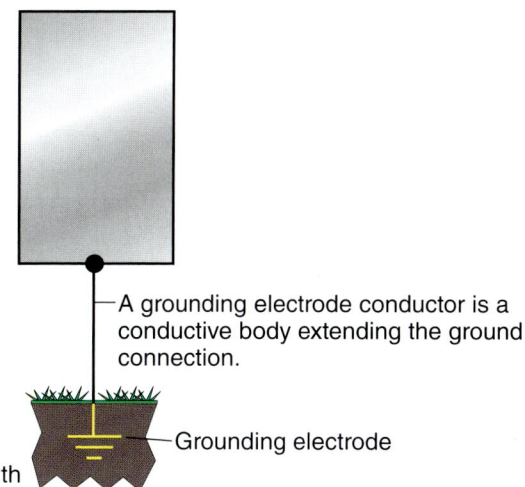

FIGURE 1-1 Grounded means connected to the ground (the Earth) directly or through a conductive body that extends the grounding connection.

FIGURE 1-2 When equipment and systems are connected to the Earth, the Earth serves as a conductor.

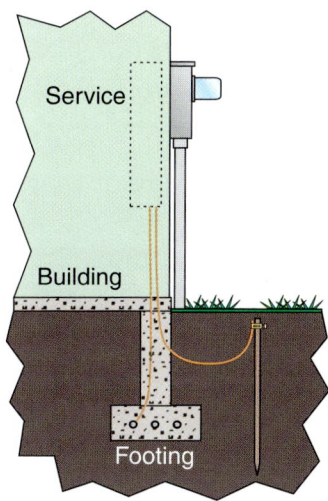

FIGURE 1-3 Buildings are constructed on solid foundations. The grounding electrode system is the foundation of a safe electrical system.

In order to accomplish grounding of equipment or systems, a connection to the planet Earth has to be made. This connection is made through grounding electrodes. Grounding is a function that results in conductive equipment parts or systems that are connected to ground, the Earth. **See Figure 1-4.** The definitions of grounding and bonding terms in Article 100 provide a common means of communication when applying the rules to installations and systems.

Bonding Concepts

When there is bonding, a connection is involved. The function of bonding is the process of connecting together. When there is bonding, two or more conductive objects become one electrically, connected together. **See Figure 1-5.**

> **Bonded (Bonding).** Connected to establish electrical continuity and conductivity.[3]

Bonding minimizes potential (voltage) differences between conductive parts that are bonded together. Bonding is a function that results in the *connection* of conductive parts. Therefore, when electrical equipment is grounded, bonding is occurring and it accomplishes the grounding (the connection to the Earth).

Bonding is a function that results in non–current-carrying conductive parts being connected together electrically, establishing continuity between them and in many instances adequate conductivity for current. Bonding safeguards against possible shock hazards and ensures effectiveness of a ground-fault current path for overcurrent device operation. Ineffective bonding connections could result in arcing or sparking being released at loose bonding connections under fault conditions, which can be a fire hazard.

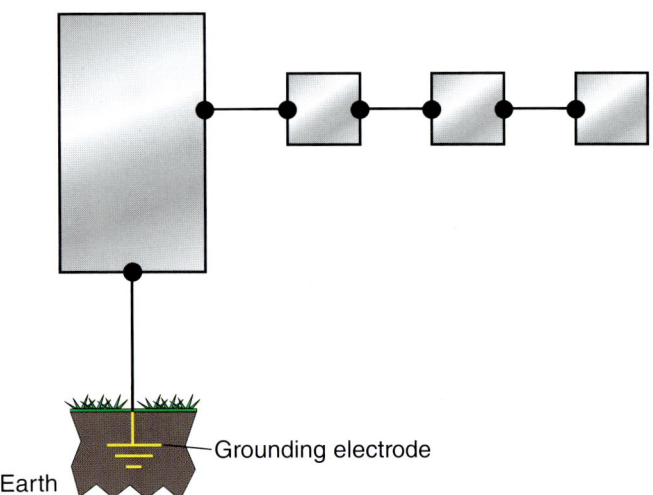

FIGURE 1-4 Grounding is an ongoing process of connecting systems and equipment to the ground (Earth). Bonding is inherent to the grounding process.

Minimizing Shock Hazards

Grounding and bonding are the safety circuits installed with electrical systems. Electrical grounding and bonding are two protective processes that happen simultaneously. These two processes not only provide essential protective functions for equipment and property, but also reduce shock hazards. The process of grounding a system or electrical equipment reduces potential differences between electrical

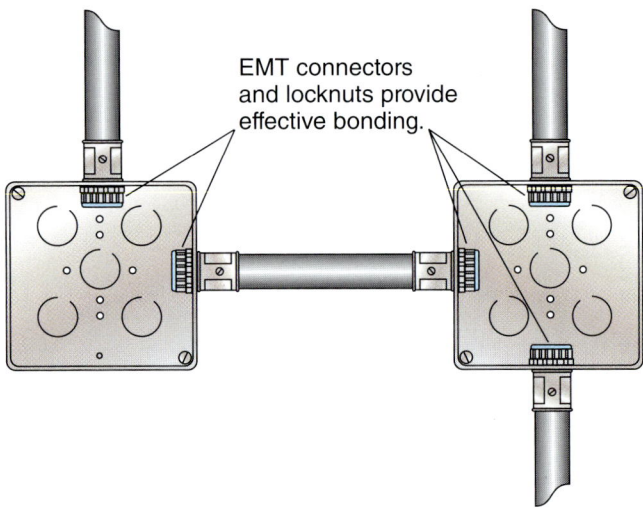

FIGURE 1-5 Bonding is a process that establishes electrical continuity and conductivity between conductive parts.

systems and equipment and the earth. See **Figure 1-6.** Bonding is the process of connecting conductive objects together so they become one electrically, thus minimizing and eliminating potential differences while at the same time reducing possibilities of electric shock. Historically, grounding and bonding have been regarded as the primary method of protecting systems, property, and personnel from electrical shock hazards and fire, while offering significant advantages in systems operation. Over the years there have been many debates about whether greater protection is achieved by grounding systems and equipment as compared to not grounding or isolating systems. Contrary to the beliefs of many, grounding only provides part of the protective functions necessary for equipment and systems designed and installed according to the *NEC*. While grounding provides degrees of protection described above, the *NEC* strongly emphasizes the critical role of an effective ground-fault current path in electrical safety systems, and how it relates to sure operation of overcurrent protective devices. This path for ground fault current is essential in grounded systems. During a ground fault event, current returning to the source is present in multiple paths such as equipment grounding conductors, the Earth, conductive parts of equipment, through a person, or any combination of ground-fault return paths. **See Figure 1-7.**

Path for Current through the Human Body

When a person becomes a pathway for electrical current, the person will incur an electrical shock. Three conditions are required to result in an electrical shock:

1. There must be a contact point on the body for entry of the circuit current.
2. The current must exit through a second contact point.
3. There must be a voltage to force the current across the human body.

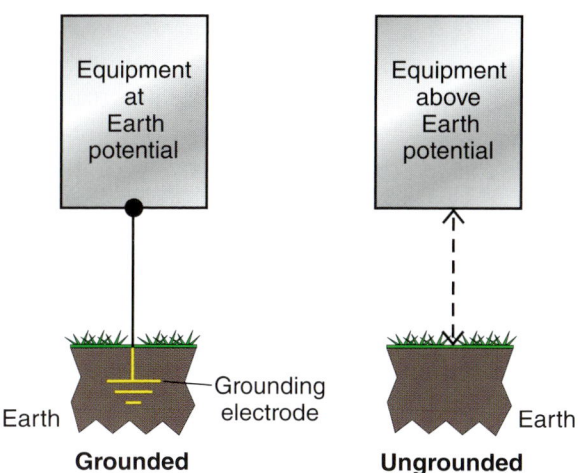

FIGURE 1-6 Grounding reduces potential differences to the ground in normal operation and during ground-fault events.

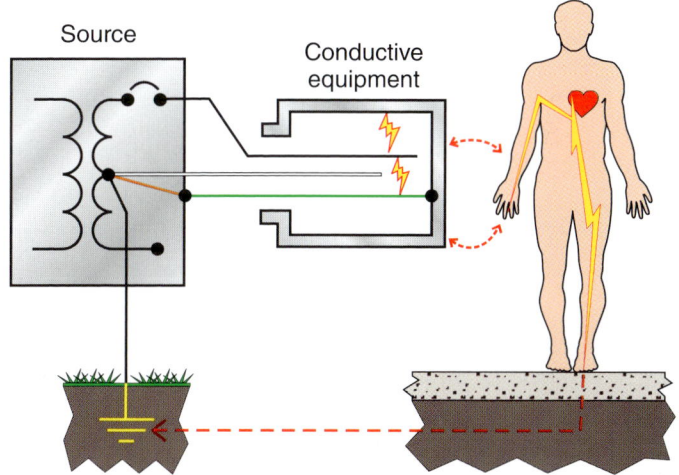

FIGURE 1-7 Ground-fault current tries to return to the source over any available conductive path.

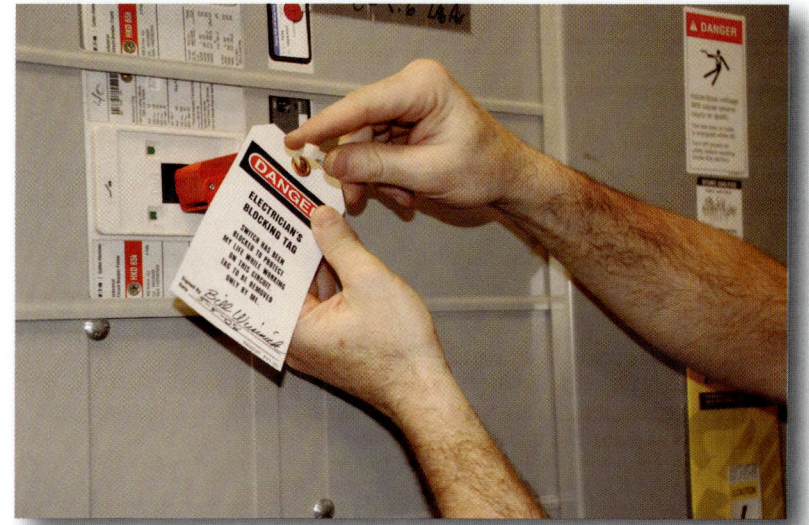

Protection from shock and other electrical hazards is achieved by establishing an electrically safe work condition in accordance with NFPA 70E Standard for Electrical Safety in the Workplace.

The severity of the shock may result in the person being seriously injured or even electrocuted. How serious the electrical shock injury generally depends on the:

1. amount of current (amperes);
2. duration (time) the current is present;
3. path the current takes through the body. **See Figure 1-8.**

Circuit frequency can also affect the severity of an electrical shock a person receives. The amount of damage an electric shock will inflict on the human body is influenced by several factors, such as moisture levels, temperatures, and location on the body where the contact occurs. A person may become involved in an electrical circuit or provide a pathway for current in a circuit through ground in one of two ways: he or she might be (1) in series contact with the electrical circuit or (2) in parallel contact with the circuit. For current to be present in a circuit, the circuit must be complete. In series contact, the person is the only current path to the ground and the EGC is not involved in the circuit. Contact can occur in many ways. Current will take any and all paths available to return to its source. **See Figure 1-9.** Electrical shock can occur when the human body is in series with electrical current or when it is in parallel with the current returning to the source.

Series Circuit through the Human Body

The severity of shock is significant when the human body is in series with the complete electrical circuit. When contact with energized parts occurs, and the body is also in contact with another conductive path to the source, complete current for the circuit can be present in series across the body. The body is a very good conductor. The epidermis (skin) provides a protective covering and acts similar to insulation, except that it is conductive. Current will be present through the body as long as there is contact. As an example, electric shock or electrocution can occur when series contact is made by inserting a conductive object into an energized receptacle while the body is also in contact with another grounded surface and fault current pathway to the source, such as a grounded metal baseboard heater. For this reason the *NEC* requires tamper-resistant receptacles in dwelling units and other occupancies where unsuspecting children are predominantly present. The current in a series circuit through the body is usually not enough to trip an overcurrent device, thus severe shock or electrocution is possible. **See Figure 1-10.**

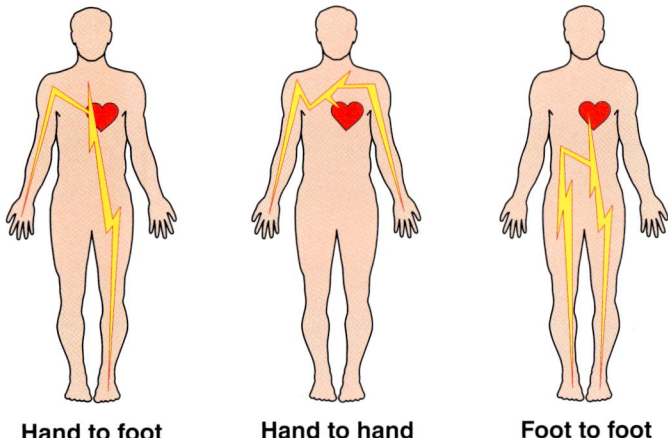

FIGURE 1-8 A person in the electrical circuit becomes a conductive path for current. Shown here are the typical paths for current through the human body.

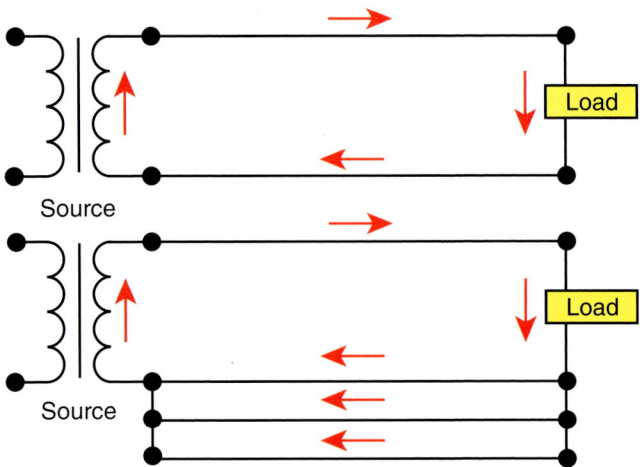

FIGURE 1-9 Current will divide over any and all paths available when returning to the source.

Parallel Circuit through the Human Body

An electrical shock can also be severe or deadly when the human body is in parallel with ground fault current. When the body is in parallel with ground fault current, the current will divide over the body and the other conductor paths that are common to the source. Impedance in each parallel path will offer opposition to current in the circuit. An example is insulation failure on internal wiring of a machine that has loose or ineffective equipment grounding conductor connections resulting in the machine frame being energized. If a person touches the energized machine and another conductor path common to the source simultaneously, there are parallel paths for fault current; the body provides one of those fault current paths. The high impedance connections and current dividing over multiple paths can prevent the fault current from reaching the instantaneous trip level which is generally much higher than a standard 15- or 20-ampere circuit breaker. In this example, high impedance in the ground-fault current path results in ineffective protection provided by the grounded system. In addition to poor or loose connections, circuit length and conductor size can contribute to impedance in equipment grounding conductors. Proper connections in the effective ground-ground fault current path are critical in three-wire, grounded systems. The effective ground-fault current path must be of low impedance, have ample capacity for any fault current imposed, and effectively operate an overcurrent device if a ground fault occurs. A report from the Research Institute and the Illinois Institute of Technology (IITRI) provides information about determined "Let-Go" current thresholds for humans and children which contributed to the 4 to 6 mA current levels for Class A GFCI protection. Expanded *NEC* requirements for ground-fault circuit interrupters have enhanced protection for people against electrical shock and electrocution. A Consumer Product Safety Commission (CPSC) paper entitled "Three-Wire Grounding Systems vs. GFCI" describes the relationship of the impedance ratio of line circuit to ground circuit and to shock current levels.

Grounding and Bonding Concepts Together

Grounding and bonding circuits are safety circuits for electrical services, feeders, and branch circuits in today's wiring systems. The *Code* provides performance language to assist users by describing what grounding and bonding must accomplish, as provided in Section 250.4. Rules in the *Code* that are performance-based make many other prescriptive requirements easier to apply and follow.

> An effective gounding and bonding system depends on the integrity of many series connections, which must be properly made and maintained.

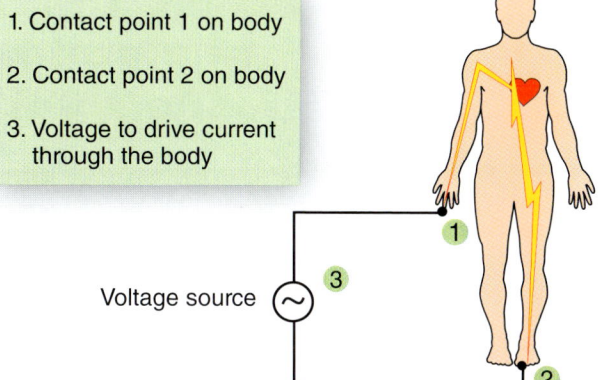

FIGURE 1-10 Current in series with the body usually causes the most severe injuries and can often result in electrocution.

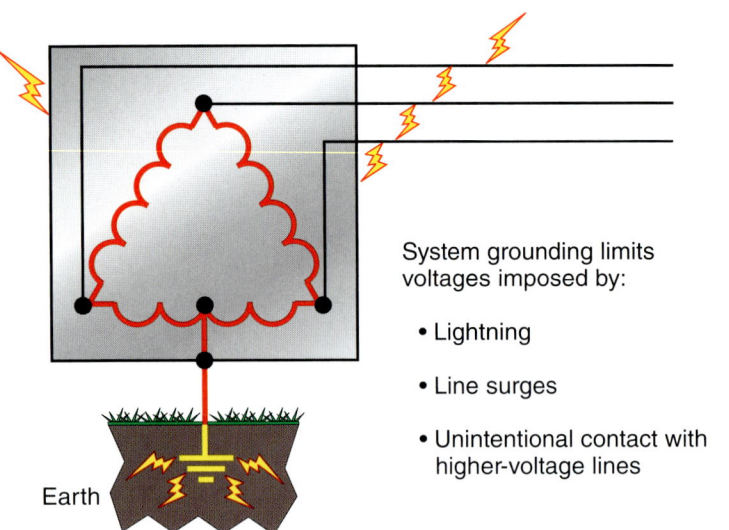

FIGURE 1-11 The graphic shows system and equipment grounding and how grounding can limit voltages imposed by lightning and line surges and unintentional contact with higher-voltage lines.

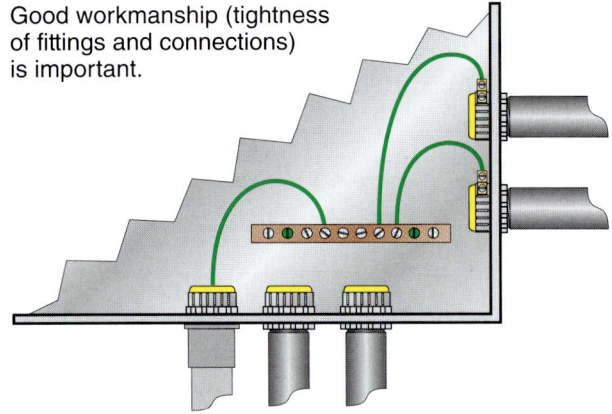

FIGURE 1-12 The figure shows bonding functions that effectively establish electrical continuity and conductivity.

REMEMBER! Good workmanship is essential for establishing thorough electrical continuity and conductivity through effective bonding. This means fittings and connections should be made up tight to perform in normal and under adverse conditions such as during system ground faults. The National Electrical Contractors Association (NECA), a standards development organization accredited by the American National Standards Institute (ANSI), has developed many National Electrical Installation Standards (NEIS) that clarify what constitutes good workmanship in the multiple aspects of the electrical contracting business.

Electrical systems are grounded to limit potentials (voltages) imposed by lightning, line surges, or unintentional contact with lines of higher-voltage lines. **See Figure 1-11.** Grounding also stabilizes voltages during normal operation. Electrical equipment is grounded to establish the same or close to the same potential between the equipment and the ground (Earth).

When electrically conductive parts are effectively bonded together, electrical continuity and conductivity are established. **See Figure 1-12.** An important part of the grounding and bonding scheme is constructing an effective ground-fault current path.

Care must be taken to ensure that good electrical connections are made for all the conductors including but not limited to the service grounding scheme. In addition, Section 110.12 of the *Code* requires that electrical conductors and equipment be installed in a manner that is workmanlike. Any worthy training program must include strong emphasis on craftsmanship and training persons in their field in a way that results in a skilled and confident workforce.

Performance *Code* Language

Part I of Article 250 provides general requirements for grounding and bonding. In the 1999 edition of the *NEC*, Article 250 was revised in a unique fashion to include text that tells what grounding and bonding must accomplish in electrical wiring. Prior to that edition of the *Code*, there were only fine print notes (now informational notes) that informed users about grounding performance. Section 250.4 was a huge step in helping to improve users' understanding of the *Code* requirements related to this subject. These general provisions tell users how grounding and bonding perform in grounded systems and ungrounded systems. In other words, these provisions explain what is anticipated when grounding and bonding meet the minimum requirements in the *NEC*. The performance language must be analyzed closely to develop a strong understanding of what must happen electrically.

This will simplify an electrician's understanding of the other prescriptive requirements in Article 250 and throughout the *Code*.

In the simplest form, grounding is made up of two concepts: system grounding and equipment grounding. **See Figure 1-13.**

Systems are either grounded or ungrounded. Grounded systems include one conductor that is intentionally connected to the ground. Other than at the point of grounding for the system, grounded system conductors and EGCs are generally kept separate from each other. A clear differentiation between these two systems must be made. In the simplest form, bonding is made up of two concepts: connecting together for continuity and connecting together for conductivity. Bonding must fulfill both of these functions. Bonding that is effective results in effective paths for ground-fault current and minimizes potential differences between conductive parts and equipment that is required to be bonded. **See Figure 1-14.**

It is important for workers to realize the performance aspects for grounding and bonding as they are constructing electrical systems in the field. With a thorough understanding of what grounding and bonding are intended to accomplish for electrical systems equipment, the *NEC* prescriptive requirements are much more understandable and more easily applied in design and installation. The actual text from the *Code* is provided, followed by a commentary that provides further clarification and examples of what is intended by each of these provisions.

Grounded Electrical Systems

Section 250.4(A) provides general performance requirements that apply to systems that are grounded. A grounded electrical system includes a conductor that is intentionally grounded. Systems can be grounded solidly or through an intentional resistance or impedance.

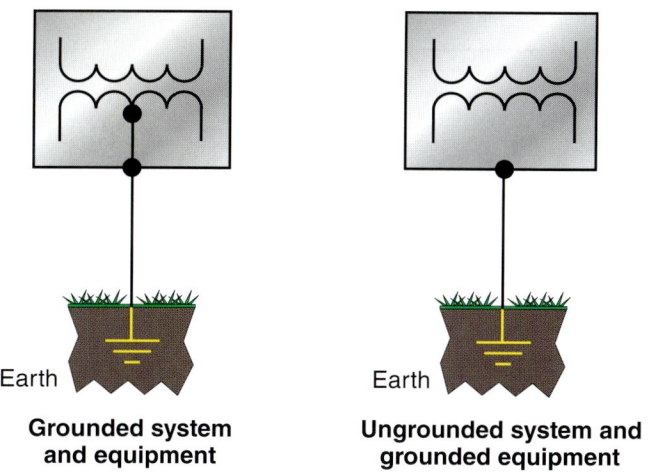

FIGURE 1-13 System grounding and equipment grounding are shown on the left side of the illustration, while only the equipment is grounded on the right.

FIGURE 1-14 Bonding is accomplished by connecting conductive parts or materials together mechanically and electrically using suitable fittings, hardware, or wire-type bonding jumpers.

> **250.4(A) Grounded Systems**
> **(1) Electrical System Grounding.** Electrical systems that are grounded shall be connected to earth in a manner that will limit the voltage imposed by lightning, line surges, or unintentional contact with higher-voltage lines and that will stabilize the voltage to earth during normal operation.[4]

When an electrical system or source is grounded, a voltage-to-ground reference is established for the system's ungrounded conductors. **See Figure 1-15.** The grounded conductor of the system is forced to take on the same potential as the ground (Earth).

Grounded systems include one conductor that is intentionally connected to the ground (Earth) and also connected to grounded electrically conductive parts. **See Figure 1-16.** System grounding helps limit voltages above ground potential, such as those caused by surges, lightning, or contact with higher-voltage systems or lines. Because the system is connected to ground, when these types of events occur in systems that are grounded, the potentials on the grounded conductor and grounded equipment are kept at or close to the ground potential for the duration of the event. This reduces flashover possibilities and shock hazards through the grounding process.

Two purposes are served by system grounding: the system is connected to ground, and the voltage to ground is stabilized during normal operation. If a grounded system is compared to one that is ungrounded, the stabilization effect of grounding systems becomes clear. Ungrounded systems are vulnerable to surges and other events that can cause the system output voltages to rise and fall. Systems that are connected solidly are stabilized to a more constant voltage-to-ground output during normal load conditions. Ungrounded systems do not offer the same stability and are subject to rise and fall conditions because the system floats above ground and not common to it, other than through induction and capacitive coupling.

An important consideration for limiting the imposed voltage is routing of bonding and grounding conductors in a manner that ensures that they are not longer than necessary to complete the connection without disturbing the permanent parts of the installation and that unnecessary bends and loops are avoided.

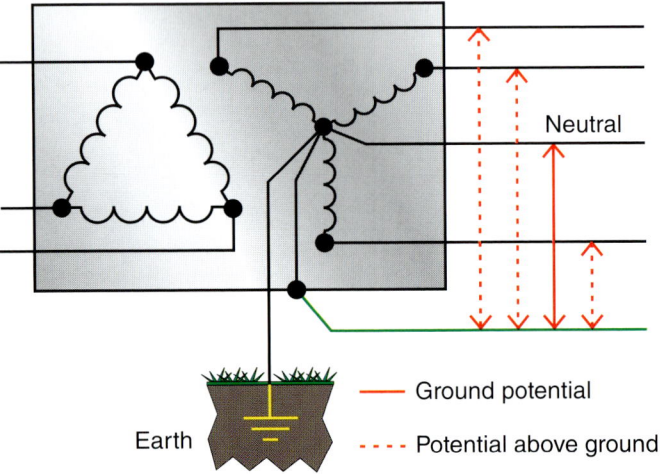

FIGURE 1-15 System grounding creates a reference to the Earth, and one conductor becomes common to the Earth.

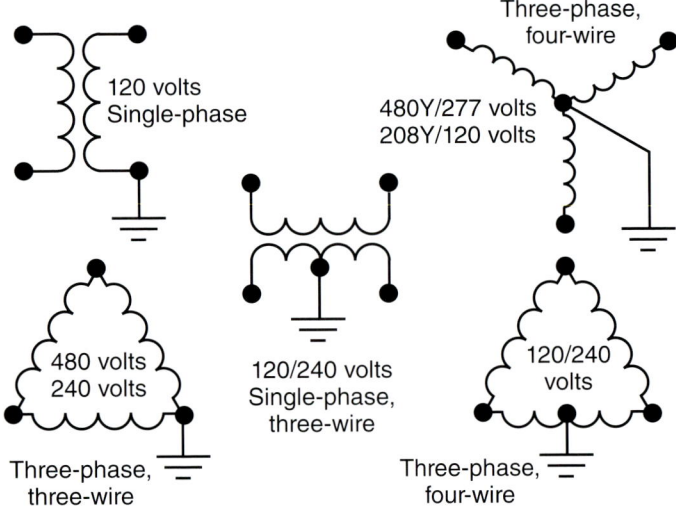

FIGURE 1-16 Common system voltages are shown where the grounded conductor is a neutral in some cases and where the grounded conductor is a phase conductor.

> **250.4(A) Grounded Systems**
> **(2) Grounding of Electrical Equipment.** Normally non–current-carrying conductive materials enclosing electrical conductors or equipment, or forming part of such equipment, shall be connected to earth so as to limit the voltage to ground on these materials.[5]

Grounding of electrical equipment involves connecting equipment to the ground (Earth), either directly or through a conductive body that extends the connection to the ground. **See Figure 1-17.** A conductive body extending the ground connection can be a bonding conductor, a grounding electrode conductor, or an EGC.

If the *Code* provides a requirement to ground equipment, it generally implies that there will be either a direct connection to the Earth or a connection through an EGC. The *Code* refers to "normally non–current-carrying metal parts enclosing electrical conductors or equipment"; this includes raceways, boxes, wireways, panelboards, switchboards, motor control centers, and so forth—all conductive parts that are not intended to carry any current during normal operation. During ground-fault events, these conductive paths will experience a high amount of fault current for a short time until the fuse or circuit breaker protecting the affected circuit can clear. EGCs are installed to connect equipment to the ground where required by the *Code*. The grounded equipment is held at or close to the ground (Earth) potential, because during normal operation and ground-fault conditions, grounding works to keep the potential from rising above ground until the overcurrent device operates and opens the circuit. The performance anticipated from equipment grounding is to connect equipment to the ground and keep it at or close to ground potential during normal conditions. This minimizes potential shock hazards.

The second performance function accomplished by equipment grounding is bonding. The function of bonding is inherent to the act of equipment grounding because bonding is the process of connecting together to establish continuity.

The last performance characteristic of equipment grounding is to serve as effective ground-fault current paths to facilitate overcurrent protective device operation. The term *ground fault* is used in the performance language of Section 250.4.

> **Ground Fault.** An unintentional, electrically conducting connection between an ungrounded conductor of an electrical circuit and the normally non–current-carrying conductors, metallic enclosures, metallic raceways, metallic equipment, or earth.[6]

> **250.4(A) Grounded Systems**
> **(3) Bonding of Electrical Equipment.** Normally non–current-carrying conductive materials enclosing electrical conductors or equipment, or forming part of such equipment, shall be connected together and to the electrical supply source in a manner that establishes an effective ground-fault current path.[7]

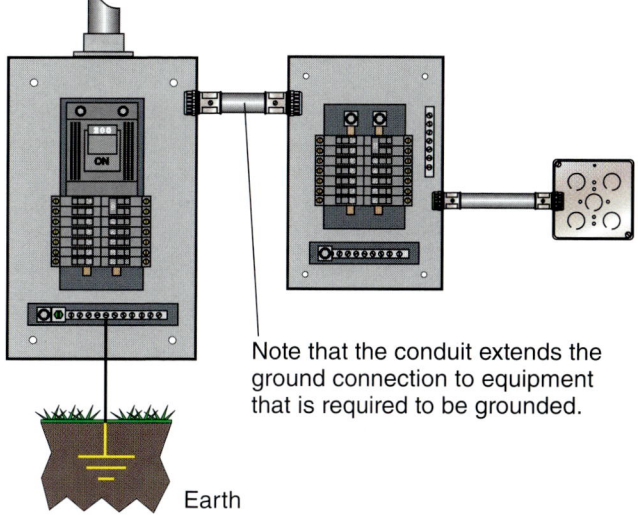

Note that the conduit extends the ground connection to equipment that is required to be grounded.

Earth

FIGURE 1-17 The phrase *grounded equipment* means the equipment is connected to the ground (Earth).

Bonding is the process of connecting two or more things together. In the electrical world, the two or more things connected together are normally non–current-carrying conductive parts in the electrical installation. **See Figure 1-18.** For example, when a length of metal conduit is attached to a metal box with two locknuts, the function of bonding is occurring.

Once this connection process is complete, the two conductive parts become one electrically and are considered bonded. Bonding establishes little or no potential differences between conductive parts, which result in the continuity aspect of a bonding function. Bonding is also required in many *Code* rules where necessary to conduct safely any imposed fault current. Sometimes bonding is required for shock protection only, such as when equipotential bonding grids are required for swimming pools or similar aquatic features.

> **250.4(A) Gounded Systems**
> **(4) Bonding of Electrically Conductive Materials and Other Equipment.** Normally non–current-carrying electrically conductive materials that are likely to become energized shall be connected together and to the electrical supply source in a manner that establishes an effective ground-fault current path.[8]

The *Code* provides some bonding requirements for conductive materials foreign to the electrical equipment and materials. Examples of such conductive materials within buildings or structures are structural building steel and metal piping systems. **See Figure 1-19.**

The bonding required by Section 250.4(A)(4) is intended to establish a connection to the ground from the bonded material, thus minimizing possible potentials above ground on these materials should they come in contact with electrical circuits. Bonding these materials also provides a direct path back to the source from materials that are "likely to become energized." This term in the *Code* requires careful judgment. Bonding is a function that happens at each connection of conductive materials or equipment. For bonding to be effective and perform properly in ground-fault conditions, each fitting, bushing, connector, coupling, and so forth, should be made up tight to keep impedance low if a fault occurs over the conduit, enclosures, and fittings. The definition of the term *bonded (bonding)* clarifies that it is a connection that establishes continuity and conductivity between conductive components.

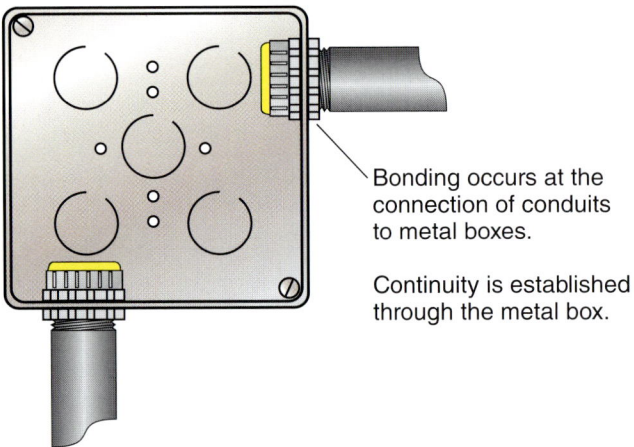

FIGURE 1-18 Bonding establishes continuity and conductivity between conductive parts.

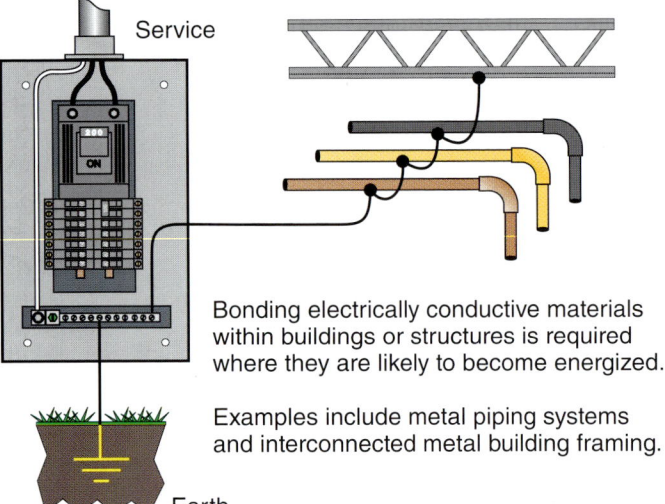

FIGURE 1-19 Metal piping systems and structural steel can become energized during abnormal circumstances such as a ground fault. Bonding provides a deliberate ground fault current path from such conductive materials.

The effectiveness of all bonding connections is directly related to workmanship. Loose fittings and poorly installed wiring methods that are not supported properly can affect bonding connections in metallic wiring methods installed for services, feeders, and branch circuits. Good workmanship is essential in all aspects of electrical installations.

> **250.4(A) Gounded Systems**
> **(5) Effective Ground-Fault Current Path.** Electrical equipment and wiring and other electrically conductive material likely to become energized shall be installed in a manner that creates a low-impedance circuit facilitating the operation of the overcurrent device or ground detector for high-impedance grounded systems. It shall be capable of safely carrying the maximum ground-fault current likely to be imposed on it from any point on the wiring system where a ground fault may occur to the electrical supply source. The earth shall not be considered as an effective ground-fault current path.[9]

This performance requirement addresses installing electrical materials that form part of or all of an effective ground-fault current path. An important aspect of this requirement is that it is related to the manner in which the materials are installed. Because the effective ground-fault current path is intentionally installed, its effectiveness is measurable based on meeting minimum requirements of the *NEC* and being installed in workmanlike fashion. An effective ground-fault current path can be a wire-type, cable-type, or in the form of a conductive metal path such as a cable tray, conduit, tubing, wireways, or others that qualify for this use. This path for fault current is relied on for fast, effective activation of overcurrent devices during a ground-fault event. Loose fittings, locknuts, or other fasteners can decrease the effectiveness of this path and increase impedance. Note that this path has to meet specific criteria that are directly related to its performance. It must be of the lowest possible impedance. This will vary because of different lengths, sizes, and other installation characteristics of branch circuits and feeders. This path also must be capable of carrying the maximum fault current likely to be imposed. In ground-fault conditions, the source will feed into a fault as much fault current as the system is capable of delivering. **See Figure 1-20.**

These heavier levels of current can cause significant damage to conductive paths or components of the circuit that are not sufficient. Impedance is opposition to current in an AC circuit, so keeping the impedance as low as possible means that most current will be present under ground fault conditions allowing fast operation of fuses or circuit breakers.

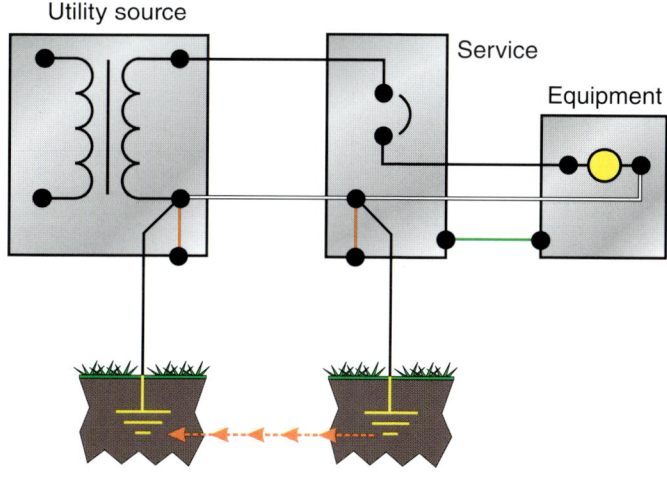

Normal Current Path

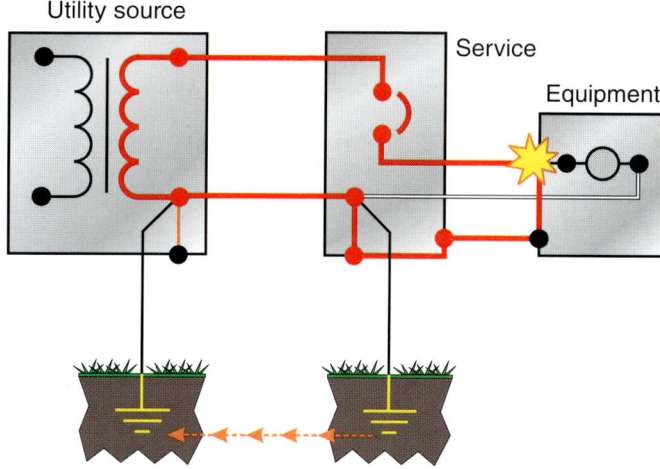

Effective Ground-Fault Current Path

FIGURE 1-20 Effective ground-fault current paths facilitate overcurrent device operation.

The Earth is not permitted as an effective ground-fault current path. This restriction of using the Earth for this purpose is necessary because although the Earth is conductive, it offers variable levels of high opposition to current. The Earth should not be depended on for any steady-state current, so obviously the Earth can never be depended on for fault current levels or to serve as an effective ground-fault current path.

> **Effective Ground-Fault Current Path.**
> An intentionally constructed, low-impedance electrically conductive path designed and intended to carry current under ground-fault conditions from the point of a ground fault on a wiring system to the electrical supply source and that facilitates the operation of the overcurrent protective device or ground-fault detectors on high-impedance grounded systems.[10]

Ungrounded Electrical Systems

Section 250.4(B) provides general performance requirements that apply to systems that are not grounded. Ungrounded systems are those that are not connected solidly or through any impedance to the Earth. Ungrounded systems do not include a conductor that is intentionally grounded. **See Figure 1-21.** The ungrounded conductors operate without a solid voltage-to-ground reference.

Voltage readings to ground from the phase conductors of an ungrounded system are typically the result of distributed leakage capacitance and can vary depending on circuit lengths, the size of conductors, and so forth. Although the system is ungrounded, the functions of grounding and bonding are still necessary for the equipment enclosing circuit conductors supplied from these systems. The following includes the *Code* performance language for ungrounded systems and is followed by commentaries that break the provisions down into simple concepts.

> **250.4(B) Ungrounded Systems**
> **(1) Grounding Electrical Equipment.**
> Non–current-carrying conductive materials enclosing electrical conductors or equipment, or forming part of such equipment, shall be connected to earth in a manner that will limit the voltage imposed by lightning or unintentional contact with higher-voltage lines and limit the voltage to ground on these materials.[11]

The performance language in Section 250.4(A)(1) is related to the grounding requirements for equipment installed for an ungrounded system. The non–current-carrying parts of raceways, equipment enclosures, and so forth, must be grounded (connected to the Earth) in such a way as to limit overvoltages caused by lightning or unintentional contact with higher-voltage lines. Grounding equipment essentially places these conductive parts at or close to the same potential as that of the Earth. **See Figure 1-22.**

Therefore, if abnormal events should cause a voltage rise on these parts, the rise will be consistent with the rise in potential of the Earth. The objective is to keep the non–current-carrying conductive parts at the same potential as that of the Earth.

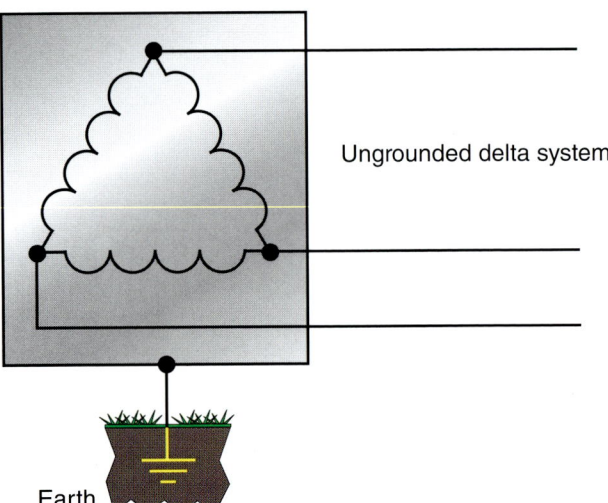

FIGURE 1-21 Ungrounded electrical systems have no solid system connection to Earth.

That is what grounding is all about. Grounding minimizes potential differences between the Earth and equipment in normal operation and during abnormal events such as ground faults. Where equipment is grounded, potential differences are reduced for the duration of time it takes overcurrent protection to react. The equipment grounding requirements in Section 250.4(B)(1) can be compared with those in Section 250.4(A)(2). The electrical performance expectations are essentially the same.

> **250.4(B) Ungrounded Systems**
> **(2) Bonding of Electrical Equipment.**
> Non–current-carrying conductive materials enclosing electrical conductors or equipment, or forming part of such equipment, shall be connected together and to the supply system grounded equipment in a manner that creates a low-impedance path for ground-fault current that is capable of carrying the maximum fault current likely to be imposed on it.[12]

When an electrical system is not grounded, there is no intentional connection between the system conductors and the Earth. When the supply conductors from ungrounded AC systems are installed in grounded metal raceways and enclosures, the effects of capacitance coupling are present, creating a potential (voltage) difference between them. This potential difference can appear as voltage-to-ground readings but is the result of leakage capacitance. An example of an ungrounded AC system is a single-phase, two-wire 480-volt system. **See Figure 1-23.**

Although this coupling (charging) voltage is present, it offers little or no effect in the event of a ground fault on the system. Instead, a ground fault from any of the ungrounded system conductors will accidentally and ineffectively ground the system. The word *ineffectively* is used because these events are typically ground faults, which are an indication of a pinched wire or other similar failure of insulation creating this condition. This differs greatly from the intentional system connection to the ground for solidly grounded electrical systems. Thus, a first ground fault on an ungrounded system is usually an intermittent and unstable connection that will manifest through intermittent minor arcing to eventually fusing to the conductive raceway or enclosure. This condition should be identified and annunciated to qualified persons by a ground detection system. If a second phase-to-ground fault were to develop on a different phase than the first fault, a short circuit and ground-fault condition results simultaneously.

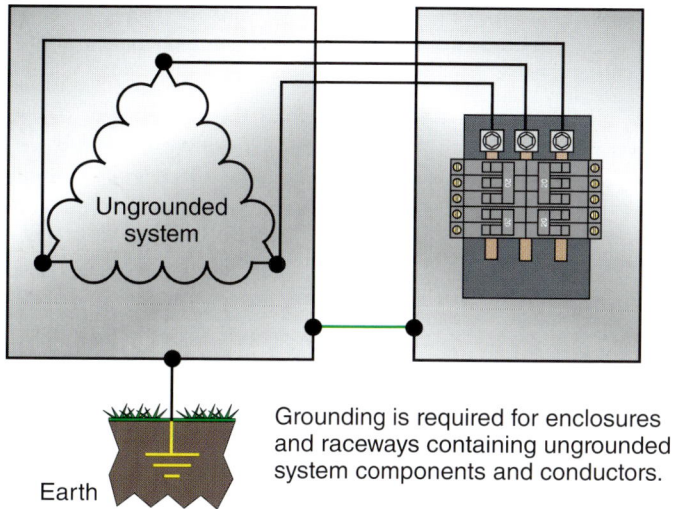

FIGURE 1-22 Raceways and electrical enclosures are examples of electrical equipment required to be grounded.

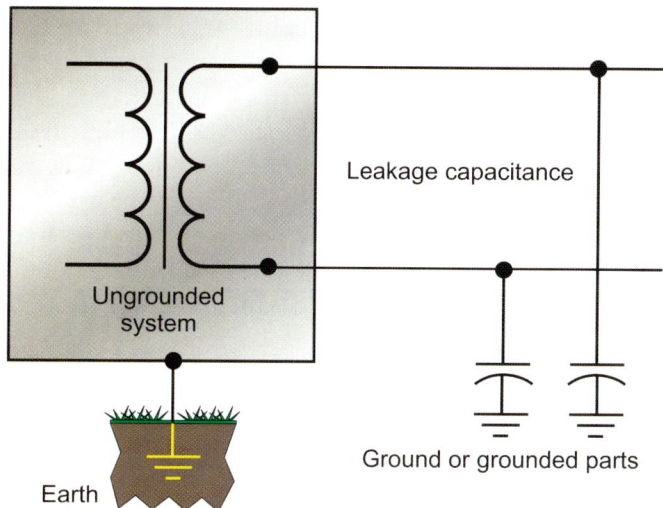

FIGURE 1-23 Ungrounded AC systems produce distributed leakage capacitance.

Heavy levels of fault current are then introduced into the system and feed into the faults until an overcurrent device opens the circuit. Bonding functions for metal equipment enclosures and metallic raceways are important in overall safety. Connections to these conductive enclosures have to be made up tight and workmanlike to perform effectively in ground-fault conditions. The bonded conductive parts and equipment must be able to maintain continuity and conductivity for any value of current imposed on them, including the highest value of fault current the system is capable of delivering. Bonding is an essential function of electrical safety and how the safety circuits operate.

imposed on them. Once again, failure to bond these conductive parts makes it possible for them to become energized, resulting in shock hazards and other possible fire hazards.

> **250.4(B) Ungrounded Systems**
> **(3) Bonding of Electrically Conductive Materials and Other Equipment.** Electrically conductive materials that are likely to become energized shall be connected together and to the supply system grounded equipment in a manner that creates a low-impedance path for ground-fault current that is capable of carrying the maximum fault current likely to be imposed on it.[13]

> **250.4(B) Ungrounded Systems**
> **(4) Path for Fault Current.** Electrical equipment, wiring, and other electrically conductive material likely to become energized shall be installed in a manner that creates a low-impedance circuit from any point on the wiring system to the electrical supply source to facilitate the operation of overcurrent devices should a second ground fault from a different phase occur on the wiring system. The earth shall not be considered as an effective fault-current path.[14]

> **Ground Fault Current Path.** An electrically conductive path from the point of a ground fault on a wiring system through normally non–current-carrying conductors, equipment, or the earth to the electrical supply source.[15]

This section deals with bonding of electrically conductive materials that are likely to become energized; this is not necessarily limited to electrical materials. Such conductive materials could include metal piping systems, metallic building frames, or other electrically conductive building components. Bonding of this material is necessary to put these parts at the same potential as the electrical grounding and bonding scheme of the service or system and to ensure a path for fault current to the system source. A low-impedance path for fault current is as important for ungrounded systems as it is for systems that are grounded. Bonding these conductive materials together and to the grounded equipment enclosures minimizes potential differences and provides a low-impedance path for any fault current likely to be

The performance requirements in Section 250.4(B)(4) deal with the path for fault current that is necessary for wiring methods and equipment installed for ungrounded systems. Once again, a first phase-to-ground fault condition will ground the system—but not effectively because these are usually loose unintentional connections or contacts between any of the ungrounded phase conductors supplied by the system. The requirements for the path for fault current are very similar to the performance requirements for the effective ground-fault current path in grounded systems. If this condition develops, the faulted phase conductor is forced to take on the same potential as that of the ground because all wiring methods and equipment are grounded. In recent *Code* cycles, a fairly new requirement for ground detection systems was added.

The *NEC* requires a ground detection system so when a phase-to-ground fault event happens on an ungrounded system an indication is provided. This alerts qualified persons of this condition so that they can fix the problem before a second phase-to-ground event on another phase develops. **See Figure 1-24.**

The informational note following Section 250.4(B)(4) clarifies that a second phase-to-ground event is indeed a ground fault, even though it would simultaneously be a phase-to-phase short circuit condition as well. This is all the more reason for quick response to annunciation of a ground fault by ground detection systems installed on ungrounded systems.

Another important characteristic of a ground fault on ungrounded systems is that the path for fault current usually includes many conductive bodies other than those installed specifically as the ground-fault current path for the system. A ground fault on ungrounded systems as well as grounded systems introduces current in multiple paths, such as combinations of EGCs, metallic wiring methods, conductive electrical equipment enclosures, and other electrically conductive materials, including conductive piping, structural steel framing, metal ducts, cable shields, and even the ground (Earth) itself. This is the reason that the word *effective* is not used for this fault current path constructed for ungrounded systems. In a first phase-to-ground fault, effective performance to facilitate overcurrent device operation is not the objective, as compared with the effective ground-fault current path constructed for grounded systems.

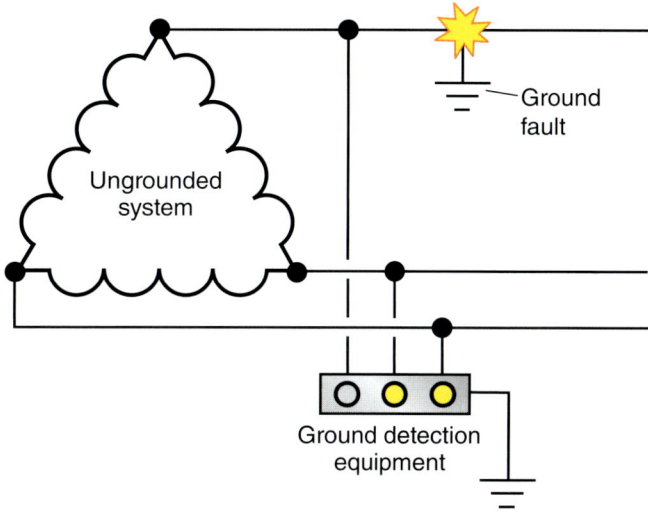

FIGURE 1-24 A ground detection system used in an ungrounded system provides annunciation of a first phase-to-ground fault condition on the system.

Summary

Electrical grounding and bonding provide essential safety for persons and property. The performance criteria of these important electrical functions have been analyzed and broken down into simple, yet thorough descriptions to provide ease of comprehension. The concepts of grounding and bonding are simple, yet many have made them more complicated than necessary. Grounding involves connecting electrical systems and equipment to the Earth. Bonding is the process of establishing and maintaining effective continuity and conductive connections between metallic parts. Grounding and bonding are actions that happen simultaneously in all electrical installations. The electrical safety system is heavily reliant upon effective grounding and bonding of equipment and systems.

References

Permission: Section 250.4 references and text from the National Electrical Code NEC-2011 are printed with permission from the National Fire Protection Association (NFPA), Quincy, MA 02169. This reprinted material is not the official position of the NFPA on the subjects covered in this chapter, which is represented only by the entire standard NFPA 70-2011.

1. NFPA 70 National Electrical Code 2011, Article 100 (National Fire Protection Association, Quincy, MA 2010), p. 70–29.
2. NFPA 70 National Electrical Code 2011, Article 100 (National Fire Protection Association, Quincy, MA 2010), p. 70–29.
3. NFPA 70 National Electrical Code 2011, Article 100 (National Fire Protection Association, Quincy, MA 2010), p. 70–29.

4. NFPA 70 National Electrical Code 2011, Section 250.4(A)(1) (National Fire Protection Association, Quincy, MA 2010), p. 70–101.
5. NFPA 70 National Electrical Code 2011, Section 250.4(A)(2) (National Fire Protection Association, Quincy, MA 2010), p. 70–101.
6. NFPA 70 National Electrical Code 2011, Article 100 (National Fire Protection Association, Quincy, MA 2010), p. 70–29.
7. NFPA 70 National Electrical Code 2011, Section 250.4(A)(3) (National Fire Protection Association, Quincy, MA 2010), p. 70–101.
8. NFPA 70 National Electrical Code 2011, Section 250.4(A)(4) (National Fire Protection Association, Quincy, MA 2010), p. 70–101.
9. NFPA 70 National Electrical Code 2011, Section 250.4(A)(5) (National Fire Protection Association, Quincy, MA 2010), p. 70–101.
10. NFPA 70 National Electrical Code 2011, Section 250.2 (National Fire Protection Association, Quincy, MA 2010), p. 70–100.
11. NFPA 70 National Electrical Code 2011, Section 250.4(B)(1) (National Fire Protection Association, Quincy, MA 2010), p. 70–101.
12. NFPA 70 National Electrical Code 2011, Section 250.4(B)(2) (National Fire Protection Association, Quincy, MA 2010), p. 70–101.
13. NFPA 70 National Electrical Code 2011, Section 250.4(B)(3) (National Fire Protection Association, Quincy, MA 2010), p. 70–103.
14. NFPA 70 National Electrical Code 2011, Section 250.4(B)(4) (National Fire Protection Association, Quincy, MA 2010), p. 70–103.
15. NFPA 70 National Electrical Code 2011, Section 250.2 (National Fire Protection Association, Quincy, MA 2010), p. 70–101.

Review Questions

1. Which of the following best defines the term *ground*?
   a. The frame of an automobile
   b. The frame of an airplane
   c. The Earth
   d. An accidental or unintentional contact between an ungrounded conductor and the Earth or another grounded object
2. _____ means connected to establish electrical continuity and conductivity.
   a. Grounded (grounding)
   b. Bonded (bonding)
   c. Guarded (isolated)
   d. Insulated
3. Which of the following *NEC* terms is best defined as connected (connecting) to ground or to a conductive body that extends the ground connection?
   a. Bonded
   b. Bonding
   c. Earthing (earthed)
   d. Grounded (grounding)
4. _____ is not considered an effective ground-fault current path.
   a. Ground
   b. Equipment grounding conductors
   c. Equipment bonding jumpers
   d. Grounded conductors
5. Which of the following is not a conductor in the grounding and bonding scheme for an ungrounded system?
   a. Grounding electrode conductor for grounding equipment enclosures and conductive parts
   b. Equipment grounding conductor
   c. Bonding conductors or jumper
   d. System grounded conductor
6. When a conductive object or system is connected to ground, it is not always connected to the Earth.
   a. True   b. False
7. Which of the following are characteristics that are necessary for an effective ground-fault current path?
   a. It is intentionally constructed.
   b. It must have low impedance.
   c. It must be an electrically conductive path.
   d. All of the above.
8. Which of the following terms is best defined as not connected to ground or to a conductive body that extends the ground connection?
   a. Grounded
   b. Ungrounded
   c. Guarded (isolated)
   d. Impedance grounded

9. Electrical systems are grounded (connected to the Earth) in a manner that will accomplish which of the following?
   a. Limiting the voltage imposed by lightning
   b. Limiting voltages imposed by line surges or unintentional contact with higher-voltage lines
   c. Stabilizing the voltage to Earth during normal operation
   d. All of the above
10. The Earth is in the electrical grounding circuit.
    a. True
    b. False
11. Which of the following *NEC* terms means connected to ground without inserting any resistor or impedance device?
    a. Solidly grounded
    b. Resistance grounded
    c. Impedance grounded
    d. Grounded
12. _____ refers to normally non–current-carrying conductive materials enclosing electrical conductors or equipment, or forming part of such equipment, that shall be connected to the Earth so as to limit the voltage to ground on these materials.
    a. Grounding electrical equipment
    b. Bonding electrical equipment
    c. Electrical system grounding
    d. None of the above
13. A ground-fault current path for a grounded system or an ungrounded system could include which of the following?
    a. Metal piping systems
    b. Equipment grounding conductors
    c. The Earth itself
    d. All of the above
14. Effective ground-fault current paths perform all but which of the following functions?
    a. They provide a low-impedance circuit.
    b. They facilitate the operation of fuses, circuit breakers, or ground detection system for high-impedance grounded systems.
    c. They require the Earth to perform its intended functions.
    d. They are capable of safely carrying the maximum ground-fault current likely to be imposed on them from any point on the wiring system where a ground fault may occur to the electrical supply source.
15. The Earth must never be depended on to function as an effective ground-fault current path.
    a. True
    b. False
16. A grounded electrical system always includes a neutral conductor that is intentionally connected to ground.
    a. True
    b. False
17. Grounding places a conductive object at or as close to the Earth's potential as possible.
    a. True
    b. False
18. Bonding electrically conductive equipment together establishes electrical continuity and conductivity to the electrical supply source in a manner that establishes an effective ground-fault current path.
    a. True
    b. False
19. The process of grounding and bonding reduces potential differences between conductive parts and the Earth, thereby reducing shock hazards.
    a. True
    b. False
20. Ground faults occur when electrical current in a circuit returns through which of the following paths?
    a. Through the equipment grounding conductor
    b. Through conductive material other than the electrical system ground (metal, water pipes, and so forth).
    c. Through a person to ground or through a combination of ground return paths
    d. All of the above
21. For a person to receive an electric shock, which of the following conditions must exist?
    a. There must be a contact point on the body for entry of the circuit current.
    b. The current must exit through a second contact point on the body.
    c. There must be a voltage to force the current across the human body.
    d. All of the above.
22. The severity of an electrical shock a human will receive is related to all of the following except _____.
    a. Length of time the current passes through the body
    b. Amount of current that passes through the body
    c. The path the current takes through the body
    d. The type of overcurrent device installed in the system

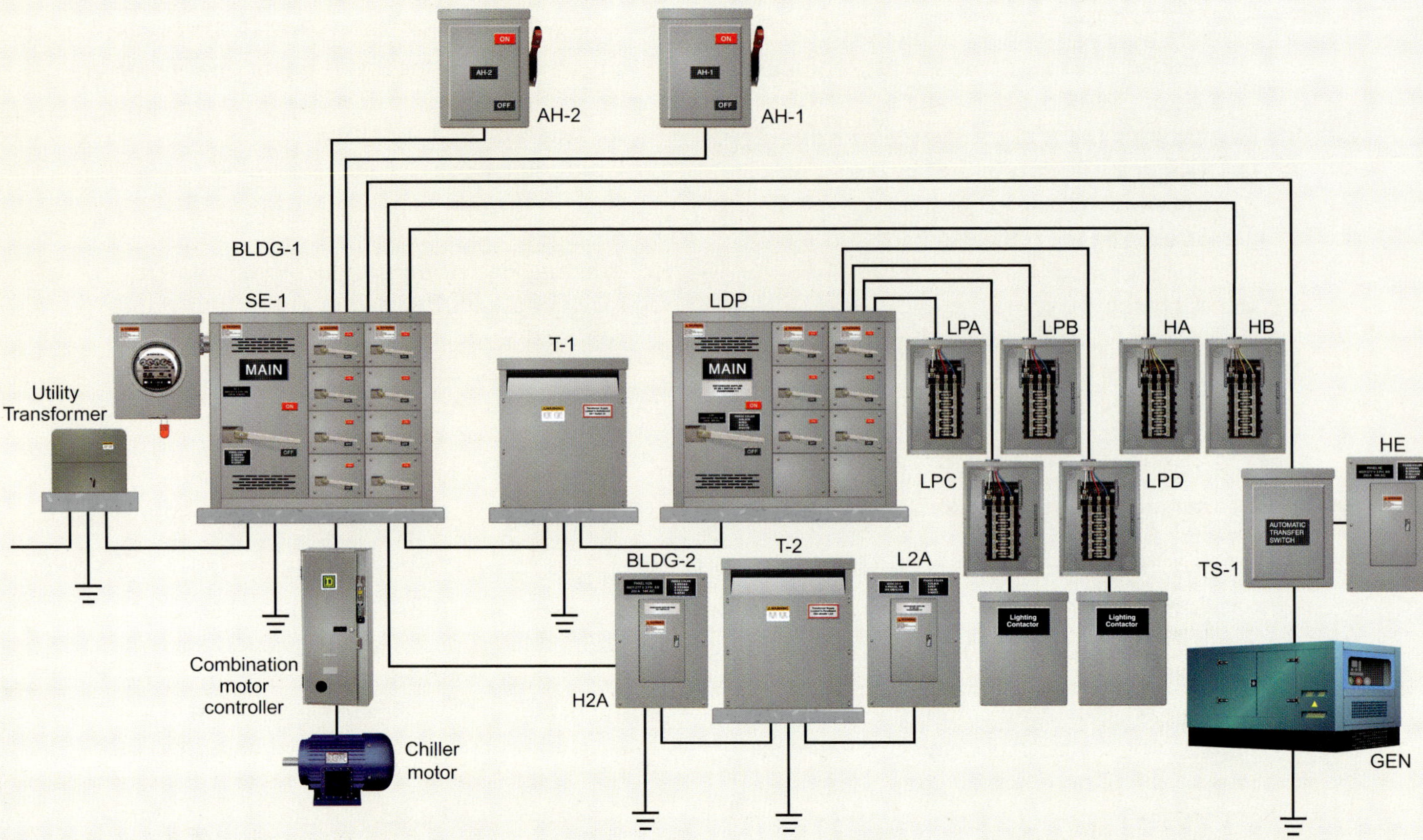

CHAPTER 2

Circuit Basics and Overcurrent Protection

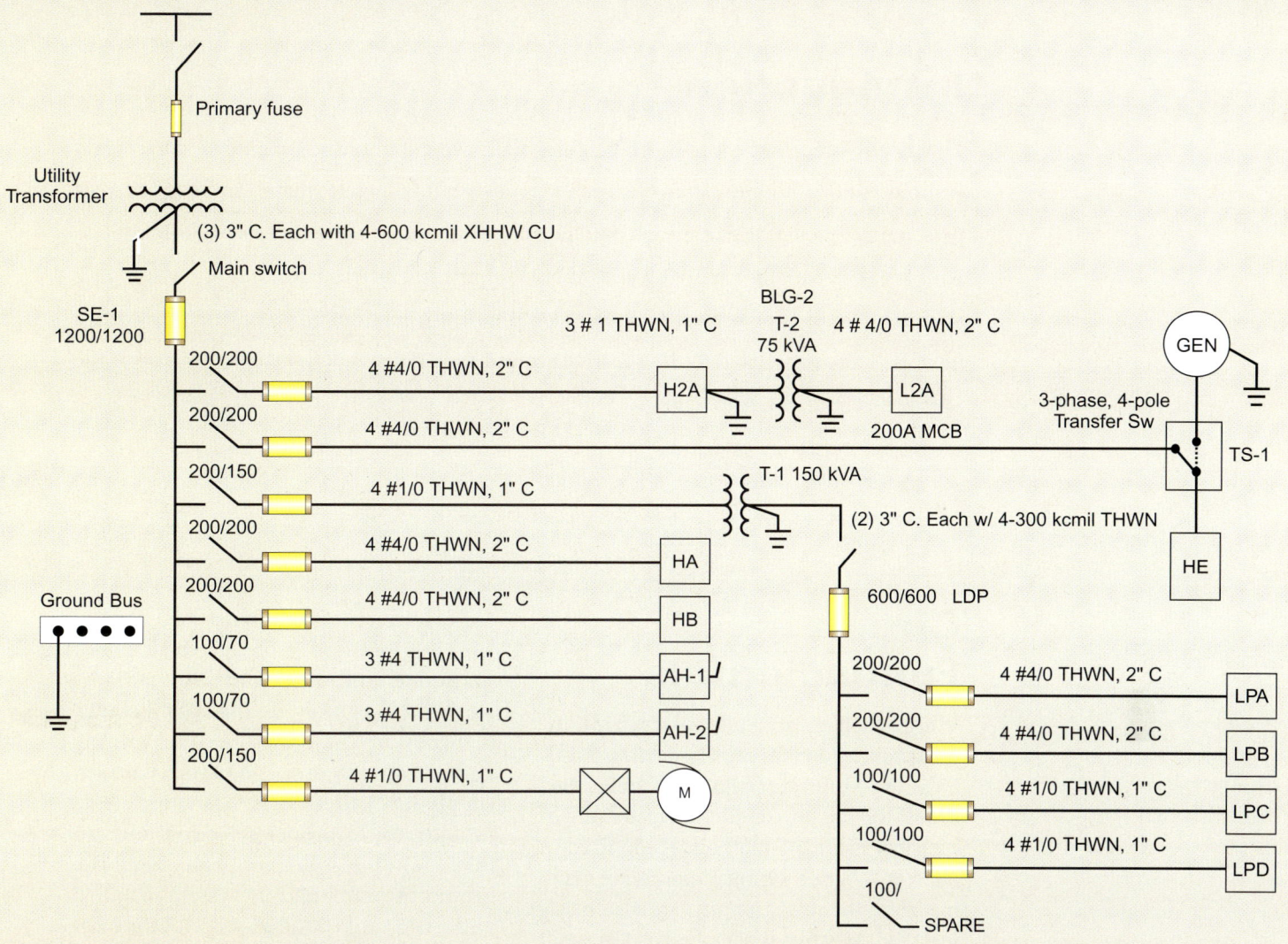

Objectives

- Understand the relationship between normal electrical circuits and grounding and bonding circuits in an electrical system
- Understand and apply circuit fundamentals to grounding and bonding and the working relationship to effective overcurrent device operation
- Understand voltage, current, resistance, and impedance in electrical circuits and how to apply Ohm's law to electrical circuits
- Know the importance of adequate equipment short circuit current ratings and proper selection of fuses and circuit breakers with adequate interrupting ratings
- Understand the importance of conductor insulation integrity, equipment grounding conductor capacity, and not exceeding the maximum withstand ratings of electrical conductors

Outline

Circuit Fundamentals

Ohm's Law

Opposition to Current in Circuits

Current in Circuits (Normal and Fault Current)

Overcurrent Protection Basics

Amperes Operate Overcurrent Protective Devices

Time and Current

Safety by System Design

Equipment Grounding Conductor Capacity

Introduction

Electrical circuits and systems operate normally when circuits are complete, intact, and without fault conditions. If an electrical circuit is not complete, there can be no current present in that circuit. This means both normal circuits and safety circuits of electrical systems are necessary for supplying utilization equipment. Electrical grounding and bonding should be thought of as electrical safety circuits. The operation of overcurrent protective devices is directly related to the effective ground-fault current path installed as part of the overall electrical safety system.

Circuit Fundamentals

When installers are building electrical infrastructures on the premises, within buildings, and within supplying structures, they install the circuit conductors using one of a variety of wiring methods recognized in the *NEC®*. While wiring methods are being installed, the safety circuits are also being installed. Workers must take great care not only when wiring is performed but also when it comes to the grounding and bonding circuits associated with the system. **See Figure 2-1.**

Electrical workers must have a good working knowledge of electrical theory to understand how and why circuits work the way they do. This knowledge is necessary to assist in determining circuit capacity and conductor sizes for supplying feeders and branch circuits in addition to proper selection and application of overcurrent protection. The grounding and bonding system installed with the ungrounded conductors for services, feeders, and branch circuits is also a network of electrical circuits that must be complete and effective to operate overcurrent protection in the event of a ground fault on the system. There is an important working relationship between the grounding and bonding circuits and the overcurrent protection provided in electrical systems. Electrical circuits in their basic form include three components: voltage, resistance, and current. Ohm's law and basic electrical theory are electrical principles that apply to the grounding and bonding circuits as well as to the current-carrying conductors of the circuits.

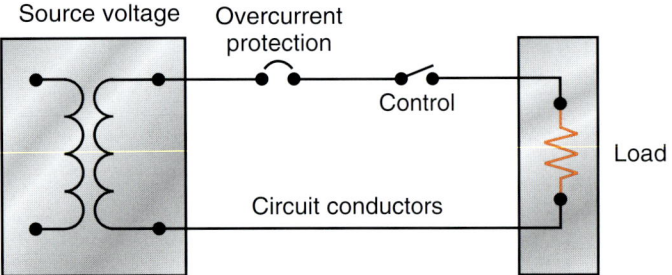

Current can only be present if voltage is applied and the electrical circuit is complete. This circuit is not complete, so there is no current.

FIGURE 2-1 A basic electrical circuit includes a source, circuit conductors, overcurrent protection, usually some method of control, and a load.

Ohm's Law

George Simon Ohm discovered the relationship between voltage, current, and resistance and developed Ohm's law.

The relationship between voltage, resistance, and current expressed in Ohm's law can be seen in a pie chart. **See Figure 2.2.**

Voltage (Electromotive Force)

A volt is a unit of electrical pressure or force. In the electrical field, voltage is the electromotive force that pushes current through resistance. **See Figure 2-3.** The voltage pushes the electrons of a circuit through the wires or other conductive components of the circuit.

Voltage can be compared to water pressure. As the pressure in the water pipe increases, the flow increases. Likewise, the stronger the voltage, the more easily current is forced through the conductors of a circuit. However, although voltage is the force that causes current in a circuit, voltage does not *flow* through a circuit; it is present in circuits. Because voltage in electrical circuits is the *electromotive* force in the circuit, it is expressed using the letter E in Ohm's law. Note that voltage is measured in volts, abbreviated V.

Ohms (Resistance)

An ohm is a unit of electrical resistance in the circuit. More specifically, it is the amount of resistance that allows 1 ampere of current when 1 volt is applied in the circuit. **See Figure 2-4.** The symbol used to represent ohms is omega (Ω); note that this is the unit of measure. In Ohm's law, the letter R is used to denote resistance, which is measured in ohms.

Resistance in an electrical circuit can also be compared to fluid piping systems. The larger the diameter of pipe is, the less resistance there is to fluid flow. In electrical circuits, the larger the conductor (wire) is, the less resistance there will be in the circuit. On the other hand, smaller conductors in a circuit offer greater resistance in that circuit. Note that this basic concept does not account for any load resistance. This concept is important to remember for professionals in the electrical field. It is also important for other electrical industry professionals such as designers, engineers, and inspectors, among others, to apply these basic principles in the field.

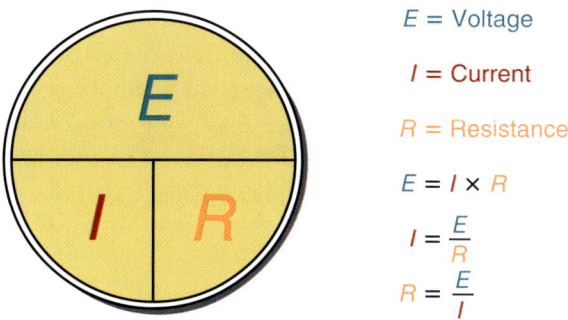

E = Voltage

I = Current

R = Resistance

$E = I \times R$

$I = \dfrac{E}{R}$

$R = \dfrac{E}{I}$

FIGURE 2-2 Basic Ohm's law formula is expressed in a simple pie chart configuration.

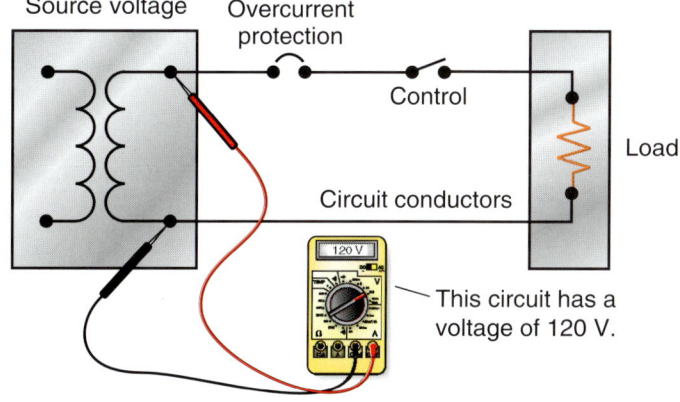

FIGURE 2-3 The voltage of an electrical circuit is the amount of pressure of electromotive (E) force in the circuit. This pressure is measured in volts.

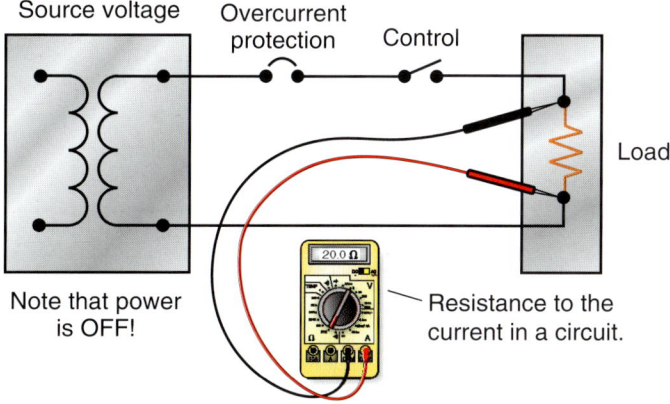

FIGURE 2-4 An ohm is a unit of electrical resistance in the circuit. Resistance to current in a circuit rises as the ohms of the circuit increase.

Amperes (Current)

The ampere, or amp, is the unit of measure used for the current in an electrical circuit. **See Figure 2-5.** The unit of amperes is represented by the symbol A. An amp is defined as the flow of 1 coulomb per second. Because the amount of amperes in a circuit is referred to as the *intensity of current* in the circuit, the symbol I is used in electrical formulas using Ohm's law.

Ohm discovered this law of proportionality and concluded that it takes 1 volt to force 1 ampere through a resistance of 1 ohm. Another way of looking at this mathematical relationship is that the current in a direct current (DC) circuit is directly proportional to the voltage applied in a circuit, and current in a circuit is inversely proportional to the resistance in that circuit. In any mathematical equation, symbols are used to represent the variables. For voltage (electromotive force), the symbol E is used. For current (intensity), the symbol I is used. For resistance (opposition), the symbol R is used. Based on the relationship reviewed earlier, the following are three simple forms of Ohm's law:

$$E = I \times R$$
(Voltage = Current × Resistance)

$$I = \frac{E}{R} \quad \left(\text{Current} = \frac{\text{Voltage}}{\text{Resistance}}\right)$$

$$R = \frac{E}{I} \quad \left(\text{Resistance} = \frac{\text{Voltage}}{\text{Current}}\right)$$

These simple forms of Ohm's law apply to both alternating current (AC) and DC circuits consisting of only resistive loads. They form the practical basis for simple electrical calculations. These three components of electrical circuits provide the basis for electrical calculations when any two of these values are known. As an example, if the voltage of a circuit is 480 volts, and the current of the circuit is 40 amperes, what is the resistance of the circuit? Because the problem is solving for resistance, the following formula is used:

$$R = \frac{E}{I}$$
$$= \frac{480 \text{ V}}{40 \text{ A}}$$
$$= 12 \text{ }\Omega$$

Another example supposes that an electrical circuit had a voltage of 120 volts and a resistance of 40 ohms. How many amperes would be present in this circuit? Because the problem is solving for current (amperes), The following formula is used:

$$I = \frac{E}{R}$$
$$= \frac{120 \text{ V}}{40 \text{ }\Omega}$$
$$= 3 \text{ A}$$

The last example of applying basic Ohm's law solves for voltage of a circuit when current and resistance are known. If the resistance of a circuit is 60 ohms and the current of the circuit is 4 amperes, then solving for voltage, Ohm's law is applied as follows:

$$E = I \times R$$
$$= 4 \text{ A} \times 60 \text{ }\Omega$$
$$= 240 \text{ V}$$

These formulas are valuable in determining current in electrical circuits. It is helpful when selecting sizes of circuit conductors and applying overcurrent protective devices (fuses and circuit breakers) in circuits. Applying Ohm's law results in solving for one component of the electrical circuit when two other components are known. **See Figure 2-6.**

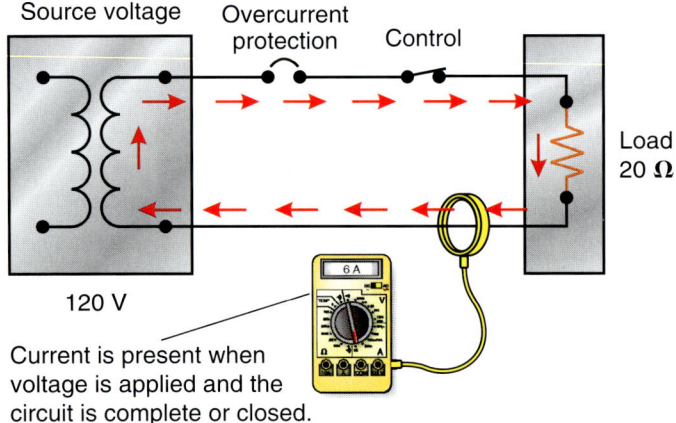

This circuit has a current of 6 A.

Current is present when voltage is applied and the circuit is complete or closed.

FIGURE 2-5 An ampere is the measure used for current in an electrical circuit.

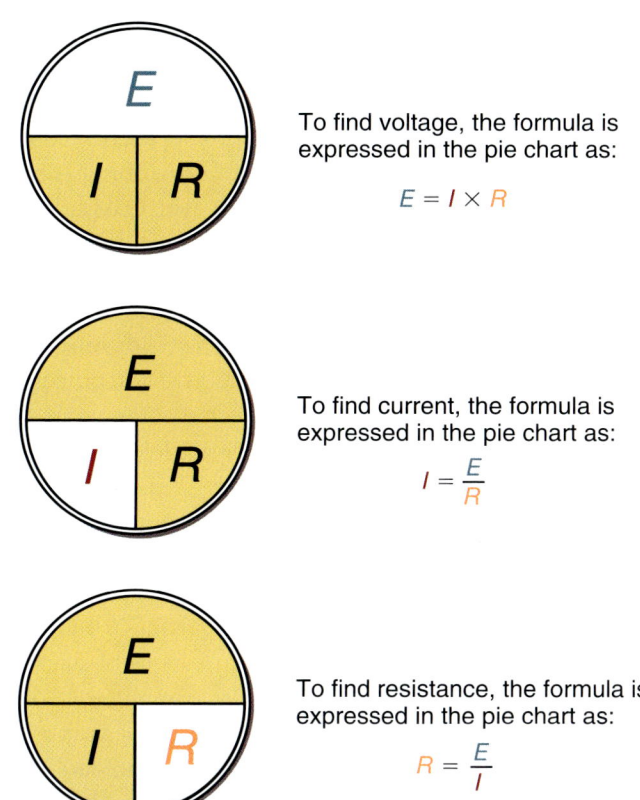

To find voltage, the formula is expressed in the pie chart as:

$$E = I \times R$$

To find current, the formula is expressed in the pie chart as:

$$I = \frac{E}{R}$$

To find resistance, the formula is expressed in the pie chart as:

$$R = \frac{E}{I}$$

FIGURE 2-6 Finding values of voltage, current, and resistance in an electrical circuit is accomplished by inserting two of the known values.

Another valuable set of formulas is provided in Watt's Wheel. In this collection of formulas, the watt (power of a circuit) is included in addition to volts, amperes, and resistance. Understanding how to apply Watt's Wheel can help solve for amperes, voltage, resistance, and wattage of a circuit. The symbol used for wattage in electrical formulas is P. Just as with Ohm's law, if any two values of a circuit are known, the third can be calculated by using the appropriate formula from Watt's Wheel. **See Figure 2-7.**

Referring to the wheel, the three basic formulas for determining power (watts) in a circuit are as follows:

$W = I \times E$
(Power = Current × Voltage)

$W = I^2 R$
(Power = Current × Current × Resistance)

$W = \dfrac{E^2}{R}$

$\left(\text{Power} = \text{Voltage} \times \dfrac{\text{Voltage}}{\text{Resistance}}\right)$

A good example of using watts in an equation is when the wattage and voltage of a circuit are known. An electric heater circuit is a good example. Normally the heater nameplate includes watts, current, and voltage rating. Applying the math, the current can be determined if voltage and wattage are known. The heater is rated at 5000 watts, and the circuit voltage applied is 277 volts.

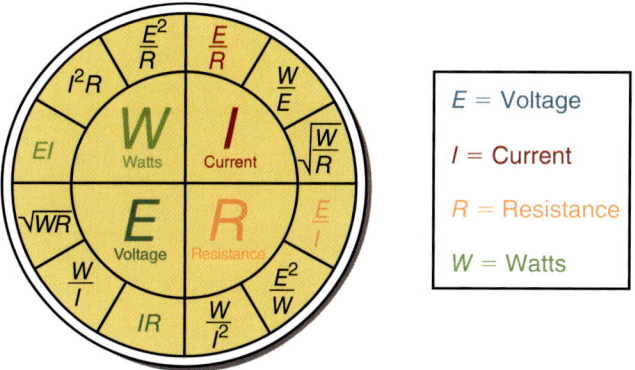

FIGURE 2-7 Watt's Wheel is a formula chart for finding values of current, voltage, resistance, and power (watts) in electrical circuits.

An equation solving for amperes using Watt's Wheel looks like:

$$I = \frac{W}{E}$$

$$= \frac{5000 \text{ W}}{277 \text{ V}}$$

$$= 18.05 \text{ A}$$

Because it is possible to solve for current in a circuit when watts and volts are known, correct circuit conductors and overcurrent protection can be designed and installed in electrical wiring systems.

Opposition to Current in Circuits

Up to this point, the opposition to current in a DC circuit (resistance) has been expressed. Although this is true for DC circuits, in AC circuits, the total opposition to current is known as the *impedance* of a circuit. Impedance of an AC circuit includes the resistance, inductance, and capacitance. The letter Z is used to represent impedance in electrical formulas:

$$Z = \sqrt{R^2 + (X_L - X_C)^2}$$

Where
Z = impedance
R = resistance
X_L = inductive reactance
X_C = capacitive reactance

The opposition to current in an AC circuit includes resistance, capacitive reactance, and inductive reactance.

The AC circuit has different operating characteristics than a DC circuit primarily because the voltage and current are changing in amplitude and direction from the 0 point in the waveform. For example, in a 60-Hz (cycle) circuit, the voltage and current change amplitude and direction 120 times per second. In a 400-Hz (cycle) circuit, the voltage and current change amplitude and direction 800 times per second. As current moves through a circuit conductor, a magnetic field is produced in a concentric and circular form around the conductor. This field is present and constantly changing direction through each cycle of the waveform. This changing of amplitude and direction causes the magnetic field to change with the change in current direction. This component of an AC circuit is the magnetic field component that results from inductive reactance.

The other component present in nearly all electrical circuits is capacitance. A capacitor, in its simplest form, is two conductors separated by insulation. From that basic description, it is easy to understand that when electrical circuits are installed, there is capacitance built into the circuits. So the impedance of current in an AC circuit is the combination of resistance, inductance, and capacitance. The subjects of inductive reactance and capacitive reactance are not covered in this text, but there are excellent training resources available that provide detailed information about these subjects. Of most importance is that the total opposition to current in an AC circuit includes a resistive component or components, as well as inductive and capacitive components.

The *NEC* uses the term *impedance* in relation to grounding and bonding in various ways. A common mention of impedance in the *Code* is when a requirement refers to a low-impedance path for ground-fault current. The idea expressed in this phrase is keeping the opposition to current as low as possible so that the maximum amount of fault current operates the overcurrent device in fault conditions. The amount of impedance will vary from circuit to circuit based on the different physical characteristics of each circuit, such as wire size, terminations, and circuit length. All of these circuit characteristics contribute to the overall impedance in an AC circuit. Maintaining low impedance in electrical circuits is important. The *NEC* includes two specific rules that help keep impedance low during normal operation through a process that results in a canceling effect.

Sections 250.134(B) and 300.3(B) both require all conductors of a circuit to be installed together in the same raceway, cable, or trench. **See Figure 2-8.** This requirement also includes the equipment grounding conductors (EGCs) because in ground-fault or short circuit conditions, the impedance of the circuit has to be as low as possible to result in fast operation of fuses or circuit breakers protecting the circuit.

This theory was proven through test modeling and experiments performed by R. K. Kaufmann, an engineer at General Electric. Kaufmann discovered that when AC circuit conductors are not installed close together, inductive reactance increases the circuit impedance, thus having a negative effect on overcurrent device operation. By keeping conductors close together, the effects of inductive reactance and circuit impedance are minimized. Also, a canceling effect happens in circuits because the supply and return side of the circuit are close together, the magnetic fields moving in opposite direction cancel each other. If the conductors were to be separated, inductive reactance would increase, which would increase the impedance in the circuit. This is the reason the *Code* requires conductors of an AC circuit to be routed together. This subject is explained in more detail in Chapter 8, which covers requirements and installation methods for EGCs.

Current in Circuits (Normal and Fault Current)

For current to be present in a circuit, the circuit must be complete. If a circuit is not complete, there can be no current. **See Figure 2-9.** This is true not only for the normal circuit wiring but also for the grounding and bonding circuits. Without complete grounding and bonding paths in the circuit wiring, the protective system is compromised. Current will always return to its source. The source could be a generator, transformer, photovoltaic array, battery, and so forth.

When a circuit is interrupted, the current returning to the source is also interrupted. Another important point here is that all current follows this principle. In other words, normal current will always return to its source and fault current will also return to its source. This concept is important to understand because it helps clarify the principles of operation for overcurrent protective devices such as circuit breakers and fuses.

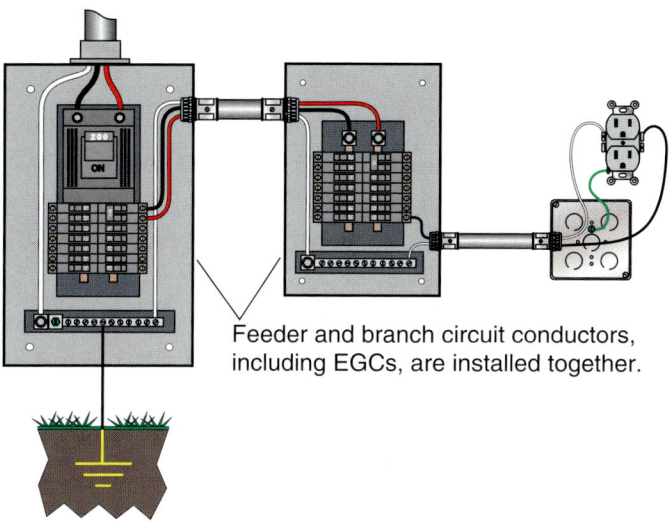

FIGURE 2-8 All conductors of AC circuits are generally required to be installed together in the same raceway, cable, or trench. This includes the ungrounded conductors, the grounded conductor, and EGC of the circuit. [*NEC* 300.3(B)].

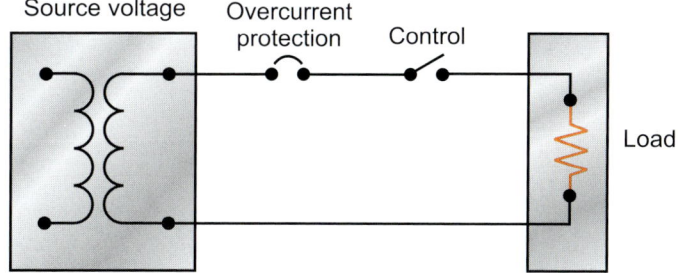

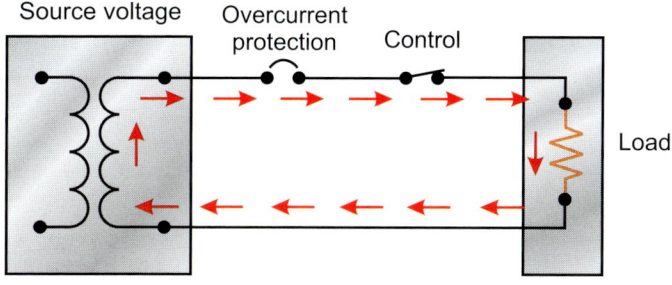

FIGURE 2-9 Current is present in an electrical circuit only when a source voltage is applied and the circuit is complete. This concept applies to the grounding and bonding circuits in addition to the normal current-carrying circuits.

The path that current will take in a circuit is related to how many paths are available that are electrically connected to the source. Current will take any and all paths available to return to its source. As for the "path of least resistance" concept, current will divide over all parallel paths in the circuit. The amount of current in each separate conductive path is related to the impedance (opposition to current) in each of the paths involved. **See Figure 2-10.**

The paths with lower impedance will carry more current than the paths with greater impedance. Lower impedance in a circuit results in higher current in the circuit. The higher the impedance in a circuit is, the more opposition to current there will be and thus the lower the amount of current present in the circuit. This has a direct effect on how overcurrent protective devices operate. This is the principle upon which the requirement to install EGCs is based. Providing a reliable low impedance path for ground fault current by installing an EGC is essential for safety electrical systems. Remember, the Earth offers significant opposition to current and is never to be considered an effective path, even though some current may attempt to return to the source through the Earth.

Overcurrent Protection Basics

A key factor in protecting electrical equipment and persons from electrical hazards is the proper application of overcurrent protective devices, fuses, and circuit breakers. Section 110.10 of the *NEC* requires the total circuit impedance, overcurrent protective devices, the component short circuit current ratings, and other circuit characteristics to be selected and coordinated so that the circuit protective devices can effectively respond and operate in a fault event. Furthermore, they must do so without causing extensive damage to the electrical equipment or conductors. Overcurrent protective devices such as fuses and circuit breakers must be selected to ensure that the short circuit current rating of any system component is not exceeded should a short circuit or high-level ground fault occur. **See Figure 2-11.**

Electrical system components include the wire, bus structures, switching, contactors, starters, overcurrent protective and disconnection devices, industrial control panels, and distribution equipment, all of which may have limited short circuit ratings. The short circuit current rating is the amount of short circuit current a component is rated to withstand without producing an explosion, fire, or shock hazard. If a short circuit or ground-fault event exceeds a component's short circuit current ratings, the component could be severely damaged or destroyed. Just providing overcurrent protective devices with sufficient interrupting ratings in accordance with *NEC* Section 110.9 does not necessarily ensure adequate short circuit protection for all components in the system.

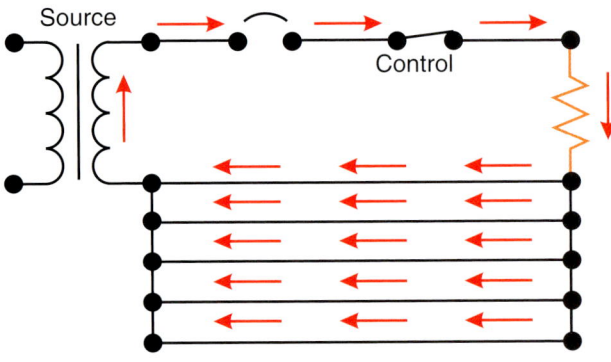

FIGURE 2-10 Current will divide over all paths (including the Earth) while returning to its source. The amount of current in each path depends on the amount of impedance in that path.

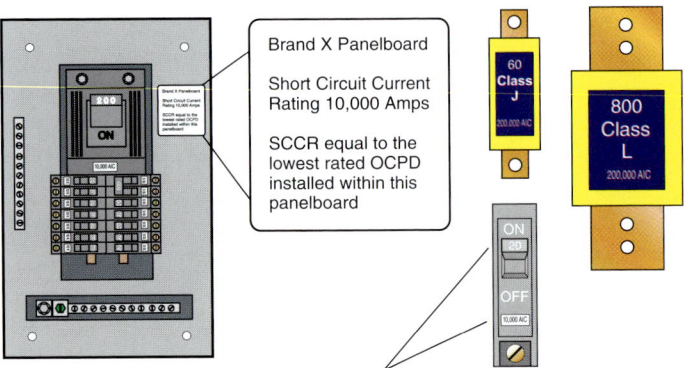

Fuses and circuit breakers are identified with normal current ratings and interrupting current ratings.

FIGURE 2-11 Equipment and overcurrent protective devices must have adequate normal current ratings, short circuit current ratings, and interrupting current ratings.

The interrupting rating ensures only that the overcurrent device can safely open the circuit if applied within its rating. In cases in which the available short circuit current exceeds the short circuit current rating of an electrical component, the overcurrent protective device selected may be able to be used to limit the let-through energy to within the rating of that electrical component. The available short circuit current (often referred to as *available fault current*) of a system is the maximum current that a system is capable of delivering at any point on a wiring system. The level of available fault current in a system is typically highest at the source. The amount of available fault current can usually be obtained by the serving utility company, or it can be calculated conservatively by using an infinite bus value. This value must be known to properly calculate the amount of available fault current at any point on the electrical system, from the source, to the furthest outlet. Knowing that all AC circuits introduce impedance in the circuit, the amount of available fault current will be reduced by the impedance of the circuit as circuits get longer and smaller when installed. **See Annex D.**

The effective paths for ground-fault current are protective safety circuits that facilitate overcurrent protective device operation during abnormal events such as ground faults and short circuits; these safety circuits are essential to protecting equipment and conductors of the system. The following definitions are helpful to understand the path for short circuit and ground-fault current during these types of events in an electrical system.

Ground Fault. An unintentional, electrically conducting connection between an ungrounded conductor of an electrical circuit and the normally non–current-carrying conductors, metallic enclosures, metallic raceways, metallic equipment, or earth.[1]

Interrupting Rating. The highest current at rated voltage that a device is identified to interrupt under standard test conditions.[2]

The term *short circuit* is not defined in the *NEC* but rather is simply described as an unintentional conducting connection between two or more ungrounded conductors of a circuit or between any of the ungrounded conductors of the circuit and the grounded parts of equipment or the EGC. A ground fault is considered one type of short circuit. **See Figure 2-12.** An overload (low-level overcurrent condition) is not considered a ground fault or short circuit.

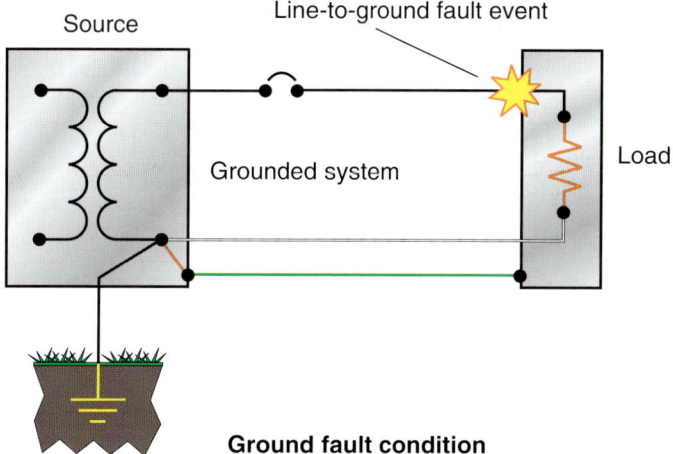

Ground fault condition

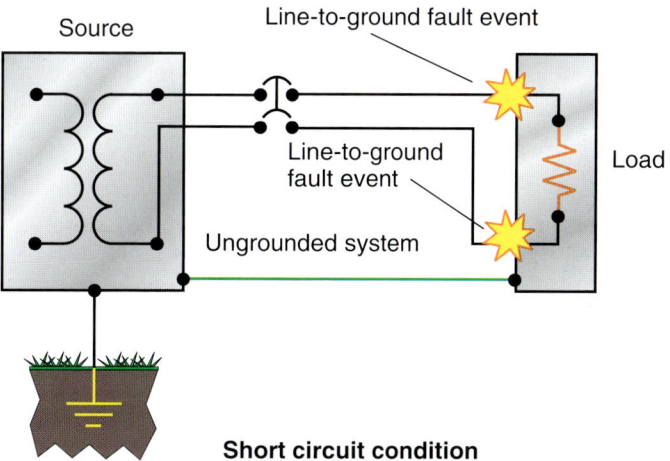

Short circuit condition

FIGURE 2-12 The comparison of a ground-fault condition and a short circuit condition shows both types of abnormal events that cause activation of overcurrent protective devices.

30 APPLIED GROUNDING AND BONDING

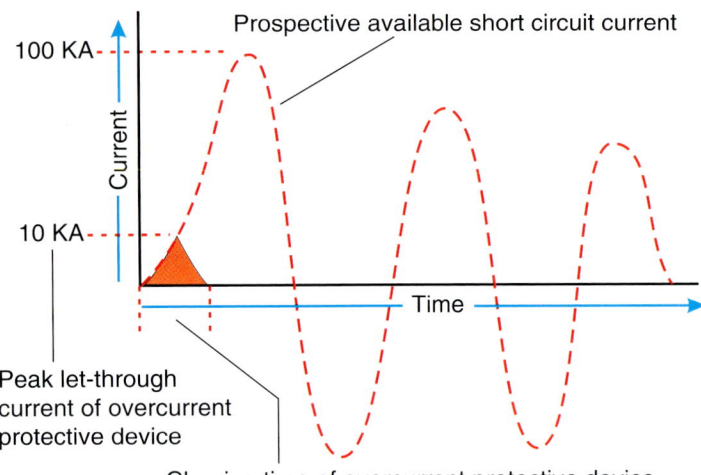

FIGURE 2-13 High short circuit current forces quick overcurrent device operation.

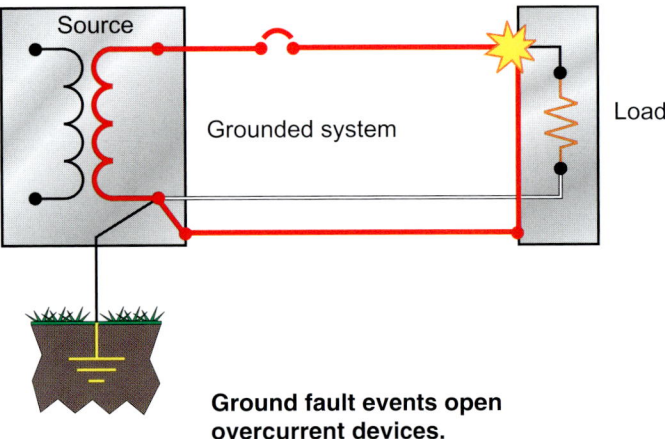

Ground fault events open overcurrent devices.

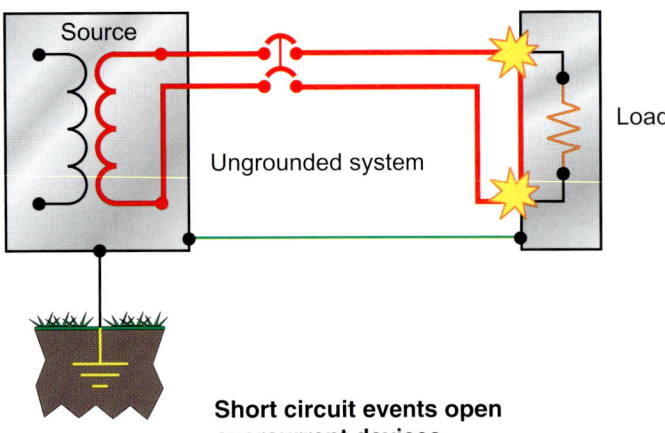

Short circuit events open overcurrent devices.

FIGURE 2-14 Overcurrent devices operate quickly to open short circuit and ground-fault current in circuits. Overload conditions cause much slower operation of overcurrent devices.

Amperes Operate Overcurrent Protective Devices

A simple way to look at how circuit breakers and fuses function is to consider how they respond to current. Normal current or amperes in a circuit will not cause a fuse or circuit breaker to open. Normal current is load current that is below the rating of the fuse or circuit breaker protecting the circuit. During normal operation, overcurrent protective devices, applied in open air, will carry current up to their rating indefinitely without causing the device to open. Because overcurrent protective devices are installed within enclosures, the *NEC* requires overcurrent devices to be sized at 125% of the continuous load, unless the assembly and overcurrent protective device are listed for 100% operation of its rating. This accomplishes the 80% limitation of continuous loads imposed by product standards.

Overcurrent protective devices also have a fusing or a tripping point. The amount of current that will operate an overcurrent protective device is inversely proportional to the amount of time. In other words, more amperes should, in theory, operate overcurrent protective devices quicker. The higher the level of current is, the quicker the overcurrent device will react to the event. **See Figure 2-13.** This operating characteristic of overcurrent protective devices is often referred to as inverse time operation. Many overcurrent devices provide current-limiting characteristics that provide enhanced equipment protection.

In a short circuit event, there is very little impedance or opposition to the maximum amount of current the electrical system is capable of supplying into the fault. A short circuit condition is where a fault happens between two or more ungrounded circuit conductors or any ungrounded conductor and the grounded conductor (usually a neutral).

These unintentional events typically cause an overcurrent device to operate quickly because of the rapid increase of available fault current through the device. **See Figure 2-14.**

In a ground-fault event, the fault is between any ungrounded circuit conductor and the Earth, EGC, enclosing metal raceway, or other grounded metal equipment. The greatest amount of damage often occurs in the first half cycle of time during a short circuit or ground fault. Excessive heat as a result of the high current can cause rapid deterioration of insulation or annealing of the conductor; it can even cause conductors to melt or vaporize in some cases. An effective ground-fault current path is integral to all electrical designs and installations to ensure fast operation of overcurrent protective devices during ground-fault or short circuit conditions. **See Figure 2-15.** This is imperative for safety and is a requirement of the *Code* as provided in Section 110.10.

The *Code* indicates that the overcurrent protective device has to function quickly enough to prevent extensive damage from occurring. Although *extensive damage* is not defined in the *Code,* it can be surmised that the equipment should be able to safely handle the fault and clear without resulting in a fire or shock hazard. Fuses and circuit breakers are directly related to overall equipment short circuit current ratings. The *Code* requires fuses and circuit breakers in all electrical systems be applied within their ratings. The short circuit current rating of equipment is the maximum amount of short circuit current that a component or assembly can safely withstand. Failure to ensure that the available fault current is within the short circuit current ratings of equipment and interrupting rating of overcurrent protective devices can result in severe damage or total component destruction in ground-fault or short circuit conditions.

Time and Current

Fuses and circuit breakers provide overcurrent protection for conductors and equipment. The general requirements for sizing overcurrent protective devices are provided in Section 240.4. The two concerns in electrical equipment and conductor protection designs are *how much current* the equipment and conductors can handle and for *how long.*

> Although *extensive damage* is not defined in the *Code,* it can be surmised that the equipment should be able to safely handle the fault and clear without resulting in a fire or shock hazard.

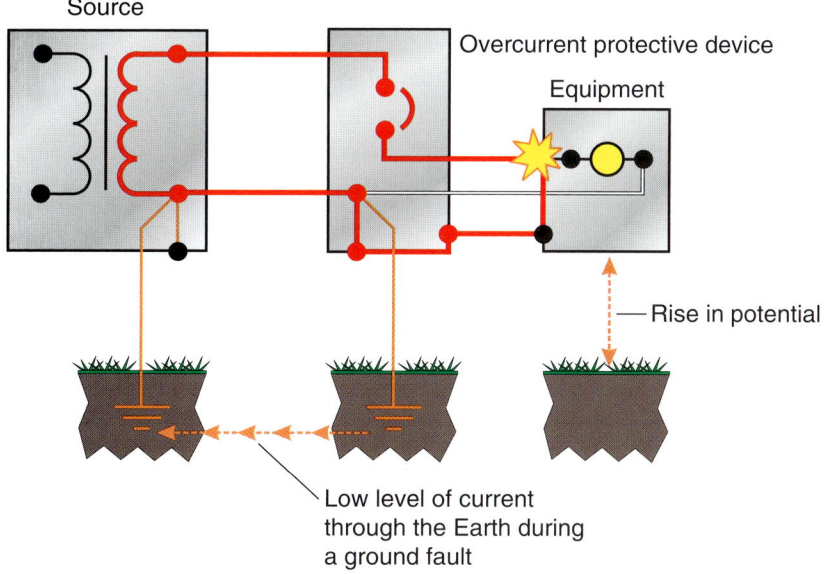

FIGURE 2-15 Effective ground-fault current path ensures fast overcurrent device operation.

Manufacturers of fuses and circuit breakers provide important data related to the operating time and current characteristics of these devices. Each overcurrent protective device has unique time and current graphs that are based on device type and size. This literature is usually in the form of a table or graph known as the *time current curves*. These graphs indicate how long it takes the device to open or clear the overcurrent condition under various levels of overload and short circuit current. **See Figure 2-16.**

The time current curves for fuses and circuit breakers are relatively simple to read. The time is shown on the graph's vertical axis, and the amount of current is provided along the horizontal axis. For circuit breakers, the vertical line in the graph represents the instantaneous trip point. The region to the left of this line is the overload region, and the region to the right is the short circuit (or instantaneous) region. The width of the time current curve indicates the manufacturing tolerance; that is, the circuit breaker clears within that band. The ranges shown in the graph indicate the manufacturer's test values or proven tolerances of the device. Of course, this range varies based on the circuit breaker type and rating. The time for a circuit breaker to function is on the right or left side of the graph, and the multiples of current levels are shown at the bottom of the chart. For lower-rated circuit breakers, the instantaneous trip range is reached sooner than for larger devices. For example, using the circuit breaker chart, look at the operating characteristics of a 20-ampere circuit breaker. **See Figure 2-17.** Now compare these operating characteristics to a 100-ampere circuit breaker. **See Figure 2-18.** The amount of time (on the left side of the chart) is greater for the larger overcurrent device to react to a short circuit or ground fault in a circuit.

The operating characteristics of fuses differ slightly from those of circuit breakers. Circuit breakers require a mechanical action to open, whereas fuses open because of melting or fusing of the element. A time current curve for fuses is a bit different in appearance in the short circuit region due to the current-limiting ability of the fuse, but the information conveyed is generally the same. In a fuse time current curve, the time is shown on the vertical axis and the amount of current is provided along the horizontal axis. Looking at a 20-ampere and a 50-ampere Class RK1 fuse, it is clear that a current of 200 amperes will open the 20-ampere fuse in 1.7 seconds and the 50-ampere fuse in 55 seconds. **See Figure 2-19.** The fuse time current curves show various Class RK1 fuse ampere ratings that help determine the clearing times based on the levels of current. **See Figure 2-20.** Understanding the information in overcurrent device time current curves is very helpful in determining whether conductors and equipment are properly protected.

Safety by System Design

Overcurrent protection is required to protect conductors and equipment from extensive damage. *NEC* Section 110.10 provides the general language that addresses this issue. Another method of achieving additional safety in electrical systems is to reduce let-through current so that incident energy (see side bar) is less during a short circuit or ground-fault event. Current-limiting overcurrent devices can provide such protection (see side bar).

Incident Energy. The amount of energy impressed on a surface, a certain distance from the source, generated during an electrical arc event. One of the units used to measure incident energy is calories per centimeter squared (cal/cm^2).[3]

Current-Limiting Overcurrent Protective Device. A device that, when interrupting currents in its current-limiting range, reduces the current flowing in the faulted circuit to a magnitude substantially less than that obtainable in the same circuit if the device were replaced with a solid conductor having comparable impedance.[4]

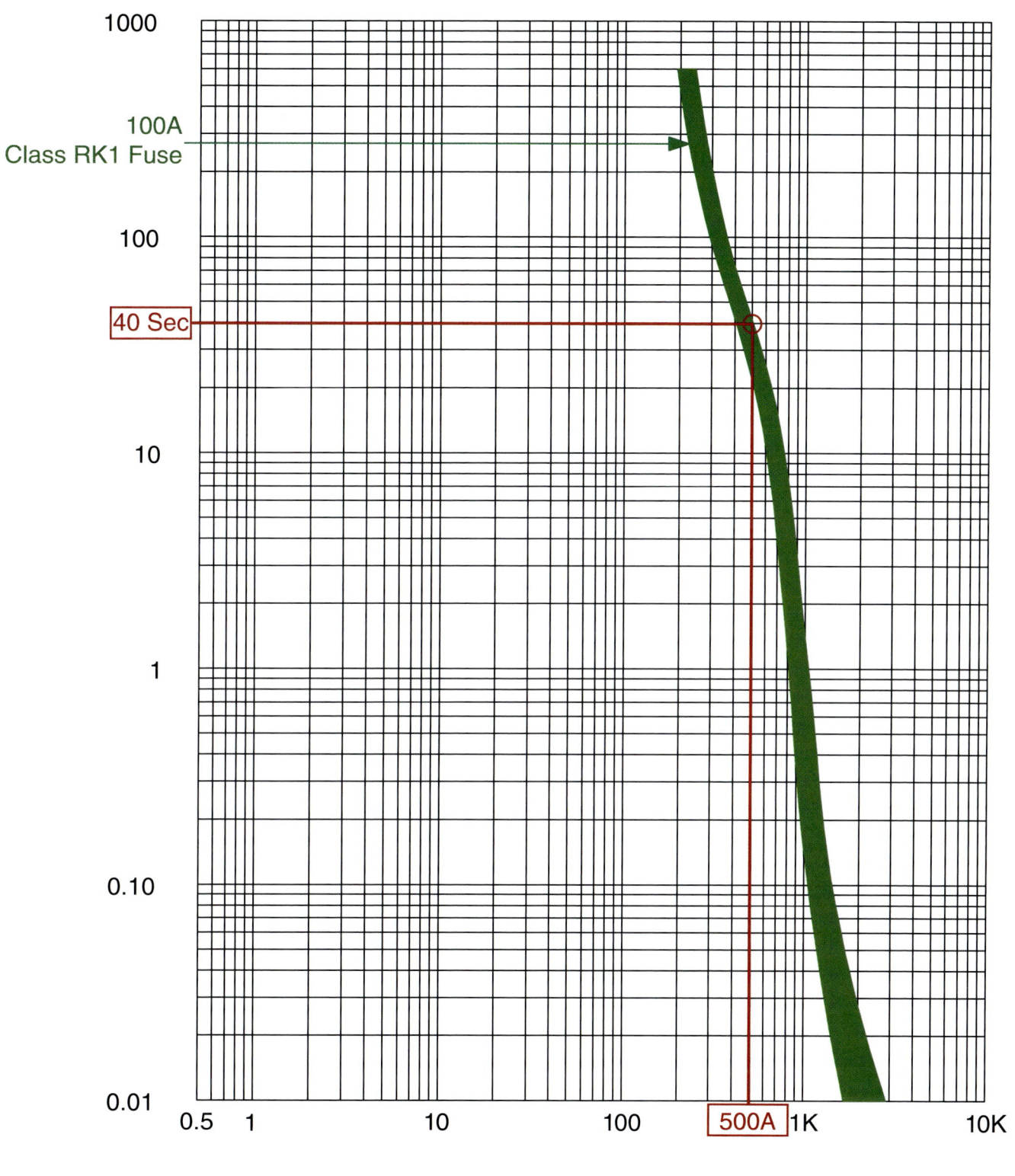

FIGURE 2-16 The time current curve for a 100-ampere RK1 fuse shows operating characteristics (in time and current) of this fuse in a graphic form. (Courtesy of Cooper Bussmann)

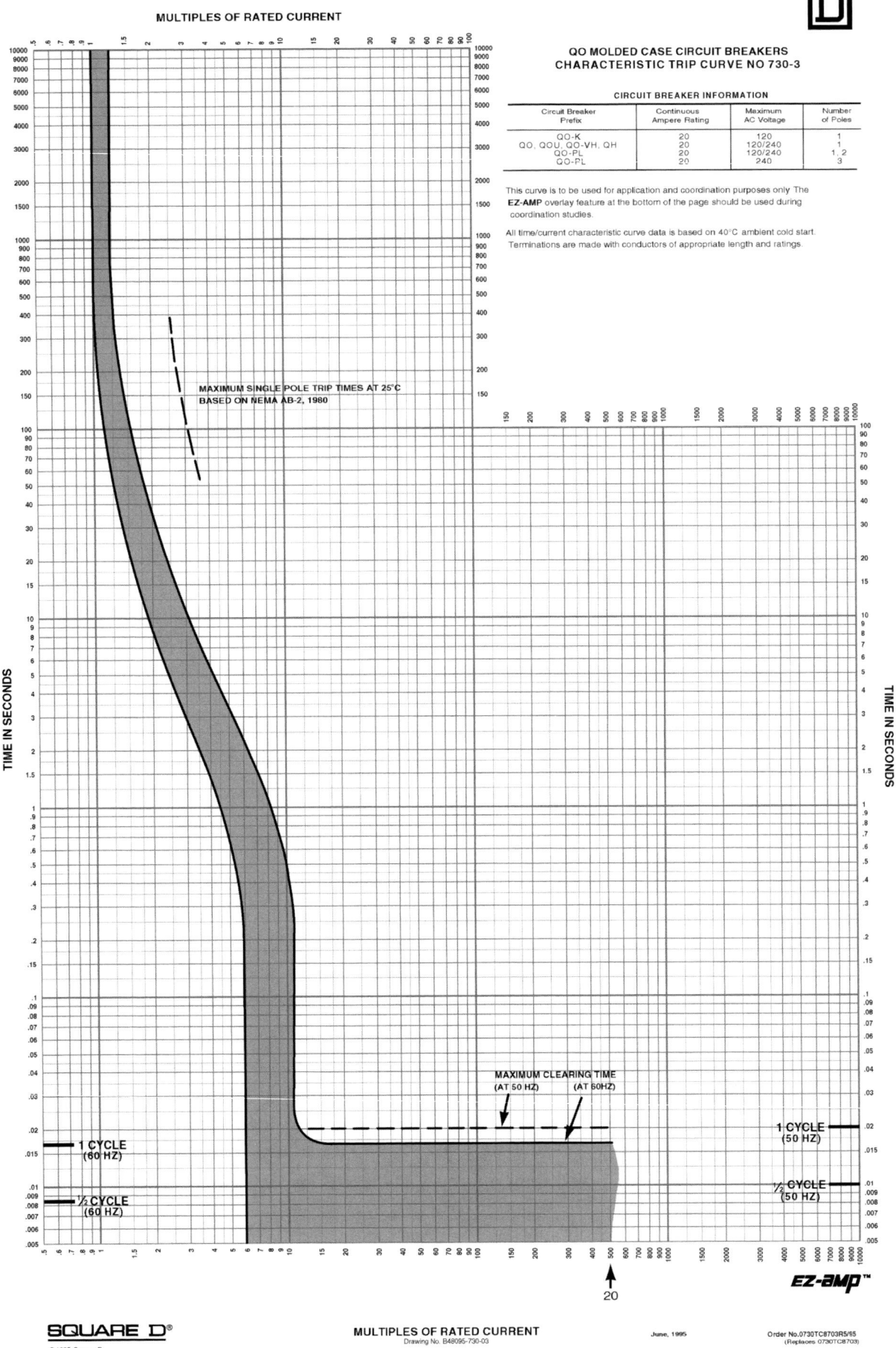

FIGURE 2-17 The time current curves for 20-ampere circuit breakers show operating characteristics (in time and current) of this circuit breaker in a graphic form. (Courtesy of Schnieder Electric Square D Company)

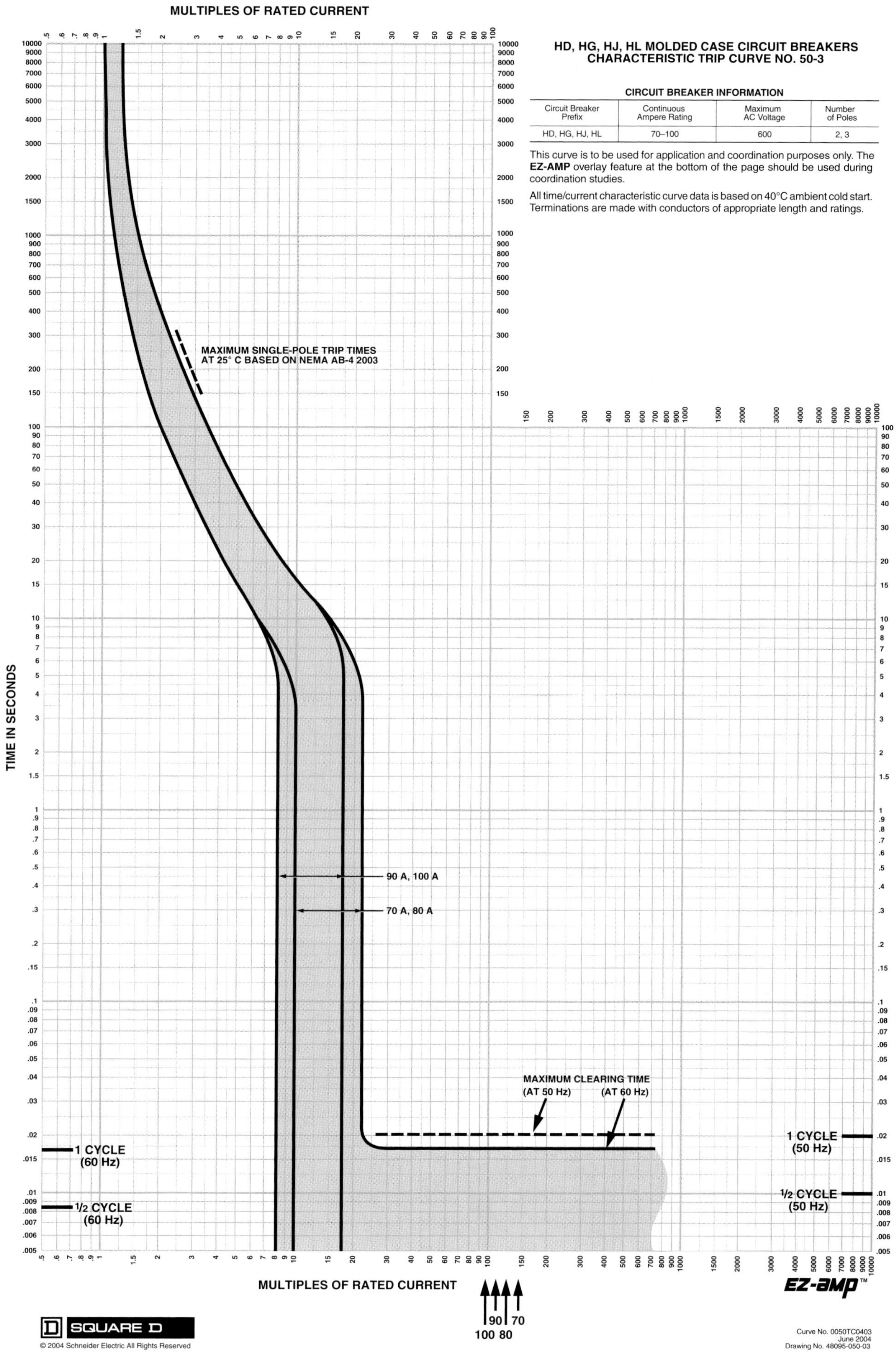

FIGURE 2-18 Time current curves for 100-ampere circuit breaker show operating characteristics (in time and current) in a graphic form. (Courtesy of Schnieder Electric Square D Company)

APPLIED GROUNDING AND BONDING

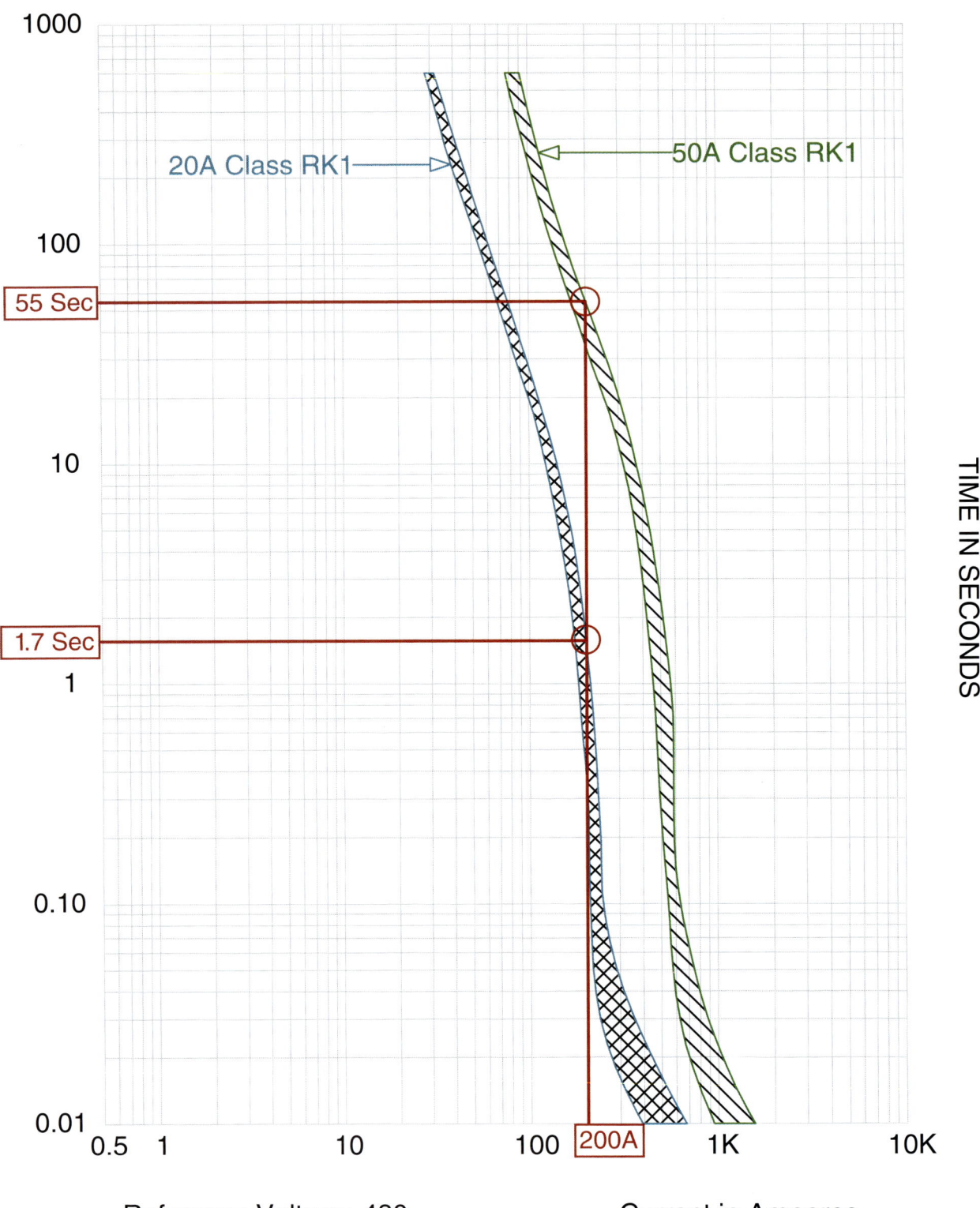

FIGURE 2-19 The fuse curve chart for a 20-ampere and a 50-ampere RK1 fuse shows operating characteristics (in time and current) of these fuse in a graphic form. (Courtesy of Cooper Bussmann)

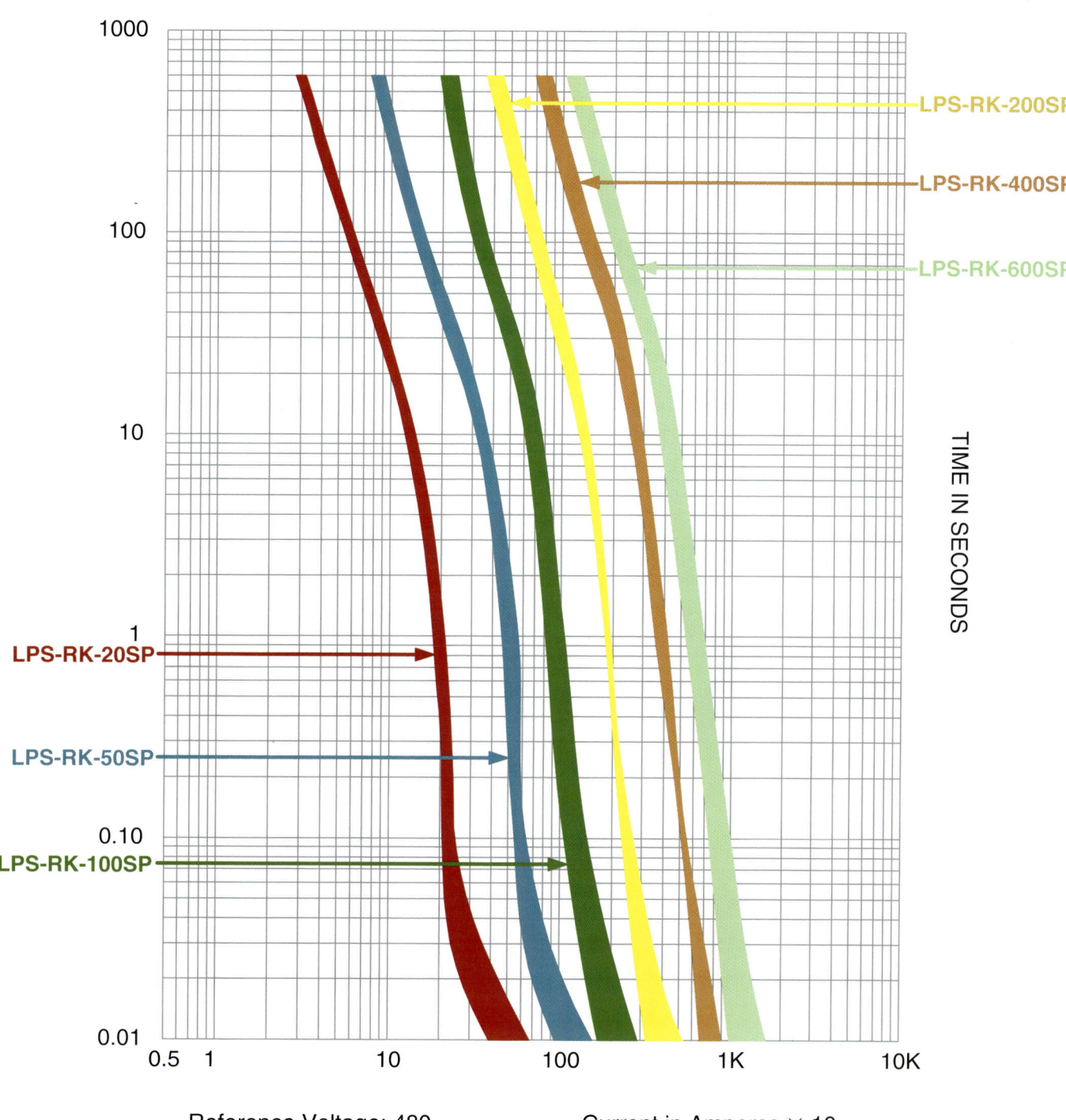

FIGURE 2-20 The fuse time current curve chart for various RK1 fuses shows operating characteristics (in time and current) of these fuse in a graphic form. (Courtesy of Cooper Bussmann)

Protecting Conductor Insulation

Power distribution systems continue to increase in capacity to handle demand. Where the transformer kilovolt ampere rating (kVA) is high, the amount of available fault current is often of very high magnitude. Ground faults and short circuits on systems with high levels of available fault current may cause serious damage to conductor insulation or to the conductor itself. The *Code* requires conductors and completed wiring installations to be free from short circuits, ground faults, or any ground connections other than as required or permitted by specific *NEC* provisions. Conductor insulation prevents the flow of electricity between points of different potential in an electrical system. Failure of the insulation system is one of the most common causes of problems in electrical installations, in both high-voltage and low-voltage systems.

Insulation tests on new or existing installations can determine the quality or condition of the insulation of conductors and equipment before it is energized. Good work practices include verifying circuit integrity before energizing any electrical circuits. This is usually accomplished using a suitable meter that verifies conductor insulation dielectric integrity. **See Figure 2-21.**

Common causes of insulation failures are heat, moisture, dirt, and physical damage occurring during and after conductor installation. Insulation can also fail due to chemical effects, exposure to sunlight, and excessive voltage stresses. Insulation integrity must be maintained during overcurrent conditions as well. Conductor damage can also include annealing and melting or vaporizing the metal. Conductors that are subjected to too much current for too long can become annealed, which softens the properties of the conductor metal. Annealing of conductors can result in loose connections, which can then result in thermal conditions that damage the components or in a poor conducting path. Melting or vaporizing conductors may result in a serious fire hazard or arc flash hazard.

> Insulation tests on new or existing installations can determine the quality or condition of the insulation of conductors and equipment before it is energized.

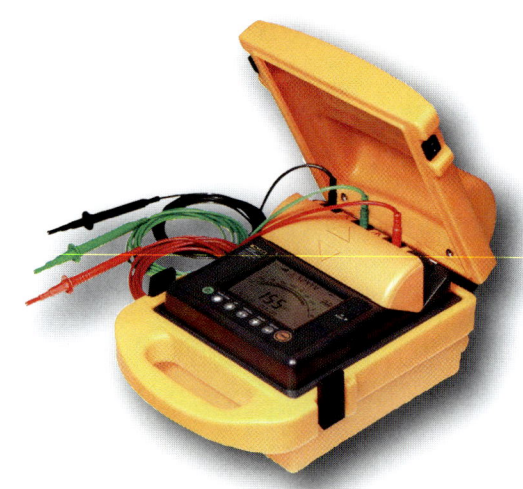

Courtesy of Fluke Corporation

Megohmmeter Model 1550B

Hand-Held Megohmmeter Model 1587

FIGURE 2-21 Megohmmeters are used to determine conductor insulation integrity prior to energizing electrical circuits.

In addition, an annealed or opened (due to melting or vaporizing) EGC can result in a shock hazard because there is a poor ground-fault return path or no effective path at all. Persons have been killed due to inadequate ground-fault return paths when a phase conductor energized normally non–current-carrying parts.

Properly sized conductors can carry current continuously and withstand overcurrents for some amount of time. The *NEC* provides ampacity tables in Article 310 for the permissible conductor current ratings (normal load current). These tables provide current values that conductors can carry indefinitely. (There are applicable conductor ampacity correction and adjustment factors that may be applicable for certain conditions.) In a ground-fault or short circuit event, the time a conductor can withstand higher levels of current is significantly reduced. The greater the overcurrent is, the less time a conductor can safely withstand the current. A good set of guidelines addressing conductor insulation abilities has been established through solid engineering research by the Insulated Cable Engineers Association (ICEA). The ICEA developed this information as a means to prevent damage to conductor insulation. The ICEA created a chart that shows that currents present for the times indicated in the graph produce maximum safe operating temperatures for each conductor size. The short circuit current, conductor cross-sectional area, and overcurrent protective devices in the circuit should be applied in a manner that does not exceed the maximum short circuit times in the ICEA chart. **See Figure 2-22.**

Short circuit protection is especially important for EGCs because reduced sizing is permitted by Table 250.122. Wire-type EGCs must be selected and applied in system designs in a manner that does not leave them vulnerable to high current levels for periods that exceed their withstand ratings. For insulated EGCs, a simple formula can be helpful to verify the amount of current a conductor can safely withstand. For every 42.25 circular mils of area, an insulated conductor can safely carry 1 ampere for 5 seconds. The measure of heat energy developed in a circuit during a short circuit or ground-fault event is characterized in formulas as I^2t. This formula is simply the current (I) squared multiplied by the time (t), in seconds. Note that the type of thermal insulation on a conductor has an impact on the true value of current it can safely withstand; it is not the same for all conductor insulation types. When this method of calculating conductor withstand values is used, the insulation type and conductor size should be provided in the formula.

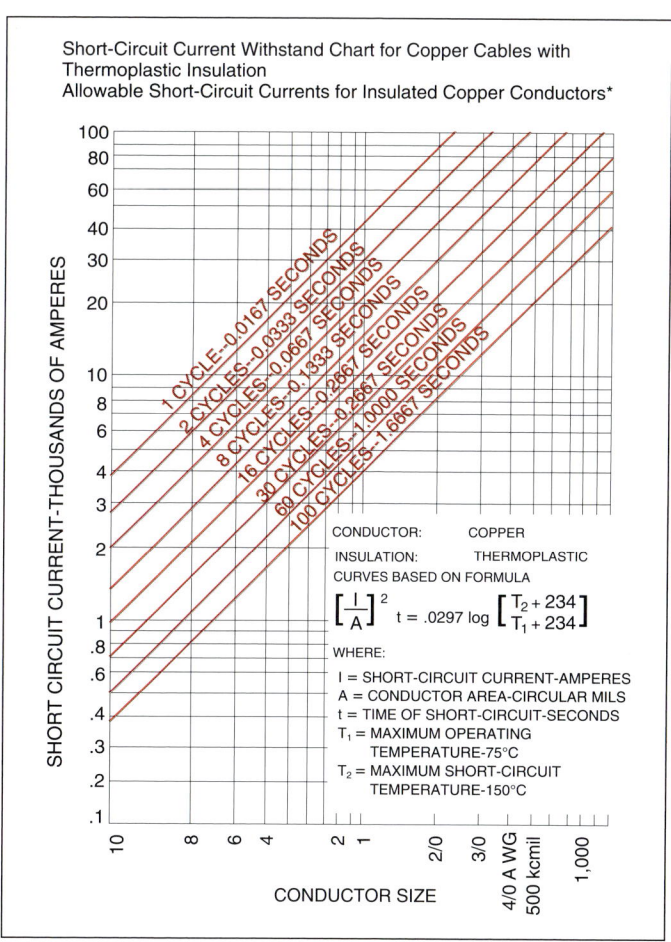

FIGURE 2-22 The *Short-Circuit Current Withstand Chart for Copper Cables with Thermoplastic Insulation* provides information about the maximum current handling capabilities for various sizes of conductors over a duration of time. (Printed with permission from the Insulated Cable Engineers Association [ICEA].)

Protecting Wire-Type Equipment Grounding Conductors

It is essential for safe electrical systems that the integrity of conductors be protected in electrical wiring systems. This requirement applies to the ungrounded circuit conductors, grounded conductors, and the protective EGCs of the wire type. Improper sizing of wire-type EGCs and/or use of the improper type of overcurrent protective devices can result in their annealing, melting, or vaporizing before the circuit overcurrent device can clear the fault. EGCs are, in some installations, much smaller than the associated circuit conductors, usually about 25% the size of the circuit conductor. Selecting the proper overcurrent device is an essential step in providing fast clearing times that protect EGCs within their withstand capabilities. Consideration must always be given to the size of wire-type EGCs, their withstand ratings, the magnitude of ground-fault currents, and the operating characteristics of the overcurrent protective devices. Protective devices that do not operate fast enough might leave EGCs vulnerable to severe damage during a ground-fault event. The solutions are to either select faster overcurrent protective devices or increase the size of the EGC.

Equipment Grounding Conductor Capacity

The *NEC* provides the minimum requirements, meaning at least that much must be done to comply. *NEC* Table 250.122 provides minimum sizes for wire-type EGCs. Safety concerns are apparent when minimum sizes for EGCs are analyzed. Even though an EGC meets the size required by Table 250.122, other factors must be considered in determining full compliance with Section 110.10. Sections 250.4(A)(5) and 250.4(B)(4) provide performance criteria for EGCs. **See Figure 2-23.** A note at the bottom of Table 250.122 refers to those sections.

This note is a mandatory requirement, as compared with advisory notes provided elsewhere throughout the *Code*. Wire-type EGC withstand capacity is the concern here. The issue of protecting EGCs was recognized and specifically addressed by Eustace Soares in the 1960s. Soares explored these issues and provided engineering solutions to conductor withstand rating issues in his book titled *Grounding Electrical Distribution Systems for Safety*.

The values in Table 250.122 are *minimum* values and their validity and capacity often have to be verified. This is an essential step in good engineering practices related to designing power systems. In some cases, the minimum sizes in Table 250.122 have to be increased because of the available short circuit current and the current-limiting abilities of overcurrent protective devices selected in the design. The best engineering practices include performing calculations of available short circuit current levels at various points on the electrical system. The circuit overcurrent protective device must be able to clear a short circuit or ground-fault event without the EGCs being damaged. The EGC must not have insulation damage, and the conductor must not be annealed, melted, or vaporized after a fault occurs.

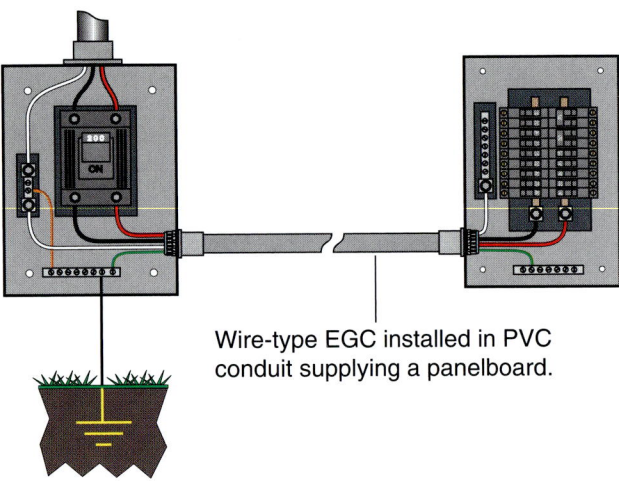

Wire-type EGC installed in PVC conduit supplying a panelboard.

FIGURE 2-23 The minimum sizes required for wire-type EGCs must be able to effectively perform during ground-fault conditions. They must be able to withstand the fault current for the amount of time it takes the overcurrent device to open.

Good design principles include consulting the fuse or circuit breaker manufacturer's literature to obtain important operating characteristics of the overcurrent protective devices installed in the system. The let-through energy levels for these devices should be compared with the withstand ratings of the EGCs. Whenever the conductor withstand ratings are exceeded, the EGC must be increased to ensure adequate capacity to perform as an effective ground-fault current path. The key word in this term is *effective,* which is measurable as it relates to size, capacity, and other physical and installation characteristics of this current path. The path for current must be effective during normal operation and during abnormal events such as ground faults.

Summary

For current to be present in a circuit, the circuit must be complete. Current will always try to return to its source. Amperes operate overcurrent protective devices. Protection of conductors and equipment requires good engineering and design practices. For an effective ground-fault current path to provide the intended protective function, electrical circuit characteristics must be carefully analyzed and applied in systems. The total circuit impedance, overcurrent protective devices, component short circuit current ratings, and other circuit characteristics must be selected and coordinated so that the circuit protective devices can effectively respond and operate during a ground-fault or short circuit event—and they must do so without causing extensive damage to the electrical equipment or conductors, including the EGCs.

References

1. NFPA 70 National Electrical Code 2011, Article 100 (National Fire Protection Association, Quincy, MA 2010), p. 70–29.
2. NFPA 70 National Electrical Code 2011, Article 100 (National Fire Protection Association, Quincy, MA 2010), p. 70–29.
3. NFPA 70E Standard for Electrical Safety in the Workplace 2009, Article 100 (National Fire Protection Association, Quincy, MA 2010), p. 12.
4. NFPA 70E Standard for Electrical Safety in the Workplace 2009, Article 100 (National Fire Protection Association, Quincy, MA 2010), p. 10.

Review Questions

1. A unit of electrical pressure in an electrical circuit is the _____ of the circuit.
   a. Amperage
   b. Resistance
   c. Reactance
   d. Voltage

2. Resistance or opposition to current in a DC circuit is expressed in _____.
   a. Amperes
   b. Watts
   c. Ohms
   d. Volts

3. For current to be present in an electrical circuit, the circuit has to be _____.
   a. Open
   b. Closed or complete
   c. Incomplete
   d. Energized

4. Using Ohm's law, if a circuit has an applied voltage of 480 volts and a resistance of 40 ohms, how much current will be present in the circuit?
   a. 12 A
   b. 192 A
   c. 0.083 A
   d. 24 A

5. The total opposition to current in an AC circuit is known as the _____ of the circuit.
   a. Resistance
   b. Current
   c. Inductance
   d. Impedance

6. The unit of electrical power is expressed in _____.
   a. Watts
   b. Volts
   c. Amperes
   d. Resistance

7. If a circuit has an applied voltage of 120 volts and the current in the circuit is 30 amperes, how many ohms of resistance are present in the circuit?
   a. 3600 Ω
   b. 40 Ω
   c. 4 Ω
   d. 90 Ω

8. An electric heating element has a resistance of 16 ohms and is connected to a voltage of 240 volts. How much current is present in the circuit?
   a. 15 A
   b. 19.20 A
   c. 0.066 A
   d. 7.5 A

9. If a 5-kW heater is connected to a 480-volt circuit, how much current will the heater draw?
   a. 96 A
   b. 0.096 A
   c. 10.4 A
   d. 20.8 A

10. In a 60-Hz (cycle) circuit, the voltage and current change amplitude and direction ____ times per second.
    a. 60
    b. 120
    c. 240
    d. 400

11. The impedance of an AC circuit includes which of the following electrical characteristics?
    a. Resistance
    b. Inductive reactance
    c. Capacitive reactance
    d. All of the above

12. Keeping the opposition to current as low as possible in electrical bonding systems is necessary so that a maximum amount of fault current causes rapid operation of overcurrent devices in fault conditions.
    a. True
    b. False

13. Generally, the *NEC* requires all conductors of AC circuits, including the equipment grounding conductors, to be run together in the same raceway, cable, or trench.
    a. True
    b. False

14. The current in an electrical circuit will take which of the following paths to return to its source?
    a. The path of least resistance
    b. The path of most resistance
    c. All paths available, including the Earth
    d. Only the Earth (ground)

15. Electrical equipment and overcurrent protective devices such as fuses and circuit breakers must be selected to ensure that the short circuit current rating of any system component is not exceeded should a short circuit or high-level ground-fault event occur.
    a. True
    b. False
16. An unintentional, electrically conducting connection between an ungrounded conductor of an electrical circuit and the normally non–current-carrying conductors, metallic enclosures, metallic raceways, metallic equipment, or Earth best defines which of the following?
    a. A short circuit
    b. An overload condition
    c. An overcurrent condition
    d. A ground fault
17. The term _____ refers to the highest current at rated voltage that a device is identified to interrupt under standard test conditions.
    a. Voltage rating
    b. Short circuit current rating
    c. Interrupting rating
    d. Withstand rating
18. If a circuit is capable of delivering 22 000 amperes of fault current, what is the minimum short circuit current rating required for the equipment, including any overcurrent device installed in the equipment?
    a. 10 000 A
    b. 22 000 A
    c. 42 000 A
    d. 5000 A
19. The load (current) on a circuit breaker is generally not permitted to exceed what percentage of the rating of the circuit breaker rating?
    a. 100%
    b. 125%
    c. 150%
    d. 80%
20. The two main concerns in electrical equipment and conductor protection designs are how much current the equipment and conductors can handle and for how long.
    a. True
    b. False
21. The *Code* does not require conductors and completed wiring installations to be free from short circuits, ground faults, or any ground connections other than as required or permitted by specific *NEC* provisions.
    a. True
    b. False
22. Testing a wire for insulation integrity is typically accomplished by use of which of the following?
    a. A voltage tester
    b. A wattage meter
    c. A megohmmeter
    d. A ground resistance tester
23. The Insulated Cable Engineers Association (ICEA) demonstrated that for every 42.25 cm of area, an insulated conductor can safely carry 1 ampere for ___ second(s).
    a. 10
    b. 1
    c. 5
    d. 60
24. All equipment grounding conductors must provide an effective ground-fault current path to perform effectively.
    a. True
    b. False

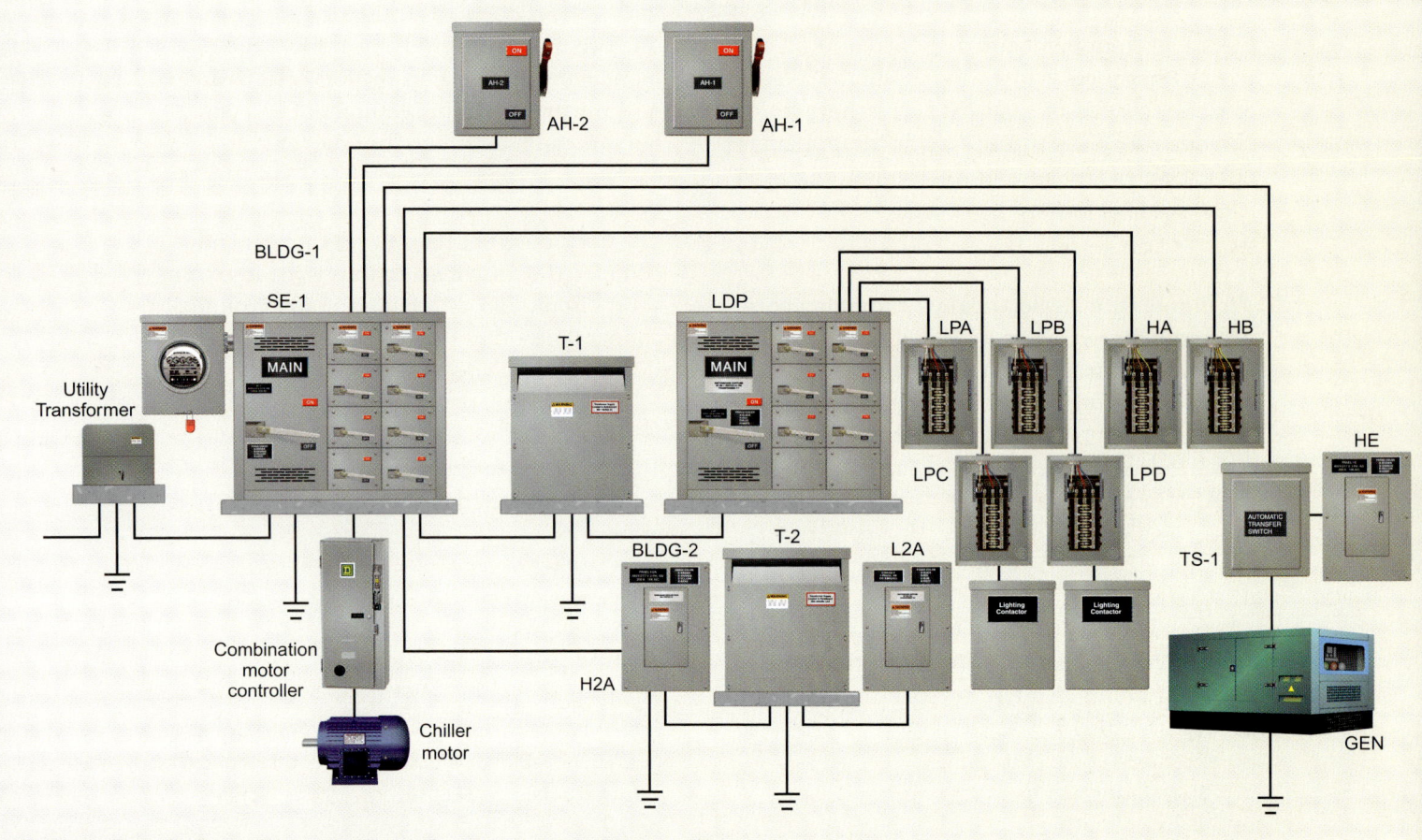

CHAPTER 3

Using the National Electrical Code®

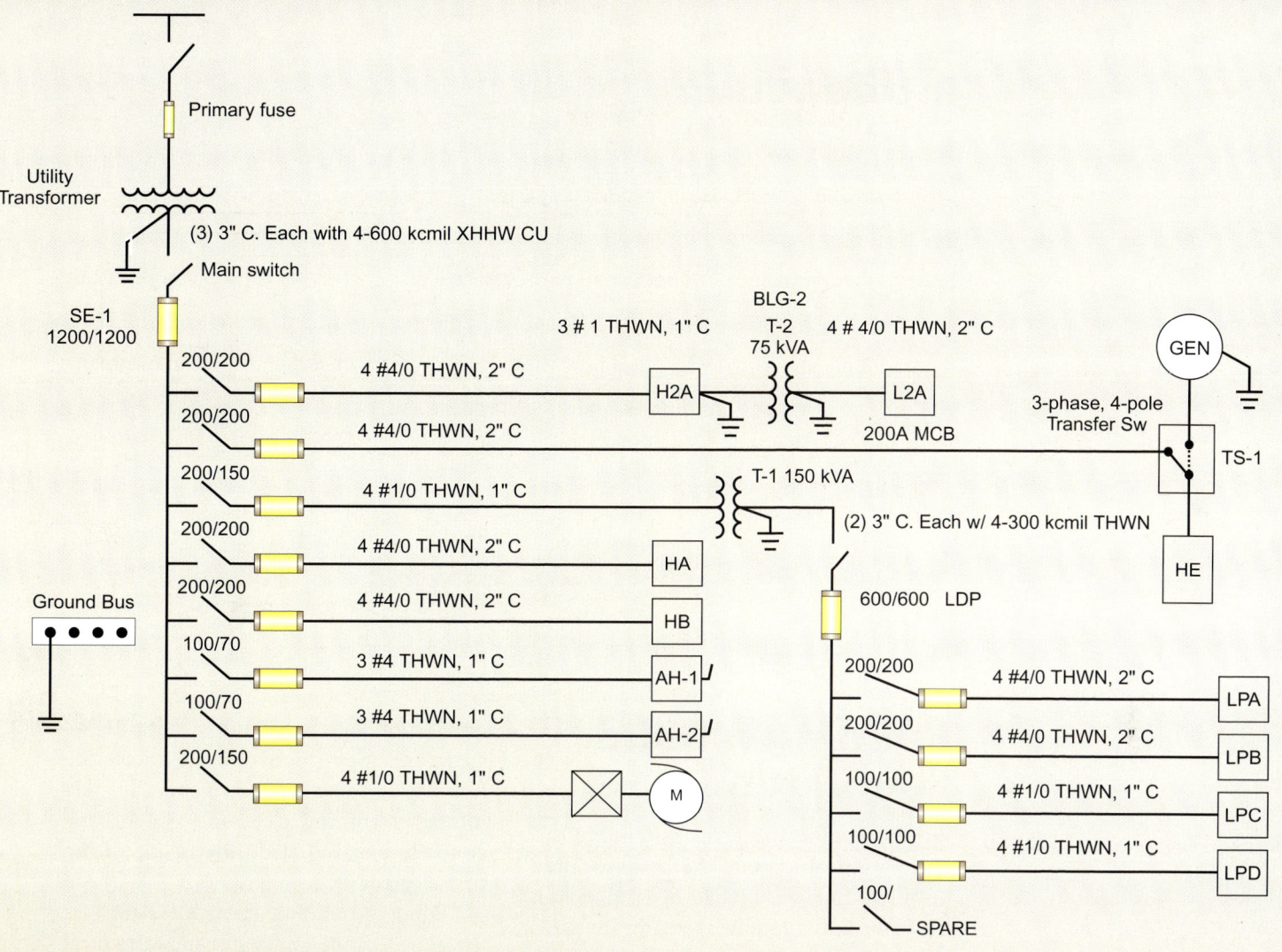

Objectives

- Understand the purpose of the *NEC* and the arrangement of Article 250
- Understand key defined grounding and bonding terms in Articles 100 and 250.2
- Determine key requirements in the *NEC* that relate specifically to electrical grounding and bonding
- Distinguish performance requirements from prescriptive requirements in the *NEC*
- Understand the how Chapters 5 though 7 of the *NEC* can modify or amend the general requirements in the *NEC*
- Understand the *NEC* tables related to grounding and bonding conductor sizing

Outline

Building a Solid Foundation
NEC Arrangement and Application
Enforcement and Approvals
Requirements, Exceptions, Alternatives, and Information
Permissive *Code* Language
Explanatory Information
Use of Defined Terms
Code-Making Panel Responsibilities
NEC Article 100—Grounding and Bonding Terms
Section 250.2 Definitions
Article 200—Identification and Use of Grounded Conductors
Article 250—Arrangement and Use
Part I—General
Table 250.66—Sizing Grounding Electrode Conductors
Table 250.122—Sizing Equipment Grounding Conductors
Special Occupancies, Equipment, and Conditions

Introduction

The purpose of the *National Electrical Code®* (*NEC;* NFPA 70) is the practical safeguarding of persons and property from hazards that arise from the use of electricity. **See Figure 3-1.** The *NEC* provides the minimum safety requirements, meaning that it is the *least* that must be done for compliance. An important part of electrical construction is proper application of *Code* rules to installations and systems. It is important to understand all general *NEC* requirements and know when and how these rules are modified due to special occupancies, special or unique conditions, and special equipment.

Building a Solid Foundation

From the time the project is being designed and blueprints are being drafted until the final receptacle is installed on the project, the requirements in the *NEC* apply. Buildings and structures must have a solid foundation on which they are built. Footings and foundations must have enough structural integrity to support all loads. Likewise, the grounding electrode system for a building serves as the foundation of the electrical system supplying the wiring system of the structure. **See Figure 3-2.** Since the footing or foundation is generally part of the grounding electrode system, the concrete-encased electrode is one of the first electrodes installed.

Other grounding electrodes, such as metal water pipe electrodes and structural metal building frames, are also grounding electrodes that are inherent to buildings. These electrodes are formed or installed as construction progresses.

NEC Arrangement and Application

The *NEC* is an installation standard that contains requirements that apply to installing electrical wiring and equipment. It consists of an introduction and nine chapters. Each chapter in the *NEC* contains articles that are further broken down into rules in the form of sections, subdivisions, and lists. Also included are exceptions to rules, informational notes, and rules in tabular form. Article 90 serves as the introduction and provides users with essential information about how the rules apply to electrical installations.

FIGURE 3-1 The purpose of the *Code* is to protect persons and property from hazards of electricity use.

The very first provision in the *NEC* indicates that the purpose of the guidelines is the practical safeguarding of persons and property from hazards arising from the use of electricity. Section 90.2 provides a scope of the *Code*, clarifying what is covered and what is not covered by *NEC* rules. Section 90.3 is an important provision that guides users in the correct application of the stated rules and how they must be applied.

Chapters 1 through 4 of the *NEC* have general application, meaning these rules apply to all electrical installations. Chapters 5, 6, and 7 include rules for special occupancies, special equipment, and other special conditions. The provisions in Chapters 5, 6, and 7 modify or amend the requirements in Chapters 1 through 4. Chapter 8 is not subject to the general requirements of the other chapters except where the other rules are referenced from within Chapter 8.[1] Chapter 9 of the *NEC* includes tables that are used in applying the other requirements of the *Code*. **See Figure 3-3.** The Annexes in the back of the *NEC* contain information that is not mandatory.

There are many grounding and bonding requirements contained in Chapter 5 of the *NEC* that are more restrictive than the general rules in contained in Article 250. For example, Section 250.118(5) recognizes flexible metal conduit as an acceptable equipment grounding conductor (EGC) if all the conditions indicated in Sections 250.118(5)(a) through (d) are met. What this means is that flexible metal conduit can be installed and used as an EGC without installing a wire-type EGC with the short run of flexible metal conduit. *NEC* Chapter 5 includes requirements for special occupancies and often modifies the general requirements of the *Code* to be more restrictive. As an example, Article 501 covers installations in Class I locations where explosion hazards exist. Section 501.30(B) specifically restricts flexible metal conduit and liquidtight flexible metal conduit from being used as an EGC and bonding means.

Chapters 1 through 4 of the *NEC* apply generally to all electrical installations and are often modified or amended by the special requirements in Chapters 5, 6, or 7.

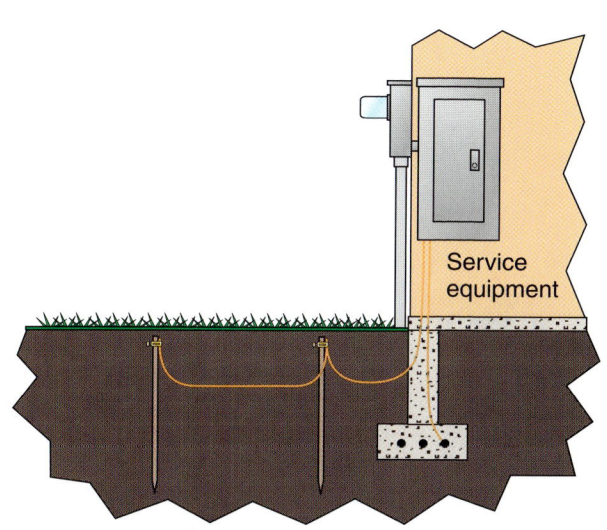

FIGURE 3-2 The foundation of the electrical system is the grounding electrode system. Grounding electrodes suitable for use in the grounding electrode system are provided in Section 250.52(A).

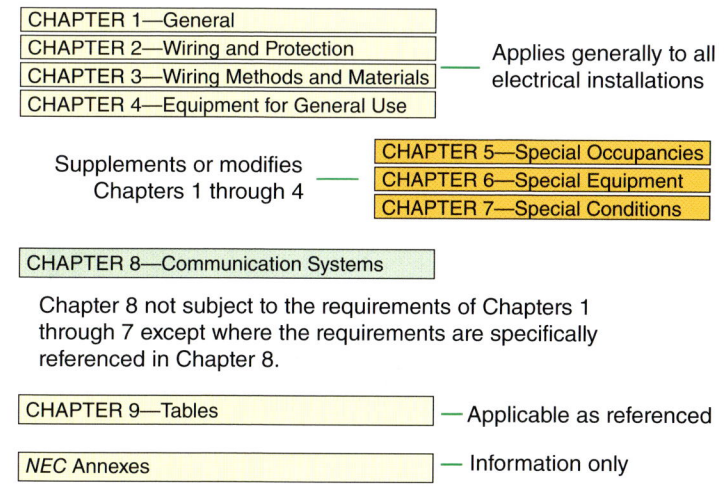

FIGURE 3-3 Section 90.3 provides the arrangement and application requirements for the *NEC*.

The result is a requirement to install a wire-type equipment bonding jumper in accordance with Section 250.102, which allows the jumper to be placed on either inside the conduit or outside the conduit. **See Figure 3-4.**

The reason for this specification is to ensure that flexible metal conduit is not depended on for carrying ground-fault current in a Class I, Division 2 location. Both the conduit and equipment bonding jumper are used, meaning the flexible metal conduit does not act as the sole effective path for ground-fault current that could result in arcing or sparks in the classified location during ground-fault conditions. This could become an ignition source in an explosive atmosphere such as gasoline fumes or vapors. Special consideration is necessary for grounding and bonding of exposed non–current-carrying metal parts of equipment, such as metal exteriors of motors, fixed or portable lamps, luminaires, enclosures, and raceways. These parts must have an effective mechanical and electrical connection in order to reduce the possibility of arcs or sparks caused by ineffective or poor grounding and bonding methods.

Enforcement and Approvals

Governing bodies have the authority to enforce the provisions of the *NEC* when it is adopted into law by a particular jurisdiction (Section 90.4). The authority having jurisdiction (AHJ) is defined in Article 100 as a person or organization that is responsible for enforcing the *Code* and issuing approvals of installations and equipment covered by the rules of the *NEC*. The AHJ is also responsible for interpreting the requirements and granting special permission as provided in some of the rules. It is important that workers establish a good working relationship with inspection authorities. This requires effective communication about how the rules will be applied to any given aspect of the installation. Being proactive is always the best approach when unsure of how a particular *NEC* requirement will be applied to an installation or system. Inspection authorities typically require electrical installations to be inspected and approved before they are concealed by building finishes. This requirement can vary among jurisdictions, so the inspection and approval procedures required in the jurisdiction where the project is being built should always be verified.

> Section 501.30(B) specifically restricts flexible metal conduit and liquidtight flexible metal conduit from being used as an EGC and bonding means.

Requirements, Exceptions, Alternatives, and Information

To effectively apply the requirements in the *NEC*, the difference between mandatory requirements and permissive and informational provisions must be understood. A misunderstanding of how exceptions apply can lead to misapplication of mandatory rules. The mandatory requirements in the *Code* are characterized by the use of *shall* or *shall not*.

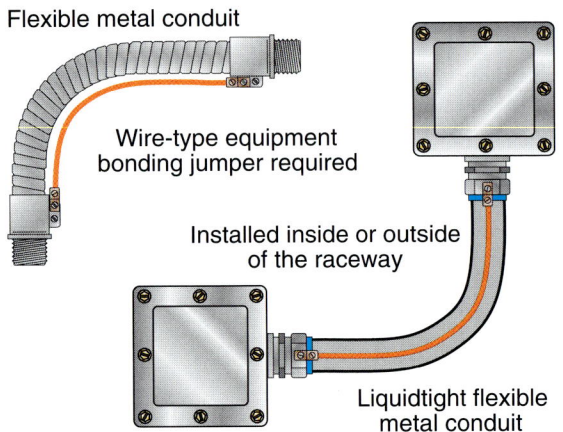

FIGURE 3-4 Section 501.30(B) modifies the general requirements in Section 250.118.

> **250.68 (A) Accessibility.** All mechanical elements used to terminate a grounding electrode conductor or bonding jumper to a grounding electrode shall be accessible.[1]

This is a general requirement that must be followed. If there were no exceptions following this requirement, all connections to grounding electrode would be required to be accessible. Knowing how impractical this would be, it makes sense that exceptions to the rule are included.

Exceptions to Rules

Exceptions to requirements modify only the rule they immediately follow unless stated differently within the exception. That is, the exception may indicate that it applies to more than the rule it follows.

> **250.68 (A) Accessibility.** All mechanical elements used to terminate a grounding electrode conductor or bonding jumper to a grounding electrode shall be accessible.
>
> *Exception No. 1: An encased or buried connection to a concrete-encased, driven, or buried grounding electrode shall not be required to be accessible.*[1]

An exception to a rule can relax the requirement in the rule. For example, Exception 1 after the preceding rule relaxes the requirement for accessibility for direct buried connections or connections that are encased in concrete. Because these types of connections are highly unlikely to be disturbed or affected once buried or once concrete is poured, this exception provides a practical exemption to the general requirement. **See Figure 3-5.** For the exception to apply, however, the connection means must be suitable for these locations. The means of connection are generally required to be evaluated and listed by a qualified electrical testing laboratory to indicate this suitability.

There are also mandatory exceptions that use the terms *shall* or *shall not*. An example of a mandatory exception follows Section 230.95. This exception indicates that ground-fault protection of equipment shall not apply to a service disconnect for a continuous industrial process where a nonorderly shutdown introduces additional or increased hazards. In these cases the *Code* recognizes a condition that introduces greater risks to persons or property.

Permissive *Code* Language

Permissive provisions in the *NEC* are characterized by the use of phrases such as *shall be permitted* or *shall not be required*. Permissive rules are options or alternative methods of achieving equivalent safety; they are not requirements. A closer review of permissive terms is essential because permissive rules are often misinterpreted.

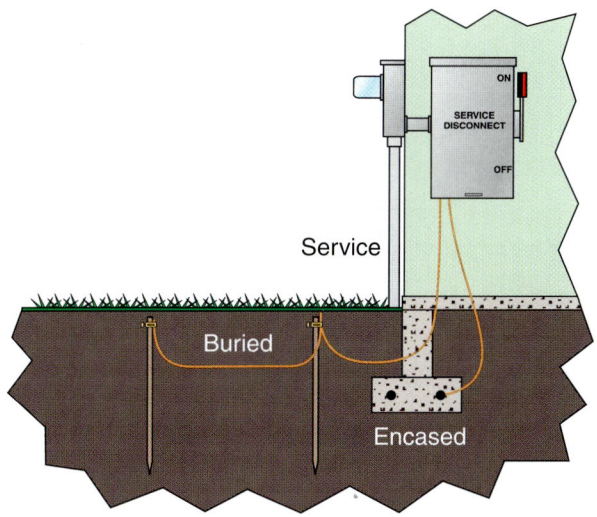

FIGURE 3-5 Exceptions modify only the *NEC* rule they follow, unless stated differently within the exception. Exception 1 to Section 250.68 relaxes an accessibility requirement for grounding electrode conductor connections. In this illustration, the connections are direct buried and encased in concrete, so Exception 1 applies.

> The inspector (AHJ) has the responsibilities of approving installations, interpreting the *NEC* rules, and granting special permission.

FIGURE 3-6 Permissive requirements often include AHJ acceptability. The authority having jurisdiction interprets and enforces the *NEC* requirements, issues approvals, and can grant special permission.

Table 250.122 Minimum Size Equipment Grounding Conductors for Grounding Raceway and Equipment (in part)

Rating or Setting of Automatic Overcurrent Device in Circuit Ahead of Equipment, Conduit, etc., Not Exceeding (Amperes)	Size (AWG or kcmil)	
	Copper	Aluminum or Copper-Clad Aluminum*
1000	2/0	4/0
1200	3/0	250
1600	4/0	350
2000	250	400
2500	350	600
3000	400	600
4000	500	750
5000	700	1200
6000	800	1200

Note: Where necessary to comply with 250.4(A)(5) or 250.4(B)(4), the equipment grounding conductor shall be sized larger than given in this table.

*See installation restrictions in 250.120.

FIGURE 3-7 Reproduction of Table 250.122 (in part). The note following Table 250.122 is a mandatory requirement. Reprinted with permission from NFPA 70–2011, *National Electrical Code*®, Copyright © 2010, National Fire Protection Association, Quincy, MA 02169. This reprinted material is not the complete and official position of the NFPA on the referenced subject, which is represented only by the standard in its entirety.

The qualifying feature of permissive rules involves determination of the AHJ as indicated in Section 90.4. **See Figure 3-6.** The AHJ has the final approving authority and responsibility. When in doubt about a permissive requirement or provision in the *Code*, consult the AHJ.

Explanatory Information

Explanatory material is provided in the form of informational notes, which were called *Fine Print Notes* (FPN) in earlier *NEC* editions. The notes are informative only and are not enforceable as requirements. Often, informational notes clarify the requirement that precedes the note. Examples of informational notes are references to other standards, references to related sections of the *Code*, or information related to a rule. Informational notes should not be confused with notes following tables, which are applicable requirements. A good example of this is found in the notes following Table 300.5 and Table 1 in Chapter 9. Another good example of a note to an *NEC* table is found following Table 250.122. **See Figure 3-7.**

Brackets containing section references to another NFPA document serve as a guide to indicate the source of the extracted informational text. Standards referred to in the brackets are for reference only and are not intended to make the referenced standard applicable as requirements. Examples of bracketed information can be found in Articles 500 and 517, but bracketed information is not limited to just those articles.

Use of Defined Terms

Defined terms help clarify the meaning of a rule in which the term appears. To fully understand how the *NEC* applies to grounding and bonding systems, a clear method or language of communication has to be established.

Article 100 contains several defined terms used in the rules. The *NEC Style Manual* indicates that when a term appears in more than two articles of the *Code,* it should be defined in Article 100. When a term is unique to a specific article in the *Code* and it is defined, the definition appears in the ".2" section of that article. For example, Section 250.2 includes the definition for the term *effective ground-fault current path* because it is used only within Article 250. **See Figure 3-8.**

However, the term *grounded (grounding)* is used in multiple articles; therefore, it is defined in Article 100. It is important to develop an understanding of defined grounding and bonding terms. In the 2008 *NEC* cycle, NFPA implemented a process change that resulted in a shift in responsibilities for definitions of terms related to grounding and bonding. *Code*-Making Panel 1 no longer has responsibility for the technical meaning of these terms. *Code*-Making Panel 5 has been assigned this responsibility, and they oversee proposals and comments for new grounding and bonding terms or revisions to existing terms; this group is also responsible for creating and revising grounding and bonding requirements every cycle. Terms used in more than one article are still located in Article 100.

Another important part of this concept is that where a defined grounding or bonding term is used within a rule, it should be used in a manner consistent with how the term is defined. The point here is to emphasize that using terms as they are defined in the *NEC* promotes effective and accurate application of requirements.

Code-Making Panel Responsibilities

There are 19 *NEC* technical "subcommittees" that are referred to as *Code*-Making Panels. *Code*-Making Panel 5 is responsible for definitions of terms related to grounding and bonding and for Articles 200, 250, 280, and 285. In the 2008 *NEC* cycle the Technical Correlating Committee (TCC) assigned a Task Group to work on revising grounding and bonding terms and to verify appropriate use of these terms throughout the *Code* to help clarify their meaning in the rules.

For example, consider the following wording:

Example 1: A grounding conductor is required to be grounded to the equipment grounding terminal bar of a panelboard.
Example 2: The ECG is required to be connected to the equipment grounding terminal bar of the panelboard.

The first example is incorrect. The correct wording is that used in the second example. The difference is how the terms *grounding* and *grounded* are used. Article 100 defines these terms as "connected (connecting) to ground or to a conductive body that extends the ground connection." Therefore, the first example does not make technical sense, even though many people in the field communicate this way.

It is important to use the language the *NEC* uses with specific respect to meaning of defined terms. This promotes a clear and consistent understanding of the requirements and which rules apply to various portions of the electrical system.

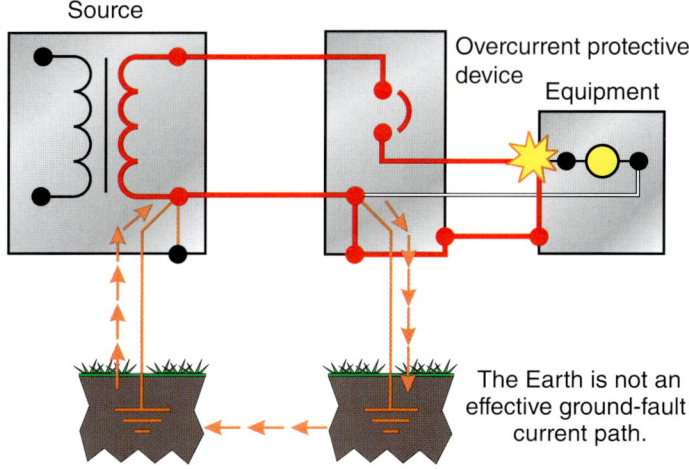

FIGURE 3-8 The term *effective ground-fault current path* is defined in Section 250.2.

Slang terms have different meaning to people and may cause confusion. For example, some people refer to a bonding jumper as a *bond*. The term *bonding jumper* is defined in the *NEC*, whereas the term *bond* is not; therefore, *bond* should not be used. The following definitions from Article 100 are included to provide readers with a general understanding of their general meaning.

NEC Article 100—Definitions of Grounding and Bonding Terms

Article 100 of Chapter 3 in the *NEC* provides the definitions of common terms related to grounding and bonding. Some definitions from the *NEC* are included here and followed by a helpful comment to describe the implied meaning of each term. Within the *NEC*, the definitions are usually presented in the simplest form. The rule in which the term is used often indicates what is intended to be accomplished and also clarifies the intended meaning of the rule with the support of its definition. Common definitions of grounding and bonding terms follow:

> **Bonded (Bonding).** Connected to establish electrical continuity and conductivity.[2]

The term *bonded* (*bonding*) means connecting conductive objects together. Once this is accomplished, the potential differences between them are minimized and they become electrically one through the bonding process. What the bonding is intended to accomplish depends on the rule in which the term is used. Bonding at service equipment establishes electrical continuity and conductivity and ensures a path for fault current likely to be imposed. The bonding required by Section 680.26 (equipotential bonding for swimming pools and similar installations) is concerned with establishing an equipotential bonding grid that has little or nothing to do with fault current or facilitating operation of overcurrent protective devices. This is why the term is defined generally in Article 100 and what is intended to be accomplished by the bonding is specifically addressed in the rule where the term is used.

> **Bonding Conductor or Jumper.** A reliable conductor to ensure the required electrical conductivity between metal parts required to be electrically connected.[3]

Bonding conductors or jumpers are typically short conductors that are installed to establish required conductivity between conductive parts. **See Figure 3-9.** A bonding jumper can serve to complete a conductive path to the Earth for grounding functions or can serve as a path for fault current, or for both purposes.

An example of a bonding jumper is the conductor installed from a bonding/grounding bushing that is used to provide a conducting connection between a raceway to the enclosure the raceway is connected to.

> **Bonding Jumper, Equipment.** The connection between two or more portions of the equipment grounding conductor.[4]

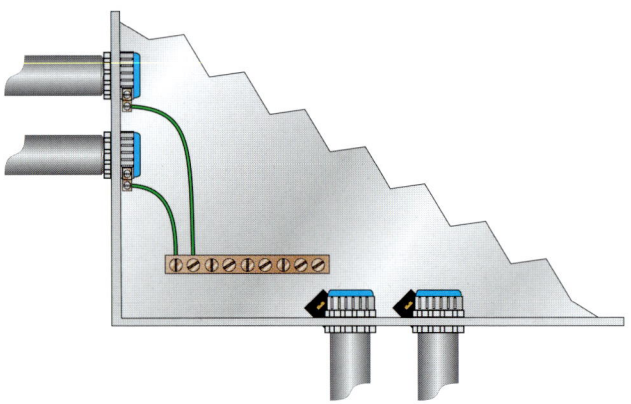

FIGURE 3-9 Bonding jumpers are reliable conductors that establish the required electrical continuity between metal parts required to be electrically connected.

An equipment bonding jumper is usually a short conductor that establishes a connection between two or more portions of an EGC. For example, if an expansion fitting is used across a structural steel expansion joint in a building, an equipment bonding jumper is often installed to ensure bonding across the special fitting that is designed to expand and contract as the building moves. **See Figure 3-10.**

If rigid metal conduit (RMC) is used for the circuit crossing the expansion joint, the integrity of the EGC path must be assured by installing an equipment bonding jumper. This jumper joins the two lengths of RMC that serve as an EGC for the contained circuit.

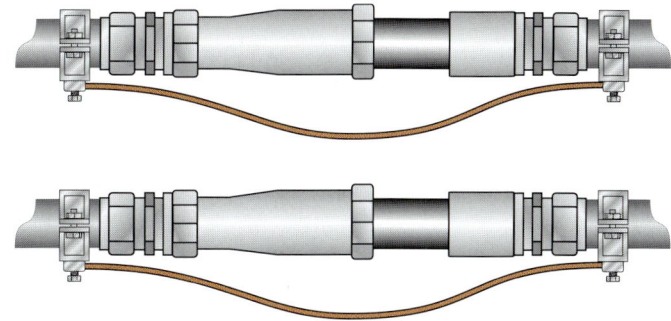

FIGURE 3-10 Equipment bonding jumpers are required to be installed with conduit expansion fittings. Listed expansion fittings should always be installed in accordance with the manufacturer's installation instructions.

Bonding Jumper, Main. The connection between the grounded circuit conductor and the equipment grounding conductor at the service.[5]

The main bonding jumper is a conductive path between the grounded conductor bus (usually a neutral bus) and both the enclosure and EGC at the service equipment. Main bonding jumpers are installed only in the service disconnecting means enclosure. **See Figure 3-11.**

Main bonding jumpers are made of copper or other corrosion-resistant material and can be a wire, bus, screw, or similar suitable conductor.

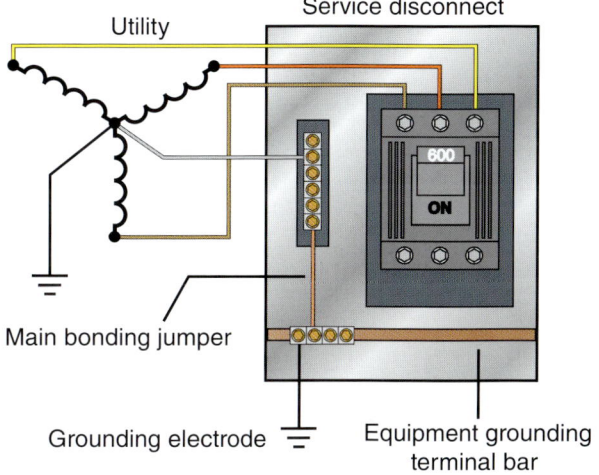

FIGURE 3-11 The main bonding jumper connects the grounded conductor to the EGC(s) at the service.

Ground. The earth.[6]

The term *ground* simply means the Earth. **See Figure 3-12.** When this term is used in the *NEC*, the planet Earth is what is implied.

Because the term *ground* is defined as the Earth, the connection implied in this definition is a connection to the Earth, or to a conductive body that extends the Earth connection, rather than a conductive body serving in place of the Earth.

FIGURE 3-12 Grounding requires a connection to the planet Earth.

The Earth is a conductive body, but it is a generally considered a poor conductor, especially when compared with typical conductor materials such as aluminum and copper. The resistivity of soil varies regionally and seasonally and thus has differing conductive properties. Earth that is normally moist and not subject to large temperature differentials can have a very low resistivity, whereas sandy or rocky Earth offers higher levels of resistivity. Obviously, the changes in seasons and fluctuations of moisture levels change the Earth's conductivity characteristics. The *Code* specifically prohibits the Earth from being used as an effective path for fault current because it is not dependable. It is a serious error to depend on the Earth as the sole return path for ground-fault current. Although the Earth is in the grounding circuit, it performs functions other than facilitating overcurrent protective device operation.

A ground fault is an event that is unintentional and usually causes a fuse or circuit breaker to operate. This type of event is usually caused by failure of insulation. During a ground-fault or short circuit condition, the supply system is under severe stress. The amount of current depends on the impedance of the circuit and the fault current available in the system. A ground-fault event can produce an arcing condition that can cause severe injury or death. This is another reason that the definition indicates that these events are unintentional.

Grounded (Grounding). Connected (connecting) to ground or to a conductive body that extends the ground connection.[8]

The term *grounded (grounding)* clearly describes what is intended to be accomplished through the process of grounding. Since the term *ground* is defined as the Earth, a connection to the planet is implied. This is usually accomplished through one or more grounding electrodes. Conductive bodies that extend the ground connection can be a grounding electrode conductor or EGC, or other conductive material in contact with the planet Earth, such as building steel or a metal water pipe. **See Figure 3-13.**

Ground Fault. An unintentional, electrically conducting connection between an ungrounded conductor of an electrical circuit and the normally non–current-carrying conductors, metallic enclosures, metallic raceways, metallic equipment, or earth.[7]

Grounded, Solidly. Connected to ground without inserting any resistor or impedance device.[9]

When a system or conductive object is solidly grounded, it is connected to the Earth by a conductive path (usually a grounding electrode conductor) that offers little or no resistance or impedance device in the path to Earth. **See Figure 3-14.**

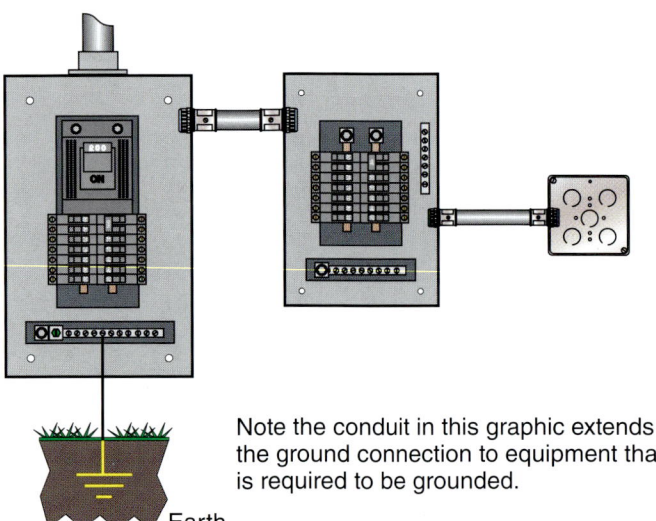

Note the conduit in this graphic extends the ground connection to equipment that is required to be grounded.

Earth

FIGURE 3-13 Grounding electrical equipment means ensuring that the equipment is grounded by connection to a conductive body that extends the ground connection.

An example of solid system grounding is a transformer or generator producing a system secondary or output voltage that is required to be grounded in accordance with Section 250.30(A).

An example of grounding that is not solid is a high-impedance neutral grounded system. The system is still connected to ground (Earth), but only through an impedance device or resistor.

> **Grounded Conductor.** A system or circuit conductor that is intentionally grounded.[10]

Systems that are grounded include one conductor that is intentionally connected to ground either solidly or through an impedance device. **See Figure 3-15.** A grounded conductor is one that connected to the ground (Earth).

Although many grounded conductors are system-neutral conductors, not all grounded conductors are neutrals. **See Figure 3-16.** Sometimes a phase conductor of a three-phase, three-wire delta system is grounded creating a corner-grounded delta system. All the rules for grounded conductors apply regardless of whether the grounded conductor is a system neutral or a phase conductor.

> **Grounding Conductor, Equipment (EGC).** The conductive path(s) installed to connect normally non–current-carrying metal parts of equipment together and to the system grounded conductor or to the grounding electrode conductor, or both.
>
> Informational Note No. 1: It is recognized that the equipment grounding conductor also performs bonding.
>
> Informational Note No. 2: See 250.118 for a list of acceptable equipment grounding conductors.[11]

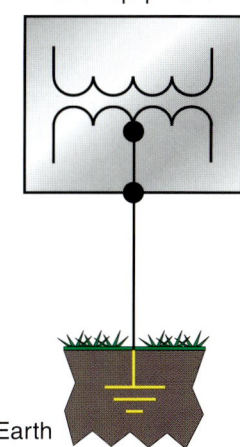

FIGURE 3-14 A solid connection to the ground with no intentional impedance in the path is shown for the system and equipment. The system and equipment are at or close to Earth potential.

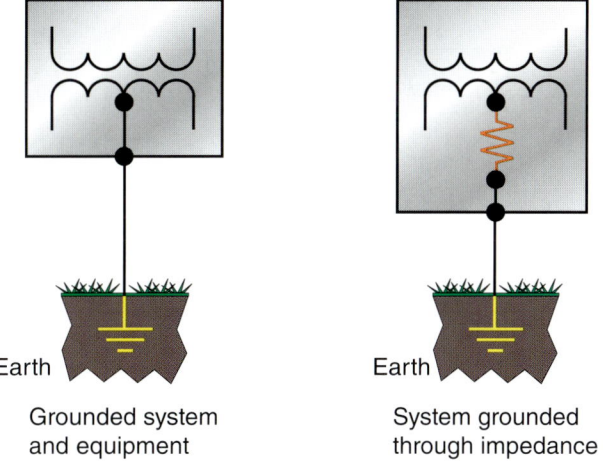

FIGURE 3-15 Grounded conductors of systems are connected to the Earth either solidly (left) or through an impedance device (right).

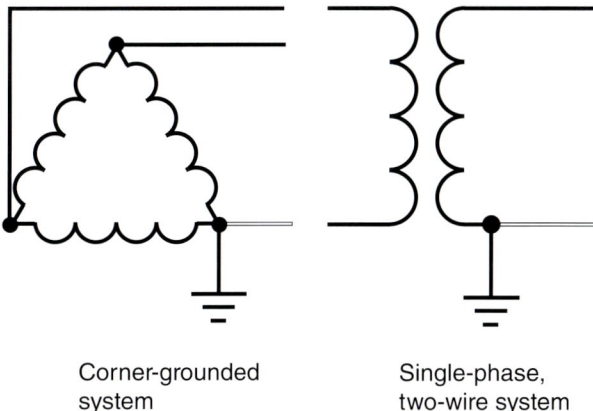

FIGURE 3-16 Most neutrals are grounded conductors, but not all grounded conductors are neutrals. Rules for grounded conductors, such as in Sections 240.22 and 200.6, apply to all grounded conductors, not just neutrals.

An EGC performs multiple functions through a single conductive path. The EGC serves as a conductive path to the ground (Earth), it performs bonding functions, and it facilitates overcurrent protective device operation when functioning as an effective ground-fault current path during ground-fault events. **See Figure 3-17.** The types of EGCs are listed in Section 250.118. EGCs can be in the form of a wire-type conductor, a metal raceway, cable tray, or other electrically conductive path recognized in Section 250.118. EGC installations must meet the provisions in Section 250.120.

> **Grounding Electrode.** A conducting object through which a direct connection to earth is established.[12]

Grounding electrodes are conductive objects in contact with the ground (Earth). For an object to qualify as a grounding electrode, there must be an Earth connection, the direct contact with the Earth must be maintained, and the object must be conductive. **See Figure 3-18.** Electrodes identified in Section 250.52(A) are acceptable as grounding electrodes. Grounding electrodes have little or no effect on the operation of overcurrent protective devices.

A grounding symbol is often included on construction blueprints. It is generally intended to mean a grounding electrode or grounding electrode system, not just a single ground rod. **See Figure 3-18.**

> **Grounding Electrode Conductor.** A conductor used to connect the system grounded conductor or the equipment to a grounding electrode or to a point on the grounding electrode system.[13]

Grounding electrode conductors are conductive bodies that extend the ground connection. They maintain potential differences at a minimum between the conductive objects and the Earth electrode to which they are connected. **See Figure 3-19.** Grounding electrode conductors are not intended to facilitate operation of overcurrent protective devices, but they do perform grounding and bonding functions. The grounding electrode conductor performs bonding between the Earth and grounded conductors and other conductive parts required to be grounded.

> **Bonding Jumper, System.** The connection between the grounded circuit conductor, and the supply-side bonding jumper, or the equipment grounding conductor, or both at a separately derived system.[14]

The system bonding jumper is located at a separately derived system and functions in similar fashion to that of the main bonding jumper. The term was defined to distinguish it from a main bonding jumper, which, by definition, is located only at service equipment.

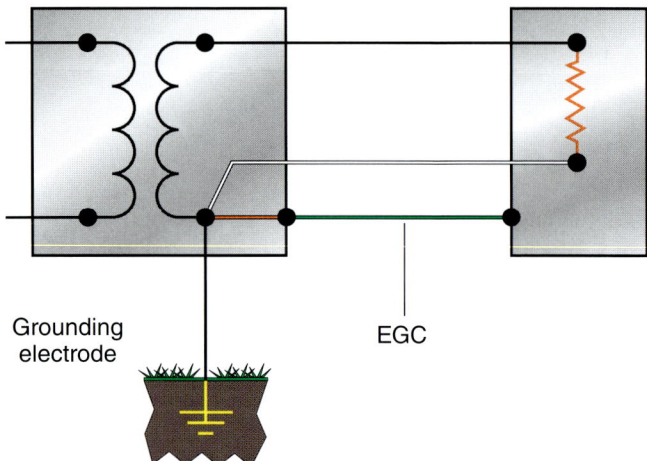

FIGURE 3-17 EGCs perform three essential functions. They (1) ensure the equipment is grounded, (2) establish bonding, and (3) provide an effective ground-fault current path.

The system bonding jumper is also addressed in Section 250.28 in regard to sizing and physical characteristics. System bonding jumpers are made of copper or other corrosion-resistant material and can be a wire, bus, screw, or similar suitable conductor.

Section 250.2 Definitions

There are some defined terms provided in Section 250.2 of Article 250. The following is a list and brief description of each term defined in Section 250.2. Again, these terms are defined here in 250.2 rather than in *NEC* Article 100 since they are unique to Article 250.

> **Effective Ground-Fault Current Path.** An intentionally constructed, low-impedance electrically conductive path designed and intended to carry current during ground-fault conditions from the point of a ground fault on a wiring system to the electrical supply source and that facilitates the operation of the overcurrent protective device or ground-fault detectors on high-impedance grounded systems.[16]

> **Bonding Jumper, Supply-Side.** A conductor installed on the supply side of a service or within a service equipment enclosure(s), or for a separately derived system, that ensures the required electrical continuity between metal parts required to be electrically connected.[15]

A supply-side bonding jumper is typically installed on the line or supply side of an overcurrent device at the service equipment or at a separately derived system. If these supply-side bonding jumpers are of the wire type, they must be sized according to the requirements in Section 250.102(C), which bases the size on the circular mil area of the largest ungrounded service-entrance conductor or largest ungrounded conductor of a separately derived system. Supply-side bonding jumpers need to be a bit more robust than a load-side bonding jumper, and have the capacity to carry fault current for the duration of time it takes an overcurrent device to open and clear the event. This overcurrent device could be on the supply side of the utility transformer or on the primary side of a transformer separately derived system.

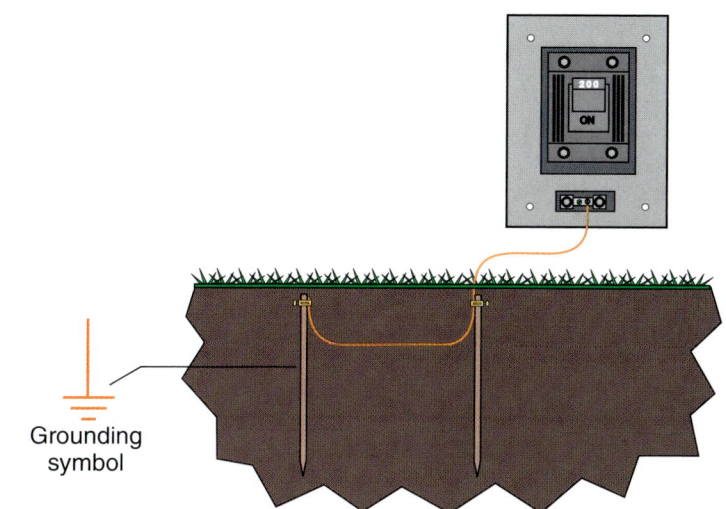

FIGURE 3-18 Grounding electrodes are conducting objects that establish a direct Earth connection. Note that the grounding symbol applies to all of the grounding electrodes specified in Section 250.52(A).

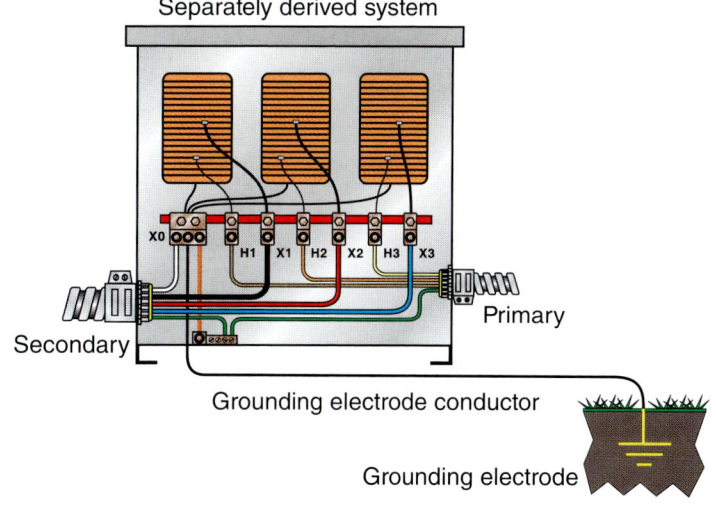

FIGURE 3-19 Grounding electrode conductors are conductive bodies that extend the ground connection.

The term *effective ground-fault current path* is also used in the performance provisions of Section 250.4 in Article 250. The purpose of an effective ground-fault current path is to carry heavy levels of fault current during ground-fault conditions for as long as it takes the overcurrent protective device to open the circuit.

Effective ground-fault current paths are built in the field by electrical installers, so their effectiveness can be evaluated and measured as part of engineering design or inspection. **See Figure 3-20.** An example of an effective ground-fault current path is a length of RMC installed for a branch circuit. Since RMC is recognized as an EGC in Section 250.118, the integrity of the conduit to serve in this fashion is related to adequate and effective support, tightness of fittings, and proper sizing. This is why the definition of the term indicates that this path is intentionally constructed. The effective ground-fault current path is usually installed simultaneously and in conjunction with conductors for feeders and branch circuits. Workmanship is important in constructing an effective ground-fault current path.

> **Ground-Fault Current Path.** An electrically conductive path from the point of a ground fault on a wiring system through normally non–current-carrying conductors, equipment, or the earth to the electrical supply source.[17]

A ground-fault current path(s) will be any and all conductive paths that fault current will be present in as it is returning to its source. Ground-fault current paths can consist of combinations of EGCs, metallic raceways, metallic cable sheaths, electrical equipment, and any other electrically conductive material such as metal water and gas piping, steel framing members, stucco mesh, metal ducting, reinforcing steel, shields of communications cables, and the Earth itself. The difference between ground-fault current paths and effective ground-fault current paths is that the effective ground-fault current paths are intentionally constructed for fault current-carrying purposes, whereas the ground-fault current path includes any conductive path between the source and the point of the ground fault on a system.

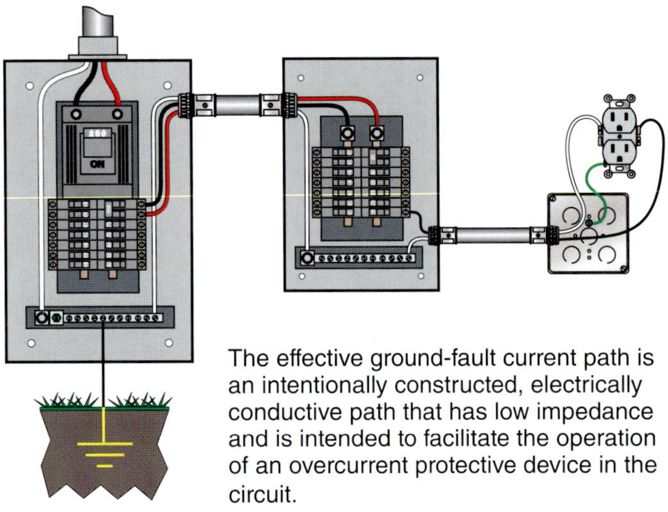

Article 200— Identification and Use of Grounded Conductors

Article 200 provides information regarding the use and identification of grounded conductors. In general, grounded conductors are identified using the colors white or gray. **See Figure 3-21.** Conductors larger than 6 AWG (American wire gauge) are permitted to be identified by a continuous white or gray insulation color or by three continuous white stripes along the entire length of a conductor insulation any color other than green; alternatively, a distinctive white or gray marking can be used at the conductor terminations.

The effective ground-fault current path is an intentionally constructed, electrically conductive path that has low impedance and is intended to facilitate the operation of an overcurrent protective device in the circuit.

FIGURE 3-20 Effective ground-fault current paths are constructed in the field by electrical installers.

The marking is typically applied in the field and must encircle the conductor or insulation. The *Code* does not specify how the marking should be affixed to the conductor. Many electrical workers use a stripe of vinyl marking tape close to where the conductor terminates. A good practice is to use enough marking tape to ensure that the conductor is readily identified within the enclosure where it is occupying enclosure space with ungrounded conductors and EGCs. This helps differentiate them from one another. Article 200 also provides rules related to the terminal identification on electrical devices such as the rules in Sections 200.9 and 200.10. These rules indicate that the terminal on a device intended for the grounded conductor has to be substantially white in color so as to readily distinguish the grounded conductor terminal from the terminals for ungrounded conductors. The terminal for a grounded conductor is often identified using a silver or chrome-colored screw or terminal.

Article 250 — Arrangement and Use

The scope of Article 250 does not specify different voltages, meaning that the rules in the article apply to installations at all voltage levels. Article 250 consists of 10 parts that are identified by the Roman numerals I through X. Each part addresses specific requirements for grounding and bonding at various points in an electrical installation. Figure 250.1 is a schematic of the article and can assist users in quickly and accurately finding the appropriate rules.

Part I — General

Part I includes rules related to the system, circuit, and equipment grounding. Here the determinations can be made as to whether a system, equipment, or circuit is required to be grounded, permitted to be grounded, or is not permitted to be grounded.

Section 250.1 is the scope of Article 250 and provides information about the general requirements for grounding and bonding and specific rules as provided in Sections 250.1(1) through (6). These sections address the following throughout the article:

- Systems, circuits, and equipment that are required, permitted, or not permitted to be grounded, and which circuit conductor must be grounded
- Location of grounding conductor connections to systems, sizes and types of grounding and bonding conductors, and the methods of grounding and bonding
- Conditions that permit guards, isolation, or insulation as alternatives to grounding

Following is a description of each part of Article 250:

Part I — General
Part II — System Grounding
Part III — Grounding Electrode System and Grounding Electrode Conductor
Part IV — Enclosure, Raceway, and Service Cable Connections
Part V — Bonding
Part VI — Equipment Grounding and Equipment Grounding Conductors
Part VII — Methods of Equipment Grounding
Part VIII — Direct-Current Systems
Part IX — Instruments, Meters, and Relays
Part X — Grounding of Systems and Circuits Over 1 kV

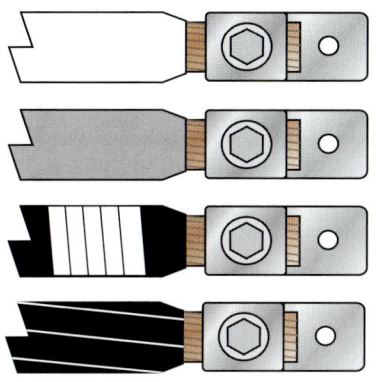

FIGURE 3-21 Identification requirements for grounded conductors are provided in *NEC* Section 200.6.

Part I of Article 250 includes all general provisions that apply in addition to the requirements in the subsequent parts of the article. An important feature of Part I is Figure 250.1, which immediately follows Section 250.1. Figure 250.1 serves as a road map or blueprint of Article 250. A closer review of Figure 250.1 reveals that it is presented in an order similar to the order of the various steps of installation in constructing grounding and bonding systems. An important feature of this figure is that bonding functions are common to all parts of the article. **See Figure 3-22.** The box on the right of the figure contains "Part V Bonding," and bonding is inherent to all other parts. This training material is crafted to parallel how projects are built and the specific aspects of grounding and bonding are covered in detail in these steps.

Part I of Article 250 provides general requirements for grounding and bonding and includes some important performance language. Parts II, VIII, and X contain the requirements for system grounding, which includes systems that supply services and separately derived systems installed on the premises. Part II of Article 250 is used to make the determination as to when these systems are required to be grounded and which rules apply. Once this is known, the rules that follow apply to constructing the appropriate grounding and bonding circuits. Part III of the article is used to determine the grounding electrode system and includes rules for installing grounding electrodes and grounding electrode conductors. When a building is being constructed from the ground up, grounding electrodes are often formed as part of the construction process. Examples include the foundation or footing, a metallic underground water piping system supplying the building, structural metal frames that qualify as grounding electrodes, and so forth. This is one of the first tasks electrical workers perform on a construction site. Part IV or the article is all about enclosure, raceway, and serviced cable grounding.

Table 250.66—Sizing Grounding Electrode Conductors

Table 250.66 is titled "Grounding Electrode Conductor for Alternating-Current Systems." This table is also used for sizing various other conductors in the grounding and bonding scheme. When used for sizing a grounding electrode conductor, the maximum size required is 3/0 copper or 250 kcmil aluminum or copper-clad aluminum conductor.

1. Table 250.66 is applied using sizes of the largest ungrounded service-entrance conductor or largest ungrounded conductor of a separately derived system.
2. This table is not based on overcurrent device ratings, in contrast to Table 250.122.

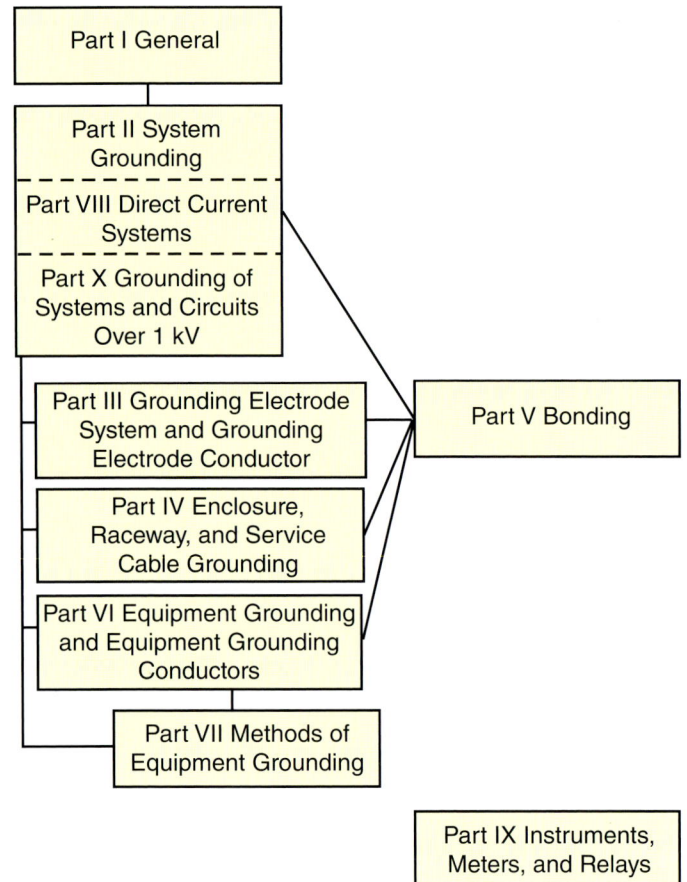

FIGURE 3-22 The order of Article 250 parallels the order of many construction projects. Shown here is a reproduction of Figure 250.1 from the *NEC*.[18]

Sizing grounding electrode conductors requires the use of Table 250.66 and Section 250.66 to determine the minimum sizes. Grounding electrode conductors need not be larger than the sizes in that table. **See Figure 3-23.**

When Table 250.66 is used for sizing equipment bonding jumpers (supply side), grounded conductors (minimum), or main bonding jumpers/system bonding jumpers, the minimum sizes in the table apply. There is one exception: When the size of the largest ungrounded service conductor or largest derived system phase conductor exceeds 1100 kcmil copper or 1750 kcmil aluminum, then the 12.5% rule has to be applied. Table 250.66 includes copper conductor sizes and aluminum or copper-clad aluminum sizes. The correct column must be used when determining the minimum size conductor required for an installation. Selecting the incorrect size can compromise safety.

When an engineering design specifies grounding electrode conductors larger than those provided in Table 250.66, the design specification usually takes precedence. The best practice is to install electrical equipment and conductors in accordance with engineering designs and in full compliance with applicable national and local codes.

Table 250.122— Sizing Equipment Grounding Conductors

Table 250.122 is titled "Minimum Size Equipment Grounding Conductors for Grounding Raceway and Equipment." The key feature of this table is that it allows efficient determination of minimum sizes of wire-type EGCs. **See Figure 3-24.** Various factors, such as voltage drop and high available fault currents, can affect the minimum sizes required for wire-type EGCs.

The note at the bottom of Table 250.122 provides users with a valuable reminder that designs must meet the minimum sizes in the table, but in some cases, the size needs to be larger to ensure that an effective path for ground-fault current is achieved. Using this table is fairly simple and is related to the size of fuse or circuit breaker protecting the circuit. For example, if a 1000-ampere feeder is installed in polyvinyl chloride (PVC) conduit and requires a wire-type EGC, the EGC must not be smaller than 2/0 copper or 4/0 aluminum or copper-clad aluminum, based on the values in the table.

Table 250.66 Grounding Electrode Conductor for Alternating-Current Systems (in part without notes)

Size of Largest Ungrounded Service-Entrance Conductor or Equivalent Area for Parallel Conductors (AWG/kcmil)		Size of Grounding Electrode Conductor (AWG/kcmil)	
Copper	Aluminum or Copper-Clad Aluminum	Copper	Aluminum or Copper-Clad Aluminum
2 or smaller	1/0 or smaller	8	6
1 or 1/0	2/0 or 3/0	6	4
2/0 or 3/0	4/0 or 250	4	2
Over 3/0 through 350	Over 250 through 500	2	1/0
Over 350 through 600	Over 500 through 900	1/0	3/0
Over 600 through 1100	Over 900 through 1750	2/0	4/0
Over 1100	Over 1750	3/0	250

FIGURE 3-23 Reproduction of NEC Table 250.66.[19] Table 250.66 is used for sizing grounding electrode conductors. Reprinted with permission from NFPA 70–2011, *National Electrical Code*®, Copyright © 2010, National Fire Protection Association, Quincy, MA 02169. This reprinted material is not the complete and official position of the NFPA on the referenced subject, which is represented only by the standard in its entirety.

Table 250.122 Minimum Size Equipment Grounding Conductors for Grounding Raceway and Equipment (in part)

Rating or Setting of Automatic Overcurrent Device in Circuit Ahead of Equipment, Conduit, etc., Not Exceeding (Amperes)	Size (AWG or kcmil)	
	Copper	Aluminum or Copper-Clad Aluminum*
15	14	12
20	12	10
60	10	8
100	8	6
200	6	4
300	4	2
400	3	1

Note: Where necessary to comply with 250.4(A)(5) or 250.4(B)(4), the equipment grounding conductor shall be sized larger than given in this table.

*See installation restrictions in 250.120.

FIGURE 3-24 Reproduction of NEC Table 250.122.[20] Table 250.122 is used for sizing wire-type equipment grounding conductors. Reprinted with permission from NFPA 70–2011, *National Electrical Code*®, Copyright © 2010, National Fire Protection Association, Quincy, MA 02169. This reprinted material is not the complete and official position of the NFPA on the referenced subject, which is represented only by the standard in its entirety.

The minimum-size copper EGC for an 80-ampere branch circuit is 8 AWG. The minimum-size aluminum EGC for a 400-ampere feeder is 1 AWG. Table 250.122 includes copper conductor sizes and aluminum or copper-clad aluminum sizes. The correct column must be used when determining the minimum size required for an installation. Selecting the incorrect size can compromise safety and the integrity of an effective ground-fault current path.

The *NEC* includes a grounding symbol in Sections 250.126 and 406.9. **See Figure 3-25.** These figures are in the form of informational notes because it is understood that there are other methods of identifying equipment grounding terminals on devices such as receptacles and switches.

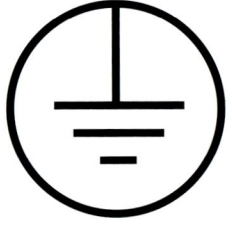 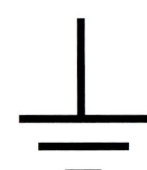

Informational Note Figure 250.126   **Grounding Symbol without Circle**

FIGURE 3-25 Informational Note Figure 250.126 provides an example of a symbol used to identify the grounding termination point for an EGC.[21]

Part X of the article provides the rules addressing specific requirements that apply to systems greater than 1000 volts. The general requirements in preceding parts of the article apply unless specifically addressed within Part X. A good sample application of this concept is sizing EGCs for medium-voltage feeders. Section 250.190 indicates that EGCs that are not an integral part of a cable assembly must not be smaller than 6 AWG copper or 4 AWG aluminum or copper-clad aluminum. Other than those conditions, the EGCs for these feeders are sized per the guidelines in Table 250.122. The purpose of medium- and high-voltage cable tape shields and stranded shields that are part of the assemblies (Section 310.6) is to address voltage stresses to insulation. The point here is that EGCs must have sufficient capacity as required in Sections 250.122 and 250.4.

Special Occupancies, Equipment, and Conditions

Section 90.3 indicates that Chapters 5, 6, and 7 of the *NEC* modify or supplement the general provisions in Chapters 1 through 4. Often the grounding and bonding rules in these later chapters are more restrictive than the general requirements. All general rules in Chapters 1 through 4 apply *in addition to* any modifications required by Chapters 5, 6, or 7.

Example Modification from Article 517 in Chapter 5

Sections 517.13(A) and (B) address branch circuits serving patient care locations and require two separate EGC paths in the form of (1) a suitable metallic raceway or cable armor and (2) a contained insulated copper EGC. **See Figure 3-26.** This provides multiple EGC paths to ensure there is effective grounding and bonding and ensure overcurrent protective device operation in the event of a ground fault on circuits serving these locations. This is a modification of the general equipment grounding requirements contained in Part VI of Article 250. These modifications are more restrictive than the general rules and affect branch circuits in health care facilities.

Example Modification from Article 600 in Chapter 6

Section 600.7(B)(7) permits a bonding conductor greater than or equal to 14 AWG copper for neon transformers that are supplied by 20- or 30-ampere branch circuits. **See Figure 3-27.** Normally these bonding conductors would be based on the rating of the fuse or circuit breaker protecting the branch circuit supplying the equipment, which would be 12 AWG or 10 AWG copper, respectively.

Because a transformer is supplying these signs and outline lighting systems, the secondary circuits supplying the luminous tube(s) are high voltage. This results in secondary currents that are only in the milliamp range.

In addition to the low current on the secondary side, these circuits for sign and outline lighting systems are required to be equipped with transformers that provide secondary-circuit ground-fault protection. The 14 AWG bonding conductor is required to keep conductive parts on the secondary side of such systems at or as close to ground (Earth) potential as possible.

Example Modification from Article 770 in Chapter 7

Section 770.100 requires non–current-carrying conductive members of optical fiber cables to be grounded by the methods specified in 770.100. This modifies the rules in Article 250.[22]

Chapter 8

Section 800.100(A)(3) indicates that the bonding conductor or grounding electrode conductor for communications systems and equipment shall not be smaller than a 14 AWG copper conductor. This is a modification to the normal use of Table 250.66 for sizing grounding electrode conductors for electrical services or separately derived systems. Article 800 provides the specific sizing and installation requirements for grounding electrode conductors for communications systems.

Chapter 8 covers communications systems installations and equipment. This chapter is not subject to the requirements in Chapters 1 through 7 except where the rules are referenced from Chapter 8.

The grounding and bonding rules for these limited-energy systems and equipment provide different safety benefits. Although the grounding and bonding concepts are the same, what is intended to be accomplished differs for limited-energy communications systems. These grounding and bonding schemes are not intended to facilitate overcurrent protective device operation. The requirements deal more with connecting equipment and cable shields to the ground (Earth) and bonding to minimize potential differences. The grounding and bonding rules in Chapter 8 provide safety from events that can cause a voltage rise on these systems.

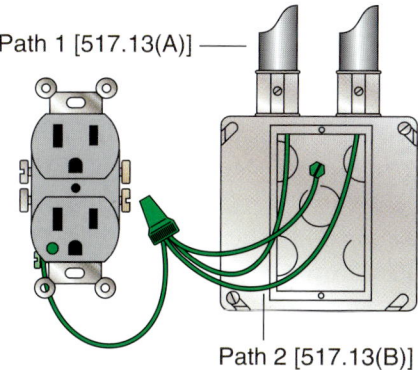

FIGURE 3-26 Branch circuits serving patient care locations require two EGC paths.

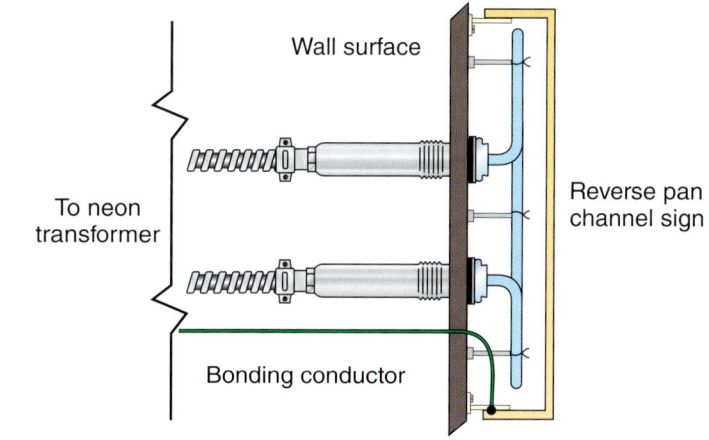

FIGURE 3-27 A 14 AWG copper conductor is required for bonding metal parts associated with a neon outline lighting system high-voltage secondary circuit.

Summary

A thorough working knowledge of the *NEC* is essential to fully understand electrical grounding and bonding and how the rules apply to installations and systems. It is equally important to understand defined terms related to grounding and bonding and an overview of the arrangement of Article 250. The *NEC* serves as the basis for the information in this training material and should be used in conjunction with this textbook. Proper use and application of the *NEC* is essential not only for students of the *Code* but for seasoned professionals as well.

References

1. NFPA 70 National Electrical Code 2011, Section 250.68 (National Fire Protection Association, Quincy, MA 2010), p. 70–115.
2. NFPA 70 National Electrical Code 2011, Article 100 (National Fire Protection Association, Quincy, MA 2010), p. 70–26.
3. NFPA 70 National Electrical Code 2011, Article 100 (National Fire Protection Association, Quincy, MA 2010), p. 70–26.
4. NFPA 70 National Electrical Code 2011, Article 100 (National Fire Protection Association, Quincy, MA 2010), p. 70–26.
5. NFPA 70 National Electrical Code 2011, Article 100 (National Fire Protection Association, Quincy, MA 2010), p. 70–26.
6. NFPA 70 National Electrical Code 2011, Article 100 (National Fire Protection Association, Quincy, MA 2010), p. 70–29.
7. NFPA 70 National Electrical Code 2011, Article 100 (National Fire Protection Association, Quincy, MA 2010), p. 70–29.
8. NFPA 70 National Electrical Code 2011, Article 100 (National Fire Protection Association, Quincy, MA 2010), p. 70–29.
9. NFPA 70 National Electrical Code 2011, Article 100 (National Fire Protection Association, Quincy, MA 2010), p. 70–29.
10. NFPA 70 National Electrical Code 2011, Article 100 (National Fire Protection Association, Quincy, MA 2010), p. 70–29.
11. NFPA 70 National Electrical Code 2011, Article 100 (National Fire Protection Association, Quincy, MA 2010), p. 70–29.
12. NFPA 70 National Electrical Code 2011, Article 100 (National Fire Protection Association, Quincy, MA 2010), p. 70–29.
13. NFPA 70 National Electrical Code 2011, Article 100 (National Fire Protection Association, Quincy, MA 2010), p. 70–29.
14. NFPA 70 National Electrical Code 2011, Article 100 (National Fire Protection Association, Quincy, MA 2010), p. 70–26.
15. NFPA 70 National Electrical Code 2011, Section 250.2 (National Fire Protection Association, Quincy, MA 2010), p. 70–100.
16. NFPA 70 National Electrical Code 2011, Section 250.2 (National Fire Protection Association, Quincy, MA 2010), p. 70–100.
17. NFPA 70 National Electrical Code 2011, Section 250.2 (National Fire Protection Association, Quincy, MA 2010), p. 70–101.
18. NFPA 70 National Electrical Code 2011, Figure 250.1 (National Fire Protection Association, Quincy, MA 2010), p. 70–101.
19. NFPA 70 National Electrical Code 2011, Table 250.66, (National Fire Protection Association, Quincy, MA 2010), p. 70–115.
20. NFPA 70 National Electrical Code 2011, Table 250.122, (National Fire Protection Association, Quincy, MA 2010), p. 70–125.
21. NFPA 70 National Electrical Code 2011, IN Figure 250.126, (National Fire Protection Association, Quincy, MA 2010), p. 70–125.
22. NFPA 70 National Electrical Code 2011, Section 770.100, (National Fire Protection Association, Quincy, MA 2010), p. 70–661.

Review Questions

1. The purpose of the *NEC* requirements is to provide for _____ of persons and property from hazards that arise from the use of electricity.
   a. Complete protection
   b. Isolation
   c. Practical safeguarding
   d. Exclusion

2. Which of the following can be compared with the foundation or footing of a building or structure?
   a. Equipment grounding system
   b. Overcurrent protective device system
   c. Signal reference grid structure
   d. Grounding electrode system

3. Which of the *NEC* chapters modify the general requirements in the first four chapters?
   a. Chapters 5, 6, and 7
   b. Chapters 5 through 8
   c. Chapter 9
   d. Chapter 8

4. Chapter 8 is not subject to any of the requirements provided elsewhere in the *NEC*.
   a. True
   b. False
5. The _____ has the authority for making interpretations and enforcing *NEC* requirements.
   a. Contractor
   b. Engineer
   c. Owner
   d. Inspector or authority having jurisdiction
6. Exceptions to rules in the *NEC* modify only the rule they follow, except if stated differently in the exception.
   a. True
   b. False
7. Mandatory requirements in the *NEC* are characterized by the use of which of the following terms?
   a. *May*
   b. *Can*
   c. *Shall* or *shall not*
   d. All of the above
8. Informational notes that follow rules in the *NEC* are enforceable as requirements as determined by the authority having jurisdiction.
   a. True
   b. False
9. Permissive provisions in the *NEC* are characterized by the use of which of the following phrases?
   a. *Shall not be required* or *shall be permitted*
   b. *Shall not*
   c. *Shall be required*
   d. *May be required*
10. The *NEC Style Manual* indicates that if a term in the *NEC* appears in more than two articles, it qualifies for a definition in which of the following?
    a. The articles where the term appears
    b. Article 90
    c. Article 100
    d. Table of Contents
11. Definitions of terms related to grounding and bonding and used only in Article 250 are located in _____.
    a. Article 100
    b. Article 200
    c. Chapter 9
    d. Section 250.2
12. A conducting object through which a direct connection to the Earth is established best defines which of the following?
    a. Equipment grounding conductor
    b. Main bonding jumper
    c. The Earth
    d. Grounding electrode
13. Article 200 of the *NEC* provides rules for which of the following in the electrical system?
    a. Grounded conductors
    b. Bonding jumpers
    c. Equipment grounding conductors
    d. Grounding electrodes
14. Article 250 in the *NEC* provides the requirements for grounding and bonding and is made up of how many parts?
    a. 6
    b. 2
    c. 8
    d. 10
15. Part X of Article 250 provides requirements for _____.
    a. Grounding of systems and circuits over 1 kV
    b. Methods of equipment grounding
    c. Direct-current systems
    d. Enclosure, raceway, and service cable connections
16. Which of the following provides a graphic road map or blueprint of the arrangement of Article 250 in the *NEC*?
    a. Table of Contents
    b. Figure 250.122
    c. Table 250.1
    d. Figure 250.1
17. Which *NEC* table provides information about sizing wire-type equipment grounding conductors?
    a. Table 250.66
    b. Table 250.122
    c. Table 250.1
    d. Table 8, Chapter 9
18. *NEC* Table 250.66 is used for sizing grounding electrode conductors.
    a. True
    b. False
19. Which of the following is covered by Chapter 5 of the *NEC*?
    a. Health care facilities
    b. Agricultural facilities
    c. Hazardous (classified) locations
    d. All of the above
20. Chapter 6 of the *NEC* provides rules that address special equipment such as elevators, cranes, and electric signs and outline lighting systems.
    a. True
    b. False
21. Article 250 includes conditions that permit guards, isolation, or insulation as alternatives to grounding.
    a. True
    b. False

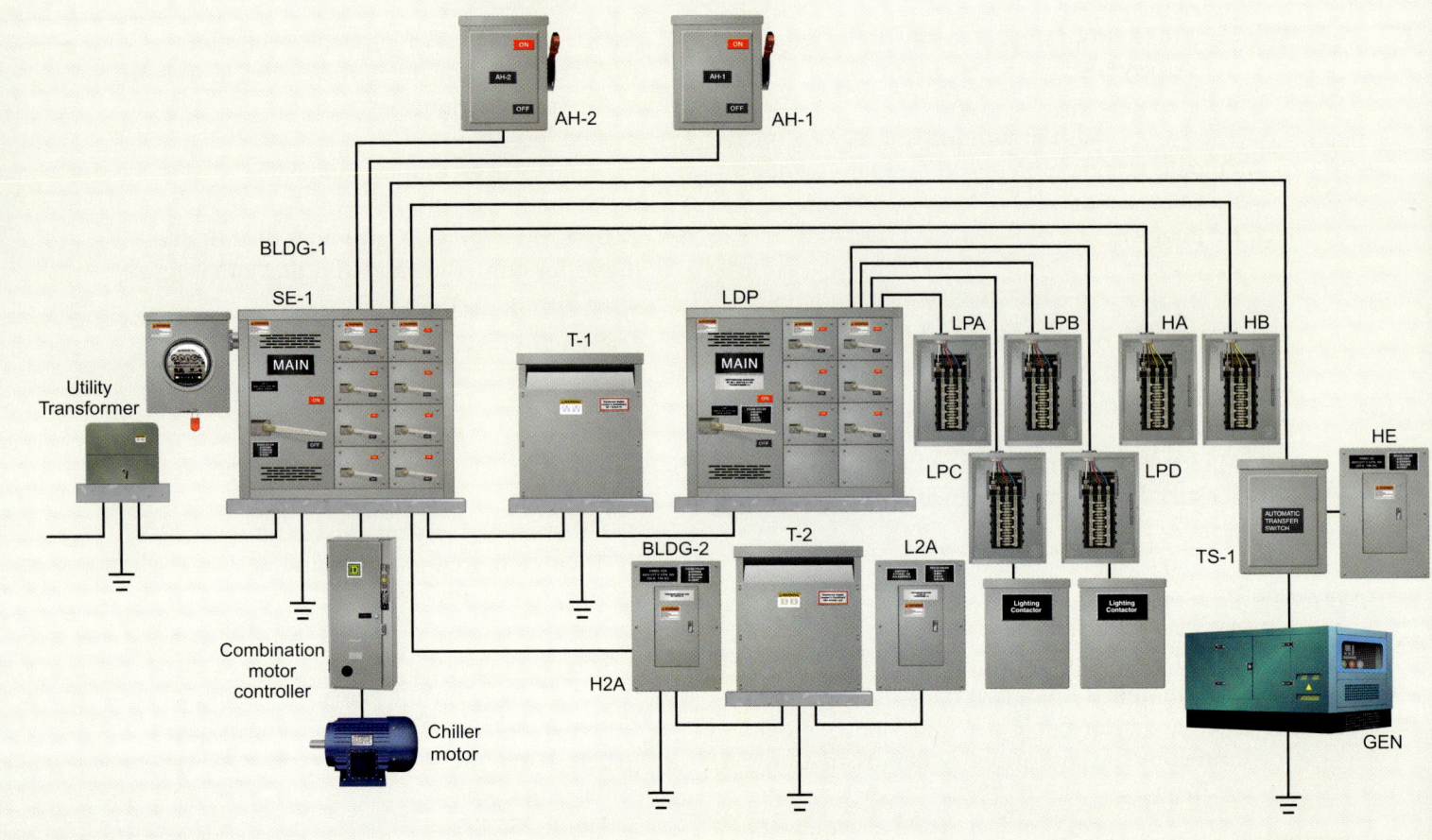

CHAPTER 4

Grounding Electrodes and the Grounding Electrode System

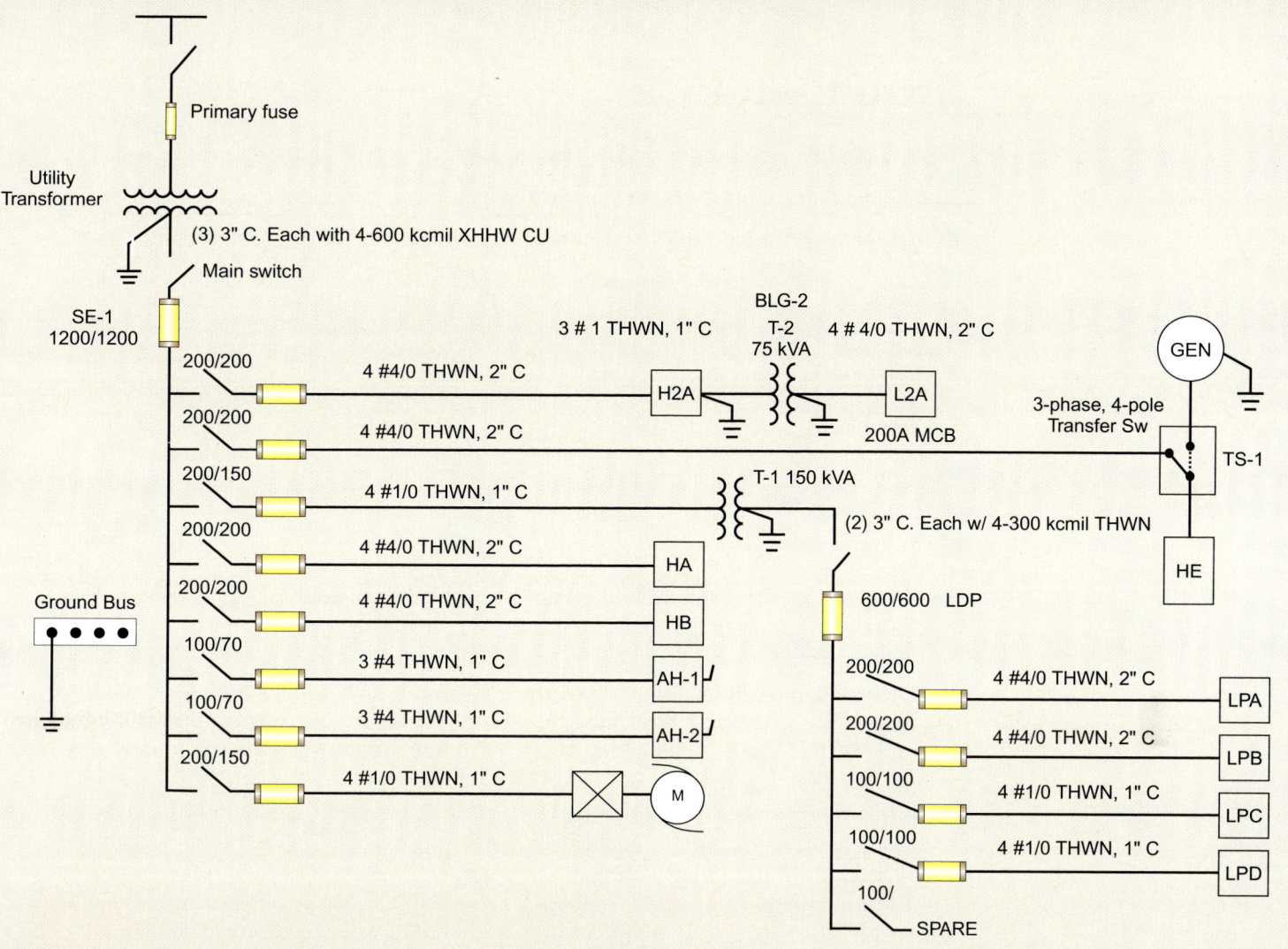

Objectives

- Understand what constitutes a grounding electrode by definition and understand the purpose served by grounding electrodes
- Determine the requirements for a grounding electrode system at a building or structure supplied by electrical services or feeders
- Determine grounding electrodes that are inherent to the construction of a building or structure
- Identify the types of grounding electrodes acceptable in accordance with the *NEC*
- Determine proper installation requirements for grounding electrodes that must be installed
- Understand the role of a lightning protection system in relation to a grounding electrode system for the power service or system of a building

Outline

Grounding Electrode Defined

Purpose and Performance of Electrodes

Grounding Electrode System Requirements

Establishing a Grounding Electrode System

Mandatory Grounding Electrodes

Types of Grounding Electrodes

Grounding Electrode Installation Requirements

Lightning Protection Fundamentals

Introduction

When a construction project begins, one of the first steps in the process is to install a footing and foundation system to support the building or structure. Because footings effectively connect the building to the Earth and provide a foundation of support, they often provide excellent grounding electrodes for electrical services and for systems installed later during construction. The *NEC®* requires that all grounding electrodes at each building or structure served be connected together to form a grounding electrode system. **See Figure 4-1.**

Footings can be very simple and small in size or very complex and massive depending on the size of the structure to be built. Footings and foundations serve as the means of connecting or attaching the building to the Earth. Engineering and designs for building footings and foundations are extremely important, as the weight of the structure must be adequately supported for the life of the building.

The structure must have a solid foundation on which it can be built. As the footing and foundation and various underground systems are being installed during the early phases of construction, grounding electrodes are often established because they are inherent to the construction process; they are part of the building.

Electrical workers are usually on the job during the installation of footings and foundations. Some tasks performed by electricians during this stage of the project are conduit sleeves, large underground conduit for services, feeders and raceways for smaller underground conduit systems, and equipment for branch circuits installed in the building slabs on grade.

The electrical service and system-grounding electrodes provide the foundation for the electrical wiring system in the building.

Existing buildings already have footings and foundations, so the grounding electrodes should already be connected to existing services, separately derived systems, and other equipment. The grounding electrode system must be used when remodeling occurs in existing buildings, for existing equipment, and existing systems.

Grounding Electrode Defined

The term *grounding electrode* was first defined in the 2005 *NEC* and has been revised to include conductive objects or materials that qualify as grounding electrodes and describe how electrodes function. An important feature of this definition is that the electrode is in direct contact with the Earth, making a connection.

Multiple grounding electrodes bonded together form a system of grounding electrodes working together to perform a common function.

FIGURE 4-1 A grounding electrode system is required at each building or structure with each grounding electrode connected to form a grounding electrode system.

This means that the electrode qualifies without a wire connected to it. **See Figure 4-2.** The term *grounding electrode* is defined as follows:

> **Grounding Electrode.** A conducting object through which a direct connection to earth is established.[1]

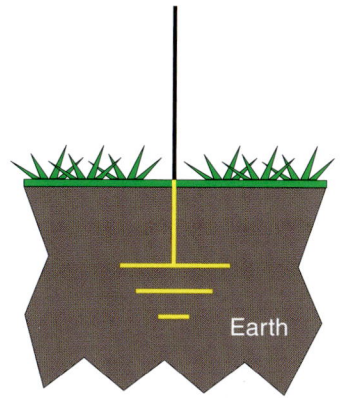

FIGURE 4-2 A grounding electrode is a conductive object through which a direct connection to Earth is established. Additional details are provided in Section 250.52(A).

It is clear from this definition that a grounding electrode not only makes the connection to ground, but it also maintains this connection. A ground rod is not a grounding electrode until it is driven into the Earth to complete the connection. A metal water pipe is not a grounding electrode unless at least 10 feet of the piping is in direct contact with the Earth. The same concept of direct connection to the Earth is found for all grounding electrodes covered by Section 250.52(A). Foundations and footings are usually in direct contact with the Earth and qualify as grounding electrodes. If a vapor barrier exists between a concrete footing and the Earth, it isolates the direct connection, interfering with the effectiveness of the footing as an electrode. When the direct connection between an electrode and the Earth is impaired in this fashion, the footing or foundation does not meet the definition of a grounding electrode. In these cases the footing cannot be used as an electrode because the direct connection to Earth is nullified. It is important to be familiar with construction specifications and local requirements to determine if such vapor barriers are a building code requirement to determine the electrodes that must be used to form the grounding electrode system. Another example of concrete that is isolated from the Earth is if a foundation wall is coated with asphalt sealants or coatings used as vapor and moisture barriers. This affects its ability to establish and maintain a direct connection to the Earth and the ability to perform as an electrode. The concept is simple. A grounding electrode is the connection to the Earth through a conductive material.

Purpose and Performance of Electrodes

The purpose of a grounding electrode is to function as the connection between grounded electrical systems and equipment and the Earth. **See Figure 4-3.** The basic performance of grounding electrodes is provided in the definition of this term.

The *Code* does not provide details about the effectiveness or resistance of a grounding electrode or grounding electrode system, other than for a single rod, pipe, or plate electrode. The *Code* also does not address long-life or end-of-life expectancy of grounding electrodes.

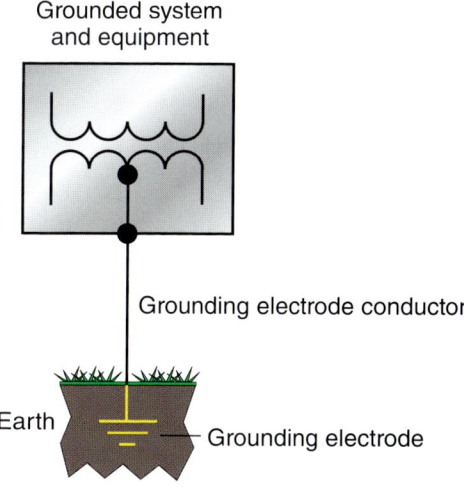

FIGURE 4-3 Grounding electrodes connect systems and equipment to ground. A grounding electrode conductor serves as the path that extends the ground connection.

Grounding electrode performance is essential as long as electrical services and systems are energized. The National Fire Protection Association (NFPA) recently conducted a national electrical grounding research project that analyzed the effects of soil and environmental conditions on grounding electrode integrity and performance. The study focused on the effectiveness of the various grounding electrodes addressed in the *NEC*. The study included periodic resistivity testing for the duration of the study, and the test results were recorded. Some grounding electrodes in the study performed better than others and maintained more consistent resistance measurements over time. This study was conducted using electrodes that were not connected to electrical power sources, so the influence of connected voltages and line surges and other events were not included in the study as normally would be expected for grounding electrodes connected to electrical systems. The test was static only and measured periodic resistivity values that could indicate deterioration effects of grounding electrodes. As the study concluded, some of the test sites were removed, and the electrodes were removed and examined. The findings revealed that some held up physically and electrically very well while others had severe deterioration, indicating a shorter life expectancy. The type of soil conditions in a geographical area appeared to have an influence on the life and effectiveness of the various types of electrodes. This demonstrated that certain types of grounding electrodes seem to be more suited for some locations than others. The Earth is a conductor, but its conductivity is variable and affected by the influence of nature. The connection to the Earth is better when the ground is moist or wet and tends to be less effective in dry or rock soil conditions. It stands to reason that the resistance between the Earth and a grounding electrode varies depending on geographical location; seasonal conditions; mineral content of the soil; and other influencing factors, such as ambient temperatures and Earth temperatures. If more than a single grounding electrode is installed and used, it is treated by the *NEC* as a grounding electrode system. **See Figure 4-4.** Electrodes that form a system benefit from multiple Earth connection characteristics and locations for the same structure.

It is reasonable to conclude that a system of multiple grounding electrodes should perform better than a single grounding electrode. This is among the reasons that any and all grounding electrodes at a building or structure must be interconnected to serve as a system of electrodes for the life of the building or structure. The grounding electrode establishes and maintains a connection to the Earth. It has little or no effect in facilitating overcurrent protective device operation. The Earth is not suitable as an effective ground-fault current path. Many have believed that the Earth was necessary in the safety system to cause overcurrent protective devices to open in ground-fault conditions. It should be understood that the Earth is not to be used as a path for normal current or as an effective ground-fault current path to carry any level of fault current necessary to cause fuses and circuit breakers to open.

> The resistance between the Earth and a grounding electrode varies depending on geographical location; seasonal conditions; mineral content of the soil; and other influencing factors, such as ambient temperatures and Earth temperatures.

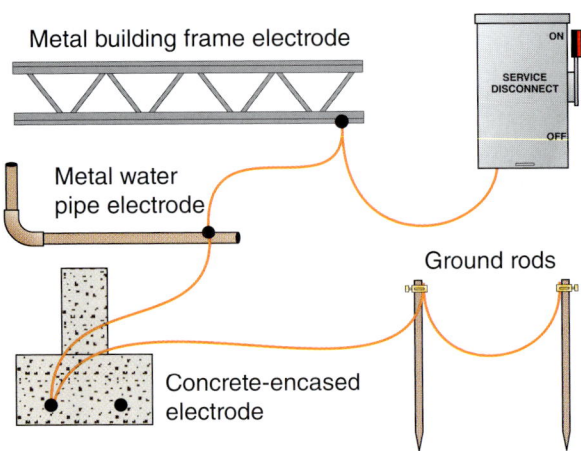

FIGURE 4-4 Multiple electrodes create a grounding electrode system.

Grounding Electrode System Requirements

The *Code* provides rules that state which electrodes have to be used for grounding electrode systems. These requirements are not optional. Grounding electrodes and the grounding electrode system are covered in Part III of Article 250. The first sentence of Section 250.50 outlines the general requirement that all grounding electrodes at a building or structure be used to form a grounding electrode system. **See Figure 4-5.** This section addresses each building or structure served. If a building or structure has no electrical system, there is no *Code* need for a grounding electrode. A network of grounding electrodes for a lightning protection system may be installed on a structure that is not supplied by an electrical power service or system, such as a barn.

The grounding electrodes addressed in Section 250.52(A) include those that are inherent to building construction and electrodes that workers must install. Examples of grounding electrodes that electricians install are ground rods, plates, ground rings, and chemically enhanced electrodes. Footings, building steel, metallic underground water pipes, and local underground conductive systems or structures are examples of grounding electrodes that are inherent to many types of building construction. Section 250.50 is very clear. If there are any electrodes present, they must be used to form the grounding electrode system.

Establishing a Grounding Electrode System

What is a system? A *system* is as an assembly or combination of entities that form together in a complete and functional arrangement. From a simple point of view, a grounding electrode system includes multiple conductive objects (grounding electrodes) connected together to form one collective connection to the Earth through several contact points. There is a bonding requirement in Section 250.50 that is an important aspect of how the grounding electrode system should be developed on the project. As defined, bonding can be accomplished by a direct connection or by other connections that establish continuity and conductivity. Section 250.53(C) addresses bonding jumpers or conductors that are used to connect grounding electrodes together to form a grounding electrode system. **See Figure 4-6.** The bonding jumpers used to form a grounding electrode system must meet the installation requirements in Section 250.64. These bonding jumpers must be sized using Section 250.66, and the connections must meet the rules for grounding electrode conductor connections in Section 250.70.

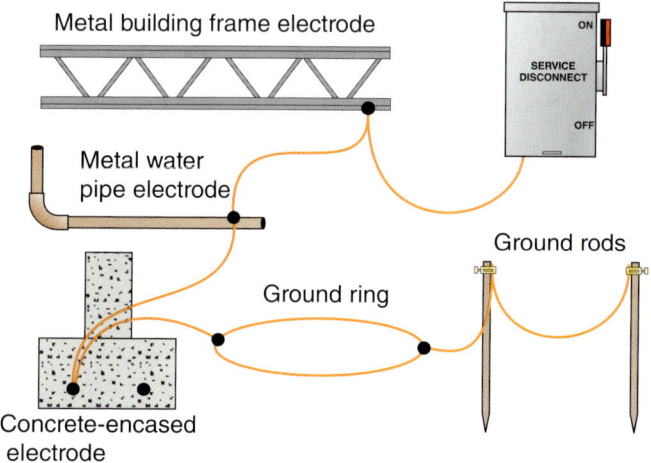

FIGURE 4-5 All grounding electrodes at a building or structure must be bonded together and used as the grounding electrode system.

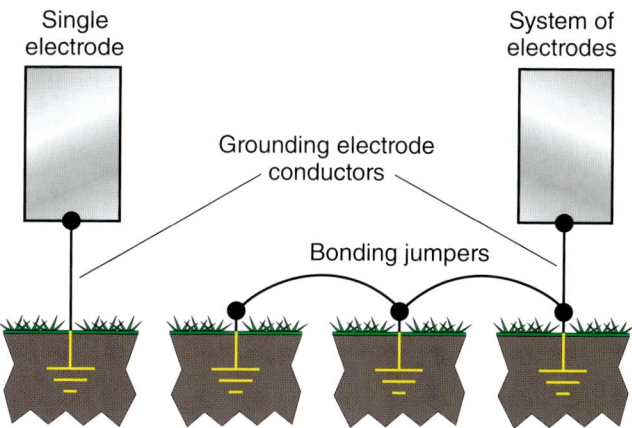

FIGURE 4-6 Bonding jumpers connect grounding electrodes together to form a grounding electrode system.

The bonding jumper size depends on its location in the system and which grounding electrodes it is bonding together. For example, if 500 kcmil ungrounded service conductors are installed at the service and the first electrode connected is an underground metal water pipe electrode, a 1/0 copper grounding electrode conductor is required. If there is a jumper installed between a ground rod and the water pipe, the bonding jumper is required to be sized no smaller than 6 AWG copper or 4 AWG aluminum or copper-clad aluminum. Two factors govern the sizing for grounding electrode conductors and bonding jumpers used to form a grounding electrode system. First, the size of the largest ungrounded supply conductor is used for sizing the grounding electrode conductor and the bonding jumpers. The type of grounding electrodes between which the bonding jumper is installed drives the minimum bonding jumper size.

Mandatory Grounding Electrodes

Grounding electrodes are often installed during building construction. The *Code* offers no option to choose among these grounding electrodes for convenience or any other reason—if the electrodes are present, they must be used in the system. Where there are no grounding electrodes inherent to the construction of a building and the building is supplied with electrical power, a grounding electrode must be installed. The *Code* does not require that more than one grounding electrode be installed to form a system of installed electrodes where none are present, only that a grounding electrode be installed. To that end there are a few choices for installing a grounding electrode such as rod, pipe, or plate electrodes; ground rings; or other listed electrodes. There are various grounding electrode products that are evaluated and listed to UL 467. An example of a listed grounding electrode is a listed chemical electrode assembly. As with any listed product, it is important to follow the manufacturer's installation instructions.

Types of Grounding Electrodes

The types of grounding electrodes recognized in the *NEC* are provided in Section 250.52(A). This section is intended to provide the details and descriptions of each grounding electrode recognized for use by the *NEC*. The list of grounding electrodes in 250.52(A) provides guidance for installers as to the electrode content for the entire grounding electrode system. Specific installation requirements for grounding electrodes are provided in Section 250.53.

Metal Underground Water Pipe Electrodes

The water pipe electrode is described in Section 250.52(A)(1). This electrode has a long history in the *NEC* and is one of the first grounding electrodes to be required for use. A metal water pipe electrode has to have a minimum of 10 feet of metal underground water piping in direct contact with the Earth. **See Figure 4-7.** This requirement does not mention depth, only that direct contact must exist. In the past, the water pipe electrode was the water service piping system routed to a building or structure, which was required by the water service provider to be buried at a minimum depth. That depth was typically based on the frost line and for protection from physical damage. Metal water pipe electrodes can be present outside the building perimeter or inside the footprint of the structure.

The *NEC* prohibits the use of an above ground metal interior water piping system that extends more than 5 feet from the point of entrance into the building to connect grounding electrodes with the grounding electrode conductor.

This limits the length of water piping that can be used as a grounding electrode conductor or conductive path to the electrode. Replacements using nonmetallic piping or fittings can interrupt the electrical continuity of the metal water piping. An exception relaxes this restriction for industrial, commercial, and institutional buildings or structures in which conditions of maintenance and supervision ensure qualified people service the installation. If these conditions are met, the interior metal water piping located more than 5 feet from the point of entry is permitted if it is exposed for the entire length. In recent years, nonmetallic types of piping have replaced metal water piping systems. Once this evolution was recognized, the *Code* included a requirement for any metal water pipe electrode to be supplemented by another grounding electrode. **See Figure 4-8.** The idea is to have a backup in case the original water pipe electrode were to be removed or replaced with a nonmetallic pipe, thus eliminating the grounding connection for electrical services and systems at the building or structure served.

Building or Structure Metal Frame Electrodes

The metal building frame is often a good grounding electrode. Section 250.52(A)(2) states that a metal building frame has to meet specific criteria to qualify as a grounding electrode. The metal frame of the building or structure has to be connected to the Earth by one or both of the following methods:

1. At least one structural metal member that is in direct contact with the Earth for 10 feet or more, with or without concrete encasement. **See Figure 4-9.**
2. The anchor bolts securing the structural steel columns are connected to a concrete-encased electrode that complies with Section 250.52(A)(3) located in the support footing or foundation. The hold-down bolts shall be connected to the concrete-encased electrode by welding, exothermic welding, the usual steel tie wires, or other approved means.[2]

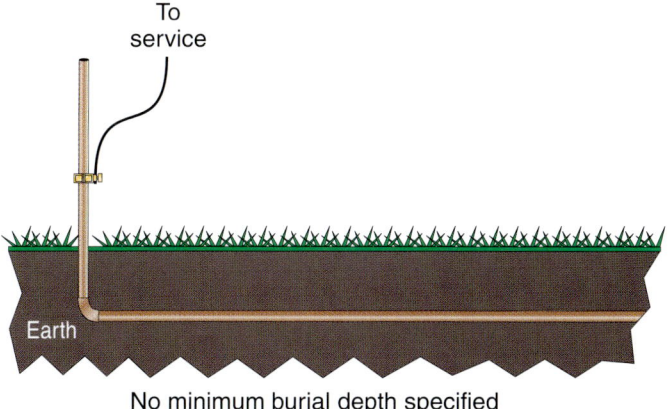

FIGURE 4-7 A water pipe can serve as grounding electrode if a minimum of 10 feet of conductive metal piping is in direct contact with the Earth.

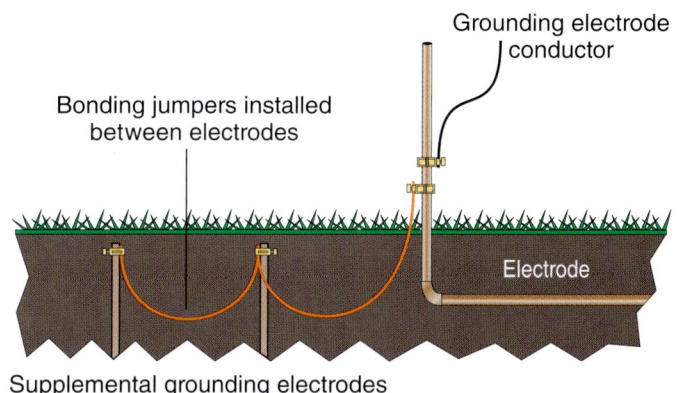

FIGURE 4-8 Water pipe electrodes have to be supplemented by at least one additional electrode.

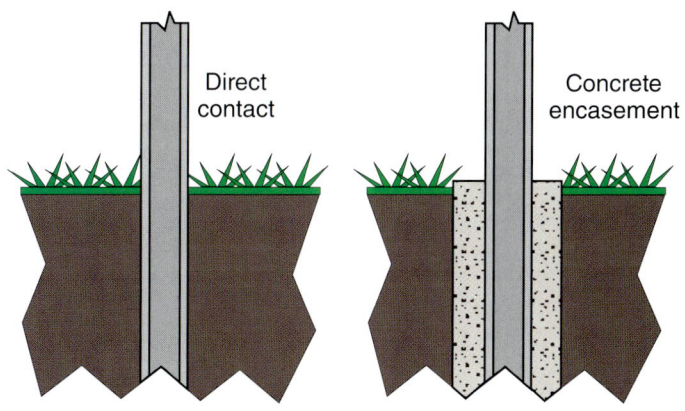

FIGURE 4-9 Metal building frames can serve as grounding electrodes according to Section 250.52(A)(2)(1). The building structure metal frame electrode must be in direct contact with the Earth for 10 feet or more, with or without concrete encasement.

The definition of *grounding electrode* clarifies that the metal frame has to be either in direct contact with the Earth or be connected through an anchor bolt system that is connected electrically to a concrete-encased electrode. Under these two conditions, the metal building frames provide a conductive path to the Earth. Structural metal frame electrodes are fairly reliable because they are unlikely to be disturbed for the life of a building. Section 250.68(C) addresses metal building frames and metal water piping systems that are not electrodes but are conductive paths to grounding electrodes that perform in similar fashion to grounding electrode conductors.

Concrete-Encased Electrodes

Concrete-encased electrodes are described in Section 250.52(A)(3). Section 250.50 requires at least one concrete-encased electrode, described in Section 250.52(A)(3), be included in the grounding electrode system for buildings or structures. This rule applies to all buildings and structures with a foundation or footing having 20 feet or more of ½ inch or larger electrically conductive reinforcing steel or 20 feet of a minimum 4 AWG bare copper conductor encased in concrete to form the electrode. If a concrete-encased electrode is not present at the building or structure supplied, it is not required that wire be used to form one, but it is an option. **See Figure 4-10.**

The *NEC* describes the concrete-encased electrode as bare or zinc galvanized or other electrically conductive coated steel reinforcing bars or rods of not less than ½ inch in diameter, installed in one continuous 20-foot length, or if in multiple pieces, connected together by the usual steel tie wires, exothermic welding, welding, or other effective means to create a 20-foot or greater length. A concrete-encased electrode can also be constructed using 20 feet or more of bare copper not smaller than 4 AWG. Note that the 20 feet of conductive rods or bare wire used in a concrete-encased electrode only establishes the connection to the concrete. The combination of the concrete and the conductive component serve as the grounding electrode as clarified in the definition of the term.

Section 250.50 mandates the use of all grounding electrodes to form the grounding electrode system. This includes all concrete-encased electrodes present at the building or structure. An exception to Section 250.50 relaxes this mandatory requirement for existing buildings and structures in which reaching the concrete-encased electrode could damage the structural integrity of the building or otherwise disturb the existing construction. Because the installation of the footings and foundation is one of the first elements of a construction project and in most cases has been completed by the time the electric service is installed, this rule necessitates an awareness and coordinated effort on the part of designers and the construction trades to ensure that the concrete-encased electrode is incorporated into the grounding electrode system during the placement of rebar and concrete footings.

The concrete-encased electrode has proved that it offers optimal performance and longevity. The footing or foundation of any building will typically be there as long as the building is.

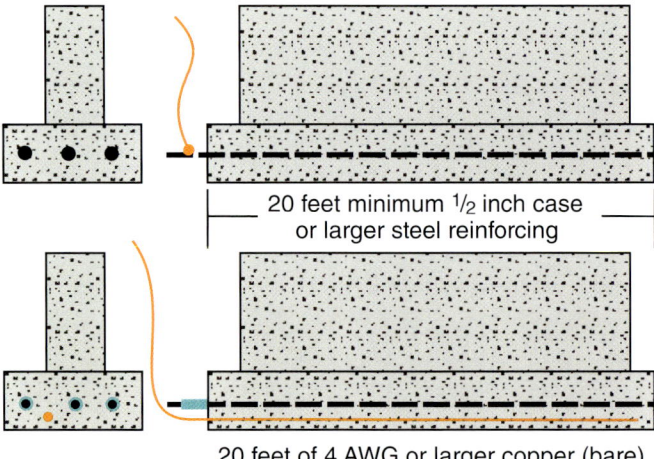

FIGURE 4-10 Concrete-encased electrodes must be included in the grounding electrode system for buildings or structures with concrete footings or foundations that provide not less than 20 feet of reinforcing steel or bare copper wire.

Because all of the rebar in the bottom perimeter of the building footing is usually tied together with tie wires, the electrode acts like a ground ring, only it has much more surface area in the connection to the Earth. The footing is present around the bottom of the building perimeter, which means there is ground (Earth) contact from concrete-encased electrodes. Concrete retains moisture and is continuously absorbing moisture through the bottom of the footing. This keeps the connection between the footing and the Earth effective. The footing of a building is also typically the largest grounding electrode in each building.

The findings of Herbert G. Ufer in the 1940s and 1950s proved the effectiveness of concrete-encased grounding electrodes. Ufer was a vice president and engineer at Underwriters Laboratories who assisted the U.S. military with ground-resistance problems at installations in Arizona. The military required low resistance (5 ohms or less) to ground connections for lightning protection systems installed at its ammunition and pyrotechnic storage sites at the Navajo Ordnance Depot in Flagstaff and Davis-Monthan Air Force Base in Tucson. Ufer developed the initial design for a concrete-encased grounding electrode that consisted of ½-inch reinforcing bars 20 feet in length placed within and near the bottom of 2-foot-deep concrete footings for the ammunition storage buildings. Test readings over a 20-year period revealed steady resistance values of 2 to 5 ohms, which satisfied the specifications of the U.S. government at that time. This work eventually resulted in what we know today as the concrete-encased electrode in the *NEC*. The slang term for this electrode is *UFER* after the engineer who created it as a solution to a significant grounding problem discovered by the U.S. military. More details about the research and findings of Herbert G. Ufer are provided in an article he wrote in October 1964. **See Annex A.**

Ground Ring Electrodes

Ground ring electrodes are described in Section 250.52(A)(4). Ground rings are electrodes that must be installed by electrical workers. The ground ring electrodes are not inherent in the construction of a building. An interesting feature of the ground ring requirement is that it must circle the entire building or structure and be not less than 20 feet. **See Figure 4-11.**

This can result in an extensive amount of work for installers when a ground ring electrode is specified as part of the grounding electrode system. A ring electrode has to be sized not smaller than 2 AWG copper, but it can be larger. Aluminum or copper-clad aluminum is not permitted as a ground ring because of the restrictions in Section 250.52(B). Aluminum is vulnerable to corrosion and deterioration when in contact with the soil. Obviously, an electrode that deteriorates and is lost to corrosive influences will be ineffective. The ring electrode is required to be buried not less than 2½ feet below grade level. This involves trenching and installing the bare conductor in contact with Earth. The ground ring should not be confused with a counterpoise system installed for lighting protection systems. Grounding electrodes for building electrical services and systems cannot be used for lightning protection and vice versa.

FIGURE 4-11 Ground ring electrodes have to circle the entire building or structure, be a minimum 2 AWG copper, and be buried a minimum of 30 inches below grade.

Ground ring installation requirements in the *Code* specify a minimum depth, but do not provide a minimum distance from the exterior walls of a building. Good design practices specify installing it as close as practical to the building yet outside the drip-line from the roof overhang and, when practical, below the permanent moisture level of the Earth, which can vary depending on geographical area. The depth of the building footing is usually a good depth to strive for when installing a ground ring, with the minimum depth at not less than 2½ feet.

Rod and Pipe Electrodes

Rod and pipe electrodes are described in Section 250.52(A)(5) and are required to be at least 8 feet in length. Rod and pipe electrodes must consist of the following materials:

- Pipe or conduit used as grounding electrodes must not be smaller than metric designator 21 (trade size ¾). If the electrode is made of steel, it has to have an outer surface that is galvanized or otherwise metal coated for corrosion protection. A pipe as covered in this section is not the metal underground water piping system previously discussed.

- Stainless steel grounding electrodes and copper- or zinc-coated steel electrodes have to be at least 15.87 mm (⅝ inch) in diameter, unless they are listed by a qualified electrical testing laboratory and are sized not less than 12.70 mm (½ inch) in diameter.[3]

Plate Electrodes

Plate electrodes are described in Section 250.52(A)(7). A plate electrode must have not less than 2 square feet of surface contact between the plate and the soil. A 1-foot square plate with two sides in contact with the Earth accomplishes this. Plate electrodes made of bare or conductively coated iron or steel must be at least ¼ inch in thickness. **See Figure 4-12.** Electrodes of nonferrous metal such as cooper or brass must be at least 0.06 inch in thickness.

Other Local Metal Underground Systems or Structures

The *Code* recognizes local metal underground systems or structures as grounding electrodes in Section 250.52(A)(8). Examples of these types of electrodes are metal piping systems, conductive underground tanks, and underground metal well casings that are not bonded to a metal water pipe. The underground structure or system mentioned in this provision has to meet the criteria of what constitutes a grounding electrode. It must be a conductive object that is in direct contact with the Earth through an established and maintained connection. Any conductive coatings that encapsulate such structures can negate or nullify the effectiveness of such structures as grounding electrodes. The decision to use these types of electrodes should include careful analysis and investigation into the long-term effectiveness of this grounding electrode.

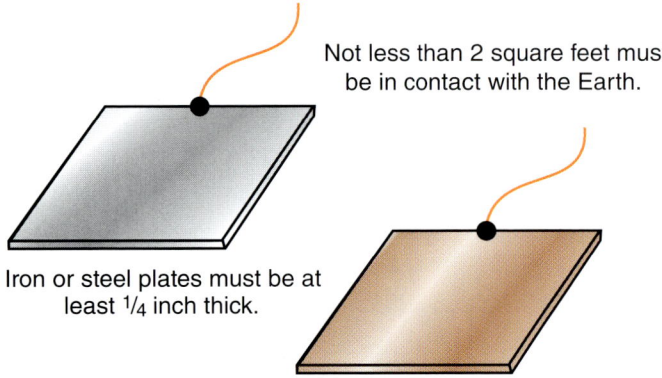

FIGURE 4-12 Plate electrodes must be buried a minimum of 30 inches deep.

Grounding Electrode Installation Requirements

Section 250.53 provides grounding electrode installation rules that apply to those electrodes that are installed by electricians and are not typically inherent in construction, as is the case with building steel, underground metal water piping systems and concrete-encased electrodes. The grounding electrodes addressed in this rule include rod, pipe, and plate electrodes; ground rings; and other listed grounding electrodes such as chemical rods and enhanced grounding systems often specified for information technology equipment rooms.

Rod and Pipe Electrode Installation

Rod, pipe, and plate electrodes are covered in Section 250.52(A)(5). These are all grounding electrodes that must be installed if necessary. If installers choose to use rod, pipe, or plate electrodes because there are no electrodes present for use, a single electrode is required to be supplemented by an additional electrode of a type specified in Sections 250.52(A)(2) through (A)(8). The supplemental electrode is permitted to be connected to the rod, pipe, or plate electrode; the grounding electrode conductor; the grounded service-entrance conductor; a nonflexible grounded service raceway; or any grounded service enclosure. **See Figure 4-13.**

There is an exception that permits using a single rod, pipe, or plate grounding electrode when the single electrode has a resistance to Earth of 25 ohms or less. Demonstrating the resistance to ground of a single electrode is possible by specialized Earth electrode testing methods. The requirement for installing two electrodes reflects how installers are typically handling this requirement in the field. The *NEC* has been improved to require the two installed electrodes as a first choice rather than installing the second when the 25-ohm resistance is exceeded. If a single electrode meets the 25-ohm requirement, there is no need for any additional electrodes to be installed. The resistivity of a single grounding electrode such as a rod, pipe, or plate is determined by testing, although the *NEC* does not directly state that testing is required.

Rod and pipe electrodes also have to be embedded below permanent moisture level when possible. This provides the most effective performance of the electrode. Frost lines can vary geographically, and although this is not typically an issue in the South, it can present electrode connection problems between the Earth and the electrode. Where the Earth freezes, the connection between the Earth and electrode is affected. Rod, pipe, or plate electrodes must be clean and free of nonconductive coatings such as paint, enamel, and other substances that affect the connection between the Earth and the electrode.

> Rod and pipe electrodes also have to be embedded below permanent moisture level when possible. This provides the most effective performance of the electrode.

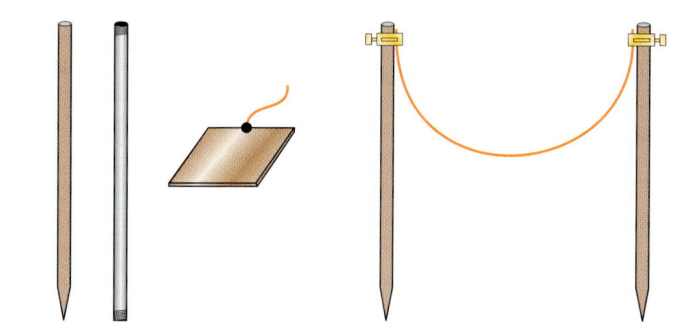

FIGURE 4-13 A single rod, pipe, or plate electrode must be supplemented by an additional electrode. The additional electrode can be any electrode specified in Sections 250.52(A)(2) through (A)(8).

Rod and pipe electrodes have to be driven into the Earth to establish the best contact. When a grounding rod or pipe electrode is installed, it is required to be in direct contact with the Earth for a distance not less than 8 feet. **See Figure 4-14.** Sometimes workers have difficulty driving an 8-foot or 10-foot ground rod or pipe electrode, but the solution is not to cut the exposed portion of the rod off. This results in less than what is required by the *NEC* minimums for electrode contact with the Earth. The *Code* recognizes this challenge and offers alternatives that can be used when the electrode cannot be driven because of rock bottom. The first step in this hierarchy of alternatives is to install the electrode at an angle not more than 45 degrees from vertical.

Often this will solve the installation problems, but, if not, the *Code* offers, as a third and last resort, the option of laying the rod or pipe electrode in a trench not less than 30 inches deep. Installers should be aware that laying the rod or pipe electrode in a trench is only a last resort as indicated by the hierarchy arrangement of Section 250.53(G).

Soil Resistivity and Ground Resistance Testing

There are conditions and installations that require low grounding electrode resistances. The overall resistance is the connection between the electrode and the soil itself. **See Figure 4-15.** The overall resistivity has a lot to do with the connection to the Earth. Resistivity values differ from location to location. In some areas of the world, the soil is rich with mineral content and stays relatively moist throughout the year, keeping the resistivity values low. This provides for low resistivity values between the grounding electrode and the surrounding Earth.

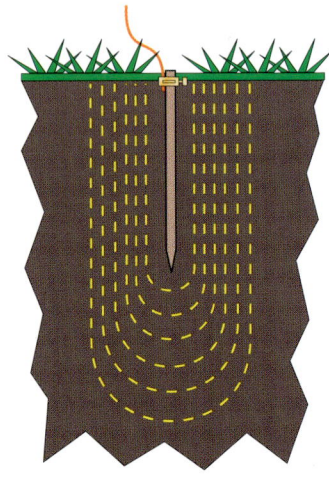

FIGURE 4-15 A single rod electrode connected to the Earth. Resistance in the connection to Earth is a variable. Soil resistivity values differ from location to location—dry and rocky soils tend to have greater soil resistivity, and moist mineral rich soils typically have less resistivity.

In other areas the soil may be sandy or very dry and rocky. These conditions can present challenges in establishing and maintaining good, effective contact between the electrode and the Earth. There are some solutions that can help improve soil resistivity values. One common solution is to increase the size or length of the grounding electrode. Rod or pipe electrodes can be installed deeper into the Earth, which tends to lower the resistance of the grounding electrode connection to the Earth. Special grounding electrode couplings are available for lengthening rod-type electrodes. Installing multiple grounding electrodes is another effective method of lowering the resistance in the grounding electrode connection to Earth.

1. A rod shall have at least 8 feet in contact with the Earth and be driven to a depth of not less than 8 feet.

2. Where rock bottom is encountered, the rod can be driven at a 45-degree angle from vertical.

3. If the rod still cannot be driven at the 45-degree angle, burial at not less than 30 inches shall be permitted.

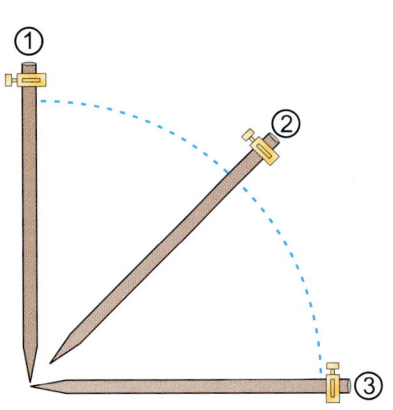

FIGURE 4-14 Rod and pipe installation must meet *Code* requirements or use an alternative method of installation.

If multiple grounding electrodes are used to form a grounding electrode system, the resistance values in the connection to Earth are typically lower because of the benefits of a system of electrodes working together to maintain effective contact with the Earth. Another effective method for lowering the Earth resistance is to treat the soil with suitable chemicals. See Figure 4-16. Special listed chemical electrodes are available for this use. Using chemicals to treat the Earth for resistivity improvement can cause increased effects of corrosion on other metal in the Earth in close proximity.

Chemically treated soil can also be a concern for the environment. Check local environmental regulations in any locations where these types of electrodes are specified in the design or installed by choice. Earth resistance testing applies the basic principles of Ohm's law. Ohm's law is the relationship between voltage, current, and resistance in a basic circuit. Applying voltage and measuring the resultant current can determine a resistance value. Ground resistance can be measured using separate voltage supply, a volt meter, and an ammeter. There are instruments and equipment available that provide electrode-testing functions in a single unit. One common method of Earth resistance testing is the three-point method, sometimes called the "fall-of-potential" or three-terminal grounding resistance testing method. This method uses two test electrodes in addition to the electrode being tested. The testing is required to be performed according to the test instrument manufacturer's instructions. The basic procedure is to ensure that the electrode being tested is not connected to the building electrode system. One test electrode should be installed approximately 100 feet from the electrode under test. The second test electrode should be installed approximately 62 feet from the electrode under test. This test method is also commonly referred to as the 62% test method. The two test electrodes should be in line with the electrode under test. The test leads of the resistance tester need to be connected to the grounding electrode under test and the other temporary testing electrodes. The Earth resistance reading should be measured and recorded. Then the test electrode should be moved in the middle one way or the other in 10-foot increments until the resistivity values are basically the same. This means the plateau area of the test has been determined. If the plateau reading cannot be determined, the test rod 100 feet from the electrode being tested needs to be installed at a greater distance. The testing procedure should then be repeated. All instructions provided by the test equipment manufacturer should be followed.

> An effective method of lowering Earth resistance is to treat the soil with chemical additives and use special listed electrodes designed for this purpose.

Courtesy of Harger Lightning and Grounding

FIGURE 4-16 Soil can be chemically treated to improve resistivity values.

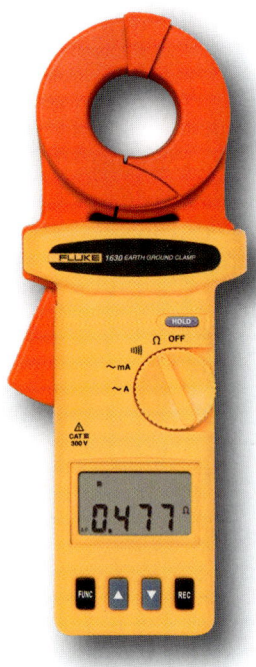

Courtesy of Fluke Corporation

FIGURE 4-17 A clamp-on style electrode resistance meter measures the resistance between a grounding rod and the Earth by transmitting a signal through a grounding electrode conductor and using the receiver of the instrument to establish the resistance of the grounding electrode conductor or system.

There are also clamp-on instruments that can provide ground resistance readings. **See Figure 4-17.** These instruments incorporate a transmitter and receiver and are designed to measure circuit current, resistance, and leakage current. A clamp-on ground resistance meter measures the resistance between a grounding rod and the Earth by transmitting a signal through a grounding electrode conductor and using the receiver of the instrument to establish the resistance of the grounding electrode conductor or system. These can also be used to determine soil resistivity.

Soil resistivity is measured to determine the type of soil conditions, the corrosive effects in the soil, and the best type of grounding electrode system to install. A four-point testing procedure can also be used to measure soil resistivity.

Plate Electrode Installation

When workers install plate electrodes, the installation has to provide a depth of not less than 2½ feet. The connection between the plate and the grounding electrode conductor will have to be buried in the soil, which drives the requirement that the connection means be listed as suitable for direct burial applications. The connection means also has to be compatible with the plate metal and the grounding electrode conductor. Some plate electrodes are available with a lead already connected by means of exothermic welding. **See Figure 4-18.** The exothermic welding process is also a popular means of connecting grounding electrode conductors to buried ground rods. See Figure 4-19.

Electrode Spacing Requirements

If rod, pipe, or plate electrodes are installed, each electrode must be spaced a minimum 6 feet from another electrode of another grounding electrode system. **See Figure 4-20.** This requirement includes those ground terminals (grounding electrodes) for lightning protection systems.

Courtesy of Thomas and Betts

FIGURE 4-18 Exothermic welding connections are often used to connect grounding electrode conductors to plate electrodes.

The purpose of the spacing requirement is to reduce the effects of overlapping spheres of influence associated with each electrode.

Ideally, the spacing distance between these installed electrodes should be not less than twice the depth of the electrode, which for 8-foot rod and pipe types is 16 feet. The *NEC* requires a minimum spacing distance of not less than 6 feet but indicates that the paralleling efficiency of rods is increased by spacing them twice the length of the longest rod. When two or more grounding electrodes are bonded together, such as two ground rods that are installed 6 feet apart, they are considered to be a single grounding electrode system made up of two grounding electrodes.

Water Pipe Electrode Supplement and Bonding Jumpers

Section 250.53(D) indicates that metal underground water pipe electrodes be supplemented by an additional electrode of a type specified in Sections 250.52(A)(2) through (A)(8). A common practice in the field is to install a ground rod to serve as this supplement even though any of the electrodes specified in 250.52(A)(2) through (8) could be used. When the supplemental electrode is a rod, pipe, or plate type, the installation has to meet the provisions in Section 250.53(A), meaning two such electrodes have to be installed unless a single electrode has a resistance of 25 ohms or less. **See Figure 4-21.**

FIGURE 4-19 Exothermic welding connections can also used to connect grounding electrode conductors to ground rod electrodes.

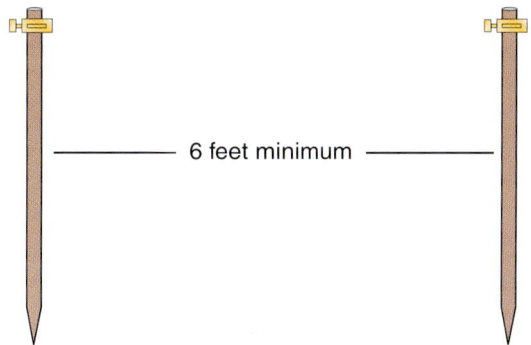

FIGURE 4-20 The *Code* sets grounding electrode spacing requirements. The paralleling efficiency of rod electrodes is typically increased by spacing them not less than twice the length of the longest rod installed.

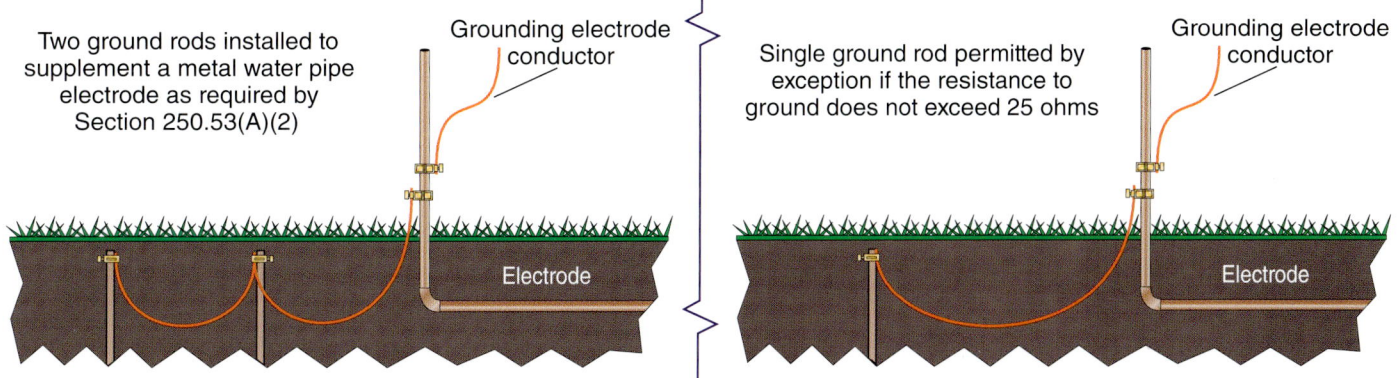

FIGURE 4-21 A water pipe electrode is supplemented by two rod electrodes unless the resistance to ground for a single rod does not exceed 25 ohms.

The supplemental electrode is permitted to be connected to the grounding electrode conductor, the grounded service-entrance conductor, the nonflexible grounded service raceway, or any grounded service equipment enclosure.

When a metal water pipe electrode is used, the continuity of the grounding path or bonding connection to the piping must not depend on water meters, filters, or similar equipment. Bonding jumpers might need to be installed around such equipment. The bonding jumper is required to be installed with enough length to allow for servicing the equipment in the water line without removing the jumper and disrupting the grounding electrode system for the service.

Auxiliary Grounding Electrodes

Auxiliary electrodes are those that are installed by choice and not to meet a requirement in the *NEC*. Auxiliary grounding electrodes are sometimes specified by electrical equipment manufacturers, electrical designers, and facility owners. It should be understood that the auxiliary grounding electrode is a connection to the Earth that is usually in close proximity to the equipment it supplements. The *Code* does not specify a distance from the equipment the auxiliary electrode supplements. An important aspect of the auxiliary grounding electrode installation is that it is connected to equipment that is also connected to an equipment grounding conductor (EGC). A good example of commonly installed auxiliary electrodes is when grounding electrodes are installed at lighting pole bases in parking lots. **See Figure 4-22.** Engineers and designers often specify these auxiliary electrodes to help dissipate lightning events into the Earth at the point of a strike. Another common use for auxiliary electrodes is for certain types of electrical machines where a manufacturer specifies a local grounding electrode connection in addition to the required EGC.

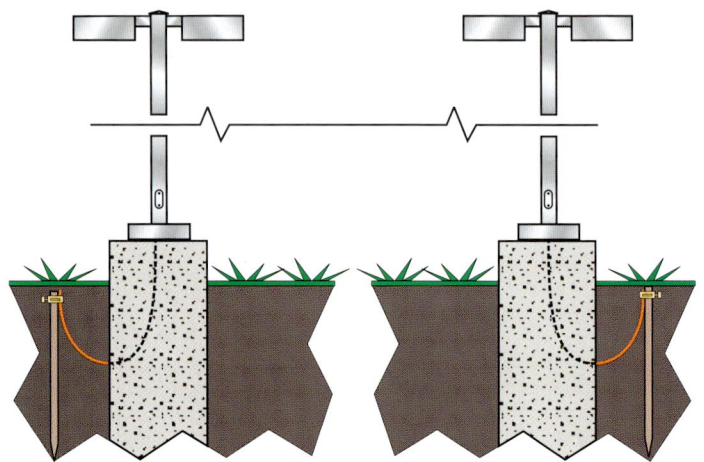

FIGURE 4-22 Auxiliary grounding electrodes supplement the required EGC of the circuit supplying equipment.

FIGURE 4-23 Gas piping systems are not permitted as grounding electrodes, but they are required to be bonded.

Material Not Permitted as Grounding Electrodes

Some conductive materials are not compatible with the Earth and are vulnerable to corrosion and deterioration. Section 250.52(B) provides restrictions for using aluminum materials and metal gas piping systems as grounding electrodes. **See Figure 4-23.**

Obviously there are concerns about corrosion for grounding electrodes that could result in deterioration of the electrode over time. Life expectancy of grounding electrodes is not addressed in the *Code*. However, experience has resulted in the prohibition of metals that are susceptible to corrosion when in contact with the Earth, such as aluminum. The other item addressed in this section is metal gas piping systems. These systems are not permitted to be used as grounding electrodes. This restriction is in parallel with the same restriction in Section 7.13.3 of NFPA 54 *National Fuel Gas Code*. Installers should note that the restriction of these systems is for use as grounding electrodes, but each code (NFPA 70 and NFPA 54) requires these metal piping systems to be bonded. The restrictions in 250.52(B) are very clear that metal gas piping is not permitted for use as a grounding electrode. Typically, gas service utilities require a dielectric fitting that creates isolation between the customer side of a gas meter and the portion of the metal gas system that is on the supply side of the meter and in contact with the Earth.

Grounding Electrodes and Utility Services

The *NEC* permits multiple services and systems to be installed for buildings or structures under specific conditions. Smaller buildings, such as dwelling occupancies, are typically supplied with only one service from a utility. Large buildings often have needs that force the design to include more than one service. Some of the factors that result in more than one service being installed on a building or structure are provided in Section 230.2(A). When an AC system or service is connected to a grounding electrode in or at a building or structure, the *NEC* requires the same electrode be used for grounding conductor enclosures and equipment installed at that building or structure. When more than one service, feeder, or branch circuit supplies a building, they must be connected to the same grounding electrode(s). The reason is to ensure that the same grounding potential is established among all grounded systems and equipment installed in a single building or structure. **See Figure 4-24.** This *NEC* requirement can be satisfied by bonding two or more grounding electrodes or grounding electrode systems together as specified in Sections 250.50 and 250.58.

It is recognized that two or more individual electrodes that are effectively bonded together are considered a single electrode and meet the common grounding electrode requirements in the *NEC*. An example of this type of installation is when two services of different voltage characteristics supply a single building or structure. If a 480Y/277-volt, 3-phase, 4-wire service and a 120/240-volt, 3-phase, 4-wire delta-connected service supply the same building, they must be grounded using the same grounding electrode system.[4]

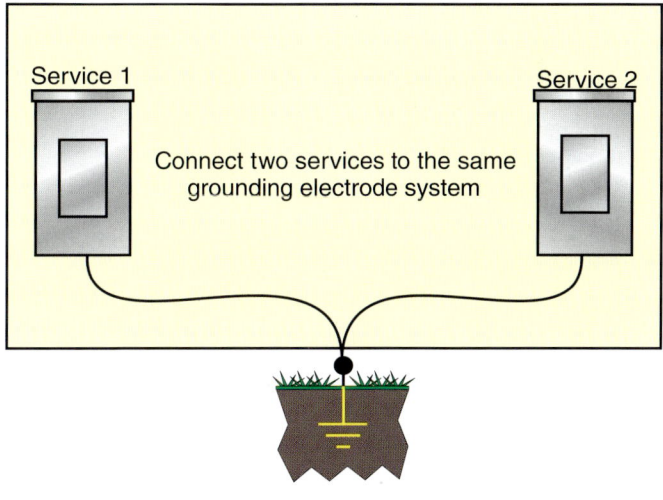

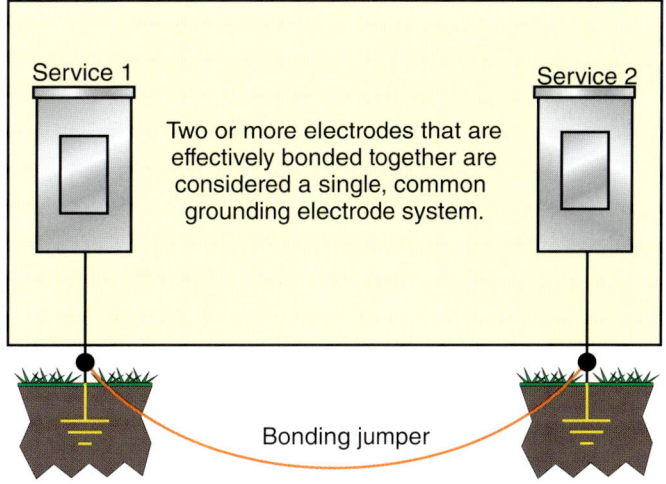

FIGURE 4-24 When more than one service is installed on a structure, the same grounding electrode must be used.

Bonding of Lightning Protection Systems to Service Electrode System

Section 250.60 restricts using a lightning protection system as a grounding electrode for electrical power systems. The requirement to bond the two systems together must be adhered to, as required in Section 250.106. **See Figure 4-25.** This rule specifies that the grounding electrode system (grounding network) of the lightning protection system be bonded to the electrical service grounding electrode system, but it does not specify a size for the bonding conductor.

The size of the bonding conductor is typically determined by the requirements in NFPA 780 and is related to the size of the lightning protection system conductors. NFPA 780 *Standard for Installing Lightning Protection Systems* provides the requirements for lightning protection systems installed on buildings or structures. Bonding may be necessary between the lightning protection system and the electrical system based on proximity and whether separation between the systems is through air or building materials. The lightning protection standard contains important detailed information on grounding, bonding, and side-flash distance from lightning protection systems. It is important to remember the scope of the *NEC* and its limitations to protect people and property from hazards associated with the use of electricity. Lightning is an uncontrollable force and is not directly covered by the rules in the *NEC*, other than where specific interconnections and bonding are required between the systems covered in NFPA 70 and 780.[5]

Lightning Protection Fundamentals

Lightning strikes can cause damage to buildings or structures and the electrical wiring systems installed within those buildings. Lighting protection systems consist of a low-impedance network of strike termination devices that are suitably connected to a special grounding electrode system installed to dissipate lightning into the Earth. Lightning protection systems are an effort to divert direct or indirect lightning strikes around the building or structure and equipment so as to be safely dissipated into the Earth. See Figure 4-26.

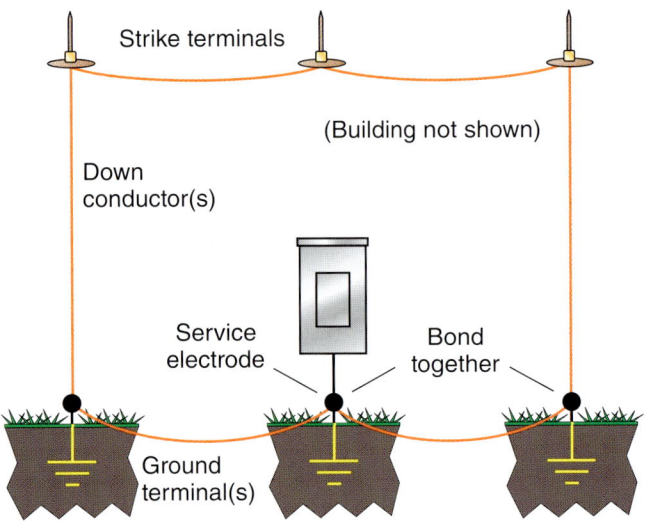

FIGURE 4-25 The electrode system for the electrical supply to a building has to be bonded to the ground network of a lightning protection system.

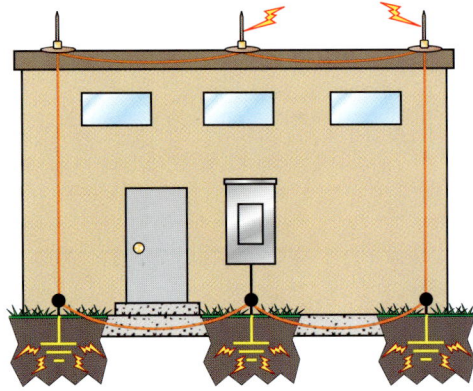

FIGURE 4-26 Lightning protection systems provide a path to Earth for lightning strikes.

Purpose of a Lightning Protection System

Lightning protection systems provide a deliberate pathway to ground for lightning. The lightning protection system must be capable of dissipating the high energy as effectively and directly to ground (Earth) as possible.

The installation of a lightning protection system is no guarantee that equipment inside a building or structure or the building itself will not be damaged by a lightning strike. The system provides a plan to provide a reasonable degree of protection from these events. Some have false beliefs that by installing a lightning protection on a building everything inside is protected from lightning. A lightning protection system is the best-known form of protection from these natural events, but in reality, there are no guarantees. Lightning is an unpredictable force that is continuously being studied to understand. The purpose of NFPA 780 *Standard for Installation of Lightning Protection Systems* is to provide safeguarding of people and property from hazards arising from lightning exposure. Lightning protection systems do not prevent lightning strikes, nor do they attract lightning from distances greater than the conventional attractive area of a building or structure.

> **Lightning Protection System.** A complete system of strike termination devices, conductors (which could include conductive structural members), grounding electrodes, interconnecting conductors, surge protective devices, and other connectors and fittings required to complete the system.[6]

As can be seen in this definition, there are several parts that are necessary to construct a lightning protection system. **See Figure 4-27.**

A closer look at this definition reveals several concepts related to how the system is intended to perform. First, the system intercepts a lightning strike through the strike termination device. Then the force is diverted down to the Earth. This process is accomplished by the specific down conductors or structural components of the building that serve as down conductors. It is important that bonding is provided from conductive parts on or within the building or structure to reduce flash-over possibilities during the event. The grounding electrodes of a lightning protection system provide the dissipation safely into the ground (Earth). And lastly, effectively applied surge protective devices handle any unwanted transient surges attempting to enter the building via the electrical supply system.

System Components

Lightning protection systems consist of several components that make up the entire system. Starting from the top of a structure and working toward the ground, the system includes a strike termination network (usually air terminals or lightning rods), a down conductor network, a grounding terminal or grounding electrode network, an equipotential bonding network, and appropriate surge protection. **See Figure 4-28.**

Courtesy of Harger Lightning and Grounding

FIGURE 4-27 Basic lightning protection system parts are strike termination devices, down conductors, grounding electrodes, interconnecting conductors, and surge protective devices.

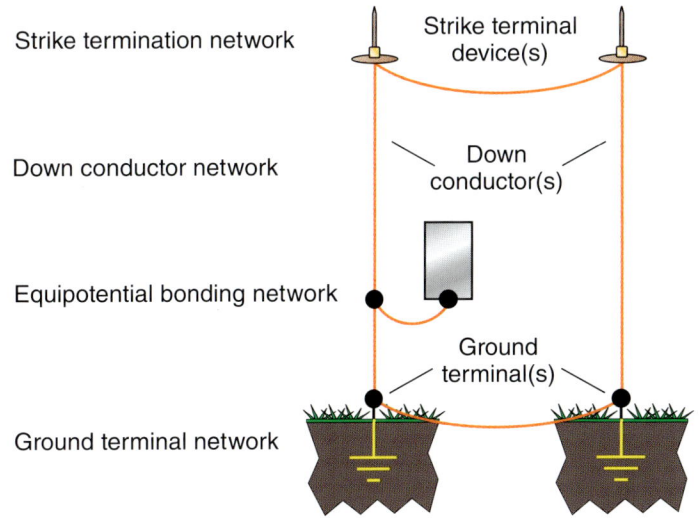

FIGURE 4-28 A lightning protection system consists of several networks.

There are two classes of materials and components used in lightning protection systems. Class II materials and components must be used in buildings taller than 75 feet, and Class I materials are used on structures 75 feet tall or less. Class I materials are typically smaller and lighter than the Class II materials. Copper or aluminum conductors can be used, but attention must be given to areas subject to corrosive influences. The conductors used for lightning protection systems have a finer woven stranding characteristic than those typically used for electrical wiring systems.

Installation Methods and Criteria

In general, each strike termination device must be provided with two separate paths (down conductors) to the ground. **See Figure 4-29.** A main conductor has to be used between strike termination devices and for the down conductors. NFPA 780 provides the minimum size required for main conductors of roof conductors and down conductors.

A low-impedance path is necessary in the down conductor network to reduce opposition in this path to Earth. Care should be taken to minimize the number of bends and to ensure that any necessary bends are long radius (as gradual as possible). A bending radius must never be less than 8 inches. Sharp bends invite flash-over possibilities. If the voltage of a strike exceeds the breakdown voltage of air space between a down conductor and another conductive object, a side flash can occur during a lightning strike. NFPA 780 Section 4.21.2 provides a formula that simplifies the calculation required to determine the probability of side flash. Network conductors on the roof, in addition to the down conductors of the system, must be securely fastened at appropriate intervals. A minimum of two down conductors is required for any application. For structures exceeding 250 feet in perimeter, additional down conductors must be installed. Down conductors should be placed at the corners of a building or structure and be separated as widely as possible in between. Good designs locate down conductors away from public areas or provide suitable protection around the down conductor in these areas. The ground network of a lightning protection system provides a low-impedance connection to the Earth. This low-impedance connection through multiple ground terminals helps minimize peak voltages that would be present on the system during an event. The ground network must be designed and installed so as to reduce the possibility of step and touch potentials. The grounding electrode network of the system is the largest contributor to the level of overvoltage that will appear on the system if the ground connection is not effective. An ineffective grounding network will create opposition to the quick and safe dissipation of a lightning strike. Each down conductor must be terminated to a grounding electrode dedicated to the lightning protection system. The grounding electrodes can be copper, copper-clad steel, or stainless steel types. Electrodes of the ground network can be rods, rings, plates, radials, and concrete-encased electrodes. Connections to the electrodes have to be made by exothermic welding, bolting, brazing, or high compression connections listed for the application. Grounding clamps listed for direct burial application are permitted. **See Figure 4-30.**

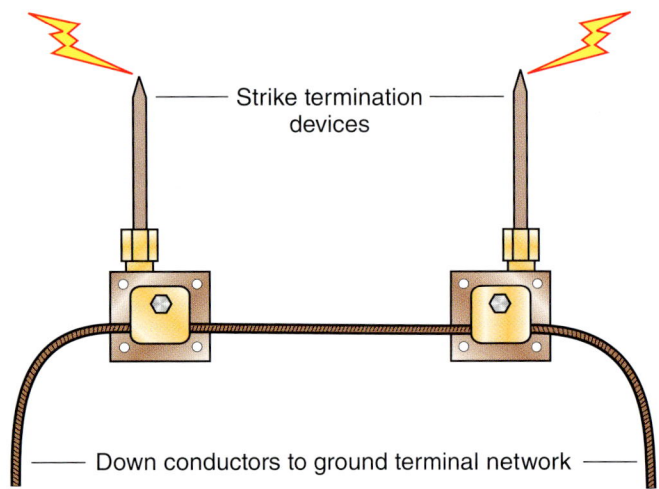

FIGURE 4-29 Two paths are required from air strike termination devices.

When possible, the electrodes should be installed below the frost line. As previously reviewed, the ground network of the lightning protection system must be bonded to the grounding electrode system for the power service supplying the building or structure.

Surge Protective Devices

Line surges can occur for a number of reasons. Among these are direct lightning strikes to incoming lines, or strikes in close proximity to the building that will cause a rise in potential on the incoming lines. This includes the electrical service lines and any limited energy lines such as telephone systems, antenna systems, or broadband communications systems. Other means may need to be taken for protecting internal electrical and electronic systems against lightning electromagnetic pulses. Potential equalization can be achieved at electrical services by installing suitable listed surge protective devices and spatial shielding. All lightning protection systems require protection against incoming surges. **See Figure 4-31.** The degree of surge protection is related to the size of the equipment and system being protected.

Quality Control for Lightning Protection Systems

The Underwriters Laboratories (UL) Master Label program and the Lightning Protection Institute (LPI) Certified System program provide quality assurances for installed lightning protection systems. Each program requires using components that are certified to UL 96. UL issues a master label for systems and components inspected on completion of the installation. **See Figure 4-32.** The components of the system have to be listed to UL 96, and the installation has to meet the requirements in UL 96A. Master Label certificates have to be renewed every five years or if the building changes structurally.

Courtesy of Harger Lightning and Grounding

FIGURE 4-30 Lightning protection conductors use a variety of connection methods.

FIGURE 4-31 Type 2 surge protection is installed at the service on the load side of an overcurrent protective device.

FIGURE 4-32 A UL Master Label provides quality assurance for lightning protection systems.

LPI Certified Systems offer similar quality control assurances. LPI Certified Systems have to be inspected to verify conformance with LPI 175, NFPA 780, and UL 96A as applicable. There are lightning protection installers that have been trained to evaluate and apply appropriately designed systems to any building or structure. Electrical contractors often offer this service as part of their contracting business, or they may subcontract to organizations that specialize in lightning protection system installation and certification. More information about lightning protection system installations and the requirements for such systems are available in NFPA 780 *Standard for the Installation of Lightning Protection System,* UL 96A *Installation Requirements for Lightning Protection Systems*, and LPI 175 *Lightning Protection Institute Standard of Practice*. For information about listed products suitable for use in lightning protection systems, refer to UL 96 *Lightning Protection Components*. **See Annex C.**

Summary

Grounding electrodes have an important role in the grounding and bonding scheme. A grounding electrode is a conductive object that establishes and maintains a connection to Earth. Grounding electrode systems are multiple electrodes used together to form a connection to Earth through multiple Earth connection points. Any and all grounding electrodes that are present at a building or structure served have to be included in the grounding electrode system. Grounding electrodes connect systems and equipment to the Earth and function to maintain those conductive parts at or as close to Earth as possible. Multiple grounding electrodes are recognized for use by the *Code*. A grounding electrode system has to be established at each building or structure served by electricity. The details and descriptions of grounding electrodes are provided in the *NEC* as well as applicable electrode installation rules for those that are installed by electrical workers. Grounding electrodes have little or no effect in the operation of fuses or circuit breakers. While the Earth is a conductor, it is not suitable for use as an effective ground-fault current path. Lightning protection systems must be installed in accordance with specific requirements of NFPA 780 and bonded to the grounding electrode system used for the electrical service serving the building or structure.

References

1. NFPA 70 National Electrical Code 2011, Article 100 (National Fire Protection Association, Quincy, MA 2010), p. 70–29.
2. NFPA 70 National Electrical Code 2011, Section 250.52(A)(3) (National Fire Protection Association, Quincy, MA 2010), p. 70–111.
3. NFPA 70 National Electrical Code 2011, Section 250.52(A)(5) (National Fire Protection Association, Quincy, MA 2010), p. 70–112.
4. NFPA 70 National Electrical Code 2011, Section 250.58 (National Fire Protection Association, Quincy, MA 2010), p. 70–113.
5. NFPA 70 National Electrical Code 2011, Section 250.60 (National Fire Protection Association, Quincy, MA 2010), p. 70–113.
6. NFPA 780 Standard for Installation of Lightning Protection Systems 2011, Section 3.3.19 (National Fire Protection Association, Quincy, MA 2010), p. 7.

Review Questions

1. The Earth is best defined in the *NEC* by which of the following?
   a. Third planet from the sun
   b. Has only one moon
   c. Ground
   d. Dirt
2. Which is best defined as a conducting object through which a direct connection to Earth is established?
   a. An equipment grounding connection
   b. A system bonding connection
   c. A grounding substitute
   d. Grounding electrode
3. As buildings or other structures are constructed, there are often grounding electrodes that are inherent to the building or structure when it is complete.
   a. True
   b. False
4. A footing or foundation of a building or structure is typically a grounding electrode that must be used for the building or structure served by electrical power.
   a. True
   b. False
5. If present, which of the following have to be used to form a grounding electrode system at a building or structure?
   a. Underground metal water pipe
   b. Structural metal building frame
   c. Ground rods
   d. All of the above
6. For a new construction, a concrete-encased electrode is always required to be used in the grounding electrode system.
   a. True
   b. False
7. The purpose of a grounding electrode is to function as the connection between grounded electrical systems and equipment and the Earth.
   a. True
   b. False
8. The *Code* requires that more than one grounding electrode be installed to form a system of installed electrodes when none are present.
   a. True
   b. False
9. Which of the following is not permitted for use as a grounding electrode?
   a. Metal underground water piping
   b. Metal underground structures
   c. Underground metal gas piping
   d. Galvanized piping systems
10. A metal underground water pipe has to have a minimum of _____ in contact with the Earth to qualify as a grounding electrode.
    a. 8 feet
    b. 6 feet
    c. 10 feet
    d. 20 feet
11. The *NEC* does not specify a burial depth for an underground metal water pipe electrode.
    a. True
    b. False
12. Aluminum piping is permitted to be used as a grounding electrode only if other electrodes specified in 250.52(A) are not available for use.
    a. True
    b. False
13. The *NEC* prohibits the use of an interior metal water piping system that extends more than _____ from the point of entrance into the building to interconnect grounding electrodes and the grounding electrode conductor.
    a. 2 feet
    b. 5 feet
    c. 10 feet
    d. 20 feet
14. A structural metal building framing member qualifies as a grounding electrode if it is in direct contact with the Earth for 10 feet or more, with or without concrete encasement.
    a. True
    b. False
15. Structural metal building frames can serve as grounding electrodes even if the anchor bolts securing the structural steel columns are not connected to a concrete-encased electrode that complies with 250.52(A)(3) located in the support footing or foundation.
    a. True
    b. False
16. The *NEC* describes the concrete-encased electrode as one or more bare or zinc galvanized or other electrically conductive coated steel reinforcing bars or rods of not less than _____ in diameter, installed in one continuous _____ length, or if in multiple pieces connected together by the usual steel tie wires, exothermic welding, welding, or other effective means to create a 20 foot or greater length.
    a. $3/8$", 10 feet
    b. $1/2$", 20 feet
    c. $5/8$", 30 feet
    d. $1/2$", 50 feet

17. A concrete-encased electrode can also be constructed using 20 feet or more of bare copper not smaller than _____ AWG.
    a. 1
    b. 2
    c. 3
    d. 4
18. A concrete-encased grounding electrode can be in a horizontal or vertical orientation as long as there is a minimum of _____ of contact with the Earth in either orientation.
    a. 5 feet
    b. 10 feet
    c. 20 feet
    d. 30 feet
19. Who was responsible for the development of the concrete-encased grounding electrode?
    a. Thomas Edison
    b. George Simon Ohm
    c. Herbert G. Ufer
    d. Eustace Soares
20. Ground ring electrodes have to encircle the building or structure completely and cannot be smaller than _____.
    a. 1 AWG copper
    b. 2 AWG copper
    c. 3 AWG copper
    d. 4 AWG copper
21. What is the minimum burial depth required for a ground ring electrode installed around a building?
    a. 1 feet
    b. 2 feet
    c. 2.5 feet
    d. 3 feet
22. When pipe or conduit is installed as a grounding electrode, it must not be smaller than which trade size?
    a. ½ inch
    b. ¾ inch
    c. 1 inch
    d. 1½ inches
23. A ground rod has to be driven to a depth of not less than _____.
    a. 4 feet
    b. 6 feet
    c. 8 feet
    d. 10 feet
24. A ground rod can be buried in a 2½-foot-deep trench only if rock bottom is encountered, and it cannot be driven either vertically or at an angle not exceeding 45 degrees from vertical.
    a. True
    b. False
25. As a general rule, when installing ground rod electrodes, two must be installed always unless a single ground rod provides not more than 25 ohms resistance to ground.
    a. True
    b. False
26. A plate electrode is required to have not less than _____ in contact with the Earth.
    a. 1 square foot
    b. 3 square feet
    c. 2 square feet
    d. 4 square feet
27. Soil resistivity is measured to determine the type of soil conditions, the corrosive effects in the soil, and the best type of grounding electrode system to install.
    a. True
    b. False
28. Rod and pipe electrodes have to be a minimum of _____ in depth and embedded below permanent moisture level, where practicable.
    a. 5 feet
    b. 10 feet
    c. 8 feet
    d. 2½ feet
29. If rod, pipe, or plate electrodes are installed, each electrode must be spaced a minimum distance of _____ from another electrode of another grounding electrode system.
    a. 6 feet
    b. 8 feet
    c. 5 feet
    d. 16 feet
30. A complete system of strike termination devices, conductors (which could include conductive structural members), grounding electrodes, interconnecting conductors, surge protective devices, and other connectors and fittings required to complete the system best defines which of the following?
    a. Lightning protection system
    b. Grounding electrode system
    c. Equipment grounding system
    d. Surge protection system

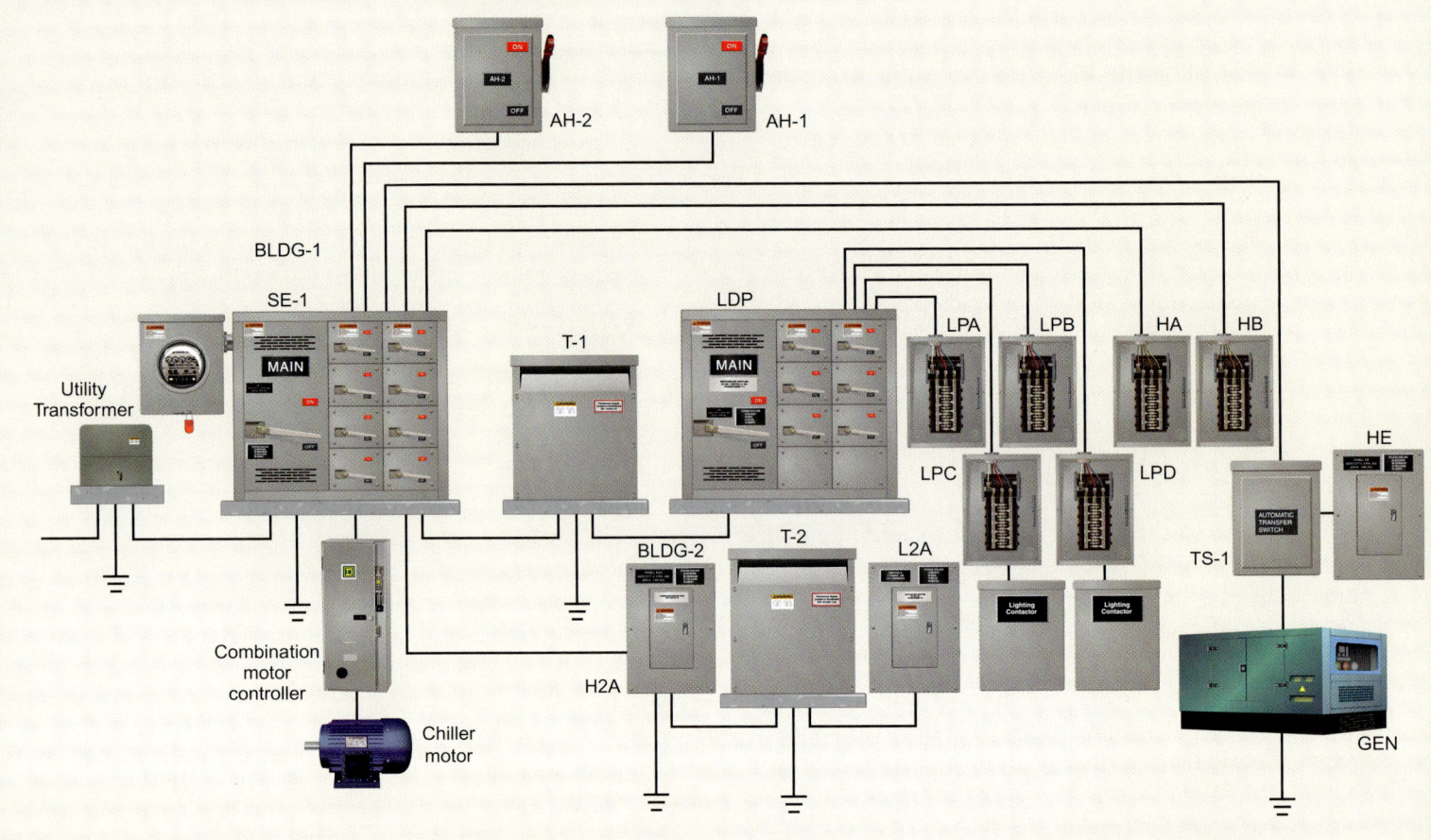

CHAPTER 5

Requirements for Grounded Conductors at Services

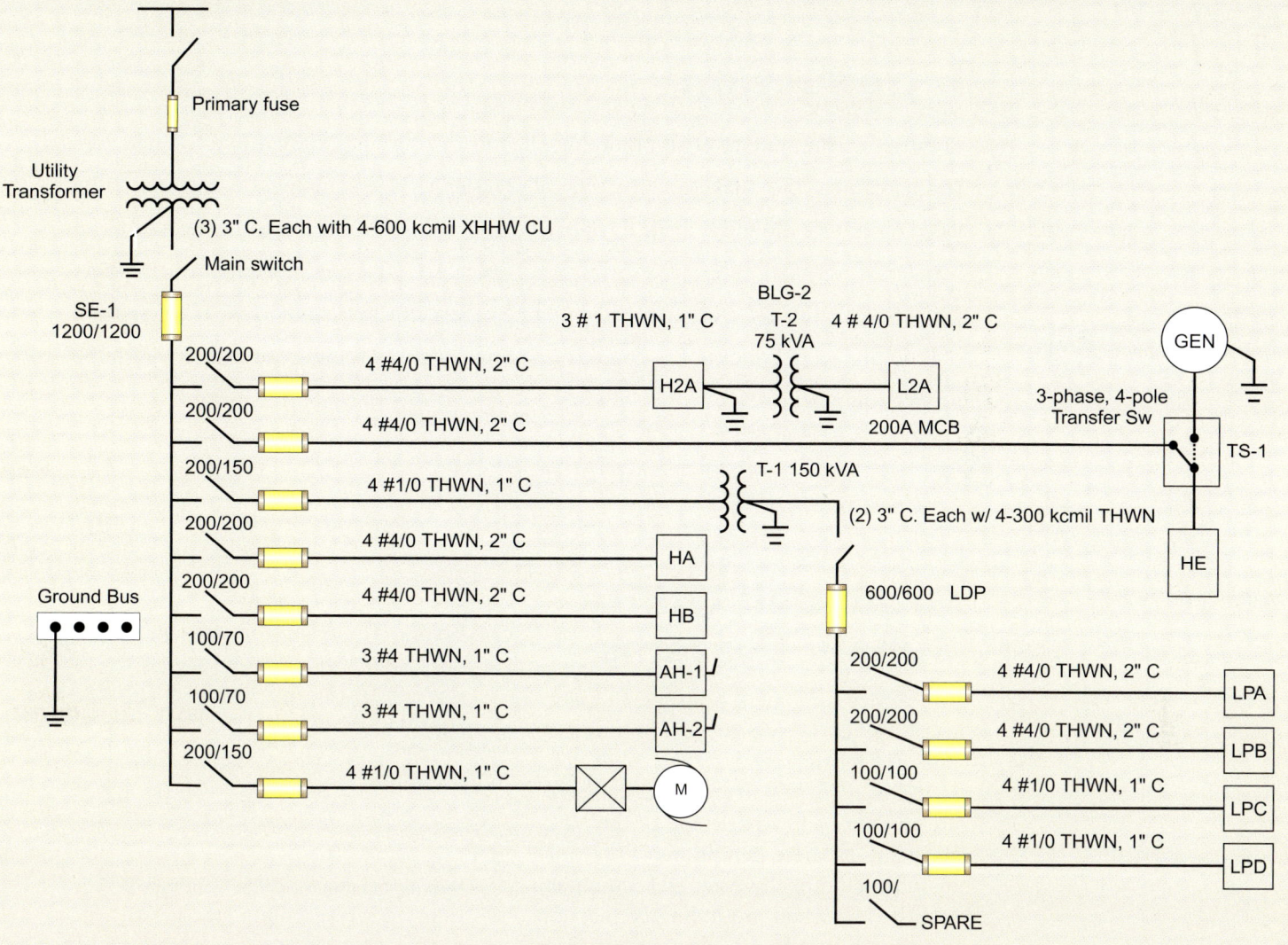

Objectives

- Understand the roles of the grounded conductor at the service equipment
- Determine the required grounding electrode conductor connection location(s) at the service equipment and outside the building or structure served
- Understand the installation and sizing requirements for the grounded conductor at services supplied by grounded systems
- Understand the physical characteristics and sizing requirements for main bonding jumpers in service equipment
- Understand the purpose of and location of the grounded conductor disconnecting means (neutral disconnecting link) in service equipment enclosures

Outline

Grounded Utility Supply Systems
First Line of Defense
Grounding Scheme for Services
Grounded Conductor Routing and Connections
Main Bonding Jumpers in Service Equipment
Dual-Fed Service Equipment
Minimizing Impedance in Service Grounded Conductors
Functions (Purposes) of the Grounded Service Conductor
Grounded Neutral Conductor
Requirements for Service Equipment (Listing)
Grounded Conductor (Neutral) Disconnect Requirement for Sevices
Requirements for Service Supplied by Ungrounded Systems
Marking Equipment for Ungrounded Systems
Grounding of Service Raceways and Enclosures

Introduction

Premises wiring systems are typically supplied from a serving utility through conductors and equipment that make up an electrical service. Service equipment is required to be listed for service use and is made up of equipment enclosures that contain switches, circuit breakers, fuses, and other accessories. **See Figure 5-1.** The service equipment is where the service conductors supplying the building or structure are connected. The first point of grounding and bonding for a premises wiring system typically occurs at or within the service equipment.

Grounded Utility Supply Systems

The service equipment is installed by electrical workers fairly early during construction. The terms *service, service conductors, service equipment,* and *grounded conductor* are all defined in Article 100 of the *NEC*. Referring to these definitions is essential in understanding this electrical equipment and the important grounding connection points associated with electrical services supplying the premises.

Service. The conductors and equipment for delivering electric energy from the serving utility to the wiring system of the premises served.[1]

Service Conductors. The conductors from the service point to the service disconnecting means.[2]

Service Equipment. The necessary equipment, usually consisting of a circuit breaker(s) or switch(es) and fuse(s) and their accessories, connected to the load end of service conductors to a building or other structure, or an otherwise designated area, and intended to constitute the main control and cutoff of the supply.[3]

Grounded Conductor. A system or circuit conductor that is intentionally grounded.[4]

FIGURE 5-1 Service equipment is typically the first point of grounding after the utility grounding connection outside the building or structure.

Premises wiring systems are generally supplied by utility systems that are grounded. **See Figure 5-2.** Where the service is supplied from a grounded system, the service conductors routed to the premises wiring equipment and systems must include a conductor that is grounded. The grounded conductor has to be electrically connected to the service equipment, as opposed to being connected through induction or magnetic coupling. The term *electrically connected* implies that there is current-carrying capacity by such systems that are electrically connected (Sections 200.2 and 200.3).

Utilities have their own rules for grounding electrical systems that supply customer-owned premises wiring. The most commonly used method of grounding for these systems is to solidly ground them, meaning no intentional impedance or resistance is used in the grounding circuit. System grounding methods and requirements for wiring systems on the load side of the service point are not the responsibility of the serving utility. Most utilities will provide service using only grounded systems. A few areas in the country still serve ungrounded systems, but those are becoming uncommon. It is always best to verify with the serving utility what their grounding requirements are for services and service equipment. Many utilities publish electrical service requirements that should be used in conjunction with the rules in the *NEC* when locating, installing, and connecting electrical service equipment.

First Line of Defense

A premises wiring system must generally be grounded, so the grounded system must include a grounded service conductor in addition to the ungrounded conductors that are connected to the service.

Grounded conductors are required to be brought to the service equipment.

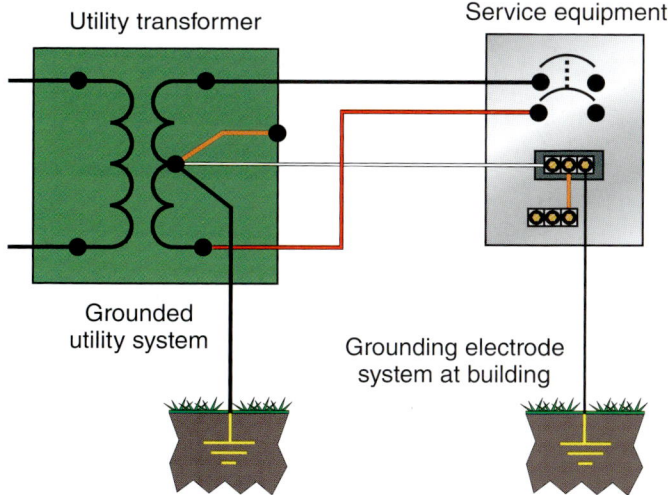

FIGURE 5-2 Utility service is typically delivered as a grounded system.

The grounded system is typically a pad-mounted transformer or one or more transformers mounted on a utility pole. **See Figure 5-3.** This is the usual location for the first system grounding connection, either at the transformer pad or at the base of the pole. The most common grounding electrode used to connect the system to Earth is a ground rod, but it could be another type. In most areas, the serving utility is responsible for this system grounding connection. This grounding connection is typically outside the footprint of the building or structure served.

This grounding is a first line of defense or protection for premises wiring systems served by grounded utility sources. This grounding connection provides a level of protection from unintentional contact with higher-voltage lines, line surges, overvoltages from lightning events, and so forth. Although the *NEC* addresses this grounding connection requirement for outdoor transformers supplying services, this grounding is usually the utility's responsibility. It is a good practice to verify the outside grounding location and method used by the serving utility. The outside electrode is a requirement that must be met by either the utility company or the installer, depending on the requirements in the area served. Safety is enhanced by establishing a path to the ground (Earth) outside the building or structure in addition to the grounding electrode conductor connection installed for the service, which is usually inside the service equipment enclosure. When transformers supplying the service are located outside buildings, Section 250.24(A)(2) requires at least one additional grounding connection from the grounded system conductor to a grounding electrode. This connection can be made at the transformer or at another point outside the building. **See Figure 5-4.**

Grounding Scheme for Services

The grounding requirements on the load side of the service point are usually the responsibility of the electrical contractor and are accomplished when installing the service equipment. Installers must verify that the equipment has markings indicating it is either "suitable for use as service equipment" or "suitable for use only as service equipment."

Courtesy of IBEW Local 26 Training Center

FIGURE 5-3 A pole-mounted utility transformer can be a single transformer supplying a single-phase, three-wire, 240/120-volt system, or a bank of transformers supplying a three-phase, four-wire 208Y/120-volt system; both are typically grounded using an electrode (ground rod) installed at the base of the pole.

The *Code* requires a grounding electrode conductor connection to be made at an accessible point that can be anywhere from the load side of the service drop or lateral up to the service equipment enclosure. **See Figure 5-5.** This long-standing requirement actually permits the grounding connection to be made at locations such as at the service weather head or in an enclosure used as service equipment. The term *accessible* is defined in Article 100 and as used in this rule, it relates to the ability to be accessed after installation. Locations such as at the service head, at meter socket enclosures, in a wireway, in an auxiliary gutter, or within a service disconnecting means enclosure all meet the accessibility requirement.

Meter socket enclosures are usually sealed by the serving utility to restrict access to unmetered conductors. This does not make the inside of meter socket enclosures inaccessible. They can be accessed by legal means in coordination with the serving utility. Some areas have a local restriction prohibiting the grounding electrode conductor connection in a meter enclosure, but this is not a national requirement or restriction. As far as the *NEC* is concerned, a meter socket enclosure is an accessible location to make such a grounding electrode conductor connection. Service equipment typically contains a terminal bar and lug that are intended for the grounding electrode conductor connection. The most common location for this connection is within listed service equipment enclosures. This location is the point of convergence for 4 conductors in the grounding scheme for grounded premises wiring systems. The four conductors that must be connected together within the service equipment are the grounded service conductor, the main bonding jumper, the grounding electrode conductor, and the equipment grounding conductor (EGC), which could be the equipment enclosure. **See Figure 5-6.**

The conductor used to accomplish the grounding is the grounding electrode conductor. This conductor connects the EGC, the service equipment enclosure, and the grounded conductor to the ground (Earth) through a grounding electrode or grounding electrode system.

FIGURE 5-4 A pad-mounted utility transformer is typically grounded using an electrode installed at the pad for the transformer (usually a ground rod).

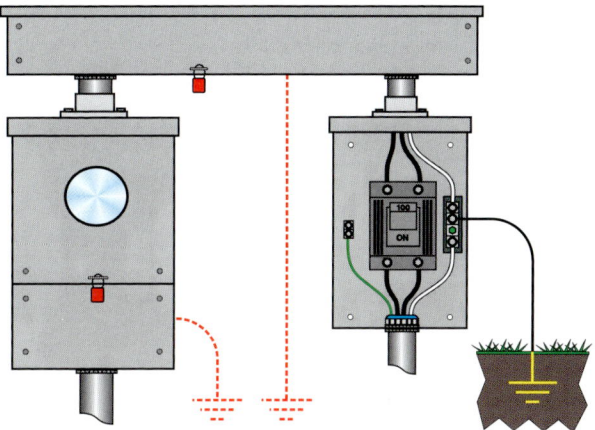

FIGURE 5-5 The grounding electrode conductor connection can be made at any accessible location.

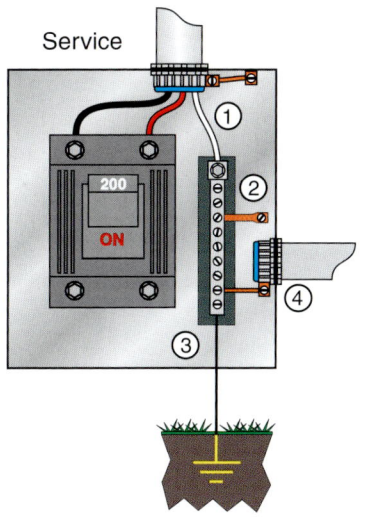

FIGURE 5-6 Four conductors are connected together in the service equipment. The connections are made at (1) the grounded conductor, (2) the main bonding jumper, (3) the grounding electrode conductor, and (4) the EGC.

98 APPLIED GROUNDING AND BONDING

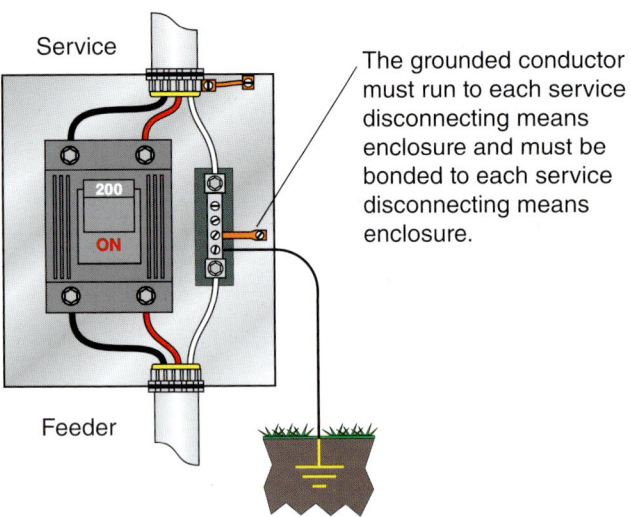

FIGURE 5-7 A grounded conductor is required to be routed to the service disconnecting means enclosure and bonded to it.

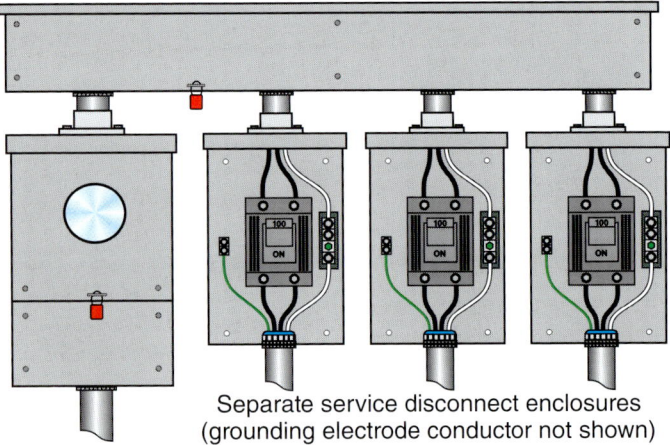

FIGURE 5-8 A main bonding jumper (MBJ) is required in each separate service disconnecting means (MBJ shown in the form of a screw).

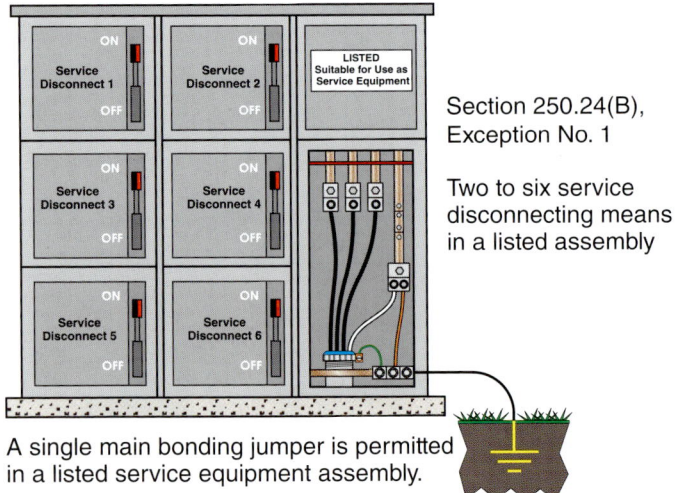

FIGURE 5-9 Multisection service switchboard assemblies are permitted to be equipped with a single main bonding jumper, by exception.

When a grounding electrode conductor is connected to the grounded conductor within a service equipment enclosure, it is required to connect to the grounded (neutral) conductor terminal bus. If a main bonding jumper (wire or busbar type) is installed between the grounded conductor busbar and the equipment grounding terminal bus in service equipment, the grounding electrode conductor is permitted to be connected to the equipment grounding terminal bar. This type of arrangement is often provided in larger switchboards or equipment containing larger equipment ground-fault protection features.

Grounded Conductor Routing and Connections

The *Code* includes a rule that requires the grounded conductor to be routed to the service equipment and connected to the equipment enclosure. **See Figure 5-7.** This rule is one of paramount importance in the grounding and bonding scheme for service equipment. This requirement applies to services operating at less than 1000 volts.

The means of connecting the grounded conductor to the service equipment enclosure is typically through a conductor identified in the *Code* as a main bonding jumper. Whether the service is supplied using an underground lateral or is supplied through a set of service conductors in a riser up to a weather head, as is done when the grounded conductor arrives in the service disconnection means enclosure, it has to be bonded to the equipment enclosure using a main bonding jumper or by direct connection to service equipment that is labeled "suitable for use only as service equipment." The main bonding jumper requirement applies to each service disconnecting means on the premises served, whether the service disconnecting means is a single main or is a group of service disconnects as permitted in Section 230.71. **See Figure 5-8.**

An exception recognizes that multiple service disconnects could be installed in a single assembly enclosure. In this case, the equipment has to be listed as service equipment and must include a grounded conductor terminal bus and a single main bonding jumper to connect the grounded conductor to the equipment enclosure. This exception applies to multisection switchboards that include two to six switches or circuit breakers as the service disconnecting means. In this case the *Code* relaxes the requirement for a grounded conductor being brought to each of the service disconnects; however, it requires the service grounded conductor to be connected to a listed service equipment assembly enclosure, rather than in each of the disconnects contained in the assembly. **See Figure 5-9.**

A similar exception applies to the main bonding jumper when multiple service disconnects are included in a listed service equipment assembly. In this case a main bonding jumper is not required in each disconnect because a single main bonding jumper is provided in the listed assembly.

Main Bonding Jumpers in Service Equipment

The term *main bonding jumper* is defined in Article 100 as the connection between the grounded conductor and the EGC at the service. **See Figure 5-10.** By definition, it is clear that this connection is made only at the service.

Section 250.28 provides information related to physical characteristics, connections, and the sizing of main bonding jumpers. Main bonding jumpers must be copper or other corrosion-resistant material and can be in the form of a screw, bus, wire, or other suitable conductor. **See Figure 5-11.**

The main bonding jumper in some equipment is a screw that connects the grounded conductor terminal bar to the service equipment enclosure. When the main bonding jumper is a screw, it must have a green finish that is visible after it is installed. **See Figure 5-12.**

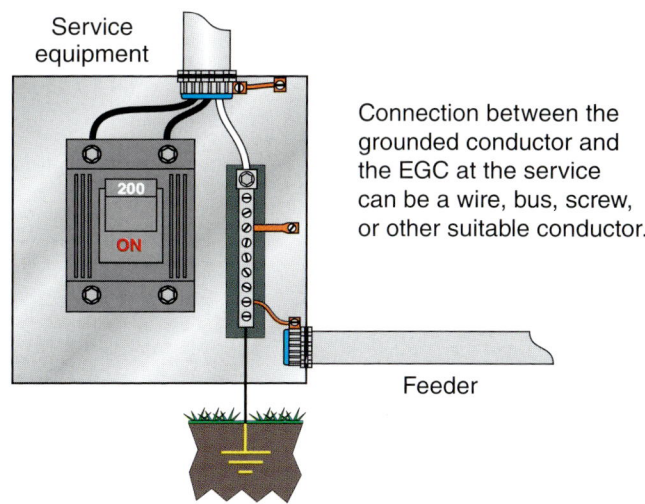

FIGURE 5-10 The main bonding jumper connects the grounded conductor to the EGC at the service.

Courtesy of IBEW Local 26 Training Center

FIGURE 5-11 The main bonding jumper can be in the form of a busbar in listed equipment.

Courtesy of IBEW Local 26 Training Center

FIGURE 5-12 A main bonding jumper in the form of a screw is required to be identified using the color green.

The color green helps installers and inspectors readily identify the main bonding jumper among many other terminal screws that may also be provided on the grounded conductor terminal busbar in the service equipment enclosure. It is fairly common for a screw-type main bonding jumper to be provided in service equipment rated up to 225 amperes. The main bonding jumper in service equipment with larger ratings is usually in the form of a busbar or wire. **See Figure 5-13.** When a main bonding jumper is installed, connections must meet the applicable requirements in Section 250.8. If a main bonding jumper is a wire type, it would be sized using Table 250.66; if the service conductor size exceeds 1100-kcmil copper or 1750-kcmil aluminum or copper-clad aluminum, it would be sized using the 12.5% rule. When a service includes multiple service disconnect enclosures, a main bonding jumper is required in each such enclosure. The size for wire-type main bonding jumpers in each enclosure must be in accordance with Section 250.28(D)(1), which bases size on the largest ungrounded service conductor serving that individual enclosure.

Use Table 250.66 or the 12.5% rule as appropriate for this sizing requirement. If the ungrounded service-entrance phase conductors and the wire-type main bonding jumper are made of different materials—for example, one is copper and one aluminum—the minimum-size main bonding jumper is based on the assumed use of phase conductors of the same material that have an ampacity equal to that of the installed ungrounded service-entrance phase conductors. For example:

1. Service size: 400 amperes
   Aluminum service-entrance conductor size: 750 kcmil
   Aluminum main bonding jumper: 3/0; copper main bonding jumper: 1/0
2. Service size: 800 amperes
   Copper service-entrance conductor size: two 600-kcmil conductors or one 1200-kcmil conductor
   Aluminum main bonding jumper: 250 kcmil; copper main bonding jumper: 3/0

Main bonding jumpers supplied with listed service equipment, such as switchboards and panelboards, can be installed without calculation of size. The manufacturer has built the equipment to meet or exceed the requirements in the applicable product safety standards, which includes grounding and bonding provisions.

Dual-Fed Service Equipment

When service equipment in a single enclosure or a group of separate enclosures is fed from two sources (dual-fed service), the grounding electrode conductor connection is permitted to be by a single grounding electrode conductor. Service equipment arranged in this fashion is typically equipped with a tie breaker. This single connection is permitted to be made at the tie point of the grounded conductor terminal bars supplied by separate power sources. **See Figure 5-14.**

Minimizing Impedance in Service Grounded Conductors

The grounded conductor for a service must be routed with its associated ungrounded conductors. **See Figure 5-15.**

FIGURE 5-13 The main bonding jumper can be a wire type.

This is also a general requirement in the *NEC,* as covered in Section 300.3(B). The reason for routing with the phase conductors is to keep the impedance of the circuit as low as possible during normal operation and during abnormal events such as ground faults or short circuits.

Separation of the grounded service conductor from its associated ungrounded conductors increases the effects of inductive reactance, thus increasing circuit impedance. This goes against the philosophies of intentionally constructing an effective ground-fault current path as required in Section 250.4(A)(5). For this reason, the *Code* requires the grounded service conductor to be routed with the ungrounded phase conductors. **See Figure 5-16.**

Functions (Purposes) of the Grounded Service Conductor

The grounded conductor at the service provides two essential functions for the premises wiring system. The first function is to serve as a current-carrying conductor for the load supplied. The grounded conductor of a service is usually a neutral conductor, but it may also be a phase conductor depending on the type of system. A corner-grounded delta system is an example of a system with a grounded phase conductor and no grounded neutral conductor. The grounded neutral conductors typically carry the maximum unbalanced neutral current to the system neutral point. During normal operation the grounded (neutral) conductor is carrying current that returns to the source windings. The same load profile characteristic applies to grounded phase conductors, except the grounded phase conductor typically carries the same current as the ungrounded phase conductors, as would be the case for a three-phase motor supplied by a corner-grounded system. Section 250.24(C)(3) requires the grounded conductor of a three-phase, three-wire, delta-connected service to have an ampacity not less than that of the ungrounded conductors of this service.

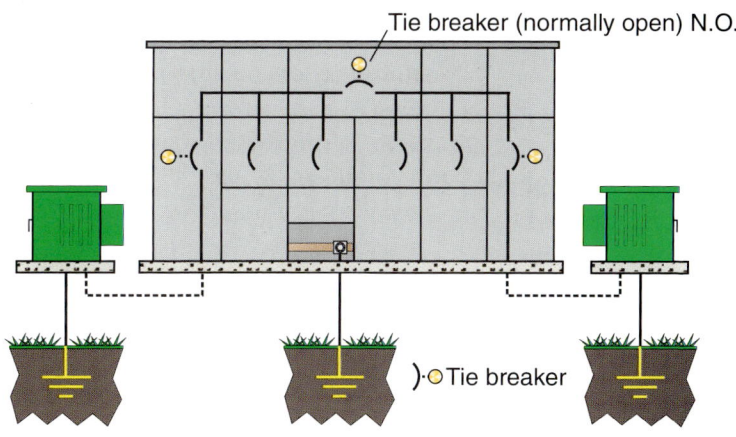

FIGURE 5-14 A single grounding electrode conductor connection is permitted at the tie point for the grounded conductor(s) of dual-fed (double-ended) services.

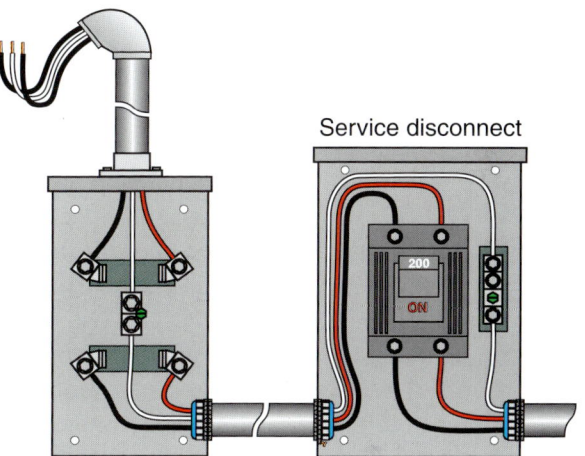

FIGURE 5-15 Route the grounded service conductor with its associated ungrounded service conductors.

FIGURE 5-16 The grounded conductor is required to be routed with the service-entrance conductors.

The second essential function of the grounded conductor is to perform as an effective ground-fault current path during ground-fault events at the service or at any point on the load side of the service equipment. **See Figure 5-17.**

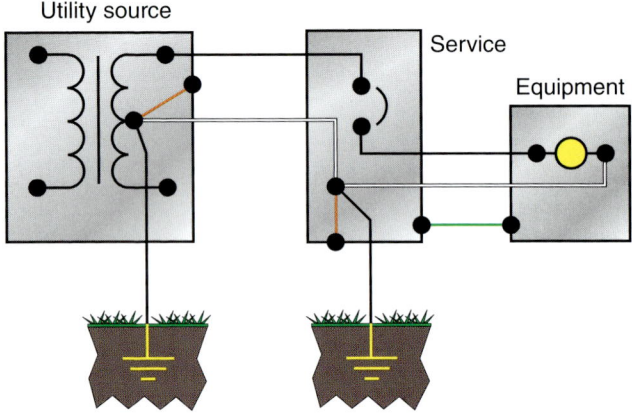

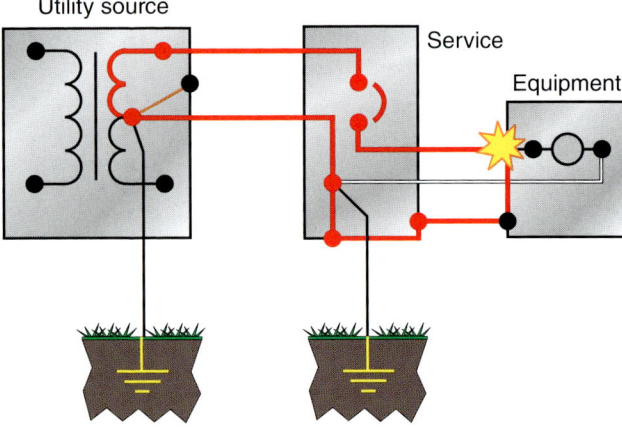

FIGURE 5-17 Grounded conductors brought to the service equipment enclosure typically perform two essential functions.

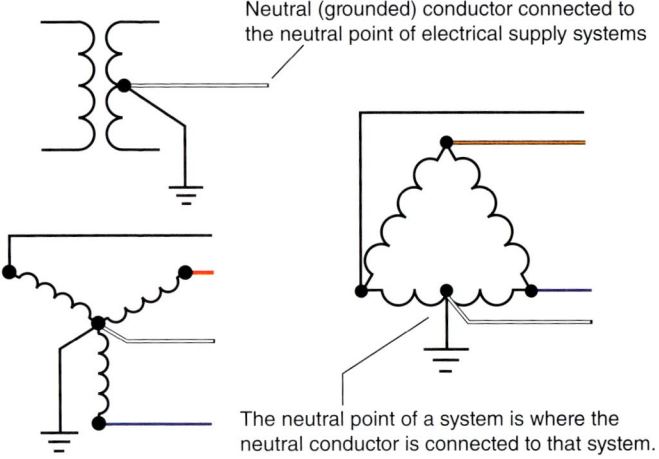

FIGURE 5-18 The neutral conductor is connected to the neutral point of the system.

The grounded conductor at the service is used for the intentionally constructed, low-impedance, effective ground-fault current path addressed in Section 250.4(A)(5). This is an essential element of effective overcurrent protective device operation for any ground fault that occurs on the load side. This is why Section 250.24(C) requires that the grounded conductor be brought to a service disconnecting means enclosure and bonded to the enclosure.

Grounded Neutral Conductor

The term *neutral* can relate to either a conductor or a system connection point. The terms *neutral conductor* and *neutral point* are defined in Article 100. These definitions help distinguish between what constitutes system neutrals and the conductors to which they are connected. **See Figure 5-18.** System neutral conductors are usually grounded, but not all grounded conductors are system neutrals. At the neutral point of the system, the vectorial sum of the nominal voltages from all other phases within the system that utilize the neutral, with respect to the neutral point, is zero potential. **See Figure 5-19.**

> **Neutral Conductor.** The conductor connected to the neutral point of a system that is intended to carry current under normal conditions.[5]

FIGURE 5-19 The neutral conductor is connected to the neutral point of a separately derived system (transformer).

> **Neutral Point.** The common point on a wye-connection in a polyphase system or midpoint on a single-phase, 3-wire system, or midpoint of a single-phase portion of a 3-phase delta system, or a midpoint of a 3-wire, direct-current system.[6]

Grounded Conductor Sizing Requirements

The grounded service conductor is generally a current-carrying conductor during normal conditions but during abnormal conditions, it must be capable of carrying fault current. Therefore, it must meet minimum sizing requirements to ensure adequate capacity to serve both functions. The neutral conductor for services or feeders must have adequate capacity for the load served, as indicated in Section 220.61. Table 250.66 is also used to determine the minimum size required for service grounded conductors for fault current. This rule applies whether there is a load on the neutral or not. **See Figure 5-20.** The 12.5% requirement is applicable for larger services that are served by service conductors exceeding 1100-kcmil copper or 1750-kcmil aluminum or copper-clad aluminum. **See Figure 5-21.** The minimum size of the grounded conductor must never be smaller than the required grounding electrode conductor for the service. If a three-phase, three-wire, corner-grounded delta service is supplied, the grounded conductor must be sized the same as the ungrounded phase conductors. **See Figure 5-22.**

The conductor must be sized this way because the current on the grounded conductor will be the same as the current on the ungrounded phase conductors during normal operation, compared with a grounded neutral conductor that may carry only the maximum unbalanced return current from associated phase conductors sharing this neutral conductor. This is usually the case for neutral current associated with a multiwire branch circuit or a feeder that includes a neutral conductor.

When the 12.5% rule is applied, Table 8 in Chapter 9 of the *NEC* has to be used. This table provides conductor properties and includes the information for converting sizes such as 2/0 or 4/0 to circular mils (see the third column).

Size of Largest Service-Entrance Conductor (Copper)	Minimum Size of Grounded Conductor (Copper)	Minimum Size of Grounded Conductor (Aluminum)
1 AWG	6	4
4/0 AWG	2	1/0
500 kcmil	1/0	3/0
750 kcmil	2/0	4/0

FIGURE 5-20 The minimum sizes required for grounded conductors are established using Table 250.66.

Size of Largest Service-Entrance Conductor (Copper)	Calculated Value (Copper)	Minimum Size of Grounded Conductor (Copper)
1200 kcmil	150 kcmil	3/0 copper (167,800 circular mils in Table 8)
1800 kcmil	225 kcmil	250 kcmil
2500 kcmil	312,500 cm	350 kcmil
3000 kcmil	375 kcmil	400 kcmil

FIGURE 5-21 The 12.5% rule must be used for sizing the grounded conductor where the size of the largest ungrounded service conductor exceeds 1100-kcmil copper or 1750-kcmil aluminum or copper clad aluminum.

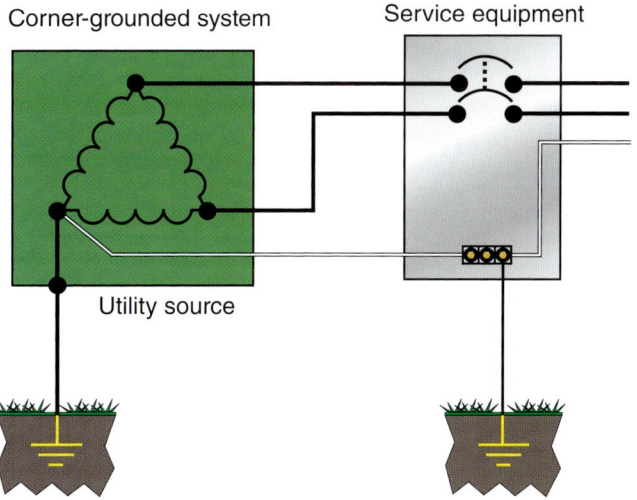

FIGURE 5-22 Grounded conductor of a corner-grounded system must be the same size as the ungrounded system conductors.

Once the 12.5% rule is applied, the minimum size required is determined by referencing *NEC* Table 8 and rounding to the next higher size than the kcmil value determined by the calculation.

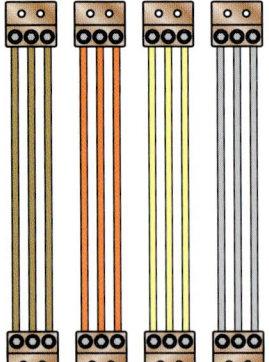

Conductors in parallel must:

1. Be the same length
2. Be of the same conductor material
3. Have the same type of insulation
4. Be terminated in the same manner
5. Be the same size
6. Not be smaller than 1/0 (in general)

FIGURE 5-23 There are specific rules for parallel conductor installations provided in 310.10(H).

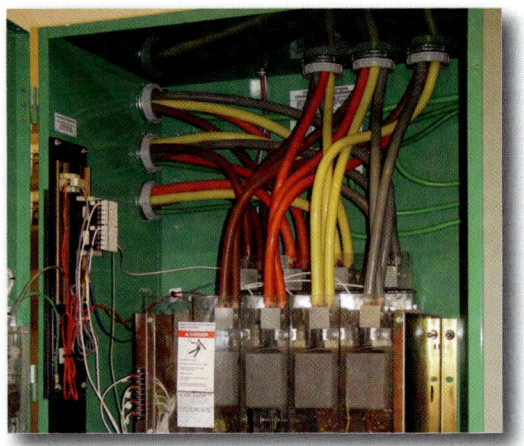

FIGURE 5-24 Conductors are permitted to be installed in parallel arrangements according to Section 310.10(H).

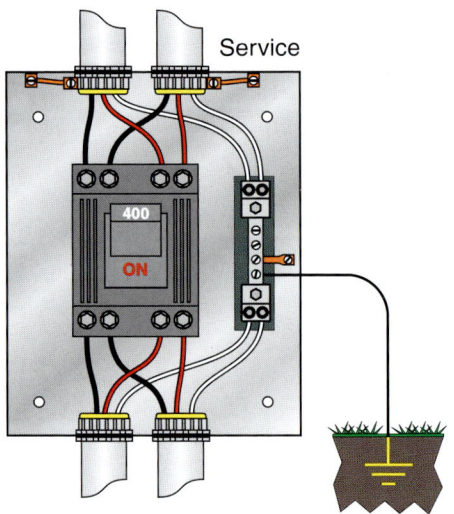

FIGURE 5-25 Use Table 250.66 to determine the appropriate size; use the 12.5% rule for larger services. Grounded service conductors cannot be smaller than 1/0 for parallel runs.

Grounded Conductor Sizing for Parallel Installations

The general requirements for conductors installed in parallel are provided in Section 310.10(H) of the *NEC*. **See Figure 5-23.** Conductors in parallel generally must be the same size, same length, and same conductor material; must have the same insulation type; and must be terminated in the same manner. **See Figure 5-24.**

Parallel conductors are generally not permitted to be installed in sizes smaller than 1/0. If a service is supplied by service conductors that are installed in a parallel arrangement, the rules in Section 250.24(C)(2) apply. The minimum size must be capable of carrying the normal neutral load current anticipated and be capable of carrying the maximum fault current anticipated during a ground-fault event. The minimum-size grounded service conductor for parallel arrangements is based on the total circular mil area of all the ungrounded service conductors in parallel. Once the total circular mil area is determined for the largest ungrounded service-entrance conductor, Table 250.66 is then used to determine the minimum-size grounded service conductor required to be installed in each conduit. **See Figure 5-25.**

The 12.5% requirement applies to parallel service conductor arrangements that are larger than 1100-kcmil copper or 1750-kcmil aluminum or copper-clad aluminum and is used to determine the minimum-size grounded conductor. These sizing requirements anticipate that all ungrounded conductors and grounded conductors are installed in the same raceway, as would be the case for a wireway or auxiliary gutter installation. When service conductors are installed in parallel using multiple raceways, which is fairly common, the minimum-size grounded conductor in each raceway is based on the size of the ungrounded service conductor in each raceway but must not be smaller than 1/0 per the requirements in Section 310.10(H).

Grounded Conductor (Load-Side Use)

The grounded conductor carries current during normal operation and when functioning as an effective ground-fault current path during a ground-fault event. For this reason, the *Code* generally restricts grounding conductor connections only to those required or permitted on the line side of and up to the service disconnecting means enclosure. This restriction applies on the load side of the grounding point at the service equipment, or the point where the main bonding jumper connection is made in the service equipment. Section 250.24(A)(5) restricts load-side grounding connections to the grounded conductor; essentially connections on the load side of the service disconnecting means are not permitted. **See Figure 5-26.** The same restriction is included for separately derived systems, as provided in Section 250.30(A). The informational notes following Sections 250.24(A)(5) and 250.30(A) indicate a few installations for which using the grounded conductor for grounding is permitted.

Load-side grounding of the grounding conductor is generally prohibited to minimize the paths that current can divide between while returning to the source windings. Once conductors leave the service equipment enclosure in the form of feeders or branch circuits, installers must not connect the grounded (usually a neutral) conductor to ground or to grounded metal enclosures. Doing so creates multiple paths for current to return to the source. This is often referred to in the field as *parallel paths for neutral current*. The goal is to isolate neutrals from the ground and grounded parts everywhere but where the system is initially grounded. **See Figure 5-27.**

If this rule is not complied with, objectionable current is introduced on conductive parts, raceways, and equipment that are not intended to carry current during normal operation. This condition is also a leading cause of many power quality problems experienced in systems today. There are very few allowances for such load-side grounding connections to the grounded conductor of services and premises wiring systems. Obviously, these are permitted for separately derived systems because a new system grounding point is established. There are also allowances for use of the grounded conductor for grounding equipment such as for existing ranges and dryer installations as provided in Section 250.140. Using the grounded conductor for grounding at separate buildings or structures was recognized in the *NEC* prior to the 2008 edition. However, the trend has been to migrate away from using the grounded conductor for equipment grounding purposes on the load side of the service disconnect or the load side of a grounding point for a separately derived system.

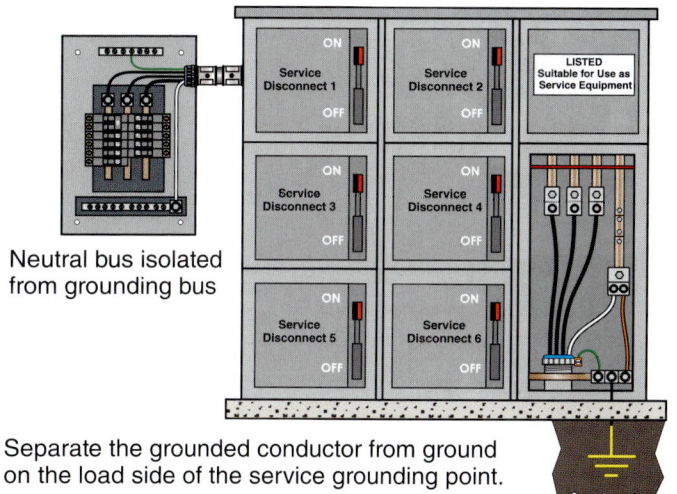

FIGURE 5-26 Generally, no load-side grounding connections of grounded conductor are permitted.

FIGURE 5-27 Separation (isolation) is generally required between the grounded conductor on the load side of the point of grounding at the service disconnecting means or separately derived system.

106 APPLIED GROUNDING AND BONDING

Electrical workers should exercise care in the connections of grounded (neutral) conductors and respect the performance concepts contemplated by keeping the neutrals and grounding connections separated in wiring installations.

Grounded Conductor Identification

There are *NEC* requirements that apply specifically to grounded conductors, whether they are grounded neutral conductors or grounded phase conductors. One such requirement is identification. Grounded conductors have to be identified according to Section 200.6. **See Figure 5-28.** This rule generally indicates that grounded conductors in sizes 6 AWG and smaller must be identified using the colors white or gray. **See Figure 5-29.** In certain conditions, the grounded conductor at a service could be a bare conductor, such as in service-entrance (SE) cable assemblies.

It is recognized that in larger sizes, conductor insulations have been less readily available in these colors. The *Code* therefore amended this general requirement by allowing sizes larger than 6 AWG to be identified using any of the following three methods:

1. A continuous white or gray outer finish
2. Three continuous white stripes along its entire length on other than green insulation
3. A distinctive white or gray marking at terminations that encircles the conductor[7]

Wire manufacturers have made it more convenient for electrical contractors to obtain conductors with insulations in a variety of colors. In item 3, the allowance is for applying identification in the form of a distinctive white or gray marking at the terminations.

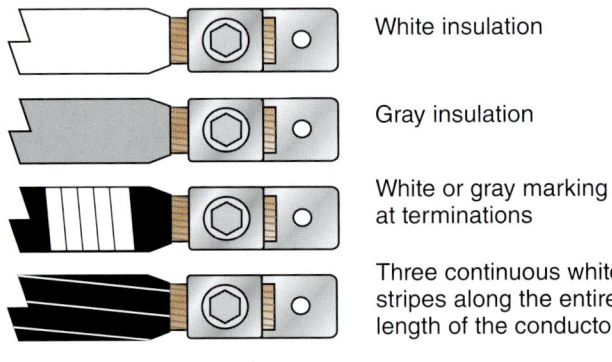

White insulation

Gray insulation

White or gray marking at terminations

Three continuous white stripes along the entire length of the conductor

FIGURE 5-28 Generally Section 200.6 of the *NEC* requires grounded conductors be identified in one of the four ways shown.

FIGURE 5-29 Grounded conductors in sizes 6 AWG and smaller are identified using white conductor insulation.

Note that the *Code* does not specify how far from the terminations the markings must be or the amount of marking that must be used. This requirement is usually satisfied when workers apply a sufficient amount of white or gray vinyl marking tape close to where the terminal is located. **See Figure 5-30.**

It is important that this marking completely encircle the conductor. This provides reasonable assurances that the marking (tape) will remain on the grounded conductor after the installation is complete and will be visible. Marking tape is a typical method of accomplishing identification of conductors larger than 6 AWG.

Requirements for Service Equipment (Listing)

Service equipment must be identified for use as service equipment. These requirements are found in Section 230.66. Equipment that is suitable for use as service equipment has been manufactured and evaluated to meet product safety standards. Some equipment that could carry the identification as suitable for service use are switchboards, enclosed switches, panelboards, motor control centers, and power outlets, among others. The product standards apply to these types of equipment. **See Figure 5-31.**

FIGURE 5-30 Identification of grounded conductors larger than 6 AWG is permitted at the termination points.

> Electrical equipment manufacturers clearly identify equipment that is suitable for use as service equipment or equipment that is suitable for use only as service equipment.

Type of Service Equipment	Applicable Product Safety Standard
Switchboards	UL 891 Deadfront Switchboards
Panelboards	UL 67 Panelboards
Service power outlets	UL 231 Power Outlets
Enclosed switches	UL 98 Enclosed Switches
Motor control centers	UL 845 Motor Control Centers

FIGURE 5-31 Some UL product standards include information related to products that are suitable for use as service equipment.

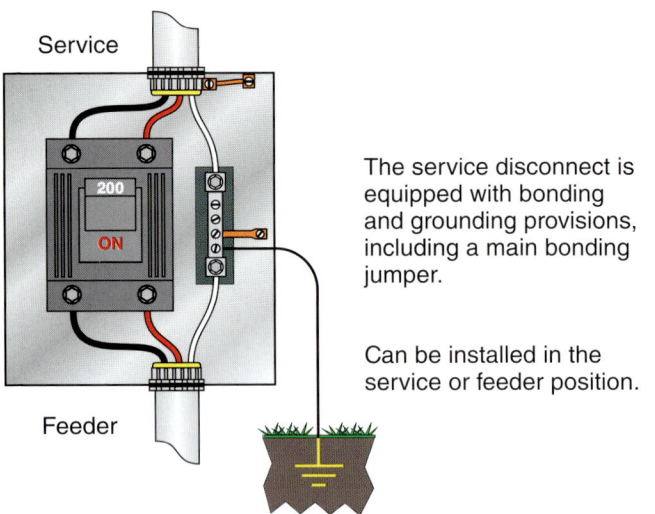

FIGURE 5-32 Equipment that is suitable for use as service equipment includes a main bonding jumper.

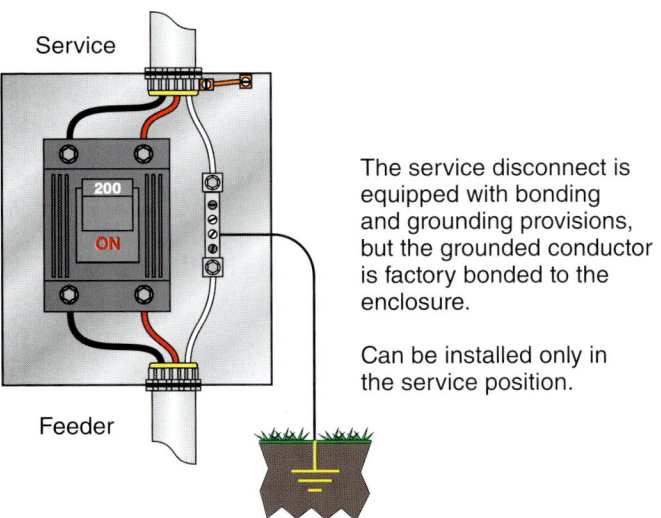

FIGURE 5-33 Equipment that is suitable for use only as service equipment does not include a main bonding jumper.

FIGURE 5-34 Equipment that is suitable for use as service equipment can be used in the service position or on the load side of the service disconnecting means.

The designation of being suitable for service equipment use identifies the equipment for use in the service position of premises wiring installations. It does not limit the use of this equipment to only service use, as could be the case for service equipment marked "suitable for use only as service equipment." Equipment that is suitable for service equipment use has been evaluated for short circuit current capabilities; it includes provisions for connection of a grounding electrode conductor, a neutral disconnecting means (link or terminal), and a main bonding jumper, to name a few.

Because this equipment includes grounding and bonding capabilities, the requirements in Section 250.24(C) are satisfied by installing the grounding and bonding connections using the means provided with the equipment. **See Figure 5-32.** The main bonding jumper in listed service equipment is usually a bus, screw, strap, wire, or other conductive material that has been evaluated by a qualified electrical testing laboratory for size and performance.

FIGURE 5-35 Equipment that is suitable for use only as service equipment does not include a main bonding jumper and should generally be used only in the service position.

This means the main bonding jumper in listed service equipment can be installed without calculating size. **See Figure 5-34.** This has been done by the manufacturer.

If equipment is marked "suitable for use only as service equipment," it has a grounded conductor terminal busbar that is bolted or otherwise bonded to the enclosure and no main bonding jumper is provided. **See Figure 5-33.** This type of equipment is suitable for use only in the service position because the grounded conductor cannot be isolated from the enclosure. **See Figure 5-35.**

If a service is constructed using a wireway or an auxiliary gutter arrangement, then the main bonding jumper of a wire type would be sized using Table 250.66; alternatively, the 12.5% rule must be used if the service conductor size exceeds 1100-kcmil copper or 1750-kcmil aluminum or copper-clad aluminum.

Grounded Conductor (Neutral) Disconnect Requirement for Services

An important requirement for service equipment is that provisions for disconnecting the grounded conductor (usually a neutral) are required by *NEC* Section 230.75. Listed service equipment includes such provisions either in the form of a busbar or terminal. **See Figure 5-36.**

In smaller service equipment such as a panelboard, the means to disconnect the grounded conductor is the terminal to which it is connected. In larger equipment such as a switchboard or motor control center, the means to disconnect a grounded conductor is a busbar that is identified as the "neutral disconnect link" in accordance with the applicable product safety standard. **See Figure 5-37.** The product standard requires that the enclosure identify which vertical section the grounded conductor (neutral) disconnect link and main bonding jumper are located in. **See Figure 5-38.**

FIGURE 5-36 A neutral disconnect link is typically in the form of a busbar in larger service equipment.

FIGURE 5-37 Vertical sections of switchboards have to be identified on the vertical section in which the main bonding jumper and neutral disconnect link are located.

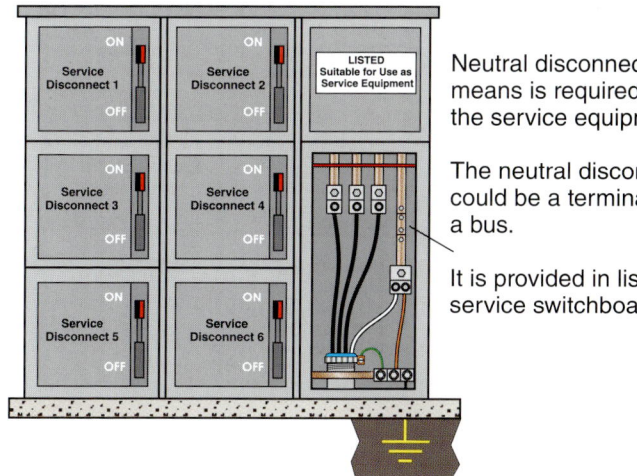

FIGURE 5-38 Neutral disconnect links are often a removable portion of bus installed in the neutral (grounded conductor) busbar of the service equipment.

The neutral disconnect link provides a means for disconnecting and isolating the neutral bus from the equipment grounding terminal bus and the enclosure. It is important that a main bonding jumper be connected on the supply side of the neutral disconnect link so that if the link is removed for any reason, the grounded conductor remains grounded and connected to the service equipment enclosure. In larger equipment that has ground-fault protection, the neutral disconnect link is an important part of the equipment ground fault protection (EGFP) performance testing that is required in various *Code* rules that address ground-fault protection for equipment such as those in Articles 215, 230, 240, and 517.

Requirements for Services Supplied by Ungrounded Systems

Some utilities will serve a premises wiring system from an ungrounded source, but this practice is uncommon. Utilities understand the safety benefits of supplying services with systems that are grounded. However, the *NEC* still includes rules for services supplied by ungrounded utility sources in Section 250.24(E). Although the system is not grounded, grounding and bonding requirements are included for metallic raceways and enclosures that contain conductors and equipment used with ungrounded systems. If a service is supplied from an ungrounded utility source, the *Code* requires that the metal enclosure have a grounding electrode conductor connection to a grounding electrode or grounding electrode system as provided in Part III of Article 250. **See Figure 5-39.** Even though a system could be supplied ungrounded, grounding non-current-carrying metallic equipment is still necessary for many of the same reasons grounding is required for enclosures containing conductors supplied from grounded systems.

The grounding electrode conductor size is determined by using Table 250.66, which bases sizing on the size of the largest ungrounded service-entrance conductor. The general performance requirements for ungrounded systems are covered in Section 250.4(B), which provides all grounding and bonding functions except for system grounding. No system conductor is connected to ground other than through magnetic coupling effects and distributed leakage capacitance.

The grounding electrode conductor installed for a service that is supplied by an ungrounded system must be connected to the metal service enclosure at any accessible point from the load end of the service drop or lateral to the service disconnecting means. The marking requirements in Section 230.66 are still applicable for these services, except there will not be a grounded conductor supplied with the utility service conductors. This means that this equipment must be suitable for use as service equipment. The grounding electrode conductor must be installed according to applicable installation rules in Section 250.64.

Another important requirement for services supplied by ungrounded systems is ground detection. Ground detection equipment must be provided for ungrounded systems as required in Section 250.21(B).

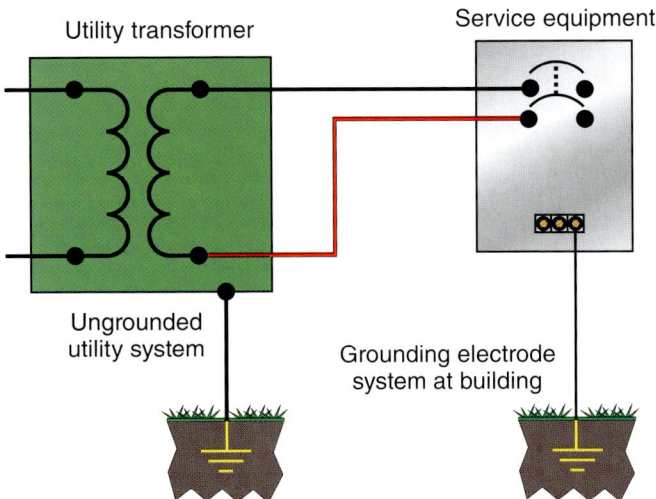

FIGURE 5-39 Equipment grounding is required for services supplied by an ungrounded system or supply source.

This equipment can either be part of the service equipment or be field installed. The ground detection sensing devices must be installed as close as practicable to the service or source. This requirement ensures that ground detection monitoring remains active even if feeder or branch circuits are disconnected. Connection of ground detection systems at points downstream from the source increases the possibility of monitor loss resulting from opening of feeder or branch circuits.

Marking Equipment for Ungrounded Systems

Section 250.21(C) provides a general requirement that ungrounded systems be marked. **See Figure 5-40.** Section 408.3(F)(2) requires switchboards and panelboards containing ungrounded electrical systems to be legibly and permanently field marked as follows[8]:

Caution: Ungrounded System Operating _____ Volts Between Conductors

This marking must be suitable for the environment in which it is installed. The installer is responsible for adding this marking and inserting the correct operating voltage of the system. As an example, if the system were an ungrounded 240-volt delta system, the marking would read, "Caution: Ungrounded System Operating at 240 Volts Between Conductors." If it were an ungrounded 480-volt delta system, the marking would read, "Caution: Ungrounded System Operating at 480 Volts Between Conductors."

This marking requirement for switchboards and panelboards is an improvement in safety for workers. Ungrounded electrical systems are less common than grounded systems, especially for premises wiring inside of buildings or structures. This warning on equipment identifies the unique voltage characteristics of ungrounded systems to raise awareness of the need to apply caution when testing for voltages and troubleshooting such systems.

Grounding of Service Raceways and Enclosures

Section 250.80 provides requirements for grounding metal raceways and enclosures for service conductors and equipment. They must be connected to the grounded system conductor if the service is supplied by a grounded electrical supply system; otherwise, they must be connected directly to a grounding electrode if the service is supplied by a system that is ungrounded.

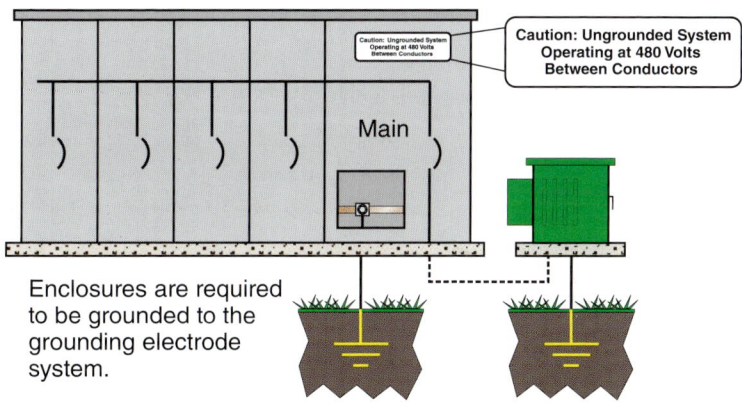

FIGURE 5-40 The marking requirement applies to ungrounded system equipment enclosures, including switchboards and panelboards.

Important Information: Ground detectors are required to be installed for ungrounded systems and the detection-sensing devices must be installed as close as practicable to the service or source location.

The *Code* relaxes the requirement (by exception) for grounding metal elbows installed in a run of polyvinyl chloride (PVC) conduit or reinforced thermosetting resin conduit (RTRC) as long as the metal elbow is isolated from possible contact by a minimum cover of 18 inches to any part of the elbow. **See Figure 5-41.**

This is usually accomplished by a concrete encasement or by the minimum required burial depths of most utility company service laterals. When the sheath or armor of a continuous underground metal-sheathed service-entrance cable system is connected to the supply system grounded conductor, the metal sheath or armor is not required to be connected to the grounded conductor at the service equipment enclosure installed at the building or structure. The same requirements apply to underground service raceways that contain a metal-sheathed or armor cable that is connected to the grounded system conductor on the supply end. In this case the metal sheath or armor is not required to be connected to the grounded conductor at the building or structure service equipment. The sheath or armor is already grounded at one end and is permitted to be insulated from the interior metal piping or raceway systems. See Section 250.84.

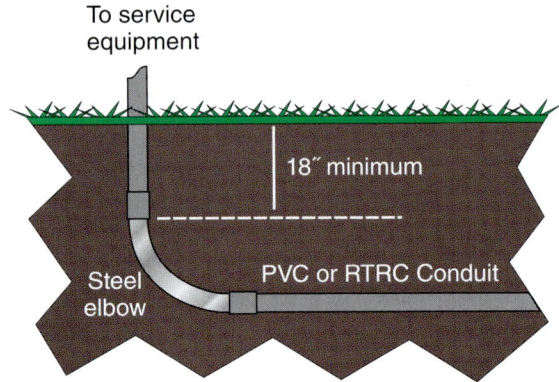

FIGURE 5-41 Grounding is not required for metal elbows buried not less than 18 inches to any part of the elbow and is isolated from possible contact by persons.

Summary

When a utility transformer is located outside a building, an additional grounding connection is required at the transformer or elsewhere outside the building. A grounded conductor is required to be brought to each service disconnecting means enclosure and connected to the enclosure. Equipment used as service equipment is required to be identified for such use. The minimum size required for grounded (often neutral) service conductors at the service cannot be smaller than that of the required grounding electrode conductor. The grounded conductor minimum size is determined based on the load served using Section 220.61; at a minimum it is sized using Table 250.66 or the 12.5% rule as required for larger services. The grounded conductor is an essential part of the effective ground-fault current path back to the source for grounded systems. Main bonding jumpers are another essential component in the effective ground-fault current path. Main bonding jumpers provided with listed service equipment can be used without calculation of size. Although services are generally supplied by grounded utility sources, when an ungrounded system is used, there is an equipment marking requirement. The equipment enclosures for ungrounded systems are required to be grounded using a grounding electrode conductor that connects to a grounding electrode in accordance with Part III of *NEC* Article 250.

References

1. NFPA 70 National Electrical Code 2011, Article 100 (National Fire Protection Association, Quincy, MA 2010), p. 70-31.
2. NFPA 70 National Electrical Code 2011, Article 100 (National Fire Protection Association, Quincy, MA 2010), p. 70-31.
3. NFPA 70 National Electrical Code 2011, Article 100 (National Fire Protection Association, Quincy, MA 2010), p. 70-32.
4. NFPA 70 National Electrical Code 2011, Article 100 (National Fire Protection Association, Quincy, MA 2010), p. 70-29.
5. NFPA 70 National Electrical Code 2011, Article 100 (National Fire Protection Association, Quincy, MA 2010), p. 70-30.
6. NFPA 70 National Electrical Code 2011, Article 100 (National Fire Protection Association, Quincy, MA 2010), p. 70-30.
7. NFPA 70 National Electrical Code 2011, Section 200.6 (National Fire Protection Association, Quincy, MA 2010), p. 70-46.
8. NFPA 70 National Electrical Code 2011, Section 250.21(C) (National Fire Protection Association, Quincy, MA 2010), p. 70-104.

Review Questions

1. The first point of grounding for a building or structure supplied by a utility service is usually within the service equipment enclosure(s).
   a. True
   b. False

2. Premises wiring systems generally are required to be supplied by a system that is _____ and includes a grounded conductor.
   a. Ungrounded
   b. Grounded
   c. Impedance grounded
   d. Reactance grounded

3. In addition to the grounding connection at the service, one additional grounding connection from the grounded system conductor to a grounding electrode must also be made at the transformer or at another point outside the building.
   a. True
   b. False

4. The equipment used on the load side of the service disconnecting means enclosure such as switchboards and panelboards have to be either suitable for use as service equipment or suitable for use only as service equipment.
   a. True
   b. False

5. When a service is supplied by a grounded system, the _____ is required to be routed from the utility supply system or source to the service disconnecting means enclosure and bonded to the service disconnecting means enclosure.
   a. Equipment bonding jumper
   b. Grounded conductor
   c. Grounding electrode
   d. Equipment grounding conductor

6. The grounding electrode conductor connection has to be made at an accessible point located anywhere from the load side of the service drop or lateral up to the service equipment enclosure. Which of the following locations are not acceptable for this grounding electrode conductor connection?
   a. Directly buried in the Earth
   b. Connected in meter socket enclosure
   c. Connected in the service disconnecting means enclosure
   d. Connected at an auxiliary gutter

7. If equipment is identified as suitable for use only as service equipment, it typically includes a terminal or bus for the grounded conductor that is fastened directly to the enclosure; there is no main bonding jumper.
   a. True
   b. False

8. The *NEC* requires the grounded conductor to be routed to the service disconnecting means enclosure and bonded to that enclosure. For equipment that is identified as suitable for use as service equipment, this is typically accomplished using a(n) _____.
   a. Bonding conductor
   b. Grounding conductor
   c. Equipment grounding conductor
   d. Main bonding jumper

9. The connection between the grounded conductor and the equipment grounding conductor at the service best defines which of the following components of the grounding and bonding system?
   a. System bonding jumper
   b. Equipment bonding jumper
   c. Main bonding jumper
   d. Neutral disconnect link

10. The main bonding jumper must be copper or other corrosion-resistant material and can be in which of the following forms?
    a. A screw
    b. A bus
    c. A wire or other suitable conductor
    d. Any of the above

11. A main bonding jumper in listed service equipment can be used without calculation of size because it is provided by the manufacturer and the equipment is suitable for use as service equipment.
    a. True
    b. False

12. When a main bonding jumper is a wire, it has to be sized using _____; alternatively, the 12.5% rule can be used when the size of the largest ungrounded service-entrance conductor exceeds 1100-kcmil copper or 1750-kcmil aluminum or copper-clad aluminum.
    a. Table 250.122
    b. Table 250.66
    c. Table 8 in Chapter 9
    d. Table 310.16

13. If the largest ungrounded service-entrance conductor for a service is 750-kcmil copper, what is the minimum size required for a wire-type main bonding jumper for the service?
    a. 2/0 AWG copper
    b. 1/0 AWG copper
    c. 3/0 AWG copper
    d. 1 AWG copper
14. If the total circular mil area of the largest ungrounded service-entrance conductor for a service is 2000-kcmil copper, what is the minimum size required for a wire-type main bonding jumper for the service?
    a. 4/0 AWG copper
    b. 250-kcmil copper
    c. 350-kcmil copper
    d. 500-kcmil copper
15. The minimum-size grounded conductor for a service supplied by a grounded system shall not be smaller than the minimum size required for a(n) _____.
    a. Equipment grounding conductor
    b. System bonding jumper
    c. Ungrounded conductor
    d. Grounding-electrode conductor
16. What is the minimum size required for the grounded conductor of a three-phase, three-wire, corner-grounded delta system that supplies a grounded service?
    a. The grounded conductor cannot be smaller than the required grounding electrode conductor.
    b. The grounded conductor must have an ampacity not less than the ungrounded conductors supplied by the system.
    c. The grounded conductor cannot be smaller that the equipment grounding conductor.
    d. The grounded conductor can be 12.5% or the largest ungrounded service conductor supplied by the system.
17. _____ is defined as the common point on a wye-connection in a polyphase system or midpoint on a single-phase, three-wire system; the midpoint of a single-phase portion of a three-phase delta system; or a midpoint of a three-wire, direct-current system.
    a. Neutral point
    b. Neutral conductor
    c. Phase conductor
    d. Grounded conductor
18. The grounded conductor at the service carries normal line-to-neutral load current and functions as an intentionally constructed, low-impedance, effective ground-fault current path, as required in Section 250.4(A)(5).
    a. True   b. False
19. Parallel service conductors are generally not permitted in sizes smaller than _____.
    a. 2 AWG
    b. 1 AWG
    c. 3/0 AWG
    d. 1/0 AWG
20. Grounded service conductors larger than 4 AWG are required to be identified using which of the following methods?
    a. A continuous white or gray outer finish
    b. Three continuous white stripes along its entire length on other than green insulation
    c. A distinctive white or gray marking at terminations that encircles the conductor
    d. Any of the above
21. An important requirement for service equipment is that provisions for disconnecting the grounded conductor (usually a neutral) are provided within the service equipment enclosure.
    a. True   b. False
22. The grounding electrode conductor for ungrounded services and grounded services is generally required to be sized based on the size of the largest ungrounded service-entrance conductor using _____.
    a. The rating of the overcurrent device in the service disconnect
    b. Table 250.122
    c. Table 8 in Chapter 9
    d. Table 250.66

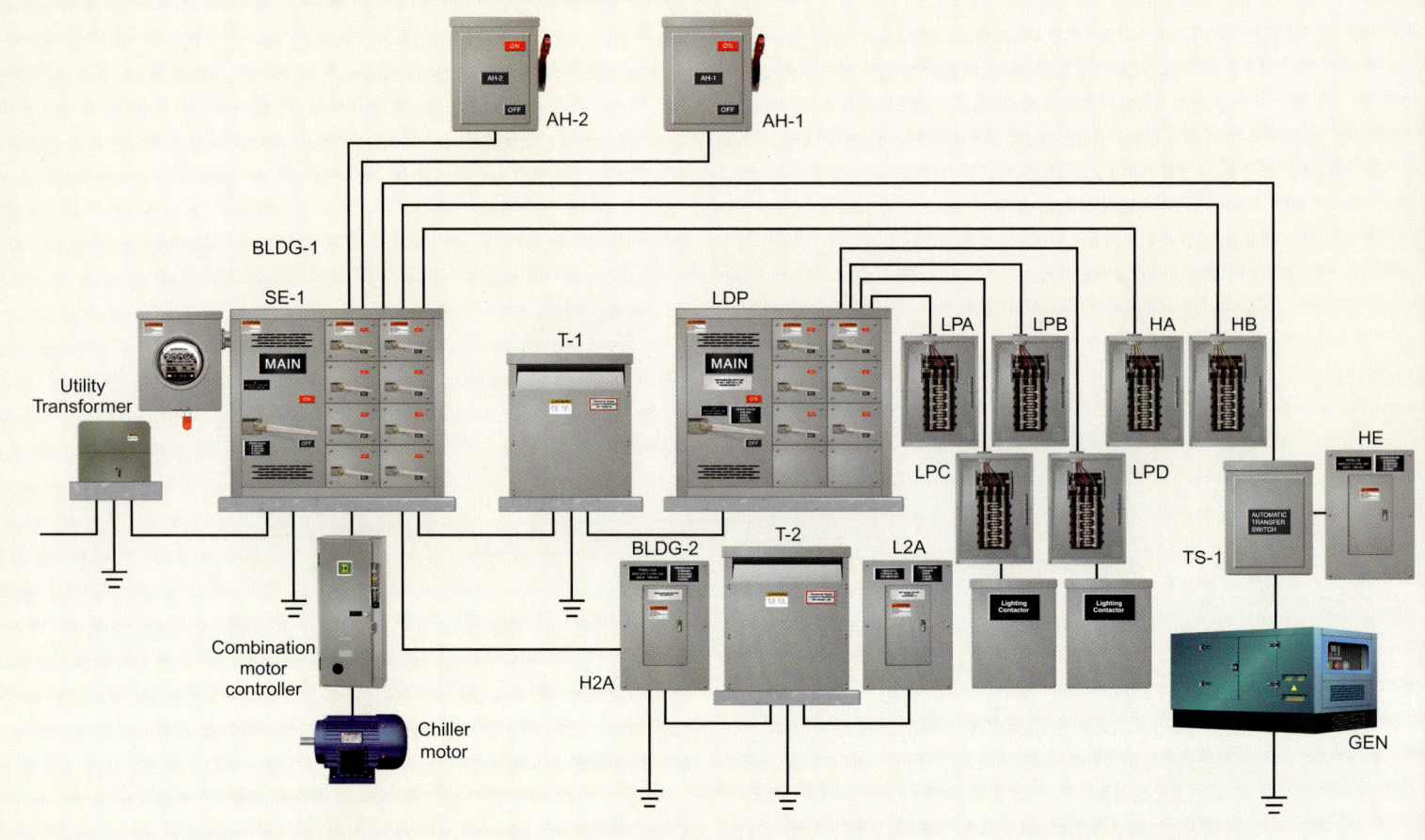

CHAPTER 6

Grounding Electrode Conductors

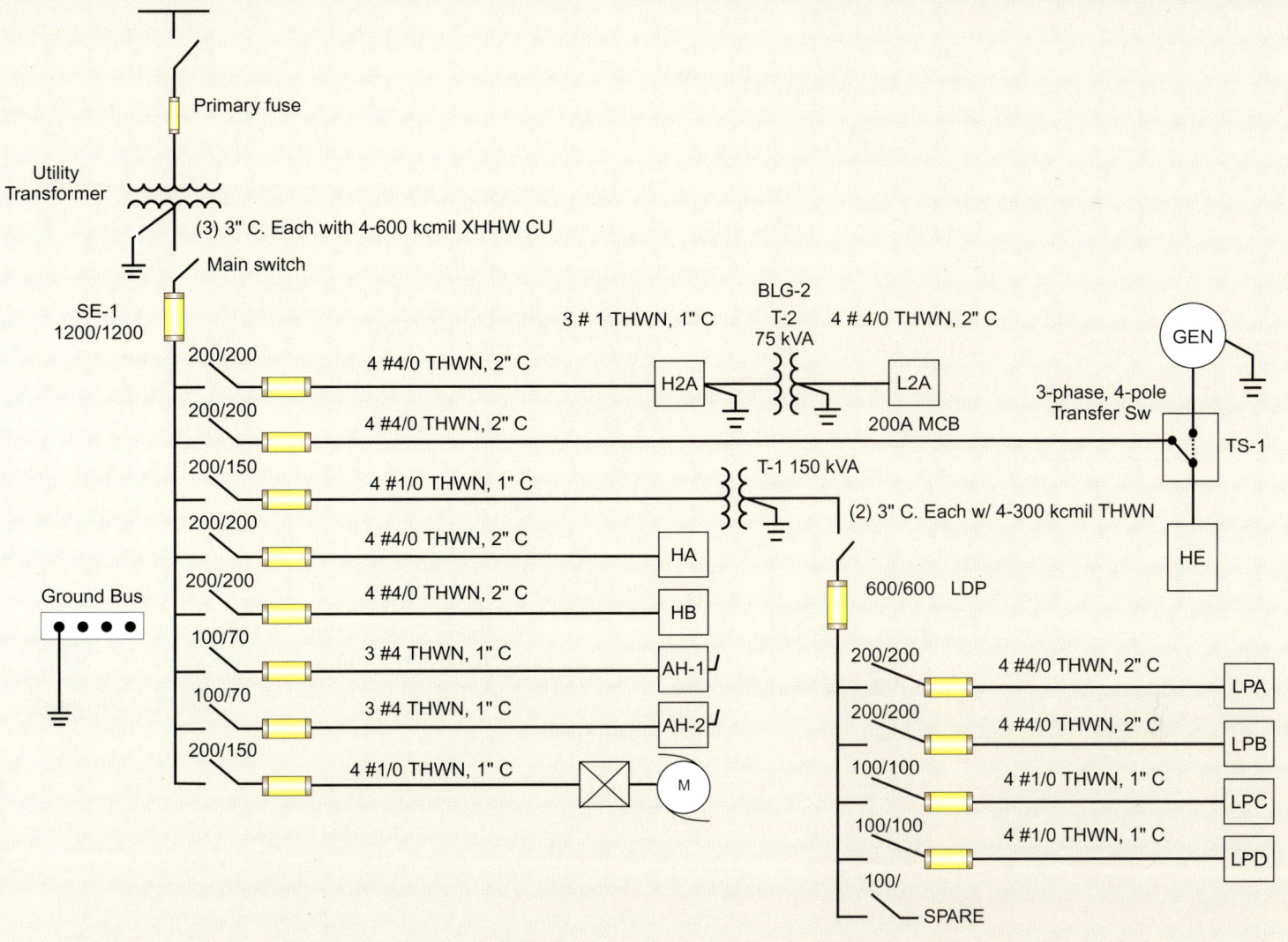

Objectives

- Understand the role of the grounding electrode conductor
- Determine the grounding electrode conductor connection locations at a service
- Understand the material permitted for a grounding electrode conductor
- Understand requirements for grounding electrode conductor installations and determine minimum sizes required for grounding electrode conductors
- Understand the requirements for grounding electrode conductor connections
- Determine the requirements for protecting grounding electrode conductors from physical damage and the effects of magnetic fields

Outline

The Path to Ground

Purpose of Grounding Electrode Conductors

Current in Grounding Electrode Conductors

Grounding Electrode Conductor Material

Sizing Grounding Electrode Conductors

Grounding Electrode Conductors for DC Systems

Grounding Electrode Conductor Installation

Effectiveness (Integrity) of the Grounding Path

Grounding Electrode Conductor Connection Locations

Grounding Electrode Conductor Connections

Magnetic Field Concerns

Introduction

Where service equipment and other equipment and systems required to be grounded are installed on the premises, a path to the ground (Earth) must be established. The grounding electrode conductor provides the conductive path to ground for grounded electrical systems and other equipment that must be grounded as required by the *NEC*®. There are specific installation requirements that apply to grounding electrode conductor installations. These requirements relate to how this conductor must perform and how it must be protected from possible damage.

The Path to Ground

This conductive path is known as a grounding electrode conductor primarily because one end of this conductor is typically connected to a grounding electrode or other conductive object that completes a path to the Earth. **See Figure 6-1.** Grounding electrode conductors are required for services, separately derived systems, and limited-energy systems such as communications systems and information technology networks.

Grounding electrode conductors provide the connection between the Earth and the object or conductor that is required to be connected to the ground. These conductors are conductive bodies that extend the ground connections indicated in the definition of *grounded (grounding)*.

> **Grounding Electrode Conductor.** A conductor used to connect the system grounded conductor or the equipment to a grounding electrode or to a point on the grounding electrode system.[1]

This definition describes a conductor that performs grounding functions and how it relates to the systems or equipment it is used with. Grounding electrode conductors are not limited to use with just services or separately derived power systems. The grounding functions required for low-voltage and limited-energy systems require a connection to the ground (Earth), and that connection is accomplished by installing a grounding electrode conductor. The sizing requirements and installation rules for grounding electrode conductors installed and used with limited-energy systems are specified within Chapters 7 and 8 of the *NEC*. This provides the relief from having to size these grounding electrode conductors according to Table 250.66, which for these limited-energy systems, would be excessive.

FIGURE 6-1 Grounding electrode conductors are generally connected to grounding electrodes.

Purpose of Grounding Electrode Conductors

An important step in grounding systems and equipment is the installation of grounding electrode conductors. This phase of construction of the grounding system is usually completed after the service equipment or separately derived system is installed and the service or feeder conductors are connected to the equipment. Grounding electrode conductors complete a conductive path to the ground (Earth) from systems and equipment that are required to be grounded or are otherwise grounded by choice. This conductor serves to establish and maintain an equal potential between the Earth and the equipment or system conductor being grounded, or as close to the same potential as electrically possible. **See Figure 6-2.**

Grounding electrode conductors carry varying amounts of current during normal operation, although ideally there would be no current. The amount of current present in a grounded conductor is usually low during normal operation. **See Figure 6-3.** During a ground-fault event, the amount of current in the grounding electrode conductor can increase for a short period. This current will be present only for the time it takes an overcurrent protective device to open the faulted circuit. The amount of current in the grounding electrode conductor during a ground-fault event is typically low because of the high amount of impedance between the fault location and the source grounding electrode.

The Earth is a fault current path that offers substantial impedance for current. For this reason, the purpose of the grounding electrode conductor is not to facilitate operation of the overcurrent protective device. Even though grounding electrode conductors are in the grounding and bonding circuit, the grounding electrode conductors are not intended as effective ground-fault current paths as clarified in Section 250.4. Grounding electrode conductors are connected to electrodes placing Earth in the return path for ground-fault current. The Earth cannot ever be depended on to serve as an effective ground-fault current path. The grounding electrode conductors are intended to experience lower levels of ground-fault current for the duration of a ground-fault condition due to multiple paths over which the current will divide, including the Earth. This accounts for the differences in sizing requirements for grounding electrode conductors as compared with equipment grounding conductors (EGCs). Two different conductors are used in the grounding and bonding scheme, and they each have different roles.

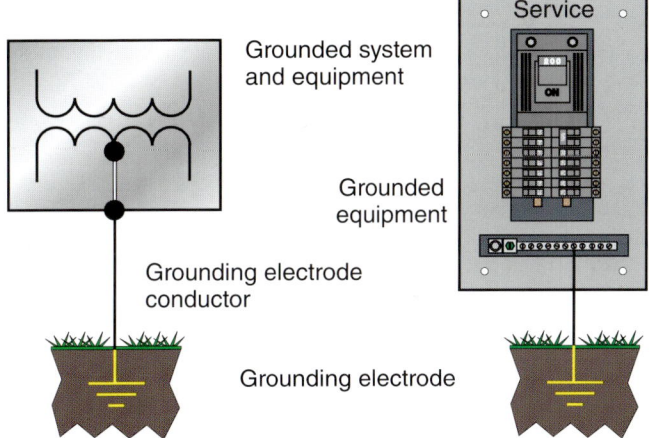

FIGURE 6-2 Grounding electrode conductors connect equipment and systems to the Earth.

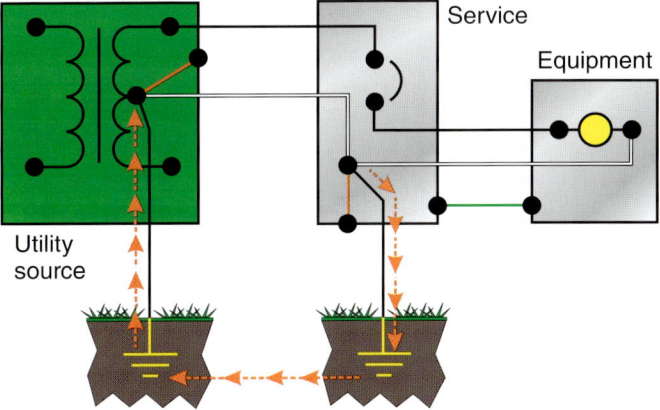

FIGURE 6-3 There is usually a small amount of current present in grounding electrode conductors during normal system operation.

Current in Grounding Electrode Conductors

The ground (Earth) is part of electrical circuits or equipment that is grounded, and current will be present in a circuit as long as the circuit is complete. Current will also be present in any and all paths available between the source and point of use or ground-fault condition. The amount of current present in each path is directly related to the amount of impedance in each path. The Earth is a conductor, but it is a very large "semi-conductive" body and the resistance (impedance) values of its conductivity vary greatly both geographically and seasonally. These are the main reasons that the amount of current present in grounding electrode conductors is typically low during normal operation and can rise and fall during ground-fault events on the system. The amount of current a grounding electrode conductor handles is limited because of the high level of impedance in the path between the service or system grounding point and the point of grounding at the source. This is one of the reasons that when a grounding electrode conductor is connected to a single rod, pipe, or plate electrode, it never has to be larger than a 6 AWG (American wire gauge) copper conductor.

The amount of current the grounding electrode conductor has to carry during normal operation and under abnormal conditions such as ground faults is kept low because of the high impedance in that path through the Earth. **See Figure 6-4.** The components in this path are usually the grounding electrode conductor (at the service or system), the grounding electrode, the Earth (between the two grounding connections), the electrode (at the system or source grounding point), and the grounding electrode conductor (at the system or source grounding point).

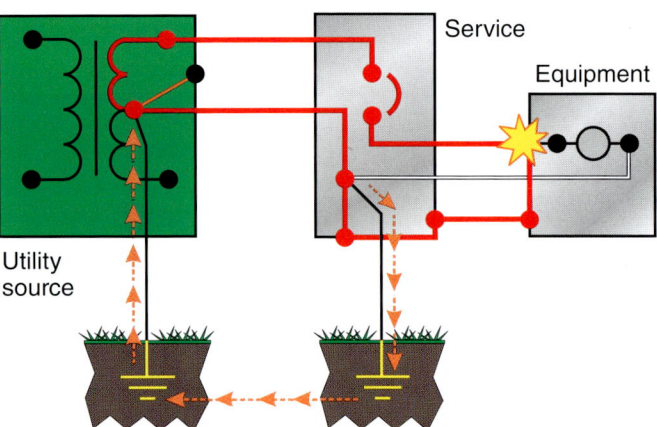

FIGURE 6-4 Current is usually relatively low in grounding electrode conductors during a ground-fault condition.

Grounding Electrode Conductor Material

Grounding electrode conductors are required to be made of copper, copper-clad aluminum, or aluminum. Copper-clad aluminum was used years ago in limited amounts as an effort to reduce electrical material costs during wartime periods, but is less common in electrical installations today because copper and aluminum conductors are available for this use. Copper-clad conductors are those manufactured with a thin copper coating over an aluminum conductor core.

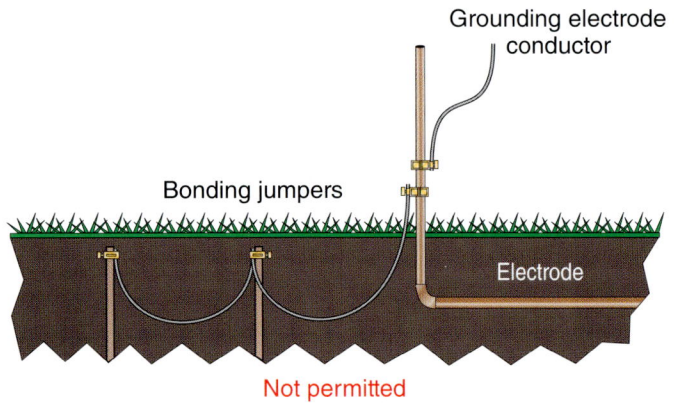

FIGURE 6-5 Aluminum grounding electrode conductor connections to electrodes are not permitted within 18 inches of the Earth outside.

Copper-clad and aluminum conductors are more vulnerable to the effects of corrosion and deterioration from Earth contact or contact with masonry. The most commonly specified and installed grounding electrode conductors today are insulated or bare copper. The *Code* includes installation and sizing requirements for aluminum and copper-clad aluminum grounding electrode conductors. One such installation restriction is the prohibition of terminating grounding electrode conductors within 18 inches of the Earth outside. **See Figure 6-5.** Other limitations for bare copper-clad and aluminum grounding electrode conductors are that they are not permitted to be in direct contact with the Earth or masonry and cannot be subjected to other corrosive conditions.

Insulated copper-clad or aluminum grounding electrode conductors offer protection from the previously mentioned corrosive conditions. A marking, such as a "W," should be used in the suffix of the insulation type to indicate suitability for wet locations.

Sizing Grounding Electrode Conductors

Sizing grounding electrode conductors is covered in Section 250.66 of the *NEC*. The size of a grounding electrode conductor for a service or separately derived system is based on the size of the largest ungrounded supply conductor. **See Figure 6-6.** The grounding electrode conductor is never sized based on an overcurrent protective device of a service or system.

Section 250.66 requires that grounding electrode conductors installed for services or separately derived systems and those installed at separate buildings or structures be sized according to the values in Table 250.66. This sizing rule applies to both grounded systems and ungrounded alternating current (AC) systems. Sections 250.66(A), (B), and (C) permit a reduction in size of the grounding electrode conductor where it is the sole connection to any of the electrode types identified in those subdivisions. For example, Section 250.66(A) indicates that when the grounding electrode conductor is connected to a rod, pipe, or plate electrode, that portion of the grounding electrode conductor that is the sole connection to the electrode is not required to be larger than a 6 AWG copper conductor or 4 AWG aluminum conductors. **See Figure 6-7.**

FIGURE 6-6 Grounding electrode conductors at services are generally sized according to Table 250.66 based on the circular mil area of the largest ungrounded service-entrance conductor.

A grounding electrode conductor connected to a single rod, pipe, or plate electrode does not have to be larger than a 6 AWG copper conductor.

Must be a sole connection to the ground rod, pipe, or plate.

FIGURE 6-7 A sole connection of a grounding electrode conductor to a single ground rod is addressed in Section 250.66(A).

Section 250.66(B) indicates that when the grounding electrode conductor is connected to a concrete-encased electrode, the portion of the grounding electrode conductor that is the sole connection to the electrode is not required to be larger than a 4 AWG copper. **See Figure 6-8.**

Section 250.66(C) indicates that when the grounding electrode conductor is connected to a ground ring, the portion of the grounding electrode conductor that is the sole connection to the electrode is not required to be larger than the ground ring. **See Figure 6-9.** The minimum-size conductor permitted for use as a ground ring electrode is 2 AWG copper, as provided in Section 250.52(A)(4).

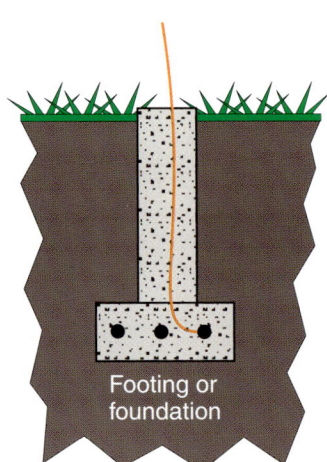

FIGURE 6-8 A sole connection to a concrete-encased electrode is addressed in Section 250.66(B).

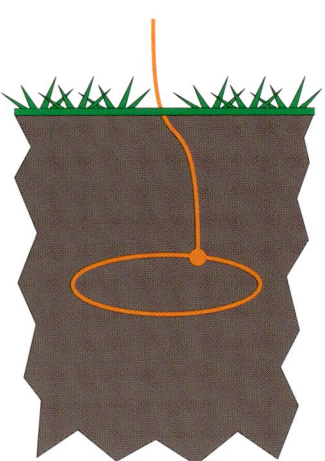

FIGURE 6-9 A sole connection to a ground ring electrode is addressed in Section 250.66(C).

Using Table 250.66

Table 250.66 is titled "Grounding Electrode Conductor for Alternating-Current Systems." Sizing grounding electrode conductors for direct current (DC) systems is covered in Section 250.166. Table 250.66 is easy to use and is based on conductor sizes rather than sizes of overcurrent protective devices. To use Table 250.66, first the size must be determined (in AWG or kcmil) of the largest ungrounded supply conductor or equivalent area for parallel conductors. Then the range should be located of conductor sizes that the supply conductor falls within, in column 1 or 2 depending on whether they are copper or aluminum. Last, column 3 or 4 should be looked at, depending on whether the conductor is copper or aluminum, to determine the grounding electrode conductor size. **See Figure 6-10.**

As an example, if a 400-ampere, three-phase, four-wire, 480Y/277 volt service was supplied by four 750-kcmil aluminum conductors (XHHW), the minimum-size copper grounding electrode conductor must be 1/0 copper or 3/0 aluminum. If an 800-ampere, three-phase, four-wire, 480Y/277 volt service was supplied by two sets of four 750-kcmil aluminum conductors (XHHW) in parallel, the copper grounding electrode conductor must not be smaller than 2/0 copper or 4/0 aluminum. When parallel conductors are installed, add the circular mil area of all conductors installed in parallel to make up one phase. In this case, two 750-kcmil service conductors is the equivalent of 1500 kcmil aluminum and requires a grounding electrode conductor not smaller than 2/0 copper or 4/0 aluminum.

Smaller grounding electrode conductors are permitted when they are the sole connection to rod, pipe, plate, concrete-encased, and ground ring electrodes as detailed in Sections 250.66(A), (B), and (C). The grounding electrode conductor never has to be sized larger than the values provided in Table 250.66 unless the engineering design specifies larger sizes. Sometimes grounding electrode conductors are installed in lengths that are excessive (200 to 300 feet for example).

In these cases, the size can be evaluated and possibly increased in size to compensate for the additional length. The *Code* does not address this issue, but in the 1950s, Eustace Soares did extensive work on the subject of sizing grounding electrode conductors to compensate for the effects of voltage drop. The concerns are obviously about the withstand capabilities of the grounding electrode conductors when installed in excessive lengths. The length of grounding electrode conductors and how length relates to performance are design issues that should be considered in the planning and design phases of projects.

Grounding Electrode Conductors for DC Systems

Grounding for DC systems is covered in Part VIII of Article 250. This part includes rules specific to DC system grounding requirements and must be applied in addition to the other parts of Article 250. A DC system is required to be grounded as follows:

1. Two-wire DC systems supplying premises wiring operating at greater than 50 volts but not exceeding 300 volts are required to be grounded (see three exceptions to 250.162(A)].
2. All three-wire DC systems are required to be grounded by connecting the neutral conductor of the system to ground.[2]

The connection point of the grounding electrode conductor for a DC system is covered in Section 250.164. If the DC source is not located on the premises, this connection shall be made at one or more supply stations. The grounding electrode conductor connection is not permitted to be made at individual services or at any point on the premises wiring if the source is not located on the premises. If the DC system (source) is located on the premises, the grounding electrode conductor connection has to be made:

1. At the source
2. At the first system disconnecting means or overcurrent protective device; or
3. By other means that provides equivalent protection and utilizes listed and identified equipment[3]

Table 250.66 Grounding Electrode Conductor for Alternating-Current Systems (in part)

Size of Largest Ungrounded Service-Entrance Conductor or Equivalent Area for Parallel Conductors (AWG/kcmil)		Size of Grounding Electrode Conductor (AWG/kcmil)	
Copper	Aluminum or Copper-Clad Aluminum	Copper	Aluminum or Copper-Clad Aluminum
2 or smaller	1/0 or smaller	8	6
1 or 1/0	2/0 or 3/0	6	4
2/0 or 3/0	4/0 or 250	4	2
Over 3/0 through 350	Over 250 through 500	2	1/0
Over 350 through 600	Over 500 through 900	1/0	3/0
Over 600 through 1100	Over 900 through 1750	2/0	4/0
Over 1100	Over 1750	3/0	250

1. Determine the size of the largest ungrounded service-entrance conductor
2. Proceed down the column to the range where the size is included
3. Proceed right horizontally to the copper column to determine the minimum size grounding electrode conductor required

FIGURE 6-10 Figure is a reproduction of *NEC* Table 250.66 (in part). Table 250.66 of the *NEC* is used to determine the appropriate size for grounding electrode conductors.

The minimum-size grounding electrode conductor for a DC system is generally required to be not less than the sizes indicated in Sections 250.166(A) and (B).

When the DC system consists of a three-wire balancer set or a balancer winding with overcurrent protection, as provided in Section 445.12(D), the grounding electrode conductor shall be sized no smaller than the neutral conductor and can never be smaller than 8 AWG copper or 6 AWG aluminum. **See Figure 6-11.** When the DC system is other than as described in Section 250.166(A), the grounding electrode conductor cannot be smaller than the largest conductor supplied by the system and can never be smaller than 8 AWG copper or 6 AWG aluminum.

When the grounding electrode conductor for a DC system is a sole connection to a rod, pipe, or plate electrode, the grounding electrode conductor does not have to be sized larger than 6 AWG copper or 4 AWG aluminum. If the grounding electrode conductor is a sole connection to a concrete-encased electrode, it never has to be larger than 4 AWG copper. If the grounding electrode conductor for a DC system is a sole connection to a ground ring, it never has to be sized larger than the size of the ring electrode. These provisions for sole connections of grounding electrode conductors installed for DC systems are identical to the allowances in Sections 250.66(A), (B), and (C) for AC systems. The installation requirements in Section 250.64 apply to grounding electrode conductors for DC systems. The requirement in Section 250.160 indicates that the other parts of the article have general application in addition to the rules in Part VIII unless a rule is specifically intended to apply to AC systems.

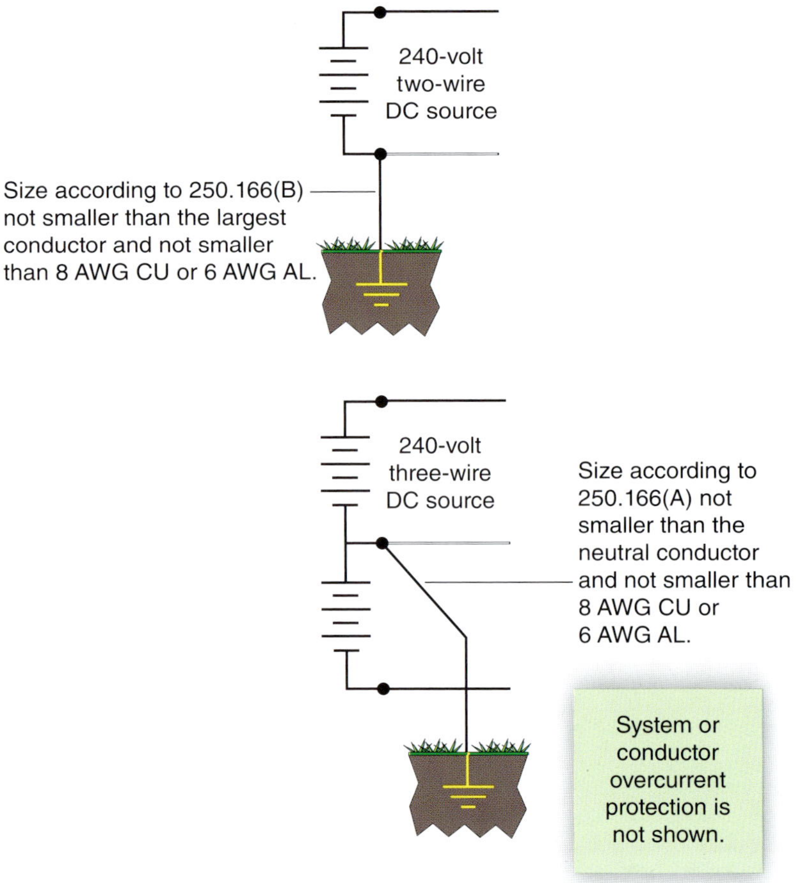

Grounding Electrode Conductor Installation

Section 250.50 requires all grounding electrodes present at a building or structure be connected together to form a grounding electrode system. There are only two exceptions to this rule. The first exception, which is not written in the *Code*, is if a building or structure had no power supplying it. There is no requirement for a grounding electrode system for that building or structure. The second exception follows Section 250.32 and relaxes the grounding electrode requirement for structures supplied by a single branch circuit that includes an EGC. An example of such an installation is a lighting pole in a parking lot. Typically an auxiliary grounding electrode is installed for these poles anyway, but it is not a *Code* requirement. It is a design consideration that is often included in plans and specifications. Another example is a detached garage or storage building supplied by only one branch circuit. An individual branch circuit or a multiwire branch circuit can qualify for the relief of this exception.

FIGURE 6-11 The grounding electrode conductor for a DC system is sized using Section 250.166.

Workers installing grounding electrode conductors must understand several important performance characteristics and requirements specified in Part III of Article 250. The grounding electrode conductor is the conductive path to the ground for equipment and systems. Specific rules apply to installations of the grounding electrode conductor. No identification requirement for grounding electrode conductors exists. They can be insulated covered or bare, but they are not required to be identified using the color green or green with yellow stripes as is required for EGC. There also is no prohibition against using the color green for identifying the grounding electrode conductor—it just is not a requirement. Verify any specific identification requirements that may be included in the project specifications. The installation requirements for grounding electrode conductors in the *NEC* are concerned with sizing requirements, connections, protection of the conductor from physical damage, and protection from external influences such as the effects of magnetic fields.

Securing and Protecting from Physical Damage

When grounding electrode conductors are installed exposed, they are required to be securely fastened to the surface on which they are placed. This means that they should be treated in similar fashion to securing a flexible wiring method such as any of the nonmetallic or metallic cable assemblies. If a grounding electrode conductor is installed in a conduit or other armor, the conduit or armor also has to be secured to the surface. Grounding electrode conductors in sizes 4 AWG or larger must be protected where installed in locations that expose them to physical damage. **See Figure 6-12.** Various protection methods can also be used, such as installing in the covered partitions of a building or burial in the Earth.

The *Code* also recognizes that if grounding electrode conductors that are 6 AWG or larger are installed such that they are not exposed to potential physical damage, they can be run without being placed in a raceway or armor if they are securely fastened. The authority having jurisdiction is typically responsible for judging whether a grounding electrode conductor is exposed to potential damage. Grounding electrode conductors also can be installed on or run through wood framing members. Physical protection in this case is usually inherently provided by this location within the construction framing. Other installation locations that provide the necessary physical protection also can be selected for grounding electrode conductors without the use of conduit or armor.

Grounding electrode conductors are required to be installed in a continuous length without a splice or joint unless splices are made using irreversible compression connectors or the exothermic welding process. **See Figure 6-13.**

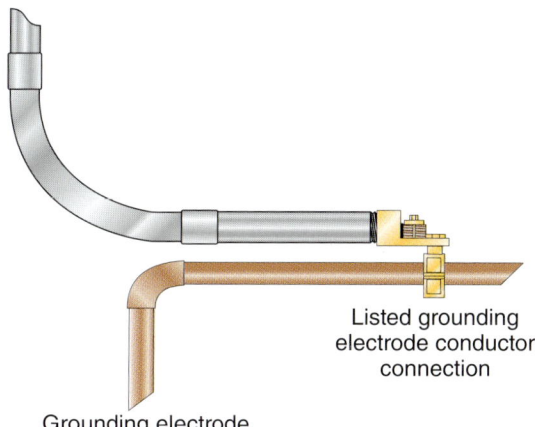

FIGURE 6-12 Grounding electrode conductors must be protected from physical damage.

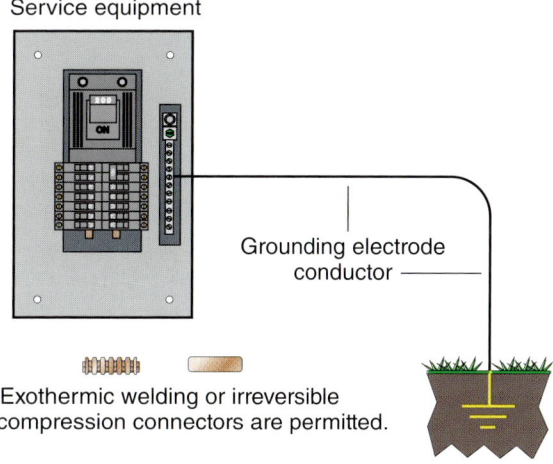

FIGURE 6-13 Grounding electrode conductors generally have to be installed without a splice or joint. Wire-type grounding electrode conductors are generally required to be installed without splices or joints other than as permitted in Section 250.64(C)(1).

126 APPLIED GROUNDING AND BONDING

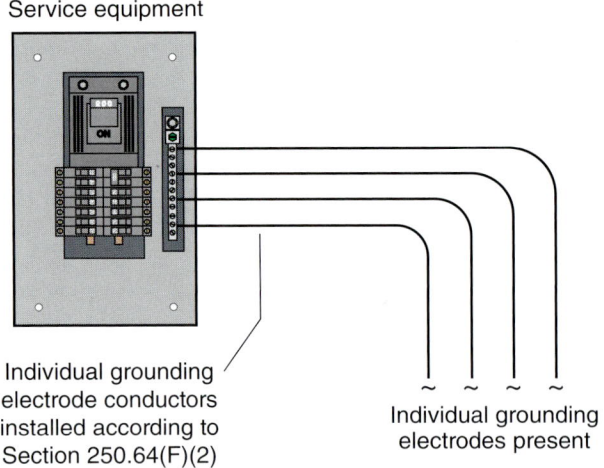

FIGURE 6-14 Grounding electrode conductors routed from individual electrodes to the service equipment.

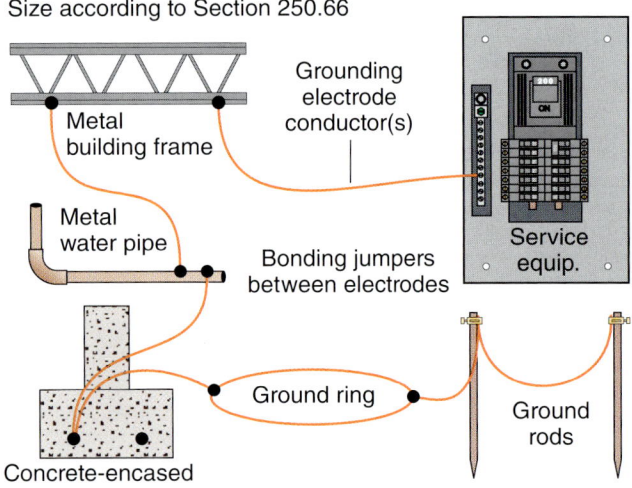

FIGURE 6-15 Table 250.66 is used for sizing the grounding electrode conductor and the bonding jumper(s) of a grounding electrode system, unless using the provisions of Sections 250.66(A), (B), or (C).

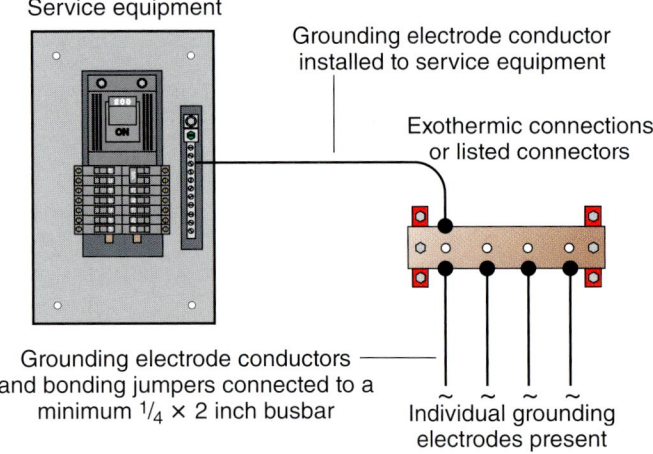

FIGURE 6-16 Grounding electrode conductors and bonding jumpers are permitted to be connected to a common busbar.

This is the general rule in Section 250.64(C). The exothermic welding process is an acceptable method of splicing grounding electrode conductor together or making a connection to a grounding electrode.

The irreversible compression connectors have to be listed as grounding and bonding equipment. This means they have been evaluated for the anticipated performance criteria in accordance with UL 467, *Grounding and Bonding Equipment*. Busbars are also permitted to be connected together in order to form grounding electrode conductors.

Methods of Installing Grounding Electrode Conductors

Individual grounding electrode conductors can be installed from each electrode back to the service, or a grounding electrode conductor can be run to one or more electrodes and bonding jumpers can interconnect the rest of the electrodes to form the grounding electrode system. **See Figure 6-14.** Section 250.64(F) describes installation methods that can be used when constructing the electrode system.

This section also provides a sizing requirement that relates to where the grounding electrode conductor or bonding jumper is connected in the grounding electrode system. When multiple grounding electrodes form a system of electrodes, the size required for the grounding electrode conductor has to be for the largest grounding electrode conductor connected in the system. For example, if a grounding electrode conductor is installed from a service to a concrete-encased electrode and a bonding jumper is installed from the concrete-encased electrode to a water pipe electrode, then both would have to be sized using Table 250.66 based on the requirement for connection to the water pipe. **See Figure 6-15.**

Grounding electrode conductors are permitted to be run to any electrode in the grounding electrode system if the other grounding electrodes are connected using bonding jumpers as indicated in Section 250.64(F). These bonding jumpers must be installed in accordance with Section 250.53(C), which indicates that sizing must be per Section 250.66 and the connections have to be made according to the requirements in Section 250.70. **See Figure 6-16.** Sometimes it is convenient and practical to install multiple grounding electrode conductors to each electrode of the system. Section 250.64(F)(2) recognizes this practice, and the same sizing requirements apply to each individual grounding electrode conductor.

Another method of interconnecting grounding electrodes to form the grounding electrode system is to connect the grounding electrode conductor(s) and bonding jumper(s) from each electrode to a copper or aluminum busbar. The busbar must be at least ¼ × 2 inch cross-sectional area. **See Figure 6-17.** These connections, as shown in Figure 6-16, must be listed and the busbar must be securely fastened to the structure in an accessible location.

When installing aluminum busbars, installers must follow the restrictive criteria in Section 250.64(A), which prohibits terminating aluminum conductors within 18 inches of the Earth. In this case the busbar is aluminum and must meet the same requirement for the same concerns about corrosive influences.

Connection Point for Grounding Electrode Conductors

A single grounding electrode conductor is permitted to be run to the grounding electrode system. **See Figure 6-18.** A common method of installing grounding electrode conductors is to run them directly from the service equipment enclosure to the grounding electrode system without a splice or joint.

FIGURE 6-17 Grounding electrode conductors and bonding jumpers for grounding electrode systems are permitted to be connected by attachment to copper or aluminum busbars.

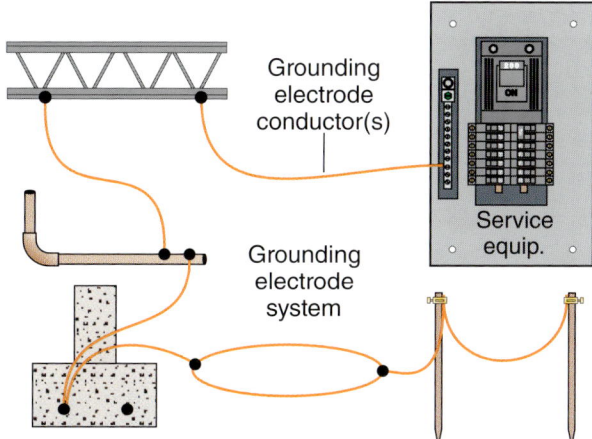

FIGURE 6-18 A single grounding electrode conductor is permitted to be run to a grounding electrode system. It is usually connected at the service equipment enclosure.

Section 250.24(A) indicates that the grounding electrode conductor connection to the system may be made at any point from the load end of the service drop or lateral up to the service equipment. Typically, a specific termination means is provided within the listed service equipment, such as within a single unit of service equipment or in assemblies such as switchboards, panelboards, motor control centers, and other equipment that is suitable for use as service equipment. **See Figure 6-19.**

Methods of Connection at Service Disconnecting Means

Services must be provided with a service disconnecting means to disconnect all ungrounded circuit conductors at a building or structure served. The service disconnecting means can be either a single (main) service disconnect in a panelboard, motor control center (MCC) or switchboard, or a separate individual disconnecting means enclosure (fused switch or enclosed circuit breaker). The service disconnecting means can also be up to six disconnects installed in a single enclosure assembly that is suitable for use as service equipment, or up to six disconnects (each suitable for use as service equipment) can be installed in separate enclosures and grouped as required in Section 230.71(A).

Single Disconnecting Means

For installations of service equipment that include only a single service disconnecting means either as a separate enclosure or as part of an assembly that is suitable for use as service equipment, a single grounding electrode conductor can be installed from the grounded conductor terminal bus in the equipment to the grounding electrode system.

FIGURE 6-19 A grounding electrode conductor terminal is often provided on the terminal bar for the grounded conductor within the listed service equipment.

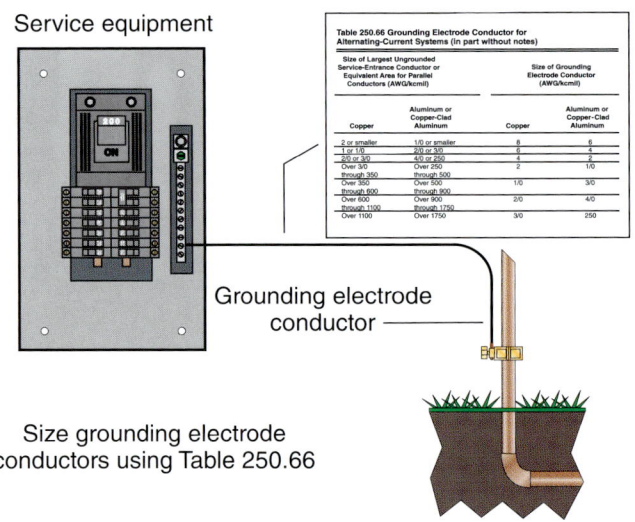

FIGURE 6-20 The general sizing rule for grounding electrode conductors is provided in Section 250.66.

The minimum size of the grounding electrode conductor is determined by using Table 250.66; sizing is based on the size of the largest service-entrance conductor. For example, if the size of the largest ungrounded service-entrance conductor for a 400-ampere service is 500 kcmil copper, then the minimum-size grounding electrode conductor is 1/0 copper or 3/0 aluminum or copper-clad aluminum. **See Figure 6-20.**

Remember that Section 250.66 permits smaller sizes to be used for certain types of electrodes. **See Figure 6-21.**

Multiple Service Disconnecting Means in Separate Enclosures

A popular method of installing grounding electrode conductors at services is by using the common grounding electrode conductor tap concept. This concept is typically used where multiple service disconnecting means are installed in separate enclosures as permitted by Section 230.71(A). In these types of installations a single, large common grounding electrode conductor is installed and multiple grounding electrode conductor taps are connected to the common grounding electrode conductor. **See Figure 6-22.**

The term *common grounding electrode conductor* refers to the conductor that is not permitted to have any splices or joints as described in the general rule in Section 250.64(D)(1). The term *grounding electrode conductor tap* is not a defined term in the *NEC* and is therefore not subject to this same restriction. The connections of the grounding electrode conductor taps to a common grounding electrode conductor must be made using exothermic welding or by listed grounding and bonding connections. A ¼ × 2 inch aluminum or copper busbar may also be used with exothermic or listed connections. **See Figure 6-23.**

The minimum size of the common grounding electrode conductor is based on the sum of the circular mil area of the largest ungrounded service-entrance conductor(s).

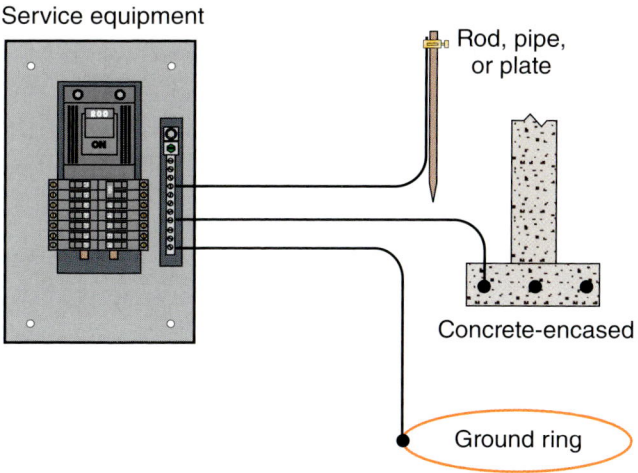

FIGURE 6-21 Grounding electrode conductors that are sole connections can be sized using Section 250.66(A), (B), or (C).

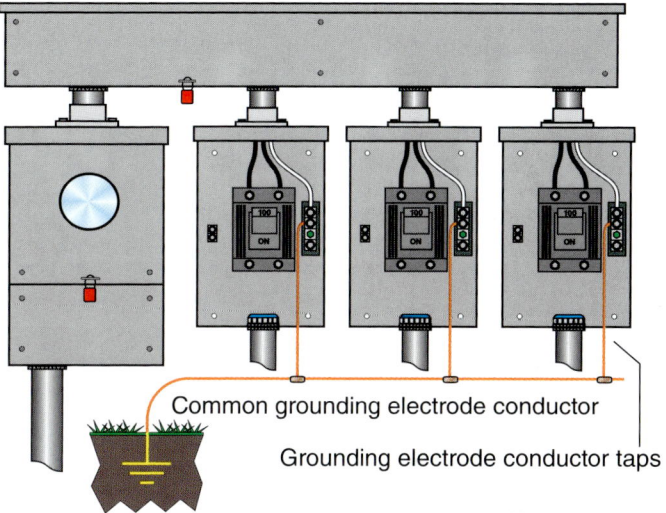

FIGURE 6-22 Grounding electrode conductor taps are permitted to be installed to separate service disconnects.

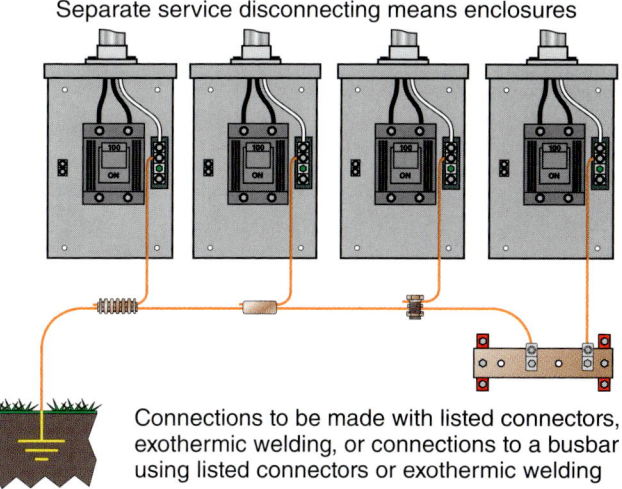

FIGURE 6-23 Grounding electrode conductor taps must be connected to a common grounding electrode conductor in a manner in which the common grounding electrode conductor remains without a splice.

When the service-entrance conductors connect directly to a service drop or service lateral, the common grounding electrode conductor shall be sized in accordance with Table 250.66, Note 1. The grounding electrode conductor tap then is extended to the inside of each service disconnecting means enclosure. Grounding electrode tap conductors must be sized using Table 250.66; sizing is based on the size of the largest phase conductor serving each service disconnecting means enclosure. **See Figure 6-24.**

The common grounding electrode conductor from which the taps are made is sized using Table 250.66; sizing is based on the sum of the cross-sectional areas of the largest ungrounded service-entrance conductors or equivalent cross-sectional area for parallel service-entrance conductors that supply the multiple separate service disconnecting means. **See Figure 6-25.** The common grounding electrode conductor tap concept is popular because it eliminates the difficulties in looping a single larger grounding electrode conductor from one enclosure to another where terminations and space within the enclosure might not meet the equipment listing requirements.

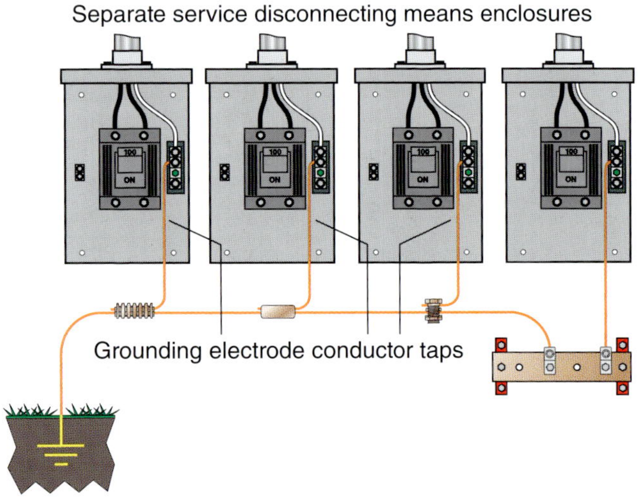

Individual Grounding Electrode Conductors to Multiple Service Disconnects

Another popular method of installing grounding electrode conductors for individual separate service disconnecting means enclosures is to run a grounding electrode conductor from the grounding electrode to each separate enclosure.

For example, a service could be installed using up to six separate service disconnecting means enclosures on the same building with a separate riser from each disconnect.

FIGURE 6-24 Use Table 250.66 to size grounding electrode conductor taps. The size to be used is based on the largest service conductor in each separate service disconnect.

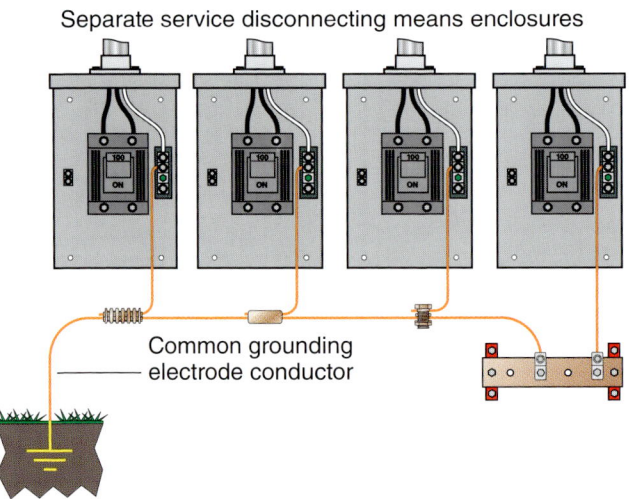

FIGURE 6-25 Use Table 250.66 to size the common grounding electrode conductor at services. The size to be used is based on the sum of the circular mil area of the largest service conductor.

Grounding Electrode Conductors

The size of the grounding electrode conductor run to each individual service disconnect enclosure is determined using Section 250.66 and is based on the largest ungrounded service conductor in each separate enclosure. **See Figure 6-26.** In these cases the same grounding electrode system is inherently used for all service disconnects on the same building or structure.

Effectiveness (Integrity) of the Grounding Path

Section 250.68 generally requires that the mechanical connections to a grounding electrode be accessible. **See Figure 6-27.**

There are some exceptions to Section 250.68 that address conditions or installations that relax this grounding electrode conductor connection accessibility requirement. **See Figure 6-28.**

1. Grounding electrode connections are buried, such as those used for rods, pipes, or plate electrodes
2. Concrete-encased electrodes where the connection is encased in concrete
3. If exothermic welding is used as the connection means or if the connections are made by irreversible compression connectors to fireproofed structural metal[4]

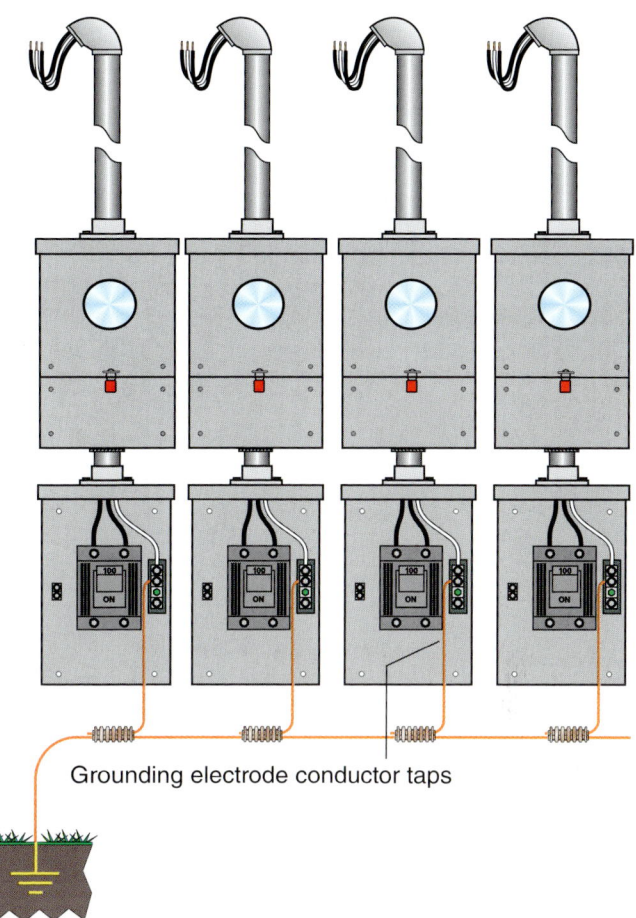

FIGURE 6-26 Individual grounding electrode conductors run to separate service disconnects are sized according to Section 250.66. The size to be used is based on the largest service conductor in each separate service disconnect.

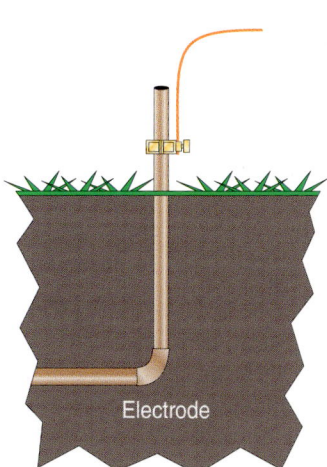

FIGURE 6-27 Grounding electrode conductor connections are generally required to be accessible.

FIGURE 6-28 Grounding electrode conductor connections are permitted to be buried in fireproofing material on building steel electrodes.

132 APPLIED GROUNDING AND BONDING

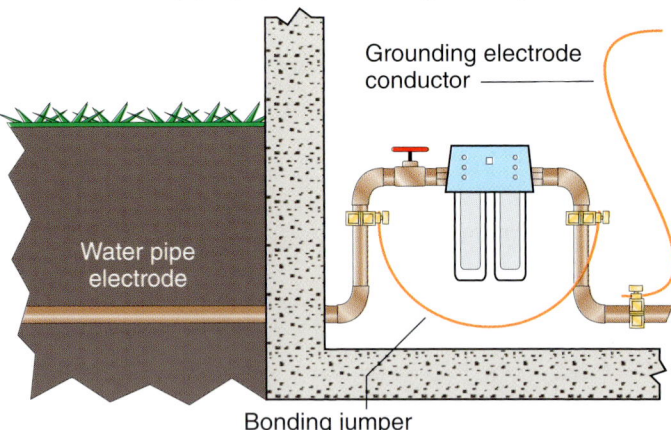

FIGURE 6-29 The integrity of grounding connection must be ensured by installing a bonding jumper around any insulating joints or equipment likely to be removed for replacement or repairs.

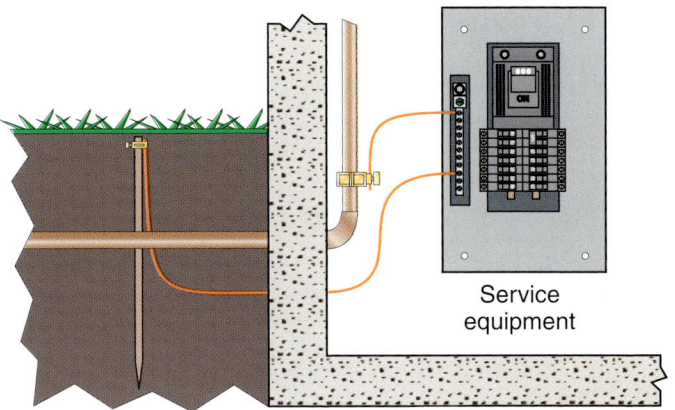

FIGURE 6-30 The grounding electrode conductor connection to water pipe electrode must be made within 5 feet of the water pipe point of entry at the building or structure served.

The connections of grounding electrode conductor(s) or bonding jumper(s) to a grounding electrode must ensure an effective path to the ground (Earth). When necessary to ensure the grounding path for a metal piping system grounding electrode, bonding must be provided around any insulated joints and around equipment that is likely to be disconnected or removed for repairs or replacement.

Bonding jumpers must be of sufficient length to permit removal of such equipment while retaining the integrity of the grounding path. **See Figure 6-29.** An example of an installation where a bonding jumper is required to ensure the grounding path is at a water meter or other equipment such as a water softener that is installed in series with a metal water piping system that is used as a grounding electrode.

Grounding Electrode Conductor Connection Locations

The connections of grounding electrode conductors can be made at a variety of locations throughout the building or structure. When a metal water pipe is used as a grounding electrode, the grounding electrode conductors or bonding jumpers from a grounding electrode system are permitted to be made to the water piping at any point on the exposed piping from the point of entrance up to a distance of 5 feet as it enters the building or structure. **See Figure 6-30.** The reason for the restriction is to limit the length of piping that must function as a grounding electrode conductor.

Section 250.68(C)(1) also prohibits the use of that portion of the interior metal water piping system that extends more than 5 feet beyond the point of entrance into the building to be used as a conductor to interconnect grounding electrodes.

This is because of concerns over the use of nonmetallic piping or fittings causing an interruption in the electrical continuity of the metal water piping.

There is an exception for commercial, industrial, and institutional facilities. This exception allows the grounding electrode conductor connection or connections of bonding jumpers of the grounding electrode system to be made at locations farther than 5 feet from the point where the water piping enters the building or structure if the following specific restrictive conditions are:

1. Conditions of maintenance and supervision ensure qualified persons service the installation.
2. The entire length of the piping is exposed other than where it passes perpendicularly through walls or floors.[5]

Buildings or structures often include conductive paths that connect to grounding electrodes. These conductive paths actually perform the function of a grounding electrode conductor, but they are not wire-type conductors. A couple of examples are interconnected metal building frames and metal water piping systems that do not qualify as grounding electrodes by definition or by the details and descriptions provided in Section 250.52(A). Previous editions of the *Code* referred to these conductive objects as grounding electrodes, but they do not meet the definition of the term *grounding electrode* and they may function as grounding electrode conductors in this case. Section 250.68(C) addresses locations of connections permitted for grounding electrode conductors and bonding jumpers of a grounding electrode system. **See Figure 6-31.**

This section includes provisions that recognize these conductive paths that are ultimately connected to grounding electrodes but that are not electrodes by definition. Metal building frames that are electrically connected to a grounding electrode or system of grounding electrodes are permitted to function as conductive paths to ground as provided in either (1), (2), or (3) below.

(1) By connecting the structural metal frame to the reinforcing bars of a concrete-encased electrode as provided in 250.52(A)(2) or ground ring as provided in 250.52(A)(4) or,
(2) By bonding the structural metal frame to one or more of the grounding electrodes as defined in 250.52(A)(5) or (A)(7) that comply with 250.53(A)(2)[6] or,
(3) By other approved means of establishing a connection to earth[6]

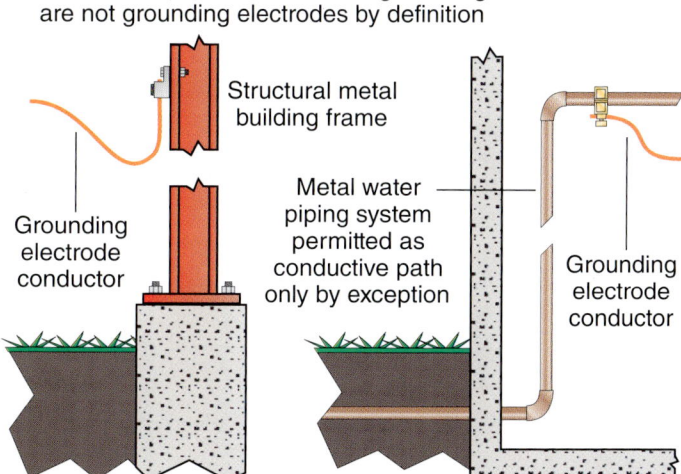

FIGURE 6-31 Building steel and metal water piping systems often perform as conductive paths to the grounding electrode system, but are not grounding electrodes by definition.

Grounding Electrode Conductor Connections

Section 250.70 provides information about acceptable methods of grounding electrode conductor connections. **See Figure 6-32.** Connections of grounding electrode conductors or bonding jumpers of the grounding electrode system can be made using one of the following methods:

1. Listed lugs
2. Exothermic welding processes
3. Listed pressure connectors
4. Listed clamps
5. Or other listed means[7]

The integrity of the connection to a grounding electrode system is important for safety and proper operation of the electrical system. Obviously, solder is not permitted to be relied on for grounding electrode conductor connections because the integrity of soldered connections can be negatively affected by heat caused by the current during a ground fault event. Listed grounding clamps must be compatible with the grounding electrode conductor material and must be suitable for the type of piping or material to which they are connected. This is important because dissimilar metals can cause a deterioration effect that can compromise the connection.

Listed lugs used for grounding electrode conductor connections.

Exothermic welding process used for grounding electrode conductor connections.

Listed pressure connectors used for grounding electrode conductor connections.

Listed clamps used for grounding electrode conductor connections.

Courtesy of Thomas and Betts

FIGURE 6-32 Grounding electrode conductor connections can be made using a variety of methods and materials that are listed (except exothermic welding) and identified for this purpose.

For grounding electrode conductors that are connected to rod, pipe, or plate electrodes, the connection device has to be listed for direct burial. Clamps that are listed for direct burial are also suitable for encasement in concrete as indicated in category KDER of the *UL General Information Directory for Electrical Equipment* (White Book). **See Figure 6-33.**

Grounding or bonding clamps must be listed and are typically suitable for connecting only one conductor. **See Figure 6-34.**

More than one grounding electrode conductor is permitted to be used with a single clamp where the clamp is identified for multiple conductors. The following provides four acceptable methods for connecting grounding electrode conductors or bonding jumpers to grounding electrodes:

1. A pipe fitting or plug that screws into a pipe or pipe fitting
2. A listed bolted clamp of cast bronze or plain malleable iron
3. A listed sheet metal strap type for indoor communications purposes
4. An equally approved and substantial means[8]

Listed grounding clamps for grounding electrode conductors used with communications systems or other limited-energy systems are available in a strap-type configuration for indoor use only. **See Figure 6-35.** These clamps must be listed and must have a rigid metal base and strap that encircles the piping system and is not likely to stretch after installation.

These types of grounding electrode conductor connections are not permitted for use on power system installations. Category KDSH in the *UL General Information Directory for Electrical Equipment* provides additional information about communications systems grounding electrode conductor connection devices. Grounding clamps are required to be protected from physical damage unless the fittings (clamps) are approved for use without additional protection.

> More than one grounding electrode conductor is permitted to be used with a single clamp where the clamp is identified for multiple conductors.

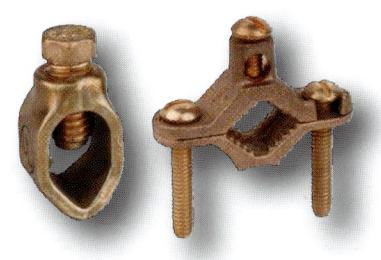

Courtesy of Thomas and Betts

FIGURE 6-33 A ground clamp listed for direct burial is also suitable for concrete encasement.

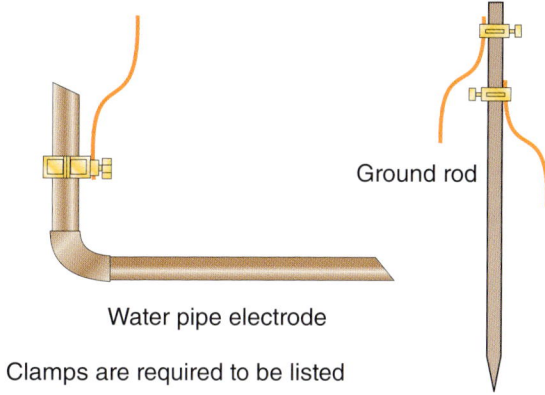

FIGURE 6-34 Grounding clamps are generally limited to connection of one grounding electrode conductor unless identified for more than one.

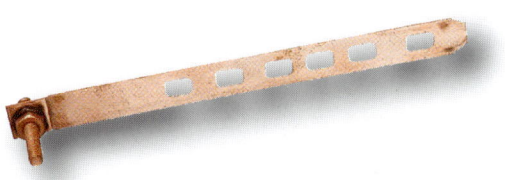

Courtesy of Thomas and Betts

FIGURE 6-35 Ground clamps are specifically listed for use with communications system grounding electrode conductors and bonding conductors.

Otherwise, the grounding clamp must be protected using metal, wood, or other equivalent materials as described in Section 250.10. **See Figure 6-36.**

Magnetic Field Concerns

Grounding electrode conductors must be protected if subject to physical damage. In addition to concerns about physical damage from external causes, grounding electrode conductors can be affected by magnetic fields. This is a not a concern for grounding electrode conductors that are not installed in ferrous (magnetic) metal raceways as a means for protection against physical damage. However, if a grounding electrode conductor is installed in a ferrous metal raceway, the raceway has to be electrically continuous from the point of attachment to the cabinet or equipment to the grounding electrode and must be securely fastened to the ground clamp or fitting. Ferrous metal raceways are those with iron or steel content such as rigid metal conduit (RMC), intermediate metal conduit (IMC), or electrical metallic tubing (EMT). These conduits and tubing have a magnetic property that reacts to rising and falling magnetic fields present in AC systems. During a ground-fault event, the current in a grounding electrode conductor can be relatively high for the duration of the event.

The strength of the magnetic field will also increase in direct proportion to the amount of current in the conductor. In many cases the magnetic lines of force in the conductor are induced into the conduit enclosing the grounding electrode conductor; they can even surpass the saturation point of the steel raceway. At the point where the grounding electrode conductor exits the conduit, the magnetic lines of force generated by the fault current in the conductor will try to be induced on the end of the conduit creating a saturation point that exceeds the capacity of the conduit. The steel conduit in this instance is acting like a steel core of a coil to concentrate the magnetic lines of force. This creates what the industry refers to as a *choke effect*. Specific bonding requirements are necessary for ferrous metal raceways that contain grounding electrode conductors.

The Choke Effect

Ferrous metal raceways are bonded to the contained grounding electrode conductor to reduce the effects of magnetic fields that are present while the system is energized and in use. The grounding electrode conductor for an AC system or service is an alternating current carrying conductor. There are varying amounts of current in a grounding electrode conductor during normal operation. This current can also rise and fall significantly depending on events such as ground faults, short circuits, or line surges. As the current rises and falls, the magnetic field of the contained conductor gets larger and smaller accordingly. This means the stresses on the contained grounding electrode conductor increase and decrease as the current rises and falls. Because the ferrous metal raceway is enclosing this conductor, there is an inductive reactance between the ferrous metal raceway and the contained grounding electrode conductor. This inductive reactance is one component of impedance and actually impedes current in the grounding electrode conductor. The magnetic field and the capacitance results in a coupling effect between the current in the conductor and the surrounding ferrous metal raceway.

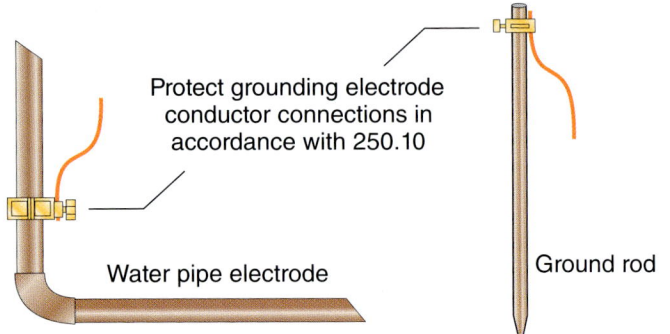

FIGURE 6-36 Protection is required for grounding electrode conductor connections unless the clamp or connection is approved for use without protection from damage.

In actuality, the majority of the current would be present in the ferrous raceway rather than the contained grounding electrode conductor. This condition is often referred to as the *choke effect* because this condition is actually the restriction of a grounding electrode conductor from performing its function.

Bonding Ferrous Metal Enclosures for Grounding Electrode Conductors

Ferrous metal enclosures for grounding electrode conductors that are not physically continuous from cabinets or equipment to the grounding electrode are required to be made electrically continuous by bonding each end of the raceway to the contained grounding electrode conductor. **See Figure 6-37.** This action puts the contained grounding electrode conductor in parallel with the enclosing ferrous metal raceway so that the two work together when the current in grounding electrode conductors rises and falls in response to various events occurring on the system.

The methods required for bonding each end of the raceway are provided in Sections 250.92(B)(2) through (B)(4). **See Figure 6-38.** These methods apply to all intervening ferrous raceways, boxes, and enclosures containing the grounding electrode conductor. If a bonding jumper is used to accomplish this bonding to intervening metal raceways and enclosures, the size of the bonding jumper must not be smaller than the required enclosed grounding electrode conductor. If a raceway is used as protection for the grounding electrode conductor, the installation rules in Chapter 3 of the *NEC* are applicable for such raceways. Nonferrous conduits such as brass and aluminum types do not have a magnetic content and do not have to be electrically continuous. Another important installation requirement to consider is the conduit and tubing fill percentage restrictions in Table 1 of Chapter 9 of the *NEC*. If a single conductor (grounding electrode conductor in this case) is installed in conduit or tubing, the percentage of fill must not exceed 53%. **See Figure 6-39.**

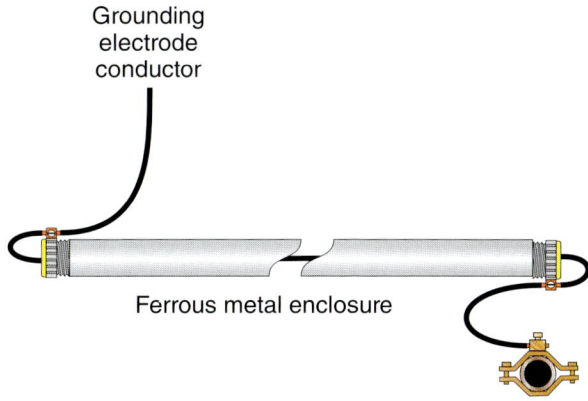

FIGURE 6-37 When ferrous metal raceways are not physically continuous from the cabinet or equipment to the grounding electrode, bond both ends of the ferrous metal raceways that enclose grounding electrode conductors.

FIGURE 6-38 Bond both ends of ferrous metal raceways that enclose grounding electrode conductors.

Table 1 Percentage of Cross Section of Conduit and Tubing for Conductors

Number of Conductors	All Conductor Types
1	53
2	31
Over 2	40

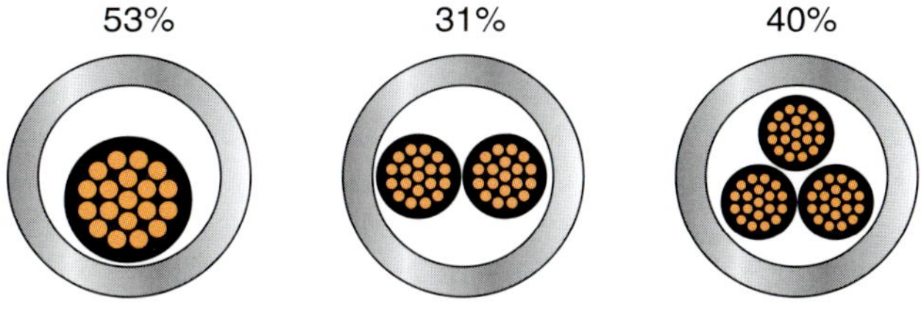

Reproduction of Table 1, Chapter 9 (without notes to tables)

FIGURE 6-39 Table 1 in Chapter 9 limits a single conductor to fill not more than 53% of the raceway. This reproduction does not include the notes to the table.

For smaller grounding electrode conductors installed in conduit or tubing, it is relatively easy to use a bonding busing on the ends of the raceway and form the grounding electrode conductor back through it to complete the bonding connection. However, with larger grounding electrode conductors, it is more difficult to meet these bonding requirements. Because the bonding must be accomplished using the same size jumper as the grounding electrode conductor, two methods can be used for larger conductors and conduit combinations. One method is to use bonding bushings at each end of the raceway and use an irreversible compression connector to crimp on a short bonding jumper that connects to the busing lug. **See Figure 6-40.**

Another method is to use bonding fittings manufactured specifically for this purpose. **See Figure 6-41.** In this case the hub-type fitting is installed on the end(s) of the conduit or tubing, and it includes a conductor clamp for completing an effective bonding connection between the hub and the grounding electrode conductor. Remember that per Section 250.64(E), the bonding at each end has to be accomplished using a conductor that is the same size as the grounding electrode conductor contained in the raceway.

The requirement to bond the contained grounding electrode conductor at both ends reduces these effects by putting the contained conductor in parallel with the ferrous metal raceway enclosing it. In this case the ferrous metal raceway and the contained grounding electrode conductor work together. Testing data have demonstrated that for practical purposes, the impedance of an AC grounding electrode conductor enclosed in a ferrous metal raceway when the conduit is bonded at both ends is approximately equal to the impedance of the conduit itself. Another simple solution to avoid the choke effect in grounding electrode conductors is to use Schedule 80 polyvinyl chloride (PVC) conduit as a means of providing physical protection. Be sure that the type of construction for the project does not restrict the use of nonmetallic (plastic) wiring methods. Because a grounding electrode conductor carries varying amounts of current during normal operation, it is not permitted to function simultaneously as an EGC, per Section 250.121.

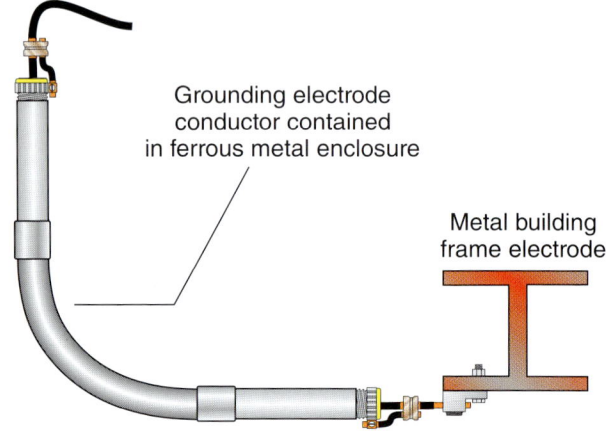

FIGURE 6-40 Grounding electrode conductors are required to be bonded to enclosing ferrous metal conduit.

Courtesy of Thomas and Betts

FIGURE 6-41 A conduit hub fitting can be used for bonding grounding electrode conductor to the ferrous metal raceway that contains it.

Summary

The grounding electrode conductor is an important component of the grounding and bonding system. This conductor performs the grounding function and maintains the connection to the Earth for systems and equipment. The grounding electrode conductor carries varying degrees of current during normal operation and does not perform as an effective ground-fault current path. Grounding electrode conductor sizing is generally determined by using Table 250.66, unless the grounding electrode conductor is installed as a sole connection, as covered in Section 250.66(A), (B), or (C). Connections must be generally accessible unless they are buried, encased in concrete, or placed under fireproofing materials. The grounding electrode conductor is generally required to be installed in a continuous length without a splice and in a manner that protects it from physical damage; otherwise, physical protection must be provided. Grounding electrode conductors can be copper, aluminum, or copper-clad aluminum. Aluminum and copper-clad aluminum grounding electrode conductors are not permitted to be terminated within 18 inches of the Earth. The grounding electrode conductor is not permitted to function as both an EGC and a grounding electrode conductor.

Metal water piping systems can be used as a conductive path to the grounding electrode system in accordance with Section 250.68(C)(1) Exception.

Structural metal building framing can be used as a conductive path to the grounding electrode system in accordance with Section 250.68(C)(2).

References

1. NFPA 70 National Electrical Code 2011, Article 100 (National Fire Protection Association, Quincy, MA 2010), p. 70–29.
2. NFPA 70 National Electrical Code 2011, Section 250.164(A) (National Fire Protection Association, Quincy, MA 2010), p. 70–128.
3. NFPA 70 National Electrical Code 2011, Section 250.164(B) (National Fire Protection Association, Quincy, MA 2010), p. 70–128.
4. NFPA 70 National Electrical Code 2011, Section 250.68(A) (National Fire Protection Association, Quincy, MA 2010), p. 70–115.
5. NFPA 70 National Electrical Code 2011, Section 250.68(C)(1) (National Fire Protection Association, Quincy, MA 2010), p. 70–116.
6. NFPA 70 National Electrical Code 2011, Section 250.52(A)(2) (National Fire Protection Association, Quincy, MA 2010), p. 70–111.
7. NFPA 70 National Electrical Code 2011, Section 250.70 (National Fire Protection Association, Quincy, MA 2010), p. 70–116.
8. NFPA 70 National Electrical Code 2011, Section 250.70 (National Fire Protection Association, Quincy, MA 2010), p. 70–116.

Review Questions

1. A grounding electrode conductor can be made of which of the following materials?
   a. Copper
   b. Aluminum
   c. Copper-clad aluminum
   d. All of the above

2. A conductor used to connect the system grounded conductor or the equipment to a grounding electrode or to a point on the grounding electrode system best defines which of the following?
   a. Grounded conductor
   b. Grounding electrode conductor
   c. Equipment grounding conductor
   d. Bonding jumper

3. *NEC* Table 250.66, used for sizing grounding electrode conductors, uses the largest _____ to determine the minimum-size grounding electrode conductor required for a service.
   a. Overcurrent protective device
   b. Service equipment rating
   c. Service conductor size
   d. Disconnect

4. Grounding electrode conductors provide the connection between the Earth and the object or conductor required to be connected to ground.
   a. True
   b. False

5. A grounding electrode conductor is an effective ground-fault current path that will facilitate operation of overcurrent protective devices during ground-fault conditions.
   a. True
   b. False

6. A grounding electrode conductor has little or no effect on facilitating overcurrent protective device operation.
   a. True
   b. False

7. A grounding electrode conductor is never sized based on an overcurrent protective device of a service or separately derived system.
   a. True
   b. False

8. When the grounding electrode conductor is connected to a concrete-encased electrode, the portion of the grounding electrode conductor that is the sole connection to the electrode is not required to be larger than a _____ AWG copper conductor.
   a. 1
   b. 2
   c. 3
   d. 4

9. When the grounding electrode conductor is connected to a ground ring that is a 1 AWG copper conductor, the portion of the grounding electrode conductor that is the sole connection to the electrode is not required to be larger than _____ AWG copper.
   a. 1
   b. 2
   c. 3
   d. 4

10. When the DC system (source) is located on the premises, the grounding electrode conductor connection to the DC system has to be made at which of the following locations?
    a. At the source
    b. At the first system disconnecting means or overcurrent protective device
    c. Either a or b
    d. Neither a or b

11. A grounding electrode conductor for a DC system is required to be sized using Table 250.66.
    a. True
    b. False

12. The requirements for installing grounding electrode conductors for AC systems are provided in _____ of Article 250.
    a. Part II
    b. Part III
    c. Part IV
    d. Part V

13. A grounding electrode conductor smaller than 6 AWG copper generally has to be protected by installation in a cable armor or raceway.
    a. True
    b. False

14. What is the minimum-size grounding electrode conductor for a 400-ampere service (600-kcmil copper service conductors) if the grounding electrode is a structural steel building frame?
    a. 1 AWG copper
    b. 2 AWG copper
    c. 3 AWG copper
    d. 1/0 AWG copper
15. A grounding electrode conductor is required to be identified by the color green.
    a. True   b. False
16. Grounding electrode conductors in sizes 6 AWG and larger that are installed such that they are not exposed to physical damage can be run without being placed in a raceway or armor as long as they are securely fastened.
    a. True   b. False
17. Grounding electrode conductors generally must be installed in a continuous length without a splice or joint unless splices are made using _____.
    a. Exothermic welding
    b. Irreversible compression connectors listed as grounding and bonding equipment
    c. Bars using listed connectors
    d. Any of the above
18. A grounding electrode conductor is required to be connected to a grounding electrode using which of the following methods?
    a. Listed lugs or listed clamps
    b. Listed pressure connectors
    c. Exothermic welding process
    d. Any of the above
19. If a single ground rod is installed as the grounding electrode for a service, what is the maximum size the grounding electrode conductor must be?
    a. 2 AWG copper
    b. 4 AWG copper
    c. 6 AWG copper
    d. 8 AWG copper
20. The mechanical connections to a grounding electrode are required to be accessible, unless _____.
    a. Grounding electrodes are buried, such as rods, pipes, or plate electrodes
    b. The electrodes at the connection are encased in concrete
    c. Exothermic welding is used as the connection means or the connections are made by irreversible compression connectors to fireproofed structural metal
    d. Any of the above
21. Metal building frames that are electrically connected to a grounding electrode or system of grounding electrodes are permitted to function as conductive paths to ground under which of the following conditions?
    a. By connecting the structural metal frame to the reinforcing bars of a concrete-encased electrode as provided in Section 250.52(A)(3) or ground ring as provided in Section 250.52(A)(4)
    b. By bonding the structural metal frame to one or more of the grounding electrodes as defined in Section 250.52(A)(5) or (A)(7) that comply with Section 250.56
    c. By other approved means of establishing a connection to Earth
    d. Any of the above
22. If a grounding electrode conductor is installed in a ferrous metal raceway, the raceway has to be electrically continuous from the point of attachment to the cabinet or equipment to the grounding electrode and must be securely fastened to the ground clamp or fitting.
    a. True   b. False

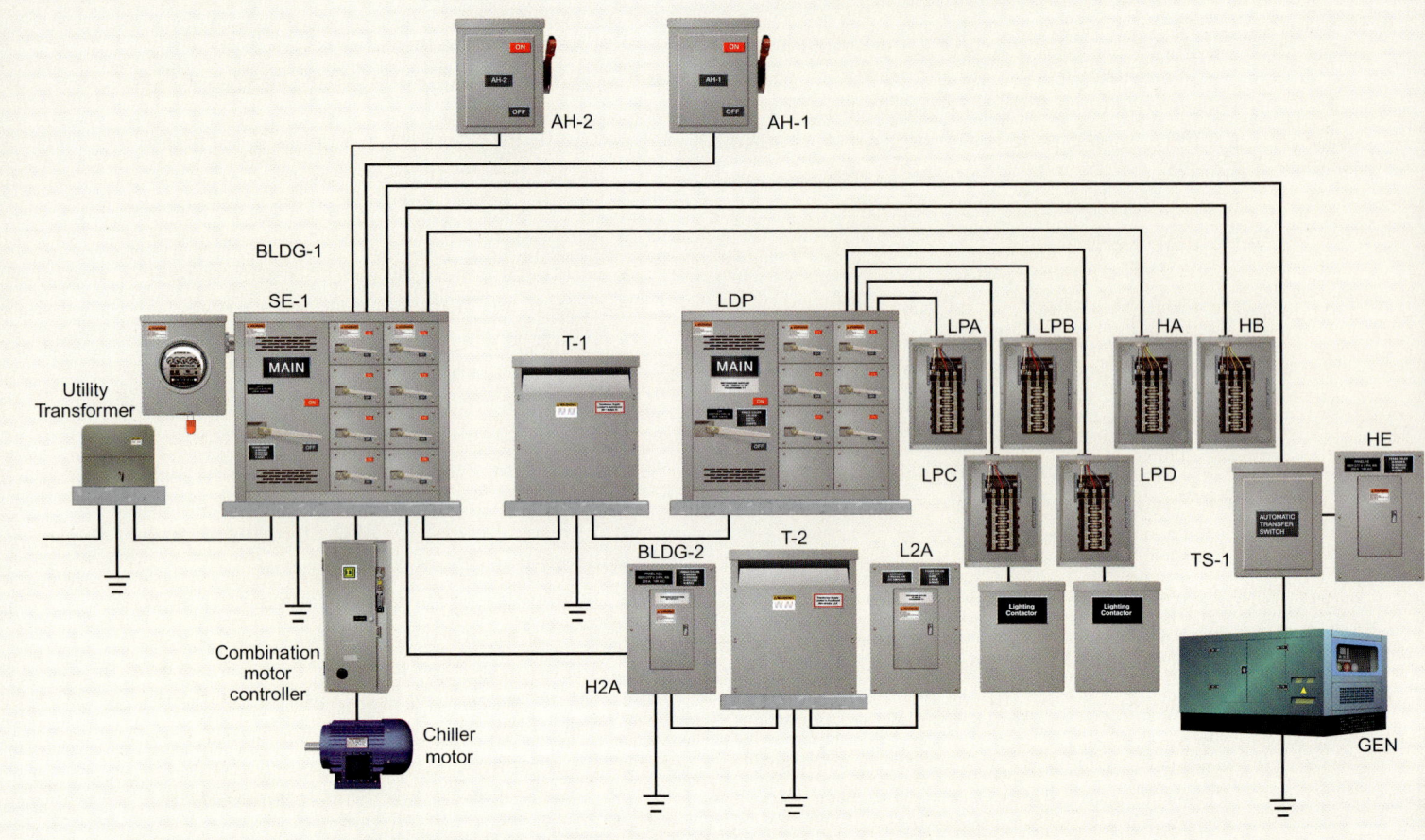

CHAPTER 7

Bonding Requirements

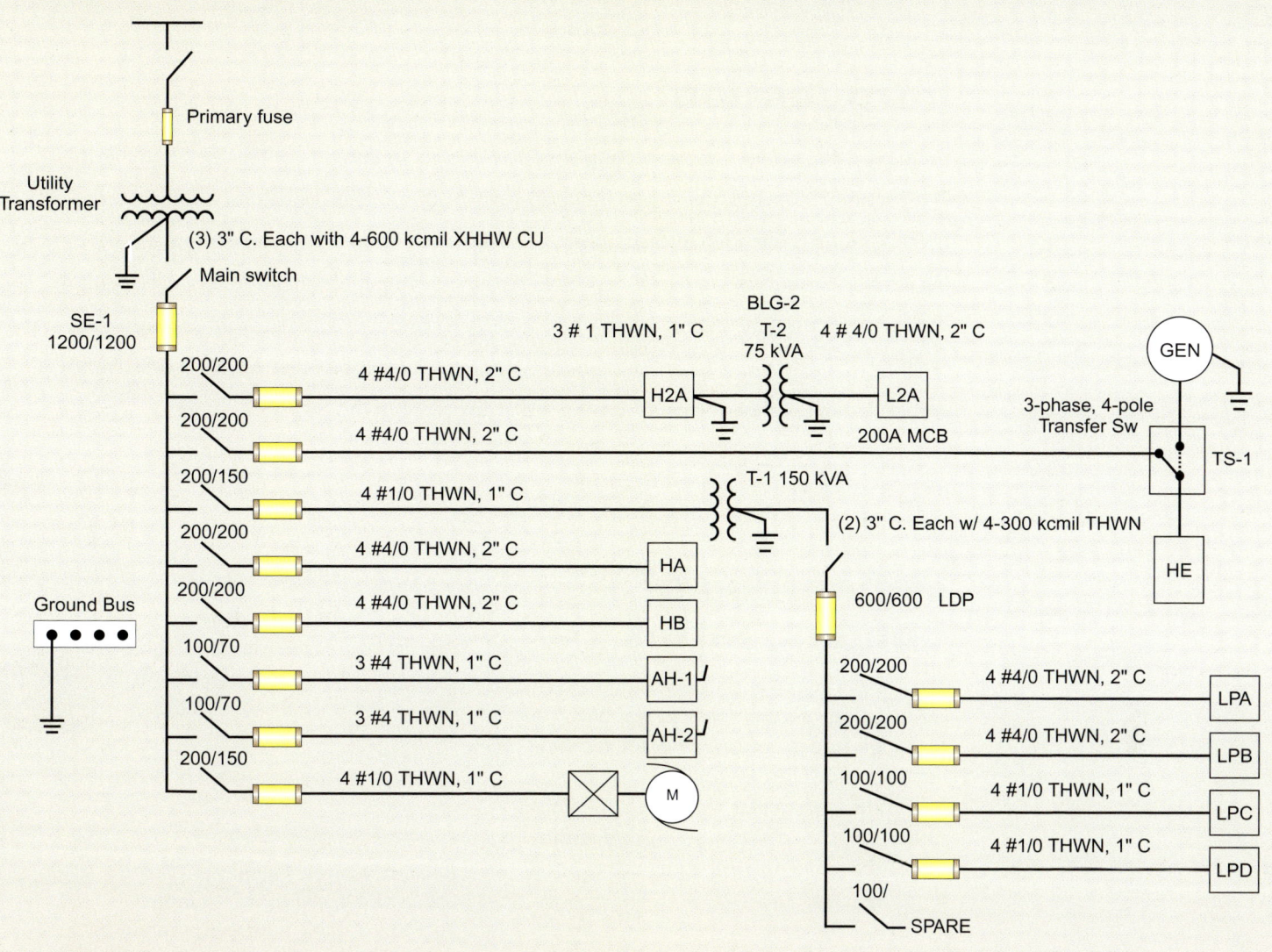

Objectives

- Understand the *NEC* definitions related to electrical bonding requirements
- Understand the requirements for bonding on the supply side of the service disconnecting means
- Determine where equipment bonding jumpers are required
- Understand the difference between supply-side and load-side bonding requirements and establish appropriate sizes for supply-side bonding jumpers and load-side bonding jumpers
- Understand the importance of bonding in the effective ground-fault current path
- Determine the requirements for bonding metal piping systems and structural metal building framing

Outline

Definitions of Bonding Terms
Bonding Performance Criteria
Maintaining Continuity
Bonding Connections (Wire-Type Conductors)
Cleaning Coated Surfaces
Bonding Jumper and Bonding Conductor Length
Equipment Bonding Jumpers (Function and Purpose)
Sizing Requirements for the Supply Side and Load Side
Service Bonding Rules (Line Side)
Reducing Washers
Boxes with Concentric or Eccentric Knockouts
Sizing Supply-Side Bonding Jumpers (Wire Types)
General Equipment Bonding Rules (Load Side)
Installation and Size of Equipment Bonding Jumpers (Load Side)
Expansion Fittings and Loose Joined Metal Raceways
Bonding Metal Piping Systems
Bonding Structural Metal Building Frames
Bonding Lightning Protection Systems

Introduction

Bonding is the process of connecting conductive parts or equipment together. **See Figure 7-1.** Section 250.4 contains important performance criteria that clearly describes what electrical bonding is intended to accomplish. Bonding methods and sizing of equipment bonding jumpers are covered in Part V of Article 250 along with requirements for bonding piping systems, structural metal building frames, and other conductive parts within or attached to buildings or structures. Electrical bonding requires connections that must be effective to ensure that optimal performance is achieved at any point in the electrical system.

Definitions of Bonding Terms

To properly apply and understand electrical bonding requirements in the *NEC*®, a thorough understanding of defined bonding terms must be developed. When a term is used in a particular *Code* rule, the definition helps clarify the intent and meaning of that rule. The following are defined bonding terms:

Bonded (Bonding). Connected to establish electrical continuity and conductivity.[1]

Bonding Conductor or Jumper. A reliable conductor to ensure the required electrical conductivity between metal parts required to be electrically connected.[2]

Bonding Jumper, Equipment. The connection between two or more portions of the equipment grounding conductor.[3]

Bonding Jumper, Main. The connection between the grounded circuit conductor and the equipment grounding conductor at the service.[4]

Bonding Jumper, Supply-Side. A conductor installed on the supply side of a service or within a service equipment enclosure(s), or for a separately derived system, that ensures the required electrical conductivity between metal parts required to be electrically connected.[5]

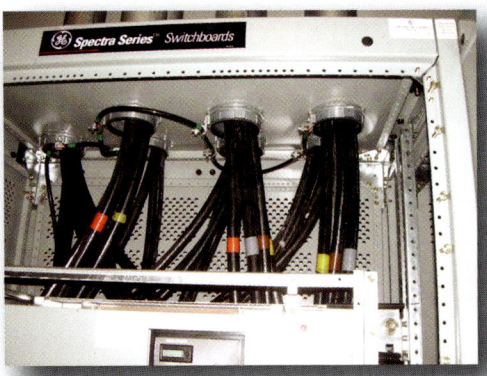

FIGURE 7-1 Bonding establishes continuity and conductivity between conductive parts of equipment required to be electrically bonded together.

Bonding Requirements 145

> **Bonding Jumper, System.** The connection between the grounded circuit conductor and the supply-side bonding jumper, or the equipment grounding conductor, or both, at a separately derived system.[6]

> **Service Point.** The point of connection between the facilities of the serving utility and the premises wiring.[7]

Bonding Performance Criteria

In electrical systems, grounding and bonding functions often go hand-in-hand and are usually happening simultaneously through a single bonding action. In other words, if a system and equipment are grounded and other conductive parts such as conduit are connected to such grounded equipment, the function of bonding is happening and grounding is accomplished simultaneously. The result is grounded and bonded equipment.

Section 250.4 provides general requirements that explain what electrical grounding and bonding is intended to accomplish and how these functions are expected to perform during normal and abnormal conditions. Section 250.4 consists of two subdivisions: (A) and (B). The difference between these two subdivisions is that (A) deals with grounding and bonding performance criteria for a grounded system and associated equipment, whereas (B) provides the performance criteria for systems that are ungrounded. **See Figure 7-2.** General performance requirements for electrical bonding are provided in both subdivisions (A) and (B) of Section 250.4. The goals of bonding are the same for grounded systems as for ungrounded systems.

The *Code* also provides bonding requirements for conductive materials foreign to electrical equipment and materials. Examples of such conductive materials within buildings or structures are structural building steel and metal piping systems. The bonding required by Section 250.4(A)(4) is intended to establish a connection to the ground from the bonded material, thus minimizing possible potentials above ground on these materials if they should come in contact with electrical circuits. Bonding these materials also provides a direct path back to the source from materials that are "likely to become energized." This phrase means "failure of insulation on," as described in the *NEC Style Manual*. This phrase in *Code* rules requires careful judgment.

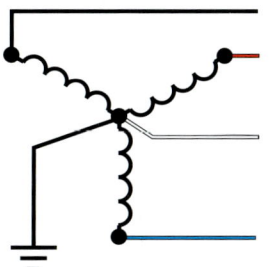 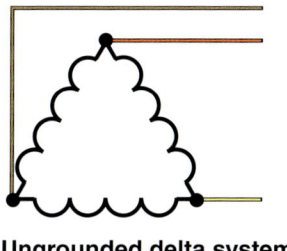

Grounded wye system **Ungrounded delta system**

FIGURE 7-2 Section 250.4(A) covers grounded systems. Section 250.4(B) covers ungrounded systems.

LEARN MORE! The National Electrical Contractors Association (NECA) publishes a family of National Electrical Installation Standards (NEIS). The *NEC* requires electrical wiring and equipment to be installed in a neat and workmanlike manner, but it leaves this requirement open-ended and subject to interpretation. The NEIS establishes a benchmark of quality in multiple aspects of the electrical contracting business, covering subjects such as installing generators, switchboards, panelboards, conduit, fire alarm systems, transformers, security systems, and many more. The National Electrical Contractors Association (NECA) publishes a family of National Electrical Installation Standards (NEIS) that help define what is meant by *good workmanship* in electrical construction. More information and the full library of NEIS is available at www.neca-neis.org.

Bonding is a function that happens at each connection of conductive materials or equipment. For bonding to be effective and perform properly during ground-fault conditions, each fitting, bushing, connector, coupling, and so forth must be made up tight to keep impedance low if a fault should occur and cause current through the conduit, enclosures, and fittings. The effectiveness of all bonding connections is directly related to workmanship. Good workmanship is essential in all aspects of electrical installations. Loose fittings and poorly installed wiring methods that are not supported properly can impair bonding connections in metallic wiring methods installed for services, feeders, and branch circuits.

It is important to understand the general intent of bonding so that the more prescriptive requirements in Article 250 are easily understood and applied to installations and systems. The specific bonding requirements are provided in Part V of Article 250. The function of bonding occurs throughout the entire electrical installation. When there is an electrical connection, bonding happens. A good example of bonding as an ongoing function is the installation of rigid metal conduit (RMC) or electrical metallic tubing (EMT) from one point to another. As the conduit is connected to an enclosure (panelboard, switchboard, box, and so forth), bonding occurs. The run of conduit is then assembled using couplings (fittings) to connect the 10-foot sections. **See Figure 7-3.**

FIGURE 7-3 Fittings (couplings) are used with EMT to connect (bond) 10-foot sections.

This is a bonding function because continuity and conductivity are established between the sections of conduit or tubing connected by listed couplings and connectors. Fittings used with electrical wiring methods, specifically metallic raceways, have been evaluated for grounding and bonding purposes. Underwriters Laboratories evaluated the effectiveness of various metallic wiring fittings for conduit and tubing and determined that the fittings typically provided satisfactory performance when good metal-to-metal contact was accomplished. **See Figure 7-4.**

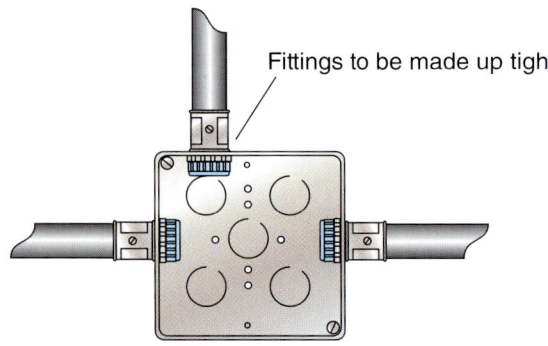

FIGURE 7-4 Listed fittings provide bonding continuity between conductive parts of equipment and raceways.

Fittings for metallic wiring methods are required to be listed, which includes being evaluated for grounding and bonding performance. The other aspect of metallic raceway fittings performing effectively is workmanship and providing proper supports for the wiring method involved. The supports must meet the minimum requirements based on the applicable wiring method article in the *NEC*; furthermore, fittings must be made up tight using suitable tools, as required in Section 250.120(A). **See Figure 7-5.**

FIGURE 7-5 Fittings for electrical conduit and tubing systems must be made up tight and installed in a neat and workmanlike fashion.

Maintaining Continuity

The process of mechanically and electrically connecting metal parts or enclosures, for conductors or components, provides bonding. This establishes the required conductivity between them. In general, bonding can be accomplished by metal-to-metal contact using suitable listed fittings or by using bonding bushings and jumpers. **See Figure 7-6.**

The primary objective of bonding is to make two conductive objects become one electrically. **See Figure 7-7.** Bonding conductive parts of an electrical system together puts the parts at the same or nearly the same potential. Section 250.90 provides a general requirement to bond where necessary to ensure electrical continuity and the capacity to conduct safely any imposed fault current. Fault current is typically at higher (heavier) levels than normal operating current of the system.

The amount of fault current that bonding connections have to endure is related to the amount of available fault current delivered by the system or serving utility. Bonding connections must endure the heavier levels of fault current for a duration long enough to cause overcurrent protective devices to operate. This requires all bonding connections in the ground-fault current path to be mechanically and electrically effective during ground-fault events. Maintaining effective bonding continuity is required from the service equipment to the farthest outlet supplied by the system. Any loose or improper bonding connections in this path can compromise the performance of the electrical safety system (the grounding and bonding circuits).

Understanding the "weakest link in the chain" concept is important. The entire electrical safety circuit is only as good as the weakest link. All bonding connections, such as locknuts, fitting setscrews, and the like must be made using suitable fittings and connections and adequate sizes of jumpers or bonding conductors to result in effective operation during both normal and ground-fault conditions. Section 250.96 requires that bonding be provided around connections of metal raceways, cable trays, cable armor, cable sheaths, enclosures, frames, fittings, and other normally non–current-carrying conductive parts used as equipment grounding conductors (EGCs) or that are otherwise in the effective ground-fault current path.

Bonding Connections (Wire-Type Conductors)

For EGCs, bonding conductors, and jumpers to perform effectively, thorough, tight connections must be made. Section 250.8 provides the acceptable methods of making grounding and bonding connections and applies generally to connections for wire-type conductors.

FIGURE 7-6 A bonding jumper is used to connect the cable tray to EMT at a transition point.

Bonding bushings and equipment bonding jumper

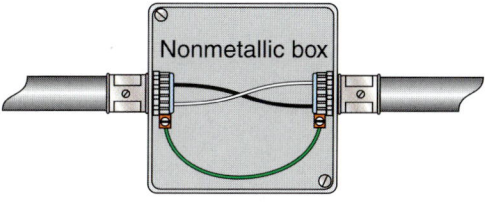

FIGURE 7-7 Bonding establishes continuity and conductivity to bring conductive parts to the same potential.

148 APPLIED GROUNDING AND BONDING

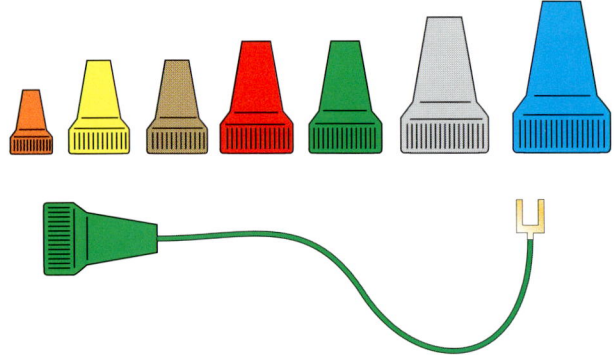

Listed wire pressure connectors are suitable for grounding and bonding connections.

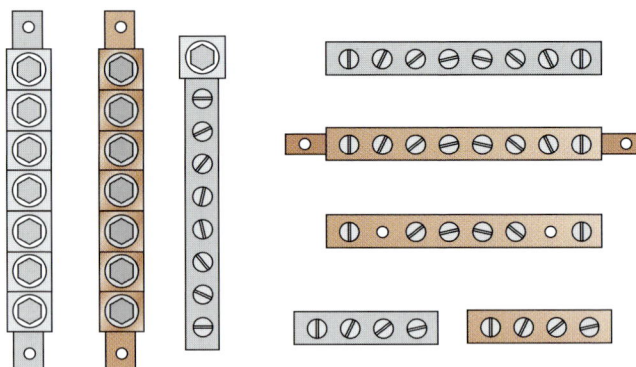

Available separately or as accessories for switchboards, panelboards, and other equipment
Equipment grounding terminal bars are often installed in panelboards and switchboards.

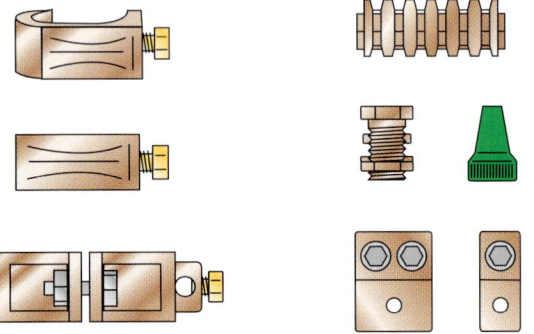

Pressure connectors and irreversible compression connectors are available and listed for use in electrical connections, and they can also be listed as grounding and bonding equipment.

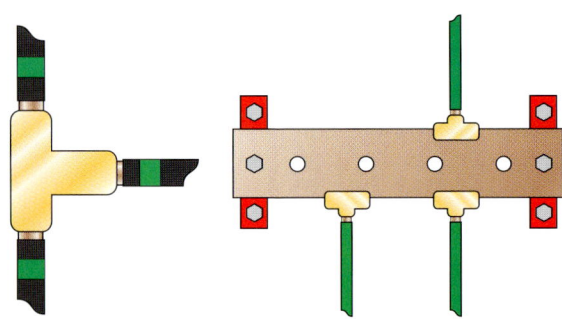

Exothermic welding processes are suitable for grounding and bonding connections. They are not required to be listed, but the manufacturer's instructions must be followed.

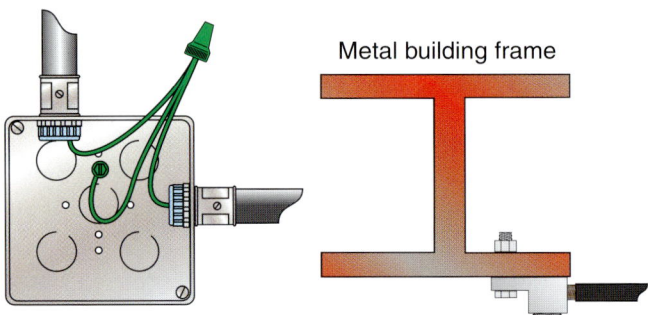

Machine screw fasteners can be used to connect (such as grounding screws) as long as no fewer than two full threads are engaged. Machine screws that are secured with a nut are also suitable for grounding and bonding connections.

FIGURE 7-8 Bonding connections can be accomplished using any of the various methods addressed in the *NEC*.

Several methods and various devices provide many choices for grounding and bonding connections. **See Figure 7-8.** The type of connection made is usually determined based on the equipment type and location in the system. The following connection means are provided in Section 250.8:

1. Listed pressure connectors
2. Terminal bars
3. Pressure connectors listed as grounding and bonding equipment
4. Exothermic welding processes
5. Machine screw fasteners engaging not less than two threads or secured with a nut
6. Thread-forming machine screws engaging not less than two threads in the enclosure
7. Connections that are part of a listed assembly
8. Other listed means

Cleaning Coated Surfaces

An important aspect of any electrical connection is the contact made between the conductive parts. For effective electrical connections, the joint or termination must offer little to no opposition in the electrical circuit. Section 250.12 of the *NEC* provides specific language that addresses cleaning of surfaces. Coated electrical products such as painted enclosures and coated raceways can introduce additional impedance in the grounding and bonding system. For this reason, the *Code* requires that coated or painted surfaces be cleaned to remove coatings such as paint, lacquer, and enamel from threads and contact surfaces. This helps ensure thorough electrical continuity. **See Figure 7-9.** To accomplish this, the coatings on some enclosures must be removed to establish metal-to-metal contact between parts required to be connected (bonded). Section 250.12 indicates that removal of paint or coatings is not necessary if the fittings used for the installation are designed to make such removal unnecessary. It is a good approach to verify this capability with the fitting manufacturer.

Obviously, if the enclosures are uncoated and metal-to-metal contact is established, the bonding required by the *NEC* is accomplished. **See Figure 7-10.** Many panelboards and other electrical enclosures are manufactured without paint or coatings that could impair proper continuity and conductivity at connection points in the system. With this type of equipment, additional steps to achieve contact between metal parts is not necessary.

Bonding Jumper and Bonding Conductor Length

The defined term *bonding conductor or jumper* addresses two concepts that accomplish the same purpose: to ensure electrical conductivity between metal parts. Bonding jumpers are generally understood to be a relatively short bonding means while bonding conductors have generally been understood to be a conductor length exceeding that of a jumper.

FIGURE 7-9 Coatings must be removed for effective bonding connections.

FIGURE 7-10 Bonding is established by tightened locknuts, which create metal-to-metal contact because the enclosure is not coated or painted.

For example, if a bonding jumper is installed from a grounded metal box to a grounding-type receptacle, as covered in Section 250.146, it is relatively short in length—about 6 inches to 10 inches. On the other hand, if a bonding jumper is installed to connect two grounding electrodes to form a grounding electrode system, as required in Sections 250.50 and 250.53(C), it will likely be longer than 6 inches to 10 inches, depending on the particular installation and the relative location of the grounding electrodes.

Equipment Bonding Jumpers (Function and Purpose)

Equipment bonding jumpers are described in the definition as the connection between two or more portions of the EGC. Bonding jumpers are connections between two conductive objects that establish continuity and conductivity between them. An example of an equipment bonding jumper installation is when it is used to connect two sections of EMT that are attached to a nonmetallic enclosure. An equipment bonding jumper is installed between the two metal conduits by using bonding bushings. **See Figure 7-11.**

Another example of an equipment bonding jumper installation is the connection between a grounded metal outlet box and a grounding-type receptacle. **See Figure 7-12.** Equipment bonding jumpers are often used for metallic conduits that emerge from the slab under an open-bottom switchboard or motor control center. The equipment bonding jumper is installed from the bonding bushing on the conduit to the equipment grounding terminal bar within the equipment enclosure. **See Figure 7-13.**

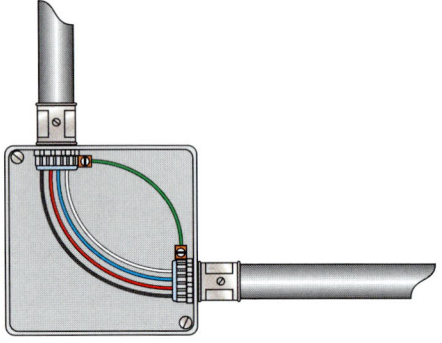

FIGURE 7-11 Bonding is accomplished by use of an equipment bonding jumper connected by bonding-type bushings. Equipment bonding jumpers are required to be installed between sections of EMT in nonmetallic boxes.

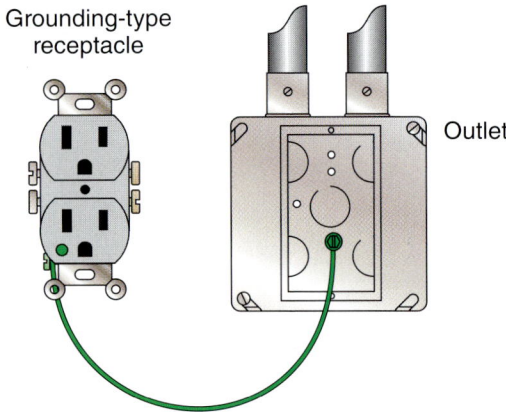

FIGURE 7-12 A short equipment bonding jumper is installed from a grounded box to a grounding-type receptacle.

FIGURE 7-13 Bonding bushings are installed on conduits entering a switchboard enclosure and bonding jumpers connect to the equipment grounding terminal bar.

Sizing Requirements for the Supply Side and Load Side

Determining sizing for bonding jumpers and bonding conductors is accomplished by following the prescriptive language in a particular *NEC* rule that provides bonding jumper or conductor sizing. The *Code* includes two general sizing requirements for these conductors and jumpers.

The applicable sizing process for each depends on whether the bonding jumper or conductor is on the supply side of a service or overcurrent protective device or if it is on the load side of an overcurrent device, as covered in Section 250.102. The definition of the term *service point* clarifies which portion of the service installation is covered by the *NEC* and which portion is the responsibility of the serving utility.

An example of supply-side bonding is often provided at the service equipment for metal conduit–containing service-entrance conductors supplying the service disconnecting means enclosure. The conductors from the service point to the service disconnect are usually not protected at their ampacity. In this case, line or supply-side bonding rules apply. Sizing bonding jumpers on the supply side of the service overcurrent protective devices is accomplished using Table 250.66 (or the 12.5% rule for larger services). Sizing is based on the size of the largest ungrounded service-entrance conductor supplying the service disconnecting means. **See Figure 7-14.** Conversely, equipment bonding jumpers for a feeder supplying a panelboard from the service disconnection means enclosure are on the load side of an overcurrent protective device. In this case the equipment bonding jumper is sized using Table 250.122 based on the rating of the overcurrent protective device for the feeder circuit. **See Figure 7-15.** Notice that *supply-side bonding jumper* is the term typically used on the line side; *equipment bonding jumper* is the term used on the load side of overcurrent protective devices.

Service Bonding Rules (Line Side)

Section 250.92 provides rules related to bonding methods for enclosures, raceways, and other normally non–current-carrying metal parts at the service and on the supply side of the service disconnecting means and overcurrent protective device. These conductive enclosures contain conductors supplied directly from a utility source that are usually not protected at their ampacity. During ground-fault conditions, these metal enclosures and raceways carry high levels of fault current for the duration of time it takes for the overcurrent protective device on the primary side of the utility transformer to open the circuit. For this reason, bonding requirements for metallic parts on the supply side of the service disconnect are more restrictive and result in more robust or strengthened bonding installations.

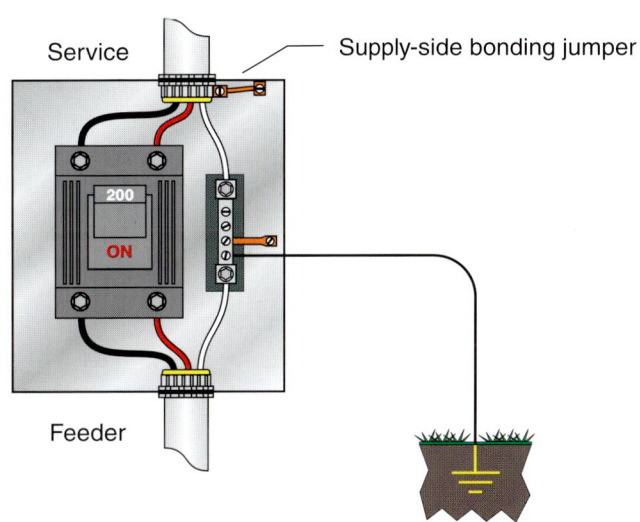

FIGURE 7-14 Supply-side bonding jumpers are sized in accordance with Section 250.102(C), specifically Table 250.66. For larger services, use the 12.5% rule to determine sizing.

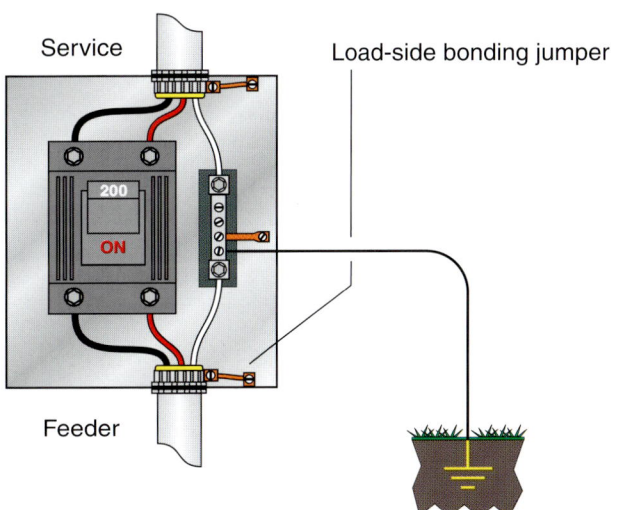

FIGURE 7-15 Load-side equipment bonding jumpers are sized per Section 250.102(D). Refer to Table 250.122.

Section 250.92 is divided into two parts. Subdivision (A) addresses the bonding requirement generally and the service parts required to be bonded; subdivision (B) describes methods by which the requirements in (A) can be accomplished. The following metal parts of equipment containing service conductors are required to be bonded together:

1. Raceways
2. Cable tray
3. Cable bus frames
4. Auxiliary gutters
5. Service cable armor or sheath

In addition, all enclosures containing service conductors, such as meter enclosures, panelboards, switchboards, boxes, and so forth, are interposed in the service raceway installation and must be bonded together. **See Figure 7-16.** Bonding the metal parts at services can be accomplished using any of the following methods:

1. Use of the grounded conductor for bonding enclosures together
2. Connections using threaded couplings or threaded hubs on enclosures if made up wrench-tight
3. Threadless couplings or connectors for metal raceways and metal-clad cables (where made up tight)
4. Other listed devices such as bonding-type locknuts or bonding bushings

It is important to bond around any compromised entries to enclosures, such as reducing washers and oversized, concentric, or eccentric knockouts. Often the process of removing concentric or eccentric knockouts results in loosening of the other concentric or eccentric rings in the punched enclosure.

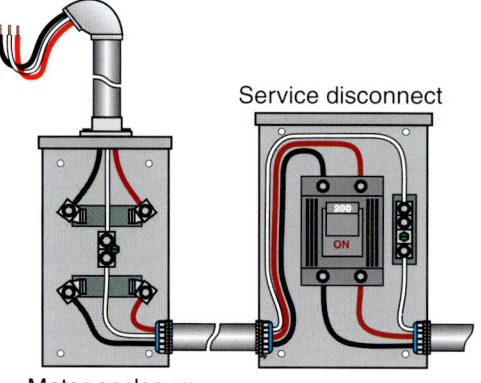

The meter enclosure and service disconnect are bonded by the grounded conductor.

Threaded hubs establish bonding between conduit and an enclosure.

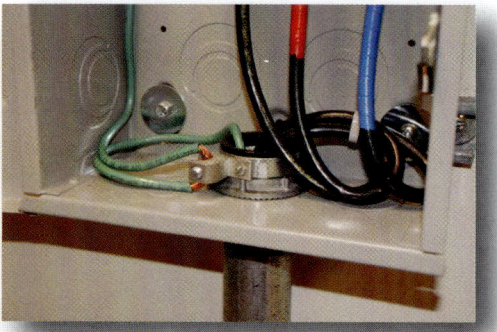

Bonding bushings can be used with bonding jumpers on the supply side or load side of the service disconnecting means.

A bonding wedge can be used for establishing a bonding connection between the conduit and an eclosure.
Courtesy of Thomas and Betts

FIGURE 7-16 The bonding required for metal raceways or enclosures containing service conductors can be accomplished by any of the methods included in Section 250.92(B) of the *NEC*.

For services, standard locknuts or bushings are not permitted as the only means to accomplish the bonding required for service raceways and enclosures. Standard locknuts, on both sides of the enclosure, are acceptable for making the mechanical connection of the service raceway to the enclosure, but they are not suitable for the heavier bonding prescribed in Section 250.92. In this case, additional bonding means such as bonding bushings and jumpers, bonding wedges, service bonding locknuts, or other methods must be used in addition to the standard locknuts.

Bonding locknuts are effective because a setscrew, in addition to the teeth of the locknut, provides an effective bonding connection between the locknut and the enclosure. Use of bonding wedges is another effective method to establish a bonding connection between a fitting or raceway and an enclosure. Bonding wedges provide two ways to ensure effective bonding between a raceway or fitting and the enclosure to which it is fastened. The first method of installing a bonding wedge is effective for bonding without the use of a bonding jumper, but only where concentric or eccentric knockouts are not encountered. The second method of installing a bonding wedge is when concentric knockouts are encountered and bonding wedges are used. In this scenario, a bonding jumper must be installed from one terminal screw on the wedge to the enclosure.

Strengthened bonding methods are also required for installations in health care facilities and hazardous locations. Section 250.100 provides the requirements for bonding metal raceways, equipment, and other enclosures installed in hazardous locations. This section requires the methods of bonding prescribed in Sections 250.92(B)(2) through (B)(4), regardless of the voltage of the circuit.

Where concentric, eccentric, or oversized knockouts are encountered in the installation, the bonding requirements in Section 250.97 apply, if the circuit voltage exceeds 250 volts phase-to-ground. In this case the electrical continuity of metal raceways and cables with metal sheaths that contain any conductor other than service conductors must be ensured by one or more of the methods specified for services in Sections 250.92(B)(2) through (B)(4). The exception to Section 250.97 relaxes this bonding requirement for installations where oversized, concentric, or eccentric knockouts are not encountered or where a box or enclosure with concentric or eccentric knockouts is listed. Listed outlet boxes have been evaluated for the bonding required by this section. If additional bonding is required, one of the following methods can be used to establish an effective bonding connection:

1. Threadless couplings and connectors for cables with metal sheaths
2. Two locknuts, on RMC or intermediate metal conduit—one inside and one outside of boxes and cabinets
3. Fittings with shoulders that seat firmly against the box or cabinet, such as EMT connectors, flexible metal conduit connectors, and cable connectors, with one locknut on the inside of boxes and cabinets
4. Listed fittings

Reducing Washers

When oversized knockouts are encountered, the common solution is to use reducing washers to connect smaller conduits, tubing, and cables into larger knockouts. **See Figure 7-17.** A common question is whether reducing washers are suitable for bonding raceways

FIGURE 7-17 Listed reducing washers are suitable for bonding when they meet listing requirements.

or cables to boxes or other enclosures. The QCRV category in the *UL General Information Directory for Electrical Equipment* (White Book) indicates that listed metal reducing washers are considered suitable for grounding in circuits over and under 250 volts when installed according to the *NEC*.

An important qualifier is that this UL information also goes on to state that reducing washers are intended for use with conductors other than service conductors, meaning they are not suitable for meeting the bonding requirements for services as required by Section 250.92. Reducing washers can be used on enclosures that do not exceed 0.053 inch in thickness. Reducing washers can be installed in boxes or other enclosures where the concentric or eccentric knockouts have been completely removed, and they are suitable for use on concentric or eccentric knockout rings that are not all removed. Be sure to remove any nonconductive coatings if reducing washers are installed in equipment that is coated with paint or other type of coating; otherwise, additional bonding is necessary to ensure a thorough bonding connection.

Boxes with Concentric or Eccentric Knockouts

Bonding around punched concentric or eccentric knockouts is not required in all cases if the box or enclosure containing the prepunched concentric or eccentric knockouts has been tested and is listed as suitable for bonding. **Figure 7-18.** Guide card information from the QCIT category of the *UL General Information Directory for Electrical Equipment* (White Book) indicates that concentric and eccentric knockouts of all metallic outlet boxes evaluated in accordance with UL 514A, Metallic Outlet Boxes, are suitable for bonding in circuits of above or below 250 volts to ground without the use of additional bonding equipment. Be sure that metal-to-metal contact is made according to *NEC* Section 250.12. For the most effective bonding connections, it is highly recommended that equipment bonding jumpers around reducing washers should be used when installed in coated metal enclosures, unless the coating is removed during installation.

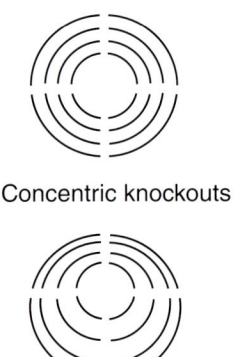
Concentric knockouts

Eccentric knockouts

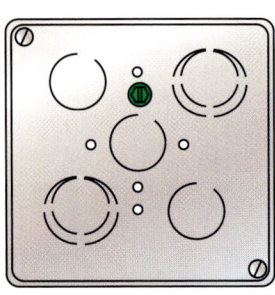
Applies only to boxes

FIGURE 7-18 Boxes with concentric or eccentric knockouts are listed for bonding in circuits over 250 volts to ground and in circuits 250 volts and less to ground.

Sizing Supply-Side Bonding Jumpers (Wire Types)

When wire-type bonding jumpers are installed on the supply side of the service disconnecting means, they must be sized according to the requirements in Section 250.102(C). Wire-type supply-side bonding jumpers are typically installed

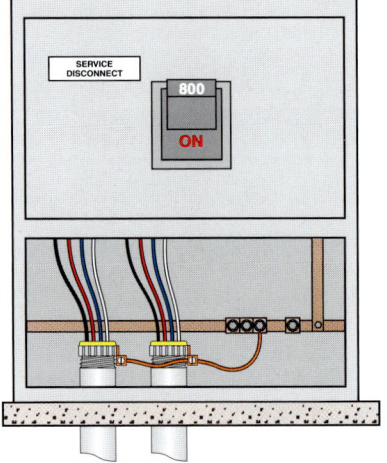

FIGURE 7-19 A single equipment bonding jumper can be installed in a "daisy-chain" arrangement. Size supply-side bonding jumpers using Table 250.66. Use the total circular mil area of the largest ungrounded service conductor in both raceways.

from bonding bushings or fittings to enclosures. They are usually short in length. The minimum size required for supply-side bonding jumpers must not be smaller than the size provided in Table 250.66, which bases sizing on the size of the largest ungrounded service-entrance conductor. If the largest ungrounded service-entrance conductor exceeds 1100-kcmil copper or 1750-kcmil aluminum or copper-clad aluminum, the minimum size required must not be less that 12.5% of the total circular mil area of the largest ungrounded service-entrance phase conductor. **See Figure 7-19.** If the ungrounded supply conductors are paralleled in two or more raceways or cables, the supply-side bonding jumper, if routed with the raceways or cables, shall be installed in parallel. The size of the supply-side bonding jumpers for each of the individual raceways in the parallel set is required to be based on the size of the largest ungrounded service-entrance conductor in each raceway or cable. **See Figure 7-20.**

If the ungrounded supply conductors and the supply-side bonding jumpers are of different materials (copper or aluminum), the minimum size of the bonding jumpers is determined by the assumed use of ungrounded conductors of the same material as the supply-side bonding jumper and with an ampacity equivalent to that of the installed ungrounded supply conductors. The ampacity equivalents can be determined from the values in Table 310.15(B)(16).

General Equipment Bonding Rules (Load Side)

Electrical bonding is also required on the load side of the service disconnect. Bonding functions occur from the point of delivery at the service equipment to the final outlets in the branch circuits. Once the service bonding is complete, bonding is required for feeders and branch circuits connected to the service equipment enclosure as well as all equipment downstream of the service equipment enclosure. The bonding requirements are a bit less restrictive on the load side of the service overcurrent device, but every bit as important.

Bonding requirements apply to metal raceways and enclosures for feeders and branch circuits. Each feeder and branch circuit is required to be provided with an EGC. This conductor can be a wire type or can be in the form of a metallic raceway or other metallic enclosure that is part of the effective ground-fault current path. An example is a feeder conduit installed from the service enclosure to a panelboard elsewhere in the building. The feeder includes an EGC that performs bonding functions by being connected at both ends of the feeder. **See Figure 7-21.**

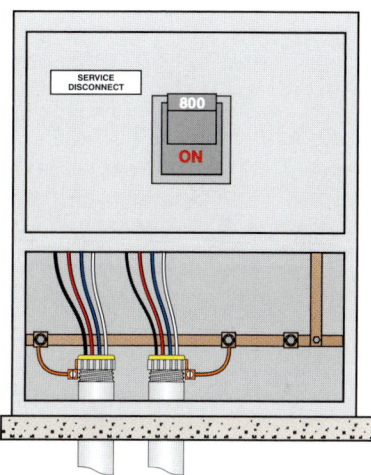

FIGURE 7-20 Individual equipment bonding jumpers can be installed individually from each raceway to the enclosure. Size supply-side bonding jumpers using Table 250.66. Use the total circular mil area of the largest conductor in each of the raceways.

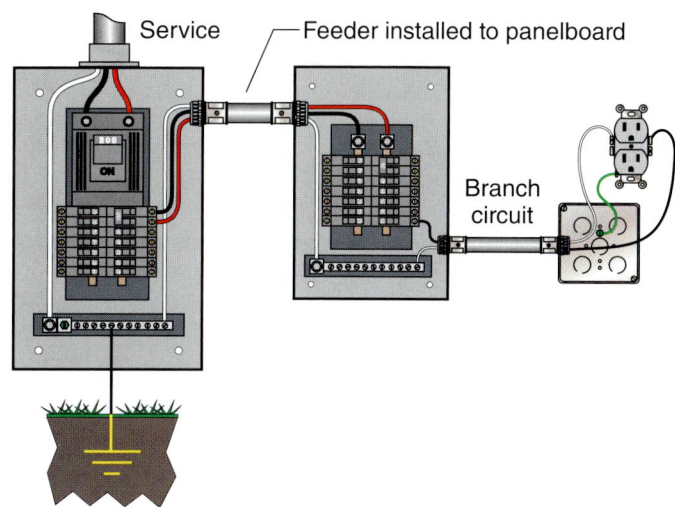

FIGURE 7-21 The feeder EGC, either a wire type or raceway, performs bonding and grounding functions.

The same is true for the EGC of branch circuits. Bonding functions are accomplished by the EGC installed from enclosure to enclosure in the entire length of the branch circuit. The definition of the term *equipment grounding conductor* includes bonding functions as indicated in the definition and the Informational Note No. 1.

FIGURE 7-22 Equipment bonding jumpers connect two or more portions of the EGC.

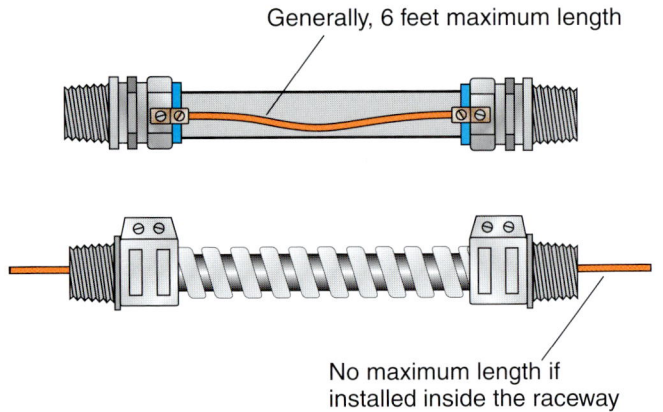

FIGURE 7-23 Equipment bonding jumpers are permitted to be installed on the outside or inside of raceways.

FIGURE 7-24 Equipment bonding jumpers are often installed around expansion conduit fittings to maintain required continuity.

Installation and Size of Equipment Bonding Jumpers (Load Side)

Equipment bonding conductors or jumpers are necessary to connect two or more portions of the EGC as indicated in the definition of *equipment bonding jumper*. **See Figure 7-22.** The general requirements for sizing equipment bonding jumpers are provided in Section 250.102.

Equipment bonding jumpers must be made of copper or other corrosion-resistant material. Bonding jumpers can be in the form of a bus, wire, or other suitable conductor. Bonding jumper connections must meet the applicable provisions in Section 250.8. Connections in the grounding electrode system, covered in Section 250.53(C), must be made in accordance with Section 250.70.

Bonding jumpers are permitted to be installed inside or outside of the raceway or enclosure. **See Figure 7-23.** If installed on the outside of the raceway, the bonding jumper or conductor is generally limited to a maximum of 6 feet in length. An exception permits longer lengths for outside pole installations.

An example of installing an equipment bonding jumper on the outside of a raceway is found in conduit runs where expansion fittings are necessary. **See Figure 7-24.** When equipment bonding jumpers or conductors are installed outside of raceways or enclosures, they are vulnerable to damage. When they are installed in locations subject to physical damage, Section 250.102(E)(3) requires that they be protected in accordance with the provisions in Sections 250.64(A) and (B).

Section 250.148 addresses required connections to boxes or enclosures.

Sizing equipment bonding jumpers on the load side of an overcurrent protective device is covered in Section 250.102(D). The size of wire-type equipment bonding jumpers or conductors is determined from Table 250.122, which bases sizing on the rating of the fuse or circuit breaker protecting the circuit in which the equipment bonding jumper is installed. An example of a requirement for sizing equipment bonding jumpers on the load side of the service disconnecting means is provided in Section 250.146. Section 250.146 requires equipment bonding jumpers from grounded metal boxes to grounding-type receptacles to be sized based on the rating of the fuse or circuit breaker protecting the branch circuit. **See Figure 7-25.**

Another example of sizing equipment bonding jumpers based on the rating of an overcurrent protective device is when conduits emerge from a slab into an open-bottom enclosure such as a switchboard. The equipment bonding jumpers provide the electrical continuity and conductivity between the exposed portion of the metal raceway and the equipment grounding terminal bus in the switchboard assembly. The equipment bonding jumper can be installed in a daisy-chain fashion or as individual jumpers from each conduit bushing to the terminal bar. **See Figure 7-26.**

Expansion Fittings and Loose Joined Metal Raceways

When expansion fittings are installed in a run of metallic raceway, the electrical continuity and conductivity must be ensured by installing an equipment bonding jumper or through other means. Listed expansion fittings are available for raceways that cross expansion joints in building construction projects. These listed fittings must be installed according to the manufacturer's installation instructions. Failure to install these fittings without the necessary amount of travel left on either side could limit movement and result in damage to the conduit system.

> The size of wire-type equipment bonding jumpers or conductors is determined from Table 250.122, which bases sizing on the rating of the fuse or circuit breaker protecting the circuit in which the equipment bonding jumper is installed.

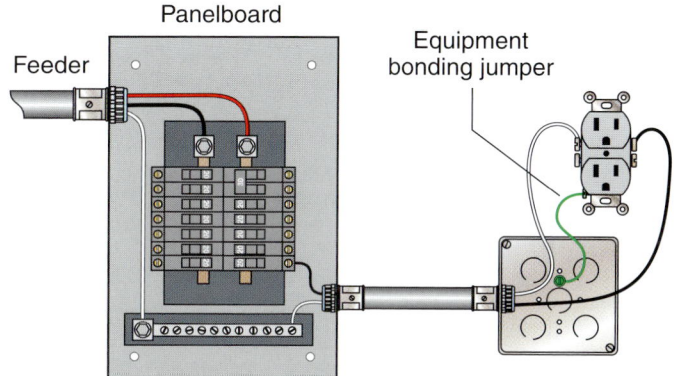

FIGURE 7-25 To size an equipment bonding jumper installed from a grounded metal box to a grounding receptacle, determine the size of the fuse or circuit breaker and then refer to Table 250.122.

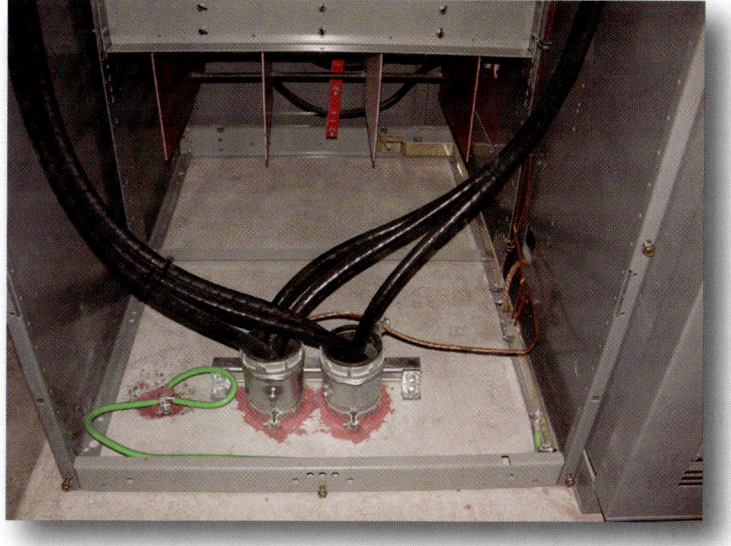

Courtesy of Jim Dollard, IBEW Local 98

FIGURE 7-26 Equipment bonding jumpers are installed from conduits in the open bottom of a switchboard to the equipment grounding terminal bar of the equipment.

These fittings often include the required wire-type equipment bonding jumpers and connections to be used across the fitting after it is inserted in the raceway. **See Figure 7-27.** Listed expansion fittings that provide the equipment bonding jumper as an internal, integral part of the overall assembly are available. **See Figure 7-28.** Installing an equipment bonding jumper on the outside is not necessary when these types are installed.

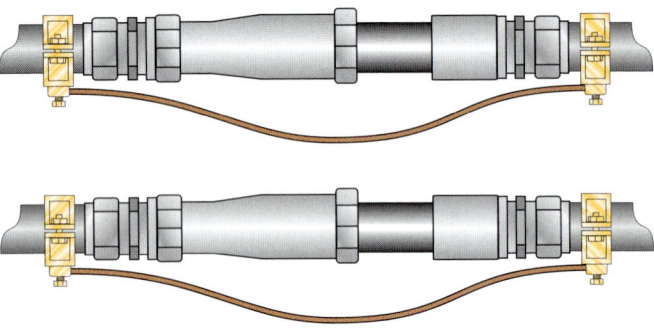

Always install listed expansion fittings in accordance with the manufacturer's installation instructions.

FIGURE 7-27 Bonding jumpers are required to be used with expansion fittings to maintain continuity across the fitting.

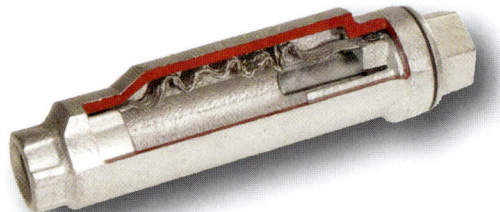

Courtesy of Thomas and Betts

FIGURE 7-28 Expansion fittings are available that include a bonding jumper that is internal to the assembly.

Bonding Metal Piping Systems

The *NEC* contains requirements for bonding metal piping systems. Metal water piping systems have one set of bonding rules, and other metal piping systems have a different set of requirements. The bonding requirements for metal piping systems are provided in Section 250.104. **See Figure 7-29.** The purpose of bonding metal piping systems is to place the metal piping at the same potential as the grounded metallic components of the electrical system and to provide a path for any fault current likely to be imposed on them.

The idea is to not leave a metal piping system isolated in a building or structure because it is possible that it could become energized and present a shock or electrocution hazard. In addition, the risk for fire or property damage is increased. Section 250.104(A) provides the bonding rules for water piping systems only. There are two key points to understand in the wording of this rule. First, the requirement applies to water piping systems, meaning the whole system is metallic. An installation of water piping that uses nonmetallic piping and short metallic sections at points of connection is not a metallic piping system; it is a nonmetallic system with short metallic nipples or short sections of metal piping installed. The second important point in this rule is that it applies to all water piping systems in or attached to the building or structure. Water piping installed inside the building or on a roof or an exterior wall is subject to the bonding requirements in Section 250.104(A). This includes potable water piping systems, sprinkler system piping, and chilled water system piping. Many make the mistake of applying this bonding rule to the only domestic potable hot and cold water systems in the building.

Sizing and Accessibility

The *NEC* requires metallic water piping systems to be bonded to the service equipment enclosure, to the grounded conductor at the service, to the grounding electrode conductor where it is of sufficient size, or to one or more of the grounding electrodes of the building grounding electrode system. The bonding jumper must not be smaller than the values listed in Table 250.66, which bases sizing on the size of the largest ungrounded service-entrance conductor. For example, if the size of the largest ungrounded service-entrance phase conductor is 4/0 copper, then a bonding jumper that is at least 2 AWG copper or 1/0 AWG aluminum is required. **See Figure 7-30.**

If a 500-ampere service is supplied with 750-kcmil copper as the largest ungrounded service-entrance conductor, the minimum-size bonding jumper for the water piping system is 2/0 copper or 4/0 aluminum. The largest size bonding jumper required by the *NEC* is 3/0 copper or 250 aluminum or copper-clad aluminum. The 12.5% requirement does not apply to water pipe bonding conductors. The points of attachment of the bonding jumper for water piping systems are required to be accessible. The phrase *accessible (as applied to wiring methods)* is defined in Article 100 as "capable of being removed or exposed without damaging the building structure or finish or not permanently closed in by the structure or finish of the building."[7]

Metal Water Piping in Multiple Occupancy Buildings

Section 250.104(A)(2) offers a water piping system bonding alternative for multiple occupancy buildings. In buildings of multiple occupancy where the metal water piping system or systems are installed in or attached to a building or structure, the metal piping system in each occupancy must be bonded. If the metal piping system for each individual occupancy is metallically isolated from all other occupancies by use of nonmetallic water piping, the metal water piping system or systems for each occupancy are permitted to be bonded to the equipment grounding terminal bus of the panelboard or switchboard enclosure that supplies the individual occupancy. This alternative is in lieu of installing a larger water piping bonding conductor to the service equipment enclosure as required by 250.104(A).

> Water piping installed inside the building or on a roof or an exterior wall is subject to the bonding requirements in Section 250.104(A).

FIGURE 7-29 Metal water piping systems are required to be bonded in accordance with Section 250.104(A).

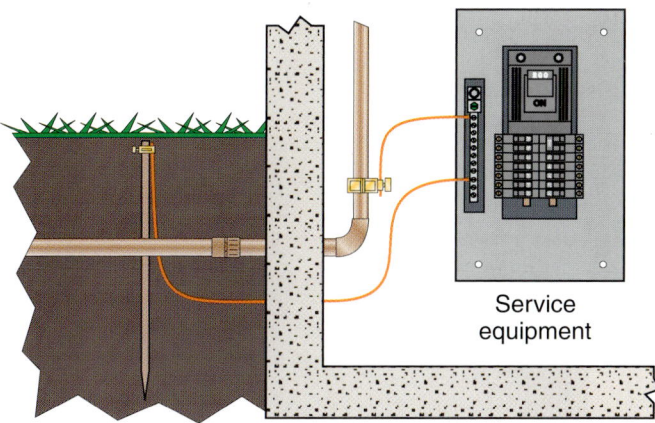

FIGURE 7-30 Size water bonding conductors using Table 250.66 based on the size of the largest ungrounded service-entrance conductor.

The bonding jumper installed for the piping system in each individual occupancy must be sized in accordance with Table 250.122, which bases sizing on the rating of the overcurrent protective device for the circuit supplying the occupancy. **See Figure 7-31.**

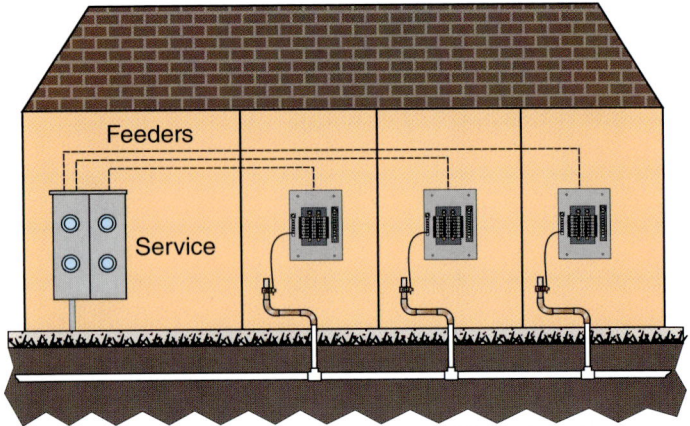

FIGURE 7-31 Water pipe bonding in multiple occupancy buildings is accomplished within each occupancy. Size bonding conductors based on the rating of the feeder overcurrent protective device for each individual unit.

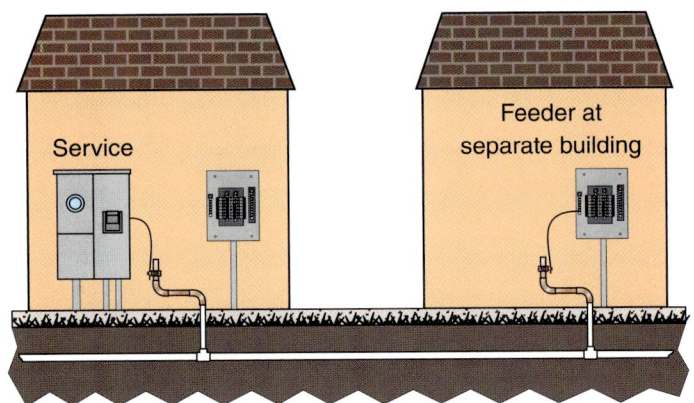

FIGURE 7-32 Size water bonding jumper(s) at separate buildings or structures using Table 250.66.

FIGURE 7-33 Bonding for other metal piping systems must be in accordance with Section 250.104(B).

An example of this type of bonding is often found in apartment complexes. The key to being able to use this sizing option and connecting to the panelboard in the apartment is that the metallic piping system within each individual unit is isolated from all other apartments or other occupancies.

Multiple Buildings or Structures Supplied by a Feeder or Branch Circuit

A metal water piping system or systems installed in or attached to a building or structure are required to be bonded to the building or structure disconnecting means enclosure where located at the building or structure, to the EGC run with the supply conductors, or to the one or more grounding electrodes used. The bonding jumpers shall be sized in accordance with Table 250.66, using the size of the largest feeder or branch circuit conductors that supply the building. The bonding jumper is not required to be larger than the largest ungrounded feeder or branch circuit conductor supplying the building or structure. **See Figure 7-32.**

Bonding Other Metal Piping Systems

There are other metal piping systems in buildings or structures that must be bonded to the electrical service supplying the building or structure under certain conditions. The requirements for bonding metal piping systems other than metal water piping systems are provided in Section 250.104(B). Some of the other metal piping systems include compressed air piping, piping systems for lubricants and fuels, and pneumatic systems for controls. One of the key drivers of this requirement is that the other metal piping systems are installed in or attached to a building or structure. The Code requires bonding only if these other metal piping systems are "likely to become energized." Likely to become energized is described in the *NEC Style Manual* as "the failure of insulation on."

The Code requires bonding if these other metal piping systems (either in or attached to the building or structure) are "likely to become energized." **See Figure 7-33.**

The bonding jumper is required to be installed and connected to the grounded conductor at the service, to the grounding electrode conductor if large enough, to any of the grounding electrodes in the grounding electrode system, or to the service equipment enclosure. At a minimum, other metal piping systems, including metal gas piping systems, are required to be bonded using a bonding jumper or conductor sized in accordance with Table 250.122, which is based on the rating of the overcurrent protective device for the circuit likely to energize the piping systems. Examples include a gas water heater and a gas furnace. If the branch circuit supplying the furnace is protected by a 30-ampere overcurrent protective device, then the 10 AWG EGC of the 30-ampere branch circuit can serve as the bonding means. In these cases the branch circuit supplying the appliance is viewed as the circuit likely to energize the piping system. **See Figure 7-34.** It is difficult to say with absolute certainty that there is no possibility of isolated piping becoming energized. Many engineering designs require a bonding jumper that is sized according to Table 250.66, which is based on the size of the largest ungrounded service-entrance conductor. This, of course, exceeds the minimum requirements in the *NEC*.

Another good example of metal piping systems that are required to be bonded is metal piping used for compressed air systems. **See Figure 7-35.** At a minimum, the EGC supplying this equipment can serve as the required bonding means because this is the circuit that would be likely to energize the piping system.

The points of attachment of the bonding jumper or conductor for the piping systems are required to be accessible, just as is required for water piping system bonding jumper connections. Some gas piping systems have been known to be vulnerable to damage from lightning. Some manufacturers of corrugated stainless steel tubing (CSST) gas piping have provided instructions for more restrictive bonding requirements than those contained in the *NEC*. Because the National Fuel Gas Code (NFPA 54) requires that manufacturer's installation instructions be followed and these bonding requirements for CSST piping systems may exceed those in the *NEC*. Prudent design dictates that a coordinated effort be made to address the bonding of CSST piping in a manner that satisfies the *NEC* (NFPA 70), the National Fuel Gas Code (NFPA 54), and the manufacturer's installation instructions. This may require coordination between the plumbing/mechanical contractor, the electrical contractor, and the inspection authority. An informational note following Section 250.104(B) references additional bonding requirements in Section 7.13 of the National Fuel Gas Code (NFPA 54). Another informational note follows Section 250.104(B) and advises about additional safety being achieved by bonding all piping and metal air ducts within the premises. This is an advisement only and not a requirement of the *NEC*.

FIGURE 7-34 A gas water heater supplied by branch circuit and EGC performs the bonding required in Section 250.104(B).

Courtesy of Cogburn Bros. Inc.

FIGURE 7-35 Metal piping systems for compressed air systems are required to be bonded in accordance with Section 250.104(B).

FIGURE 7-36 Structural metal building framing, if likely to become energized, must be bonded.

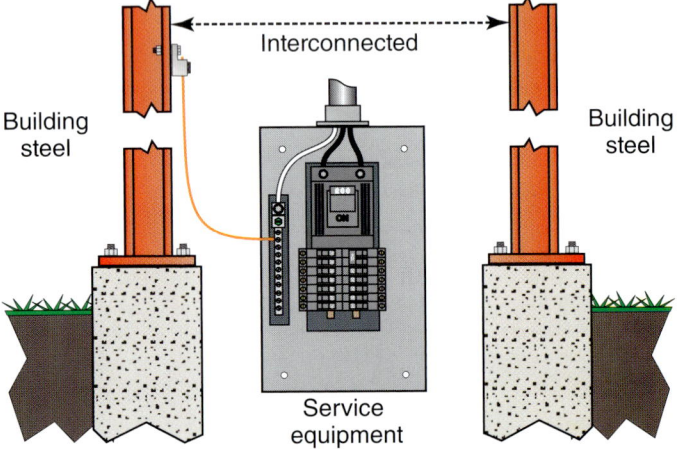

FIGURE 7-37 Bonding jumpers for interconnected structural metal building frames must be sized according to Table 250.66.

FIGURE 7-38 Coating or paint on building frames has to be removed for effective metal-to-metal contact.

The bonding connection for a metal water piping system that does not serve as a grounding electrode must be made at any accessible location where installed within or attached to a building or structure.

Bonding Structural Metal Building Frames

Any exposed structural metal that does not qualify as an electrode, is interconnected to form a building frame, and is not already intentionally grounded or bonded is required to be bonded to the electrical supply service of the building or structure. **See Figure 7-36.**

This requirement also applies only when the structural metal building framing is likely to become energized. The bonding jumper has to be connected to one of the following locations:

1. The service equipment enclosure
2. The grounded conductor at the service
3. The grounding electrode conductor, where large enough
4. The disconnecting means enclosure supplied by a feeder or branch circuit
5. Any of the grounding electrode in the grounding electrode system

The bonding jumper is required to be sized in accordance with Table 250.66, which is based on the size of the largest ungrounded service-entrance conductor supplying the service equipment. **See Figure 7-37.** The maximum size required for the structural metal frame bonding jumper is 3/0 copper or 250-kcmil aluminum or copper-clad aluminum.

Just like the points of attachment for the water piping system, the points for attachment to the building metal frame have to be accessible unless the connection is covered with fireproofing material and the installation meets the applicable provisions in Section 250.68(A) Exception No. 2. Structural metal building frame sections that are isolated by expansion joints must be bonded together using a bonding jumper meeting the sizing requirements of Section 250.102(C). Remember that the connection of the bonding jumper to the building steel must be in accordance with Section 250.8 and the metal surface must be cleaned to bare metal in accordance with Section 250.12. **See Figure 7-38.**

Bonding Lightning Protection Systems

Requirements for lightning protection systems are provided in NFPA 780 *Standard for the Installation of Lightning Protection Systems*. Section 250.106 of the *NEC* requires that the ground terminals of a lightning protection system be bonded to the building or structure grounding electrode system. This ensures that any rise or fall of potential on conductive objects in or on the building or structure will occur at the same potential and reduce flashover possibilities.

Summary

Bonding is the process of connecting conductive parts or equipment together. In electrical systems, grounding and bonding functions go hand-in-hand and usually occur simultaneously through a single bonding action. In other words, if a system and equipment are grounded and other conductive parts such as conduit are connected to such grounded equipment, the function of bonding is happening and grounding is accomplished simultaneously. The result is grounded and bonded equipment. Bonding methods and sizing of equipment bonding jumpers were covered, along with requirements for bonding piping systems, structural metal building frames, and other conductive parts within or attached to buildings or structures. Electrical bonding results in conductive parts connected together and establishing conductivity and continuity between them in a manner that functions electrically to provide safety for the overall electrical system.

References

1. NFPA 70, National Electrical Code 2011, Article 100 (Quincy, MA, National Fire Protection Association 2010), p. 70–26.
2. NFPA 70, National Electrical Code 2011, Article 100 (Quincy, MA, National Fire Protection Association 2010), p. 70–26.
3. NFPA 70, National Electrical Code 2011, Article 100 (Quincy, MA, National Fire Protection Association 2010), p. 70–26.
4. NFPA 70, National Electrical Code 2011, Article 100 (Quincy, MA, National Fire Protection Association 2010), p. 70–26.
5. NFPA 70, National Electrical Code 2011, Section 250.2 (Quincy, MA, National Fire Protection Association 2010), p. 70–100.
6. NFPA 70, National Electrical Code 2011, Article 100 (Quincy, MA, National Fire Protection Association 2010), p. 70–26.
7. NFPA 70, National Electrical Code 2011, Article 100 (Quincy, MA, National Fire Protection Association 2010), p. 70–32.

Review Questions

1. Bonding is the process of connecting conductive objects together.
   a. True
   b. False
2. Bonding establishes continuity and conductivity between conductive parts connected together.
   a. True
   b. False
3. The requirements for bonding are provided in _____ of Article 250.
   a. Part II
   b. Part III
   c. Part IV
   d. Part V
4. A(n) _____ is a reliable conductor used to ensure the required electrical conductivity between metal parts required to be electrically connected.
   a. Equipment bonding jumper
   b. Bonding conductor of jumper
   c. Main bonding jumper
   d. Equipment grounding conductor
5. A _____ is a conductor installed on the supply side of a service or separately derived system to ensure the required electrical conductivity between metal parts required to be electrically connected.
   a. Supply-side bonding jumper
   b. Load-side bonding jumper
   c. System bonding jumper
   d. Bonding busbar
6. The *NEC* defines *equipment bonding jumpers* as the connection between two or more portions of the equipment grounding conductor.
   a. True
   b. False
7. Section _____ provides rules related to bonding methods for enclosures, raceways, and other normally non–current-carrying metal parts at the service and on the supply side of the service disconnecting means and overcurrent protective device.
   a. 250.66
   b. 250.4
   c. 250.120
   d. 250.92
8. Which of the following products are not suitable for bonding on the supply side of the service disconnecting means?
   a. Threaded hubs
   b. Bonding bushings with jumpers
   c. Standard locknuts
   d. Threadless connectors made up tight
9. Which of the following metal parts of equipment containing service conductors are required to be bonded together?
   a. Raceways
   b. Cable tray
   c. Auxiliary gutters
   d. All of the above
10. All enclosures containing service conductors, such as meter enclosures, boxes, and so forth, interposed in the service raceway installation are required to be bonded in accordance with Section 250.92(A).
    a. True
    b. False
11. When concentric, eccentric, or oversized knockouts are encountered in the installation, the bonding requirements in Section 250.97 apply where the circuit voltage exceeds _____ phase-to-ground.
    a. 120 V
    b. 220 V
    c. 250 V
    d. 100 V
12. Reducing washers are suitable for bonding on the supply side of the service when they meet all listing requirements and no paint or coatings are provided on the service equipment enclosure.
    a. True
    b. False
13. What is the minimum-size supply-side bonding jumper required if installed on the line side of a 300-ampere service disconnecting means if the service conductors are sized at 400-kcmil copper?
    a. 3 AWG copper
    b. 2 AWG copper
    c. 1/0 AWG copper
    d. Parallel 2 AWG copper conductors

14. An equipment bonding jumper is the connection between two or more portions of an equipment grounding conductor.
    a. True   b. False
15. Bonding jumpers can be in the form of a(n) _____.
    a. Bus
    b. Wire
    c. Other suitable conductor
    d. All of the above
16. A metal water piping system installed in a building or structure is required to be bonded to the service equipment enclosure, to the grounded conductor at the service, to the grounding electrode conductor if of sufficient size, or to one or more grounding electrodes installed for the service and sized using _____.
    a. Table 9 in Chapter 8
    b. Table 250.66
    c. Table 250.122
    d. Section 250.122
17. When expansion fittings are installed in a run of metallic raceway, the electrical continuity and conductivity must be ensured by installing an equipment bonding jumper or through other means. This means is permitted to be either inside or outside of the raceway.
    a. True   b. False
18. The equipment grounding conductor of a circuit likely to energize a copper compressed air piping system can be used as the bonding means.
    a. True   b. False
19. What is the minimum-size equipment bonding jumper required from a grounded metal outlet box to a 40-ampere receptacle?
    a. 12 AWG copper
    b. 10 AWG copper
    c. 6 AWG copper
    d. 8 AWG copper
20. If two lengths of conduit for a 200-ampere feeder have to be bonded together, what is the minimum-size aluminum equipment bonding jumper required?
    a. 10 AWG
    b. 8 AWG
    c. 6 AWG
    d. 4 AWG
21. Points of attachment of bonding jumpers for metal water piping systems and structural metal building frames are required to be _____.
    a. Accessible
    b. Inaccessible
    c. Not accessible to the public
    d. Guarded
22. The grounded conductor is not permitted to be used for bonding between a meter enclosure and a service disconnecting means enclosure.
    a. True   b. False
23. For circuits over 250 volts to ground, bonding is accomplished when concentric, eccentric, or oversized knockouts are not encountered by which of the following means?
    a. Threadless couplings and connectors for cables with metal sheaths
    b. Two locknuts, on rigid metal conduit or intermediate metal conduit—one inside and one outside of boxes and cabinets
    c. Fittings with shoulders that seat firmly against the box or cabinet, such as electrical metallic tubing connectors, flexible metal conduit connectors, and cable connectors, with one locknut on the inside of boxes and cabinets
    d. Any of the above

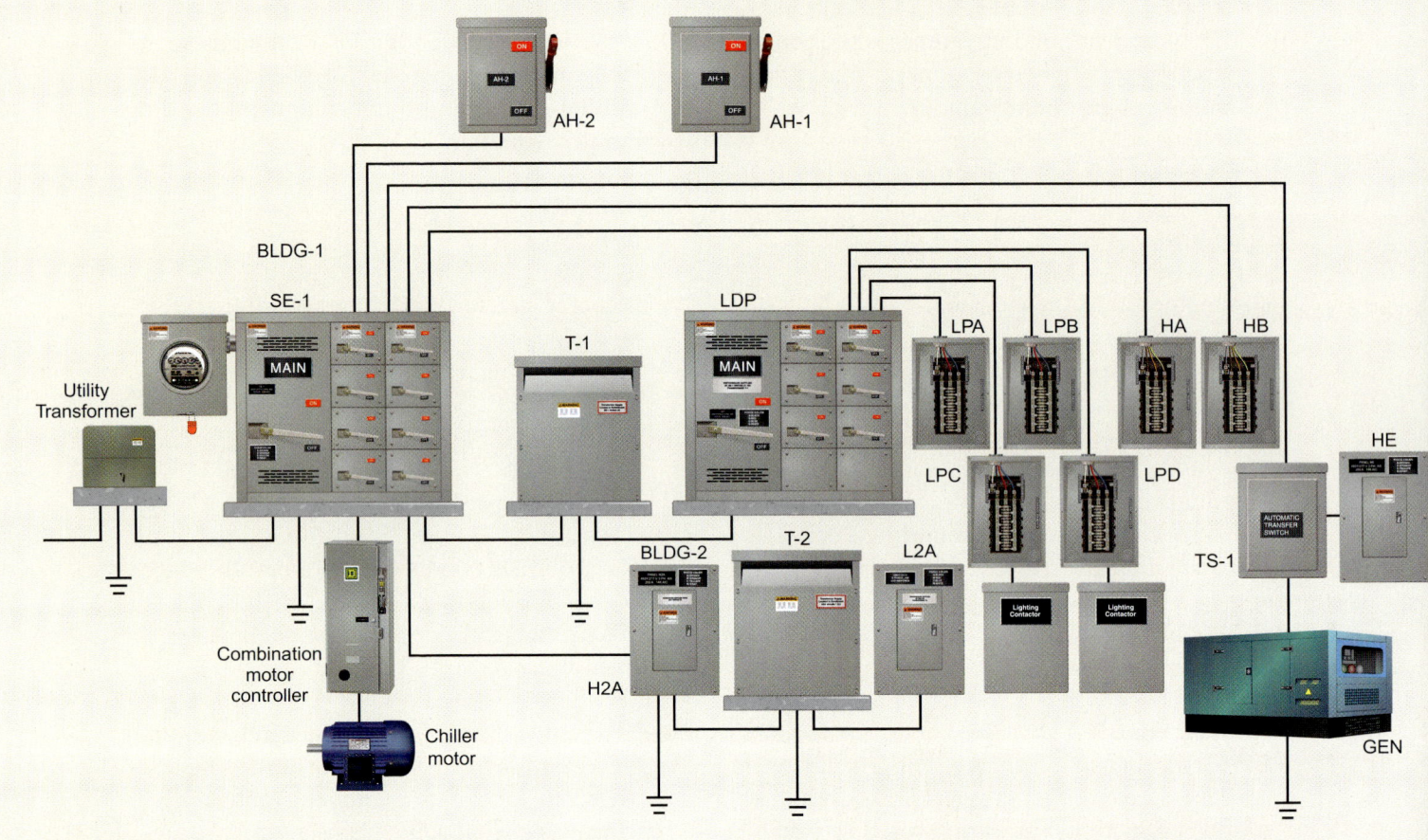

CHAPTER 8

Equipment Grounding Conductors

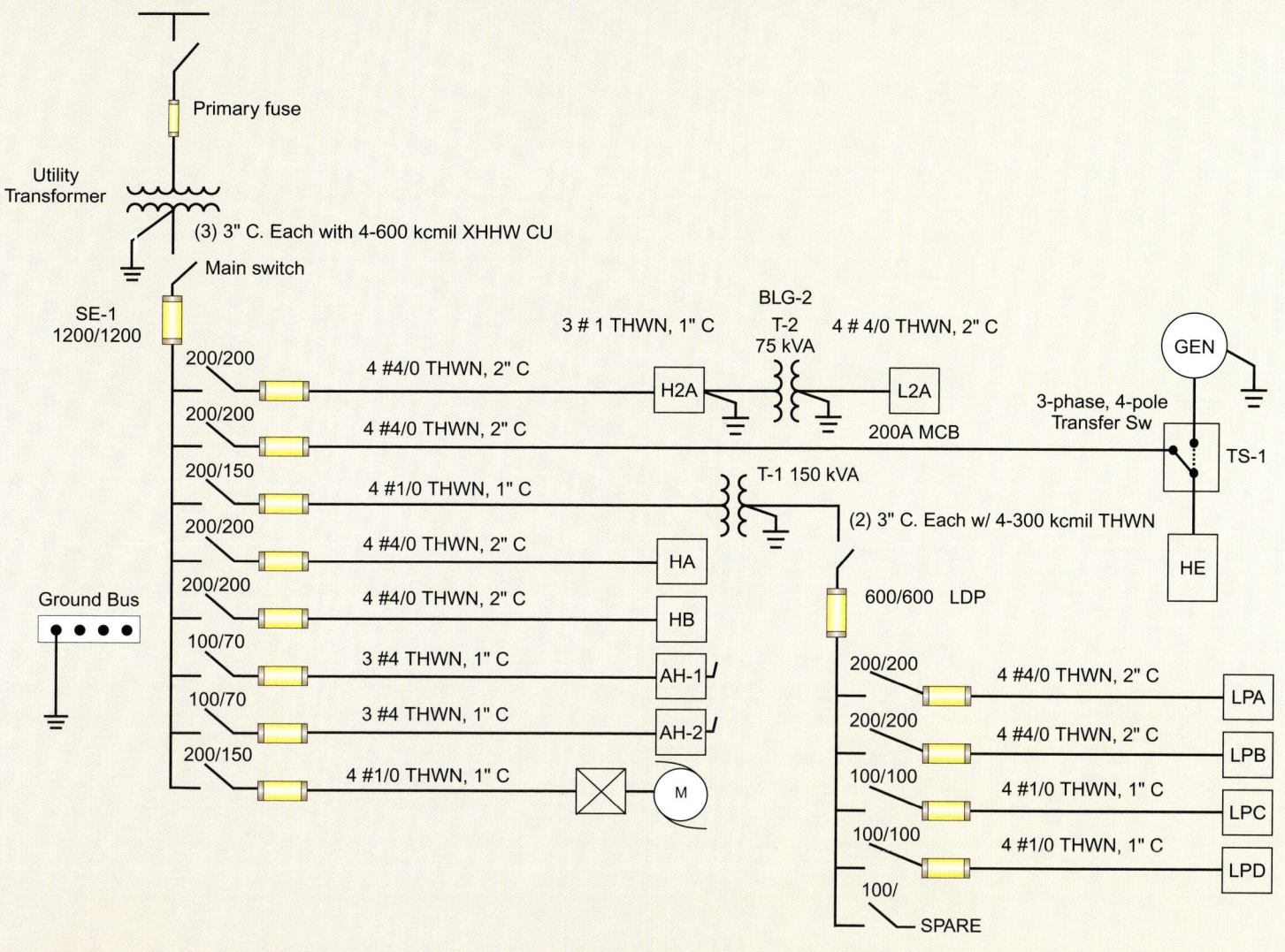

Objectives

- Understand the definition of the term *equipment grounding conductor*
- Identify the types of equipment grounding conductors recognized by the *NEC*
- Understand the purpose of the equipment grounding conductor in the electrical system
- Determine the requirements for identification and connections of equipment grounding conductors
- Understand installation requirements and determine minimum sizes for equipment grounding conductors of the wire type

Outline

Equipment Grounding Conductor Defined

Purpose of Equipment Grounding Conductors

Equipment Grounding Conductor Material

Types of Equipment Grounding Conductors

Equipment Grounding Conductor Installations

Equipment Grounding Conductor Connections

Equipment Grounding Conductor Identification

Equipment Grounding Conductor Sizing

Current in Equipment Grounding Conductors

Introduction

The equipment grounding conductor (EGC) is another important component in the grounding and bonding system. EGCs are typically installed with feeders and branch circuits of electrical systems. They perform grounding, bonding, and serve as effective ground-fault current paths. EGCs are electronically conductive paths that extend the ground (Earth) connection to equipment that is required to be grounded. The performance of EGCs is directly related to the integrity of this conductive path that is ensured through good workmanship.

Equipment Grounding Conductor Defined

The general requirements for EGCs are provided in Part VI of *NEC*® Article 250. These rules cover EGC types, identification, installation, and sizing. An EGC is required to be installed with feeders and branch circuits and can be in the form of a wire-type conductor. **See Figure 8-1.**

Wiring methods such as conduit, tubing, and cable armor can also qualify as EGCs. **See Figure 8-2.** Whether an EGC is a wire type, conduit or tubing, or cable armor, it must be able to perform as an effective ground-fault current path.

EGCs perform multiple functions, one of which is to ground equipment. To strengthen understanding of the role of the EGC in the safety system, it is important to review related *NEC* definitions.

Grounding Conductor, Equipment (EGC). The conductive path(s) installed to connect normally non–current-carrying metal parts of equipment together and to the system grounded conductor or to the grounding electrode conductor, or both. Informational Note No. 1: It is recognized that the equipment grounding conductor also performs bonding. Informational Note No. 2: See 250.118 for a list of acceptable equipment grounding conductors.[1]

FIGURE 8-1 Wire-type EGCs are connected to the equipment grounding terminal bar in an enclosure.

Grounded (Grounding). Connected (connecting) to ground or to a conductive body that extends the ground connection.[2]

Effective Ground-Fault Current Path. An intentionally constructed, low-impedance electrically conductive path designed and intended to carry current under ground-fault conditions from the point of a ground fault on a wiring system to the electrical supply source and that facilitates the operation of the overcurrent protective device or ground-fault detectors on high-impedance grounded systems.[3]

Purpose of Equipment Grounding Conductors

The EGC performs three important functions in the electrical safety system. EGCs are intended to provide a path that connects equipment to ground (the Earth), thereby performing grounding functions. **See Figure 8-3.** The EGC extends the ground connection to various points in the electrical system because it is generally installed with feeders or branch circuits. The role of grounding is to place a conductive object (equipment) at or as close to Earth (ground) potential as possible. The EGC limits voltages above ground potential on conductors and equipment enclosures during normal operation and during conditions such as a ground fault or short circuit.

Another important function performed by EGCs is bonding. In the definition, the words *connect* and *together* are used. The act of connecting together is a bonding function. For example, when an EGC is installed from one metallic outlet box to another and is connected to the metal box at both ends, the boxes not only become grounded, but also they are bonded electrically together. **See Figure 8-4.**

The third important role the EGC has is to serve as an effective ground-fault current path during abnormal events such as ground faults.

Courtesy of IBEW Local 26 Training Center

FIGURE 8-2 Electrical metallic tubing is recognized as an EGC in accordance with *NEC* Section 250.118(4).

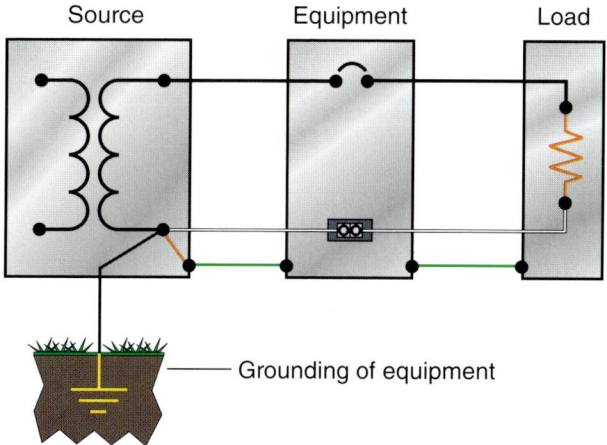

FIGURE 8-3 EGCs perform grounding functions.

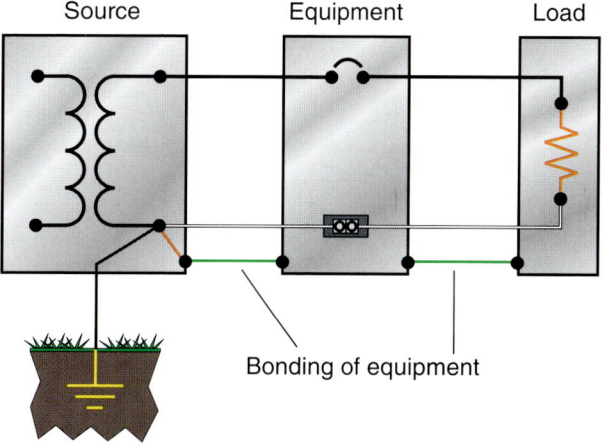

FIGURE 8-4 EGCs perform bonding functions.

The conductor must be capable of carrying the available fault current back to the source for the time it takes the overcurrent protective device to open and clear the event from the circuit. Section 250.4(A)(5) clearly provides the performance requirements and criteria for an effective ground-fault current path. Ideally there will not be a ground fault on the circuit, but insulation breakdown or failure can occur, resulting in the undesirable ground fault. Insulation can be in the form of a dielectric material or air space, such as the space between busbars in a switchboard or panelboard. Ground faults are typically unintentional. In addition to wire insulation failure, a ground fault can result from human error or accidents such as the dropping of a conductive tool in an electrical enclosure during live work. During those events, the EGC must be capable of withstanding the higher level of current to perform its all-important safety function. The functions of EGCs are quite simple. They provide grounding for equipment, perform bonding functions, and facilitate overcurrent device operation, making them an important component of the electrical safety system. **See Figure 8-5.**

Equipment Grounding Conductor Material

Section 250.118 provides the conductive materials and wiring methods that can be used as EGCs with feeders and branch circuits. List item (1) in Section 250.118 states EGCs can be made of copper, aluminum, or copper-clad aluminum material. These conductor materials can be in the form of a wire (stranded or solid) or busbar of any shape, and they can be insulated, covered, or bare.

Types of Equipment Grounding Conductors

Section 250.118 includes a list of recognized EGCs that must be run with the circuit conductors. EGCs can be in various forms. They are required to be any one or any combination of the following types. **See Figure 8.6.**

Electrical conduit and tubing are acceptable as EGCs according to Section 250.118. These wiring methods qualify as EGCs on their own without a wire-type conductor being installed. **See Annex B.** Where conduit or tubing is installed, all fittings (connectors, couplings, and locknuts, for example) must be made up tight in a workmanlike manner, and the wiring system must be secured and supported properly in accordance with the applicable rules in Chapter 3 of the *Code*.

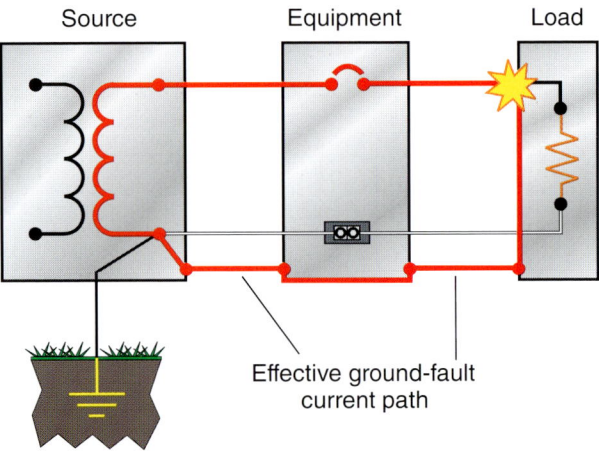

FIGURE 8-5 EGCs provide an effective path for ground-fault current and facilitate overcurrent device operation during ground-fault conditions.

A copper, aluminum, or copper-clad aluminum conductor. This conductor shall be solid or stranded; insulated, covered, or bare; and in the form of a wire or a busbar of any shape

Rigid metal conduit (RMC)

Intermediate metal conduit (IMC)

Electrical metallic tubing (EMT)

Armor of Type AC cable, as provided in Section 320.108

The copper sheath of mineral-insulated, metal-sheathed cable

Cable trays, as permitted in Sections 392.10 and 392.60

Cable bus framework, as permitted in Section 370.3

Other listed electrically continuous metal raceways and listed auxiliary gutters

Surface metal raceways listed for grounding

FIGURE 8-6 *NEC* Section 250.118 lists the types of acceptable EGCs.[4]

For example, rigid metal conduit must generally be supported at intervals not exceeding 10 feet and also must generally be secured within 3 feet of outlets, junction boxes, conduit bodies, etc. Similarly, Section .30 in each respective wiring method article provides securing and supporting requirements. **See Figure 8-7.** As an example, Section 358.30 requires that electrical metallic tubing (EMT) be installed as a complete system and securely fastened in place and supported in accordance with Section 358.30(A) and (B). The general requirement in this section calls for secure fastening of the tubing at intervals not exceeding 10 feet, and within 3 feet of each outlet box, junction box, device box, cabinet, conduit body, or other tubing termination.

Listed flexible metal conduit can serve as an EGC but in limited applications. **See Figure 8-8.** It is important to understand that the interlocking metal-tape construction of flexible metal conduit provides a high-impedance path for ground-fault current. Thus it is limited in length and amount of current when serving as an EGC. This type of wiring must meet four conditions to qualify as an EGC.

Section 250.118(5) indicates that listed flexible metal conduit is only suitable as an EGC if it meets all of the following conditions:

1. The conduit is terminated in listed fittings.
2. The circuit contained in the conduit is protected by an overcurrent device rated 20 A or less.
3. The combined length of flexible metal conduit and flexible metallic tubing and liquid-tight flexible metal conduit in the same ground-fault current path does not exceed 6 feet.
4. If used to connect equipment in which flexibility is necessary to minimize the transmission of vibration from equipment or to provide flexibility for equipment that requires movement after installation, an EGC must be installed. **See Figure 8-9.**

An example of flexible metal conduit installed where flexibility is necessary after installation is a flexible conduit "whip" that supplies chain-hung fluorescent luminaires.

FIGURE 8-7 Secure and supported metal raceway installations help provide an effective ground-fault current path.

FIGURE 8-8 Flexible metal conduit is permitted as an EGC in accordance with the conditions in Section 250.118(5)(A) through (D).

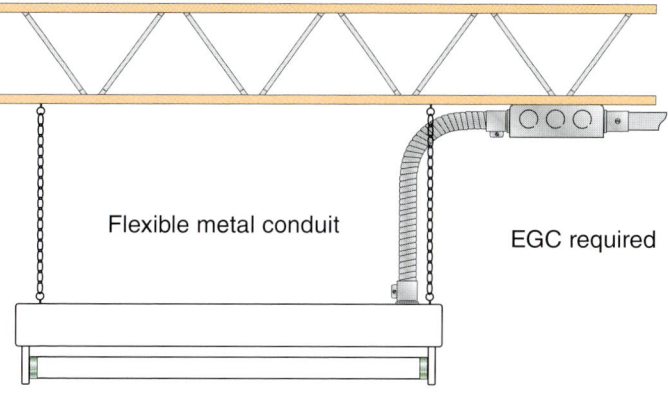

FIGURE 8-9 Listed flexible metal conduit is not permitted as EGC if flexibility is necessary after installation.

172 APPLIED GROUNDING AND BONDING

FIGURE 8-10 Listed flexible metal conduit is permitted as an EGC in accordance with the conditions in Section 250.118(5)(A) through (D).

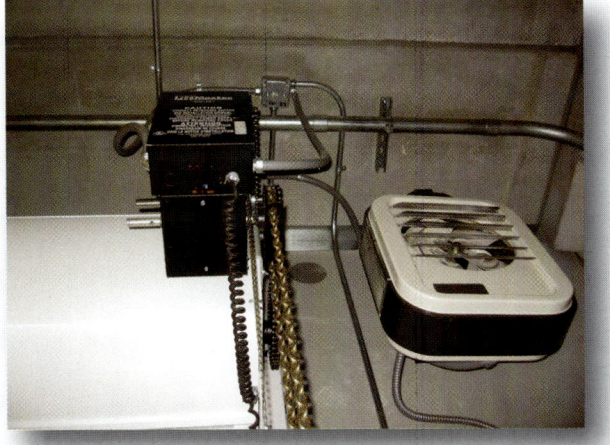

Listed liquid-tight flexible metal conduit can be used as an EGC if all the conditions of Section 250.118(6) have been met.

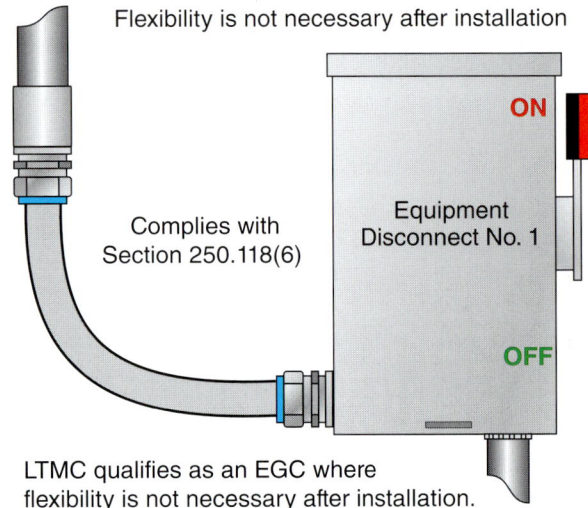

FIGURE 8-11 Listed liquidtight flexible metal conduit is suitable as an EGC when installed and used within the limitations of Section 250.118(6).

There is anticipated movement in the flexible metal conduit, so an EGC is required. An example of equipment that introduces vibration is a motor or transformer. These types of equipment are often connected using flexible metal conduit. EGCs are required in these types of installations. When there is no flexibility required after installation and the installation meets the provisions in Section 250.118(5)(A) through (D), an EGC is not required. An example of this type of installation could be a commercial garage door opener wired using flexible metal conduit. Some localities and some specifications require an EGC in any installation of flexible metal conduit. This is more restrictive than the *NEC* minimum, so it is important to verify any local electrical code requirements with the approving authority in that jurisdiction.

Listed liquid-tight flexible metal conduit can also serve as an EGC but in limited applications. **See Figure 8-10.** Understanding that the interlocking metal-tape construction of liquid-tight flexible metal conduit also provides a high-impedance path for ground-fault current is important. Thus it has restrictions that must be met to qualify as an EGC.

Section 250.118(6) indicates that listed liquid-tight flexible metal conduit is suitable for use as an EGC under the following conditions:

1. The conduit is terminated in listed fittings.
2. For trade sizes ³⁄₈ through ½, the circuit conductors contained in the conduit are protected by overcurrent devices rated at 20 amperes or less.
3. For trade sizes ¾ through 1¼, the circuit conductors contained in the conduit are protected by overcurrent devices rated not more than 60 amperes and there is no flexible metal conduit, flexible metallic tubing, or liquid-tight flexible metal conduit in trade sizes ³⁄₈ through ½ in the ground-fault current path.

4. The combined length of flexible metal conduit and flexible metallic tubing and liquid-tight flexible metal conduit in the same ground-fault current path does not exceed 6 feet.
5. If used to connect equipment in which flexibility is necessary to minimize the transmission of vibration from connected equipment or to provide flexibility for equipment that requires movement after installation, an EGC must be installed. **See Figure 8-11.**

An example of liquid-tight flexible metal conduit being used to minimize the effects of vibration is when it is used to supply air-conditioning and refrigeration equipment on a rooftop or a large dry-type transformer. **See Figure 8-12.** An example of a connection to equipment that might also require more movement after installation is a large commercial refrigerator that is "hard-wired" with liquid-tight flexible metal conduit and is on wheels so it can be rolled away from a wall for cleaning or service operations.

Flexible metallic tubing is not common anymore, but it is still recognized in the *Code* as an EGC when the tubing is terminated in listed fittings and the installation meets the following conditions:

1. The circuit conductors contained in the tubing are protected by overcurrent devices rated at 20 amperes or less.
2. The combined length of flexible metal conduit and flexible metallic tubing and liquid-tight flexible metal conduit in the same ground-fault current path does not exceed 6 feet.

Metal-clad (MC) cable is available with three types of armor: the interlocking metal-tape armor (the most common type), the corrugated-tube type, and the smooth-tube type. The physical characteristics of interlocking metal-tape–style MC cable limit its performance characteristics. **See Figure 8-13.** This type of MC cable assembly must contain an EGC. The armor of the smooth-tube type or corrugated-tube type can serve as an EGC when used with fittings listed for the particular cable. **See Figure 8-14.**

FIGURE 8-12 A dry-type transformer is wired using liquid-tight flexible metal conduit to reduce noise transmission and vibration.

FIGURE 8-13 The armor of conventional metal-clad cable Type MC (interlocking metal-tape construction) is not an effective ground-fault current path.

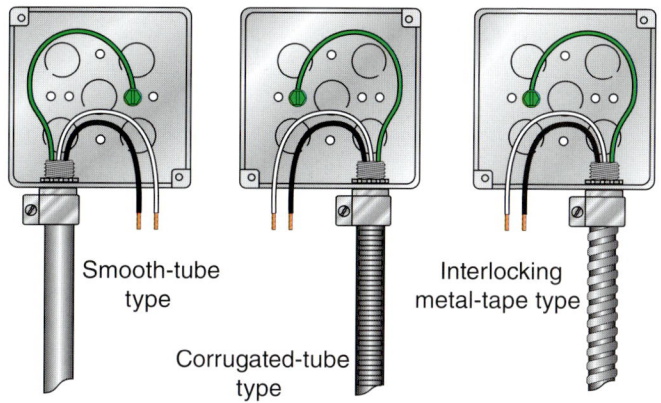

FIGURE 8-14 MC cable must be installed with listed fittings, as shown.

Metal-clad cable Type MC is also suitable for use as an EGC in accordance with any of the following:

1. It contains an insulated or uninsulated EGC in compliance with Section 250.118(1).
2. The combined metallic sheath and uninsulated equipment grounding/bonding conductor of interlocked metal-tape–type MC cable that is listed and identified as an EGC.
3. The metallic sheath or the combined metallic sheath and EGCs of the smooth or corrugated-tube–type MC cable that is listed and identified as an EGC.

There is one type of MC cable available that has interlocking metal-tape–type construction that has an armor listed as an EGC. **See Figure 8-15.** It is similar in construction to Type AC cable in that it includes a separate bare conductor (typically, a 10 AWG aluminum) within the assembly that is installed on the outside of the plastic-wrapped conductors of the cable assembly and in intimate contact with the cable armor. It should be installed according to manufacturer's instructions. The bonding conductor is in contact with armor between cable terminations.

This MC cable assembly is manufactured with and without contained insulated copper EGC, but the armor plus the bare bonding conductor serves as an EGC when installed using listed fittings. As always, to ensure compliance with Section 110.3(B) of the *Code*, follow the manufacturer's installation instructions when installing this and any other product. If this type of cable is installed in patient care locations of health care facilities, it must include an insulated copper EGC internal to the cable assembly to satisfy the requirements for two EGC paths as required in Section 517.13.

Armored-Clad Cable

Section 250.118(8) recognizes the armor sheath of armored-clad (Type AC) cables as an EGC. This cable armor qualifies as an EGC because of the bare internal bonding strip that is in intimate contact with the armor from fitting to fitting. **See Figure 8-16.**

The combination of the internal bonding strip in the assembly, together with the interlocking metal-tape–type armor, work as an effective ground-fault current path. **See Figure 8-17.** Some armored-clad cable assemblies are manufactured with an insulated copper EGC and therefore meet the criteria for isolated grounding circuits and redundant EGCs specified in Sections 517.13(A) and (B) for use in branch circuits serving patient care locations. These cable assemblies are all required to be installed using listed fittings.

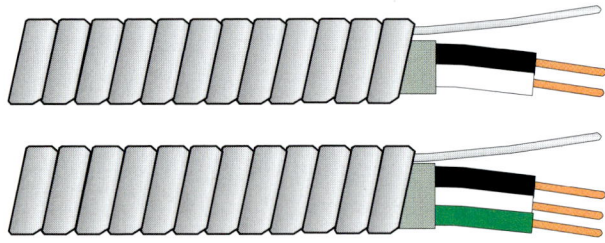

FIGURE 8-15 MC cable assembly is shown with an armor that is suitable as an EGC. The bonding conductor is in contact with armor between cable terminations.

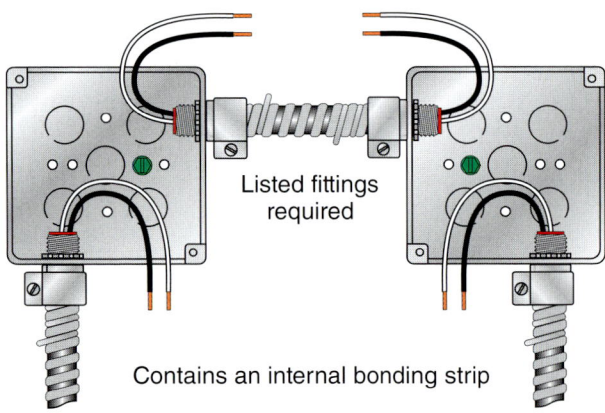

FIGURE 8-16 The armor of armored-clad cable is suitable as an EGC because of its contained internal bonding strip in contact with the armor.

Equipment Grounding Conductor Installations

Section 110.12 requires electrical conductors and equipment to be installed in a neat and workmanlike manner. This general requirement also applies to EGCs installed with feeders and branch circuits. **See Figure 8-18.** Where the wiring method, such as rigid metal conduit or electrical metallic tubing, is installed and serves as the required EGC, the installation must conform to the provisions in Section 110.12. This includes, but is not limited to, sufficient support and methods of securing the wiring method. If the conduit or tubing is secured properly, it should not be subject to movement that could compromise its performance as an effective ground-fault current path due to loosening of fittings.

Section 250.120 provides installation requirements for EGCs of all types. Essentially, whatever EGC type is installed, the applicable provisions in the *NEC* have to be met. This means that if a cable tray is the EGC, it has to meet the provisions in Articles 250, 300, and 392 relative to cable trays performing as EGCs. If a raceway such as electrical metallic tubing (EMT) is installed, it has to meet the applicable installation requirements in Articles 250, 300, and 358. The fittings and terminations used with the wiring method chosen have to be suitable for use with the type of wiring method installed. For example, fittings (connectors, couplings, and locknuts, for example) used with EMT have to be listed for use with EMT, as required by Section 358.6. **See Figure 8-19.**

FIGURE 8-17 Listed armored-clad cable Type AC is suitable as an EGC when used with listed fittings.

FIGURE 8-18 NECA-101 *Standard for Installing Steel Conduit (Rigid, IMC, EMT)* provides specific details about installing metal conduit and tubing installations.

FIGURE 8-19 Electrical metallic tubing is suitable as an EGC when used with listed fittings and properly secured and supported according to the methods required in Section 358.30.

All connections and joints have to be made tight using suitable tools. This goes back to the basic workmanship requirement in Section 110.12. It is important to tighten fittings because of the functions they are expected to perform both in normal operation and during abnormal conditions such as ground faults. Loose fittings such as set-screw couplings, connectors, and locknuts introduce impedance into the ground-fault current path and could affect quick operation of overcurrent devices. **See Figure 8-20.** Chapter 3 of the Code also includes requirements for securing and supporting conduit and other raceways that are included in Section 250.118 as EGCs. The integrity of the effective ground-fault current path established by the wiring method itself depends on effective, code-compliant support and securing of the raceway system. Workmanship is important for many reasons.

Protection from Physical Damage

An important installation requirement applies to aluminum or copper-clad aluminum conductors. These conductors are more vulnerable to the effects of corrosion and deterioration than copper conductors. For this reason, bare EGCs are not permitted to come in contact with masonry or the Earth and are not permitted where subject to corrosive conditions. These types of EGCs are not permitted to be terminated within 18 inches of the Earth. Section 250.120(C) indicates that EGCs smaller than 6 AWG are required to be protected from physical damage by a raceway or cable armor. This damage protection rule is relaxed for EGCs installed in hollow spaces of walls or partitions that protect them from physical damage. An example of this is in the rare case when EGCs are installed separately from the circuit conductors as permitted in Section 250.130(C).

Installation with Circuit Conductors

Sections 300.3(B) and 250.134(B) provide important information about installing EGCs. One of the most important is to keep the EGC as close to its associated circuit conductors as possible. This keeps the impedance values as low as possible during normal operation and during ground-fault conditions. **See Figure 8-21.** Section 300.3(B) requires all conductors of the circuit including any grounded conductor or EGC to be run in the same raceway cable or trench. When the EGC of the circuit is rigid metal conduit, intermediate metal conduit, electrical metallic tubing, or any of the other raceway types mentioned in Section 250.118, the EGC is automatically run with the circuit conductors and is integral to the wiring method.

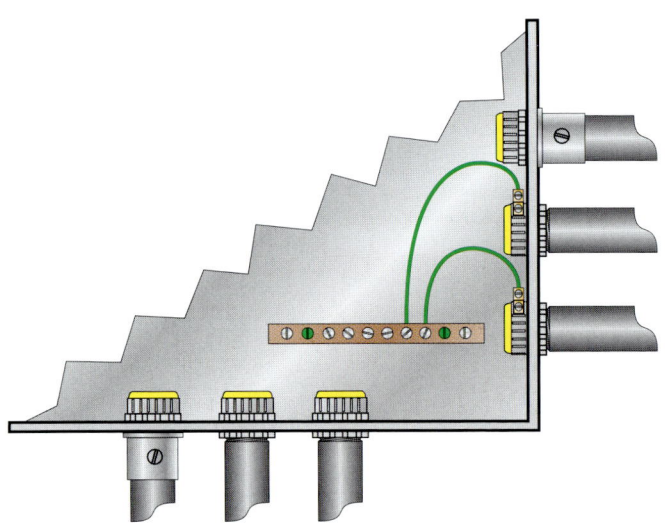

FIGURE 8-20 Tighten fittings in conduit and tubing system installations to ensure an effective ground-fault current path.

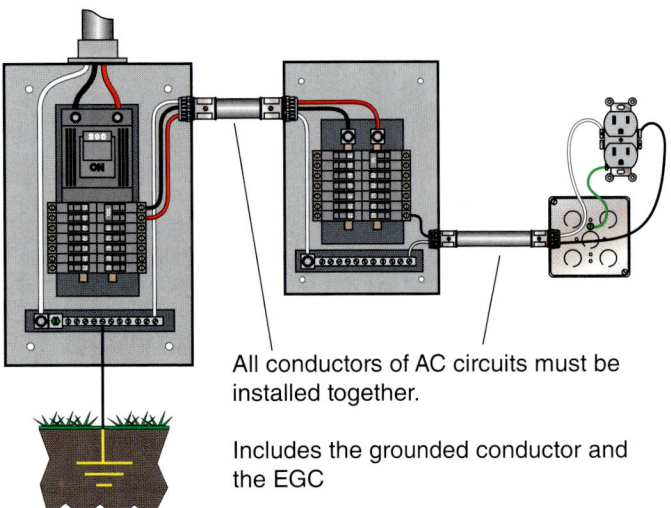

FIGURE 8-21 Install EGCs with the associated ungrounded conductors of the circuit.

Section 250.134(B) has a similar requirement that addresses equipment being connected to an EGC that is run with the circuit conductors and contained in the same raceway or cable or otherwise installed with the circuit conductors. If EGCs are separated from the associated circuit conductors, the inductive reactance of the circuit is also increased; thus circuit impedance is added. This can become problematic during ground-fault events, when the objective is to keep the impedance as low as possible to allow the highest level of fault current to quickly operate the fuse or circuit breaker protecting the circuit.

Equipment Grounding Conductor Connections

Connections of grounding and bonding conductors are covered in Section 250.8. The list of connection means is provided to clarify how EGCs must be terminated. Any method not mentioned in this list is not recognized by the *NEC*. The connections must be tight, and if the EGC is of the wire type, installers must torque connections at terminal lugs to manufacturer's requirements. **See Figure 8-22.**

Terminal lugs and equipment generally provide torque values for installers to attain the correct tightness of wire-type conductors. It is important not to undertighten or overtighten electrical connections, thus manufacturers provide specific torque values. This requirement applies not only to the ungrounded circuit conductors, but also to the grounding and bonding conductor connections. **See Figure 8-23.** Obviously, loose connections or terminal lugs in the EGC path will introduce impedance into that path.

Again, it is important to tighten all EGC connections to establish an effective path for ground-fault current. When fittings such as couplings, connectors, and locknuts are used to install wiring methods such as conduit, the fittings should be made tight for the same reason—integrity of the ground-fault current path.

> The connections must be tight, and if the EGC is of the wire type, installers must torque connections at terminal lugs to manufacturer's requirements.

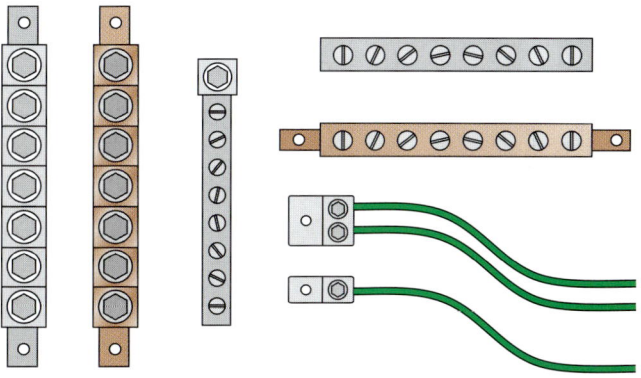

FIGURE 8-22 Terminals for grounding and bonding connections must be torqued to manufacturer's requirements.

Courtesy of Jim Dollard, IBEW Local 98

FIGURE 8-23 EGCs terminate on a grounding terminal bar in a switchboard enclosure.

Equipment Grounding Conductor Identification

The *Code* provides specific identification requirements for EGCs of the wire type. EGCs can be insulated, covered, or bare. **See Figure 8-24.** Section 250.119 recognizes a bare conductor as an EGC. An example is the bare EGC often included in nonmetallic sheathed cable or Type SE cable assemblies. If the conductor is insulated or covered, the insulation or covering is required to have a continuous outer finish that is either green or green with one or more yellow stripes.

The color green or green with one or more yellow stripes cannot be used for grounded (usually neutral) conductors or ungrounded (phase or hot) conductors. EGCs installed in raceway systems are often green for the entire length or green with one or more yellow stripes.

EGCs larger than 6 AWG must meet the applicable requirements in Sections 250.119(A)(1) and (A)(2). As indicated in (A)(1), an insulated or covered conductor larger than 6 AWG is permitted to be identified at each end and at every point where the conductor is accessible. **See Figure 8-25.** By exception, wire-type EGCs larger than 6 AWG are not required to be marked in conduit bodies that contain no splices or unused hubs. The identification means chosen must encircle the conductor and is required to be accomplished by one of the following:

1. Stripping the insulation or covering from the entire exposed length
2. Coloring the insulation or covering green at the termination
3. Marking the insulation with green tape or green adhesive labels at terminations

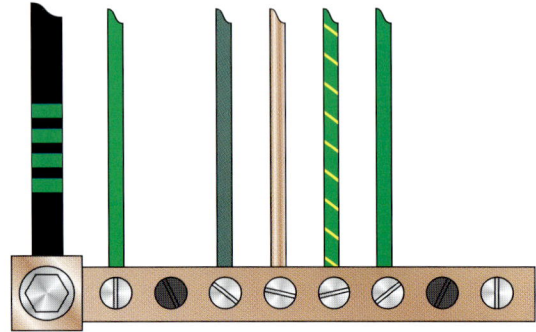

Equipment grounding terminal bar in electrical equipment

FIGURE 8-24 EGCs can be insulated, covered, or bare and must be identified by any of the methods in Section 250.119.

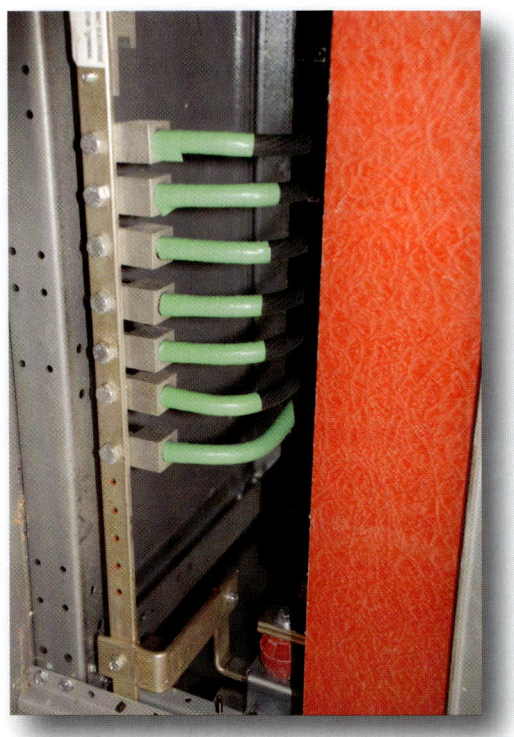

Courtesy of IBEW Local 26 Training Center

FIGURE 8-25 Green marking tape can be used at terminations to identify EGCs larger than 6 AWG.

Multi-conductor Cables

Section 250.119(B) indicates that when the conditions of maintenance and supervision ensure that only qualified people service the installation, one or more insulated conductors in a multi-conductor cable, at the time of installation, are permitted to be identified as EGCs at each end and at every point where the conductors are accessible by any one of the following methods:

1. Stripping the insulation from the entire exposed length
2. Coloring the exposed insulation green
3. Marking the exposed insulation with green tape or green adhesive labels

Sizing Criteria

Sizing requirements for EGCs of the wire type are found in Section 250.122 and Table 250.122 of the *NEC*. The minimum sizes are provided in Table 250.122 and are related to the short-time withstand capabilities of the conductor. **See Figure 8-26.** An important note follows Table 250.122 and provides an appropriate reference to Section 250.4. This is an indication to *Code* users that the minimum size required for wire-type EGCs could be larger than sizes provided in the table.

The conductor sizes in Table 250.122 of the *NEC* offer an approximate relation to the size of the overcurrent device given in the table. The I^2T values (short-time rating or withstand rating) of the EGC sizes are between 13 and 28 times their nominal continuous rating based on 1 ampere for every 42.25 circular mils of conductor. This value was used to develop the 5-second withstand rating for insulated conductors. Using this formula, a 5-second rating can be established for insulated conductors covered in the *NEC*. This is the proven conductor withstand formula developed through extensive testing and data collection by the Insulated Cable Engineers Association (ICEA). The value of 42.25 circular mils is applicable to insulated conductors. The value is 29.1 circular mils for uninsulated conductors based on the findings of the ICEA.

Example: To determine the 5-second withstand rating of a 4 AWG insulated conductor, use the values in the third column of *NEC* Table 8, Chapter 9 to determine the circular mil area of a 4 AWG conductor. The value is 41 740 cm. Take the value 41 740 cm ÷ 42.25 cm and the result is 987.9 amperes, which can be carried safely for 5 seconds by a 4 AWG conductor. Five seconds is a long time for any overcurrent device to open. Overcurrent protection typically operates in cycles or fractions of a cycle. The faster the overcurrent device operates, the more fault current the insulated conductor can carry safely without degradation or annealing of the conductor. This point emphasizes that there are factors such as voltage drop and higher amounts of fault current in systems that could result in the EGC sizes in Table 250.122 being insufficient. Good engineering designs require a careful study and selection of overcurrent protection in conjunction with sufficiently sized EGCs that will ensure fast and effective operation of overcurrent protective devices in short circuit and ground-fault conditions.

Equipment Grounding Conductor Sizing

In addition to the engineering basics of the effective ground-fault current path, the sizing rules in the *NEC* for EGCs are also important. The driving text of Section 250.122 is that the minimum size required for wire-type EGCs is not to be less than the values in Table 250.122. **See Figure 8-27.** This rule goes on to state

Table 250.122 Minimum Size Equipment Grounding Conductors for Grounding Raceway and Equipment (in part)

Rating or Setting of Automatic Overcurrent Device in Circuit Ahead of Equipment, Conduit, etc., Not Exceeding (Amperes)	Size (AWG or kcmil)	
	Copper	Aluminum or Copper-Clad Aluminum*
15	14	12
20	12	10
60	10	8
100	8	6
200	6	4
300	4	2
400	3	1

Note: Where necessary to comply with 250.4(A)(5) or 250.4(B)(4), the equipment grounding conductor shall be sized larger than given in this table.

* See installation restrictions in 250.120.

FIGURE 8-26 The minimum sizes for wire-type EGCs are provided in Table 250.122. Figure 8-26 is a reproduction of *NEC* Table 250.122 (in part). Reprinted with permission from NFPA 70-2011, *National Electrical Code*®, Copyright © 2010, National Fire Protection Association, Quincy, MA 02169. This reprinted material is not the complete and official position of the NFPA on the referenced subject, which is represented only by the standard in its entirety.

> In no case are the EGCs required to be larger than the circuit conductors supplying the equipment.

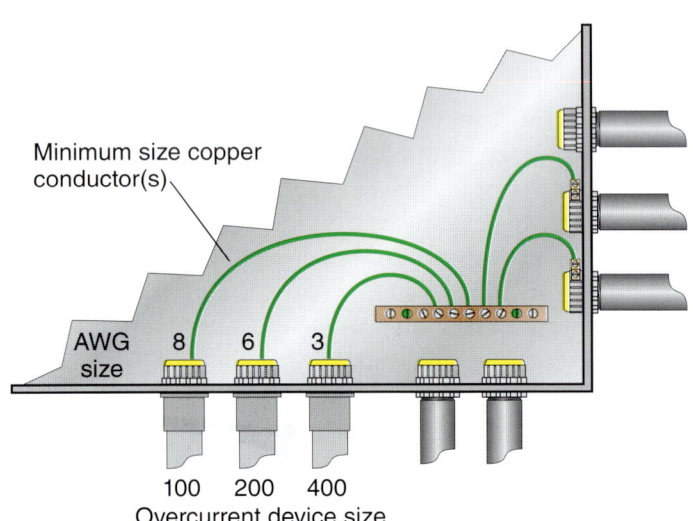

FIGURE 8-27 Wire-type EGCs must be sized according to Table 250.122.

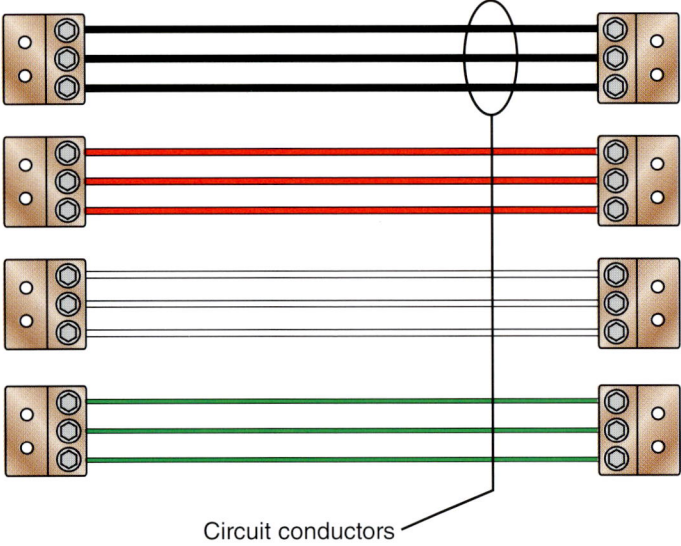

FIGURE 8-28 The circuit conductors in parallel arrangements include all the ungrounded conductors of each phase and/or grounded conductor to create the entire circuit.

that in no case are the EGCs required to be larger than the circuit conductors supplying the equipment.

It is important to mention parallel arrangements and their relationship to the circuit. The EGC is never required to be larger than the circuit conductors. Where circuit conductors are installed in parallel to create the equivalent of a larger circuit conductor, the size of the circuit conductor is the total area of all individual conductors in parallel that create the larger circuit conductor using multiple parallel paths. **See Figure 8-28.** It is important to understand that the *NEC* does not permit conductors to be installed in parallel to create an EGC. The EGC in a parallel installation must be full size.

Section 250.122 also indicates that raceways, cable sheaths, and cable trays serving as EGCs have to meet the performance requirements in Section 250.4(A)(5) or (B)(4) as applicable. For multi-conductor cable, the EGC within the assembly is permitted to be sectioned as long as the combined circular mil area of the sectioned EGC meets the size requirement in Table 250.122. Using Table 250.122 requires knowing the rating of the overcurrent device protecting the branch circuit or feeder. Once this value is known, the rating or a rating that does not exceed the value in the left column of the table should be found. The reader can then move across the table horizontally from left to right to determine the minimum size EGC expressed in AWG or circular mils. The appropriate column for aluminum as compared with copper EGCs should be used.

Examples: Several feeders and branch circuits are installed in the slab of a commercial building. All are installed in PVC conduit. Each circuit requires that an EGC be connected to the equipment. This information can be used to determine the minimum size copper and aluminum EGCs (wire type) for feeders or branch circuits protected at 20, 45, 60, 90, 110, 225, 350, and 450-amperes.

Size of Overcurrent Device	Equipment Grounding Conductor (Copper)	Equipment Grounding Conductor (Aluminum)
20	12	10
45	10	8
60	10	8
90	8	6
110	6	4
225	4	2
350	3	1
450	2	1/0

Using Table 250.122, the minimum size EGC required for a 4,000-ampere feeder can be found. Answer: The minimum size EGC for a 4,000-ampere feeder circuit is not less than 500 kcmil copper or 750 kcmil aluminum or copper-clad aluminum.

Increases in Size

EGCs are required to be increased in size proportionate to any increase of associated ungrounded conductors.

If the ungrounded conductors of a circuit are increased in size, the wire-type EGCs must also be increased proportionately according to the circular mil area of the ungrounded conductors. This can be verified by following these steps using a 400-ampere feeder as an example:

The 400-ampere feeder (420 allowable ampacity) is generally installed using 600 kcmil copper circuit conductors and a 3 AWG copper EGC. For voltage drop reasons, the 600 kcmil conductor has to be increased in size from 600 kcmil to (2) paralleled 400 kcmil copper conductors for each ungrounded phase conductor and the neutral conductor. The circular mil values are added together (400 plus 400) to result in 800 kcmil copper now required for the circuit conductors in this installation. The adjusted size (800 kcmil) is then divided by the originally required size (600 kcmil) to determine the proportionate value of circular mil area adjustment.

800 ÷ 600 = 1.3 (multiplier)

The minimum size wire-type EGC for a 400-ampere feeder is normally a 3 AWG copper, according to Table 250.122. The circular mil value of a 3 AWG conductor is 52620 as provided in *NEC* Table 8, Chapter 9. Take the value 52620 cm and multiply by 1.3 to come up with 68406 circular mils.

52620 × 1.3 = 68406 circular mils

Take this value back to *NEC* Table 8, Chapter 9 and round up to the next higher value, in the third column, to determine the minimum size EGC as adjusted proportionately to the increase in size for the ungrounded phase conductors of the circuit. The next higher circular mil value in *NEC* Table 8 is 83690. The new minimum size required for this EGC is a 1 AWG copper based on the adjustment. It is always best to perform this simple calculation to verify that the adjusted size of the EGC meets or exceeds the minimum requirements. Simply increasing the size of the EGC to the next higher size is not adequate in all cases.

Any increases in size of the ungrounded conductors of a circuit requires an increase in size of the wire-type EGC installed with that unit.

182 APPLIED GROUNDING AND BONDING

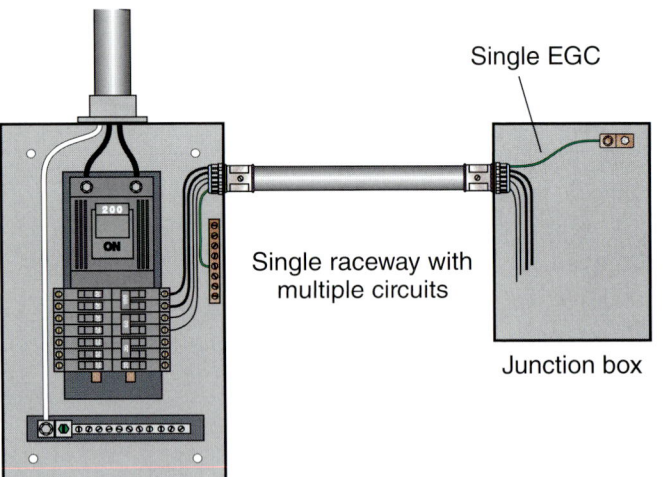

FIGURE 8-29 Only one EGC is needed when multiple circuits are combined in the same raceway.

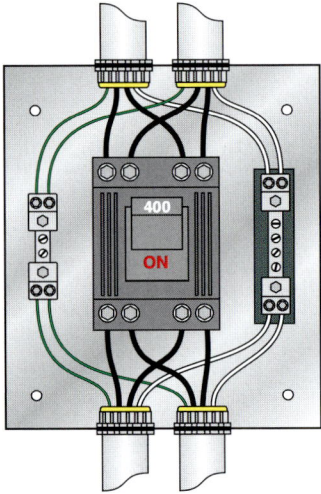

FIGURE 8-30 EGCs in parallel must comply with all of the requirements in Section 310.10(H) and must be installed in each raceway of the parallel run.

FIGURE 8-31 A single EGC is permitted with multiple circuits installed in the same cable tray.

Multiple Circuits in a Single Raceway or Cable Tray

It is common for installers to economize in construction installation methods. One way to economize is to combine electrical circuits into a single raceway rather than running individual conduits or cables. This combination offers the advantage of requiring only a single EGC in the run. **See Figure 8-29.**

When multiple circuits are installed in a single raceway, cable, or cable tray, a single EGC is permitted. The sizing requirement is based on the rating of the largest overcurrent protective device ahead of any circuit in the raceway, cable, or cable tray. When EGCs are installed in cable tray, they have to meet the requirements in Sections 250.122 and 392.10(B)(1)(c). The required ampacity adjustment factors may be a disadvantage, however, because of the number of current-carrying conductors in the same raceway.

Equipment Grounding Conductors for Parallel Runs

Many commercial and industrial electrical designs use parallel arrangements for large feeders or branch circuits to equipment. Installing parallel conductors becomes a necessity when supplying large switchboards and other large electrical equipment, simply because large single conductors are not practical, economical, or even available in many cases. Installing feeders or circuits in parallel requires compliance with Section 310.10(H), which means the conductors must be the same length, same material, and same size; have the same insulation; and be terminated in the same manner. **See Figure 8-30.**

Installing conductors in parallel for feeders means that multiple conductors are electrically connected at both ends of the circuit. When the entire parallel arrangement of conductors is installed in a single raceway, cable, or cable tray, a single EGC is permitted to be installed.

The metallic raceway or cable tray could also qualify as an EGC in accordance with Section 250.118. **See Figure 8-31.**

In this case, a wire-type EGC is not necessary to meet the minimum *NEC* requirements. There are specific rules for wire-type EGCs installed with parallel conductor installations. Section 250.122(F) states that when conductors are run in parallel, in separate raceways or cables, any wire-type EGCs are also required to be run in parallel in their respective raceway or cable. **See Figure 8-32.**

If wire-type EGCs are installed with the paralleled, ungrounded feeder conductors, the EGCs must also be installed in parallel, but they can be smaller than 1/0. This is because multiple EGCs are not being installed to share ground fault current in the event of a ground fault event, but to provide a low-impedance path regardless of where the fault occurs. They are already fully sized according to Section 250.122, as required. When parallel arrangements of conductors are installed in separate raceways such as in multiple conduits, and wire-type EGCs are required, an EGC must be installed in each of the separate raceways. **See Figure 8-33.** In this type of installation, each EGC installed in parallel is required to be sized using Table 250.122 based on the rating of the fuse or circuit breaker protecting the entire parallel set. **See Figure 8-34.**

It is important to remember Section 300.3(B) generally requires all conductors of the circuit, including the EGCs, to be installed in the same raceway, cable, or trench.

One reason for the requirement to include an EGC in each of the raceways is that the full-sized EGC prevents overloading and possible damage of a smaller, inadequately sized EGC, should a ground fault occur in one of the parallel branches. Another reason is that it keeps impedance levels low during normal operation and during ground-fault events. Installing only an EGC in one of the raceways and no EGC in the raceways of the other parallel feeder separates the EGC from its associated ungrounded conductors of the same circuit.

FIGURE 8-32 Circuits installed in parallel require that EGCs also be run in parallel.

FIGURE 8-33 Each EGC for circuits in parallel must be sized using Table 250.122.

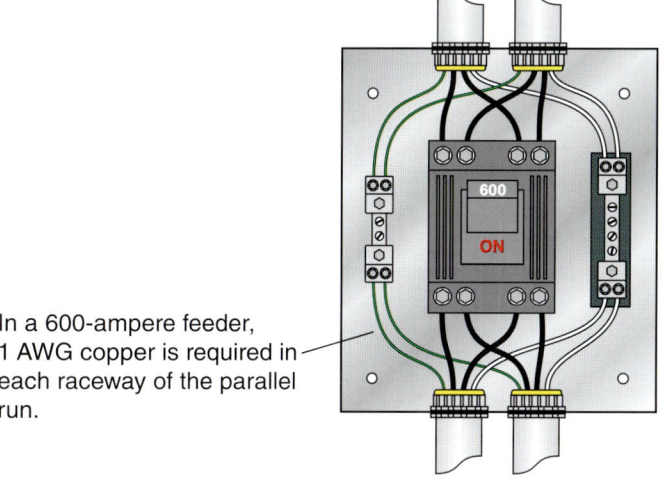

In a 600-ampere feeder, 1 AWG copper is required in each raceway of the parallel run.

FIGURE 8-34 When sizing EGCs for circuits in parallel arrangements, the minimum size must be not smaller than the sizes in Table 250.122 based on the rating of the fuse or circuit breaker.

This is a violation of Sections 250.122(F), 300.3(B), and 310.10(H). Keeping the impedance in the effective ground-fault current path as low as possible ensures fast, effective operation of fuses or circuit breakers in the case of a ground fault.

Sizing Examples

Questions about sizing EGCs for parallel installations are common. Installing EGCs in parallel arrangements is not overly complicated. The following examples address a couple of sizing exercises.

Example 1: A 4000-ampere feeder is installed in (10) PVC conduits in a parallel arrangement, each containing (4) 750 copper conductors. What is the minimum size copper EGC is required in each conduit?

Answer: A 500 kcmil copper EGC is required in each raceway based on the 4000 ampere overcurrent device in accordance with Table 250.122.

Example 2: If an 800-ampere feeder is installed in (2) raceways in a parallel arrangement, each containing (4) 750 copper conductors, what is the minimum size wire-type EGC for this circuit?

Answer: A 1/0 AWG copper EGC is required in each raceway.

Cable Assemblies in Parallel

Cable assemblies, such as MC Cable, are manufactured in standard conductor size configurations unless they are a special-order item. The EGC in a standard cable is typically sized adequately for single circuit use but might not be adequate for all parallel circuit installations. If cable assemblies are installed in large-capacity parallel circuits, it is necessary to verify that the EGC in each of the individual cables of the parallel set is sized as required by Table 250.122, based on the size of the fuse or circuit breaker protecting the entire parallel circuit. Installing cable assemblies in parallel arrangements may necessitate a special order that includes sizing the EGC in each cable large enough to allow the cable to be connected in a parallel installation. Some cable manufacturers can produce cable assemblies with larger EGCs when parallel arrangements are planned for a project.

Section 250.122(A) indicates that EGCs never have to be larger than the ungrounded conductors of the circuit. In a parallel feeder, the ungrounded conductors of the circuit include all of the conductors in parallel that are added together to make a single conductor. For example, if (4) 600-kcmil conductors are installed in parallel to create a 1600-ampere feeder circuit, the circuit conductors are 2400 kcmil. The minimum size EGC required in each raceway is 4/0 AWG copper based on the 1600 ampere overcurrent device.

Equipment Grounding Conductors for Motor Circuits

Sizing requirements for EGCs in motor circuits are provided in Section 250.122(D). The basic rule here is that the EGC (wire-type) be sized not smaller than determined by Section 250.122(A) based on the rating of the branch-circuit short-circuit and ground-fault protective device of the motor circuit as determined from Section 430.52. **See Figure 8-35.**

It is important to remember that the branch-circuit short–circuit ground fault protective device is usually sized larger to carry the starting current of the motor, which affects the size of a wire-type EGC. When an instantaneous trip circuit breaker or motor short-circuit protector is selected as the overcurrent protective device for a motor circuit, a wire-type EGC is required to be sized no smaller than provided in Section 250.122(A) using the maximum rating of a dual element time-delay fuse selected for branch-circuit short-circuit and ground-fault protection in accordance with Section 430.52(C)(1) Exception No. 1.

Equipment Grounding Conductors with Feeder Taps

When feeder taps are installed in accordance with the provisions in Section 240.21, the size of the EGC with the tap conductors must not be smaller than the size required based on the rating of the overcurrent protection for the feeder to which the tap is connected. For example, a 400-ampere feeder that is tapped by (2) 200-ampere feeders would require a 3 AWG copper EGC in each of the feeder tap raceways. **See Figure 8-36.**

Current in Equipment Grounding Conductors

In normal operation, no current should be present in the EGC. If current is present, then there probably is a noncompliant neutral-to-ground connection on the load side of the service disconnect or the load side of the first system overcurrent device of a separately derived system. Sections 250.24(A)(5) and 250.30(A) state these restrictions. EGCs should only carry current during abnormal conditions such as a ground-fault event, when heavy fault current is present in the EGC for the duration of time it takes the fuse or circuit breaker to clear the fault. The *Code* does not permit the EGC to perform double duty as a grounding electrode conductor and EGC, even though this has been practiced in the past, particularly with separately derived system installations. Section 250.121 now provides a clear restriction for this type of installation; previous editions of the *NEC* did not address the issue. The restriction clarifies that EGCs are not permitted to be used as grounding electrode conductors. The reason is that the grounding electrode conductor can carry varying levels of current during normal operation. This is unacceptable for an EGC because it results in increased objectionable current and can present a shock hazard in some conditions.

> EGCs should only carry current during abnormal conditions such as a ground-fault event, when heavy fault current is present in the EGC for the duration of time it takes the overcurrent protective device to clear the fault.

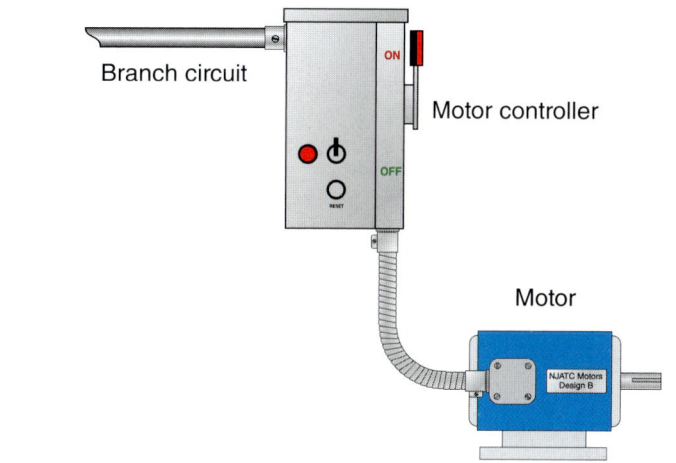

FIGURE 8-35 Size EGCs for motor circuits using Table 250.122 based on the rating of the branch-circuit short-circuit and ground-fault protective device.

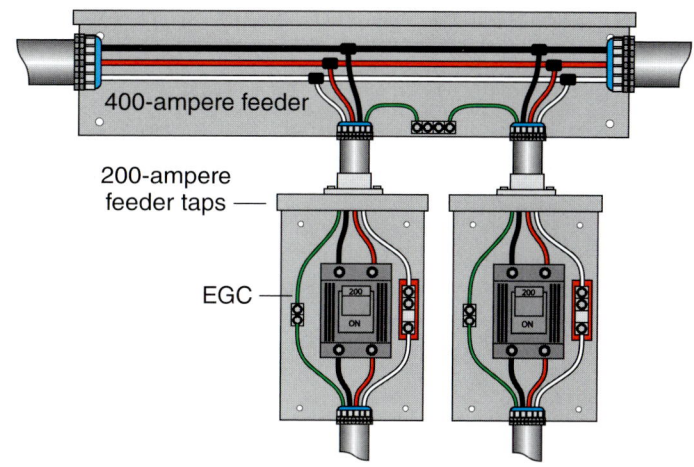

FIGURE 8-36 EGCs installed with feeder taps are sized using Table 250.122 based on the rating of the overcurrent protection on the supply side of the tap conductors.

Summary

EGCs perform three essential functions in the grounding and bonding safety system. They perform a grounding function for equipment required to be grounded; they perform a bonding function because they connect equipment together—which is bonding; and they facilitate overcurrent device operation while performing as an effective ground-fault current path during ground-fault conditions. EGCs are available in a variety of *Code*-recognized forms, including raceways, cable assemblies, cable tray, wireways, flexible metal conduit, and liquid-tight flexible metal conduit. The minimum size required for wire-type EGCs is determined using Table 250.122 on the basis of the overcurrent device protecting the circuit.

References

1. NFPA 70 National Electrical Code 2011, Article 100 (National Fire Protection Association, Quincy, MA 2010), p. 70–29.
2. NFPA 70 National Electrical Code 2011, Article 100 (National Fire Protection Association, Quincy, MA 2010), p. 70–29.
3. NFPA 70 National Electrical Code 2011, Section 250.2 (National Fire Protection Association, Quincy, MA 2010), p. 70–100.
4. NFPA 70 National Electrical Code 2011, Section 250.118 (National Fire Protection Association, Quincy, MA 2010), p. 70–122.

Review Questions

1. The conductive path installed to connect normally non–current-carrying metal parts of equipment together and to the system grounded conductor or to the grounding electrode conductor or both best defines which of the following components of the electrical grounding system?
   a. System bonding jumper
   b. Equipment bonding jumper
   c. Grounding electrode conductor
   d. Equipment grounding conductor
2. *NEC* Section 250.118 includes a list of recognized equipment grounding conductors.
   a. True   b. False
3. Equipment grounding conductors can be copper or copper-clad, but not aluminum.
   a. True   b. False
4. Equipment grounding conductors perform bonding functions, among other functions.
   a. True   b. False
5. Which of the following is not recognized by the *NEC* as a suitable equipment grounding conductor?
   a. Flexible metal conduit in lengths not longer than 8 feet
   b. Electrical metallic tubing
   c. Rigid metal conduit
   d. The armor of Type AC cable
6. Flexible metal conduit longer than 10 feet is acceptable as an equipment grounding conductor as long as the overcurrent device protecting the circuit does not exceed 20 amperes.
   a. True   b. False
7. When flexible metal conduit is installed for flexibility, a wire-type equipment grounding conductor is required.
   a. True   b. False
8. Which of the following characteristics is not related to an effective ground-fault current path?
   a. Electrically conductive
   b. Sufficient capacity for any fault current likely to be imposed on it
   c. Intentionally constructed
   d. High impedance
9. Which of the following conditions is not required for flexible metal conduit to qualify as an equipment grounding conductor?
   a. Listed fittings must be used
   b. Length is limited to 6 feet in the ground-fault return path
   c. The circuit contained is protected at no greater than 20 amperes
   d. The flexible metal conduit is installed for flexibility

10. When rigid metal conduit is installed and used as an equipment grounding conductor, all fittings and joints are required to be _____.
    a. Tested
    b. Approved
    c. Sealed
    d. Made up tight using suitable tools

11. The steel alloy sheath of mineral-insulated cable (TYPE MI) is suitable as an equipment grounding conductor.
    a. True   b. False

12. When listed liquid-tight flexible metal conduit is installed for equipment, and flexibility of the conduit is necessary after installation, an equipment grounding conductor (wire-type) must be installed in the conduit.
    a. True   b. False

13. Flexible metal conduit is never allowed as an equipment grounding conductor when flexibility is not necessary after installation.
    a. True   b. False

14. Which of the following does not qualify as an equipment grounding conductor?
    a. A copper conductor sized at 12 AWG
    b. Flexible metallic tubing 3 feet in length in which the circuit contained is protected by a 30-ampere overcurrent device
    c. Electrical metallic tubing
    d. Intermediate metal conduit

15. What table in the *NEC* is used to size equipment grounding conductors of the wire type?
    a. Table 1, Chapter 9
    b. Table 250.122
    c. Table 250.66
    d. Table 310.16

16. Table 250.122 establishes minimum sizes for wire-type equipment grounding conductors based on the size of the _____ of the circuit.
    a. Overcurrent device
    b. Largest ungrounded service conductor
    c. Overload protective device
    d. Smallest ungrounded conductor

17. An equipment grounding conductor is never permitted to perform as both the equipment grounding conductor and grounding electrode conductor simultaneously for a separately derived system.
    a. True   b. False

18. What is the minimum size wire-type equipment grounding conductor for a 4,000-ampere feeder circuit?
    a. 3/0 AWG copper
    b. 250 aluminum
    c. 400 kcmil copper
    d. 500 kcmil copper

19. What is the minimum size equipment grounding conductor for a motor circuit if the motor branch circuit short circuit and ground-fault protective device is sized at 1,200 amperes?
    a. 1/0 AWG copper
    b. 2/0 AWG copper
    c. 3/0 AWG copper
    d. 250 kcmil copper

20. If the ungrounded circuit conductors for a 400-ampere circuit are increased in size from 600 kcmil copper to 1,000 kcmil copper due to length and excessive voltage drop, what is the minimum size required for a wire-type equipment grounding conductor?
    a. 3 AWG copper
    b. 2 AWG copper
    c. 1 AWG copper
    d. 1/0 AWG copper

21. Green insulation along the entire length of a conductor is one method of identification for equipment grounding conductors of branch circuits.
    a. True   b. False

22. Table 250.122 provides the minimum sizes for wire-type equipment grounding conductors, and they may need to be larger to meet the performance requirements in Section 250.4.
    a. True   b. False

23. When a 400-ampere tap is made to a 1,000-ampere feeder, what is the minimum size equipment grounding conductor (wire-type) required to be run with the tap conductors to the equipment?
    a. 3 AWG copper
    b. 1 AWG aluminum
    c. 4/0 AWG copper
    d. 2/0 AWG copper

24. If (6) 500 kcmil copper conductors are run in parallel (per phase) as part of a 2,000-ampere feeder, what is the minimum size wire-type equipment grounding in each PVC raceway?
    a. 3/0 AWG copper
    b. 4/0 AWG copper
    c. 250 kcmil aluminum
    d. 250 kcmil copper

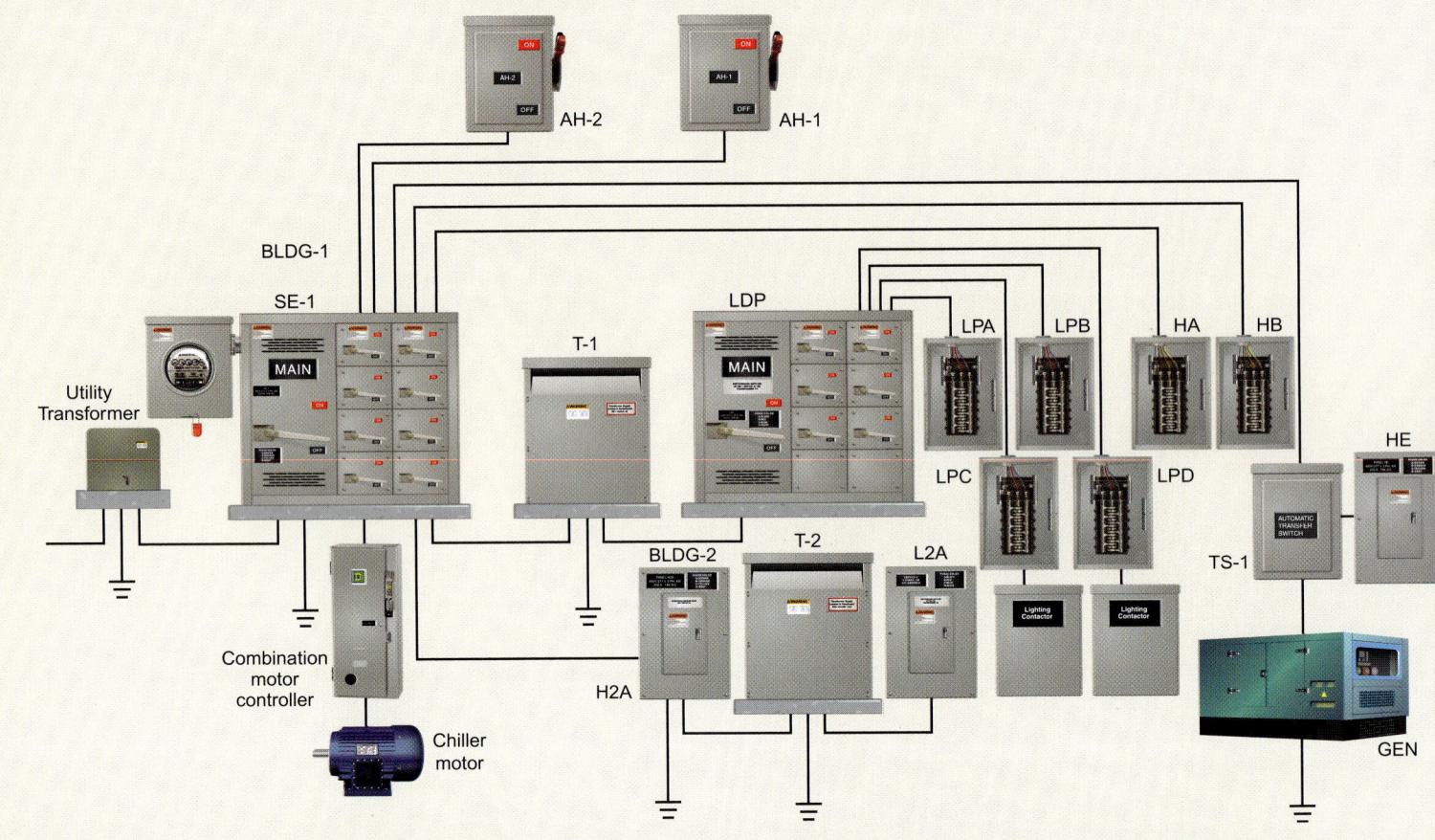

CHAPTER 9

Grounding Electrical Equipment

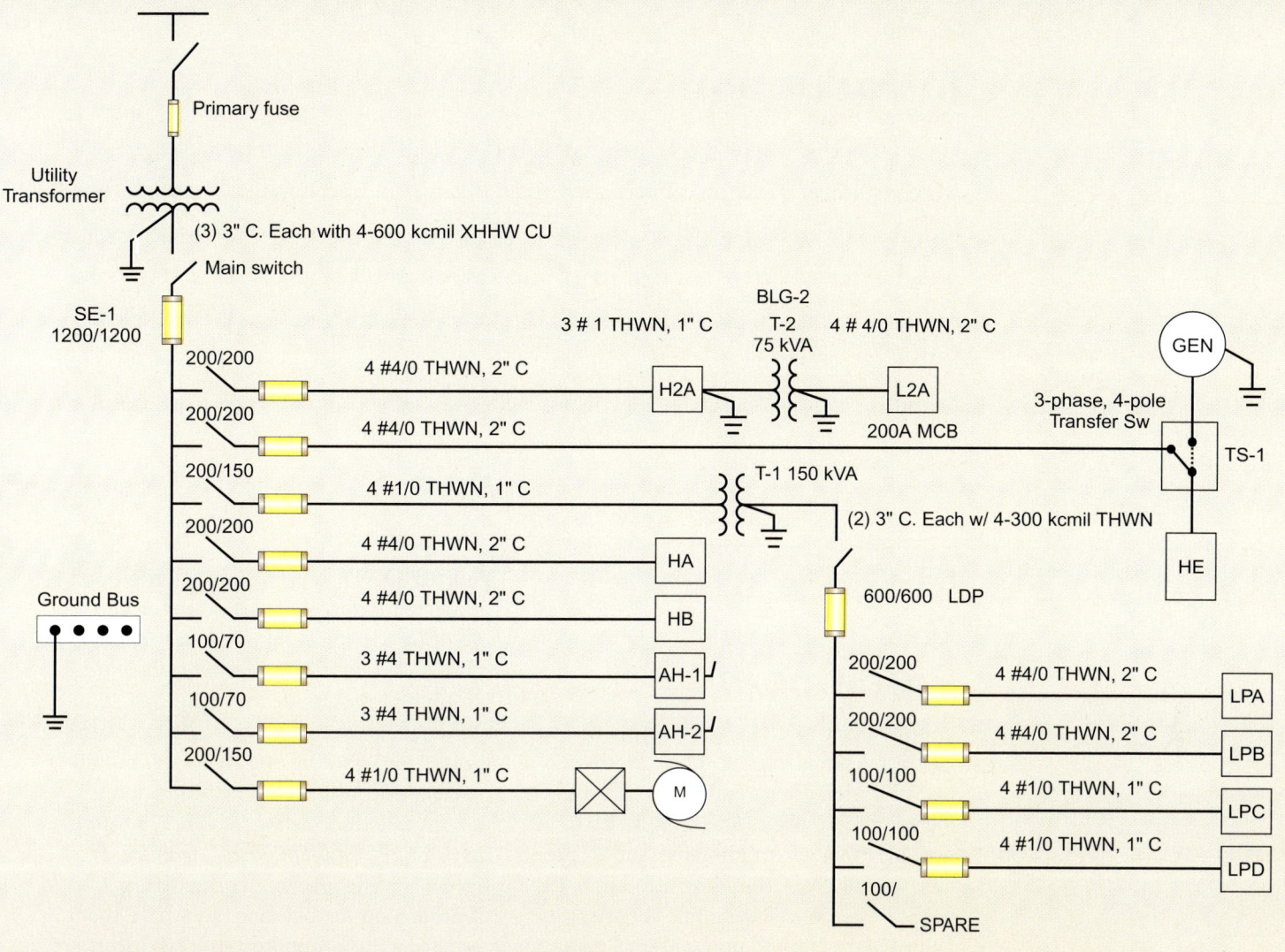

Objectives

- Understand the reasons for grounding electrical equipment
- Identify the methods of equipment grounding for feeders and branch circuits and specific conditions that provide exemptions from grounding equipment
- Understand the methods for installing equipment grounding conductors for devices such as receptacles and switches
- Understand the requirements to isolate neutrals from ground and grounded metal parts
- Determine the requirements for auxiliary grounding electrodes installed for equipment

Outline

Purpose of Grounding Equipment

General Grounding Rules for Equipment

Conductor Enclosure and Raceway Grounding Requirements

Methods of Grounding Equipment

Connections of Equipment Grounding Conductors

Receptacle Grounding Connections

Receptacle Replacements

Grounding Appliances Using the Grounded Conductor

Auxiliary Grounding Electrode Requirements

Grounding Nonelectrical Equipment

Equipment Grounded by Secure Metal Supports

Use of the Grounded Conductor for Grounding

Introduction

The *NEC*® includes several rules for grounding electrical equipment. Grounding equipment is accomplished by a direct connection to ground (Earth), or by connection to a conducting body that extends the grounding connection, or both. Specific *NEC* rules apply to fixed equipment that must be grounded with very few exceptions. Auxiliary grounding electrodes are addressed in the *NEC* as optional, but where installed, they must meet specific requirements but are not permitted as the only grounding means for equipment. Grounded conductors are permitted for grounding equipment such as appliances but only in existing installations.

Purpose of Grounding Equipment

Grounding is necessary to establish an Earth reference (connection) from connected systems and equipment. Grounding equipment places it as close to Earth as possible, thereby minimizing shock hazard possibilities. Grounding also limits the voltage to ground during line surge events, lightning events, and unintentional contact with higher-voltage lines. **See Figure 9-1.**

Equipment grounding, as required by the *NEC*, is not intended to provide total protection from lightning strikes. The purpose of the *NEC* is protection of people and property from the hazards arising from the use of electricity. Lightning is not electricity that is used; it is an unpredictable force. Grounding requirements in the *NEC* provide ancillary benefits of providing a path to ground for lightning events, but it is not the primary purpose. Lightning protection systems installed according to NFPA 780 *Standard for the Installation of Lightning Protection Systems* provide an increased level of protection for buildings that have them installed. The grounding process involves connecting a system of conductive parts to ground or to a conductive body that extends the ground connection. Grounding electrical equipment is generally accomplished by connection to an equipment grounding conductor (EGC) of a feeder or branch circuit. The Earth is not permitted as an effective path for ground-fault current.

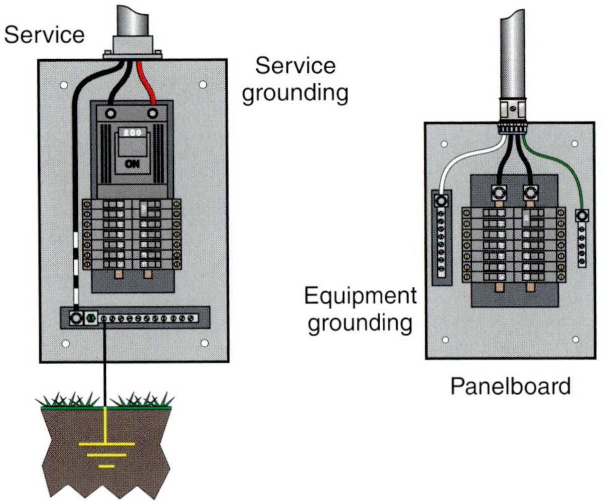

General Grounding Rules for Equipment

The general requirements for grounding equipment that is fastened in place or connected by permanent wiring methods (fixed) are provided in Part VI or Article 250. There are instances in which substitutes for grounding can be applied such as isolation, insulation, or guarding.

FIGURE 9-1 Grounding equipment places it as close to Earth as possible, reducing shock hazards.

These alternatives to grounding are addressed in Section 250.1(6) and more specifically in Section 250.110. If the equipment operates at more than 150 volts to ground, it is generally required to be grounded by connection to an EGC. **See Figure 9-2.**

Fixed equipment with exposed non–current-carrying metal parts that are likely to become energized is required to be grounded (connected to an EGC) under any number of conditions. For example, if the equipment is within 8 feet vertically or 5 feet horizontally of ground or grounded metal objects that are subject to contact by people, it must be grounded. There is a condition that relaxes the grounding requirement by isolation or elevation. If the equipment is 10 feet above the ground and not subject to contact by people, grounding is optional if the location is not wet or damp.

Another requirement for equipment grounding is that when the equipment is in a wet or damp location and is not isolated, the equipment also has to be grounded if it is in contact with other metal material. The idea is to result in all of the metal surfaces being grounded and not left isolated or floating at a potential above ground. When electrical equipment is in a hazardous (classified) location, it is required to be grounded as covered in Articles 500 through 517. When equipment is supplied by metal raceways, metal-sheathed or metal-clad cables, or another wiring method that provides an EGC, the equipment has to be grounded by connection to the EGC. **See Figure 9-3.**

Three exceptions follow the grounding requirements provided in Section 250.110. One exception relaxes the grounding requirement for frames of electrically heated appliances that are permanently and effectively insulated from ground. Another exception (Exception No. 3) is a case in which insulation is used as a substitute for grounding and offers equal and effective safety, by using double-insulated equipment.

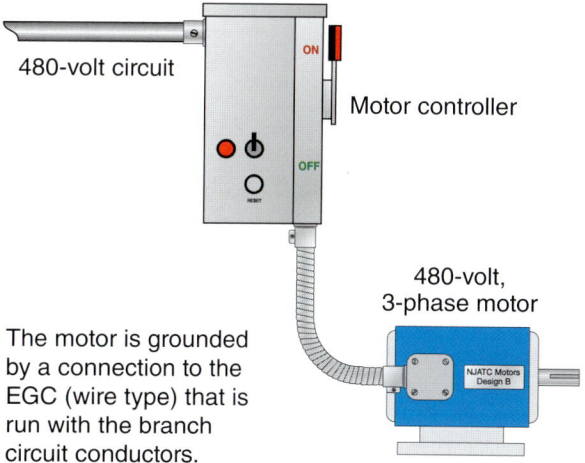

FIGURE 9-2 Equipment operating at more than 150 volts to ground is generally required to be grounded by connection to an EGC of the supply circuit.

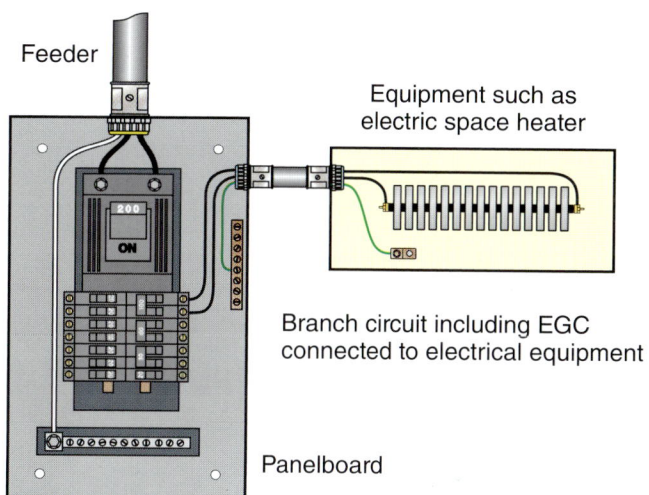

FIGURE 9-3 Equipment is grounded by being connected to an EGC run with the circuit conductors.

Wire-type EGCs are required to be sized in accordance with *NEC* Table 250.122 based on the overcurrent device rating.

Another exemption (Exception No. 2) from the grounding requirement is for distribution apparatuses such as transformers and capacitor enclosure cases mounted on wood poles and elevated to a height that exceeds 8 feet above ground. **See Figure 9-4.**

FIGURE 9-4 Grounding is not required for pole-mounted transformers or capacitors. Another relaxation of the grounding requirement applies to double-insulated equipment that is distinctly marked to indicate it employs a system of double insulation.

FIGURE 9-5 Motor frames in electrically operated pipe organs are required to be grounded.

Section 250.112 provides a list of specific equipment that is connected by permanent wiring and is required to be grounded by connection to an EGC. Equipment covered by the equipment grounding rules in Section 250.112 consists of the following:

> Switchboard frames and structures
> Enclosures for motor controllers
> Electric signs
> Luminaires (lighting fixtures)
> Pipe organs
> Elevators and cranes
> Motion picture projection equipment
> Skid-mounted equipment
> Motor frames
> Garages, theaters, and motion picture studios
> Remote-control, signaling, and fire alarm circuits
> Motor-operated water pumps and metal well casings

The exposed, normally non–current-carrying parts of equipment described in Section 250.112(A) through (M) are required to be grounded by connection to an EGC, regardless of the operating voltage applied. Switchboard frames and structures supporting switching equipment, except frames of 2-wire DC switchboards that are effectively insulated from ground, must be grounded. Section 250.112(A) clarifies that DC switchboards insulated from ground are not required to be grounded. Generator and motor frames in an electrically operated pipe organ, unless effectively insulated from ground, are required to be grounded, in addition to the motor driving the pipe organ. **See Figure 9-5.**

Motor frames are required to be grounded by connection to an EGC as provided in Section 430.242. **See Figure 9-6.**

Enclosures for motor controllers must be grounded unless attached to ungrounded portable equipment. **See Figure 9-7.** These could be part of a motor control center assembly.

Electrical equipment for elevators and cranes is required to be grounded by connection to the EGC of the supply circuit. **See Figure 9-8.**

Electrical equipment in commercial garages, theaters, and motion picture studios, except pendant lampholders, all have to be grounded except when they are supplied by circuits not more than 150 volts to ground. Electric signs, outline lighting, and associated equipment are all required to be grounded by connection to an EGC as provided in Section 600.7. **See Figure 9-9.**

FIGURE 9-6 Motor frames are required to be grounded.

FIGURE 9-8 Electrical equipment for cranes and elevators are required to be grounded.

FIGURE 9-7 Motor controllers are generally required to be grounded.

FIGURE 9-9 Electric signs and equipment associated with outline lighting are required to be grounded with the exception of equipment supplied by Class II power supplies, such as LED secondary circuits.

All motion picture projection equipment must be grounded. Equipment supplied by Class 1 circuits shall be grounded unless the equipment operates at less than 50 volts. Equipment supplied by Class 1 power-limited circuits, by Class 2 and Class 3 remote-control and signaling circuits, and by fire alarm circuits must be grounded only when system grounding is required by Part II or Part VIII of Article 250. In other words, if the supply system is grounded, the equipment it supplies has to be grounded. Luminaires (lighting fixtures) are required to be grounded as provided in Part V of Article 410. **See Figure 9-10.**

Permanently mounted electrical equipment and skids are also required to be connected to the branch circuit EGC sized as required by Table 250.122. Motor-operated water pumps, including the submersible type, are required to be grounded by connection to an EGC. When a submersible pump is used in a metal well casing, the well casing must be connected to the pump circuit EGC. **See Figure 9-11.**

FIGURE 9-10 Luminaires are required to be grounded in accordance with Part V of Article 410.

FIGURE 9-11 A submersible well pump casing must be bonded to the EGC of the branch circuit.

Conductor Enclosure and Raceway Grounding Requirements

Section 250.86 provides grounding requirements for metal raceways and enclosures other than those for service conductors. The general requirement in Section 250.86 is that these metal enclosures be grounded by connecting them to an EGC. There are three exceptions that relax this requirement.

The first exception is for metal enclosures and raceways used for conductors that extend existing installations of old knob-and-tube systems and nonmetallic sheathed cable systems installations. The following conditions have to be met to qualify for the exemption from grounding requirements:

1. The circuits do not contain or include an EGC. If they do include an EGC, grounding these raceways and enclosures is required.
2. The runs are in lengths less than 25 feet. For longer runs, the grounding requirement applies.

3. The installation is isolated and free from possible contact with ground, grounded objects, metal lath or other conductive material.
4. The installation is guarded from contact by people. If all of these conditions are met, then grounding is not required for raceways or enclosures that are installed to extend existing knob-and-tube installations or existing nonmetallic sheathed cable installations.

Another exception is for short sections of metal raceway or metal enclosures used for support or to provide protection from physical damage for cable assemblies. An example of a short section of metal not required to be grounded would be a length of conduit used to provide physical protection for UF cable emerging from the ground at a pole. **See Figure 9-12.** The first 8 feet of this UF cable must be protected from physical damage, as required by Section 300.5(D)(1).

Another example is when a metal conduit sleeve is installed for a run of nonmetallic sheathed cable that has to penetrate a masonry wall. The metal sleeve is not required to be grounded. The *NEC* falls short of addressing the term *short sections* relative to a maximum distance. This is typically left up to the judgment of the authority having jurisdiction.

Metal elbows that are isolated from public contact by a minimum of not less than 18 inches when buried or encased in not less than 2 inches of concrete are not required to be grounded. **See Figure 9-13.** See the exceptions following Section 250.86 for the actual *Code* provisions. Section 250.132 requires isolated sections of metal raceway or cable armor to be connected to an EGC if they are required to be grounded in accordance with the provisions in Section 250.134.

Methods of Grounding Equipment (Part VII of Article 250)

As the feeder and branch circuit wiring is installed on the job site, EGCs are being installed as required. Section 215.6 requires EGCs to be provided with feeder conductors if they supply branch circuits that require EGCs. Because most equipment is required to be grounded, as required by Sections 250.110 and 250.112, nearly all feeders and branch circuits have an EGC installed with the circuit conductors.

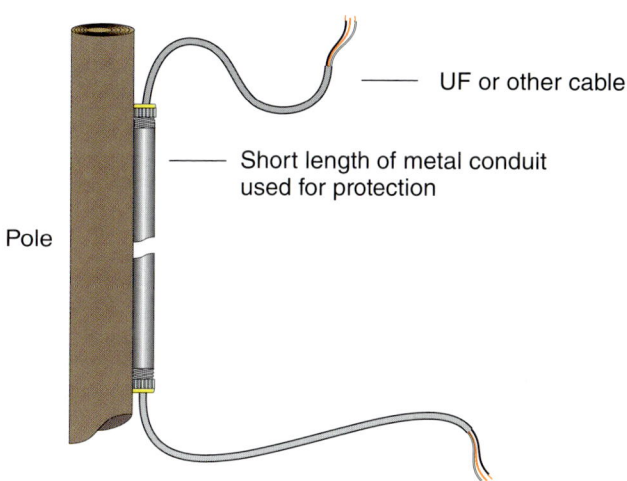

FIGURE 9-12 Short sections of metal raceways are not required to be grounded.

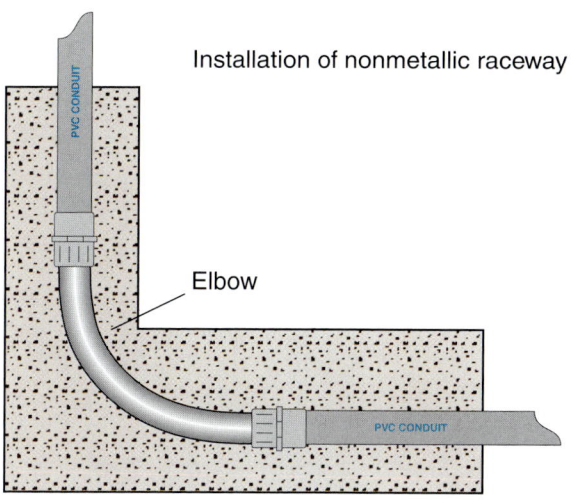

FIGURE 9-13 Metal elbows encased in not less than 2 inches of concrete and isolated from contact by people are not required to be grounded.

The requirements of Section 250.134 indicate that if equipment such as raceways and other enclosures is required to be grounded, then a method in Section 250.134(A) or (B) must be used. This rule states that equipment grounding can be accomplished by connection to any EGC type specified by Section 250.118, which could be a wire, raceway, or other type mentioned in the list. In general, the EGC has to be installed with its associated circuit conductors. It could be a separate conductor contained in the wiring method of the circuit, or it could be the wiring method itself, such as a metal raceway or cable armor assembly (Type AC cable, for example).

Two cases in which EGCs are permitted to be installed separately from the associated circuit conductors are addressed in the exceptions to Section 250.134(B). Exception No. 1 relaxes this requirement for separate EGCs installed to address grounding conductors for existing branch circuit extensions of circuits without an EGC as provided in Section 250.130(C). Exception No. 2 indicates that EGCs of DC circuits are permitted to be run separately from the associated circuit conductors.

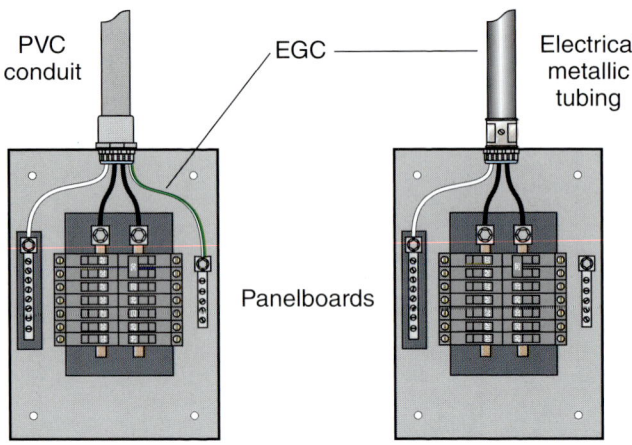

FIGURE 9-14 Section 215.6 generally requires an EGC to be installed with the feeder conductors supplying switchboards, panelboards or other distribution equipment.

Feeders and Branch Circuits

As the power circuits are distributed on the project, feeders typically are run from the service equipment to panelboards or other distribution equipment installed at various locations throughout the facility. From the panelboards or other distribution equipment such as motor control centers or other control panels, branch circuits are installed to the final outlets or connected directly to utilization equipment. As the feeder wiring is connected to the panelboard enclosure, so is an EGC, which complies with the general provisions in Section 215.6. Sometimes the EGC is a wire type, and often it is the conduit, tubing, or qualifying cable armor containing the feeder conductors. **See Figure 9-14.**

If the feeder and branch circuits connected at a panelboard enclosure are all installed using rigid metal conduit (RMC) or intermediate metal conduit (IMC), by *Code* minimum, separate (wire-type) EGCs are not required. In this case, the connection between the feeder EGC and the EGCs of the branch circuits is accomplished through the metal panelboard enclosure. **See Figure 9-15.**

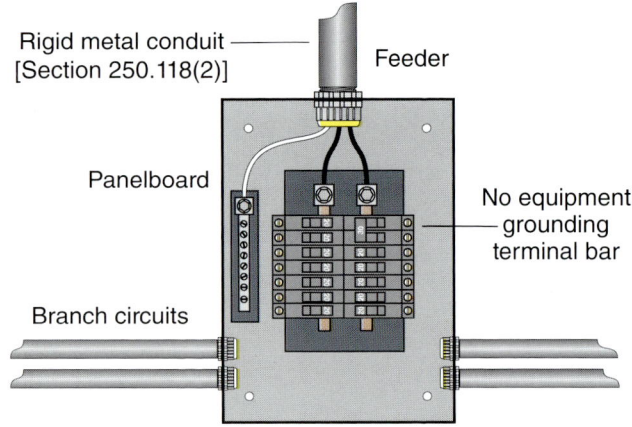

FIGURE 9-15 Rigid metal conduit qualifies as an EGC for the feeder and the branch circuits.

FIGURE 9-16 Equipment grounding terminal bars are provided by panelboard manufacturers.

It is important that suitable locknuts or other fittings be installed properly to maintain the integrity of the EGC connections and path from the feeder to the branch circuit.

Panelboard Equipment Grounding Terminal Bars

Section 408.40 requires an equipment grounding terminal bar in panelboards if nonmetallic raceways or cables are used, or where separate EGCs of the wire type are provided. The equipment grounding terminal bar is typically an accessory feature provided by the panelboard manufacturer. **See Figure 9-16.** The installation and use of an equipment grounding terminal bar for EGCs is driven by panelboards used with nonmetallic raceway or cable or when separate EGCs are provided. This terminal bar may be installed on the panelboard or its enclosure. A terminal bar assembly kit must include instructions for installation and panelboard or enclosure markings. **See Figure 9-17.**

Note that if no wire-type EGCs are present with the branch circuits or feeder, the equipment grounding terminal bar is not required or necessary. However, this is rarely the case.

Isolate Grounded Conductor from Ground

Another important requirement for panelboard wiring, on the load side of the service equipment, is that the grounded (usually the neutral) conductor of the panelboard must not be connected to ground (the enclosure). This restriction also applies to panelboards on the load side of the grounding point for a separately derived system. Sections 250.24(A)(5) and 250.142(B) provide clear direction on the requirement to isolate and separate the neutrals and EGCs. **See Figure 9-18.**

The exceptions to this general restriction are for meter enclosures and the frames of existing ranges or dryers in which the grounded conductor can be used for grounding the equipment.

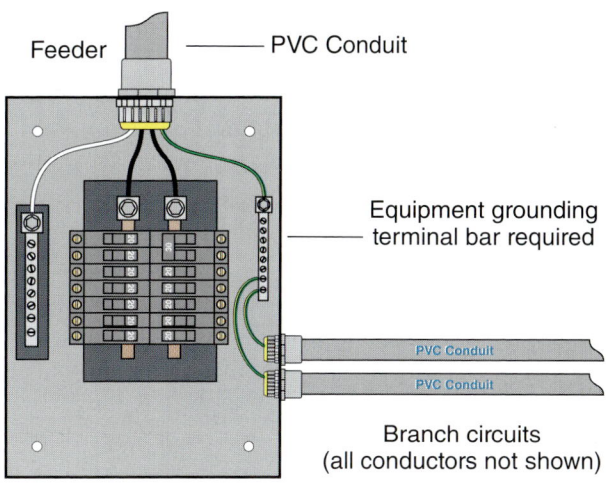

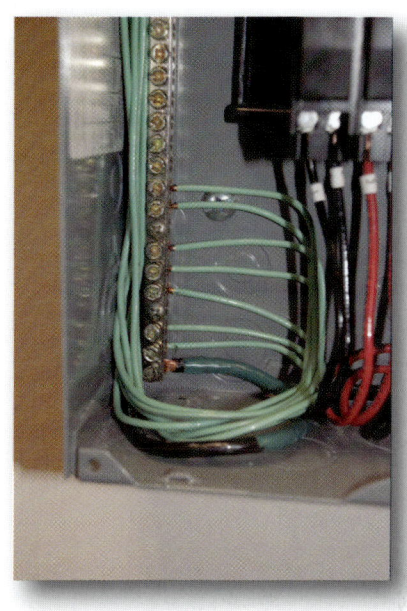

FIGURE 9-17 An equipment grounding terminal bar must be installed in panelboard enclosures where there are wire-type EGCs contained in the circuits entering the enclosure.

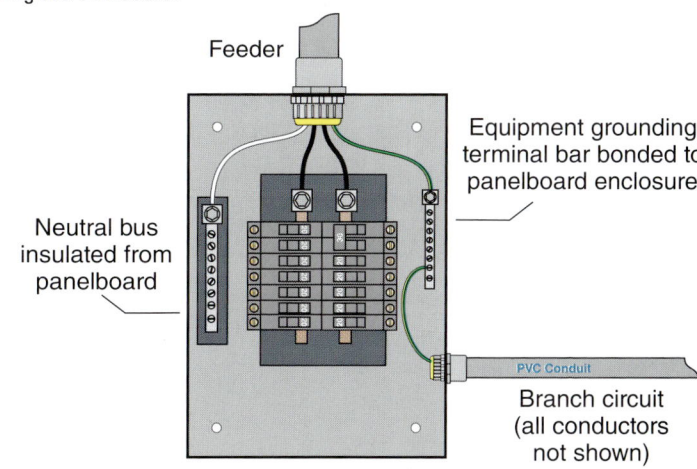

FIGURE 9-18 Grounded conductors (usually neutral conductors) and EGCs in load-side equipment generally must be isolated to prevent multiple paths for normal neutral current.

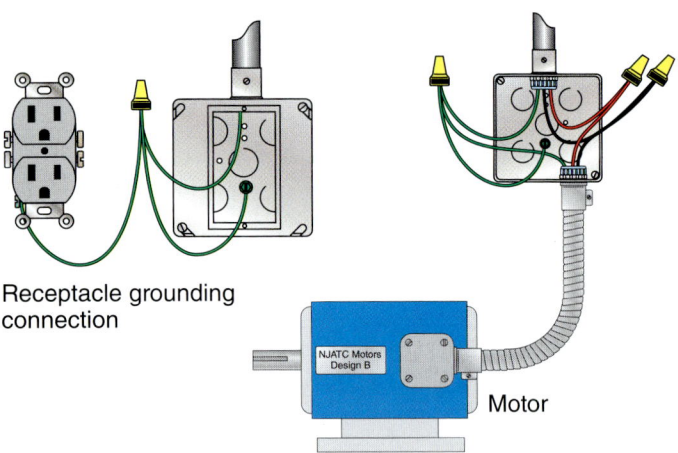

FIGURE 9-19 EGC connections can be through direct wiring connection or through a receptacle connection.

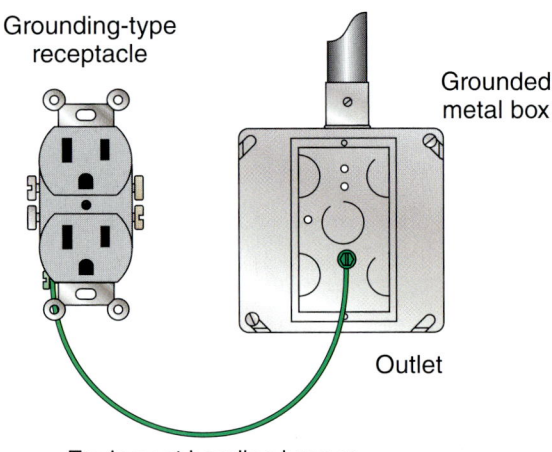

FIGURE 9-20 An equipment bonding jumper is generally required from a grounded metal box to a grounding-type receptacle.

FIGURE 9-21 An equipment bonding jumper is not required for receptacles mounted in surface boxes if metal-to-metal contact is established between the box and receptacle mounting strap.

There are also exceptions to this requirement for DC systems and electrode-type boilers that operate at more than 600 volts. The requirements for grounding this type of equipment are provided in Sections 490.72(E)(1) and 490.74.

Connections of Equipment Grounding Conductors

Once the panelboards, motor control centers, and other distribution equipment and feeders are in place, branch circuit wiring is connected and distributed to outlets and utilization equipment within the building, structure, or elsewhere on the same premises. The wiring from the final overcurrent device to the outlet(s) is the branch circuit, by definition. The outlet is a point of the electrical system at which current is taken to supply utilization equipment, by definition. Receptacles are typically installed at outlets, or utilization equipment can be directly wired to the outlets without use of a receptacle. **See Figure 9-19.** Sections 250.146 and 250.148 provide the base requirements for attachment of EGCs at outlets and other boxes.

Receptacle Grounding Connections

Section 406.4(C) requires that branch circuit wiring methods provide an EGC to which the EGC terminal of receptacles is to be connected. Section 250.146 requires an equipment bonding jumper to be installed from a grounded metal box to the grounding terminal on a grounding-type receptacle. The equipment bonding jumper has to be sized from Table 250.122 based on the rating of the fuse or circuit breaker protecting the branch circuit. **See Figure 9-20.**

Section 250.146(A) through (D) offers alternatives to this general requirement.

Section 250.146(A) deals with surface metal boxes only. If the grounded metal box is mounted on the surface and there is direct metal-to-metal contact between the mounting strap of the device and the grounding metal box, the equipment bonding jumper is not required. **See Figure 9-21.** One condition for this allowance is that at least one of the insulating washers (6-32 device screw retainers) be removed to ensure metal-to-metal contact is established.

The equipment bonding jumper is also not required for surface box and cover combinations in which the receptacle cover has provisions to securely fasten the receptacle with rivets, thread locking screws, or a screw and nut combination and the cover has a flat, nonraised portion that seats firmly against the grounding metal box. **See Figure 9-22.** These provisions apply only to surface mounted boxes.

Another alternative to installing an equipment bonding jumper from the receptacle to the grounded metal box is when self-grounding receptacles are installed. These types of receptacles have a spring tension device that maintains an effective connection between the 6-32 device mounting screw and the grounded metal box. **See Figure 9-23.**

Listed floor boxes that provide satisfactory grounding continuity between the grounding-type receptacle and the grounded metal portion of the assembly do not require an equipment bonding jumper, as provided in Section 250.146(C). **See Figure 9-24.**

Some designs specify isolated grounding-type receptacles (sometimes called a quiet ground) for the reduction of electrical noise (electro-magnetic interference) on the grounding circuit. Section 250.146(D) requires an insulated EGC be connected to a receptacle that is designed specifically to isolate the grounding terminal and mounting strap from the grounding metal box.

These receptacles are referred to as *isolated grounding-types*. **See Figure 9-25.**

FIGURE 9-22 Surface covers with a flat, nonraised portion making metal-to-metal contact on surface boxes relieve the equipment bonding jumper requirement.

FIGURE 9-23 Equipment bonding jumpers are not required when self-grounding receptacles are installed. (Drywall not shown.)

Courtesy of Thomas and Betts

FIGURE 9-24 Equipment bonding jumpers are not required when listed floor boxes are installed.

FIGURE 9-25 The mounting strap for isolated grounding-type receptacles is isolated from the grounding connection to the receptacle. (Plaster ring and drywall not shown.)

200 APPLIED GROUNDING AND BONDING

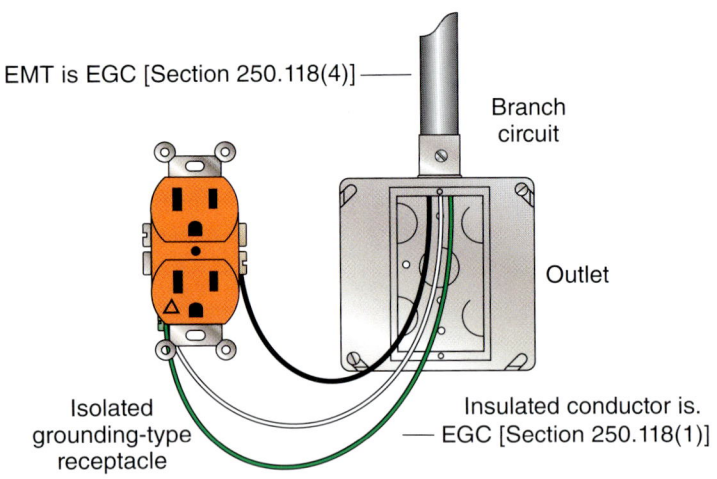

FIGURE 9-26 Installation of isolated grounding-type receptacles requires two EGCs.

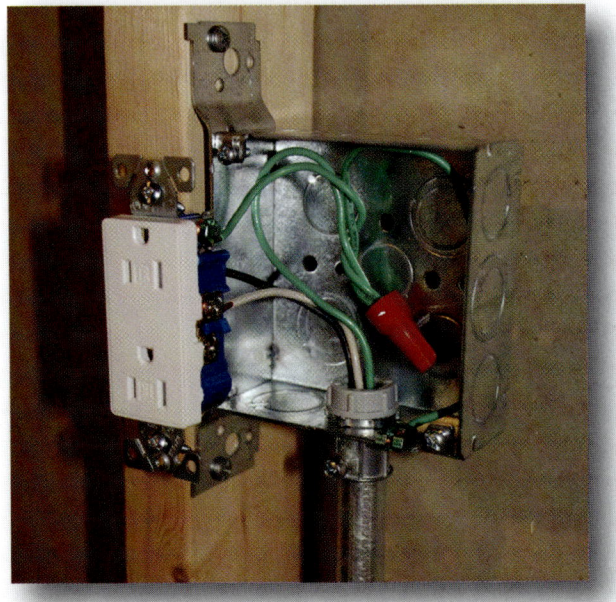

FIGURE 9-27 An equipment bonding jumper can be fastened to a grounded metal box using a grounding screw (top) or a grounding clip (bottom). (Plaster rings and drywall not shown.)

The requirements in Sections 406.3(D), 408.40 Exception, and 250.146(D) must all be followed when installing isolated grounding-type receptacles. **See Figure 9-26.** In these types of installations, the insulated EGC is run with the circuit conductors but does not connect to the grounding metal box where the receptacle is installed. Another EGC or the metal raceway (often referred to as the dirty ground) is connected to the box to accomplish the grounding required for the metal box and plate.

The isolated (insulated) EGC used with isolated grounding (IG) receptacles is permitted to pass through boxes, wireways, panelboards, and other enclosures so as to terminate at the grounding point for the applicable service or separately derived system. Installing an IG circuit and receptacle does not relieve the requirement for grounding metal boxes and or metallic portions of the branch circuit supplying the IG receptacle.

Equipment Grounding Conductor Continuity

When EGCs are spliced within a box or connected to equipment such as devices that are secured to the box, any EGCs associated with the branch circuit conductors are required to be connected to the box or within the box with suitable devices such as a grounding screw, listed grounding clip, or other equipment listed for accomplishing this grounding connection. **See Figure 9-27.**

When EGCs are spliced together in a box, the connections or splices have to meet the requirements in Section 110.14(B), except insulation of the splice is not required. This means that standard wire nuts can be used to splice EGCs together in outlet boxes. Wire nuts that are specifically listed as grounding and bonding equipment can also be used for this purpose. See the acceptable methods of connecting grounding and bonding conductors in Section 250.8.

Often the EGCs are spliced together in the box. An equipment bonding jumper is connected to the box and another equipment bonding jumper is connected to the grounding-type receptacle. **See Figure 9-28.** This method of connecting EGCs and equipment bonding jumpers at outlet boxes results in compliance with Section 250.148(C).

The arrangement and connections of EGCs in outlet boxes have to be in such a fashion that the removal of a device such as a receptacle, switch, or luminaire does not interfere with the grounding connection or continuity as required by Section 250.148(B). A common method to accomplish this is to splice all EGCs together and install an equipment bonding jumper to the box using a grounding screw or grounding clip. **See Figure 9-29.** One method for connecting EGCs in larger junction and pull boxes is to install terminal lugs or equipment grounding terminal bar for this purpose. Be sure to remove any coating such as paint to ensure good metal-to-metal contact is established as required in Section 250.12.

When EGCs are installed for branch circuits and nonmetallic boxes are used, the EGCs have to be spliced together and provisions have to be made to connect the EGC to any device or equipment that is installed in or attached to the box that requires grounding. This type of wiring at outlets is common in residential installations where nonmetallic-sheathed cable and nonmetallic boxes are installed for the branch circuits.

Grounding Snap Switches

Section 404.9(B) provides specific grounding requirements for snap switches. The grounding rules for switches are similar to those for receptacles because both of these types of equipment (devices) have a mounting strap or yoke. For snap switches, an EGC must be connected to the switch mounting strap so that the grounding is extended to a metal faceplate even if a nonmetallic faceplate is initially installed.

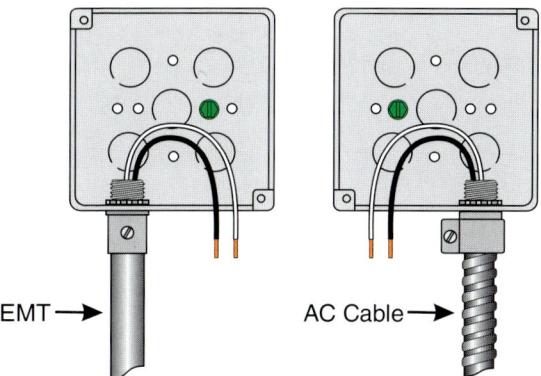

ECG CONTINUITY RELIES ON PROPER INSTALLATION OF CONNECTORS AND LOCKNUTS.

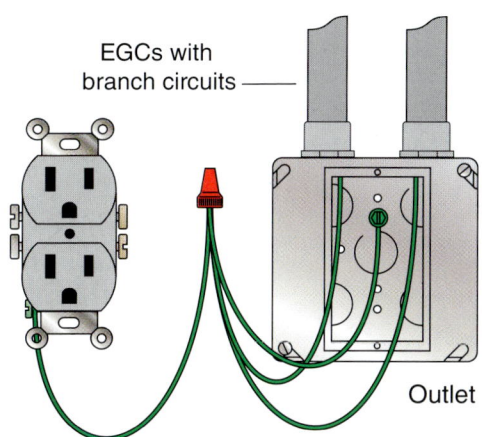

FIGURE 9-28 Continuity of EGCs is required at outlet boxes.

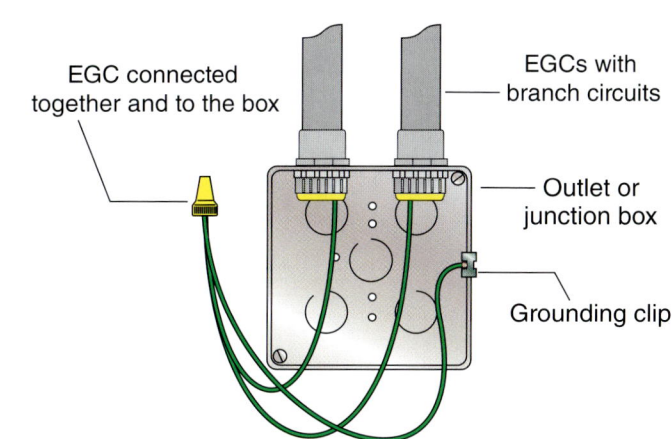

FIGURE 9-29 Continuity of EGCs is required at outlet boxes and an equipment bonding jumper is connected to the metal box.

See Figure 9-30. The mounting yoke of a snap switch can serve as an effective ground-fault current path under the following conditions:

1. The switch is mounted with metal screws to a metal box or metal cover that is connected to an EGC or to a nonmetallic box with integral means for connecting to an EGC.
2. An EGC or equipment bonding jumper is connected to an equipment grounding termination of the snap switch.

There are three exceptions that relax the snap switch grounding requirements as follows:

1. If the wiring method does not include an EGC or if no means in the enclosure exists to connect to one, then an EGC is not required for replacement snap switches. Under this condition, the faceplate and its mounting screws, if within 8 feet vertically or 5 feet horizontally from ground or exposed grounded parts, must be nonmetallic unless either the snap switch strap is nonmetallic or the circuit has GFCI protection.
2. Entirely nonmetallic (accessible parts) listed kits or assemblies with "non standard" faceplate mounting means do not require an EGC.
3. A snap switch with integral nonmetallic enclosure for NM cable that complies with Section 300.15(E) does not require an EGC.[1]

Identification of Wiring Device Terminals

Wiring devices such as receptacles and switches are generally required to provide a terminal for connecting an EGC to the device. This grounding terminal must be identified by one of the following methods:

1. A green screw with a hexagon head that is not easily removed from the device.
2. A green hexagonal nut that is not easily removed.
3. A green pressure wire connector.
4. If the connection point for the EGC to the device is not visible, then the EGC entrance hole has to be marked with the words *green* or *ground,* the letters G or GR, a grounding symbol, or otherwise identified by a distinctive green color. The installer does not have to provide this identification on devices because it is provided by the manufacturer.

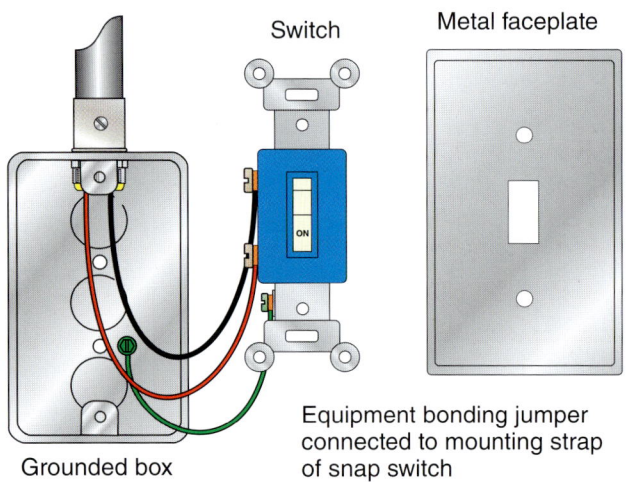

FIGURE 9-30 Grounding of snap switches must be accomplished using one of the methods provided in 404.9(B).

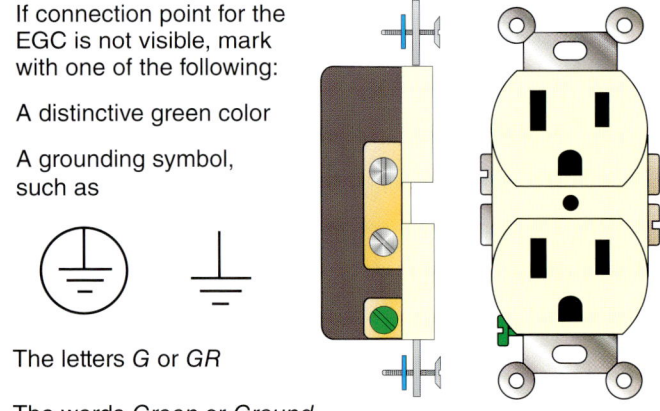

FIGURE 9-31 The grounding symbol is one way to identify the terminal for attaching EGCs to receptacles.

Informational Note Figure 406.10(B)(4) provides an example of a grounding symbol that might be used to identify a grounding terminal connection point.[2] **See Figure 9-31.**

Continuity between Attachment Plugs and Receptacles

EGC continuity is established through receptacles and attachment plugs. These are referred to in the *NEC* as "separable connections." Section 250.124 requires separable connections to provide for a "first-make, last-break" EGC connection. The first-make, last-break connection means is not a requirement for equipment connections that afford an interlock that does not allow disconnecting or connecting in the energized state. Standard listed receptacles and attachment plugs are manufactured to provide first-make, last-break of the grounding conductor connections of the devices. Automatic switches or cutouts are not permitted in the EGC of the premises' wiring unless they disconnect all power sources when operated.

Equipment Grounding Conductor Connections

EGC connections at services must be made in accordance with Section 250.130(A) or (B). EGC connections for separately derived systems are made in accordance with Section 250.30. Section 250.130(A) addresses premises wiring that is supplied by grounded systems. This rule requires that the EGCs be connected to the grounded conductor and the grounding electrode conductor at the service. **See Figure 9-32.** When the wiring system is supplied by a service that is ungrounded, the connection of the EGCs has to be made to the enclosure and grounding electrode conductor at the service. **See Figure 9-33.**

Receptacle Replacements

Grounding requirements are provided in the *Code* for situations requiring receptacle replacements. It is not uncommon to have to replace receptacles over time because of wear and tear. Many older installations were installed without EGCs. Many include an EGC. This segment reviews some of the options recognized in the *NEC* for receptacle replacements either by equipment grounding or other methods of ensuring safety.

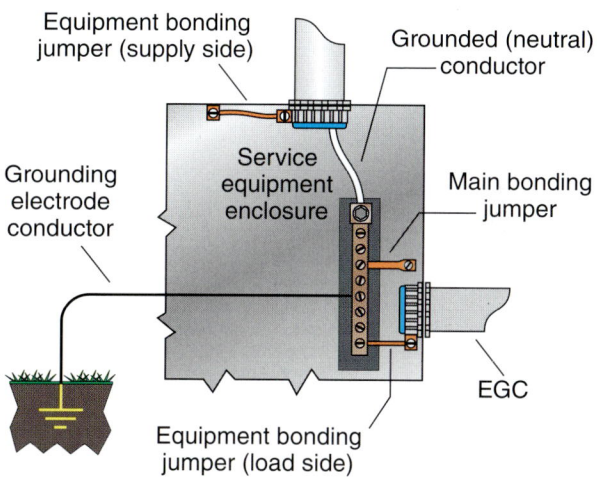

FIGURE 9-32 EGC connections are required to be made to the grounded conductor at service equipment supplied by a grounded system.

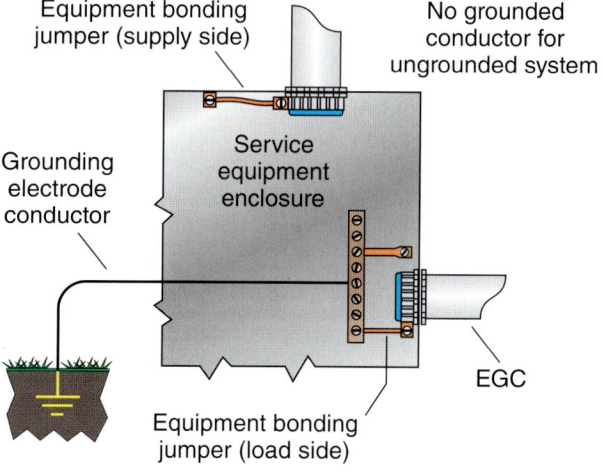

FIGURE 9-33 EGC connections are required to be made to the enclosure at service equipment supplied by an ungrounded system.

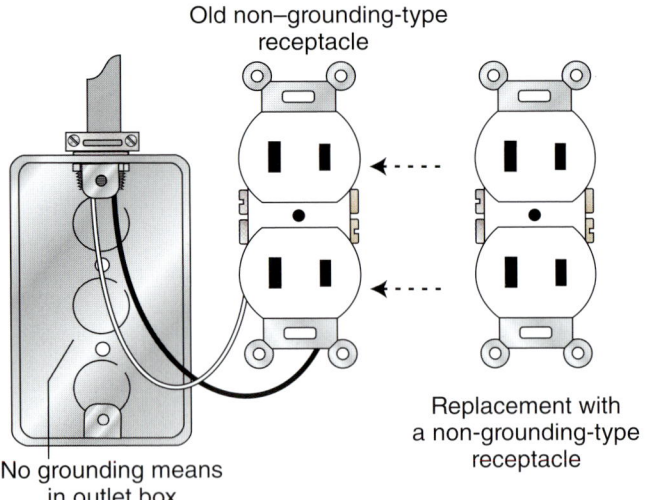

FIGURE 9-34 A non–grounding-type receptacle can replace another non–grounding-type receptacle when no equipment grounding means exist at the outlet.

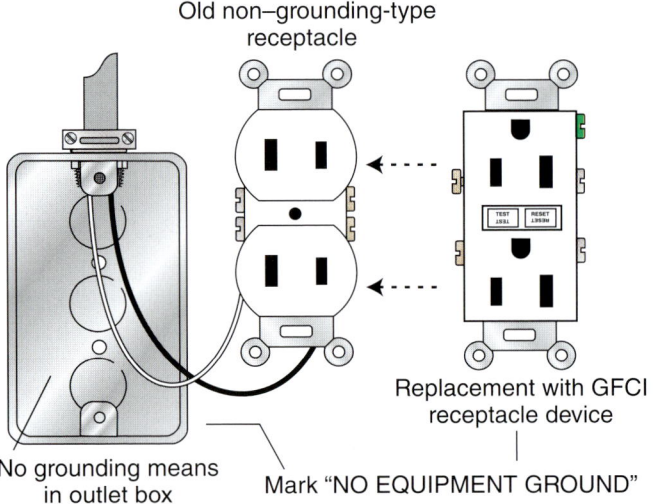

FIGURE 9-35 A GFCI receptacle is permitted to replace a non–grounding-type receptacle.

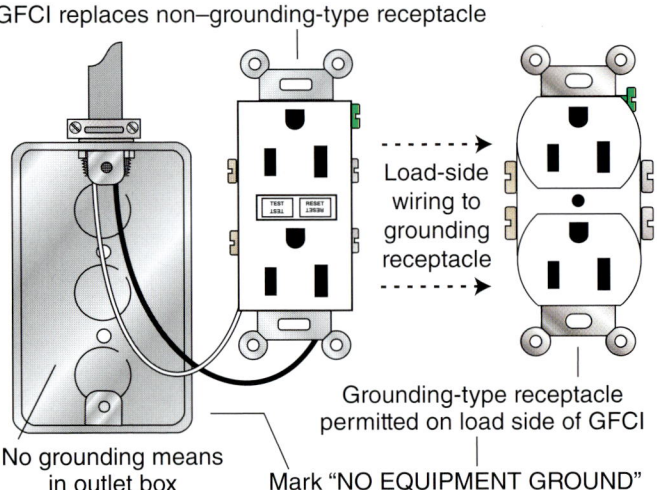

FIGURE 9-36 Grounding-type receptacles are permitted as replacements on the load side of a GFCI receptacle device if no EGC is installed at the downstream outlets and marked as such.

Replacing Grounding-Type Receptacles

Section 406.4(D)(1) indicates that grounding-type receptacles are required to be replaced only with grounding-type receptacles. Non–grounding-types are not permitted as replacements when the outlet provides an EGC.

Replacing Non–Grounding-Type Receptacles

Section 406.4(D)(2) addresses replacements of non–grounding-type receptacles and provides three alternatives. First, a non–grounding-type receptacle can be replaced with another non–grounding-type receptacle. **See Figure 9-34.** Non–grounding-type receptacles are still available for this use.

The second alternative allows a ground-fault circuit-interrupter (GFCI) receptacle device to replace a non–grounding-type receptacle. If this alternative is chosen, the GFCI device has to be marked "No Equipment Ground." **See Figure 9-35.** Using this alternative, an EGC is not permitted to be installed from the GFCI receptacle device to any outlet downstream supplied by the GFCI replacement.

The third alternative allows grounding-type receptacles to be installed as replacements for non–grounding-types on the load side of the GFCI replacement in the circuit, provided each grounding-type receptacle on the load side of the GFCI device is marked "No Equipment Ground." An EGC is not permitted to be connected between the GFCI replacement and the grounding-type receptacles installed on the load side of the GFCI downstream. **See Figure 9-36.**

Section 250.130(C) includes some criteria that can be applied for replacements using grounding-type receptacles if an EGC does not exist at the outlet. An EGC is permitted to be installed separately from the existing branch circuit wiring already in place within the building walls, ceilings, or other inaccessible locations.

See Figure 9-37. If a separate, wire-type EGC is used for any reason, such as the need for an IG receptacle, the EGC of a grounding-type receptacle or a branch circuit extension shall be permitted to be connected to any of the following:

- Any accessible point on the grounding electrode system as described in Section 250.50
- Any accessible point on the grounding electrode conductor
- The equipment grounding terminal bar within the enclosure where the branch circuit for the receptacle or branch circuit originates
- For grounded systems, the grounded service conductor within the service equipment enclosure
- For ungrounded systems, the grounding terminal bar within the service equipment enclosure[3]

Grounding Appliances Using the Grounded Conductor

The frames of wall-mounted ovens, counter-mounted cooking units, ranges, dryers, and associated outlet or junction boxes are required to be connected to an EGC and must include an insulated grounded circuit conductor if it is needed for the load, according to Sections 250.140 and 250.142. **See Figure 9-38.**

The result is a requirement for a four-wire device being installed at the outlets for this type of equipment, specifically at ranges and clothes dryers. The *NEC* relaxes this requirement (by exception) and allows the grounded conductor to be used for grounding in existing three-wire circuits supplying this type of equipment, but under the following conditions:

1. The supply circuit is 120/240-volt, single-phase, 3-wire; or 208Y/120-volt derived from a 3-phase, 4-wire, wye-connected system.

Grounding-type receptacles or branch circuit extensions may be connected to any of the following:

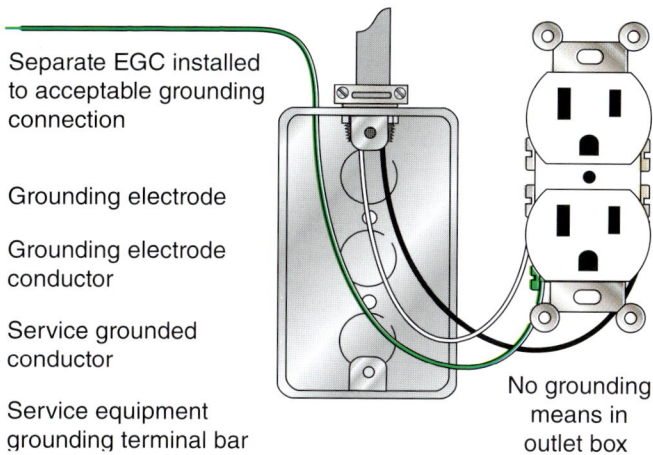

Separate EGC installed to acceptable grounding connection

Grounding electrode

Grounding electrode conductor

Service grounded conductor

Service equipment grounding terminal bar

No grounding means in outlet box

FIGURE 9-37 Installation of a separate EGC is permitted under restrictive conditions.

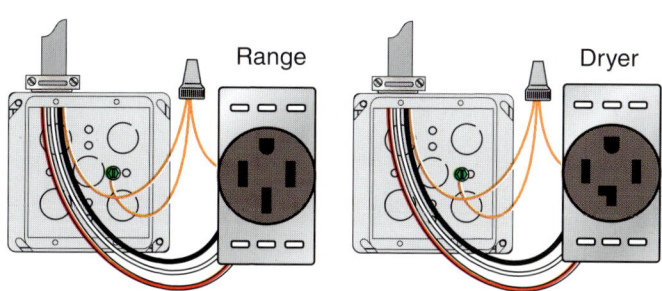

Grounding arrangement for new range and dryer circuits

An EGC and an insulated neutral conductor are required (Section 250.140)

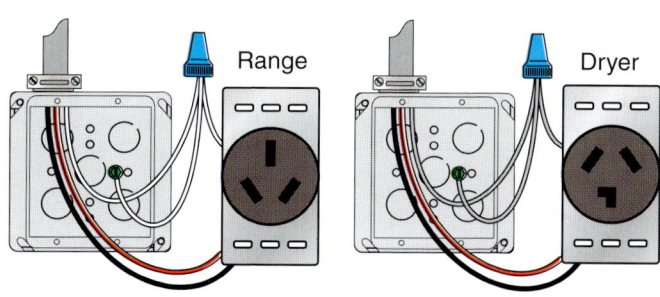

Grounding arrangement for existing range and dryer circuits

The neutral conductor is permitted for grounding the box and the receptacle in accordance with the Section 250.140 exception.

FIGURE 9-38 Four-wire branch circuits are required for wall-mounted ovens, counter-mounted cooking units, ranges, dryers, and associated outlet or junction boxes. These circuits include two ungrounded conductors, an insulated neutral conductor, and an EGC (covered, insulated, or bare).

2. The grounded conductor is not smaller than 10 AWG copper or 8 AWG aluminum.
3. The grounded conductor is insulated, or the grounded conductor is uninsulated and part of a Type SE service-entrance cable and the branch circuit originates at the service equipment.
4. Grounding contacts of receptacles furnished as part of the equipment are bonded to the equipment.[4]

The grounded conductor (neutral) of newly installed branch circuits supplying ranges and clothes dryers is no longer permitted to be used for grounding the non–current-carrying metal parts of the appliances. Branch circuits installed for new appliance installations are required to provide an EGC sized in accordance with Table 250.122 for grounding the non–current-carrying metal parts.

Auxiliary Grounding Electrode Requirements

Auxiliary grounding electrodes are often specified as a design requirement but are not required by the *NEC*. However, if an auxiliary electrode is installed, important requirements apply. Section 250.54 indicates that it is permissible to connect one or more grounding electrodes to the EGCs specified in Section 250.118. An example of an auxiliary grounding electrode installation is typically found at light-pole bases in parking lots. The EGC is installed with the branch circuit supplying the pole luminaires as a requirement, so the equipment is grounded and an effective path for ground-fault current is provided as required. An auxiliary grounding electrode is often installed and connected to the metallic pole in addition to the required EGC. **See Figure 9-39.** This allows a lightning strike to be dissipated into the Earth locally at the light pole base. The local Earth connection of this equipment also establishes an equipotential between the conductive pole and the Earth. Auxiliary grounding electrodes do not have to be bonded to the grounding electrode system for the building or structure as indicated in Section 250.50 or 250.53(C).

The auxiliary electrode is bonded to the grounding electrode system of the supply system by being connected to the EGC of the branch circuit. It is important that installations of auxiliary grounding electrodes for equipment meet the provisions in Section 250.54 and satisfy the provisions in Section 250.4(A)(5) or 250.4(B)(4). The Earth is never permitted as an effective path for ground-fault current.

Grounding Nonelectrical Equipment

The metal parts of some nonelectrical equipment are required to be connected to an EGC. For example, the frames and tracks of electrically operated cranes and hoists must be grounded by connection to an EGC of the circuit supplying them. The frames of nonelectrically driven elevator cars have to be grounded, and hand-operated metal shifting ropes or cables of electric elevators must be grounded.

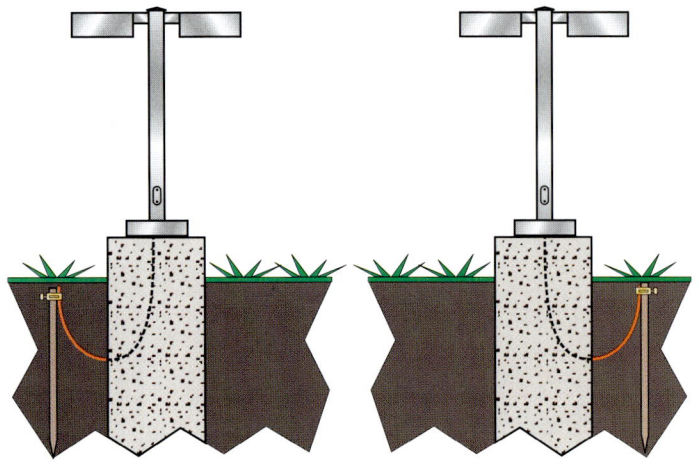

FIGURE 9-39 Auxiliary electrodes supplement the required EGC of the circuit supplying equipment.

Grounding of nonelectrical equipment provides a higher level of safety because it reduces the possibility of these metal parts remaining energized, resulting in a shock hazard, which could occur if the parts are not grounded.

Equipment Grounded by Secure Metal Supports

There are installations in which a support frame can provide the equipment grounding required by the *NEC*. For example, a common frame that supports multiple motors can be grounded by a single EGC that grounds all the motors mounted to the common metal frame or rack. **See Figure 9-40.** The EGC, if of the wire type, has to be sized based on the requirements in Table 250.122. Section 250.136 recognizes these types of installations as meeting the minimum requirements for grounding equipment.

While a structural metal building frame may qualify as an electrode, be used to interconnect electrodes, or be bonded, it is important to understand that the structural metal framing of a building or structure is not permitted to be used as the required EGC for AC circuits or equipment.

Use of the Grounded Conductor for Grounding

Up to this point, with the exception of the provisions for grounding services, the objectives on the load-side of the service disconnecting means (grounding point) are to isolate the grounded (neutral) conductor from the EGC and grounded enclosures. Section 250.142 includes a few conditions in which the grounded conductor is permitted to be used for grounding equipment, but these are restrictive conditions. On the supply side or within the enclosure of the service, a grounded (usually the neutral) conductor can be used for grounding non–current-carrying metal parts of equipment, raceways, and other enclosures. The grounded conductor can also be used for grounding equipment at locations on the supply side or within the enclosure of the main disconnecting means for separate buildings, as provided in Section 250.32(B), and at locations on the supply side or within the enclosure of the main disconnecting means or overcurrent devices of a separately derived system, as permitted by Section 250.30(A)(1).

FIGURE 9-40 Equipment grounding can be accomplished by connection to a common metal frame that is grounded.

The restrictions of using grounded conductors for grounding are supported by the rules in Sections 250.24(A)(5) and 250.142(B), which both prohibit the grounded (often a neutral) conductor from being used for grounding on the load side of the grounding point at a service or separately derived system. This is related to the objective of keeping normal load current on the path it is intended to be on and limit normal current from being imposed on the EGC or other conductive paths within the building or structure.[5]

Summary

Equipment grounding can be accomplished by a number of methods addressed in the *NEC*. Parts VI and VII include the general requirements for equipment grounding. This chapter reviewed the requirements for specific fixed equipment that must be grounded and how to make suitable grounding connections at equipment. The important aspects of equipment grounding connections at standard receptacles and switches and special receptacles such as range and dryer receptacles were reviewed. Substitutes for grounding such as isolation, insulation, or guarding were also reviewed. These alternatives to grounding are addressed in Section 250.1(6) and more specifically in Section 250.110. Grounding of equipment can be accomplished by the use of an auxiliary electrode in addition to the required EGC. Separation and isolation is required between the grounded (usually the neutral) conductor of a circuit and the EGC and grounded metal parts on the load side of the grounding point either at the service equipment or the source of a separately derived system.

References

1. NFPA 70 National Electrical Code 2011, Section 404.9(B) Exception (National Fire Protection Association, Quincy, MA 2010), p. 70–268.
2. NFPA 70 National Electrical Code 2011, Section 250.126 (National Fire Protection Association, Quincy, MA 2010), p. 70–124.
3. NFPA 70 National Electrical Code 2011, Section 250.130(C) (National Fire Protection Association, Quincy, MA 2010), p. 70–125.
4. NFPA 70 National Electrical Code 2011, Section 250.140 Exception (National Fire Protection Association, Quincy, MA 2010), p. 70–126.
5. NFPA 70 National Electrical Code 2011, Section 250.142 (National Fire Protection Association, Quincy, MA 2010), p. 70–126.

Review Questions

1. Equipment grounding places equipment at or as close to Earth as possible, thereby minimizing shock hazard possibilities and limiting the voltage to ground during line surge events, lightning events, and unintentional contact with higher-voltage lines.
    a. True
    b. False
2. Lightning protection systems installed according to NFPA 780 *Standard for the Installation of Lightning Protection Systems* provide an increased level of protection for buildings that have them installed.
    a. True
    b. False
3. Generally, if equipment operates at more than _____ volts to ground, it is required to be grounded by connection to an equipment grounding conductor.
    a. 30
    b. 50
    c. 120
    d. 150
4. Equipment is required to be grounded if it is located _____ vertically or _____ horizontally of ground or grounded metal objects that are subject to contact by people.
    a. 8 feet, 5 feet
    b. 10 feet, 6 feet
    c. 12 feet, 7 feet
    d. 15 feet, 8 feet
5. If equipment were installed 10 feet above the ground and were not subject to contact by people, grounding would always be required if the location was not wet or damp.
    a. True
    b. False
6. The *NEC* relaxes the grounding requirement for short sections of metal enclosures that are used for protecting wiring from physical damage, as long as the short sections do not exceed which of the following distances?
    a. 2 feet
    b. 4 feet
    c. 6 feet
    d. No distance is provided in the *NEC* for the term *short section*
7. When a submersible pump is used in a metal well casing, the well casing must be connected to the _____ of the pump circuit.
    a. Grounding electrode conductor
    b. System bonding jumper
    c. Equipment grounding conductor
    d. Grounded conductor
8. An exception to the grounding requirements in Section 250.86 is for short sections of raceway or metal enclosures used for support or to provide some protection from possible physical damage.
    a. True
    b. False
9. Metal elbows are not required to be grounded if they are isolated from public contact by a minimum cover of not less than 18 inches where buried or encased in not less than _____ of concrete.
    a. 2 inches
    b. 6 inches
    c. 12 inches
    d. 24 inches
10. Equipment grounding conductors are generally required to be installed with feeders of electrical systems.
    a. True
    b. False
11. The equipment grounding conductor with feeders and branch circuit conductors can be any of the types provided in which of the following *Code* rules?
    a. Section 250.110
    b. Section 250.112
    c. Section 250.118
    d. Section 250.122
12. If a panelboard is wired using rigid metal conduit for the feeder and branch circuits, and no equipment grounding conductors of the wire type are installed, an equipment grounding terminal bar is required to be installed in the panelboard enclosure.
    a. True
    b. False
13. The grounded (usually the neutral) conductor of the panelboard must not be connected to ground (the enclosure) on the load side of the service disconnecting means or on the load side of the grounding point for a separately derived system.
    a. True
    b. False
14. An equipment bonding jumper connected from a grounded metal outlet box to a grounding-type receptacle is required to be sized based on _____, using the rating of the overcurrent device ahead of the branch circuit.
    a. Table 250.66
    b. Table 1, Chapter 9
    c. Table 250.122
    d. Minimum size of the ungrounded conductors

15. An equipment bonding jumper is not required to be installed from a grounding-type receptacle to a grounded metal box mounted on the surface if which of the following conditions are met?
    a. The box and cover combination has provisions to securely fasten the receptacle to the cover with rivets, thread locking screws, or a screw and nut combination
    b. The cover has a flat, nonraised portion that sits firmly against the grounding metal box
    c. The metal box is grounded by connection to an equipment grounding conductor
    d. All of the above
16. An equipment bonding jumper is required to be installed from a self-grounding-type receptacle to a flush-mounted grounded metal outlet box.
    a. True   b. False
17. Isolated grounding receptacles are often installed for branch circuits in an effort to reduce the _____ on the grounding circuit.
    a. Static
    b. Stray current
    c. Excessive voltage
    d. Electromagnetic interference
18. An isolated grounding-type receptacle is required to have an equipment bonding jumper installed between the grounding terminal of the device and the grounded metal outlet box in which it is installed.
    a. True   b. False
19. When equipment grounding conductors are spliced within a box or connected to equipment such as devices that are secured to the box, any equipment grounding conductors associated with the branch circuit conductors are required to be connected to the box or within the box with which of the following means?
    a. A device such as a grounding screw
    b. A listed grounding clip
    c. Other equipment listed for accomplishing this grounding connection
    d. Any of the above
20. The arrangement and connections of _____ in outlet boxes has to be in such a fashion that the removal of a device such as a receptacle, switch, or luminaire does not interfere with the grounding connection or continuity.
    a. Equipment grounding conductors
    b. Grounded conductors
    c. Ungrounded conductors
    d. Grounding electrode conductors
21. The mounting yoke of a snap switch can serve as an effective ground-fault current path under which of the following conditions?
    a. The switch is mounted with metal screws to a metal box or metal cover that is connected to an equipment grounding conductor or to a nonmetallic box with integral means for connecting to an equipment grounding conductor.
    b. An equipment grounding conductor or equipment bonding jumper is connected to an equipment grounding termination of the snap switch.
    c. Either a or b
    d. Neither a nor b
22. The branch circuits installed for wall-mounted ovens, counter-mounted cooking units, ranges, dryers, and associated outlet or junction boxes are required to be connected to an equipment grounding conductor and have to include an insulated grounded circuit conductor when it is needed based on the load served.
    a. True   b. False
23. Wiring devices such as receptacles and switches are generally required to provide a terminal for connecting an equipment grounding conductor to the device. This grounding terminal can be identified by any one of the following methods *except* _____.
    a. A green screw with a hexagon head that is not easily removed from the device
    b. A green hexagonal nut that is not easily removed
    c. A green pressure wire connector
    d. A green screw that has a round head
24. Standard receptacles and attachment plugs are manufactured to provide first-make, last-break of the grounding conductor connections of the devices.
    a. True   b. False
25. Non–grounding–type receptacles are permitted as replacements when the outlet does not provide an equipment grounding conductor.
    a. True   b. False

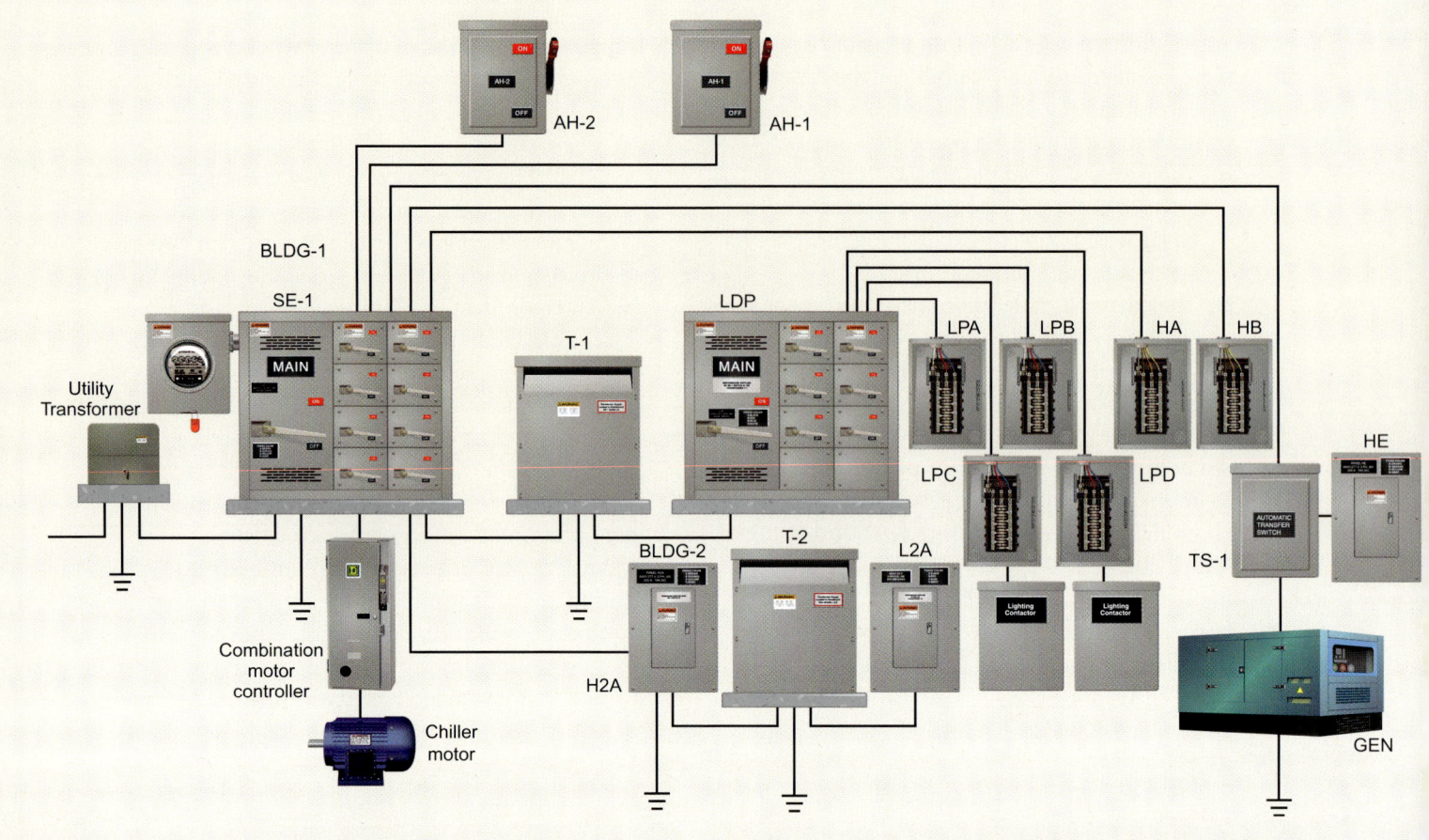

CHAPTER 10

Isolated/Insulated Grounding Circuits and Receptacles

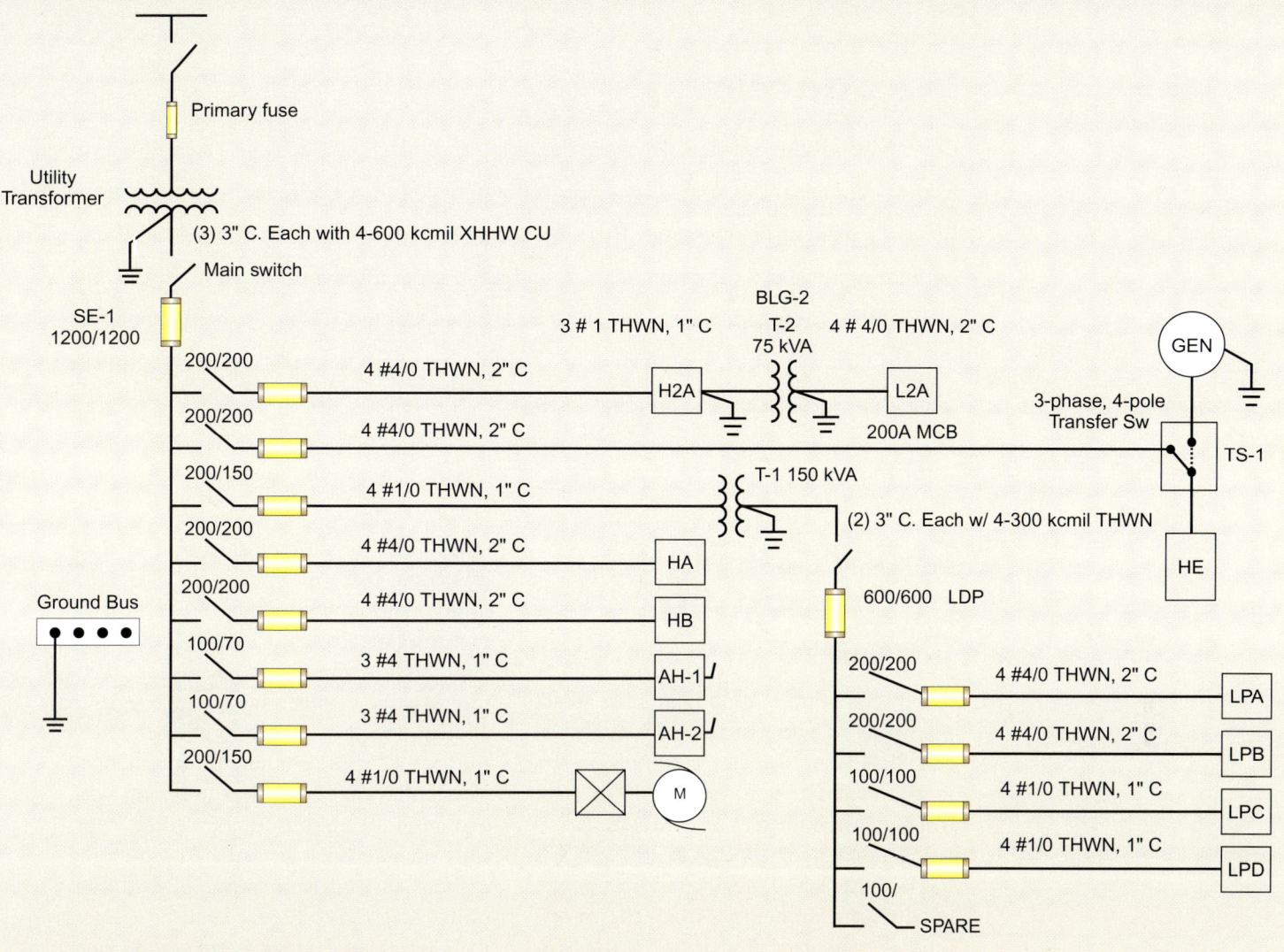

Learning Objectives

- Identify provisions in the *NEC* that address objectionable current in the grounding and bonding system
- Understand sources of electromagnetic interference (EMI) that can affect normal operation of electronic equipment
- Understand various alternatives for reducing objectionable current and noise (EMI) in the grounding circuits and review use of surge protective devices (SPDs)
- Understand installation requirements for isolated equipment grounding conductors and isolated grounding receptacles
- Determine specific grounding requirements that apply to information technology equipment and rooms and understand the purpose of signal reference structures (grids)

Outline

Electrical Noise in Grounding Circuits

Purpose of Isolated Grounding Circuits and Receptacles

Objectionable Currents in Grounding Paths

Power Quality System Grounding Analysis

Isolated Grounding Circuits

Use of Auxiliary Grounding Electrodes

Grounding and Bonding in Information Technology Centers

Signal Reference Structures (Grids)

Surge Protection

Introduction

Clean power is the key objective when designing and installing electrical systems for electronic equipment. The term *clean power* is not defined in the *Code*; neither are the terms *isolated ground* or *quiet ground*. There are wiring techniques that can be used to achieve optimal performance in equipment grounding circuits for electronic equipment while maintaining compliance with the *NEC*® safety regulations. Wiring isolated grounding circuits and receptacles for information technology equipment must never compromise safety in the grounding and bonding system.

Electrical Noise in Grounding Circuits

Some electronic equipment can react negatively to noise in the grounding circuit. The term *sensitive electronics* refers to equipment that is vulnerable to electromagnetic interference (EMI) or circulating currents, typically at low current levels. Other standards that address power quality issues sometimes use the term *susceptible equipment*. Article 647 of the *NEC* addresses wiring methods and systems for sensitive electronic equipment. A common objective in wiring systems for computer equipment is minimizing "ground loops," or circulating currents, and their effects on electronic equipment. The minimum requirements of the *NEC* must be met for safe electrical installations. Some alternative equipment grounding techniques can be applied in system designs to address concerns related to normal operation of electronic equipment. Installations of circuits for electronic equipment such as computer circuits in IT centers typically include isolated/insulated grounding circuits derived from power distribution units (PDUs). **See Figure 10-1.**

Objectionable current and *ground loops* are not defined in the *Code,* but in the IT world, these terms often refer to circulating currents through grounding paths. **See Figure 10-2.** The current in these paths is thought to be moving in circular fashion or over multiple paths while returning to the source. This circular movement of current is more commonly called a ground loop.

Ground loops, or low-level circulating current, are common when computer equipment is connected to multiple circuits supplied from different power sources and then those power sources are interconnected with shielded communication cables. Multiple equipment grounding points or paths can also cause circulating current in the frames of electronic equipment. Because electronic circuits in these types of equipment operate at low-voltage levels, typically less than 5 volts, they are more affected by electrical noise from power circuits that are installed in close proximity. Differences in potential can also cause data errors or losses in electronic equipment.

I-Stock Photo Courtesy of NECA

FIGURE 10-1 IT rooms typically use isolated grounding receptacles and circuits supplied from power distribution units (PDUs).

As potential differences attempt to equalize over multiple grounding paths, circulating currents are present.

Practical Solutions

One solution is to intentionally reduce or eliminate potential differences by isolating the grounding circuit supplying the equipment and connecting the frames of all equipment to a common signal reference grid structure. **See Figure 10-3.**

Another method for solving this problem is to supply all equipment from the same power supply or power distribution unit (PDU), often the case in IT room designs. **See Figure 10-4.** The equipment grounding means of the branch circuit wiring will help keep the ground potential the same.

This solution may be impractical when the computer is far from peripheral equipment and is supplied by a different system. Some other solutions to the problem of circulating currents are:

1. Single point grounding and use of a single power supply system
2. Modems, which are normally used as interfaces with telephone circuits
3. Fiber optic transmission over completely nonconducting paths or optical isolators
4. Interface devices (surge arresters and surge protection devices)

High-Frequency Effects (Resonance)

Another challenge in IT room installations or with electronic equipment is the high-frequency effects in grounding circuits. Avoiding resonance at high frequency is important and more challenging for IT equipment because of the higher frequencies in today's digital signaling circuits. Processor speeds have increased far beyond the megahertz range. Resonance can occur when the length of a conductor and the frequency of alternating current are in tune. This effect is similar to the principle of tuning a radio transmitter and antenna for maximum resonance and radiation.

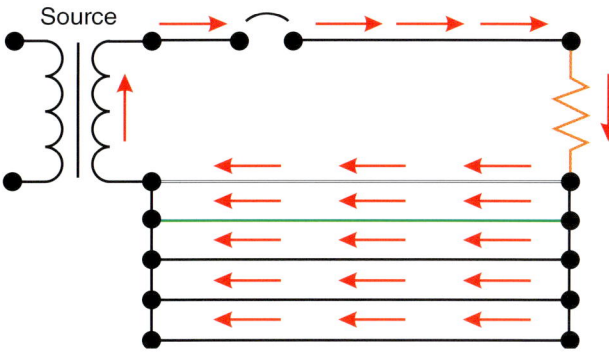

FIGURE 10-2 Circulating currents divide over all available paths to return to the source.

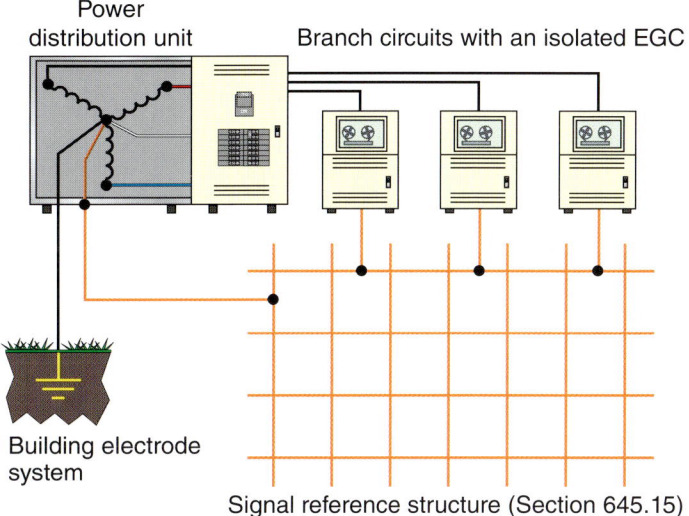

FIGURE 10-3 IT room designs typically include an isolated grounding circuit and a signal reference structure connected to the IT equipment.

Courtesy of Cogburn Brothers, Inc.

FIGURE 10-4 Power distribution units (PDUs) are often installed for supplying branch circuits and equipment in information technology rooms.

At frequencies above and below resonance, the partial resonance increases circuit impedance and thus is less effective as a constant ground potential, which is needed for reference and steady-state operation. Good engineering designs resolve this problem in IT facilities by using signal reference structures (grids) and short bonding straps, usually only a few feet in length, to equalize potential and minimize high-frequency effects in the grounding paths.

Purpose of Isolated Grounding Circuits and Receptacles

When sensitive electronic equipment is installed and used in IT rooms, for example, there are electrical circuit designs that can exceed the minimum requirements of the *NEC*. Isolated grounding circuits and receptacles are installed in an effort to reduce electrical noise that can interfere with data systems and equipment. **See Figure 10-5.**

This type of circuit design can reduce or minimize electromagnetic interference on the equipment grounding circuits by insulating the conductive paths and reducing the grounding circuit to a single insulated path that extends back to the source grounding point, usually at a service or separately derived system. **See Figure 10-6.**

Electrical noise can affect some electronic equipment and lead to data errors and sometimes data loss. In the IT world three characteristics are sought regarding electrical power supplying these systems: reliability, power quality, and minimization of electrical noise in the equipment grounding circuits or other grounding paths. When solutions to electrical noise interference problems are being sought, the minimum requirements of the *NEC* must never be compromised. Safety always comes first.

For a truly isolated EGC, the EGC must remain insulated from ground until it connects to the point of grounding at the applicable service or separately derived system. Section 250.146(D) has permissive text that allows the insulated EGC to pass through one or more panelboards or other enclosures without a grounding connection so as to terminate at the grounding point at the service or separately derived system. These provisions of the *Code* do not require the isolated grounding conductor to return all the way to the source grounding point. Similar provisions are included in Section 408.40 Exception.

Courtesy of Pass and Seymour/Legrand

FIGURE 10-5 Isolated grounding receptacles installed in a branch circuit include an isolated, insulated EGC installed in accordance with Section 250.146(D).

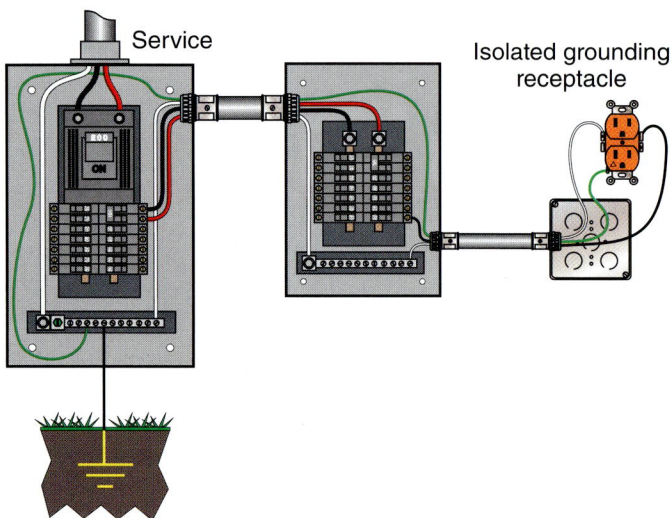

FIGURE 10-6 The isolated EGC is installed from the receptacle to the grounding point at the service. No connection is made to the intervening panelboard or any other grounded enclosure such as boxes, wireways, or conduit bodies.

From a design perspective, the optimal isolated EGC is one that is kept insulated from all other grounded conductive surfaces, which will minimize the possibility of creating a ground loop and circulating current disturbance for connected electronic equipment. Obviously it is better from a design standpoint to keep this EGC isolated (insulated) all the way back to the source or service. As an example, in a remodel project where isolated grounding circuits and receptacles are specified, the point of connection of the EGC might be at an existing panelboard or switchboard located downstream of the point of grounding at the service or separately derived system. The requirements in 250.146(D) are permissive, meaning there is a choice of where the isolated/insulated EGC connects to ground. The important factor is that isolated grounding circuits provide an effective path for ground-fault current in each completed installation.

The EGC must be an effective ground-fault current path, even if it is an isolated/insulated EGC. When isolated/insulated EGCs are installed, two separate EGC paths are necessary for the branch circuit. **See Figure 10-7.** One will serve as the required EGC of the wiring method and enclosures; the other will be the additional isolated/insulated EGC, which terminates directly on the isolated grounding receptacle, without a connection to the grounded metal outlet box. Section 250.118 provides a list of wiring methods that qualify as EGCs.

Objectionable Currents in Grounding Paths

The *NEC* addresses objectionable current in Section 250.6 and provides some alternatives to reduce objectionable current. Objectionable currents through EGCs and other grounding paths can contribute to the overall electrical noise of grounding circuit conductors and other grounding paths.

Objectionable currents in grounding circuits are often a result of improper neutral-to-ground connections on the load side of the service disconnecting means or the load side of the point of grounding for a separately derived system. **See Figure 10-8.**

Temporary current, such as that present during ground-fault events, is not considered objectionable current when applying the provisions in Section 250.6.

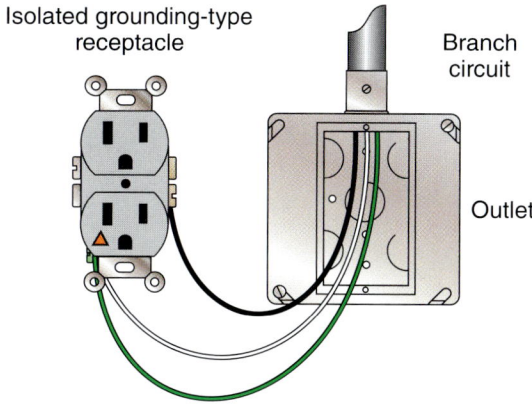

FIGURE 10-7 Two equipment grounding paths are created for isolated grounding circuits, one is the required EGC for the branch circuit, the other is the additional isolated/insulated EGC installed according to Section 250.146(D).

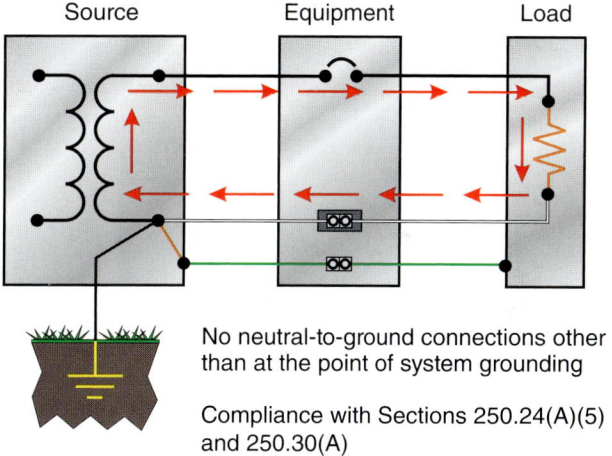

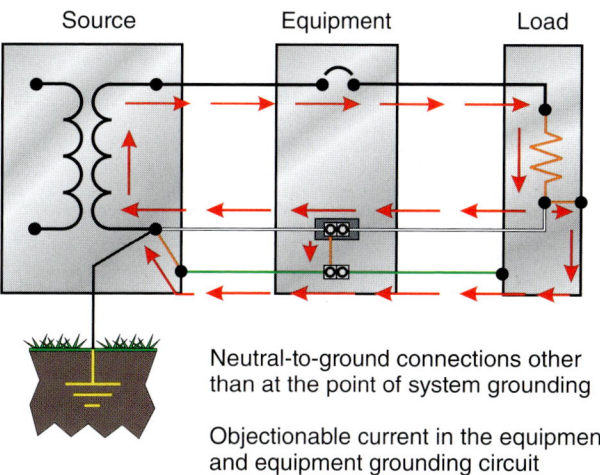

FIGURE 10-8 Improper neutral-to-ground connections create objectionable current in the grounding circuits.

Remedial alternatives for eliminating or minimizing objectionable currents are addressed in Section 250.6(B), where objectionable currents are encountered, but it is essential that the effective ground-fault current paths remain continuous and functional. If multiple grounding connections result in objectionable current, one or more of the following alterations are permitted as long as the effective ground-fault current path is not interrupted.

1. Disconnection of one or more, but not all, grounding connections
2. Change of the location of the grounding connections
3. Interruption of the continuity of the conductor or conductive path causing objectionable current
4. Other remedial action approved by the authority having jurisdiction (AHJ)[1]

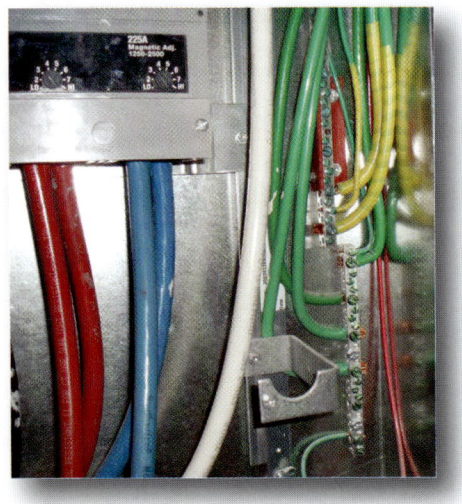

Isolated grounding conductors from IG receptacles are often terminated on an insulated terminal bar installed in a panelboard.

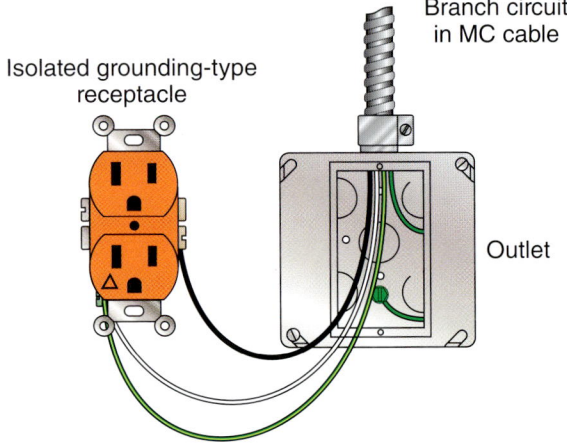

FIGURE 10-9 Identification of wire-type EGCs can be accomplished using two methods.

The alternatives provided in Section 250.6 do not allow for the removal of safety equipment grounding in compliance with *NEC* minimums. Current that introduces noise or data errors in electronic equipment is not considered objectionable current, as addressed in Section 250.6.

Power Quality System Grounding Analysis

Good power quality and effective, *Code*-compliant grounding and bonding are essential for proper electronic equipment operation. When a building power quality analysis is performed, it should always include a thorough analysis of the building grounding and bonding system. There can be no neutral-to-ground connections on the load side of the service grounding point or on the load side of the grounding point for a separately derived system, other than those few permitted exceptions in the *NEC* [Sections 250.24(A)(5) and 250.30(A)]. Neutral-to-ground connections downstream of a main bonding jumper or system bonding jumper in a separately derived system cause current in the EGC circuit(s) and in other conductive paths connected to the source. The other important part of this analysis is to determine that the grounding electrode system meets *NEC* requirements. Before power quality issues can be effectively handled by using filtering equipment, surge arresters, surge protection devices, and other remedies, the grounding and bonding system for the structure must be *Code*-compliant.

Isolated Grounding Circuits

When isolated/insulated EGCs are installed with the branch circuit, there are two EGC paths. The first path is the required EGC for safety; the next path is the desired isolated/insulated EGC for performance. The first path can be metallic conduit, tubing, cable armor, and so forth.

However, the second path must always be an insulated conductor of the wire type. Sometimes a cable assembly such as metal-clad (MC) cable will include two insulated conductors of the wire type for use with these circuits. One will generally be identified as green, and the other as green with one or more yellow stripes in accordance with Section 250.119. **See Figure 10-9.**

Isolated Ground Receptacle Wiring Rules

The *Code* addresses isolated grounding circuits and receptacles that are installed for the reduction of electrical noise. The objective of an isolated grounding circuit is to remove the possibility of circulating currents through the grounding circuit by insulating it from other grounded conductive paths between the source and the outlet connection point. Isolated ground receptacles are manufactured with a grounding terminal that is deliberately isolated from the mounting strap of the device. Isolated grounding-type receptacles must be marked with an orange triangle on the face of the receptacle. **See Figure 10-10.**

There are specific requirements for installing isolated grounding receptacles and branch circuits. Section 250.146(D) includes the permissive text that allows an insulated EGC to pass through panelboards and other enclosures without connecting to them, as long as they are terminated at the point of grounding of the circuit. The point of grounding is either at the service equipment or at a source of a separately derived system, which could be a transformer or PDU in an IT room. This is usually the point where the main bonding jumper or system bonding jumper is installed. The isolated/insulated EGC "clean ground" is installed in addition to the normally required EGC "dirty ground" for the circuit. In the completed wiring installation, two separate EGC paths are present from the outlet to the source grounding point. **See Figure 10-11.**

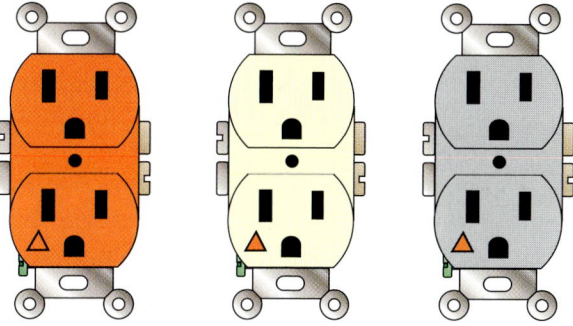

FIGURE 10-10 An orange triangle must appear on the face of isolated grounding-type receptacles. The ground terminal is isolated from the mounting yoke of the receptacle.

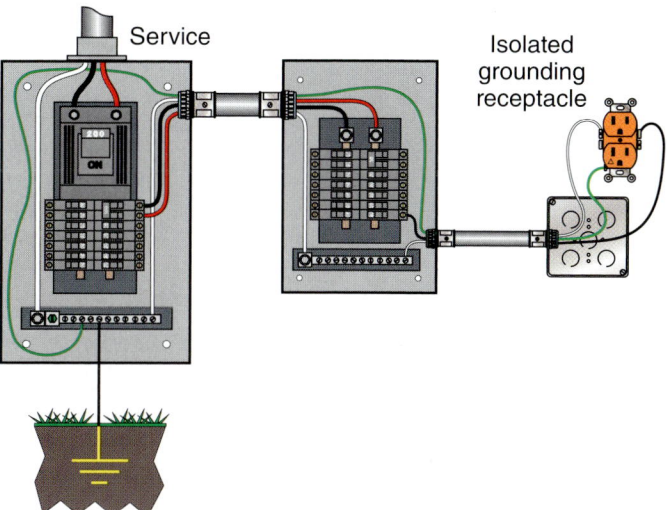

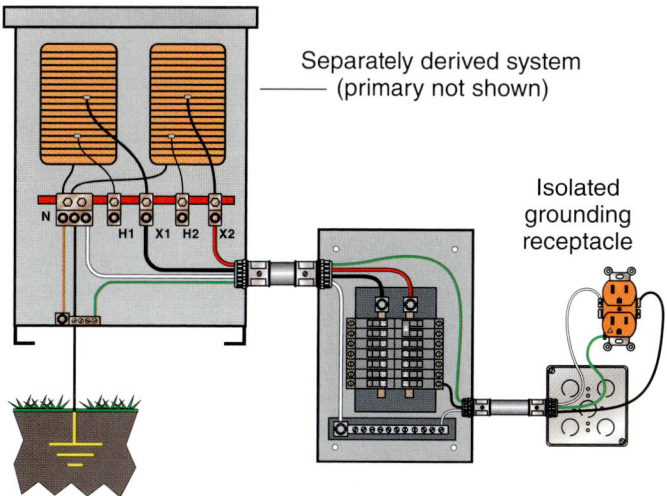

FIGURE 10-11 The isolated/insulated EGC conductor is installed from the receptacle to the grounding point at the service or separately derived system. No connection is made to the intervening panelboard or any other grounded enclosure. (Plaster rings not shown at outlet box.)

The required EGC ("dirty ground" in this case) can be achieved by installing a conduit or other raceway system that qualifies as an EGC in accordance with Section 250.118. Examples are rigid metal conduit (RMC), intermediate metal conduit (IMC), electrical metallic tubing (EMT), and armor-clad cable (Type AC). **See Figure 10-12.**

There are listed armored-clad (Type AC) and metal-clad (Type MC) cable assemblies that provide an outer armor that qualifies as an EGC when used with listed fittings. These cable assemblies may also include a separate insulated EGC (wire type) within the assembly. These cables with two EGC paths are suitable for use in isolated grounding circuits. The metallic path inherent to these wiring methods establishes the required EGC for the circuit. The isolated/insulated EGC connected to the isolated grounding-type receptacle establishes a "clean ground," which is the second EGC path. Both paths terminate together at the grounding point of either the service equipment or the grounding point at the source of a separately derived system. At the receptacle, however, the two EGCs are insulated from each other. **See Figure 10-13.**

Panelboards and Isolated Grounding Circuits

In some instances, multiple isolated grounding circuits are supplied from a single panelboard. Some designs go beyond the minimum *NEC* requirements and specify that an isolated grounding terminal bar be installed in the panelboard for connecting all isolated/insulated EGCs from such branch circuits. In such designs, typically two EGCs are run with the feeder to the panelboard. One EGC serves as the normal EGC and connects directly to the enclosure; the other isolated/insulated EGC connects to an equipment grounding terminal bar that is mounted in but isolated from the panelboard enclosure. Isolated grounding terminal bars built specifically for this purpose are available from various panelboard manufacturers. The size of both EGCs installed with the feeder is based on the overcurrent protective device ahead of the feeder supplying the panelboard; refer to Table 250.122. **See Figure 10-14.**

Sometimes a separately derived system supplies a panelboard. The same installation techniques and methods are applied to this type of installation, except the isolated/insulated EGC terminates at the grounding point for the separately derived system, usually within the transformer enclosure. **See Figure 10-15.** The *Code* does not specifically address this type of design, but it is logical to conclude it can be installed and sized the same as the normal feeder EGC for the circuit.

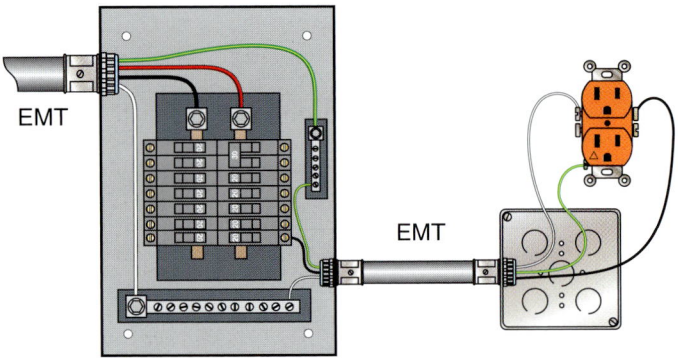

FIGURE 10-12 EMT qualifies as an EGC for branch circuits, as provided in Section 250.118(4). The insulated wire-type EGC connects directly to the isolated grounding-type receptacle. (Plaster ring not shown at outlet box.)

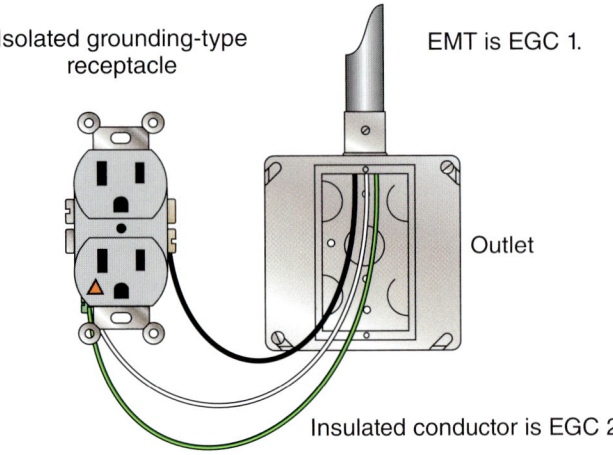

FIGURE 10-13 The isolated grounding conductor is separate from the required EGC at the isolated grounding-type (IG) receptacle.

Isolated Equipment Grounding Circuits for Equipment

The term *clean ground* often refers to the isolated EGC sometimes specified for IT equipment or receptacle outlets by the equipment manufacturer or owner. Section 250.96(B) indicates that when a reduction of electrical noise on grounding circuits is desired, an equipment (typically IT equipment) enclosure is permitted to be supplied by a branch circuit containing an insulated EGC that is isolated from metal raceways by the use of listed nonmetallic raceway fittings (polyvinyl chloride [PVC] fittings). **See Figure 10-16.** Isolation is accomplished by using a nonmetallic fitting between the conductive frame of the equipment and the grounded metal raceway, thus reducing the vulnerability to any electrical noise or circulating currents that may be present on the metallic raceway.

The *NEC* does not specifically address who determines what circumstances justify the use of isolated equipment grounding. This is a design issue and not an *NEC* requirement. Usually, the owner, design engineer, or equipment manufacturer specifies the circuit to have isolated equipment grounds. This *NEC* provision does not mandate the use of a metal raceway to supply the IT equipment. If PVC conduit is used, the insulating connector between the metal raceway and the IT equipment is not required. However, an EGC must be installed inside the PVC conduit to serve as the required equipment grounding means for the IT equipment.

Isolated Grounding Circuits in Health Care Facilities

Isolated/insulated equipment grounding circuits must provide an effective ground-fault current path in addition to providing a clean grounding connection for the equipment. Essentially, the isolated/insulated EGC serves both purposes. The required EGCs must always be in place and effective in addition to any desired isolated/insulated EGC.

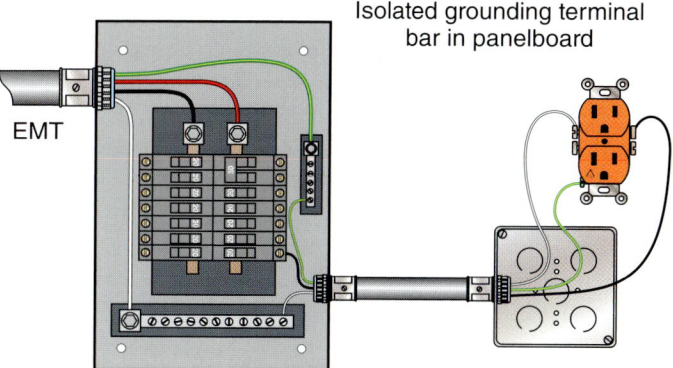

FIGURE 10-14 In this panelboard, the EMT and metallic enclosures are EGCs. The insulated EGC connects to the isolated ground receptacle. (Plaster ring not shown at outlet box.)

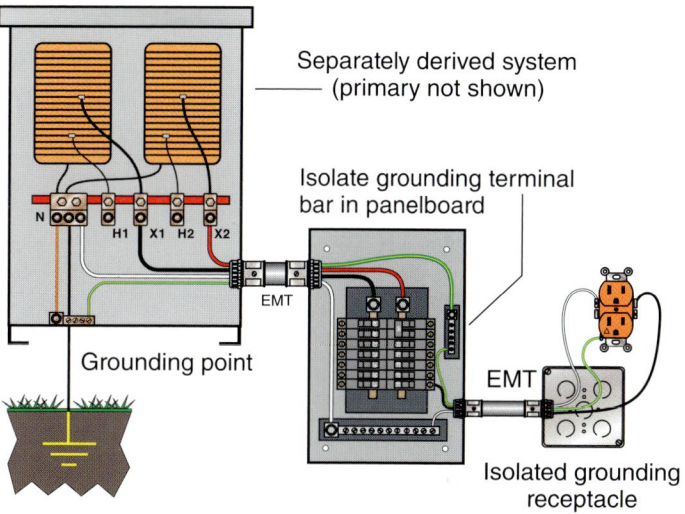

FIGURE 10-15 Some designs go beyond *NEC* requirements and specify that an isolated EGC be installed with the feeder supplying a panelboard from a separately derived system (transformer). (Plaster ring not shown at outlet box.)

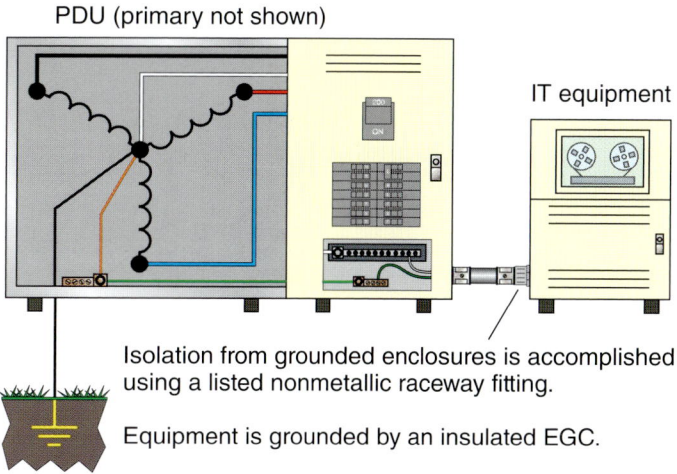

FIGURE 10-16 Some installations of isolated/insulated grounding circuits are provided for equipment that is wired directly to circuits without using an IG receptacle.

When designs call for more specialized equipment grounding means for reducing unwanted electromagnetic interference on the grounding circuit, an additional insulated EGC path is usually installed. In the patient areas of health care facilities, where isolated/insulated equipment grounding circuits were installed with the branch circuits, three EGC paths are typically installed. **See Figure 10-17.**

This is because the branch circuits serving patient care locations must meet the requirements in Sections 517.13(A) and (B) in addition to any isolated/insulated equipment grounding that may be desired over and above the required redundant grounding. The isolated EGC cannot be counted as one of the two grounding paths required by Section 517.13 because this path does not provide a functional benefit of being in parallel with the metal raceway or cable system for the branch circuit. The concept of two EGCs in branch circuits serving patient care locations is to provide redundancy in the equipment grounding circuit. A good way to approach these installations is to always strive to satisfy what is required by the minimum requirements of the *NEC* before applying any desired isolated/insulated equipment grounding circuits. Although the *NEC* no longer permits the installation of isolated grounding circuits and receptacles as restricted by Section 517.16, NFPA 99 *Standard for Health Care Facilities* still addresses installations of isolated grounding circuits and receptacles. NFPA 99 still requires periodic testing of grounding systems, which includes installations of isolated grounding receptacles and circuits in health care facilities.

Use of Auxiliary Grounding Electrodes

When isolated/insulated circuits are installed, there is often a desire to install a separate supplemental connection only to the Earth at the equipment location. This grounding electrode is known as an auxiliary grounding electrode. Installation of auxiliary grounding electrodes does not relieve the requirement for connection of an EGC. It is installed in addition to the required EGC for the branch circuit. **See Figure 10-18.**

When an auxiliary grounding electrode is installed, both the EGC and the grounding electrode conductor to the electrode must be connected to the equipment. It is a violation of the *Code* to eliminate the required EGC, and it creates a safety hazard. Sections 250.4(A)(5) and 250.54 both clearly indicate that the Earth is not permitted to be used as an effective ground-fault current path. Sometimes enhanced grounding electrode systems are installed for IT centers. These grounding electrodes cannot be isolated from the building grounding electrode system; they are in addition to and must be bonded to the building grounding electrode system. Otherwise, unwanted and unsafe differences of potential can result.

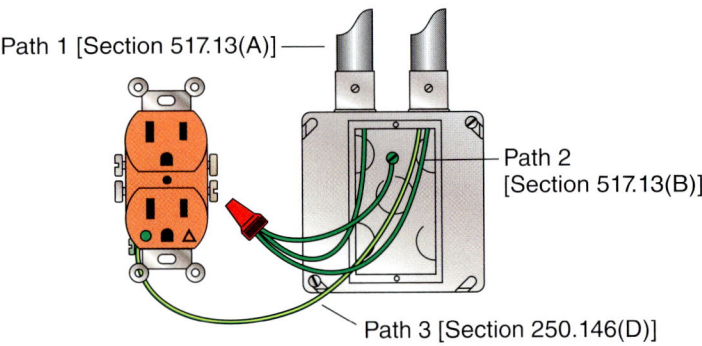

FIGURE 10-17 An isolated/insulated EGC is present in addition to the two EGCs required by Section 517.13.

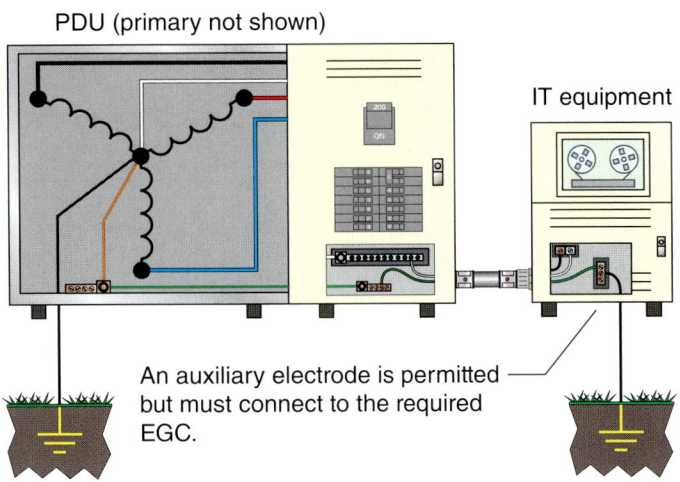

FIGURE 10-18 Auxiliary grounding electrodes must supplement the required EGC of the circuit supplying equipment.

Grounding and Bonding in Information Technology Centers

In large facilities it is common for all IT equipment, servers, data storage, backup drives, data processing, and so forth, to be located in a single room or data center within a facility. Sometimes an entire building is constructed to meet these criteria, depending on the extent of the IT needs for the particular business. Essentially an envelope within the building is being created to house all IT systems and equipment. This room is constructed to meet specific criteria to qualify as an IT room. Article 645 of the *NEC* provides specific requirements that must be met before the rules in Article 645 can be applied to an IT room. In other words, to be able to use the provisions in Article 645 all requirements in Section 645.4 have to be complied with. The following must be provided for an IT room to take advantage of the provisions in Article 645.

1. A disconnecting means according to Section 645.10 has to be provided.
2. A separate HVAC system is provided (dedicated to the room and equipment).
3. Listed communications equipment and IT equipment are installed.
4. The room is accessible to and occupied only by operators and maintainers of the equipment/system.
5. The room is separated from other occupancies in the building by fire-resistant-rated construction.
6. Only equipment and wiring associated with the operation of the IT room are located in the room.[2]

Note: See Section 645.4 for complete information about each of these six items.

Section 645.4 and NFPA 75 *Standard for the Protection of Information Technology Equipment* both provide more detailed information on the construction requirements for an IT room. If the room is not constructed to meet the criteria and items 1 through 6 are not provided, then all the rules in Chapters 1 through 4 of the *NEC* apply. The driving text in Section 645.4 is clear that Article 645 "shall apply provided all of these conditions are met."[3]

Section 645.15 includes specific grounding requirements for equipment in an IT system. The primary requirement is that all non–current-carrying metal parts of such equipment be connected to the EGC of the supply branch circuit or feeder in accordance with Article 250.[4] There is an exemption from the grounding requirement, but only where the IT equipment is double-insulated.

To minimize possible differences in potential in the grounding systems for power circuits supplying IT equipment, it is common for these centers to be equipped with a single PDU or multiple PDUs. The term *power distribution unit* is not defined in the *NEC*, but it is described in Section 645.17. **See Figure 10-19.** The PDUs that are used for IT equipment are permitted to have multiple panelboards within a single cabinet, provided that the PDU is utilization equipment listed for IT application.[5]

FIGURE 10-19 PDUs often include a single transformer and multiple panelboards in a single enclosure.

PDUs are typically built with transformers and panelboards in a single enclosure or assembly. This unit is supplied by a feeder, and branch circuits are routed to the IT equipment from the PDU. The PDUs usually provide a convenient shunt trip feature that affords easy compliance with the disconnecting means rule in Section 645.10. The power systems derived in listed IT equipment (PDUs) supplying IT systems through specially constructed receptacles and cable assemblies are not considered as separately derived for the purposes of applying the grounding requirements for separately derived systems.[6] This means that installing, wiring, and grounding these circuits and systems all must be in accordance with specific instructions provided with the IT equipment. Some of these PDUs require a connection to the building electrode system; others do not. The listed equipment provides the requirements that installers must follow for grounding of such circuits and systems. The isolated/insulated grounding circuits and receptacles are usually supplied as premanufactured "whips" by the supplier of the PDU. The grounding and bonding connections for all such circuits are made within the PDU because that is the power source. Thus, electrical noise (electromagnetic interference) in the grounding circuits supplying the IT equipment is kept to a minimum, because these circuits are relatively short and do not extend throughout the building or structure. The informational note following Section 645.15 indicates that the grounding and bonding requirements provided in the product standards that apply to listed IT equipment ensure that the rules in Article 250 are complied with.

Signal Reference Structures (Grids)

The IT term *signal reference grid* refers to a common conductive structure such as a computer floor, copper interconnected sheet strips, or a copper mesh grid installed under the raised floor. This grid provides an effective equipotential bonding structure to which all equipment can be connected. **See Figure 10-20.**

IT equipment is connected to the signal reference grid to equalize the potential between components. Sometimes the support structures for the raised floor are part of the signal reference structure. **See Figure 10-21.** All equipment in the room, including equipment that is mounted to the wall, should be connected to the grid, as should the EGC in each supply branch circuit.

Courtesy of Rick Maddox, Clark County, NV

FIGURE 10-20 Signal reference structures are typically installed in an IT room before installing the raised floor (platform).

Courtesy of Harger

FIGURE 10-21 Signal reference structures often incorporate the support frame for the raised floor in the IT room.

The *NEC* does not address this subject but, as with other bonding or grounding methods used for IT equipment, such bonding cannot substitute for the EGC required to be installed with the branch circuit conductors supplying the IT equipment. The signal reference grid can be thought of as overlaying the EGCs (sometimes referred to as *safety grounds*) that are required. **See Figure 10-22.**

The signal reference grid serves as a signal reference plane over a broad range of frequencies. Signal reference grids are also commonly referred to as *broadband grounding systems*. The grid structure minimizes potential differences and reduces, eliminates, or controls the conductors connected to computers that tend to resonate at higher frequencies. A grid provides multiple parallel conducting paths between its metal parts. If one path is a high-impedance path because of full or partial resonance, other paths of different lengths will be able to provide a lower-impedance path. A signal reference grid is commonly constructed of continuous sheet copper or aluminum or any number of pure or composite metals with good surface conductivity. Listed products are available for this purpose. **See Figure 10-23.**

Signal reference grids can also be made up of a number of conductive surfaces such as the raised floor framing, the suspended ceiling grid system, or a constructed grid under the raised platform or floor of an IT room. Grids of copper or aluminum strips are sometimes installed under a computer-raised floor. This grid provides a constant potential reference network over a broad range of frequencies from direct current (DC) to higher than 30 MHz, even in the gigahertz ranges today. Typically, the grids are made in mesh configurations using a minimum of 4 AWG copper or aluminum conductors that have been electrically joined at their intersections or by thin copper straps that are about 2 inches wide, also joined at their intersections. **See Figure 10-24.**

These constructed grids are typically placed directly on the subfloor under the IT room raised floor. Cables and conduits under the floor would normally be below the raised floor but above the grid.

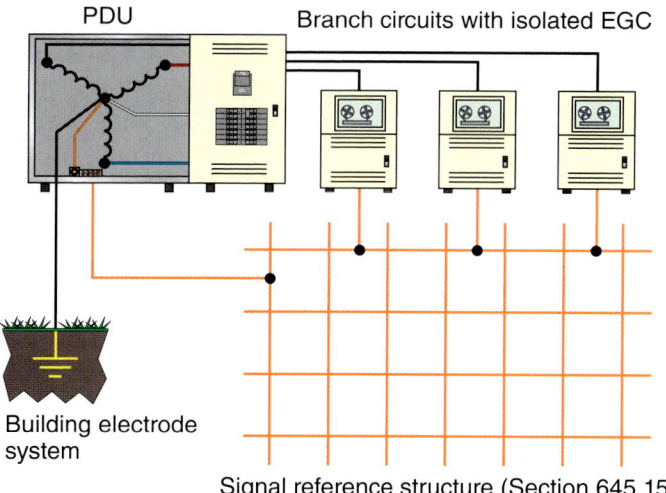

FIGURE 10-22 The signal reference structure in IT rooms must overlay the required EGCs of the branch circuits for IT equipment.

Courtesy of Harger Lightning and Grounding

FIGURE 10-23 Listed products are available for constructing a signal reference grid in an IT center.

Courtesy of Harger Lightning and Grounding

FIGURE 10-24 Signal reference structure can be in the form of overlapping of solid copper straps under the raised platform (floor) of an IT room.

The IT equipment is then connected to the grid by using short braided bonding conductors (usually 2 to 4 feet in length). This common bonding grid is *not* a substitute for the required EGC supplying the IT equipment; it is in addition to it. Section 645.15 requires the signal reference grid to be bonded to the EGC provided in the IT equipment.[7] This is accomplished at each separate IT unit in the room and, typically, at each PDU grounding point.

Surge Protection

Surge protection is often desired for IT equipment installations. These devices can provide a degree of protection against power line surges. Installations of surge protective devices (SPDs) are covered by the *NEC*; however, they are not required by the *NEC*. If surge protection is installed, it has to comply with the applicable provisions in Article 285. **See Figure 10-25.**

Unlike surge arresters rated above 1000 volts, SPDs are required to be listed. Article 280 in the *NEC* includes requirements for surge arresters rated over 1000 volts.

SPDs are not permitted for use on circuits exceeding 1000 volts and are restricted from use on ungrounded systems, impedance grounded systems, or corner grounded systems, unless they are specifically listed for use on those systems. This means they have been evaluated to applicable product safety standards. The product certification or listing also drives the requirements for installers to follow manufacturer's installation instructions, as required in Section 110.3(B). This requirement resolves many of the questions relative to what is required for SPD installation. Article 100 defines four types of surge protection listed to UL 1449 and describes their use. When an SPD is connected and used with a separately derived system, it must be connected on the load side of the first overcurrent protective device supplied by the derived system. This could be a Type 2 or Type 3 SPD because it is located on the load side of a service overcurrent protective device. The supply conductors and grounding conductors for an SPD are not permitted to be smaller than 14 AWG copper or 12 AWG aluminum conductors. The length of conductors for Type 3 SPDs is limited to 30 feet unless further restricted by the manufacturer's installation instructions. Grounding conductor connections to surge protection devices must meet the requirements in Part III of Article 250, and when the grounding conductor is installed in a ferrous metal raceway, it must comply with Section 250.64(E). This section requires the grounding conductor contained in a ferrous metal raceway to be bonded to both ends of the raceway, thus minimizing impedance and choke effects during surge events.

FIGURE 10-25 SPDs provide a level of equipment protection against line surges.

Summary

Installations of circuits for electronic equipment such as computers, servers, and so forth in information technology centers often necessitate installing isolated grounding circuits and receptacles. The minimum requirements for installing isolated grounding circuits and receptacles were covered. The *NEC* includes some alternative equipment grounding techniques that are often applied in system designs to address concerns related to normal operation of electronic equipment. Various common methods are applied to achieve optimal performance in the grounding system while meeting minimum requirements of the *NEC,* simultaneously. These two must to go hand in hand. Article 645 of the *NEC* provides specific rules for wiring information technology rooms, specifically the grounding circuits and systems for such installations. Additional information about constructing IT rooms and protecting IT equipment is contained in NFPA 75 *Standard for the Protection of Information Technology Equipment.*

Courtesy of Bill McGovern, City of Plano, TX

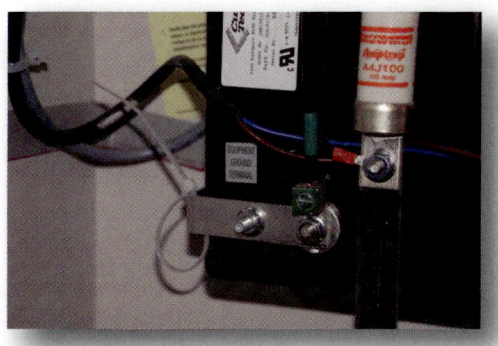

Surge protective devices (SPDs) must be installed according to the manufacturer's installation instructions.

References

1. NFPA 70 National Electrical Code 2011, Section 250.6 (National Fire Protection Association, Quincy, MA 2010), p. 70–103.
2. NFPA 70 National Electrical Code 2011, Section 645.4 (National Fire Protection Association, Quincy, MA 2010), p. 70–560.
3. NFPA 70 National Electrical Code 2011, Section 645.4 (National Fire Protection Association, Quincy, MA 2010), p. 70–560.
4. NFPA 70 National Electrical Code 2011, Section 645.15 (National Fire Protection Association, Quincy, MA 2010), p. 70–562.
5. NFPA 70 National Electrical Code 2011, Section 645.17 (National Fire Protection Association, Quincy, MA 2010), p. 70–563.
6. NFPA 70 National Electrical Code 2011, Section 645.15 (National Fire Protection Association, Quincy, MA 2010), p. 70–562.
7. NFPA 70 National Electrical Code 2011, Section 645.15 (National Fire Protection Association, Quincy, MA 2010), p. 70–562.

Review Questions

1. According to the *NEC*, which of the following is not a recognized method of reducing objectionable current in the grounding system?
   a. Disconnection of one or more, but not all, grounding connections or change of the location of the grounding connections
   b. Disconnection of the EGC of the circuit supplying the equipment
   c. Interruption of the continuity of the conductor or conductive path causing objectionable current
   d. A remedial action approved by the authority having jurisdiction (AHJ)

2. The term *ground loop* is used in the IT industry and is best associated with _____.
   a. A ground ring
   b. A grounding electrode system
   c. Circulating currents in multiple grounding paths
   d. A counterpoise system

3. Resonance can occur when the length of a conductor and the _____ of alternating current are in tune. This effect is similar to the principle of tuning a radio transmitter and antenna for maximum resonance and radiation.
   a. Wattage
   b. Amperage
   c. Voltage
   d. Frequency

4. Ground loops, or low-level circulating current, are commonly present when computer equipment is connected to multiple circuits supplied from different power sources and then those power sources are interconnected with shielded communication cables.
   a. True
   b. False

5. Which of the following are other known solutions to the problem of circulating currents in the grounding system?
   a. Installation of a single point grounding and use of a single power supply system
   b. Use Fiber optic transmission over completely nonconducting paths or optical isolators
   c. Interface devices (surge arresters and transient voltage surge suppressors)
   d. Any of the above

6. Isolated grounding circuits and receptacles are installed in an effort to reduce electrical noise that can interfere with data systems and equipment.
   a. True
   b. False

7. The insulated EGC connected to an isolated grounding-type receptacle is permitted to pass through one or more panelboards or other enclosures without a grounding connection as long as it terminates at the grounding point at the _____.
   a. Applicable service or separately derived system
   b. Ground rod
   c. Building steel
   d. Surge arrester

8. An isolated grounding-type receptacle is required to be identified by which of the following means?
   a. The color orange
   b. An orange triangle on the receptacle
   c. A green dot on the receptacle
   d. An orange faceplate

9. An EGC must be an effective ground-fault current path, even if it is an isolated/insulated EGC installed with the branch circuit.
   a. True
   b. False

10. An isolated/insulated EGC from an isolated grounding-type receptacle is permitted to be connected only to a ground rod for a true isolated grounding connection.
    a. True
    b. False

11. Neutral-to-ground connections downstream of a main bonding jumper in a service or system bonding jumper in a separately derived system can cause current in the EGC circuit(s) and over other common conductive paths to the source. They are in violation of _____.
    a. Section 250.24(A)(5)
    b. Section 250.30(A)
    c. Both a and b
    d. Neither a or b

12. When isolated/insulated EGCs are installed with a branch circuit, there will be _____ EGC path(s).
    a. One
    b. Two
    c. Three
    d. Four

13. An auxiliary grounding electrode is permitted to be the only grounding connection for electronic equipment when noise on the equipment grounding circuit is a problem.
    a. True
    b. False
14. The isolated/insulated EGC installed and connected to an isolated grounding-type receptacle is required to be identified by which of the following methods?
    a. Bare
    b. White
    c. Green or green with one or more yellow stripes
    d. Orange
15. Although the *NEC* no longer permits the installation of isolated grounding circuits and receptacles as restricted by Section 517.16, NFPA 99 *Standard for Health Care Facilities* still addresses these circuits. NFPA 99 also still requires periodic testing of grounding systems, which include installations of isolated grounding receptacles and circuits in health care facilities.
    a. True
    b. False
16. To minimize possible differences in potential in the grounding systems for power circuits supplying IT equipment, it is common for these facilities to be equipped with a single PDU or multiple PDUs.
    a. True
    b. False
17. Signal reference grids can be any number of conductive surfaces in the IT room, including _____.
    a. The raised floor framing
    b. The suspended ceiling grid system
    c. A constructed grid under the raised platform or floor of an IT room or rooms
    d. Any of the above
18. A(n) _____ is a protective device for limiting transient voltages by diverting or limiting surge current; it also prevents continued flow of follow current while remaining capable of repeating these functions.
    a. Grounding electrode
    b. Surge arrester
    c. SPD
    d. Uninterruptable power supply
19. Permanently connected SPDs intended for installation on the load side of the service disconnect overcurrent protective device, including SPDs located at the branch panel, best describes which of the following SPDs?
    a. Type 1
    b. Type 2
    c. Type 3
    d. Type 4
20. If a grounding electrode conductor for an SPD is installed in a ferrous metal raceway, the grounding electrode conductor is required to be bonded to both ends of the raceway, thus minimizing impedance and choke effects during surge events.
    a. True
    b. False

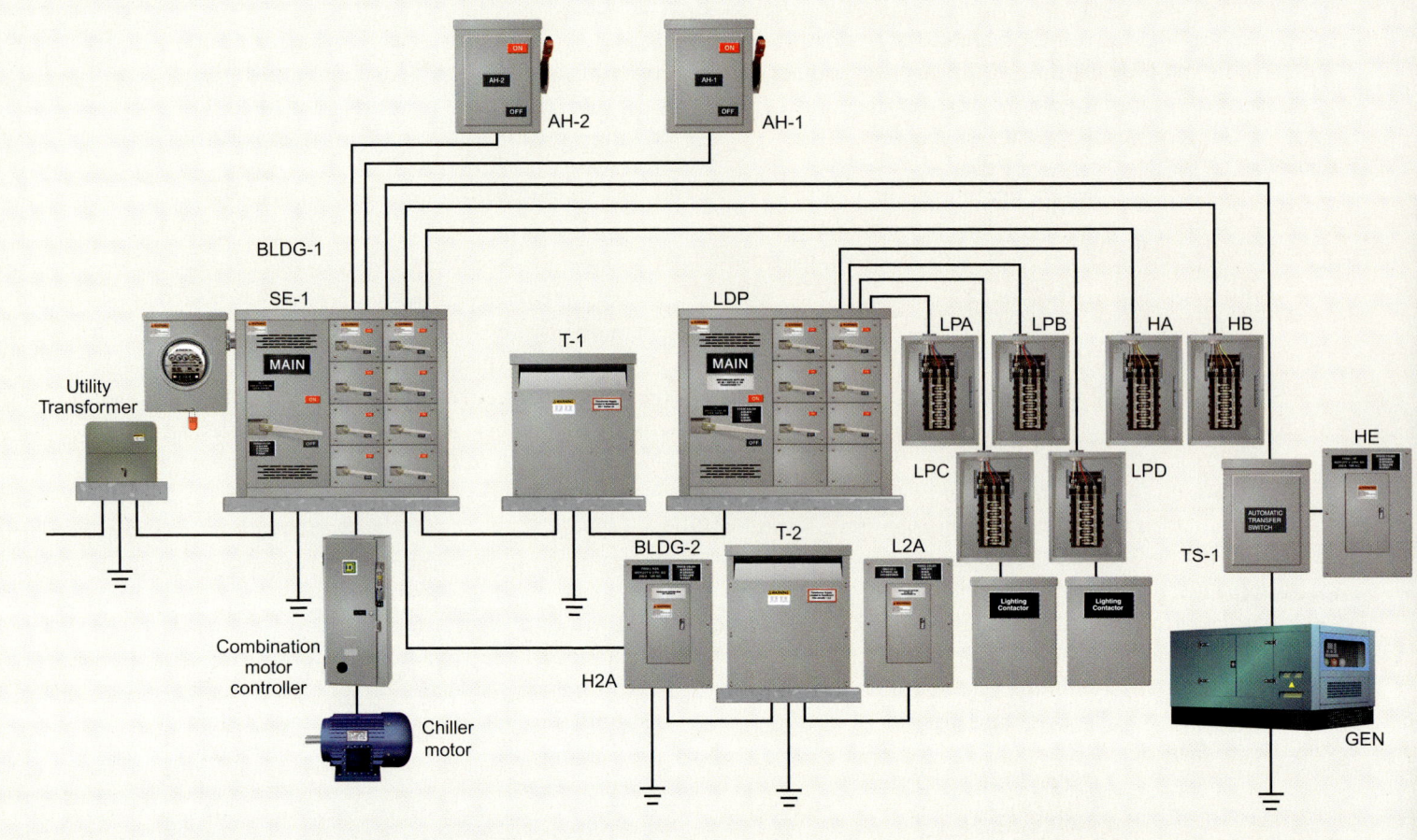

CHAPTER 11

Grounding at Separate Buildings or Structures

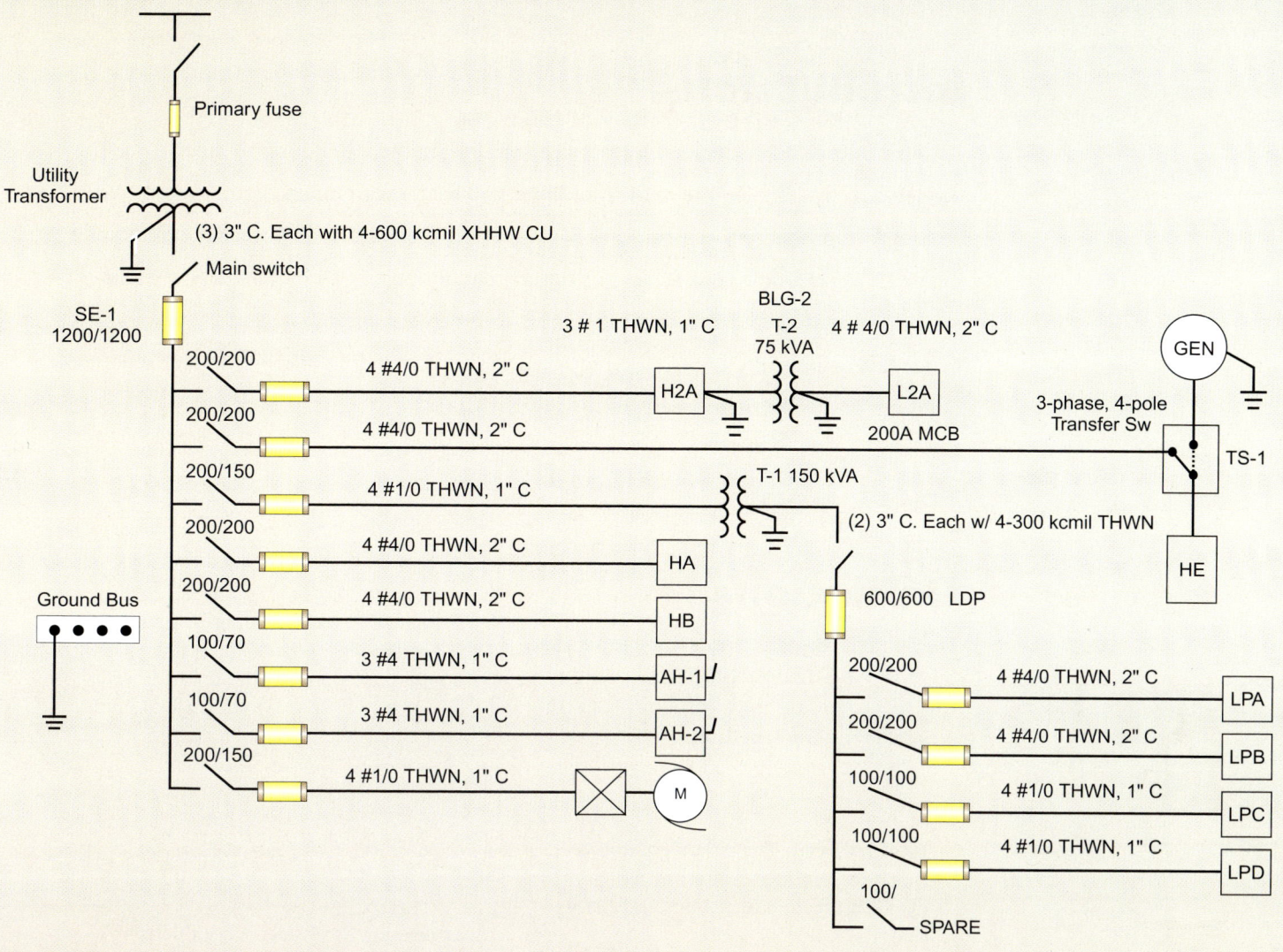

Learning Objectives

- Determine the requirements for a grounding electrode system at buildings or structures supplied by a branch circuit(s) or feeder(s)
- Understand the requirements for equipment grounding conductors to be installed with feeders or branch circuits supplying separate buildings or structures
- Understand the exception for grounded conductors used for equipment grounding at a separate building or structure supplied by a feeder or branch circuit
- Determine the bonding requirements for metal water piping systems and structural metal building framing of separate buildings or structures
- Understand the grounding and bonding rules associated with generators supplying separate buildings or structures

Outline

Definitions

Supplying Power to Separate Buildings or Structures

Purpose of Grounding and Bonding at Separate Buildings or Structures

Grounding Electrode Required

Grounding Electrode Conductor

Feeder and Branch Circuit Requirements

Ungrounded Systems Supplying Services

Supplied by a Separately Derived System

Building Disconnecting Means Requirements

Metal Water Pipe Bonding and Other Bonding

Disconnecting Means Remote from Building or Structure

Buildings or Structures Supplied by Separately Derived Systems

Buildings or Structures Supplied by an Ungrounded System

Buildings or Structures Supplied by Generators

Introduction

At some properties, a single utility service supplies multiple buildings or structures. The service directly supplies one of the buildings, and feeders or branch circuits supply other buildings from that service equipment. Alternatively, the utility service can be freestanding, such as on a pole, and feeders or branch circuits supply all buildings or structures. Specific grounding and bonding rules apply to separate buildings or structures supplied by feeders or branch circuits. The requirements for grounding and bonding at separate buildings or other structures are located in Part III of *NEC*® Article 250.

Definitions

There are a few key words and terms defined in *NEC* Article 100 that directly relate to grounding and bonding rules for separate building sructures.

Branch Circuit. The circuit conductors between the final overcurrent device protecting the circuit and the outlet(s).[1]

Branch Circuit, Multiwire. A branch circuit that consists of two or more ungrounded conductors that have a voltage between them, and a grounded conductor that has equal voltage between it and each ungrounded conductor of the circuit and that is connected to the neutral or grounded conductor of the system.[2]

Building. A structure that stands alone or that is cut off from adjoining structures by fire walls with all openings therein protected by approved fire doors.[3]

Feeder. All circuit conductors between the service equipment, the source of a separately derived system, or other power supply source and the final branch-circuit overcurrent device.[4]

Grounding Electrode. A conducting object through which a direct connection to earth is established.[5]

Grounding Electrode Conductor. A conductor used to connect the system grounded conductor or the equipment to a grounding electrode or to a point on the grounding electrode system.[6]

Structure. That which is built or constructed.[7]

Supplying Power to Separate Buildings or Structures

Many buildings or structures are supplied by power from other than a utility source or service. Examples range from a small detached garage supplied by a feeder from a house (single-family dwelling) to a university campus or industrial facility that has an outdoor substation service supplied at a higher-voltage level and 15-kV or 480-volt feeders supplying the buildings.

A service by definition is supplied from an electric utility. If the supply, such as a transformer or generator, is customer owned, it is not a service and therefore is either a feeder or branch circuit. When a building or structure is supplied by a feeder(s) or branch circuit(s), specific rules in the *NEC* have to be applied. Section 250.32 of the *NEC* provides the requirements for grounding and bonding at separate buildings or structures that are supplied by other than a service.

Purpose of Grounding and Bonding at Separate Buildings or Structures

Section 250.32 includes the grounding and bonding requirements for buildings and structures supplied by feeders or branch circuits. The purpose of grounding and bonding systems at branch circuit or feeder supplied separate buildings or structures is similar to service supplied buildings or structures. A connection to ground is required at separate buildings or structures to establish a reference to ground at that location.

Residential occupancy supplied by a utility service and two separate buildings supplied by feeders originating at the service equipment on the dwelling unit.

Industrial property often includes multiple buildings supplied by feeders from a common or multiple utility services. Separate buildings shown among the bulk storage tanks.

See Figure 11-1. This places all normally conductive non–current-carrying metal parts and other conductive materials at or as close as possible to Earth potential.

The reasons equipment and systems are grounded at separate buildings or structures are essentially the same reasons grounding is required if the building or structure is supplied by a utility service. The performance grounding requirements for equipment supplied by a grounded system are provided in Section 250.4(A)(2), and for electrical equipment supplied by an ungrounded system, the performance requirements are in Section 250.4(B)(1). These performance rules both indicate that normally non–current-carrying conductive materials enclosing electrical conductors or equipment, or forming part of such equipment, must be connected to the Earth (grounded) in a manner that will limit the voltage imposed by lightning or unintentional contact with higher-voltage lines and limit the voltage to ground on these conductive materials.[8] These concepts also apply at separate buildings or structures supplied by feeders or branch circuits.

Grounding Electrode Required

Section 250.52(A) lists grounding electrodes recognized in the *NEC*. Section 250.52(B) lists materials not permitted to be used for grounding electrodes. A grounding electrode system is generally required at a separate building or structure supplied by feeders or branch circuits. This condition is relaxed (1) if the building or structure is not supplied by electrical power or (2) if the structure or building is supplied by a single branch circuit (including multiwire branch circuit) that includes an equipment grounding conductor (EGC) for grounding non–current-carrying parts of equipment.

Section 250.32(A) provides the general requirement for a grounding electrode system in accordance with Part III of Article 250. This means that the same grounding electrode requirements in Section 250.50 must be applied to separate buildings or structures supplied by feeders or branch circuits. Use the same approach as if the building or structure were supplied by a utility service. If any of the grounding electrodes in Section 250.52(A) are present (exist) at the building or structure served, they must be bonded together to form a grounding electrode system for the separate building or structure. **See Figure 11-2.** This includes the water pipe electrodes, metal building frame electrodes, concrete-encased electrodes, and so forth, as provided in Section 250.52(A).

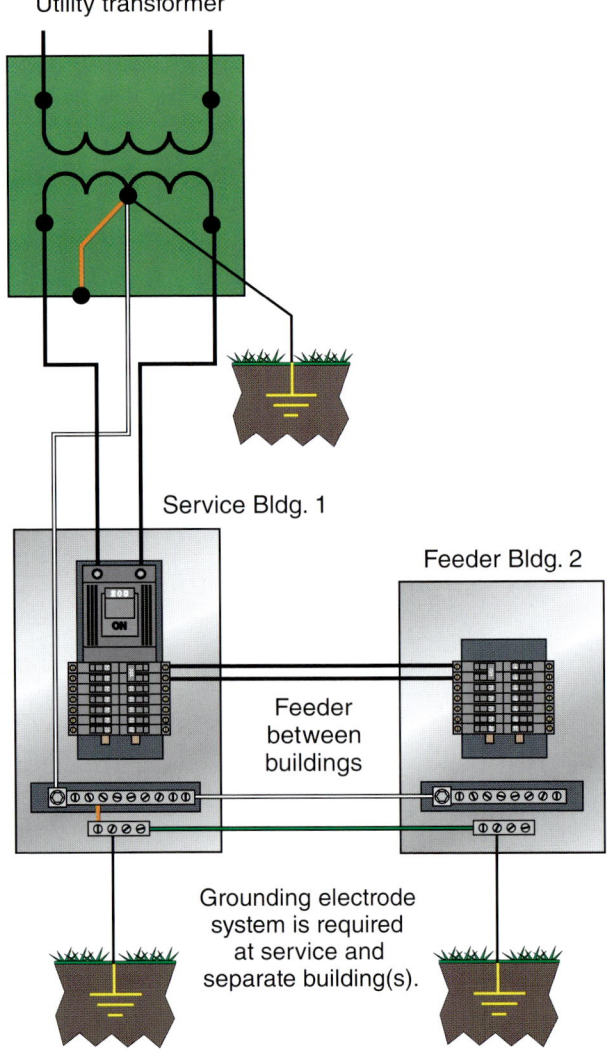

FIGURE 11-1 A grounding electrode system is generally required at buildings supplied by services and building supplied by feeders or branch circuits.

If none of these electrodes is present to form a grounding electrode system, one must be installed. The exception to the grounding electrode requirement in Section 250.32 applies to a building or structure that is supplied by a single branch circuit (either an individual or multiwire branch circuit) and that includes an EGC for grounding electrical equipment associated with it. **See Figure 11-3.** An example of a building supplied this way is a detached garage at a dwelling occupancy. The terms *multiwire branch circuit* and *structure* are defined in Article 100 and clarify the meaning of the rules in which these terms appear. An example of a structure supplied by a single branch circuit is a light pole base in a parking lot, which qualifies as a structure by definition.

Grounding Electrode Conductor

Section 250.32(E) provides requirements for the grounding electrode conductor installation at separate buildings or structures supplied by feeders or branch circuits. Part III of Article 250 provides the requirements for grounding electrode conductor installation and sizing. The grounding electrode conductor is generally sized according to Table 250.66, which bases sizing on the largest ungrounded circuit conductor supplying the building or structure. **See Figure 11-4.** For example, if a building is supplied by a 200-ampere feeder using four 4/0-copper conductors (three phases and a neutral), the minimum size required for the grounding electrode conductor is 2 AWG copper or 1/0 AWG aluminum.

If the grounding electrode is a sole connection to a rod, pipe, or plate electrode, per Section 250.66(A), the maximum size required is a 6 AWG copper or a 4 AWG aluminum conductor. If the grounding electrode conductor is a sole connection to a concrete-encased electrode, per Section 250.66(B), the maximum size required is a 4 AWG copper conductor.

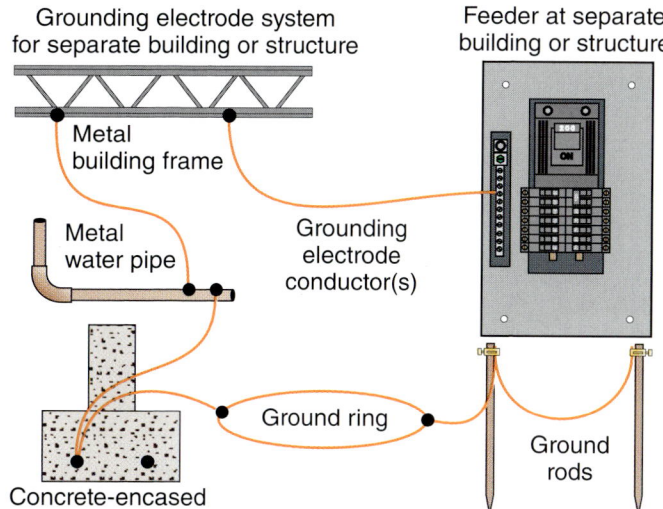

FIGURE 11-2 All grounding electrodes must be bonded together to form a grounding electrode system for the separate building or structure supplied by feeders or branch circuits.

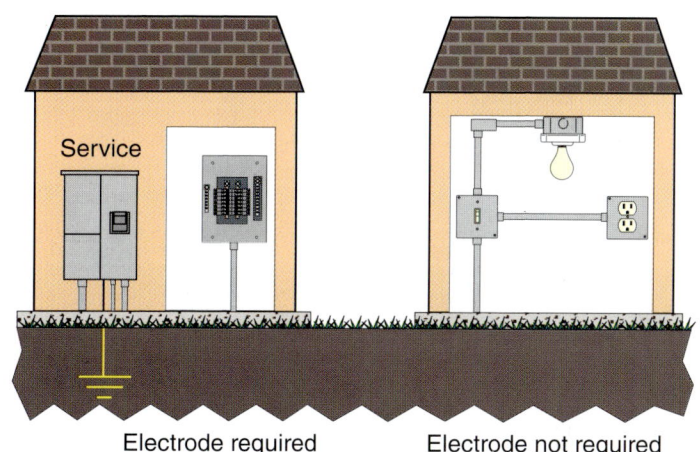

FIGURE 11-3 A grounding electrode is not required for separate structures supplied by a single branch circuit that includes an EGC. A multiwire branch circuit is also permitted.

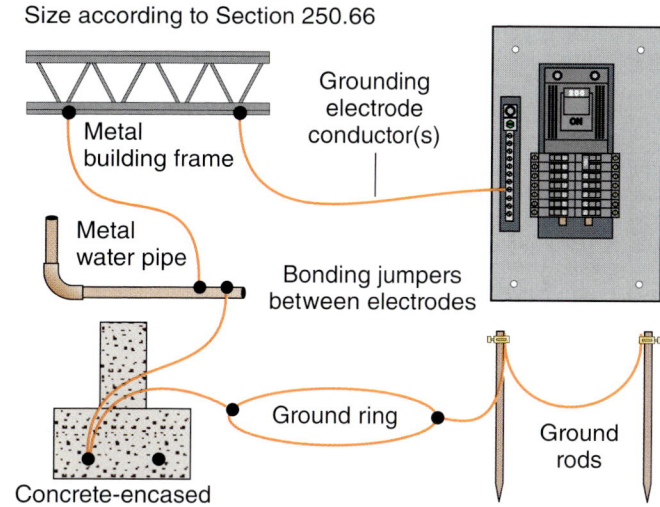

FIGURE 11-4 A grounding electrode conductor must be installed according to 250.64 and sized according to 250.66.

Section 250.66(C) indicates that if a ring electrode is installed and the grounding electrode conductor is solely connected to the ring, the size must not be smaller than the size of the ring and never smaller than 2 AWG copper. The grounding electrode conductor must be installed in accordance with the applicable requirements in Section 250.64. This includes installing them without a splice or joint, as a general rule, or meeting the requirements for splicing grounding electrode conductors if they are spliced or joined together. Protecting grounding electrode conductors against physical damage is required in accordance with Section 250.64 as applicable, and connections to grounding electrodes have to meet the requirements in Section 250.70. Basically, the grounding electrode conductor requirements for separate building or structure installations supplied by feeders or branch circuits are the same as if the building or structure were supplied by a utility service. Grounding is done the same way, except that for new installations the grounding electrode conductor is connected to the EGC, not the grounded conductor.

Feeder and Branch Circuit Requirements

Feeders are generally required to include an EGC, as indicated in Section 215.6. **See Figure 11-5.** When feeders or branch circuits are installed to supply a separate building or structure, an EGC must also be installed. The EGC can be a wire type, or it can be any wiring method in Section 250.118 that qualifies as an EGC. If the EGC is a wire type, it must be sized in accordance with Section 250.122. **See Figure 11-6.**

If a feeder to a separate building or structure is installed using rigid metal conduit (RMC), the conduit can serve as the required EGC. Note that any of the metal raceways provided in Section 250.118 that qualify as EGCs can be used as EGCs. The common metallic raceways installed for feeders supplying separate buildings or structures are rigid metal conduit (RMC) and intermediate metal conduit (IMC). Corrosion or deterioration and their effect on the life expectancy of wiring methods in contact with the Earth can be an issue in some areas. Corrosion and deterioration protection rules for wiring methods are provided in Section 300.6. Good design practices provide a wire-type EGC to ensure that the safety system is not compromised in the short or long term. Note that the *NEC* falls short of addressing specific life expectancy of grounding electrodes or other materials in contact with the Earth; however, corrosion protection is addressed in Section 300.6.

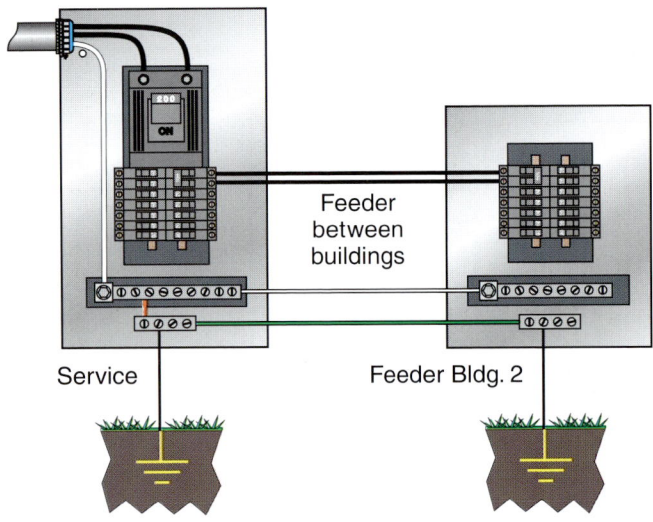

FIGURE 11-5 An EGC is generally required with a feeder or branch circuit supplying a separate building or structure.

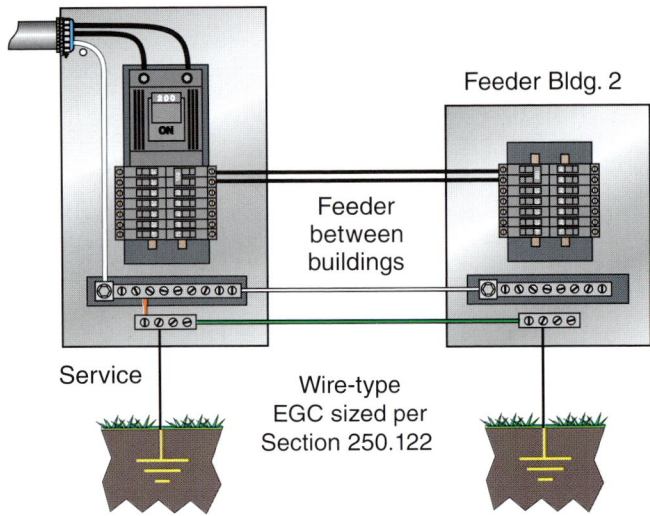

FIGURE 11-6 Wire-type EGCs must be sized according to Section 250.122.

In addition, voltage drop considerations must be applied to designs for feeders and branch circuits supplying separate buildings or structures. The EGC has to provide an effective ground-fault current path. Section 250.32(B)(1) restricts a grounded conductor from being connected to a grounding electrode or the EGC for the separate building or structure. **See Figure 11-7.**

There is an exception that allows the grounded (usually the neutral) conductor of feeders or branch circuits to be used for grounding at separate buildings or structures. This exception applies only to existing premises wiring systems and is not applicable to new installations. The conditions of the exception are as follows:

1. An EGC is not included with the supply circuit to the separate building or structure.
2. There are no common electrically continuous metallic paths between the feeder source and the destination at the building or structure served.
3. Ground-fault protection of equipment is not provided on the supply side of the feeder.[9]

For existing premises wiring systems that use a grounded conductor in this manner, the grounded conductor must not be smaller than the sizes required in Section 250.122 or 220.61 as applicable.

Ungrounded Systems Supplying Services

When an ungrounded source supplies separate buildings or structures, by feeders or branch circuits, the requirements in Section 250.32(C) apply. For an ungrounded system, an EGC meeting the requirements in Section 250.118 must be installed with the supply conductors and must be connected to the building or structure disconnecting means and to the grounding electrode system. **See Figure 11-8.**

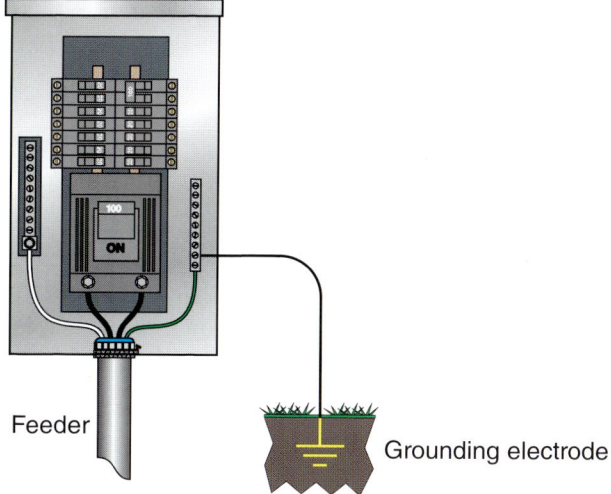

FIGURE 11-7 The grounded (usually the neutral) conductor must be isolated from grounded equipment and the EGC [Section 250.24(A)(5)]. Then, the EGC and the grounding electrode conductor must be connected to the disconnecting means enclosure.

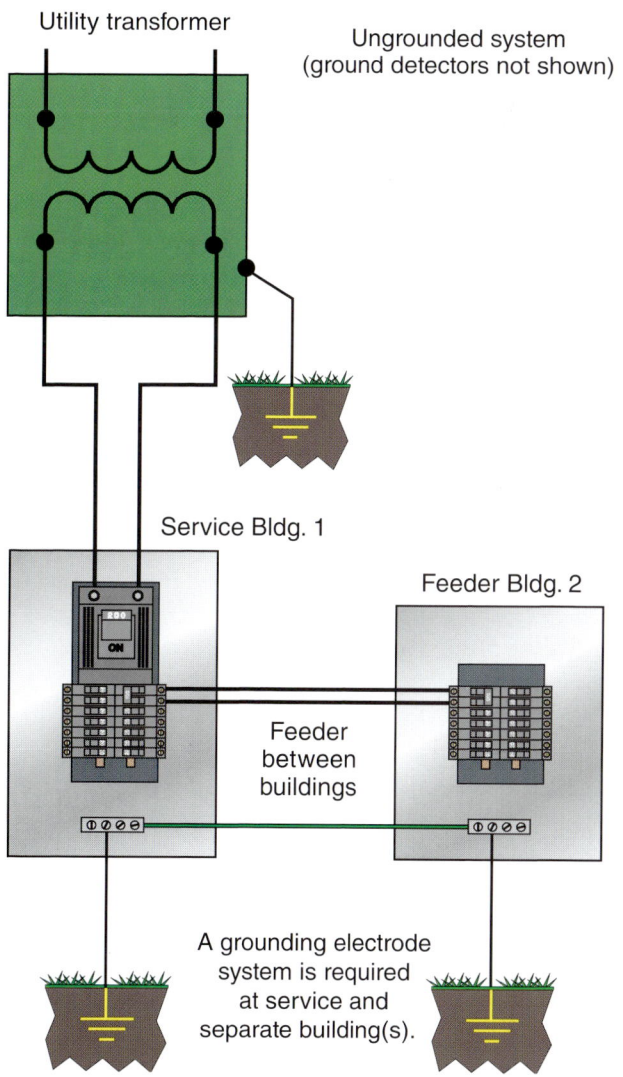

FIGURE 11-8 A grounding electrode system is required at buildings or structures supplied by services and at buildings supplied by feeders or branch circuits from an ungrounded system.

The grounding electrodes at the separate building or structure served must be connected to the building or structure disconnecting means.[10]

Supplied by a Separately Derived System

Section 250.32(B)(2) addresses buildings or structures supplied by separately derived systems, such as transformers installed outside the building or structure. When the disconnecting means is remote from those buildings or structures that are supplied by feeders or branch circuits, the rules in Section 250.32(D) must be applied.

Systems with Overcurrent Protection

Sometimes overcurrent protection is provided on the secondary side of a separately derived system, such as may be the case for unit substation equipment supplying a building or structure. There are also pad-mounted transformers that include overcurrent protection on the secondary side. **See Figure 11-9.** If overcurrent protection is provided where the conductors originate, the installation of a feeder or branch circuit to a separate building or structure has to include an EGC in accordance with Section 250.32(B)(1).

Systems without Overcurrent Protection

Often, separately derived systems, specifically transformer types, do not provide overcurrent protection on the secondary side at the transformer. Section 240.21(B)(4) permits outside feeder conductors, from a transformer secondary, to supply a building or structure; there is no tap conductor length limitation. However, once the feeder conductors arrive at the building, the requirements for overcurrent protection apply. The feeder or branch circuit supplying a separate building or structure is outside, so the rules in Article 225 also apply. The disconnecting means is required by Section 225.31 and must be located in accordance with Section 225.32, either outside or inside the building, nearest to the point of entrance of the feeder conductors. Section 240.21(B)(5)(3) requires the disconnecting means to contain overcurrent protection for the conductors; alternatively, the overcurrent protection must be located immediately adjacent to the disconnecting means. If overcurrent protection is not provided where the conductors originate, at the separately derived system source, the installation must comply with Section 250.30(A). If installed with the feeder to the building or structure, the supply-side bonding jumper shall be connected to the building or structure disconnecting means and to the grounding electrode system. **See Figure 11-10.** This means that the system bonding jumper will be connected in the source enclosure and a supply-side bonding jumper must be installed from the source enclosure to the disconnecting means.

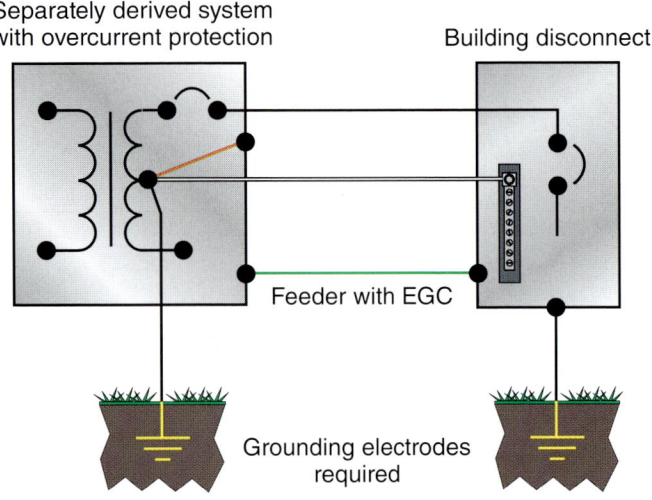

FIGURE 11-9 Where the separately derived system has overcurrent protection at the source, the feeder must include an EGC installed to the separate building or structure disconnecting means.

A grounded (usually a neutral) conductor, if installed, must be kept separate and isolated from ground and grounded parts at the building or structure disconnecting means. See Section 250.30(A).

Building Disconnecting Means Requirements

The disconnecting means installed for a feeder or branch circuit supplying the separate building or structure also has to be rated "suitable for use as service equipment," as specified by Section 225.36. **See Figure 11-11.**

This means the equipment must meet certain requirements that apply to service disconnecting means such as short circuit current ratings, normal current ratings, and overcurrent protection, and it must include grounding and bonding capabilities. Generally all grounding and bonding conductors and connections will be made in this equipment. This requirement could also be met by using a panelboard that is suitable for use as service equipment and that meets the disconnecting means requirements.

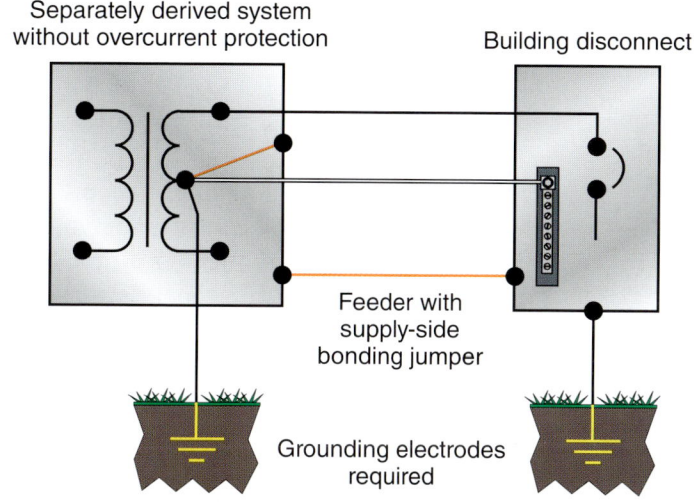

FIGURE 11-10 Where the separately derived system has no overcurrent protection at the source, the feeder must include a supply-side bonding jumper installed to the separate building or structure disconnecting means.

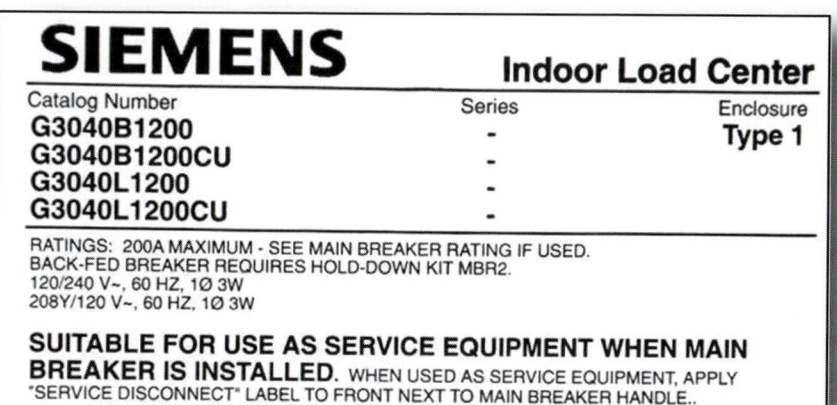
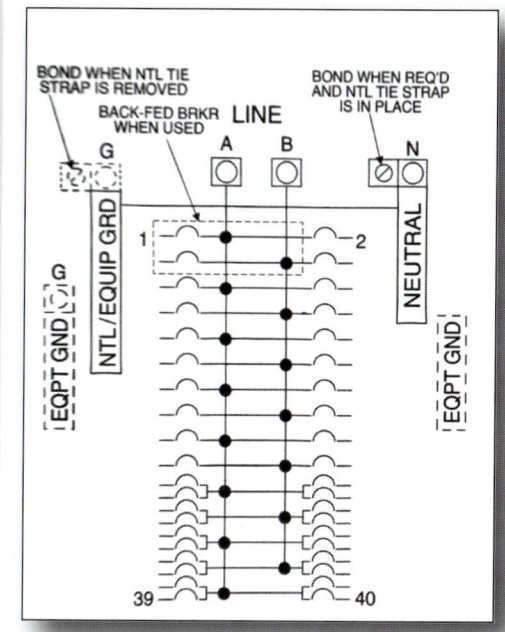

FIGURE 11-11 Some panelboards are suitable for use as service equipment and must be so identified. (Photos of actual labels within a panelboard that is suitable for use as service equipment)

See Figure 11-12. The one exception to this requirement allows a snap switch, or a set of three- or four-way switches at garages and other accessory buildings on residential property. These snap switches are exempt from the service disconnecting means rating requirement in Section 225.36.

Metal Water Pipe Bonding and Other Bonding

Bonding water piping systems at separate buildings or structures is required, especially for installations where the piping system does not qualify as a grounding electrode and is not already part of the grounding electrode system. The metal water pipe bonding at separate buildings or structures is treated the same way it is as if the separate building were supplied by a utility service and not a feeder or branch circuit. Section 250.104(A)(3) requires that metal water piping systems be installed in or attached to a building or structure supplied by feeders or branch circuits to be bonded. This includes all exterior metal water piping systems for the building or structure served. The bonding jumper can be attached to the building or structure disconnecting means enclosure, to the EGC run with the feeder conductors to the building, or to one or more of the grounding electrodes used at the separate building or structure. See Figure 11-13.

The bonding jumper must not be less than the sizes provided in Table 250.66, which are based on the size of the largest ungrounded supply conductor to the building or structure; however, the bonding jumper does not ever have to be larger than the size of the largest ungrounded feeder conductor. For example, if the building were supplied with a feeder that used four 750-kcmil copper conductors (one for each phase conductor and grounded conductor), the minimum-size bonding jumper for the water piping system is 2/0 copper or 4/0 aluminum.

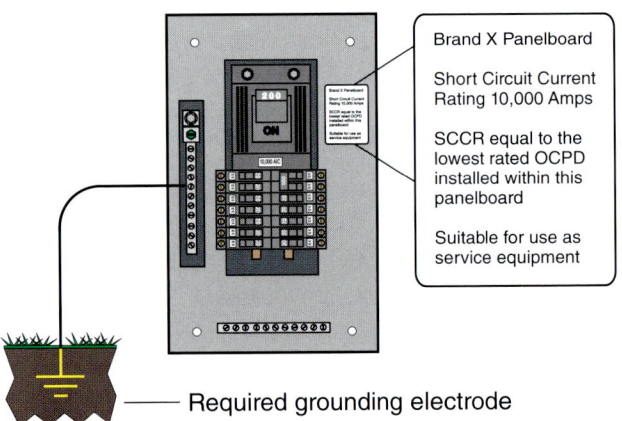

FIGURE 11-12 The building disconnecting means generally has to be suitable for use as service equipment (Section 225.36).

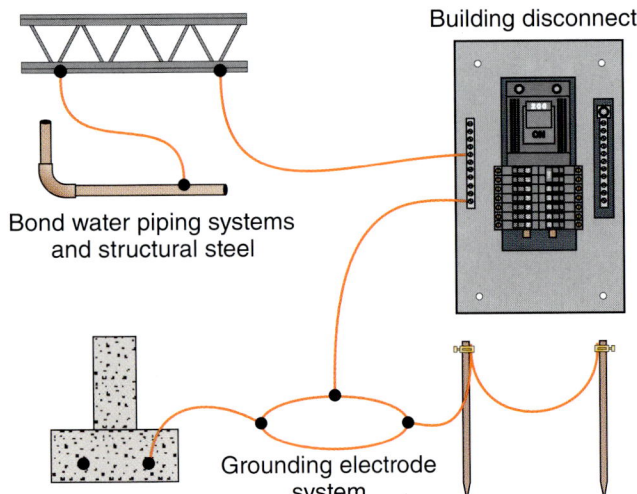

FIGURE 11-13 Bonding of piping systems and building steel is required at separate buildings or structures if they do not qualify as grounding electrodes.

Disconnecting Means Remote from Building or Structure

Feeders or branch circuits that supply a separate building or structure are required to have a disconnecting means installed either outside or inside the building or structure nearest to the point of entrance of the supply conductors (Section 225.32). There are two exceptions to this general rule that permit the building disconnecting means to be located elsewhere on the premises.

To qualify for these exceptions, restrictive conditions, such as the building has safe switching procedures, qualified persons to service the installation, and single management, must be met. See the exceptions for the full set of qualifying conditions.

There are also other circumstances that permit a feeder or branch circuit disconnect to be remote from a building or structure, such as for generators equipped with a disconnecting means and used with emergency systems or legally required standby systems. The conditions that must be met to gain relief from the building disconnect requirements are as follows:

1. The generator disconnect is readily accessible.
2. The disconnecting means is located within sight from the building it serves.
3. The disconnecting means is suitable for use as service equipment.[11]

If all of these conditions are met, the feeder disconnecting means, generally required at the building or structure, can be eliminated. The following conditions and installation requirements must be met when the required building disconnects are located remotely. These are in accordance with Section 250.32(D); Section 225.32 Exceptions 1 or 2; or, for generators, Section 700.12(B)(6), 701.11(B)(5), or 702.12.

1. There is no connection made between the grounded conductor of the feeder and the grounding electrode and EGC at the separate building or structure.
2. An EGC is installed to the separate building or structure in accordance with Section 250.32(B)(1) and is connected to a grounding electrode at the building or structure, except when an electrode is not required, as provided in Section 250.32(A) Exception.
3. There are provisions for terminating the grounding electrode conductor and EGC in a box, panelboard, or other enclosure located immediately on the outside or inside of the building or structure. A grounding terminal bar inside the enclosure is a common method for meeting this requirement.[12] **See Figure 11-14.**

Buildings or Structures Supplied by Separately Derived Systems

Many installations necessitate that buildings be supplied by separately derived systems. The conductors supplied by a separately derived system could be feeders or branch circuit conductors. Most often, the conductors connected to the secondary of a transformer supply more than a single circuit, so by definition they are feeder conductors. Many campus-style electrical systems include a separately derived system supplying separate buildings or structures. Section 250.32(B)(2) provides specific direction for installers when buildings or structures are supplied directly from a separately derived system. The first things that should be reviewed are the requirements for grounding electrodes. Each separate building or structure served must have a grounding electrode system.

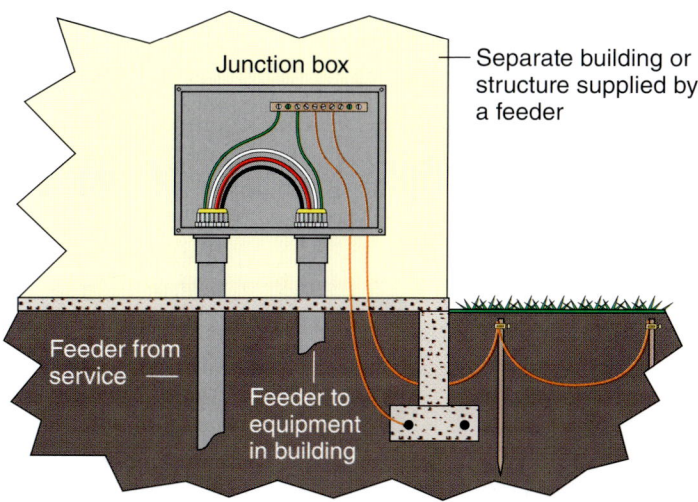

FIGURE 11-14 A grounding electrode conductor connection to the EGC of building supply circuits is permitted to be made in a junction box. The building disconnect in this figure is remote in accordance with Section 250.32(D).

If the separately derived system, such as a transformer, is installed outside the building or structure served, a grounding electrode must be installed and connected at the source location per the requirements in Section 250.30(C). Sections 250.32(B)(2)(a) and (B)(2)(b) provide two different installation requirements for separately derived systems supplying separate buildings. The first condition addresses derived systems that have overcurrent protection at the source location. An example is a unit substation that includes secondary overcurrent devices within an assembly enclosure. In these circumstances, a feeder is installed from the load side of an overcurrent protective device at the separately derived system to the building or structure served. The feeder must include an EGC of any type identified in Section 250.118. If the EGC is a wire type, then the size cannot be smaller than what is required by Section 250.122. **See Figure 11-15.**

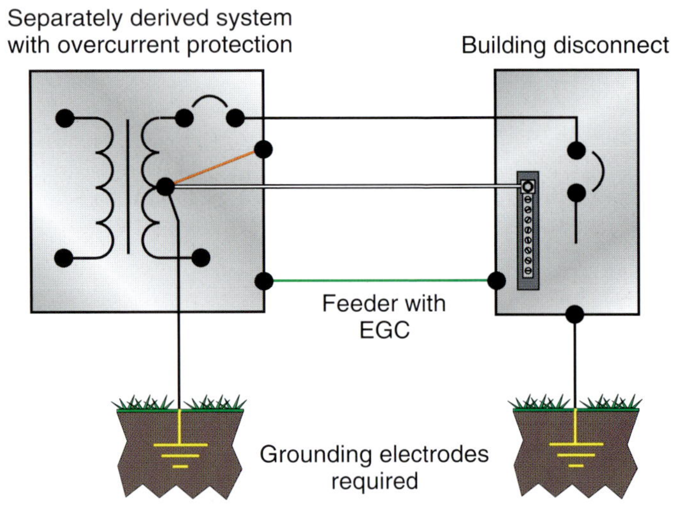

If there is no overcurrent device at the point of origin of the feeder supplied by the separately derived system, then the installation must meet all of the grounding and bonding requirements for separately derived systems as set forth in Section 250.30(A). When the feeder from the derived system includes a wire-type supply-side bonding jumper, it has to be connected to the building or structure disconnecting means and the required grounding electrode for the building or structure. **See Figure 11-16.**

FIGURE 11-15 Where overcurrent protection is provided at the origin of the feeder from a separately derived system supplying a building or structure, an EGC is required to be installed with the feeder.

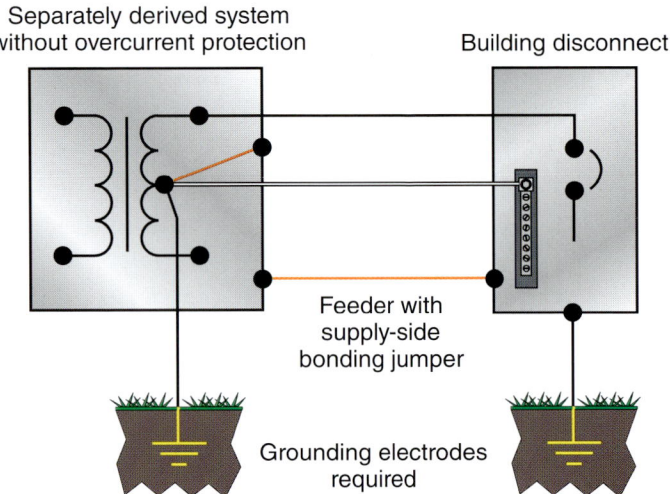

FIGURE 11-16 Where no overcurrent protection is provided at the origin of the feeder from a separately derived system supplying a building or structure, a supply-side bonding jumper is required to be installed with the feeder.

Buildings or Structures Supplied by an Ungrounded System

Although uncommon, when a building or other structure is supplied by an ungrounded separately derived system, the requirements in Section 250.32(C) have to be applied. (An ungrounded system is one with no conductor supplied by the system that is intentionally grounded, such as a three-phase, 3-wire ungrounded delta system.) Section 250.21(B) requires that ground detectors be used for this type of installation. The feeder or branch circuit from an ungrounded separately derived system that supplies a separate building or structure must include an EGC or supply-side bonding jumper with the feeder conductors to the building disconnecting means; furthermore, the EGC or supply-side bonding jumper must be connected to the disconnecting means enclosure as well as to the grounding electrode conductor.

The building or structure grounding electrode system must be connected to the building disconnecting means by a grounding electrode conductor; the size of the grounding electrode conductor is based on requirements in Section 250.66.

If overcurrent protection is provided where the conductors originate, an EGC in accordance with Section 250.118 has to be installed with the circuit conductors supplying the building or structure. If the EGC is a wire type, it has to be sized in accordance with Section 250.122, which bases sizing on the rating of the overcurrent protective device for the feeder or branch circuit supplying the building or structure served. If overcurrent protection is not provided where the conductors originate, the installation must comply with Section 250.30(B). If installed, the supply-side bonding jumper shall be connected to the building or structure disconnecting means and to the grounding electrode system. For buildings or structures supplied by ungrounded separately derived systems, a grounding electrode system is required in accordance with Section 250.32(A). **See Figure 11-17.**

Buildings or Structures Supplied by Generators

Generators are sometimes installed as an emergency system, legally required standby system, or optional standby power system for a building or structure. These installations require feeders from the generators to supply the building or structure; if they are located outside the building or structure, the requirements in Article 225 apply, specifically, Sections 225.31 through 225.39. **See Figure 11-18.**

The grounding and bonding requirements for permanently installed generators are found in Section 250.35. If the transfer equipment for the generator system switches the grounded (neutral) conductor, the generator must be grounded and bonded in accordance with Section 250.30(A).[13]

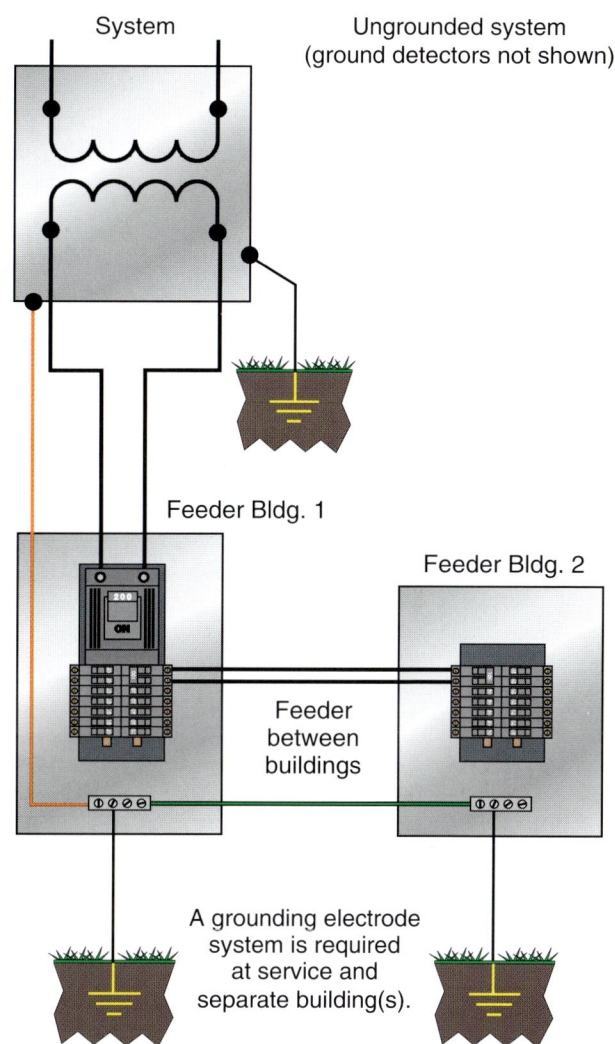

FIGURE 11-17 A grounding electrode system is required at buildings supplied by services and buildings supplied by feeder or branch circuits connected to ungrounded systems.

FIGURE 11-18 Generators often supply emergency or standby systems for separate buildings or structures.

If the generator system has transfer equipment that does not switch the grounded (neutral) conductor, then the grounding and bonding connections for the generator feeder must comply with Section 250.35(B), depending on where the first overcurrent device for the generator feeder is located. If the overcurrent device is located on the generator, then the feeder is on the load side of an overcurrent device. In this case the EGC with the feeder is sized based on the requirements in Section 250.102(D). This requires sizing based on the rating of the overcurrent protective device at the generator using Table 250.122. **See Figure 11-19.** If the first overcurrent protective device is not located at the generator, a supply side bonding jumper must be installed with the feeder up to the first system overcurrent protective device enclosure. It must be sized based on the requirements in Section 250.102(C). The bonding jumper is sized per Table 250.66 or by applying the 12.5% rule, which is based on the largest ungrounded phase conductor connected to the generator. **See Figure 11-20.**

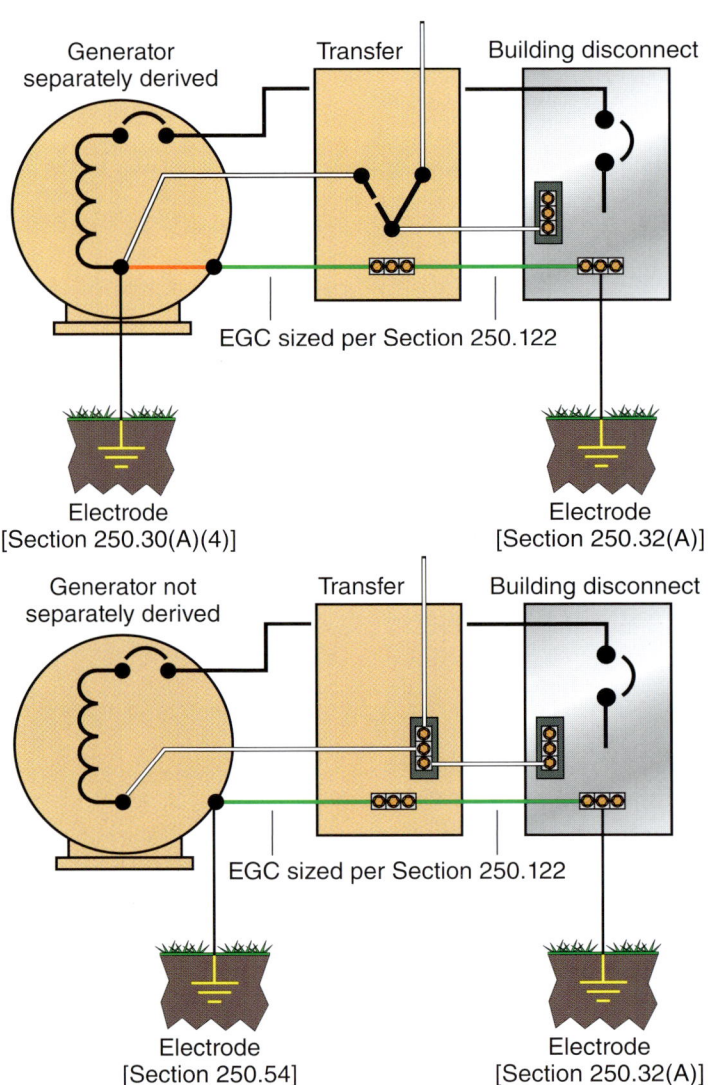

FIGURE 11-19 Where generators with an overcurrent protective device supply a separate building or structure, size the EGC according to Section 250.122. Note that the ungrounded conductor connections to the transfer switch are not shown.

Grounding at Separate Buildings or Structures **245**

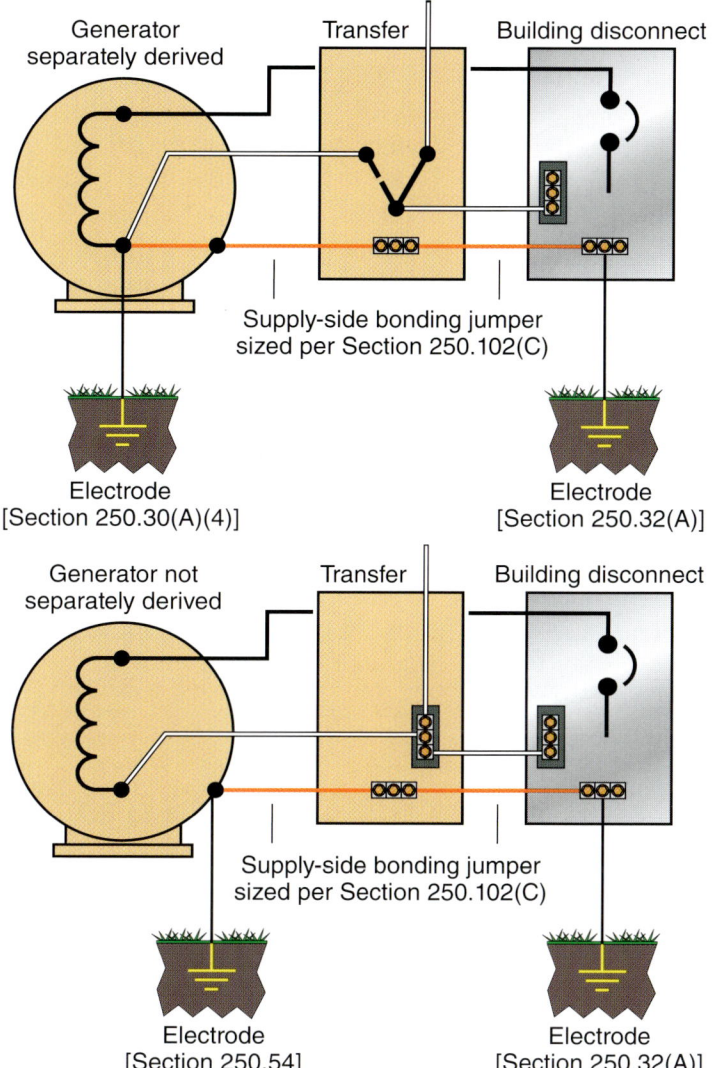

FIGURE 11-20 Where generators without an overcurrent protective device supply a building or structure, size the supply-side bonding jumper according to Section 250.102(C). Note that the ungrounded conductor connections to the transfer switch are not shown.

Summary

Electrical grounding and bonding requirements apply to buildings or structures supplied by feeders or branch circuits. The grounding and bonding performance concepts and purpose are generally the same as if the building or structure were supplied by a utility service. The requirements for grounding electrodes were reviewed, in addition to the methods for grounding at buildings supplied from grounded systems and ungrounded systems. Alternative restrictive grounding rules for existing buildings or structures were also reviewed, as were grounding rules for buildings that have a disconnecting means remote from the building or structure. Whether a building or structure is supplied by a feeder or branch circuit derived from a grounded system or ungrounded system, grounding and bonding rules apply for those separate buildings or structures. The specific grounding and bonding requirements for buildings or structures supplied by generators were also reviewed. In most cases, the grounded conductor must be separated and isolated from ground, the grounding electrode, and the EGC at separate buildings or structures, with the exception of existing premises wiring systems. These rules are not applicable if the building or structure is not supplied with electrical power.

References

1. NFPA 70 National Electrical Code 2011, Article 100 (National Fire Protection Association, Quincy, MA 2010), p. 70–27.
2. NFPA 70 National Electrical Code 2011, Article 100 (National Fire Protection Association, Quincy, MA 2010), p. 70–27.
3. NFPA 70 National Electrical Code 2011, Article 100 (National Fire Protection Association, Quincy, MA 2010), p. 70–27.
4. NFPA 70 National Electrical Code 2011, Article 100 (National Fire Protection Association, Quincy, MA 2010), p. 70–28.
5. NFPA 70 National Electrical Code 2011, Article 100 (National Fire Protection Association, Quincy, MA 2010), p. 70–29.
6. NFPA 70 National Electrical Code 2011, Article 100 (National Fire Protection Association, Quincy, MA 2010), p. 70–29.
7. NFPA 70 National Electrical Code 2011, Article 100 (National Fire Protection Association, Quincy, MA 2010), p. 70–32.
8. NFPA 70 National Electrical Code 2011, Section 250.4(B)(1) (National Fire Protection Association, Quincy, MA 2010), p. 70–101.
9. NFPA 70 National Electrical Code 2011, Section 250.32(B) Exception (National Fire Protection Association, Quincy, MA 2010), p. 70–109.
10. NFPA 70 National Electrical Code 2011, Section 250.32(C) (National Fire Protection Association, Quincy, MA 2010), p. 70–109.
11. NFPA 70 National Electrical Code 2011, Section 225.32 (National Fire Protection Association, Quincy, MA 2010), p. 70–74.
12. NFPA 70 National Electrical Code 2011, Section 250.32(D) (National Fire Protection Association, Quincy, MA 2010), p. 70–110.
13. NFPA 70 National Electrical Code 2011, Section 250.35(A) (National Fire Protection Association, Quincy, MA 2010), p. 70–110.

Review Questions

1. When a building or other structure is supplied by feeders or branch circuits, a grounding electrode system is generally required at the separate building or structure served.
   a. True
   b. False
2. A _____ is a structure that stands alone or that is cut off from adjoining structures by fire walls, with all openings therein protected by approved fire doors.
   a. Concrete pad
   b. Foundation
   c. Building
   d. Footing
3. _____ refers to all circuit conductors between the service equipment, the source of a separately derived system, or other power supply source and the final branch circuit overcurrent protective device.
   a. Branch circuit
   b. Feeder
   c. Service conductors
   d. Derived system conductors
4. An EGC is required to be installed with a new feeder supplying a separate building or structure.
   a. True
   b. False
5. If present, which of the following grounding electrodes must be used to form a grounding electrode system for a building or structure served by a feeder or branch circuit?
   a. Underground metal water pipe
   b. Structural metal building frame electrode
   c. Ground rods
   d. All of the above
6. If polyvinyl chloride (PVC) conduit is installed for a 600-ampere feeder supplying a separate building, a _____ wire-type copper EGC is required to be installed with the feeder conductors.
   a. 1 AWG
   b. 1/0 AWG
   c. 2/0 AWG
   d. 3/0 AWG

7. According to the *NEC*, a _____ is that which is built or constructed.
   a. Building
   b. Structure
   c. Footing
   d. Foundation
8. The disconnecting means installed in the feeder supplying a separate building has to be suitable for use as service equipment.
   a. True
   b. False
9. If no grounding electrodes are present for use at a separate building supplied by a feeder, which of the following applies?
   a. A grounding electrode is not required.
   b. The EGC of the feeder can serve as the only grounding means.
   c. A grounding electrode has to be installed.
   d. By exception, a ground ring electrode is required.
10. A grounding electrode is not required for a separate structure that is supplied by only one branch circuit even if the circuit does not include and EGC.
    a. True
    b. False
11. When a single multiwire branch circuit that includes an EGC supplies a separate building or structure and no grounding electrodes are present at the building or structure served, installing a grounding electrode is not required.
    a. True
    b. False
12. The grounding electrode conductor installed at a separate building or structure supplied by a feeder must be sized using _____, which bases sizing on the circular mil (cm) area of the largest ungrounded feeder conductor.
    a. Table 8, Chapter 9
    b. Table 250.66
    c. Table 250.122
    d. Table 250.4
13. A 400-ampere feeder supplies a separate building. If the grounding electrode installed at this separate building is a single ground rod that has less than 25 ohms resistance to ground, the maximum size for the copper grounding electrode conductor required as the sole connection is _____.
    a. 3 AWG
    b. 1/0 AWG
    c. 2 AWG
    d. 6 AWG
14. Section 250.32(B)(1) generally restricts any installed grounded conductor from being connected to a grounding electrode or the EGC for the separate building or structure.
    a. True
    b. False
15. The grounded (usually the neutral) conductor of feeders or branch circuits is permitted to be used for grounding at separate buildings or structures only in existing premises wiring systems and in accordance with which of the following conditions?
    a. An EGC is not included with the supply circuit to the separate building or structure.
    b. There are no common electrically continuous metallic paths between the feeder source and the destination at the building or structure served.
    c. Ground-fault protection of equipment is not provided at the service location on the supply side of the feeder.
    d. All of the above.
16. Metal water piping systems in separate buildings or structures are required to be bonded to which of the following locations?
    a. The building or structure disconnecting means
    b. The EGC run with the feeder conductors to the building
    c. One or more of the grounding electrode used at the separate building or structure
    d. Any of the above
17. When a generator for standby power supplies a separate building, a separate disconnecting means where the feeder arrives at the building or structure is not required if _____.
    a. The generator disconnect is readily accessible.
    b. The disconnecting means is located within sight from the building it serves.
    c. The disconnecting means is suitable for use as service equipment.
    d. All of the above.
18. When a building is supplied by an ungrounded system, the feeder from the ungrounded system must include an EGC or supply-side bonding jumper with the conductors to the building disconnecting means and the EGC or supply-side bonding jumper must be connected to the disconnecting means enclosure as well as to the grounding electrode conductor.
    a. True
    b. False

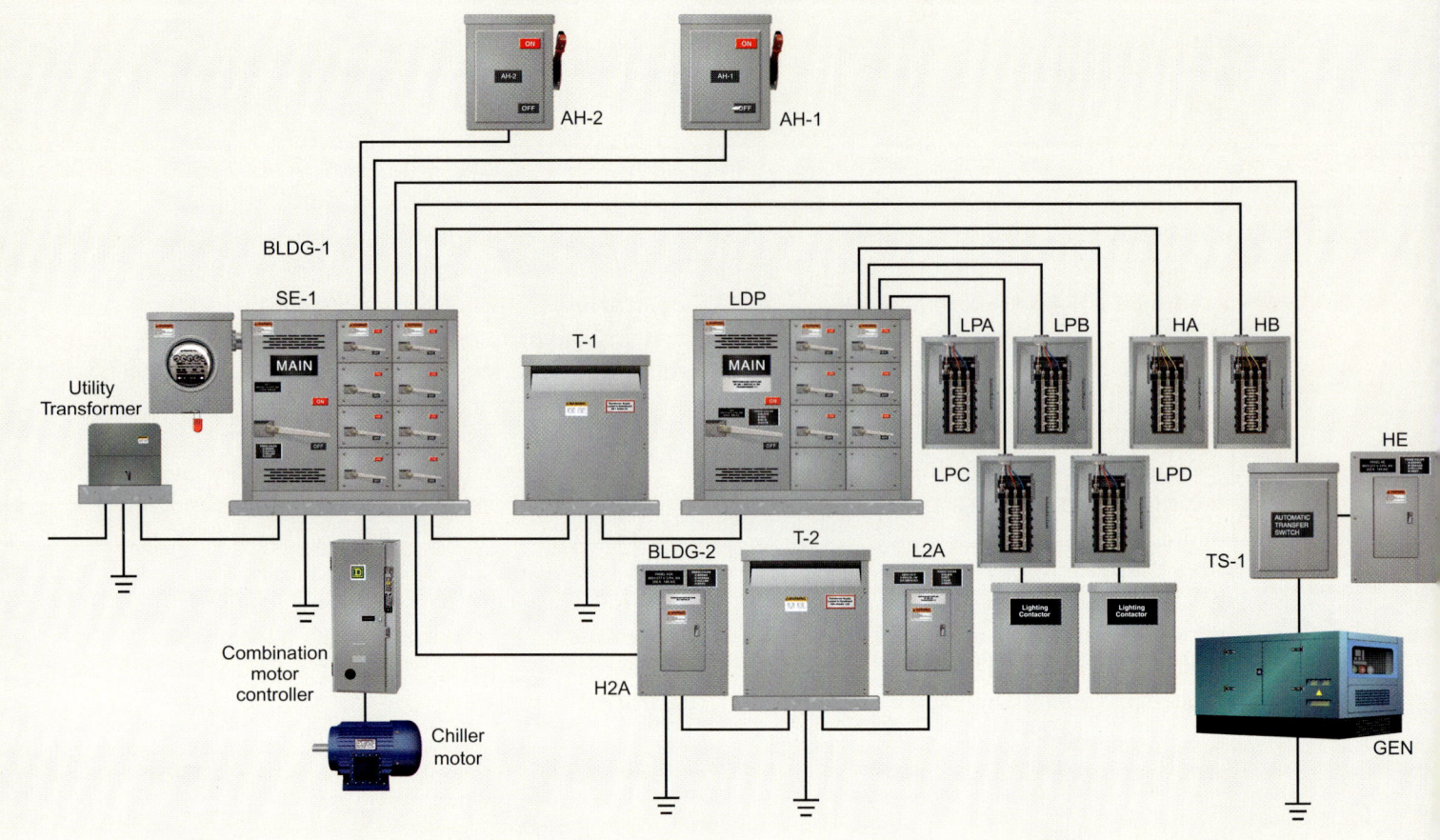

CHAPTER 12

Grounding Electrical Systems

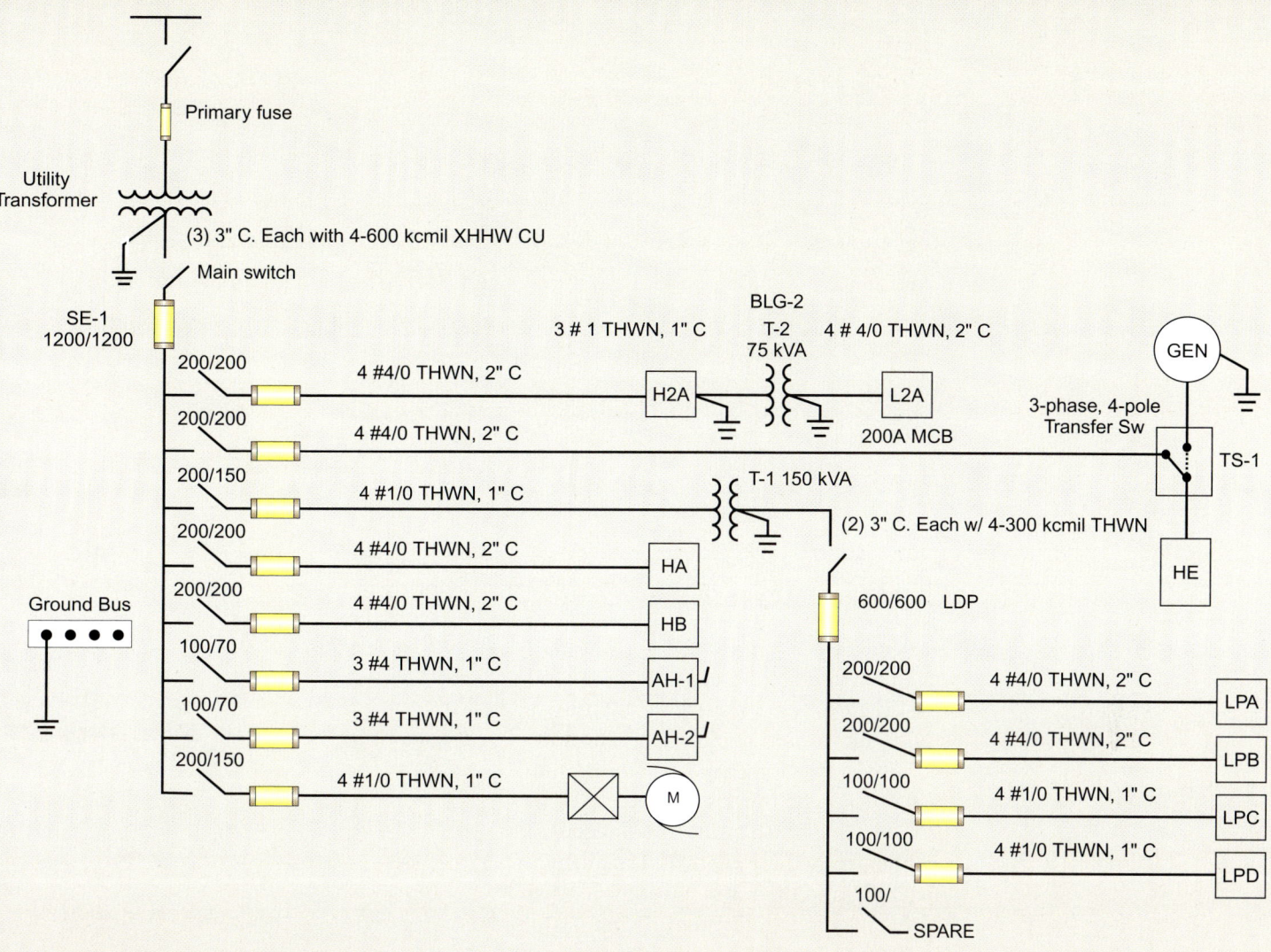

Objectives

- Understand the requirements for grounding electrical systems and differentiate between a system that is grounded and one that is not
- Determine when a system has to be grounded in comparison to when a system is permitted to be grounded by choice
- Understand when electrical systems are not permitted to be grounded and the requirements for ground detectors on ungrounded systems
- Identify specific installation rules for high-impedance grounded neutral systems
- Identify various methods for grounding electrical systems and review common voltages for grounded systems

Outline

Definitions

System Grounding

Methods of System Grounding

System Grounding Requirements

Mandatory System Grounding

Optional System Grounding

System Grounding Prohibited

High-Impedance Grounded Neutral Systems

Ungrounded Systems (Concepts)

Introduction

The *NEC®* includes many requirements for electrical system grounding, provisions for systems that are grounded by choice, and restrictions that prohibit certain systems from being grounded. Grounding methods can vary and include solid grounding, impedance or resistance grounding, grounding through surge arresters, and grounding through an inductor. Where an electrical system is not grounded, ground detectors are generally required to be installed. If system is grounded by a requirement—or if grounded by choice—all *NEC* requirements for grounded systems must be applied.

Definitions

The code rules applicable to grounding separately derived systems use several specific terms that must be fully understood for proper application. Grounding an electrical system means that one system conductor is connected to ground (the Earth) and a reference to ground from the system is established. Installing and operating an ungrounded system means no reference to ground (the Earth) from the system conductors is established other than through capacitance. Section 250.30 provides rules for separately derived systems that are grounded and those that are not grounded.

Grounded (Grounding). Connected (connecting) to ground or to a conductive body that extends the ground connection.[3]

Grounded Conductor. A system or circuit conductor that is intentionally grounded.[4]

Grounded, Solidly. Connected to ground without inserting any resistor or impedance device.[5]

Ground. The earth.[1]

Ungrounded. Not connected to ground or to a conductive body that extends the ground connection.[6]

Ground Fault. An unintentional, electrically conducting connection between an ungrounded conductor of an electrical circuit and the normally non–current-carrying conductors, metallic enclosures, metallic raceways, metallic equipment, or earth.[2]

Voltage to Ground. For grounded circuits, the voltage between the given conductor and that point or conductor of the circuit that is grounded; for ungrounded circuits, the greatest voltage between the given conductor and any other conductor of the circuit.[7]

System Grounding

System grounding is the process of establishing a connection from one conductor of the system to ground (the Earth). Therefore, when a system is grounded, one conductor of the system is connected to ground solidly; through an impedance device, resistor, or inductor; or by some other means. Grounded systems have a system grounded conductor. **See Figure 12-1.**

The system conductors are those supplied from the system, often referred to as the *source of the system*, and include all ungrounded conductors and the conductor that is grounded. For example, in a 120/240 volt, single-phase, three-wire system, there are three system conductors, two ungrounded conductors, and one grounded (neutral) conductor. **See Figure 12-2.**

In a 208Y/120 volt, three-phase, four-wire system, there are four system conductors on the secondary side, three ungrounded phase conductors, and one grounded (neutral) conductor. **See Figure 12-3.**

In a 240/120 volt, three-phase, four-wire, delta-connected system, there are four system conductors on the secondary side, three ungrounded phase conductors, and one grounded (neutral) conductor.

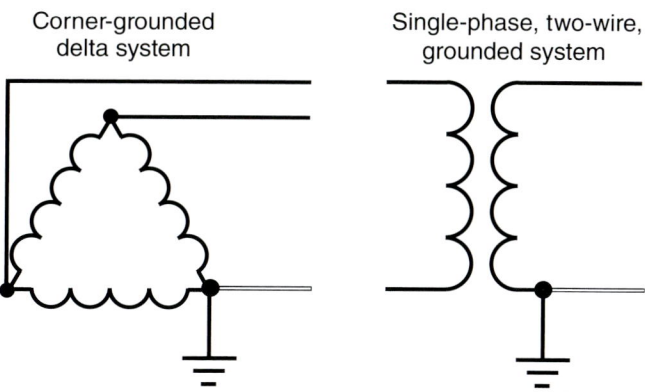

FIGURE 12-1 Grounded systems include a system conductor that is grounded either solidly or through an impedance device (solid grounding is shown).

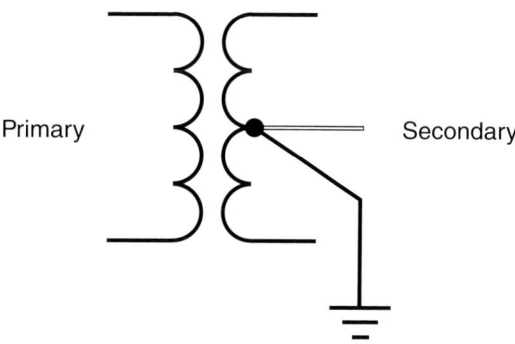

FIGURE 12-2 The grounded (neutral) conductor of a 120/240 V, single-phase, three-wire system is common to both ungrounded conductors of the system and is connected to the neutral point of the system.

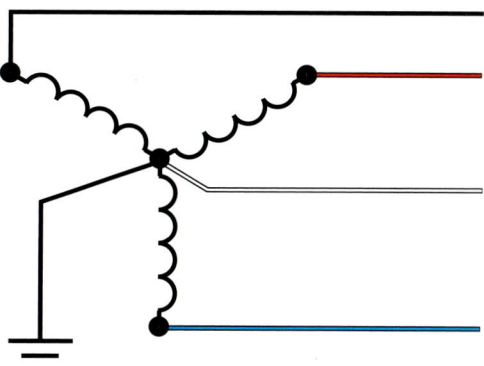

FIGURE 12-3 The grounded (neutral) conductor of a 208Y/120 V, three-phase, four-wire system is common to all three ungrounded phase conductors of the system and is connected to the neutral point of the system.

See **Figure 12-4.** This system is often referred to as a *high-leg delta system*.

If an electrical system is not grounded, there is no supply conductor of the system that is connected to the Earth. **See Figure 12-5.**

Webster's New Dictionary defines a *system* as a group of units so combined to form a whole and to operate in unison.[8] In terms of the entire electrical installation that operates as one in unison, it is commonly referred to as the electrical system of the premises. If a system is grounded, one conductor of the entire system is grounded (usually at the source), creating a grounded system for use on the premises. Common methods of producing systems on the premises include transformers, generators, photovoltaic arrays, and wind power turbines. The transformer is the most common system installed and used in premises wiring. Often, the utility supply to a commercial or industrial building is at a voltage and phase configuration such as 480Y/277-volt with three phases and four wires. To create usable voltages for 120-volt loads, transformers are usually installed. Each transformer that is installed on the premises is a new system (or source). The grounding requirements apply to each of those systems and depend on the voltage produced by the system. In other words, the level of voltage produced by an electrical system is directly related to the requirements for grounding the system or to whether the system is permitted to be operated ungrounded.

Methods of System Grounding

Electrical systems can be grounded in a variety of ways. **See Figure 12-6.** The grounding method chosen is often based on a *Code* requirement, an owner's specification, or part of an engineering design. Common grounding methods include solid grounding (connecting to ground solidly), reactance grounding (grounding through an inductor), resistance grounding (grounding through a low amount of resistance), high-resistance grounding (grounding through the highest permissible resistance), or grounding through surge arresters.

Where systems rated 600 volts and below are grounded, they are usually solidly grounded or grounded through a high-impedance device. Medium-voltage systems are grounded either solidly or through a resistor, and higher-voltage systems are typically grounded through surge arresters.

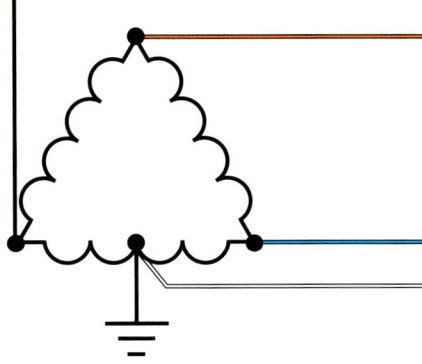

FIGURE 12-4 This 240/120 volt, three-phase, four-wire delta system with a high leg on the B phase has a grounded (neutral) conductor common to both of the ungrounded conductors of one phase winding of the system.

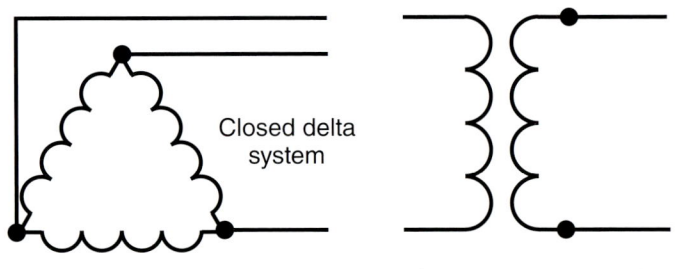

FIGURE 12-5 An ungrounded system has no conductor intentionally connected to ground.

System Grounding Requirements

Almost since the beginning of electricity use, there have been many discussions and even heated debates about the benefits of operating systems grounded as compared to ugrounded. The *NEC* rules today often make those choices. Certain electrical systems must be grounded, some systems are permitted to be grounded, and some systems are not permitted to be grounded. The *Code* provisions are broken down for each system application. Part II of Article 250 provides the requirements for electrical system grounding. The specific rules for grounding electrical systems are found in Section 250.20. Before moving on to the requirements for grounding systems, it is important to consider the purpose of grounding an electrical system. Section 250.4(A)(1) describes the purpose of system grounding and what is intended to be accomplished by grounding a system. Grounded systems are connected to the Earth in a fashion that limits voltage imposed by higher-voltage lines, line surges, lightning events, and so forth. Grounding a system also establishes a reference to the Earth from the system and stabilizes the voltage to ground during normal operation.[9] It should be understood that during abnormal events, such as a line surge or a lightning strike, the system potential and the potential on conductive enclosures of the system will attempt to rise for the duration of the event. Grounding helps keep the potential of the Earth and the connected system and equipment the same during the rise and fall of potential due to the reasons provided and as stated in Section 250.4(A)(5). A ground fault event attempts to force a rise in potential on grounded equipment and systems for the duration of the fault condition or until an overcurrent device opens the circuit. Grounding helps limit these above ground potentials during abnormal events such as ground faults or line surges and lightning events. The effective ground-fault current path provides a means to quickly facilitate overcurrent device operation, thus reducing the amount of time the abnormal condition exists.

Mandatory System Grounding

Section 250.20 includes the driving text indicating that electrical system grounding is required in accordance with Sections 250.20(A) and (B) depending on the voltage and phase arrangement of each system. If a system is not required to be grounded by the *Code* but the choice is made to ground the system, all system grounding rules in the *NEC* apply. An example of a system that is permitted to be grounded is a 480 volt, three-phase, delta-connected system.

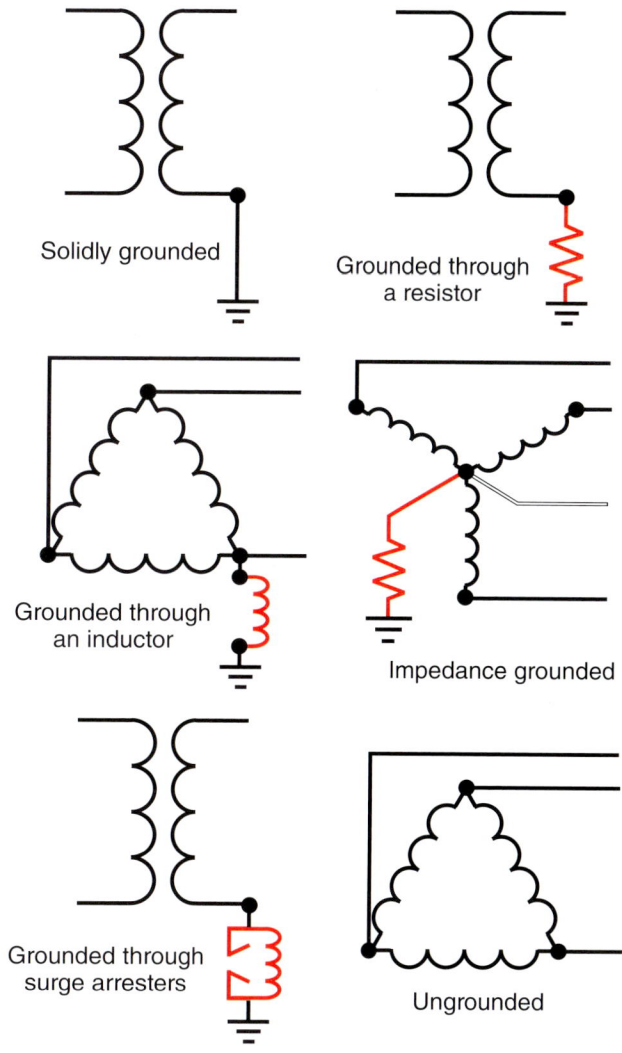

FIGURE 12-6 Systems can be grounded solidly or by other means.

Section 250.20(A) provides the requirements for grounding systems of less than 50 volts. Alternating current (AC) systems of less than 50 volts are required to be grounded under any of the following conditions:

1. Where supplied by transformers, if the supply system exceeds 150 volts to ground
2. Where supplied by transformers, if the supply system is ungrounded
3. Where installed outside as overhead conductors[10]

In item 1, if the supply of the transformer is 480 or 277 volts, the low-voltage secondary side has to be grounded. In item 2, if the supply system for the low-voltage transformer is not grounded (for example, supplied by a 480-volt, ungrounded system), the low-voltage side of the transformer has to be grounded. An example of item 3 is a low-voltage system or circuit installed as overhead conductors between buildings on a property. The reason for system grounding here is because these overhead conductors are more vulnerable to lightning. System grounding is necessary for increased safety of the system, conductors, components, and property. **See Figure 12-7.**

Section 250.20(B) addresses the grounding requirements for premises wiring and premises wiring systems of 50 volts to less than 1000 volts. Systems in this voltage range have to be grounded under any of the following conditions:

1. If the system can be grounded so that the maximum voltage to ground on the ungrounded conductors does not exceed 150 volts
2. If the system is three-phase, four wire and is wye connected and the system neutral conductor is used as a circuit conductor
3. If the system is three-phase, four wire and is delta connected in which the midpoint of one phase winding is used as a circuit conductor[11]

The preceding grounding requirements apply to many premises wiring systems installed today. In item 1, if the system can be grounded in a way that the phase-to-ground voltage is less than 150 volts, the system must always be grounded. An example of this is a single-phase, two-wire system with a 120-volt output (secondary). If one conductor or the other is grounded, the phase-to-ground voltage of the system is 120 volts, which is less than 150 volts. **See Figure 12-8.**

Here is an example of where grounding results in a voltage-to-ground greater than 150 volts. If a 240 volt, three-phase, three-wire, delta-connected transformer is grounded (corner grounded), the voltage to ground is 240 volts, which is greater than 150 volts. Therefore, grounding of this system is optional. **See Figure 12-9.**

By reviewing item 2, it is clear that this grounding requirement applies to systems such as 480Y/277 volt, three-phase, four-wire, wye-connect arrangements if the neutral of such systems is used as a circuit conductor. This is usually the case for these types of systems, so grounding is typically required.

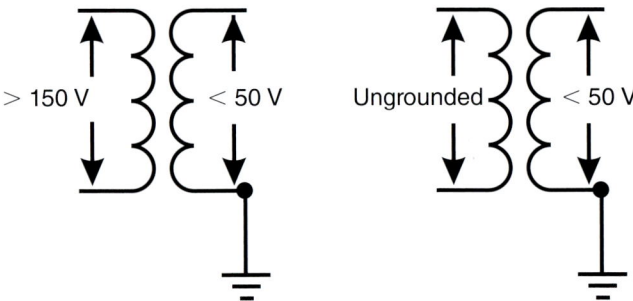

FIGURE 12-7 Systems less than 50 volts are required to be grounded in accordance with 250.20(A)(1), (2), or (3). Where conductors of systems less than 50 volts are outside buildings overhead, the systems are required to be grounded.

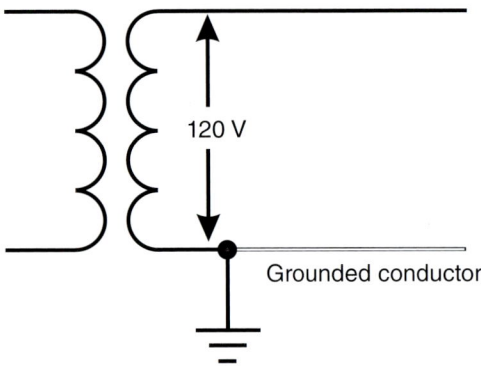

FIGURE 12-8 System grounding is required where the resulting voltage is less than 150 V to ground from any ungrounded conductors of the system.

Grounding could be an option for the 480Y/277 volt system because the voltage to ground exceeds 150 volts but only if the neutral of the system is not used as a circuit conductor. **See Figure 12-10.**

By reviewing item 3, it is clear that the system addressed in this section is a three-phase, four-wire, delta-connected system with a grounded midpoint of one phase winding in the bank. In other words, this is a high-leg, delta-connected system. The typical voltages of these systems are 240 volts from phase-to-phase around the delta and 120 volts from phase to ground on the transformer that has a grounded midpoint. From the B phase of such systems, the phase-to-ground voltage is 208 volts. The high-leg voltage results from taking the square root of three and multiplying it by the phase-to-ground voltage (120 volts). **See Figure 12-11.**

Because the grounded midpoint of the A and C phases creates a 120-volt output, the neutral created is usually used as a circuit conductor to supply 120-volt loads; thus, grounding such systems is required. Section 110.15 requires identifying the high leg of these systems with an orange marking or other effective means. An example of other effective means would be by tagging or using labels acceptable to the authority having jurisdiction. Section 408.3(E) also indicates that the high leg of a system is generally required to be the B phase except for high-leg connections in metering equipment or multi-section switchboards or panelboards that contain a metering section, according to Sections 408.3(E) and (F)(1). The delta systems shown in the figures of this textbook are closed delta systems. Be aware that open delta systems are also used; however, the kilovolt ampere (kVA) capacity of open delta systems is roughly 65% of that of closed-delta systems.

Grounded System Voltages

Systems can be solidly grounded in a few ways. How the system is grounded determines the resulting output voltage of the system. Wye-connected systems are typically grounded at the center or common point of the wye that connects all three of the phase windings.

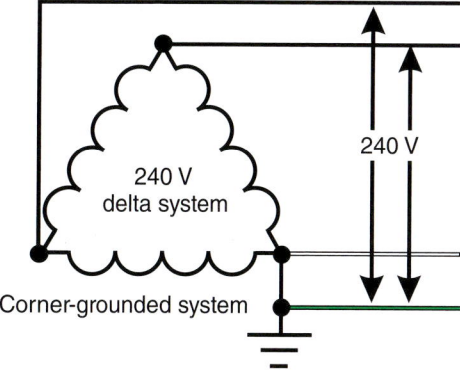

FIGURE 12-9 System grounding is not required where the phase-to-ground voltage is above 150 volts. Examples are 240 volt and 480 volt, three-phase delta systems.

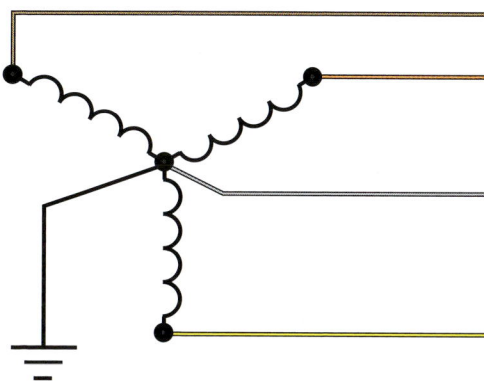

480Y/277 V, three-phase, four-wire system

FIGURE 12-10 Grounding is optional for a 480Y/277 volt system only where the neutral is not used as a circuit conductor and because the phase-to-ground voltage exceeds 150 volts.

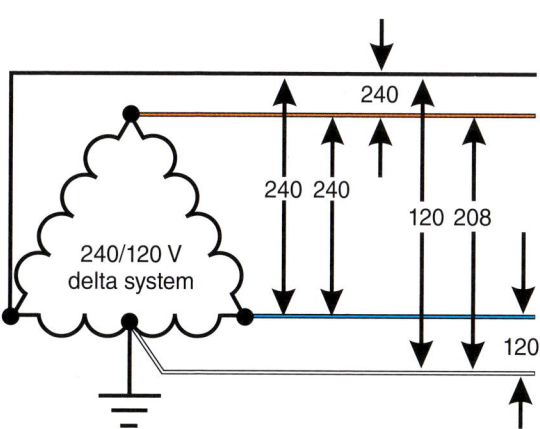

Phase to ground from the A and C phases is 120 V.
Phase to ground from the B phase is 208 V.

FIGURE 12-11 In a three-phase, four-wire, high-leg, delta-connected system, a neutral point is established between the ungrounded conductors of one transformer in the closed delta bank. This is the grounded neutral point of the system.

The terminal in a dry-type transformer with a wye-connected secondary is generally identified as the *XO terminal*. This is the center of the wye and the neutral point of the system. Typical wye-connected systems are 480Y/277 and 208Y/120 volt, three-phase, four-wire systems. **See Figure 12-12.**

Single-phase, two-wire systems that are grounded produce a phase-to-ground voltage that is the equivalent of the output voltage of the system. For example, a 120 volt, two-wire system that is grounded has a voltage of 120 volts between the output conductors and from the ungrounded conductor to ground. **See Figure 12-13.**

Delta-connected systems can be solidly grounded using a few methods. One method of grounding a three-phase, three-wire, delta-connected system is to ground one of the phase conductors. This creates a three-phase, corner-grounded system. The phase-to-phase voltage of a corner-grounded system is also the phase-to-ground voltage of that system. For example, the phase-to-phase voltage of a 480-volt delta system is 480 volts. If one phase is grounded, the phase-to-ground voltage from the other two phases is 480 volts. **See Figure 12-14.**

Delta-connected systems with the midpoint of one transformer grounded typically are used for 240/120 volt loads, as previously discussed.

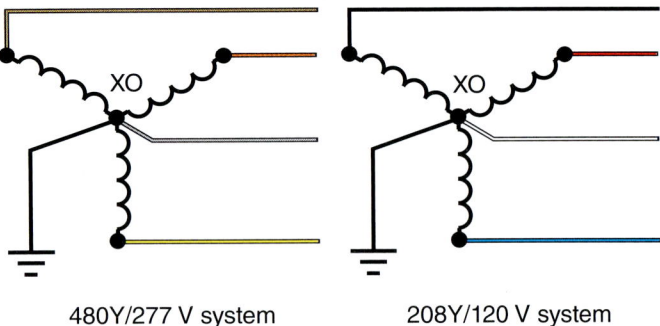

FIGURE 12-12 The neutral point of a typical three-phase, four-wire, wye-connected system is the point of grounding for such system and is usually identified as X0.

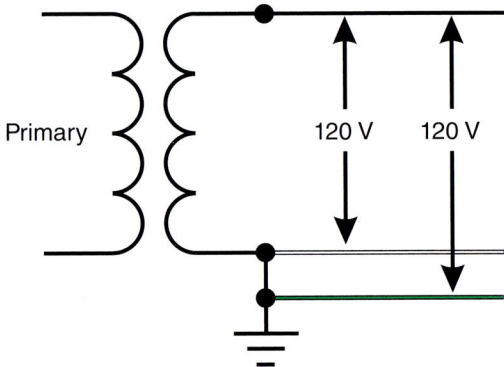

FIGURE 12-13 In a single-phase, two-wire system, the system output voltage (120) is the same as the voltage to ground from the ungrounded conductor of the system.

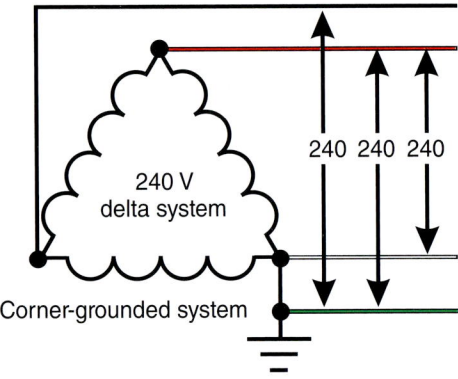

FIGURE 12-14 In a corner-grounded system, one phase of the system is grounded. Note that the grounded conductor is a phase conductor, not a neutral conductor.

Grounding Using Grounding Transformers

Delta systems can also be grounded using zigzag transformers, as covered in Section 450.5. The term *zigzag* is related to how the transformer is wound around the iron core. It is wound in one direction around the core and wound in the opposite direction back around the same core. Zigzag transformers are commonly referred to as *T-connected transformers* and are connected to three-phase, three-wire ungrounded systems to create a three-phase, four-wire distribution system or provide a neutral point for grounding purposes. **See Figure 12-15.** Such transformers must have a continuous per-phase current rating and a continuous neutral current rating. Zigzag-connected transformers are not permitted to be installed on the load side of any system grounding connection.

This method of grounding creates a ground reference for delta-connected systems and provides the ability to facilitate overcurrent device operation should a ground fault develop on any of the ungrounded conductors of such systems. The phase current in a grounding autotransformer is one-third of the neutral current. These systems should not be used to supply line-to-neutral loads.

Systems with voltages greater than 1000 volts are covered in Part X of Article 250. Where these systems are mobile or portable, such as on a trailer, they are required to be grounded. If they are permanently installed, they are permitted to be grounded as provided in Section 250.20(C). If an electrical system is grounded through an impedance device, it must be grounded in accordance with the requirements of either Section 250.36 or 250.186, depending on the system voltage.

Optional System Grounding

Section 250.21(A) provides a list of electrical systems that are permitted, but not required, to be grounded. These systems are as follows:

1. Systems exclusively for industrial electric furnaces for melting, refining, tempering, and so forth
2. Separately derived systems exclusively for rectifiers supplying only adjustable-speed industrial drives
3. Separately derived systems supplied by transformers that have a primary voltage rating of less than 1000 volts, meeting all of the following conditions:
    a. The system is used exclusively for control circuits.
    b. Qualified persons service the installation.
    c. Continuity of control power is required.
4. Other systems that are not required to be grounded in accordance with the requirements of Section 250.20(B)[12]

Typical systems permitted, but not required, to be grounded include 240 volts, three-phase, three-wire and 480 volt, three-phase, three-wire, delta-connected systems. If a three-phase, three-wire system is corner grounded, one phase conductor of the system is connected to the Earth (ground). The result is that the phase-to-ground voltage is the same as the phase-to-phase voltage. See Figure 12-16.

Grounding Separately Derived Systems

Where a system is separately derived and must be grounded because of the requirements in Section 250.20, or grounded as an option in accordance with the provisions in Section 250.21, it must be grounded in conformance with all applicable requirements in Section 250.30(A).

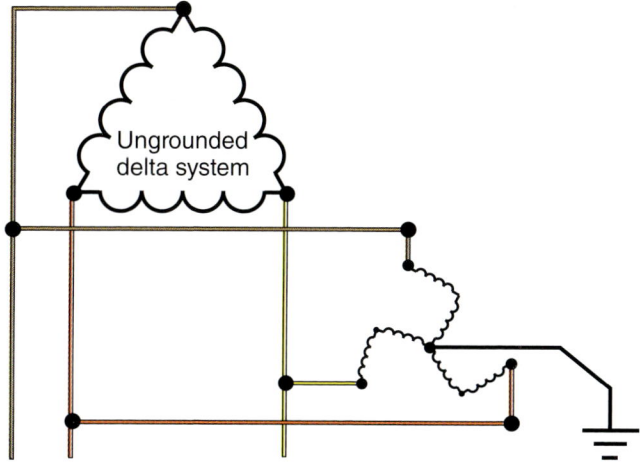

FIGURE 12-15 Grounding a delta using zigzag grounding transformers creates a grounded system (no neutral loads) and establishes effective ground-fault current path.

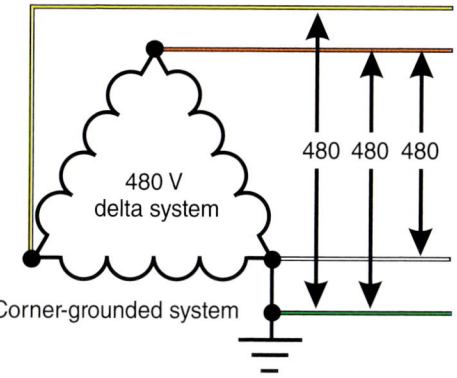

FIGURE 12-16 In a corner-grounded delta system, the phase-to-ground voltage is the same as the phase-to-phase voltage.

The requirements for grounding a separately derived system are driven by the output voltage, phase arrangement, and use of the system. Another feature for systems supplied by a generator is the switch arrangement provided in the transfer equipment. Informational Note 1 to Section 250.30 provides clear direction on grounding requirements for generators that have to be grounded in accordance with Section 250.30(A) because of the type of transfer switch applied in the design. Direct current (DC) systems are required to be grounded in accordance with Part VIII of Article 250.

Requirements for Ground Detection

For the systems addressed in Section 250.21(A), system grounding is not required. Where a system is not grounded and operates at no less than 120 volts and no greater than 1000 volts, ground detectors are required to be installed. **See Figure 12-17.** The requirement for ground detection provides the ability to monitor ungrounded systems to detect a first phase-to-ground fault on the system.

The first phase-to-ground fault will not cause overcurrent device operation, so continued service is achieved. However, it is important that those monitoring the system react to the annunciation, investigate the first phase-to-ground condition, and remove it. If the first phase-to-ground condition is not cleared and a second phase-to-ground fault develops on a different phase, the result is a simultaneous phase-to-phase short circuit and phase-to-ground fault event. This type of abnormal event can lead to significant destruction of equipment and downtime. There are some benefits of operating a system ungrounded, where permitted by the *NEC*, but it is important to monitor it and react appropriately if a phase-to-ground condition develops. Ungrounded systems are often installed and used in industrial facilities where continuity of power is desired for assembly lines and other continuous processes that would be damaged or could cause personal injury if a first phase-to-ground fault event were to result in interruption of power to the system. The choice to install and operate this type of system is determined from the nature of the process, the operational characteristics of the process and the operators/owners desired method of operation. Where ground detectors are installed on an ungrounded system, the sensors for such systems must be located as close to the supply source as possible, as indicated in Section 250.21(B)(2). Listed ground detection equipment is available for use on ungrounded systems.

According to Section 250.21(C), all enclosures containing equipment and conductors for an ungrounded system must be field-marked "Ungrounded System" at the source or at the first system disconnecting means. This marking has to be durable enough for the environment involved. There is also a specific marking requirement for switchboards and panelboards that contain ungrounded systems, as provided in Section 408.3(F)(2). This marking is required to read as follows:

Caution Ungrounded System Operating _____ Volts Between Conductors

This marking gives qualified people an additional notification of the type of system contained within the enclosure. The phase-to-ground voltage readings of such a system are not as familiar to workers as are grounded system voltage readings.

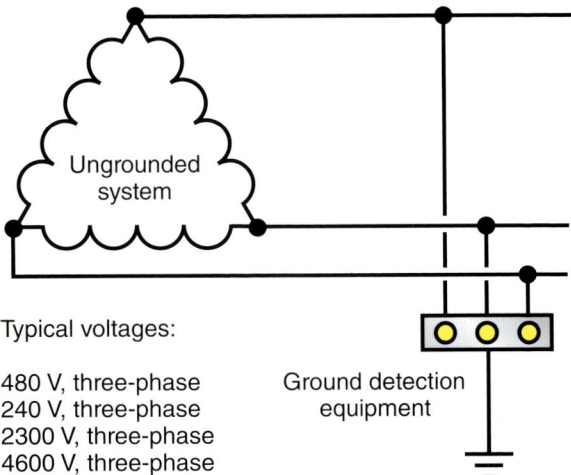

FIGURE 12-17 Ground detectors are required for ungrounded systems.

The marking requirement provides an added level of safety for personnel who must work on ungrounded systems either to troubleshoot or to add to such systems.

Conductor to be Grounded

If an AC electrical system is required to be grounded, one conductor of the system has to be connected to ground. Section 250.26 provides the information regarding which conductor of various systems is the conductor that must be grounded. **See Figure 12-18.** For a single-phase, two-wire system, either conductor supplied by the system can be grounded. This does not create a system neutral conductor; it creates a system grounded conductor. For a single-phase, three-wire system, the system neutral conductor has to be grounded. For multiphase systems having a conductor that is common to all phase conductors, the common (neutral) conductor has to be the grounded conductor. In a multiphase system that is corner or end grounded, any phase conductor can be grounded. For high-leg, delta-connected systems in which one phase winding is grounded at the midpoint to create a neutral, the neutral is required to be the grounded conductor.

Rules for the System Grounded Conductor

The *NEC* provides rules for grounded conductors. Usually, a grounded conductor is used throughout the system. This conductor must be identified according to the requirements in Section 200.6. **See Figure 12-19.** It is sometimes referred to in the field as an identified conductor, but in *Code* terms, it is a grounded conductor that is identified. The more common term used in the field is *neutral*. This is so because various other system conductors require identification. Section 200.6 has two basic identification rules. Section 200.6(A) deals with conductors in sizes 6 AWG and smaller, and Section 200.6(B) covers sizes 4 AWG and larger. The next sections review the identification requirements.[13]

Sizes 6 AWG or Smaller

Grounded (usually neutral) conductors in sizes 6 AWG and smaller must be identified generally by one of the following means:

1. Continuous white or gray insulation along the entire length of the conductor (Figure 12-19)
2. Three continuous white stripes on insulation other than green for the entire length of the conductor

See Sections 200.6(A)(4) through (8) for other specific conditions that require different identification means for grounded conductors.

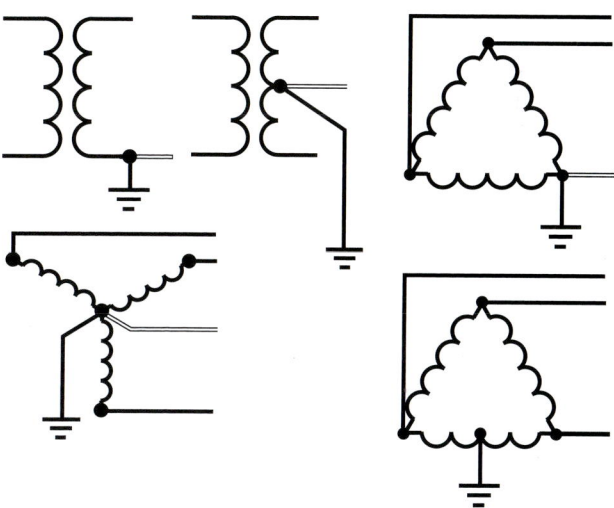

FIGURE 12-18 The conductor required to be grounded is specified in Section 250.26.

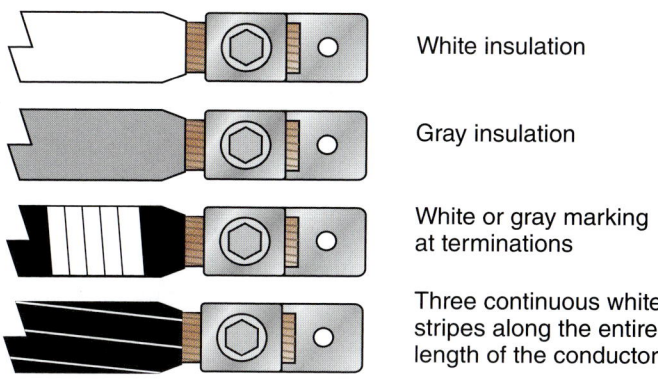

White insulation

Gray insulation

White or gray marking at terminations

Three continuous white stripes along the entire length of the conductor

FIGURE 12-19 Identification requirements for grounded conductors are provided in Section 200.6.

Sizes 4 AWG and Larger than 6 AWG

Grounded (usually neutral) conductors in sizes 4 AWG and larger than 6 AWG must be identified generally by one of the following means:

1. Continuous white or gray insulation along the entire length of the conductor
2. Three continuous white stripes on insulation other than green for the entire length of the conductor
3. Identification at the time of installation by a distinctive white or gray marking at the termination of the conductor; the marking has to encircle the conductor

There are other important *NEC* rules pertaining to grounded (usually neutral) conductors, such as the general restrictions from overcurrent protection, as provided in Section 240.22; use with multiwire branch circuits, as provided in Section 210.4; or a common neutral conductor for use with a feeder, as provided in Section 215.4(A).

Grounded separately derived systems typically include a neutral point to which a neutral conductor is connected. The terms *neutral point* and *neutral conductor* are defined in Article 100 as follows:

Neutral Point. The common point on a wye-connection in a polyphase system or midpoint on a single-phase, 3-wire system, or midpoint of a single-phase portion of a 3-phase delta system, or a midpoint of a 3-wire, direct-current system.[14]

Neutral Conductor. The conductor connected to the neutral point of a system that is intended to carry current under normal conditions.[15]

The informational note following the definition of a neutral point in Article 100 clarifies that at the neutral point of the system the vectorial sum of the nominal voltages from all other phases within the system that use the neutral, with respect to the neutral point, is zero potential. **See Figure 12-20.** These terms were added so that the appropriate conductor could be identified whenever this term is used in a requirement, such as in Sections 220.61, 250.26, and 250.36.

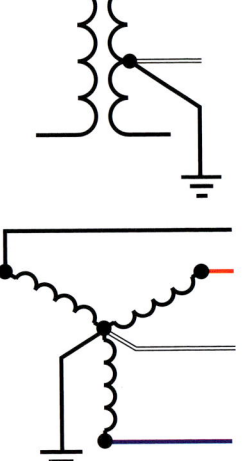
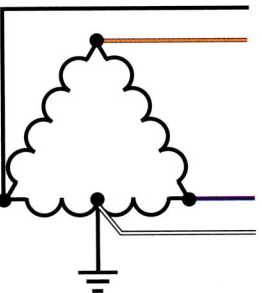

FIGURE 12-20 Neutral conductors are usually the grounded conductors of systems. The neutral conductor connects to the neutral point of electrical supply systems.

It is important to remember that the neutral conductor can be a current-carrying conductor. Many believe that, because the neutral conductor is a grounded conductor, it is safe to work on a neutral conductor while it is energized. This is a dangerous practice that has led to many serious electric shocks and electrocutions, specifically with multiwire branch circuits that share a neutral conductor. The definitions also clarify that a neutral conductor can be present at an outlet box that contains just a single 120-volt circuit supplying a receptacle. The reason is that the grounded conductor is a conductor that is connected to the neutral point of a supply system. Most neutral conductors are grounded conductors, but not all grounded conductors are neutrals. An example of a grounded conductor that is not a neutral is found in a three-phase, three-wire, corner-grounded system. The grounded conductor of this system is a grounded phase conductor, not a neutral conductor.

System Grounding Prohibited

The previous sections reviewed applicable rules for systems that are either required or permitted to be grounded. Section 250.22 provides the electrical systems that are not permitted to be grounded. Circuits for overhead cranes that operate over combustible fibers in Class III hazardous locations are not permitted to be grounded. The idea here is that a first phase-to-ground fault will not create a shower of sparks or hot particles that could cause a fire due to the accumulations of fibers on the floor below. This condition is common in textile mills because of the Class III locations associated with those manufacturing processes. Another type of system in Section 250.22 that is not permitted to be grounded is isolated power systems used in health care facilities. The requirements for isolated power systems in health care facilities are provided in Section 517.160. Section 250.22 also prohibits circuits for equipment within an electrolytic cell working zone, as provided in Article 668. Electrolytic cells are commonly used in the aluminum- and chlorine-processing industries. Secondary circuits of low-voltage lighting systems are not permitted to be grounded, as indicated in Section 411.5(A). Also, low-voltage lighting systems for underwater pool lighting supplied by isolation transformers are not permitted to be grounded. The listed transformers for these systems are of the isolation type with a grounded metal barrier between the primary and the secondary windings. Note that all of these systems are not permitted to be grounded but that the normally non–current-carrying metal parts of equipment enclosures and raceways that contain these ungrounded systems conductors and equipment are generally required to be grounded by connection to an equipment grounding conductor (EGC).

High-Impedance Grounded Neutral Systems

Section 250.20(D) provides the reference to Section 250.36 for installation of high-impedance grounded neutral systems. As previously mentioned, high-impedance grounded neutral systems have special requirements, all of which need to be followed. This type of installation requires special equipment, and various manufacturers produce the equipment designed specifically for these types of installations. What is a high-impedance grounded neutral system? This is a system in which an impedance device, usually a resistor, limits the current in a phase-to-ground fault condition to a low level. This allows the system to remain operational during the ground-fault condition. These types of systems are typically installed in industrial applications. High-impedance grounded neutral systems are permitted for use in AC systems with voltages ranging from 480 to 1000 volts, but all of the following restrictions must be met:

1. Conditions of maintenance and supervision ensure qualified persons service the installation.

2. Ground detection is provided for the system.
3. Line-to-neutral loads are not supplied by the system.[16]

Sections 250.36(A) through (G) provide the specific installation requirement for high-impedance grounded neutral systems. The grounding location has to be between the grounding electrode conductor and the system neutral point. If there is no neutral point, one has to be derived from a grounding transformer. More information about grounding transformers is provided in Section 450.5. The grounded (neutral) conductor from the neutral point of the source to where it connects to the impedance device has to be fully insulated and have an ampacity of no less than the maximum current rating of the grounding impedance. It can never be smaller than an 8 AWG copper or 6 AWG aluminum conductor. The system grounding connection shall only be made through the impedance device. **See Figure 12-21.**

The insulated neutral conductor of such systems is permitted to be installed in a separate raceway from the ungrounded conductors. Where an equipment bonding jumper is installed, it must be installed without splices and run from the first system disconnect or overcurrent device to the grounded side of the impedance device. The grounding electrode conductor must be connected from any point on the grounded side of the impedance device to the equipment grounding connection at the service equipment or first system disconnecting means. The minimum size required for the equipment bonding jumper of a high-impedance grounded neutral system has to be determined in accordance with either of the following:

1. Where the grounding electrode conductor connection is made at the grounding impedance, the equipment bonding jumper shall be sized in accordance with Table 250.66, based on the size of the service entrance conductors for a service or the derived phase conductors for a separately derived system.
2. Where the grounding electrode conductor is connected at the first system disconnecting means or overcurrent device, the equipment bonding jumper shall be sized the same as the neutral conductor in Section 250.36(B).[17]

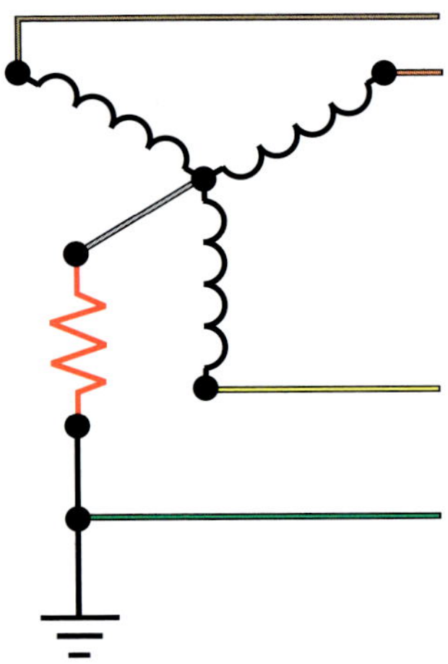

Ungrounded Systems (Concepts)

The decision to install and operate an ungrounded system is typically a combined effort that includes a design or engineering team, the owner, the operators, and sometimes the authority having jurisdiction. Common reasons for choosing to operate an ungrounded system are providing continuity of electrical operation and minimizing downtime from system outages. Common ungrounded delta systems are as follows:

240 volt, three-phase, three-wire, delta-connected

480 volt, three-phase, three-wire, delta-connected

2300 volt, three-phase, three-wire, delta-connected

4600 volt, three-phase, three-wire, delta-connected

13 800 volt, three-phase, three-wire, delta-connected

Grounding location is between the grounding electrode conductor and the neutral grounding point.

Neutral conductor has to be fully insulated.

Neutral conductor must have an ampacity no less than the maximum current rating of the grounding impedance.

Grounding connection can only be made through the grounding impedance device.

FIGURE 12-21 The neutral of a high-impedance grounded neutral system has to be fully insulated and no equipment grounding or equipment bonding connections are permitted on the supply side of the impedance device.

The disadvantages of an ungrounded system is that a first phase-to-ground fault condition can be difficult to find and can take a considerable amount of investigation and time. The voltage to ground in an ungrounded system is in theory 0 volts because there is no ground connection from any system conductor. But there is distributed leakage capacitance present throughout such systems. Phase-to-ground voltage levels that can appear result from capacitance coupling effects from the system circuits. **See Figure 12-22.**

Another important point about voltage-to-ground levels in ungrounded systems is covered in the definition of the term *voltage to ground*. The definition clarifies that the voltage to ground of a grounded system is the voltage between the given conductor and that point or conductor of the circuit that is grounded. For example, in a 120/240 volt, single-phase system, the voltage is 120 volts from any ungrounded phase conductor to ground. However, for ungrounded systems, the greatest voltage between the given conductor and any other conductor of the circuit is also the phase-to-ground voltage. For example, on a 480 volt, three-phase, three-wire, ungrounded delta system, the phase-to-phase voltage is 480 volts. This (480 volts) is also the phase-to-ground voltage for this system, based on the definition.

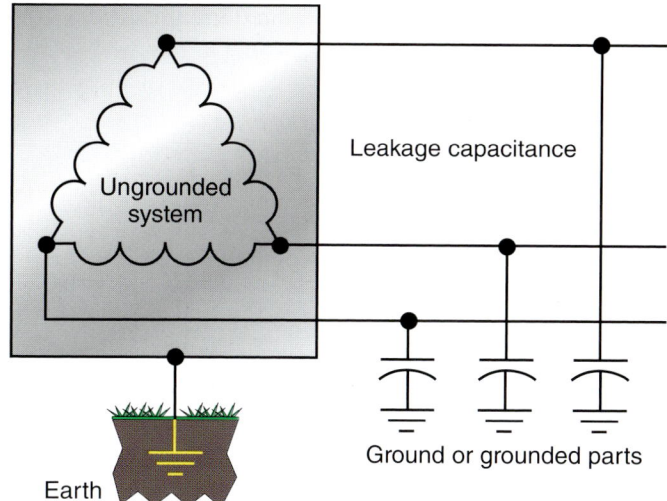

FIGURE 12-22 Phase-to-ground voltage in ungrounded systems can result from distributed leakage capacitance.

Summary

Methods of grounding can vary and include solid grounding, impedance or resistance grounding, grounding through surge arresters, and grounding through an inductor. The method of grounding is not always a matter of choice or a design consideration, but in many cases it is required by the *NEC*; then there is no choice, as the *Code* makes this decision. The *NEC* indicates the systems required to be grounded, systems permitted to be grounded, and those systems not permitted to be grounded. Ungrounded systems must generally be provided with ground detection systems.

References

1. NFPA 70 National Electrical Code 2011, Article 100 (National Fire Protection Association, Quincy, MA 2010), p. 70–29.
2. NFPA 70 National Electrical Code 2011, Article 100 (National Fire Protection Association, Quincy, MA 2010), p. 70–29.
3. NFPA 70 National Electrical Code 2011, Article 100 (National Fire Protection Association, Quincy, MA 2010), p. 70–29.
4. NFPA 70 National Electrical Code 2011, Article 100 (National Fire Protection Association, Quincy, MA 2010), p. 70–29.
5. NFPA 70 National Electrical Code 2011, Article 100 (National Fire Protection Association, Quincy, MA 2010), p. 70–29.
6. NFPA 70 National Electrical Code 2011, Article 100 (National Fire Protection Association, Quincy, MA 2010), p. 70–29.

7. NFPA 70 National Electrical Code 2011, Article 100 (National Fire Protection Association, Quincy, MA 2010), p. 70–33.
8. *Webster's New Dictionary of the English Language*, System (Popular Publishing Company, New York, NY 2002), p. 525.
9. NFPA 70 National Electrical Code 2011, Section 250.4(A)(1) (National Fire Protection Association, Quincy, MA 2010), p. 70–101.
10. NFPA 70 National Electrical Code 2011, Section 250.20(A) (National Fire Protection Association, Quincy, MA 2010), p. 70–104.
11. NFPA 70 National Electrical Code 2011, Section 250.20(B) (National Fire Protection Association, Quincy, MA 2010), p. 70–104.
12. NFPA 70 National Electrical Code 2011, Section 250.21(A) (National Fire Protection Association, Quincy, MA 2010), p. 70–104.
13. NFPA 70 National Electrical Code 2011, Section 200.6 (National Fire Protection Association, Quincy, MA 2010), p. 70–46.
14. NFPA 70 National Electrical Code 2011, Article 100 (National Fire Protection Association, Quincy, MA 2010), p. 70–30.
15. NFPA 70 National Electrical Code 2011, Article 100 (National Fire Protection Association, Quincy, MA 2010), p. 70–30.
16. NFPA 70 National Electrical Code 2011, Section 250.36 (National Fire Protection Association, Quincy, MA 2010), p. 70–110.
17. NFPA 70 National Electrical Code 2011, Section 250.36 (National Fire Protection Association, Quincy, MA 2010), p. 70–110.

Review Questions

1. The *NEC* specifies which electrical systems are required to be grounded, which systems are permitted to be grounded, and which systems are not permitted to be grounded.
    a. True
    b. False
2. The term _____ means connected to ground or to a conductive body that extends the ground connection.
    a. Grounded conductor
    b. Grounded (grounding)
    c. Bonded (bonding)
    d. Grounding electrode
3. Grounded conductors _____ or larger than 6 AWG are permitted to be identified either in the same manner as that required for conductors smaller than 6 AWG, or, at the time of installation, by a distinctive white or gray marking at the terminations.
    a. Of aluminum and smaller
    b. 4 AWG
    c. Of copper and smaller
    d. Larger
4. A 480 volt, three-phase, three-wire, delta-connected system is always required to be grounded.
    a. True
    b. False
5. Where 30-volt AC systems are installed outdoors as overhead conductors, they _____ be grounded.
    a. Are not required to
    b. Shall not
    c. Shall be permitted to
    d. Are required to
6. Which of the following conductors is not present when installing an ungrounded system?
    a. Grounding electrode conductor
    b. Equipment grounding conductor
    c. Grounded conductor
    d. Ungrounded conductors
7. Where a system can be grounded so that the maximum voltage to ground from any system ungrounded conductor does not exceed 240 volts, the system must be grounded.
    a. True
    b. False
8. The phase-to-ground voltage of a high leg in a 240/120 volt, three-phase, four-wire, delta-connected system is _____ volts.
    a. 120
    b. 208
    c. 240
    d. 480

9. Separately derived systems supplied by transformers having a primary voltage rating of less than 1000 volts are not required to be grounded if all but the following condition is met:
   a. This system is used for control circuits only.
   b. Qualified persons service this installation.
   c. Continuity of control power is required.
   d. Continuity of line power is required.
10. Where a transformer secondary is 30 volts (AC) and the primary is supplied by an ungrounded source, the secondary must be grounded.
    a. True   b. False
11. Which of the following circuits is required to be grounded?
    a. Circuits for cranes over Class III locations
    b. 120 V lighting circuits
    c. Lighting systems as provided in Section 411.5(A)
    d. Isolated power systems for health care facilities
12. A _____ system does not have a neutral point.
    a. 480 V, three-phase, three-wire, delta
    b. 480Y/277 V, three-phase, four-wire, wye-connected
    c. 208Y/120 V, three-phase, four-wire, wye-connected
    d. 120/240 V, single-phase, three-wire
13. A 240 volt, three-phase, three-wire, delta-connected system is required to be grounded.
    a. True   b. False
14. Which of the following circuits are permitted to be grounded?
    a. Circuits for equipment within electrolytic cell working zones, as provided in Article 668
    b. A 480 V, three-phase, three-wire, delta system
    c. Isolated power systems in health care facilities covered by Section 517.160
    d. Secondary circuits of lighting systems for pools, as covered in Section 680.23(A)(2)
15. A neutral conductor is defined in the *NEC* as the conductor connected to the neutral point of a system and that generally carries current under normal conditions.
    a. True   b. False
16. If grounding of a separately derived system is required by the *NEC*, it must be as specified in accordance with Section _____ of the *NEC* rules.
    a. 250.30(B)
    b. 250.30(A)
    c. 250.20(B)
    d. 250.24
17. An AC system supplying premises wiring has to be grounded where the maximum voltage to ground from any ungrounded conductor does not exceed 150 volts.
    a. True   b. False
18. A phase-to-ground fault in an ungrounded system accidentally grounds the system and activates ground detectors.
    a. True   b. False
19. If a system is not grounded, only one conductor supplied by the system is intentionally grounded.
    a. True   b. False
20. Ground detectors are not required for 240 volt, three-phase, ungrounded, delta-connected systems.
    a. True   b. False
21. An insulated grounded conductor of a system is required to be identified by any of the following methods *except* _____.
    a. Green insulation
    b. White insulation
    c. Gray insulation
    d. Three white stripes along the entire length of the conductor
22. What is the phase-to-ground voltage of the ungrounded conductors supplied by a 480 volt, three-phase, three-wire, corner-grounded delta system?
    a. 277 V
    b. 240 V
    c. 120 V
    d. 480 V
23. The voltage readings from the phase conductors of an ungrounded system are due to distributed leakage capacitance in the system.
    a. True   b. False

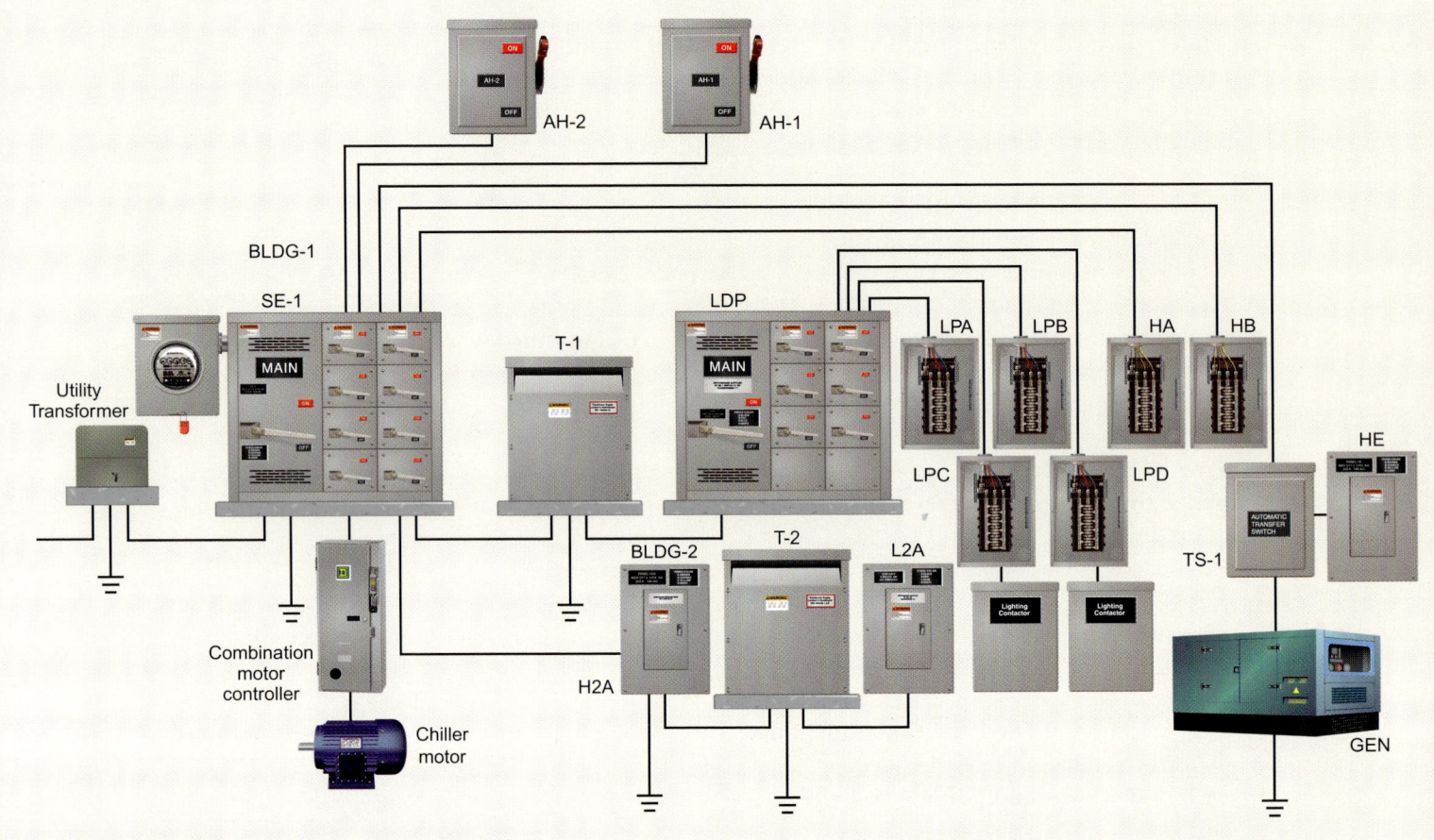

CHAPTER 13

Grounding and Bonding for Separately Derived Systems

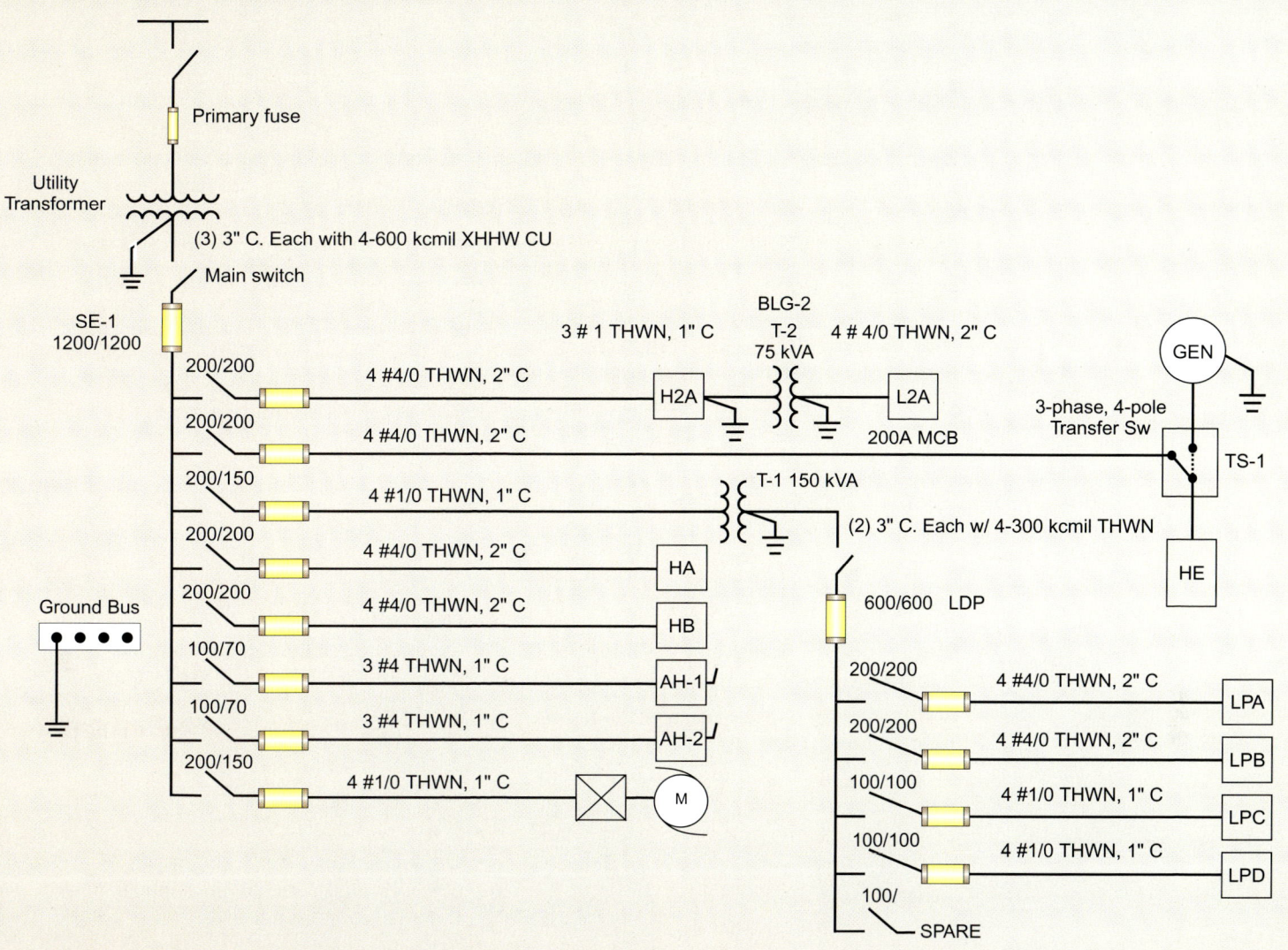

Objectives

- Understand what constitutes a separately derived system as it is defined and how to distinguish it from systems that are not separately derived
- Determine *NEC* grounding and bonding requirements that apply to separately derived systems
- Understand the application of the common grounding electrode conductor tap concept for multiple separately derived systems
- Size grounding electrode conductors, system bonding jumpers, supply side bonding jumpers, and grounded conductors for a separately derived system
- Determine how to ground and bond a generator-type separately derived system
- Understand the relationship between the transfer equipment and the system grounding requirements for generator-type separately derived systems

Outline

Definitions

Determining a Separately Derived System

Grounding Requirements

Grounded Systems

Grounding Electrodes

Bonding Water Piping and Building Steel

Outdoor Source

Ungrounded Systems

Generators and Transfer Equipment

Small Wind Electrical Systems

Grounding DC Systems

Ungrounded DC Systems

Introduction

Electrical wiring and power distribution systems for commercial, industrial, institutional, and even some residential occupancies typically include the installation and use of separately derived systems. Generally, a separately derived system is a separate power source, such as a generator, photovoltaic system, wind turbine, battery, or other source that produces electrical power. The *NEC®* provides an extensive set of rules specifically related to grounding and bonding for separately derived systems. To understand and properly apply these rules to separately derived system installations, the meaning of certain terms must be clear.

Definitions

The use of *NEC* terminology is important to determine which rules apply to installations and systems. A variety of "*Code-*defined" terms are used on the subject of grounding and bonding of separately derived systems. A common language of communication must be used when applying *Code* rules to any system or installation.

Bonded (Bonding). Connected to establish electrical continuity and conductivity.[1]

Bonding Jumper, Supply-Side. A conductor installed on the supply side of a service or within a service equipment enclosure(s), or for a separately derived system, that ensures the required electrical conductivity between metal parts required to be electrically connected.[2]

Bonding Jumper, System. The connection between the grounded circuit conductor and the supply-side bonding jumper, or the equipment grounding conductor, or both, at a separately derived system.[3]

Grounded (Grounding). Connected (connecting) to ground or to a conductive body that extends the ground connection.[4]

Grounding Electrode. A conducting object through which a direct connection to earth is established.[5]

Grounding Electrode Conductor. A conductor used to connect the system grounded conductor or the equipment to a grounding electrode or to a point on the grounding electrode system.[6]

Separately Derived System. A premises wiring system whose power is derived from a source of electric energy or equipment other than a service. Such systems have no direct connection from circuit conductors of one system to circuit conductors of another system, other than connections through the earth, metal enclosures, metallic raceways, or equipment grounding conductors.[7]

Determining a Separately Derived System

One of the keys to determining whether or not a system is separately derived has to do with it not having a "direct electrical connection" to another source. **See Figure 13-1.** Some examples of separately derived systems include generators, batteries, converter windings, transformers, solar photovoltaic systems, and wind turbine generators. Common separately derived systems installed for commercial projects include transformers and generators.

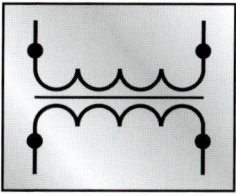
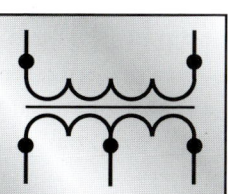

Transformers

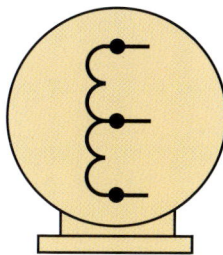

Generators

FIGURE 13-1 Separately derived systems produce electrical power and have no direct electrical connection between the circuit conductors of another supply system.

Dry-type transformers are often installed as a separately derived system in commercial and industrial installations and are rarely installed in residential applications.

Courtesy of Jim Dollard, IBEW Local 98

Generator-type separately derived systems are often installed as optional standby systems or emergency systems for commercial and industrial facilities. Generators can be separately derived power systems depending on how they are interconnected with other systems.

Some transformers do not qualify as separately derived systems because one winding in the transformer is common to both the input and the output side of the transformer. These are autotransformers addressed by requirements in Article 450. **See Figure 13-2.** Examples of autotransformers are a core and coil ballast in a fluorescent luminaire and a "buck-boost" transformer used for raising or lowering voltage levels for a particular application or single piece of utilization equipment.

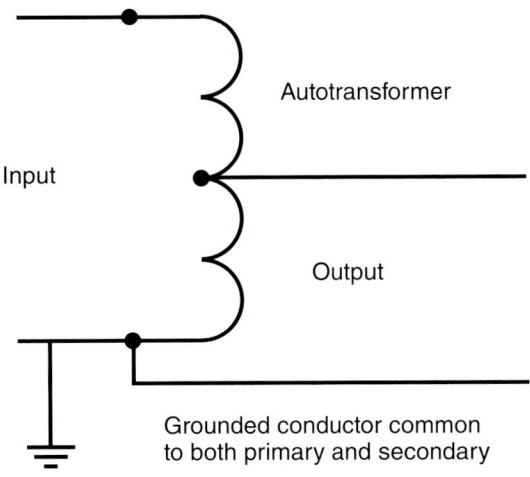

FIGURE 13-2 Autotransformers are not separately derived systems because one system or circuit conductor is common to both the input and output side of the transformer.

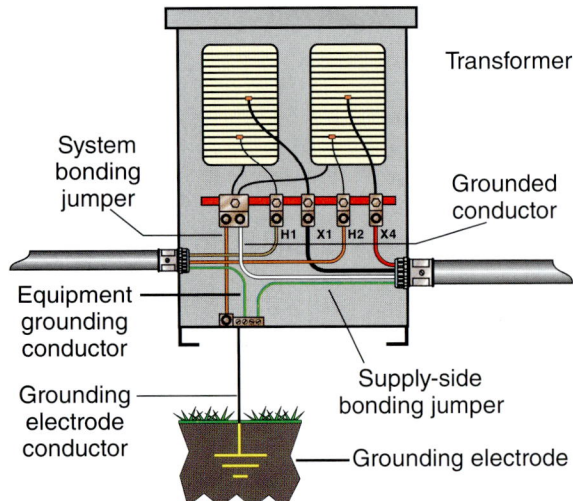

FIGURE 13-3 There are multiple grounding and bonding components necessary for transformer separately derived systems.

Grounding Requirements

Grounding a separately derived system means that the system itself will be connected to the Earth, in addition to the enclosure containing the system, if applicable. This grounding connection happens through a grounding electrode. When a system is connected to the Earth, one conductor supplied by the system is intentionally grounded. This creates a grounded conductor of this system to which all *NEC* rules for grounded conductors must apply. Section 250.30 indicates that the grounding requirements of Sections 250.30(A) and (B) apply to separately derived systems. Sections 250.20, 250.21, and 250.26 are also directly related to the requirement for grounding a separately derived system. If the system is required to be grounded because of provisions in any of these rules, the system must be grounded according to the rules in Section 250.30(A), which is one of the longest sections in the *NEC*. This chapter breaks the provisions of 250.30(A) into small pieces to clarify how to ground and bond separately derived systems. The arrangement of this chapter mostly follows the sequence of Section 250.30.

Grounded Systems

If a separately derived system is required to be grounded, based on the provisions in Section 250.20 or 250.21, the rules in Sections 250.30(A)(1) through (8) apply. There are some components that are common to the grounding and bonding scheme of a typical separately derived system (transformer type). **See Figure 13-3.**

Regardless of the type of separately derived system, the same grounding and bonding components are typically installed and thus, the same *NEC* requirements apply. Because separately derived systems are another source of power, the grounding and bonding system has similar physical and operational characteristics to those for service equipment.

The same concerns about keeping neutral current on its intended conductive path are provided in the rules for separately derived systems. It is clear that, unless otherwise permitted in Article 250, the grounded (usually a system neutral) conductor is generally not permitted to be re-grounded on the load side of the system bonding jumper. Impedance grounded neutral systems have to meet the requirements for neutral conductor connections, as provided in Section 250.36 or 250.186.

System Bonding Jumper

The system bonding jumper of a separately derived system connects the grounded conductor of the system to the enclosure, to the supply-side bonding jumper, to the grounding electrode conductor, and to an equipment grounding conductor (EGC) if there is a primary supply to a transformer-type derived system. **See Figure 13-4.**

Courtesy of Bill McGovern, City of Plano, TX

Separately derived systems can be in the form of inverter windings such as those used in micro-inverters.

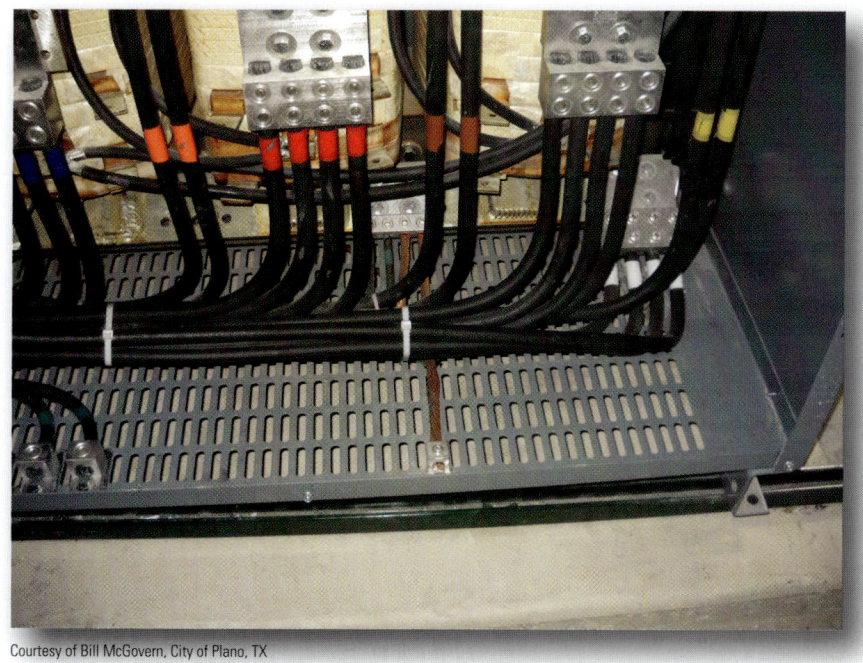

Courtesy of Bill McGovern, City of Plano, TX

FIGURE 13-4 A system bonding jumper is permitted to be installed at the dry-type transformer enclosure or the first system overcurrent device enclosure supplied by the system.

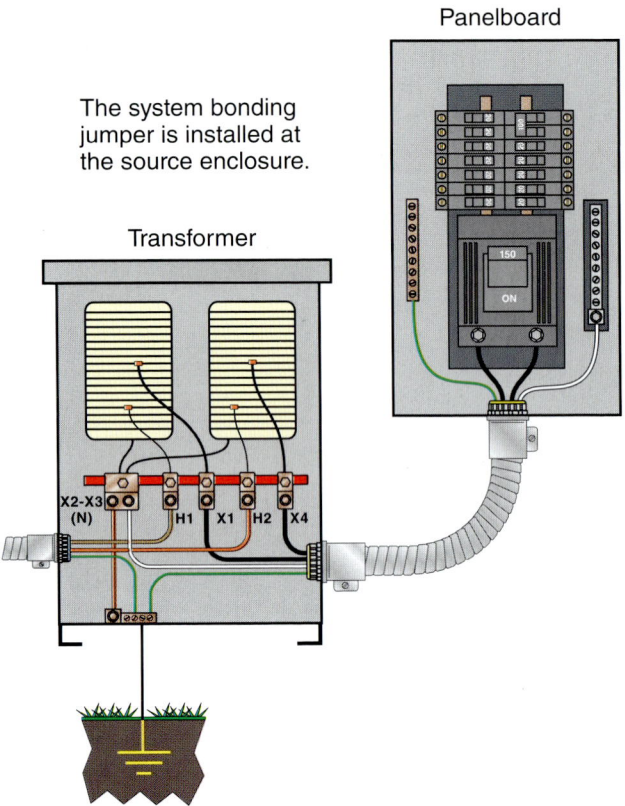

FIGURE 13-5 The system bonding jumper connects the grounded conductor to the supply-side bonding jumper at the separately derived system source enclosure.

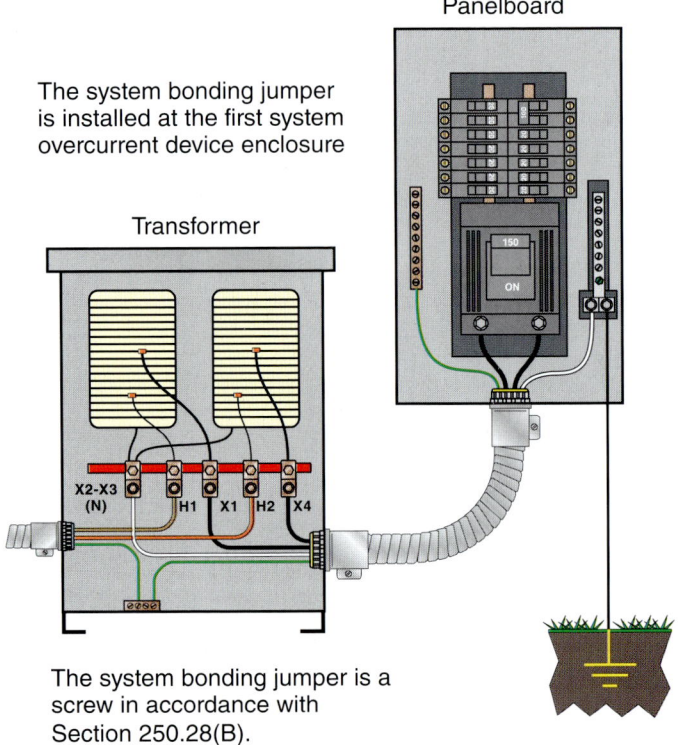

FIGURE 13-6 The system bonding jumper is permitted to connect the grounded conductor to the supply-side bonding jumper or EGC in the first system disconnecting means enclosure.

The system bonding jumper can be located at any point from the source enclosure up to the first system overcurrent device or disconnecting means.

If there is no disconnecting means or overcurrent device at the load end of the conductors supplied by the system, the system bonding jumper must be installed in the source enclosure. **See Figure 13-5.** Where the system bonding jumper is installed in the source enclosure, it must connect the grounded conductor to the supply-side bonding jumper and to the metal enclosure. If the system bonding jumper is installed at the first disconnecting means enclosure, it must connect the grounded conductor to the supply-side bonding jumper, to the disconnecting means enclosure, and to the EGC(s). **See Figure 13-6.** The *Code* also requires the system bonding jumper to remain within the enclosure in which it originates. Note that for a transformer-type separately derived system there are usually an EGC with the primary feeder connection and a supply-side bonding jumper with the secondary connection. These two are connected to the source enclosure. **See Figure 13-7.** It is important that this connection does not constitute a direct connection between the system conductors from primary to secondary.

The system bonding jumper must be installed according to Sections 250.28(A) through (D). Where a system bonding jumper is a wire type, it is sized using Table 250.66 or the 12.5% rule based on the size of the largest ungrounded derived phase conductor or conductors or the total circular mil (cm) area of all conductors connected to any one ungrounded phase at the source. **See Figure 13-8.** As an example, if the derived phase conductors connected to a transformer secondary are 750 kcmil copper, the minimum size required for a system bonding jumper is no less than 2/0 copper or 4/0 aluminum.

The term *derived phase conductors* is often used in discussions about separately derived systems. This term relates to the conductors connected to the output side of a power source that is by definition a separately derived system.

The *Code* refers to these conductors as the ungrounded secondary conductors connected to a separately derived system. The circular mil area of these conductors is used for sizing grounding electrode conductors, system bonding jumpers, and supply-side bonding jumpers that are part of a separately derived system. Table 250.66 and the 12.5% rule are used for establishing these sizes rather than Table 250.122, which is used for sizing EGCs.

Normally, there is only one system bonding jumper for a separately derived system that is grounded. By exception, a system bonding jumper is permitted at the source enclosure and the first disconnecting means enclosure if there are no parallel paths for current in the grounded conductor of the system.

An example of an installation where two system bonding jumpers would be acceptable by Section 250.30(A)(1) Exception 2 is an installation for a transformer outside a building or structure that supplies an open-bottom switchboard and has polyvinyl chloride (PVC) conduit installed between the transformer enclosure and the switchboard. **See Figure 13-9.** In this type of installation, the grounded conductor not only carries the neutral load in normal operation but also serves as an effective ground-fault current path during ground-fault conditions. This is why it is important that the minimum size for the grounded conductor, in this case, be no smaller than the system bonding jumper. Because the Earth does not qualify as an effective path for current, especially fault current, this example meets the provisions of the exception.

Supply-Side Bonding Jumper

The requirements for supply-side bonding jumpers installed for separately derived systems are provided in Section 250.30(A)(2). The conductors supplied from a transformer (separately derived system) are typically installed in accordance with Section 240.21(C) because they are transformer secondary conductors.

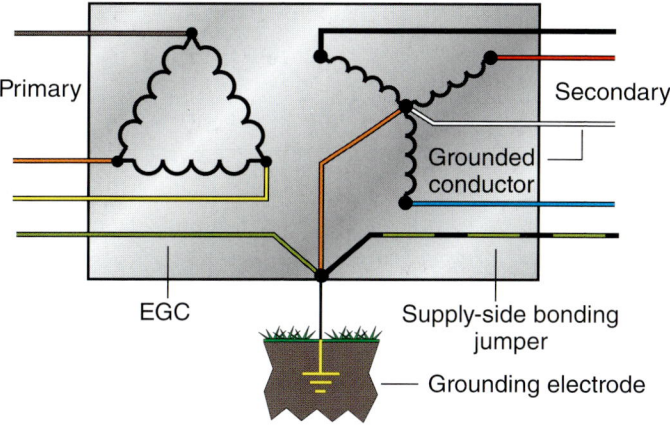

FIGURE 13-7 The EGC of the transformer primary and the supply-side bonding jumper of the transformer secondary are both connected to the source enclosure.

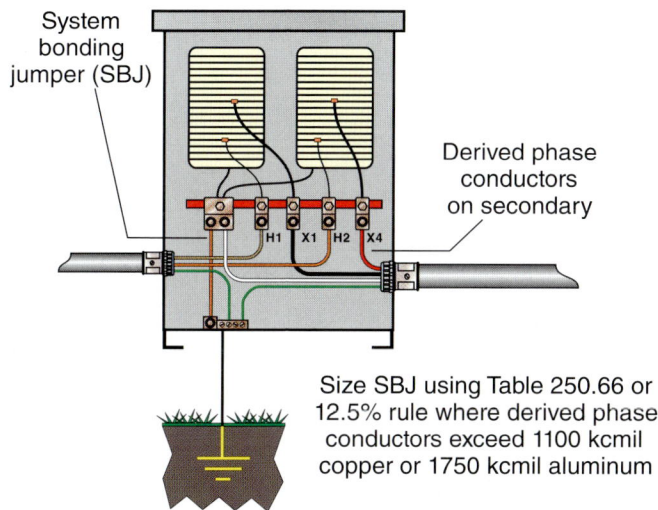

FIGURE 13-8 A system bonding jumper is sized based on the circular mil (cm) area of the largest ungrounded derived phase conductor and must remain in the enclosure from which it originates.

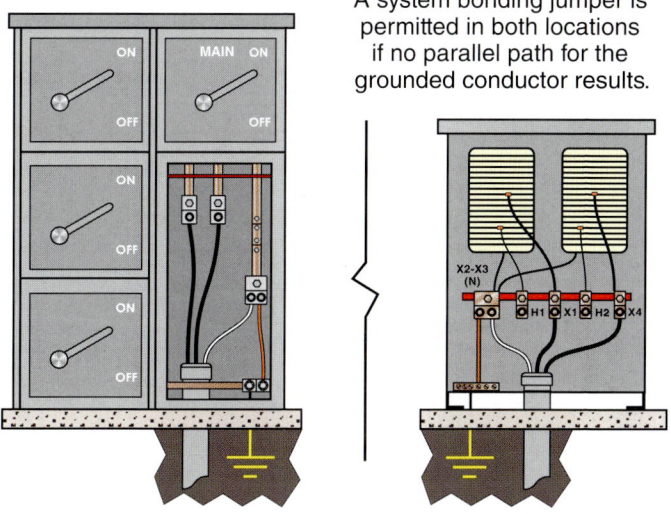

FIGURE 13-9 A system bonding jumper is permitted at the source and first disconnecting means if doing so does not create a parallel path for neutral current.

The secondary conductors are routed to the first system overcurrent device, usually in a separate enclosure, using a suitable wiring method that provides protection from physical damage. It is also acceptable to use wiring methods such as rigid metal conduit, intermediate metal conduit, or electrical metallic tubing for this type of installation, and these types of wiring methods qualify as the bonding means required between the enclosures. Common wiring methods used for connections to dry-type transformer enclosures are flexible metal conduit or liquidtight flexible metal conduit. An equipment bonding jumper is generally required for these wiring methods. The *Code* refers to this bonding jumper as a supply-side bonding jumper because it is routed with conductors on the supply side of the overcurrent device in which the transformer secondary conductors terminate. **See Figure 13-10.** The flexible wiring methods are more popular as they reduce vibration and provide flexibility during installation. A wire-type supply-side bonding jumper is required to be sized in accordance with Section 250.102(C) based on the size of the largest derived phase conductor connected to the system. **See Figure 13-11.** The supply-side bonding jumper is not required to be larger than the derived phase conductors supplied by the system.

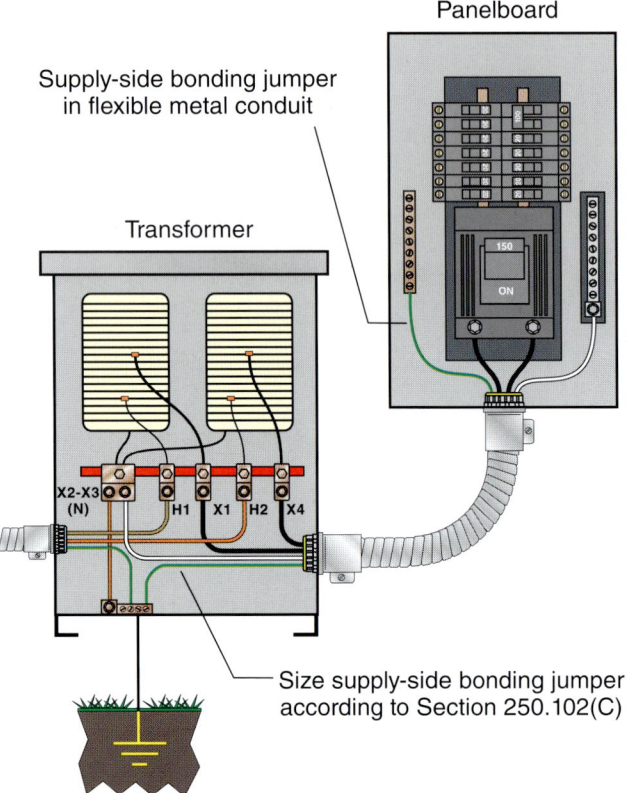

FIGURE 13-10 A supply-side bonding jumper is typically routed with derived phase conductors and installed with the derived system conductors using flexible wiring methods.

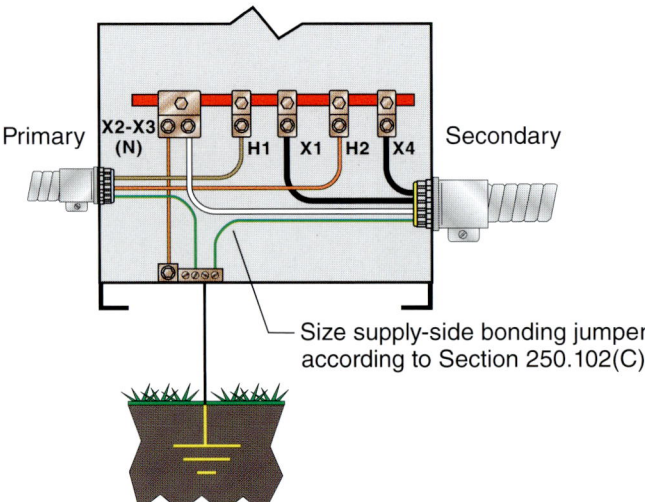

FIGURE 13-11 The supply-side bonding jumper is sized using Section 250.66 or the 12.5% rule based on the size of the largest ungrounded derived phase conductor.

Grounded Conductor Sizing

If a separately derived system is grounded, a system connection is intentionally established to ground. Grounded systems include a one-system conductor that is grounded, which could be either a system neutral or a grounded phase conductor. If a grounded conductor is necessary, such as for supplying line-to-neutral loads, then it can be installed. If there is no need for a grounded conductor, one does not have to be installed—but in that case, a supply-side bonding jumper must be provided for ground-fault current between the separately derived system and the overcurrent protection device.

The grounded conductor sizing requirements are located in Sections 220.61 and 250.30(A)(3). If the system bonding jumper is not located at the source enclosure, the grounded conductor of the separately derived system must meet specific sizing requirements. **See Figure 13-12.**

Where installed in a single raceway with the ungrounded phase conductors, the grounded conductor cannot be smaller than the required grounding electrode conductor based on Table 250.66, but it is not required to be any larger than the largest ungrounded derived phase conductor. For ungrounded derived phase conductors exceeding the values in Table 250.66, the grounded conductor cannot be smaller than 12.5% of the circular mil area of the largest ungrounded conductor or set of ungrounded conductors per phase. If the ungrounded derived phase conductors of a system are installed in a parallel arrangement, using two or more raceways, the grounded conductor must be installed in parallel in each raceway. The minimum size of the grounded conductor in each raceway is based on the circular mil area of the largest ungrounded derived phase conductor in each raceway, but each cannot be smaller than 1/0 to meet the requirements for parallel conductors in Section 310.10(H). For three-phase, three-wire, corner-grounded, separately derived systems, the grounded conductor must be no less than the ungrounded derived phase conductor size supplied by the system. As covered in Chapter 12, the grounded conductor of a high-impedance grounded neutral system has to meet the sizing requirements in Section 250.36 or 250.186.

Grounding Electrodes

Grounding a separately derived system requires a connection to the Earth through a grounding electrode. The *Code* is specific in Section 250.30(A)(4) about the electrode that must be used. An order of selection is established by the requirements in the *Code*; it is not a matter of choice in many cases. Two criteria that the grounding electrode for the system must meet are as follows:

1. It must be as near as practicable and preferably in the same area.
2. The grounding electrode be the nearest of either a metal water pipe electrode or a structural metal electrode. **See Figure 13-13.**

Courtesy of Morse Electric, Inc.

FIGURE 13-12 Grounded conductors supplied by separately derived systems must be sized according to Section 250.30(A)(3).

Grounding a separately derived system requires a connection to the Earth through a grounding electrode.

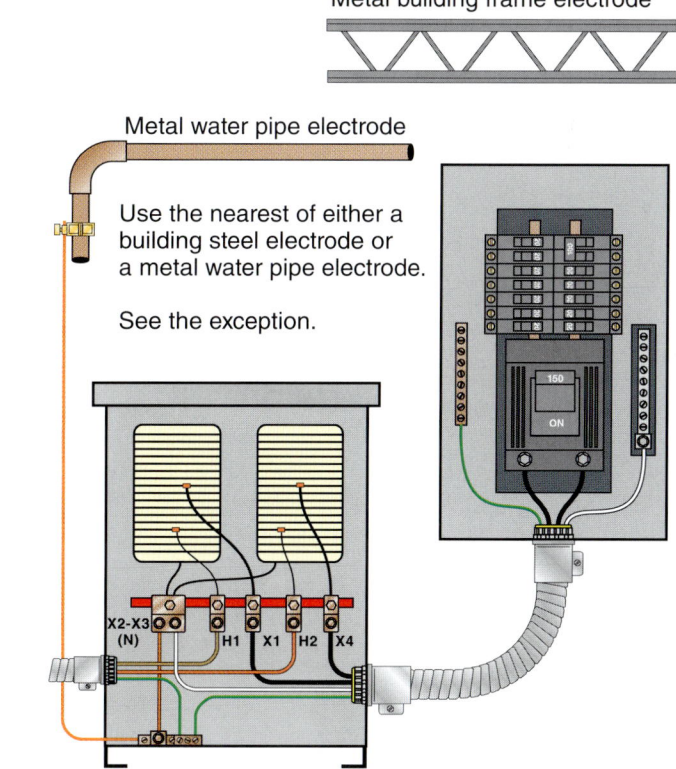

FIGURE 13-13 If present at the building or structure served, the nearest metal building frame electrode or underground metal water pipe electrode must be used as the grounding electrodes for separately derived systems.

If a separately derived system is installed in a building or structure that has a metal water pipe electrode and a structural metal frame electrode, the nearest of those two electrodes must be used for the system.

If there is no water pipe electrode or structural metal building frame electrode, then a grounding electrode must be established. By exception, any of the other grounding electrodes in Section 250.52(A) can be used in this case.

Grounding Electrode Conductor for an Individual System

Grounding separately derived systems requires installing a grounding electrode conductor. For a single separately derived system, the grounding electrode conductor generally has to be sized using Section 250.66 based on the size of the largest ungrounded derived phase conductors. **See Figure 13-14.**

The grounding electrode conductor has to connect the grounded conductor of the system to the grounding electrode described in Section 250.30(A)(4). The connection of the grounding electrode conductor to the system must be made where the system bonding jumper is installed. **See Figure 13-15.**

There is an exception for situations in which a separately derived system originates in listed equipment that is suitable for use as service equipment and a grounding electrode is connected to the equipment with a grounding electrode conductor of the minimum size required for the derived system.

Courtesy of Jim Dollard, IBEW Local 98

FIGURE 13-14 The grounding electrode conductor for a separately derived system is generally sized using Section 250.66 based on the size of the largest ungrounded phase conductor supplied by the system.

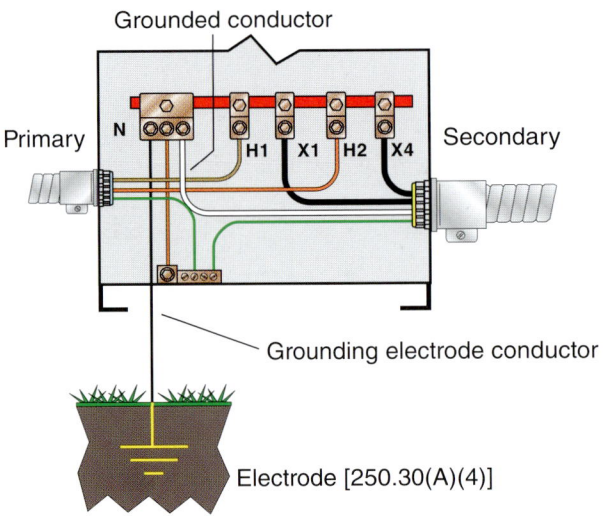

FIGURE 13-15 A grounding electrode conductor connects the grounding electrode to the grounded conductor of the system.

In cases in which the system bonding jumper is a wire or busbar, the grounding electrode conductor is also permitted to be connected to an equipment grounding terminal bar within the equipment, provided it is of sufficient size for the separately derived system. **See Figure 13-16.**

Grounding Electrode Conductor for Multiple Systems

The common grounding electrode conductor tap concept can be used for grounding multiple separately derived systems. The phrase *common grounding electrode conductor* is not defined in Article 100 or 250; however, from the description of its installation requirements in Section 250.30(A)(6), it can be surmised that this grounding electrode conductor is common to more than one separately derived system. This type of installation requires installing a single common grounding electrode conductor sized no less than 3/0 AWG copper or 250 kcmil aluminum. **See Figure 13-17.** This common grounding electrode conductor must be connected to a grounding electrode as specified in Section 250.30(A)(4). From each separately derived system, a grounding electrode conductor tap is required to be installed and connected to the common grounding electrode conductor.

Each tap conductor must be sized using Table 250.66 based on the largest ungrounded derived phase conductor(s) of the individual derived system it serves. An example is in a concrete high-rise structure with separately derived systems installed on each floor. With the closest grounding electrode being in the basement of the building, an efficient method for grounding all of the separately derived systems is to use this concept. The common grounding electrode conductor is typically run vertically through the building core electrical rooms, where the connections from multiple separately derived systems are made to it.

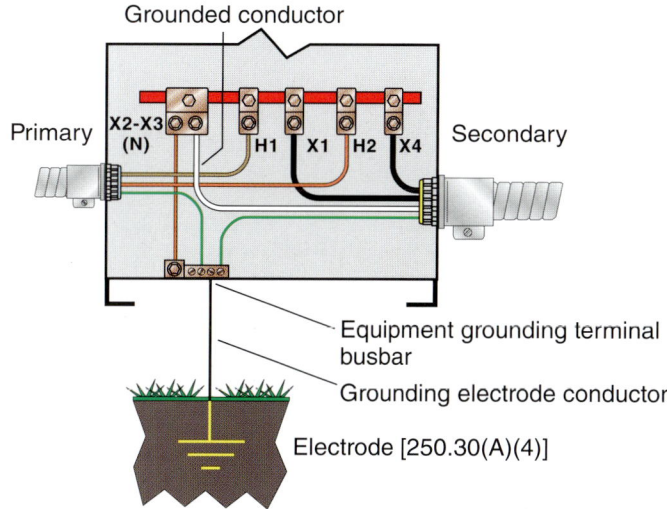

FIGURE 13-16 By exception, the grounding electrode conductor for a separately derived system is permitted to be connected to equipment grounding terminal busbar within equipment if it is of sufficient size.

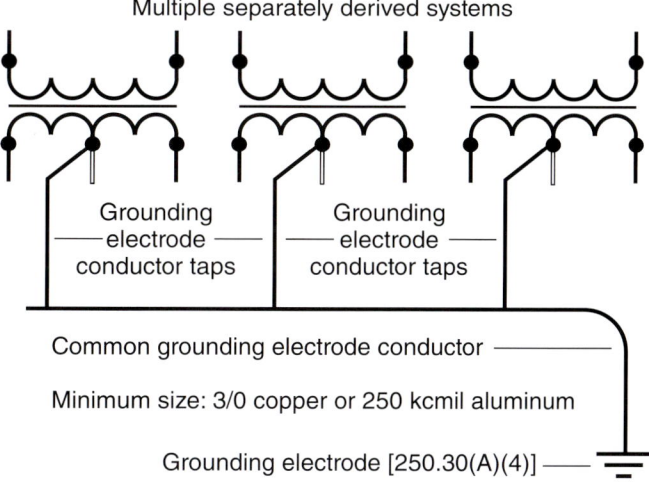

FIGURE 13-17 The common grounding electrode conductor for multiple separately derived systems is not permitted to be sized less than 3/0 AWG copper or 250 kcmil aluminum.

FIGURE 13-18 The grounding electrode conductor taps for multiple separately derived systems are permitted to be connected to a busbar to which the common grounding electrode conductor is connected (green marking tape optional).

The connections of the grounding electrode conductor tap to the common grounding electrode conductor have to be made at an accessible location. **See Figure 13-18.**

The connections can be made by an exothermic welding process, a connector listed as grounding and bonding equipment, or listed connections to a $\frac{1}{4} \times 2$ inch copper or aluminum busbar. **See Figure 13-19.** This connection between the tap and the common grounding electrode conductor has to be made in a way that the common grounding electrode conductor remains without a splice or joint. **See Figure 13-20.**

Regardless of whether a grounding electrode conductor is installed for a single separately derived system or a common grounding electrode conductor tap concept is used for multiple derived systems, the grounding electrode conductors have to be installed in accordance with Sections 250.64(A) through (C) and (E).

Bonding Water Piping and Building Steel

Section 250.104(D) provides bonding requirements for metal water piping systems and structural steel that are in the same area served by the separately derived system. Installers have to be aware that this bonding requirement is typically duplicated for each separately derived system in the building or structure.

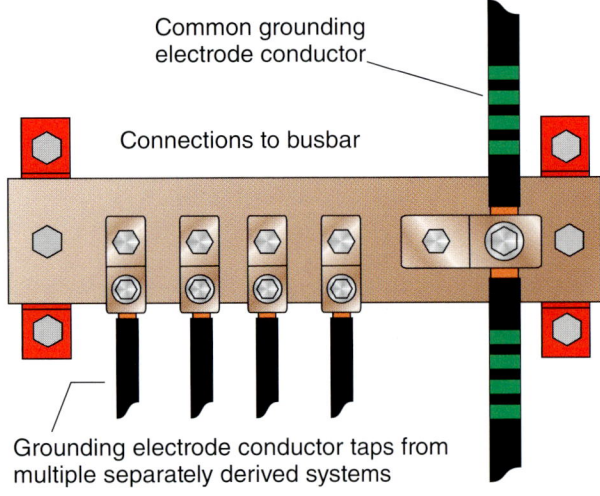

FIGURE 13-19 Connections of grounding electrode conductor taps to a busbar must meet all the requirements in Section 250.30(A)(6) and be made using a listed connection.

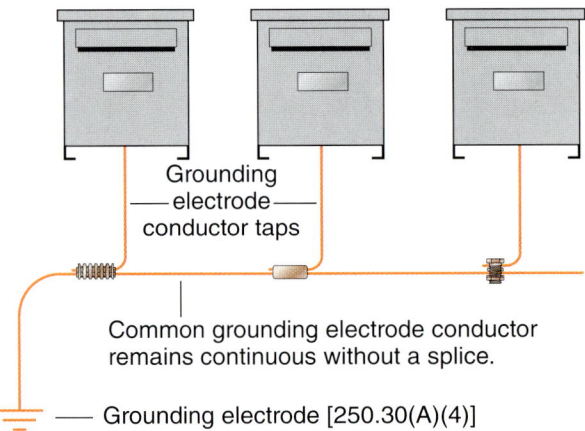

FIGURE 13-20 Connections of common grounding electrode conductor taps to the common grounding electrode conductor must be made in a manner such that the latter remains without a splice or joint other than an irreversible compression connection or an exothermic welding connection.

For example, in a high-rise building, there may be a transformer-type separately derived system installed on every floor or every other floor. The water piping or structural steel in the same area served by these systems must be bonded to each system. The purpose of this bonding requirement is to put the metal water piping and the structural metal frame at the same potential as the secondary grounding point of the separately derived system. Another important performance requirement is accomplished because this bonding jumper provides a direct path for ground-fault current back to the source windings (secondary) and should cause an overcurrent device to operate, clearing the faulted condition. Indirect paths back to the source windings would likely have more impedance due to the route of the fault current. The size of the bonding jumper for both the water piping and the structural steel is in accordance with Table 250.66 based on the largest ungrounded conductor supplied by the separately derived system. The bonding jumper has to be connected to the grounded conductor of the separately derived system, and the connection has to be made where the system bonding jumper for the derived system is located. **See Figure 13-21.**

Where a common grounding electrode conductor is installed for multiple systems that supply areas with metal water piping systems and exposed structural metal building framing, the metal water piping and structural metal frame must be bonded to the common grounding electrode conductor. The *NEC* does not specifically provide a size for this bonding jumper; however, since the minimum size for the common grounding electrode conductor is 3/0 copper, making this jumper the same size is a conservative approach.

Outdoor Source

Some premises wiring installations include separately derived systems such as generators, photovoltaic systems, or transformers installed in outdoor locations. **See Figure 13-22.** If the source is located outside the building or structure it supplies, a grounding electrode connection to the source is required at the source location outside the building or structure. The grounding electrode used for this connection to the Earth has to be in accordance with Section 250.50.

This means that if any of the electrodes in Sections 250.52(A)(1) through (7) are present for use, they have to be used for this grounding connection.

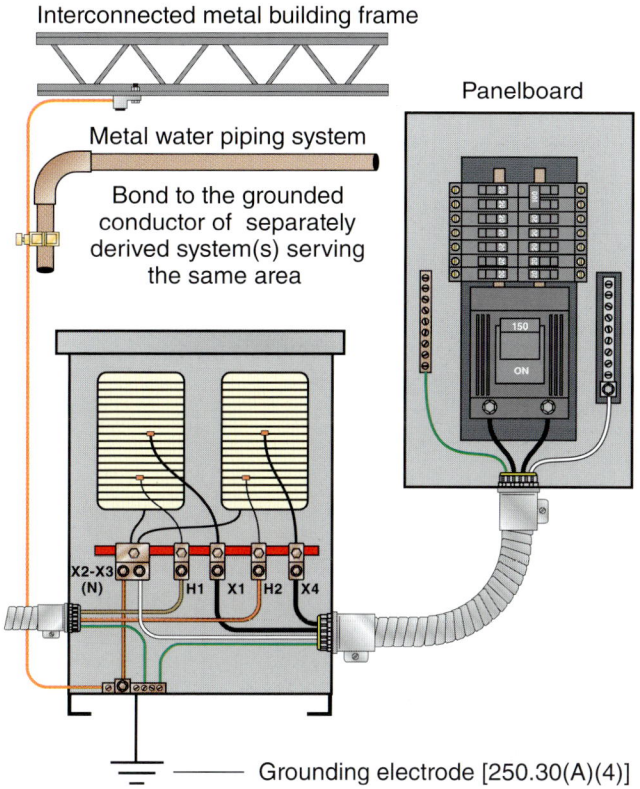

FIGURE 13-21 Metal water piping and structural metal building frames are required to be bonded to separately derived systems serving the same area.

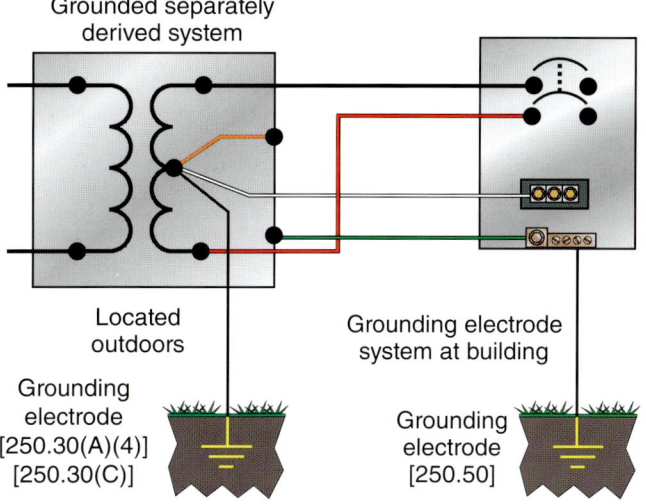

FIGURE 13-22 A grounding electrode connection is required outside for separately derived systems installed outdoors.

In addition to the required grounding connection outside, the installation has to meet all applicable requirements in Section 250.30(A) if the system is grounded or in Section 250.30(B) if the system is ungrounded. A grounding electrode conductor connection is not required at the outdoor source for a high-impedance grounded neutral system, which must be installed according to Section 250.36.

Ungrounded Systems

Ungrounded separately derived systems have to meet specific grounding requirements provided in Section 250.30(B)). First, it should be understood that these systems are installed ungrounded, operate ungrounded, and are required to have ground detectors installed in accordance with Section 250.21(B). No conductor supplied by an ungrounded system is intentionally or solidly grounded, other than an indirect connection to ground or grounded parts through distributed leakage capacitance. **See Figure 13-23.**

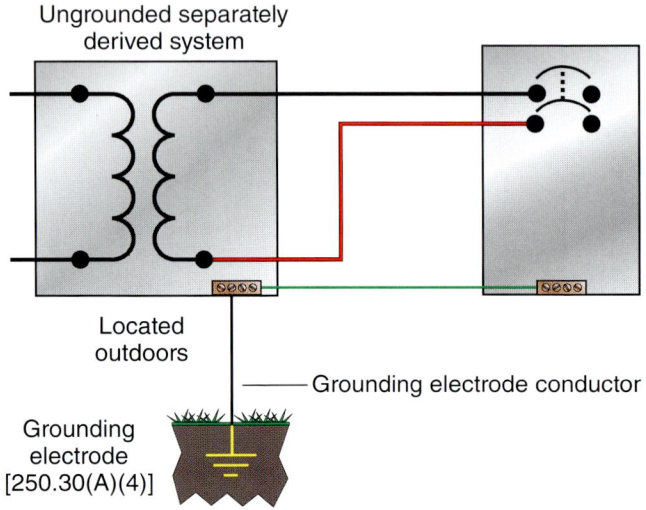

FIGURE 13-23 Grounding requirements apply to enclosures and wiring methods installed for ungrounded separately derived system conductors and equipment.

The enclosures and other normally non–current-carrying equipment are required to be connected to ground (the Earth) by a grounding electrode conductor that connects to an electrode that meets the provisions in Section 250.30(A)(4). This section is specific that the electrode for the system must be as near as practicable and preferably in the same area, and must be either a structural metal frame electrode or a metal underground water pipe electrode if present, whichever is closest to the system. If there is no water pipe electrode or structural metal building frame electrode, then an electrode must be established. By exception, any of the other grounding electrodes in Section 250.52(A) can be used in this case. The grounding electrode conductor for an ungrounded system is sized the same way the grounding electrode conductor for a grounded system is sized. Use Section 250.66 based on the size of the largest ungrounded derived phase conductors. The grounding electrode conductor must connect the metal enclosures to the grounding electrode in one of the methods specified in Section 250.30(A)(5) or (6), depending on whether it is a single system or the common grounding electrode conductor tap concept for multiple systems is used.

The grounding connection can be made at any point on the derived system from the source enclosure to the first system disconnecting means. **See Figure 13-24.**

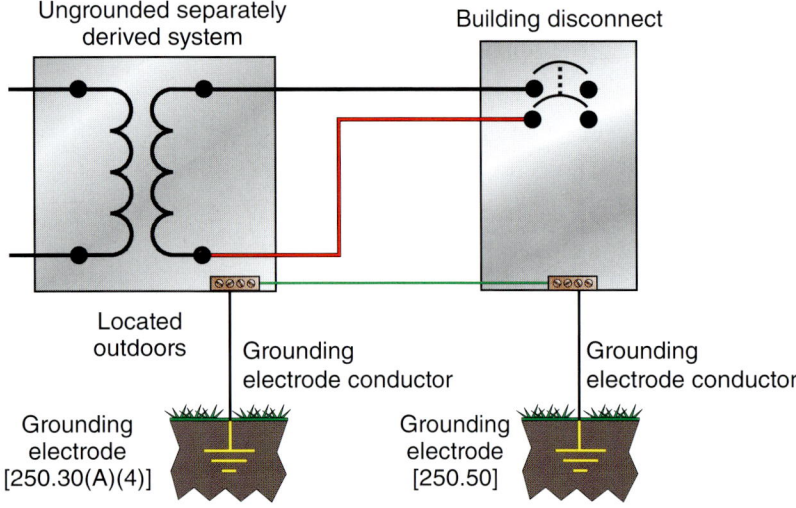

FIGURE 13-24 Grounding electrode conductor connections are required for the conductive wiring methods and equipment installed for ungrounded separately derived systems.

For ungrounded separately derived systems located outside a building or structure supplied, the grounding electrode conductor connection to the system must be made outside in accordance with Section 250.30(C). EGCs are required to be installed with the feeders and branch circuits supplied by an ungrounded system. They still have to perform protective functions as described in Sections 250.4(B)(1) through (4). This means that the protective features of a ground fault current path must be provided even though the system is installed and operated ungrounded. Section 250.30(B)(3) includes requirements to provide a supply-side bonding jumper or path between an ungrounded separately derived system and the first disconnecting means. A wire-type bonding jumper must be sized using Section 250.102(C) based on the size of the derived phase conductors. **See Figure 13-25.**

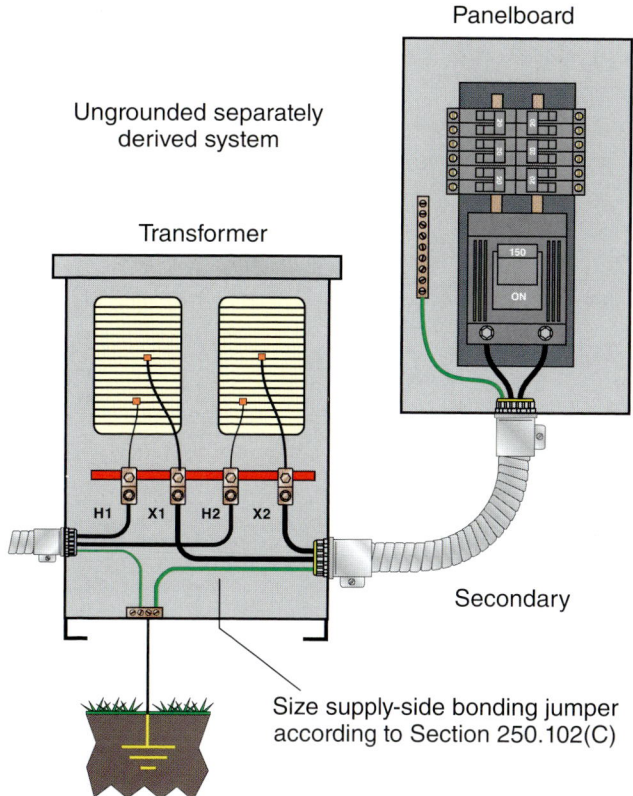

FIGURE 13-25 A supply-side bonding jumper is required for ungrounded separately derived systems.

Generators and Transfer Equipment

Generators can be separately derived power systems. How the grounding and bonding connections are made at a generator is usually determined by the type of transfer equipment. **See Figure 13-26.** There is an important informational note following Section 250.30 that describes the transfer switch and how the grounding connections should be made for the generator.

First, if a transfer switch for a generator includes a switching action in the grounded conductor, then the generator has to be grounded as a separately derived system in accordance with all applicable requirements in Section 250.30(A). The reason is that in the normal mode the grounded conductor is connected to the service grounding electrode, whereas in the standby mode the grounded conductor is switched over to the generator, which has to be grounded as a separately derived system.

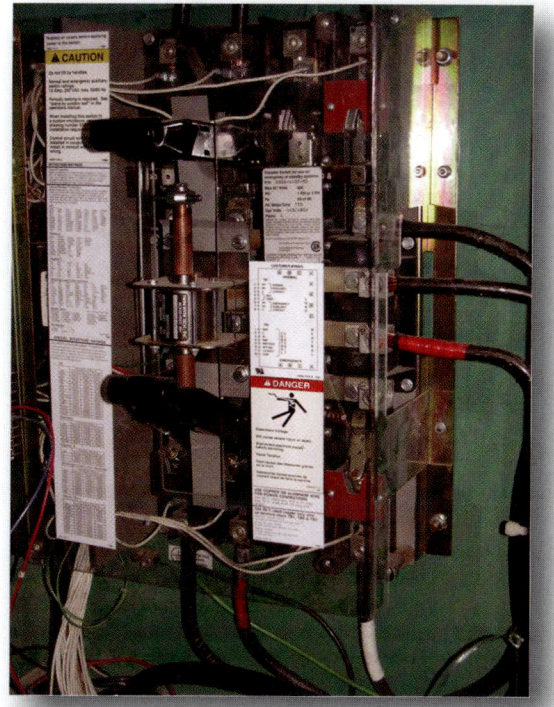

FIGURE 13-26 When transfer equipment includes a switching action in the grounded conductor, the generator must be grounded according to the requirements in 250.30(A).

The result is that in either position of the transfer switch, the system is grounded. **See Figure 13-27.** If there is no switching action in the grounded conductor by the transfer equipment, then the generator system is grounded with the transfer switch in either position. The grounding and bonding connections in this case have to meet the requirements in Section 250.35 for permanently installed generators. **See Figure 13-28.**

The performance concepts of Section 250.35 are focused on providing an effective ground-fault current path with the supply conductors of the generator to the first disconnecting means or equipment supplied. Generators grounded as separately derived systems meet this requirement when installed according to the rules in Section 250.30(A). If the generator does not include overcurrent protection, a supply-side bonding jumper must be installed between the generator equipment grounding terminal and the equipment grounding terminal bar or bus of the enclosure supplied by the system. The sizing requirements for this bonding jumper are related to the location of the first system overcurrent device. If the bonding jumper is on the supply side of the first system overcurrent device, it is sized as a supply-side bonding jumper in accordance with Section 250.102(C).

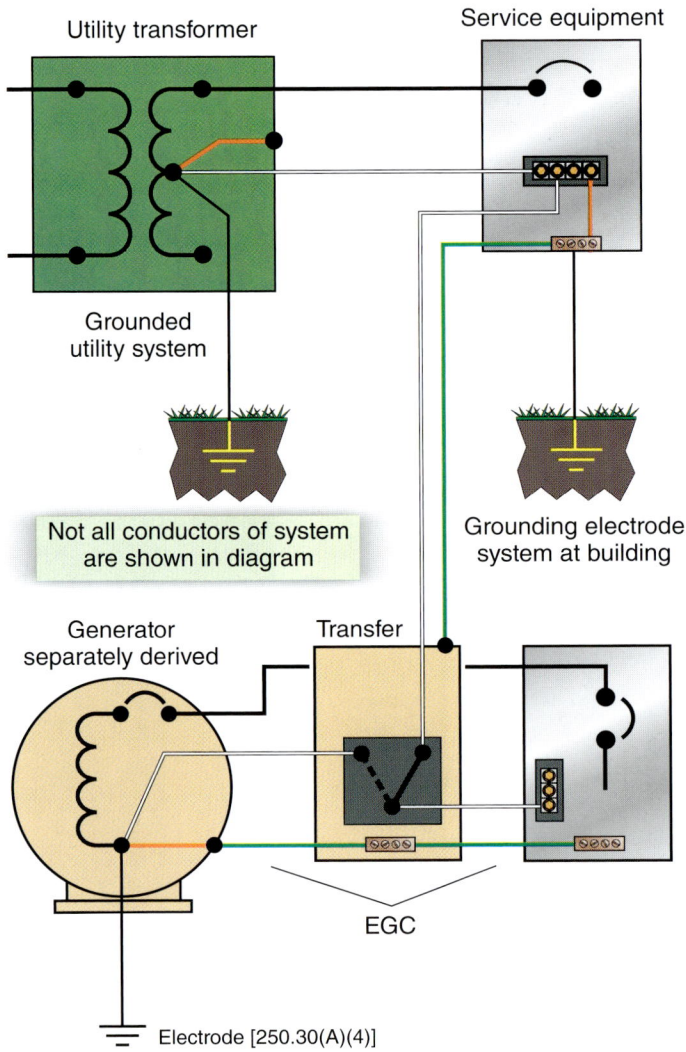

FIGURE 13-27 When the generator is grounded as a separately derived system and a transfer switch switches the grounded conductor of the system, the system remains grounded if the transfer switch is in the normal position or the standby position.

If a transfer switch for a generator includes a switching action in the grounded conductor, then the generator has to be grounded as a separately derived system in accordance with all applicable requirements in Section 250.30(A).

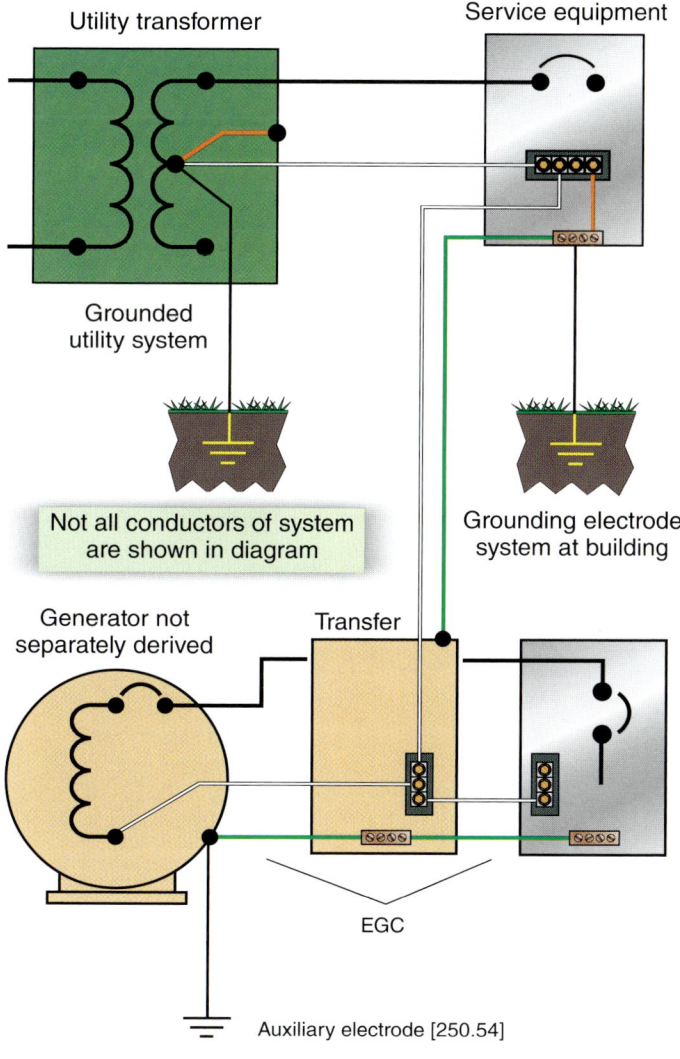

FIGURE 13-28 If a transfer switch does not switch the grounded conductor, the separately derived system remains grounded with the switch in either position. The system is grounded by the connection to the service grounded conductor.

Generators are often installed outside of buildings or structures with the feeder routed to the transfer equipment, typically inside the building or structure.

This means that the size is based on Table 250.66 or the 12.5% rule, based on the circular mil area of the ungrounded derived phase conductors supplied from the generator. **See Figure 13-29.** The conductor installed on the load side of a generator overcurrent device is an EGC because it is a load-side installation. This EGC is required to be sized using Table 250.122, based on the rating of the overcurrent device supplied. **See Figure 13-30.**

Again, the importance of these grounding and bonding connections and sizing rules is to provide an effective path for the ground-fault current back to the generator (source).

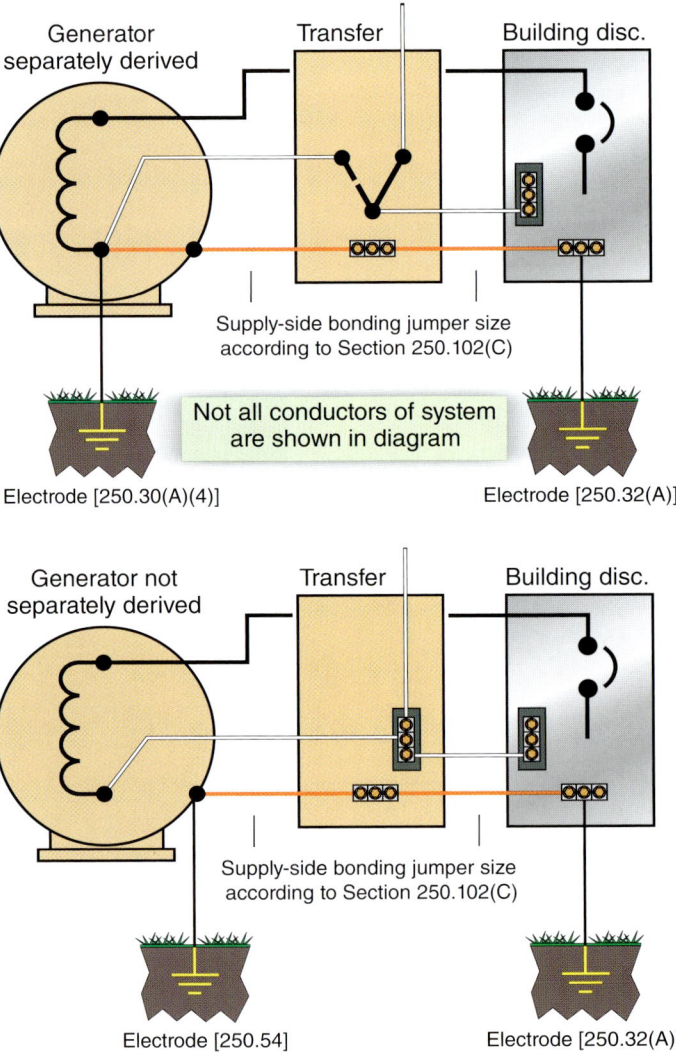

FIGURE 13-29 A supply-side bonding jumper for generators without overcurrent protection at the source must be sized according to Section 250.102(C).

Where an overcurrent device is installed at the generator, an EGC is installed with the generator conductors to the first enclosure supplied by the system. This is an EGC and is required to be sized using Table 250.122, based on the rating of the overcurrent device.

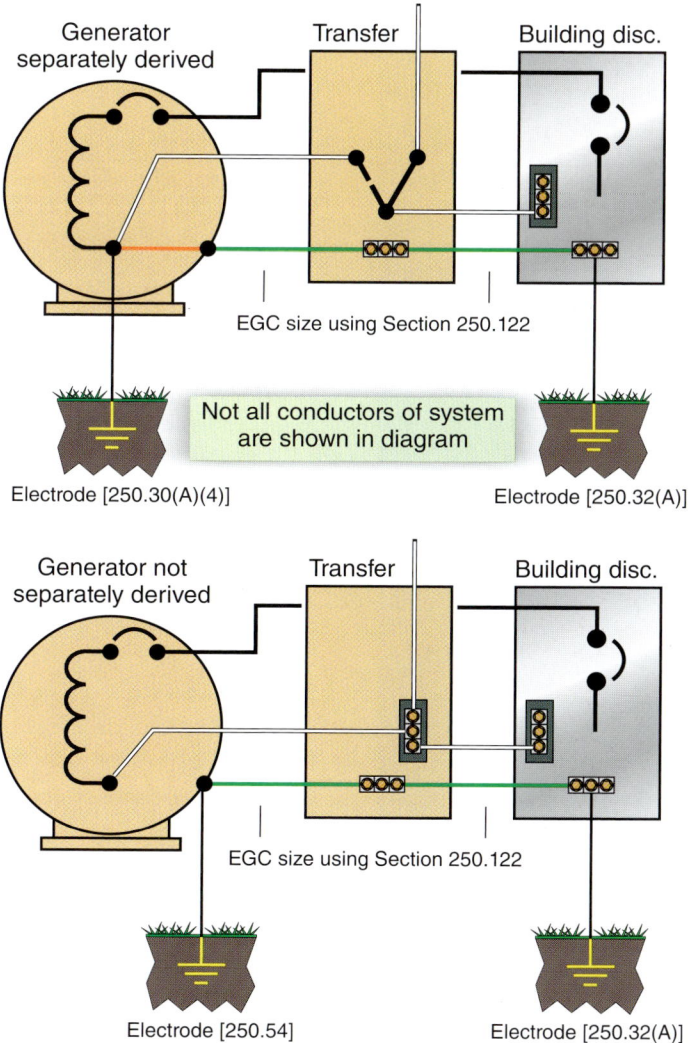

FIGURE 13-30 A load-side EGC for a generator that is provided with overcurrent protection at the source is sized based on Section 250.122.

Small Wind Electrical Systems

Article 694 of the *NEC* applies to small wind electrical systems that consist of one or more wind-driven electric generators with individual systems up to and including 100 kW. **See Figure 13-31.** Part V of Article 694 provides the grounding and bonding rules specific to these systems and associated equipment. Exposed non–current-carrying metal parts of towers, turbines, other equipment, and conductor enclosures are required to be grounded in accordance with Section 250.134 or 250.136(A), regardless of voltage. Turbine blades and tails that are not likely to become energized are not required to be grounded.

Tower Grounding

A wind turbine tower is required to be grounded with one or more auxiliary electrodes to limit voltages imposed by lightning. Auxiliary electrodes must be installed in accordance with Section 250.54. Electrodes that are part of the tower foundation and that meet the requirements for concrete-encased electrodes are acceptable. A grounded metal tower support that also qualifies as a grounding electrode according to Section 250.52 is acceptable if it meets the requirements of Section 250.136(A). This means the conductive tower can also serve as a grounding means for equipment connected to it in some cases as long as the equipment is connected to an EGC installed with the branch circuit supplying the equipment. If a tower for a wind generator is installed in close proximity to any galvanized tower anchors or foundation components, galvanized grounding electrodes must be installed. The reason is because of the electrolytic corrosion of galvanized foundation and tower anchoring components where copper and copper-clad grounding electrodes are used.

System Grounding

Turbines that drive generators create an electrical power system. Separately derived systems that are required to be grounding in accordance with Section 250.20 must also be grounded according to the requirements in Section 250.30(A).

Equipment Grounding

An EGC is required between a turbine and the premises grounding system in accordance with Section 250.110. EGC installations must be in accordance with Section 250.120, and the types of EGCs used are provided in Section 250.118. Sizing EGCs for overcurrent-protected feeders, supplied from turbine systems, must be done according to Section 250.122.

Grounding Connections

The EGC and grounding electrode conductors are required to be connected to a metallic tower by exothermic welding, listed lugs, listed pressure connectors, listed clamps, or other listed means.

Courtesy of NECA ©2010 Rob Colgan

FIGURE 13-31 Small wind electrical systems can be separately derived systems and have to be grounded according to Section 250.30(A).

Devices such as connectors and lugs shall be suitable for the material of the conductor and the structure to which they connect to ensure compatibility between the two metals. When practicable, contact between dissimilar metals in the system must be avoided to reduce galvanic action and corrosion. All mechanical terminations of EGCs and grounding electrode conductors must be accessible. Buried and concrete-encased electrode connections are generally not required to be accessible.

Auxiliary electrodes and grounding electrode conductors are recognized as lightning protection system components if they meet applicable requirements. If a tower lightning protection system grounding network is installed, the power system electrodes must be bonded to the tower auxiliary grounding electrode system. Guy wire lightning protection system grounding electrodes are not required to be bonded to the tower auxiliary grounding electrode system.

Grounding DC Systems

The current in DC systems is unidirectional, meaning the current flows in one direction. This is unlike AC, which is bi-directional, meaning AC reverses its direction of flow at regular intervals. Batteries are a common DC source. Battery systems are used for backup power systems, such as uninterruptible power supplies, and for emergency lighting backup systems. **See Figure 13-32.**

Generators and photovoltaic arrays are other sources of DC power. DC generators operate on the principle of magnetic induction. Grounding requirements for DC systems are related to the type of system and output voltages, similar to what governs AC system grounding. Grounding for DC systems is covered in Part VIII of Article 250. This part includes requirements specific to DC system grounding requirements and is required to be applied in addition to the other applicable parts of Article 250. A DC system is required to be grounded as follows:

1. Two-wire DC systems supplying premises wiring operating at a voltage greater than 50 V but not exceeding 300 V are required to be grounded. See the three exceptions to Section 250.162(A).
2. All three-wire DC systems are required to be grounded by connecting the neutral conductor of the system to ground.[8]

Where a two-wire DC system is required to be grounded in accordance with Section 250.162(A), the grounded conductor can be either output conductor, as indicated in Section 250.26. There are three alternatives that relax this grounding requirement as follows:

1. A DC system supplying only industrial equipment in limited areas and provided with a ground-fault detection system is permitted to be installed and operated ungrounded.
2. If a rectifier-derived DC system is supplied from an AC system in accordance with 250.20, it is permitted to be installed and operated ungrounded.

Courtesy of Jim Dollard, IBEW Local 98

FIGURE 13-32 DC systems are often supplied by batteries that are connected to a charging means.

3. System grounding is not required for the DC source for a fire alarm system if the maximum source circuit current is not more than 0.030 amperes as specified in Article 760, Part III.

In a three-wire DC system, the neutral conductor must be grounded. **See Figure 13-33.**

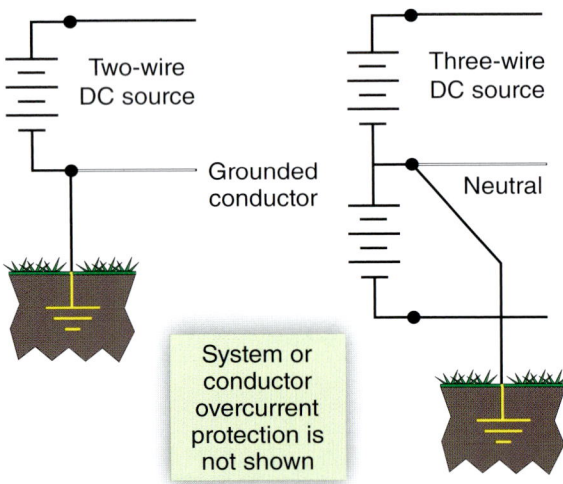

FIGURE 13-33 Grounded DC systems must be in accordance with the requirements in Part VIII of Article 250.

Point of Grounding Connection for DC Systems

The connection of the grounding electrode conductor for a DC system is covered in Section 250.164. If the DC source is not on the premises, this connection shall be made at one or more supply stations. The grounding electrode conductor connection is not permitted to be made at individual services or at any point on the premises wiring if the source is not on the premises.

When the DC system is on the premises, the grounding electrode conductor connection has to be made at one of the following locations:

1. At the source
2. At the first system disconnecting means or overcurrent protective device
3. At another location that provides equivalent protection and uses listed and identified equipment[9]

Size of a DC Grounding Electrode Conductor

The minimum size of a grounding electrode conductor for a DC system is generally required to be no less than the sizes indicated in Sections 250.166(A) and (B).

Where the DC system consists of a three-wire balancer set or a balancer winding with overcurrent protection, as provided in Section 445.12(D), the grounding electrode conductor shall be no smaller than the neutral conductor and can never be smaller than 8 AWG copper or 6 AWG aluminum. Where the DC system is other than a 3-wire balancer set, the grounding electrode conductor cannot be smaller than the largest conductor supplied by the system and can never be smaller than 8 AWG copper or 6 AWG aluminum.

Note that when the grounding electrode conductor for a DC system is a sole connection to a rod, pipe, or plate electrode, the grounding electrode conductor does not have to be sized larger than 6 AWG copper or 4 AWG aluminum. If the grounding electrode conductor is a sole connection to a concrete-encased electrode, it never has to be larger than 4 AWG copper. If the grounding electrode conductor for a DC system is a sole connection to a ground ring, it never has to be larger than the size of the ring electrode. Note that these provisions are for sole connections of grounding electrode conductors installed for DC systems and are identical to the allowances in Sections 250.66(A) through (C). The installation requirements in Section 250.64 apply to grounding electrode conductors for DC systems. This includes, but is not limited to, installing them in a continuous length without a splice or joint and protecting them where they are subject to physical damage.

Size of a DC System Bonding Jumper

Where a DC system is grounded, a system bonding jumper is required to connect the grounded conductor of the system to the EGC or EGCs.

This connection has to be made at the source or first system disconnecting means where the system is grounded. **See Figure 13-34.** The size of a DC system bonding jumper has to be no less than the system grounding electrode conductor, based on Section 250.166.

The system bonding jumper must be a copper or other corrosion-resistant conductor and can be in the form of a wire, bus, screw, or other suitable conductor. If the system bonding jumper is a screw, the screw has to be green. System bonding jumpers must be connected by using any method specified in Section 250.8.

Ungrounded DC Systems

An ungrounded system has no system conductor that is intentionally grounded. This means that the output conductors will be ungrounded. Where a DC system is a stand-alone power source, such as a battery- or engine-driven generator, the enclosures and other conductive equipment parts are required to be connected to ground (the Earth) by using a grounding electrode conductor that connects to an electrode that meets the provisions of Part III of Article 250.

The grounding requirements apply to all metal raceways, cables, and other exposed non–current-carrying metal equipment parts. The grounding connection to DC source equipment enclosures, raceways, and so forth, is permitted to be made at any point on the system from the source up to the first system disconnecting means enclosure or overcurrent device. The grounding electrode conductor installed for a DC system has to be sized based on Section 250.166. **See Figure 13-35.**

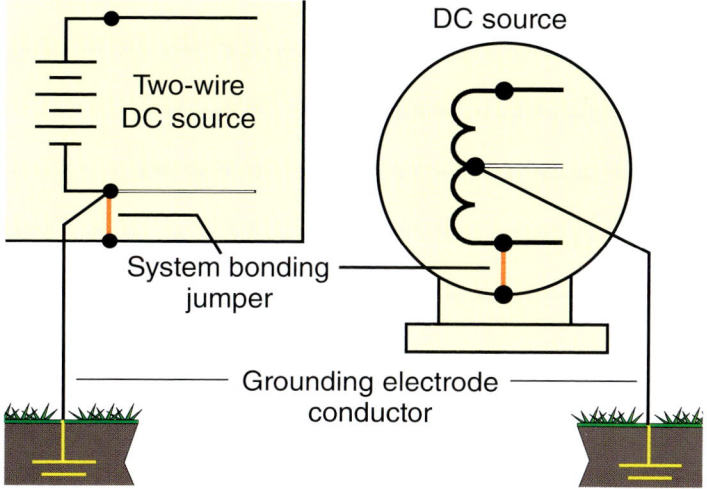

FIGURE 13-34 The DC system bonding jumper is typically installed at the source or at the first system disconnecting means enclosure.

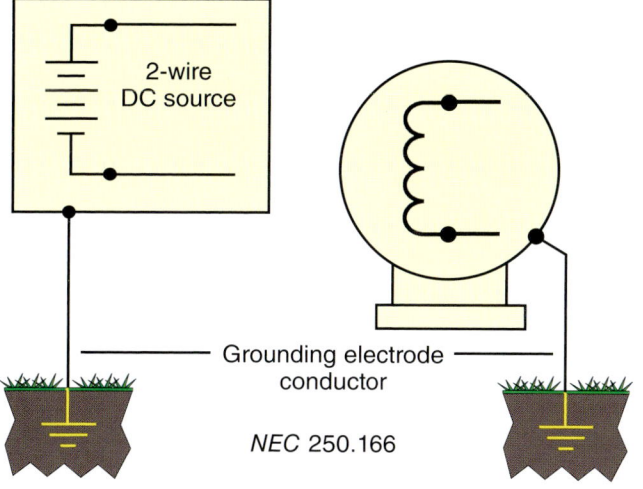

FIGURE 13-35 The grounding electrode conductor for DC power systems must be sized according to Section 250.166.

Summary

System grounding requirements are provided in Part II of *NEC* Article 250. Whether or not a system is required to be grounded is specified in Section 250.20. Separately derived systems can be in the form of a transformer, generator, inverter windings, photovolataic array, and so forth. When a system is required to be grounded, the rules in Section 250.30(A) have to be applied. There are specific requirements in Section 250.30(B) that apply to ungrounded separately derived systems. Part VIII of Article 250 provides specific rules for DC systems that must be grounded.

Courtesy of Bill McGovern, City of Plano, TX

Transformers are generally the most common separately derived systems in commercial and industrial occupancies.

Courtesy of IBEW Local 26 Training Center

Generators are grounded as separately derived systems when the transfer equipment switches the neutral conductor.

References

1. NFPA 70 National Electrical Code 2011, Article 100 (National Fire Protection Association, Quincy, MA 2010), p. 70–26.
2. NFPA 70 National Electrical Code 2011, Section 250.2 (National Fire Protection Association, Quincy, MA 2010), p. 70–100.
3. NFPA 70 National Electrical Code 2011, Article 100 (National Fire Protection Association, Quincy, MA 2010), p. 70–26.
4. NFPA 70 National Electrical Code 2011, Article 100 (National Fire Protection Association, Quincy, MA 2010), p. 70–29.
5. NFPA 70 National Electrical Code 2011, Article 100 (National Fire Protection Association, Quincy, MA 2010), p. 70–29.
6. NFPA 70 National Electrical Code 2011, Article 100 (National Fire Protection Association, Quincy, MA 2010), p. 70–29.
7. NFPA 70 National Electrical Code 2011, Article 100 (National Fire Protection Association, Quincy, MA 2010), p. 70–31.
8. NFPA 70 National Electrical Code 2011, Section 250.162 (National Fire Protection Association, Quincy, MA 2010), p. 70–128.
9. NFPA 70 National Electrical Code 2011, Section 250.164(B) (National Fire Protection Association, Quincy, MA 2010), p. 70–128.

Review Questions

1. The following electrical system or source does not qualify as a separately derived system:
   a. Generator
   b. Photovoltaic system
   c. Battery system
   d. Autotransformer

2. A premises wiring system whose power is derived from a source of electric energy or equipment other than a service best describes a _____.
   a. Separately derived system
   b. Utility power network
   c. Hydropower facility
   d. Utility generating station

3. The connection between the grounded circuit conductor and the supply-side bonding jumper, the EGC, or both at a separately derived system best defines _____.
   a. An equipment bonding jumper
   b. A main bonding jumper
   c. A system bonding jumper
   d. An EGC

4. A conductor installed on the supply side of a service, within a service equipment enclosure, or for a separately derived system that ensures the required electrical conductivity between metal parts required to be electrically connected best defines which component of the grounding and bonding system?
   a. Equipment bonding jumper
   b. Supply-side bonding jumper
   c. System bonding jumper
   d. EGC

5. Common types of separately derived systems installed in commercial and industrial applications are generators and transformers.
   a. True
   b. False

6. If there is no disconnecting means or overcurrent device at the load end of the conductors supplied by a separately derived system, the system bonding jumper has to be installed _____.
   a. In the service equipment enclosure
   b. In the first overcurrent device enclosure
   c. In the source enclosure
   d. Outside of the source enclosure on the load side of the system

7. If the system bonding jumper for a separately derived system is installed at the first disconnecting means enclosure, it must connect the grounded conductor of the system to _____.
   a. The supply-side bonding jumper
   b. The disconnecting means enclosure
   c. The EGC or EGCs
   d. All of the above

8. The system bonding jumper for a separately derived system has to be sized in accordance with _____.
   a. Sections 250.28(A) through (D)
   b. Section 250.122
   c. NEC Table 8, Chapter 9
   d. Table 310.16

9. A system bonding jumper for a separately derived system is not permitted at the source enclosure or the first disconnecting means enclosure even if there are no parallel paths for current in the grounded conductor of the system.
   a. True
   b. False

10. Where a nonmetallic raceway is installed for the secondary of a separately derived system, a supply-side bonding jumper is required to be installed and sized according to _____, based on the largest ungrounded derived system conductor.
    a. Table 250.66 or the 12.5% rule
    b. Table 310.16
    c. Table 250.122
    d. NEC Table 8, Chapter 9

11. A grounded conductor supplied by a separately derived system must be no smaller than the required _____ if there is no neutral load on the system.
    a. EGC
    b. System bonding jumper
    c. Equipment bonding jumper
    d. Grounding electrode conductor

12. A separately derived system has to be grounded by connection to the nearest of which of the following grounding electrodes?
    a. Structural metal building frame
    b. Ground rods
    c. Metal water pipe electrode
    d. Both a and c

13. The grounding electrode conductor for a separately derived system has to be connected to the grounded conductor where the _____ is installed.
    a. First overcurrent device
    b. Grounding electrode
    c. System bonding jumper
    d. Equipment bonding jumper
14. A 400 A motor control center is supplied by a separately derived system, and the largest ungrounded conductor is 600 kcmil copper. The minimum size required for the grounding electrode conductor is _____ AWG copper if the grounding electrode is a metal water pipe.
    a. 3
    b. 1
    c. 1/0
    d. 2/0
15. Where the common grounding electrode conductor tap concept is used for grounding multiple separately derived systems, the minimum size required for the common grounding electrode conductor is _____ AWG copper.
    a. 3
    b. 1
    c. 1/0
    d. 3/0
16. Where the grounding electrode tap conductors for multiple separately derived systems are connected to a common grounding electrode conductor, the minimum size required for each tap conductor is _____ AWG copper if each system secondary derived ungrounded conductor is sized at 300 kcmil copper.
    a. 2
    b. 3
    c. 4
    d. 1/0
17. The connections of grounding electrode conductor taps to a common grounding electrode conductor can be made using all but which of the following methods?
    a. Exothermic welding process
    b. A connector listed as grounding and bonding equipment
    c. A connection to a ¼ × 2 inch copper or aluminum busbar
    d. Solder connections
18. If the transfer switch for a generator system switches the grounded (neutral) conductor, the generator has to be grounded as a separately derived system.
    a. True
    b. False
19. If a generator has a 1200 A overcurrent device installed at the generator set, a minimum size of _____ AWG copper is required for the EGCs installed in a parallel set of 4-inch PVC conduits routed from the generator to the first system overcurrent device enclosure.
    a. 1/0
    b. 2/0
    c. 3/0
    d. 4/0
20. If a generator without an overcurrent device supplies a building, and a supply-side bonding jumper is installed with 400 kcmil copper feeder conductors, the minimum size required for the supply-side bonding jumper is _____ AWG copper.
    a. 3
    b. 2
    c. 1
    d. 1/0
21. Metal water piping and structural metal framing that exists in the area served by a separately derived system has to be bonded to the grounded conductor of that separately derived system. _____ is used to size this bonding jumper or conductor.
    a. Table 250.122
    b. Table 310.16
    c. Table 250.66
    d. *NEC* Table 8, Chapter 9
22. The minimum size required for a copper wire-type system bonding jumper is _____ AWG if the size of the largest ungrounded derived phase conductor connected to the system is 750 kcmil aluminum.
    a. 1/0
    b. 2/0
    c. 3/0
    d. 4/0
23. The size of a DC system bonding jumper has to be no less than the system grounding electrode conductor based on Section 250.66.
    a. True
    b. False

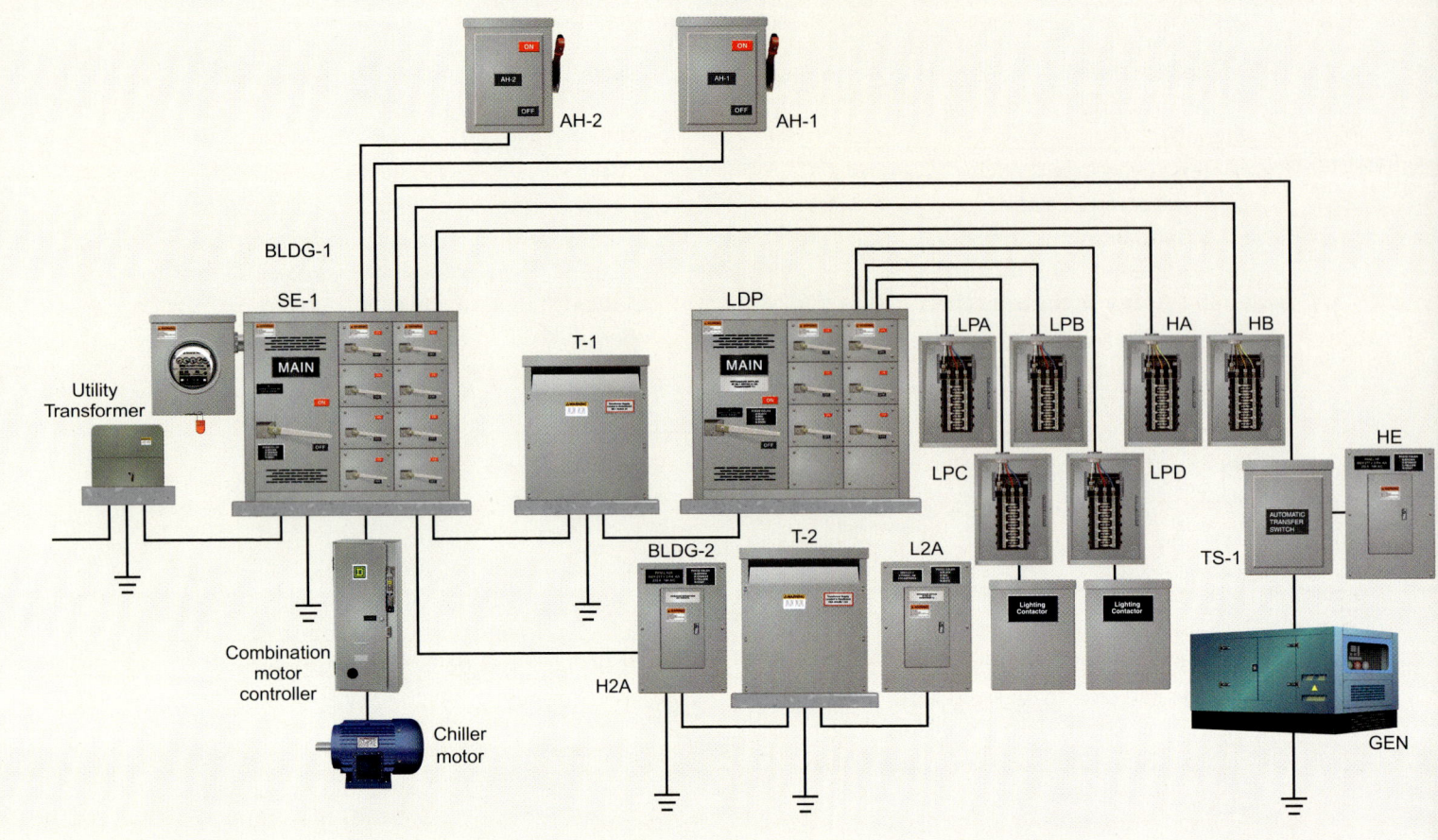

CHAPTER 14

Special Occupancies and Conditions

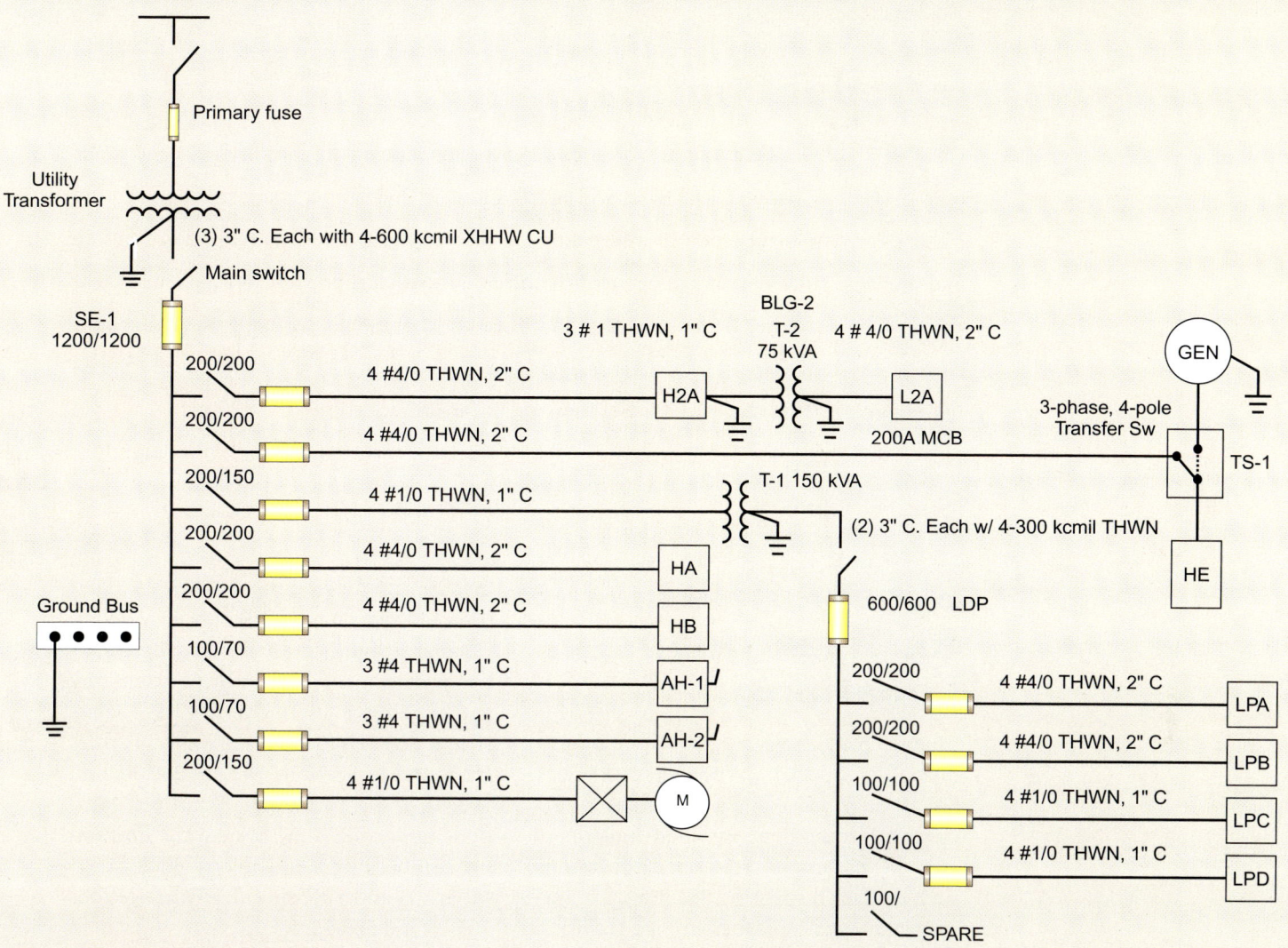

Objectives

- Understand why grounding and bonding requirements for special occupancies are often more restrictive than general *NEC* grounding and bonding requirements
- Understand the specific grounding and bonding rules for hazardous (classified) locations
- Understand the special requirements for grounding and bonding in patient care locations of health care facilities
- Understand the unique grounding and bonding problems in agricultural facilities and the rules that apply
- Determine special grounding and bonding requirements for mobile homes and manufactured homes

Outline

Special Rules for Hazardous Locations

Special Rules for Health Care Facilities

Special Rules for Agricultural Installations

Mobile and Manufactured Home Grounding and Bonding Rules

Introduction

Chapter 5 of the *NEC* provides rules for special occupancies that modify the general requirements of Chapters 1-4. Electrical installations in special occupancies often require a more restrictive approach in addressing specific concerns and can require grounding and bonding rules that exceed general requirements. It is important to establish a clear comprehension of the electrical wiring that exceeds what would normally be required for electrical systems in general locations. Examples include restrictive grounding and bonding requirements for health care facilities and hazardous locations.

Special Rules for Hazardous Locations

Chapter 5 of the *NEC* provides requirements for special occupancies. Articles 500 through 517 contain specific rules for electrical equipment installed and operated in hazardous (classified) locations such as fuel dispensing facilities, chemical plants, and bulk fuel storage facilities. **See Figure 14-1.** Chapters 1 through 4 of the *Code* apply generally to these installations but special requirements provided in Chapter 5 significantly modify these general provisions. For equipment grounding and bonding requirements in Class I, Class II, and Class III locations, the specific rules are found in Sections 501.30, 502.30, and 503.30, respectively. For hazardous locations under the Zone System the grounding and bonding rules are provided in Articles 505 and 506, specifically in Sections 505.25 and 506.25.

Special equipment is required in locations classified as hazardous. This means the equipment needs to be suitable for use in the specific location. Suitability is often established through product listing and identification. The *NEC* rules for conductive equipment, raceways, and other electrical wiring installed in these areas are more restrictive in most cases. In explosive atmospheres, it is essential to control and eliminate all possible sources of ignition, such as an electrical arcing event. A motor controller or switch that has make and break arcing contacts are good examples of electrical equipment that must be installed in suitable enclosures that effectively contain the arcing under normal conditions.

Part V of Article 250 contains general bonding requirements for electrical installations. Section 250.90 provides a performance requirement that bonding be provided to ensure electrical continuity and adequate capacity for any fault current. Effective bonding is important for safety

FIGURE 14-1 More specific and strengthened methods of bonding are required for metal raceways and enclosures in hazardous (classified) locations.

in general locations, but in hazardous locations there are increased concerns about establishing an effective ground-fault current path that will not create an explosion or fire if a ground fault were to occur. During a ground fault event, the amount of fault current is significant to the point where special bonding methods are necessary for metallic wiring methods. The reason for enhanced bonding is to address concerns about arcing and sparking at any fitting terminations. The strengthened bonding required in these locations ensures an effective path for ground fault current to facilitate fast operation of the overcurrent device protecting the circuit and equipment. Fast opening of an overcurrent device reduces possibilities of a hot spot developing on an enclosure at the point of the ground fault. Otherwise, such hot spots or heated enclosure surfaces could quickly become an ignition source if the arcing event is sustained even for a short period of time. Section 500.8(E) requires installing conduit with five full threads fully engaged and wrench-tight. The purpose is to maintain the explosion-proof integrity of enclosures and conduit system, and prevent arcing or sparking at threaded joints if fault current passes over the conduit during a ground fault event. **See Figure 14-2.**

Section 250.100 provides requirements for bonding in hazardous locations and applies regardless of the voltage of the system or circuit. This rule indicates

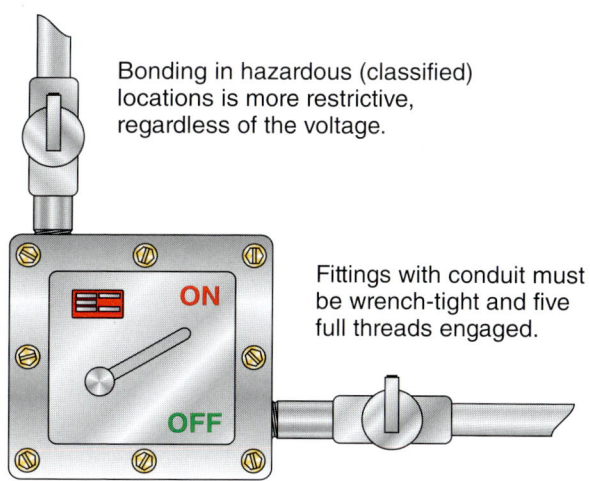

FIGURE 14-2 The threads of couplings, fittings, boxes, and so forth must have five full threads fully engaged when installed in hazardous locations.

FIGURE 14-3 Bonding in hazardous locations must be made tight using suitable wiring methods and fittings.

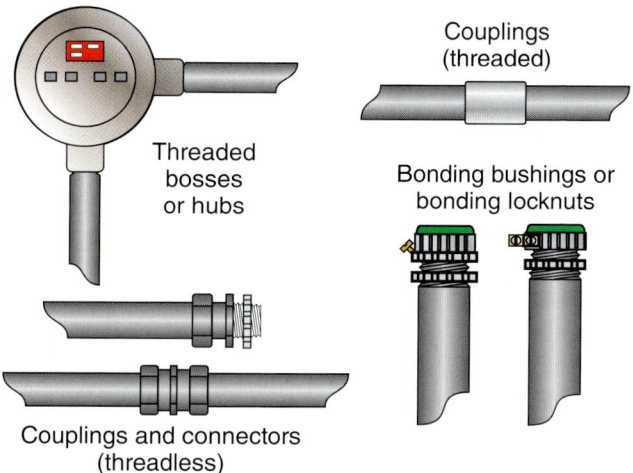

FIGURE 14-4 Any of the methods in Sections 250.92(B)(2) through (4) are acceptable for bonding in hazardous locations.

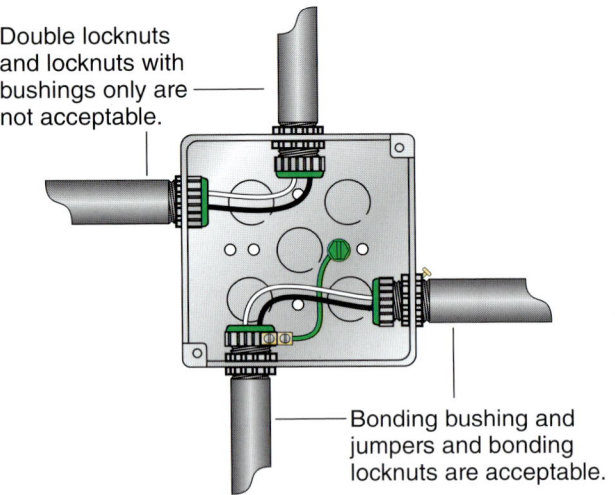

FIGURE 14-5 Bonding locknuts and bonding bushings with jumpers are acceptable methods of bonding in hazardous (classified) locations, while standard locknuts and a locknut and bushing arrangement is not.

that the non–current-carrying metal parts of equipment, raceways, and other enclosures in hazardous locations must be bonded using one of the methods provided in 250.92(B)(2) through (B)(4). **See Figure 14-3.** This enhanced bonding for metallic wiring methods is required even if an equipment grounding conductor of the wire type is installed. **See Figure 14-4.** Standard locknuts or bushings are not permitted to accomplish the bonding required for wiring in a hazardous location. Where metallic wiring connections are made at equipment such as boxes, enclosures, cabinets, and panelboards, effective bonding must be ensured around any joints in the fault current path to prevent sparking and ensure a low-impedance

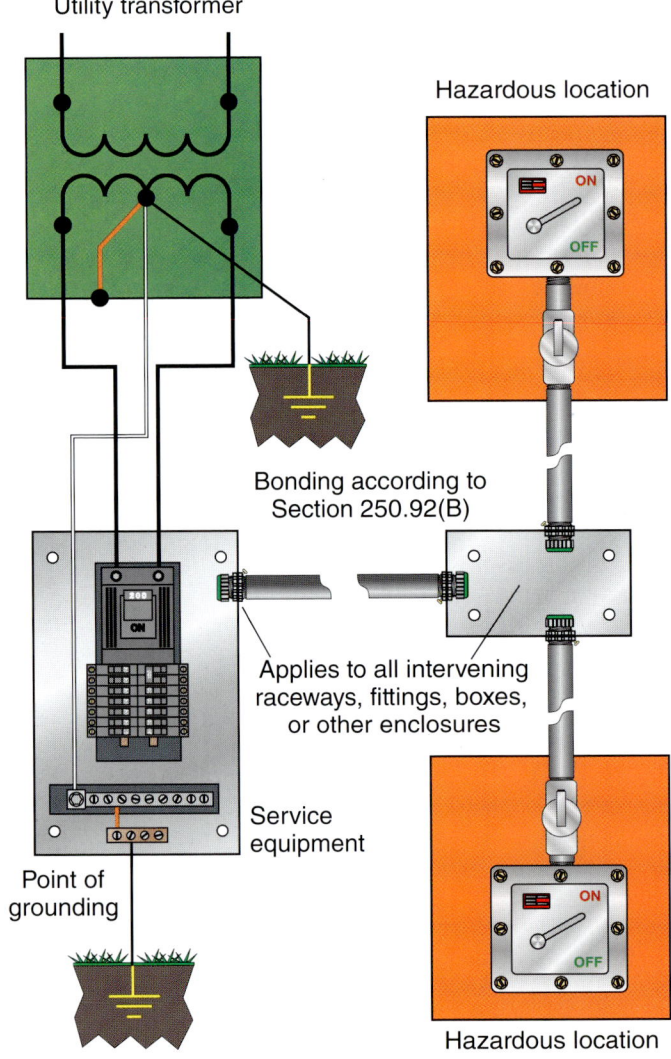

FIGURE 14-6 Bonding requirements must extend to all intervening metal raceways and back to the applicable service or separately derived system grounding point.

path for any ground fault current. Standard locknuts and bushings can be used to connect the raceways to the enclosures, but bonding continuity must be ensured around any standard locknut, bushing, or combination, using one of the methods provided in 250.92(B)(2) through (B)(4). **See Figure 14-5.**

During a ground fault, heavy levels of current will be present for the time it takes the overcurrent device to open the faulted circuit. Strengthened bonding methods are required for wiring within the hazardous locations, and are also required for the entire metallic raceway system not in the hazardous location. The bonding requirements in Chapter 5 of the *NEC* clarify that the enhanced bonding applies to all metal raceways and enclosures in the hazardous location and to all metal raceways and enclosures of the circuit run extending to the grounding point of the applicable service or derived system. **See Figure 14-6.** The "point of grounding" is typically where the service main bonding jumper is installed or where the system bonding jumper is installed for a separately derived system. This is also where the grounding electrode conductor connection is made at the service or system.

The *Code* permits flexible wiring methods such as flexible metal conduit (FMC) and liquidtight flexible metal conduit (LFMC) in Division 2 locations under restrictive conditions as provided in Sections 501.10(B) and 502.10(B). **See Figure 14-7.** LFMC is permitted in Class III locations as indicated in Section 503.10(A)(3)(2). Where FMC or LFMC is installed for flexibility in hazardous locations, a wire-type equipment bonding jumper must be installed in accordance with Section 250.102. **See Figure 14-8.** See the .30(B) section of Articles 501 through 503 and Sections 505.25(B) and 506.25(B). Bonding jumpers are often installed external to the flexible conduit installation using suitable fittings that provide an attachment lug. The equipment bonding jumpers can be installed inside or outside the conduit. **See Figure 14-9.**

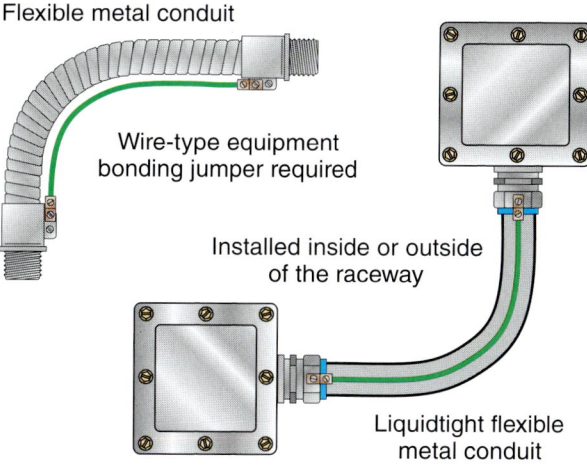

FIGURE 14-7 A wire-type equipment bonding jumper is required for flexible conduit installations and can be installed inside or outside the conduit.

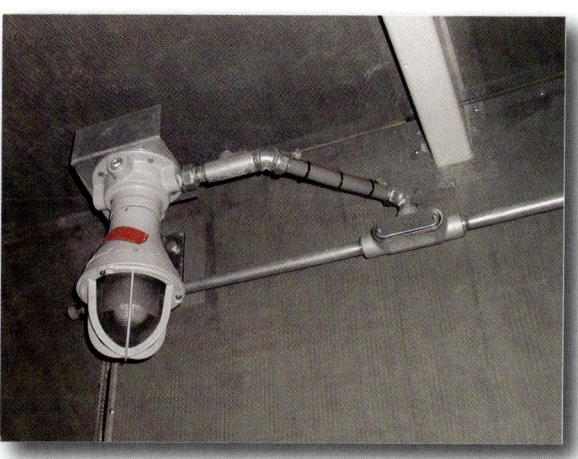

FIGURE 14-8 A wire-type equipment bonding jumper is required for LFMC installed in hazardous (classified) locations.

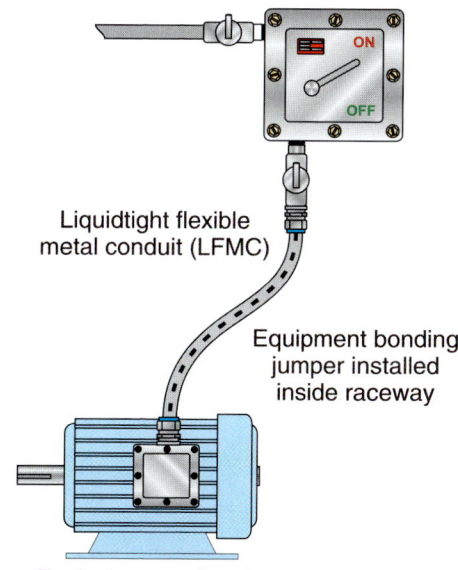

FIGURE 14-9 A wire-type equipment bonding jumper is required for liquidtight FMC installed in hazardous (classified) locations and can be installed inside or outside the raceway.

There are additional concerns about static electricity as an ignition source in hazardous locations. Static electricity problems are typically addressed using equipotential bonding and grounding techniques that reduce or eliminate potential differences in the hazardous location. In some cases, specifically in indoor applications, raising humidity levels provides an effective solution. Each static problem requires its own analysis and solution.

For information on protection against static electricity and lightning hazards in hazardous locations, see NFPA 77, *Recommended Practice on Static Electricity*; NFPA 780, *Standard for the Installation of Lightning Protection Systems*; and API RP 2003-1998, *Protection Against Ignitions Arising Out of Static Lightning and Stray Currents*. The NJATC textbook on hazardous (classified) locations provides more information about grounding and bonding and specific protection systems and methods for handling static problems in hazardous locations.

Branch circuits in patient care areas of health care facilities must provide two independent equipment grounding paths.

Special Rules for Health Care Facilities

Reducing differences of potential between conductive equipment or other objects and the Earth helps minimize shock hazards in normal circuit operation. In patient care locations, it is even more important that overcurrent devices operate quickly during ground-fault conditions to minimize the potential (voltage) on conductive parts. **See Figure 14-10.** These are the main reasons for the more restrictive requirements for equipment grounding conductors (EGCs) for branch circuits that supply patient care areas in a health care facility.

This requirement applies to all patient care areas in all health care facilities. It does not apply to the EGCs for feeders that supply the panelboards containing branch circuits for patient care locations. Electrical feeder EGCs must meet the general requirements in Section 215.6 that indicate feeders generally must include an equipment grounding condcutor. Where the feeder is installed in a metallic wiring method, such as rigid metal conduit (RMC) or electrical metallic tubing (EMT), it must comply with the rules in Section 517.19(D). Feeder grounding and bonding requirements are covered in more detail later in this chapter. For now, this chapter focuses on the rules for the branch circuits serving patient care locations.

The objectives of the two separate EGCs for branch circuits serving patient care areas required by Section 517.13 emphasize creating redundancy in the safety (grounding and bonding) components of these branch circuits. It is important to remember that many of the requirements for electrical wiring in health care facilities that are included in the *NEC* and NFPA 99, *Standard for Health Care Facilities*, focus on providing redundancy of systems. In this case, the redundancy is required to be built into the branch circuit safety equipment grounding means.

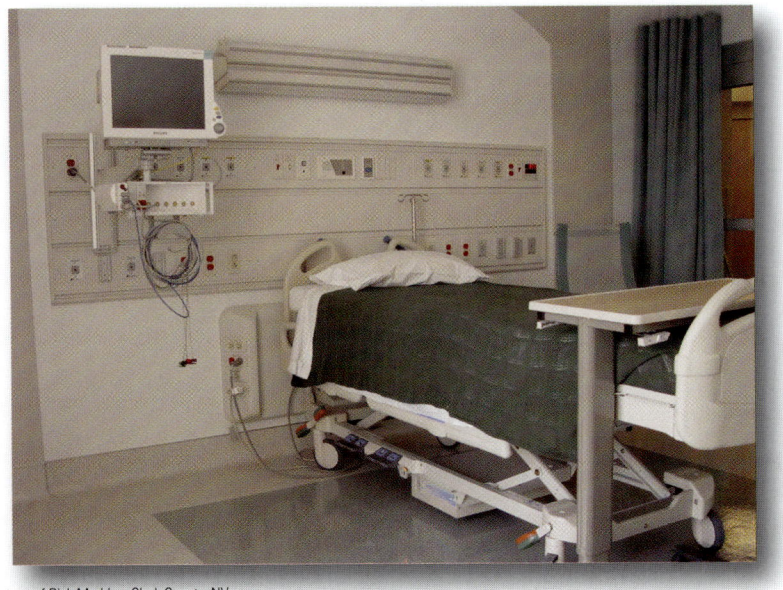

Courtesy of Rick Maddox, Clark County, NV

FIGURE 14-10 Branch circuits serving patient care locations have to meet the requirements in Section 517.13.

Grounding of Receptacles and Fixed Equipment in Patient Care Areas

The requirements for patient care area branch circuit EGCs are provided in Sections 517.13(A) and (B) of the *NEC*. It is important to understand that both subdivisions (A) and (B) must be applied together to each branch circuit serving patient care locations to meet the minimum requirements in this section.

Article 517 of the *NEC* provides specific definitions and rules that assist in determining patient care locations where this more restrictive grounding and bonding are required. **See Figure 14-11.** Branch circuits in patient care areas of health care facilities are required to provide two independent equipment grounding paths for all non–current-carrying conductive surfaces of fixed electrical equipment likely to become energized that are subject to personnel contact. **See Figure 14-12.** Section 517.13(A) includes requirements related to the type of wiring method that can be used for branch circuit conductors. Approved wiring methods for patient care areas include metallic raceway systems or cables with a metallic armor or sheath that qualifies as an EGC in accordance with Section 250.118. EMT is often specified and installed as a branch circuit wiring method in patient care areas.

It is important to recognize the types of conduit, tubing, cables, and other wiring methods that qualify as EGCs in accordance with Section 250.118. A review of this section reveals a fairly extensive list of such qualifying wiring methods: RMC, intermediate metal conduit, EMT, armored cable (Type AC), and so forth. These wiring methods inherently provide a path for ground-fault current through the raceway or metallic cable armor itself. The wiring methods listed in Section 250.118 that meet the minimum requirements in the *NEC* as acceptable EGCs are acceptable as wiring methods for branch circuits in patient care locations. Remember that the integrity of the EGC has everything to do with the workmanship providing suitable supporting and securing means, as well as proper coupling and fitting installation. The rules addressing securing and supporting requirements for wiring methods approved for patient care areas are typically located in the .30 section of each applicable wiring method article. For example, the rules for securing and supporting EMT are found in Sections 358.30(A) and (B).

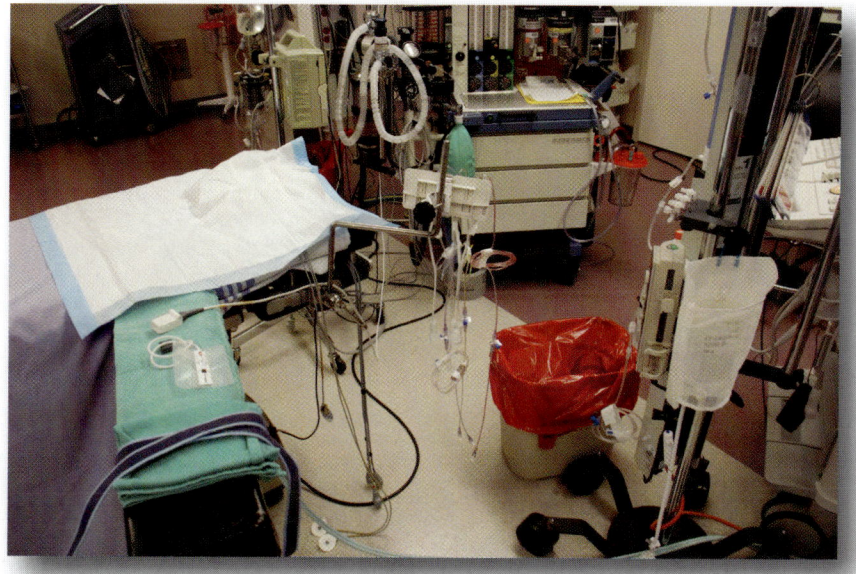

FIGURE 14-11 Branch circuits supplying an operating room in a health care facility must include EGCs that meet Section 517.13.

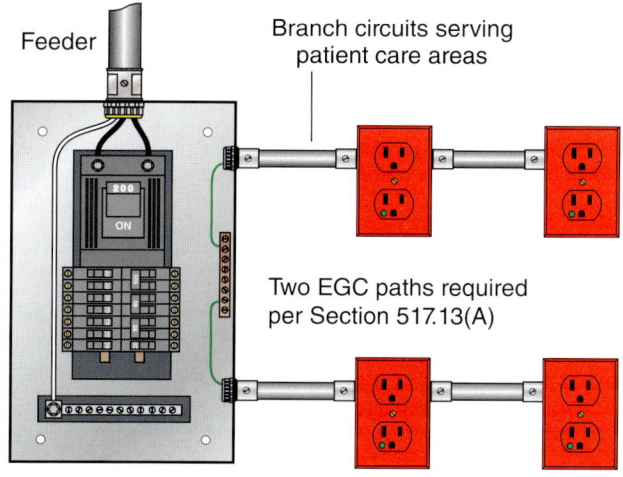

Only branch circuit EGCs are shown.

FIGURE 14-12 Two separate EGC paths are required for branch circuits serving patient care locations.

Listed flexible metal conduit (FMC) is suitable as an EGC that meets the minimum requirements of the *Code,* but it does so under more restrictive conditions. Where listed FMC is installed for branch circuit wiring in a patient care area, it not only must meet the restrictive conditions provided in Section 250.118(5) but also must include an insulated copper EGC with the branch circuit conductors. **See Figure 14-13.** This EGC, like all EGCs, must be no smaller than the sizes provided in Table 250.122 based on the rating of the overcurrent protective device ahead of the circuit.

Listed FMC might be used in patient care areas for facilitating flexible connections to equipment such as medical headwall assemblies in patient care rooms of hospitals. When this wiring method is used in patient care areas, the wiring method for the branch circuit must meet the requirements of both Section 250.118(5) and Sections 517.13(A) and (B).

Listed liquidtight flexible metal conduit (LFMC) is, with restrictions, also suitable as an EGC. LFMC must meet the following conditions to be suitable for and function as an EGC for branch circuits serving patient care locations: It must meet the restrictive conditions provided in Section 250.118(6), and it must include an insulated copper EGC with the branch circuit conductors. **See Figure 14-14.** Like listed FMC, listed LFMC might be used in a patient care area for facilitating flexible connections to equipment. When this wiring method is used in patient care areas, it must meet the requirements of both Section 250.118(6) and Sections 517.13(A) and (B).

When cable wiring methods are used for branch circuit wiring in patient care areas of health care facilities, the conductive armor must provide an effective ground-fault current path and qualify as an EGC. There must also be an insulated copper EGC included in the cable assembly if used for a patient care location. Some Type MC cable is suitable as an EGC, according to Section 250.118(10).

(10) Type MC cable that provides an effective ground-fault current path in accordance with one or more of the following:
  a. An insulated or uninsulated EGC in compliance with Section 250.118(1)
  b. The combined metallic sheath and uninsulated equipment grounding/bonding conductor of interlocked Type MC cable that is listed and identified as an EGC conductor

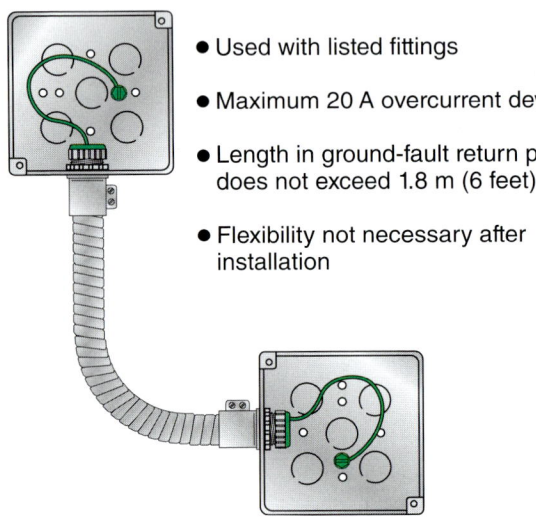

FIGURE 14-13 Listed FMC with an insulated copper EGC can be used as a branch circuit wiring method for patient care locations under restrictive conditions in Section 250.118(5).

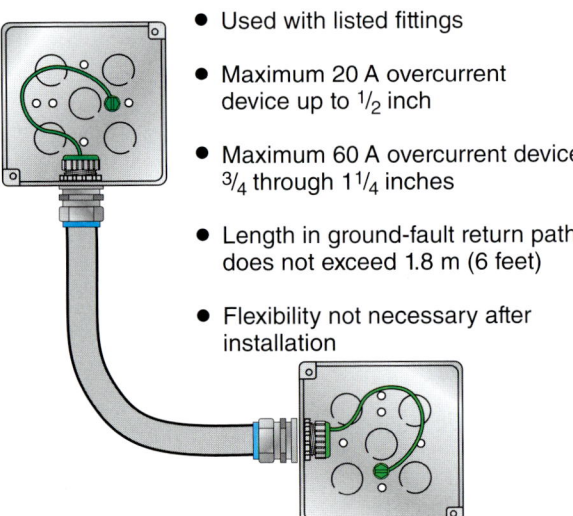

FIGURE 14-14 Listed LFMC with an insulated copper EGC can be used as a branch circuit wiring method for patient care locations under restrictive conditions in Section 250.118(6).

c. The metallic sheath or the combined metallic sheath and EGCs contained within smooth or corrugated tube–type MC cable that is listed and identified as an EGC

Only Type MC cables addressed in list items b and c are acceptable for use in a patient care area if they also contain an insulated copper EGC.

The *UL Guide Information for Electrical Equipment* under categories (PJAZ) and (PJOX) provides additional product information about Type MC cable assemblies and fittings (see Annex C of this textbook). There is at least one specific type of MC cable that has interlocking metal tape–type construction, and the armor is suitable as an EGC. It is similar in construction to standard type AC cable in that it includes a separate bare conductor within the assembly that is installed on the outside of the plastic-wrapped conductors of the cable assembly and is in intimate contact with the cable armor. **See Figure 14-15.** This type of cable is covered by Section 150.118(10)(b).

This Type MC cable has been listed and evaluated for use as an EGC when used with suitable listed fittings. This Type MC cable is manufactured with and without a contained insulated copper EGC, so it would be acceptable for use in health care facility installations only if the internal insulated EGC is present. **See Figure 14-16.** As always, to ensure compliance with Section 110.3(B) of the *Code,* follow the manufacturer's installation instructions when installing this and any other product.

Section 517.13(A) specifies metallic raceways or cable assemblies that are acceptable as EGCs in accordance with Section 250.118. This is one of two EGC paths required for branch circuits serving these locations. It is important to understand that this effective ground-fault current path in the form of a metallic wiring method is different in characteristics from the contained insulated copper conductor addressed in Section 517.13(B). So in review, Section 517.13(A) indicates that the wiring method selected for the branch circuits in patient care areas must qualify as an EGC in accordance with Section 250.118 and Section 517.13(B) requires an insulated copper EGC of the wire type.

Armored Cable (Type AC)

The armor sheath of Type AC cable is recognized by Section 250.118(8) as an EGC. Type AC cable that includes an insulated copper EGC complies with the requirements in Sections 517.13(A) and (B).

FIGURE 14-15 Type MC cable is suitable for use in patient care locations where it includes an insulated copper EGC.

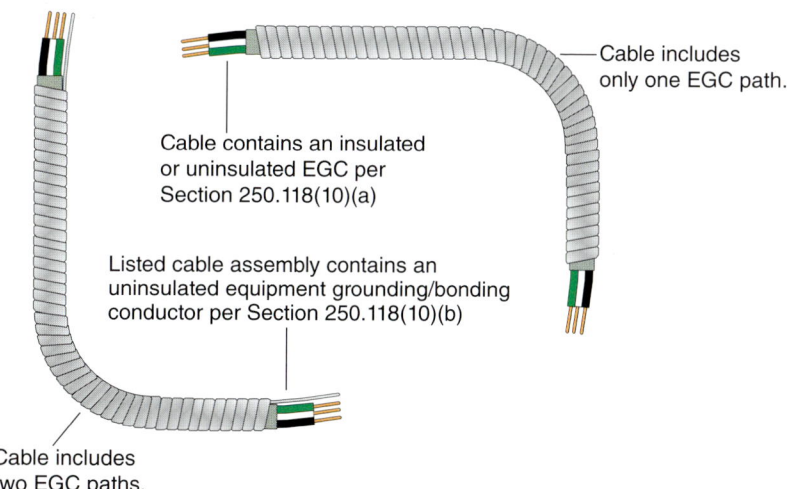

FIGURE 14-16 The armor of Type MC cable has to qualify as an EGC, and the assembly has to contain an insulated copper EGC.

This cable armor qualifies as an EGC because of the bare internal bonding strip that is in intimate contact with the armor from fitting to fitting. **See Figure 14-17.**

The internal bonding strip in the cable assembly and the interlocking metal tape–type armor work in combination as an effective ground-fault current path.

Nonmetallic types of wiring methods, such as rigid polyvinyl chloride (PVC) conduit, and cables where the outer jacket (armor) of the cable does not qualify as an EGC are not suitable for use as branch circuit wiring methods in patient care areas. The objective of this redundant EGC requirement is to ensure adequate equipment grounding in the patient care area and provide a redundant path for the ground-fault current should a ground-fault condition develop. Quick operation of branch circuit overcurrent devices is essential to safely clear the event. Remember, the rule requires the installation of two separate EGC paths to ensure that one will operate when necessary. A good approach to selecting suitable cable wiring methods that meet this more restrictive requirement is to refer to the *General Information of Electrical Equipment Directory (UL White Book)* for the specific criteria. Typically, the engineering design and correspondence with the governing body of the health care facility will have already determined which areas of the facility are patient care locations and the design team will have specified suitable wiring methods that meet the redundant equipment grounding requirements.

The fittings used with approved wiring methods are required to be listed. Listed Type AC and MC cable fittings are evaluated for use with these cable assemblies and provide a suitable connection means for establishing continuity and conductivity between the cable armor and the enclosure to which it connects. **See Figure 14-18.** Each cable must be used with fittings designed and listed for use with that particular cable. Verify with the manufacturer if there is doubt as to compatibility. Cable fittings are required to be identified for the use. This identification can be found on the product itself or on the smallest shipping carton.

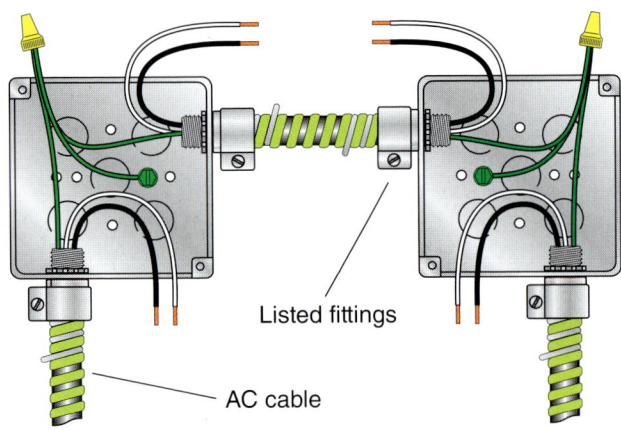

FIGURE 14-17 The armor of Type AC cable is suitable as an EGC.

FIGURE 14-18 Type AC cable is suitable for use in patient care locations when an insulated copper EGC is included in the cable assembly and listed fittings are used.

Wire-Type EGC Required

As previously stated, in addition to the wiring method qualifying as an EGC in accordance with Section 250.118, each branch circuit and outlet serving patient care areas must include an insulated copper EGC. This insulated copper conductor must be connected to the grounding terminals of receptacles and non–current-carrying conductive surfaces of fixed electrical equipment likely to become energized. This requirement applies to circuits operating at a voltage of more than 100 volts, which generally applies to all line voltage branch circuits serving patient care areas.

Notice that the equipment grounding requirements in Section 517.13(B) are yet more restrictive than what would normally be required in Section 250.118.

The requirement for an EGC that is both insulated and copper exceeds the basic provisions of Section 250.118. The sizing rules for insulated EGCs required by Section 517.13(B) are the same as those provided in Section 250.122 for all branch circuits and feeders. The size is based on the rating of the final overcurrent protective device protecting the circuit in accordance with Table 250.122. The requirements of Section 250.122 still have general application to branch circuit EGCs, including requirements for conditions where the ungrounded (phase) conductors may be increased in size to handle voltage drop conditions, and so on.

The requirements of Sections 517.13 (A) and (B) apply to the branch circuits serving patient care areas. Looking at the definition of *branch circuit*, it can be seen that two separate EGC paths are required with the wiring installed from the branch circuit panelboard (final overcurrent device) to the outlet. **See Figure 14-19.** This means cord- and plug-connected equipment is only required to have a single equipment grounding means. Metal faceplates that cover receptacles installed in these outlets are permitted to be grounded by the 6-32 fastening screw that attaches the plate to the device in accordance with Exception 1 to Section 517.13(B).

Branch Circuit. The circuit conductors between the final overcurrent device protecting the circuit and the outlet(s).[1]

Furthermore, the requirements for grounding using an insulated EGC are not necessary for luminaries or switches that are outside the patient vicinity, as indicated in Exception 2 to Sections 517.13(B). So, by the allowances in these two exceptions, it can be seen that the definitions of the terms *branch circuit* and *patient care vicinity* are important in clarifying the minimum requirements of the *Code*. Even though the term *patient care vicinity* is used in Exception 2 to Sections 517.13(B), it should be recognized that many governing bodies of health care facilities specify the location for patient care as the entire room. Design and engineering firms understand the importance of meeting the minimum requirements, but many times their designs exceed the minimum requirements in the *NEC* and other applicable standards. Be sure to maintain the integrity of the engineered design by following plans and specifications. If changes or modifications are desired or necessary, always work cooperatively with the engineering firms. Remember, the key concept to recall is that the *Code* requires two grounding paths to be installed for electrical equipment in patient care areas, assuring that at least one return path for the fault current is always present if one should fail. This is often referred to in the field as redundant grounding for patient care areas. The concept of this redundancy should be viewed from the standpoint that installation of two suitable EGC return paths is required to ensure at least one path is available with the circuit.

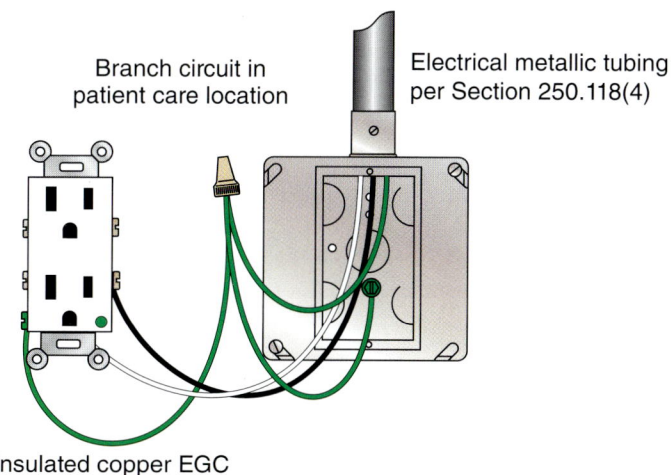

FIGURE 14-19 An insulated copper EGC is required to be connected to the grounding terminal of receptacles serving patient care locations.

Isolated Ground Receptacle Installations in Patient Care Locations

Often, medical equipment manufactures specify isolated equipment grounding circuits and isolated grounding receptacles. Reducing the amount of electrical magnetic interference (EMI) in the grounding circuit for sensitive electronics is why isolated grounding is specified. Isolated grounding introduces additional requirements for another EGC in patient care areas. In addition to the two separate EGCs, required by Sections 517.13(A) and (B), a third isolated, insulated EGC must be installed with the circuit conductors in accordance with Section 250.146(D) and the exception to Section 408.40. Section 406.2(D) requires isolated grounding-type receptacles to be identified with an orange triangle on the face of the receptacle. **See Figure 14-20.** In addition, isolated grounding receptacles used in patient bed locations of health care facilities are required to be listed as hospital-grade types.

Prior to the 2011 edition of the *NEC*, Section 517.16 and the associated fine print addressed isolated ground receptacles installed in patient care areas. It is critical to note that the isolated EGCs do not satisfy the Section 517.13(B) requirements because they do not provide the functional benefit of being connected in parallel with the raceway or cable armor at both ends. Because of this requirement, any isolated, insulated EGC connected to an isolated grounding-type receptacle must be installed *in addition to* the two EGC paths required by Sections 517.13(A) and (B) for branch circuits serving patient care areas. While the *NEC* no longer permits the installation of isolated grounding circuits and receptacles as restricted by Section 517.16, NFPA 99 *Standard for Health Care Facilities* still addresses isolated grounding circuits and receptacles. NFPA 99 still requires periodic testing of grounding systems, which include installations of isolated grounding receptacles and circuits in health care facilities.

Patient Equipment Grounding Point

An optional method of protecting patients from differences of potential and possible shock hazard is the installation of a patient equipment grounding point. This type of installation usually includes specific devices manufactured for this purpose that incorporates one or more listed grounding and bonding jacks. This type of optional protection is often part of electrical system designs and specifications for operating rooms and other critical procedure locations in health care facilities. At minimum, a 10 AWG equipment bonding jumper must be installed from the grounding terminals of all grounding-type receptacles to the patient equipment grounding point.

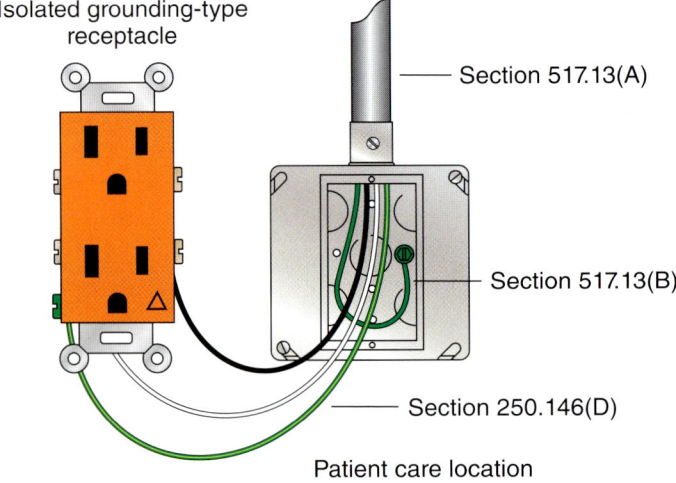

FIGURE 14-20 Isolated grounding receptacles are identified with an orange triangle on the face of the receptacle and must be connected to an insulated copper EGC installed with the branch circuit in accordance with 250.146(D).

Where an isolated power system is used, the patient equipment grounding point is typically connected to the reference grounding bar in the isolated power panel. The bonding conductor used for the equipment grounding point is permitted to be installed separately to each conductive part and receptacle grounding terminal, or it may be looped if this is convenient. See Section 517.19(C) for the actual *Code* requirements.

Isolated Power System Grounding

Isolated power systems installed in health care facilities must comply with Section 517.160. **See Figure 14-21.** These systems are required to operate ungrounded. **See Figure 14-22.** EGCs are terminated to the grounding terminal of grounding-type receptacles connected to these systems. Section 517.160(A)(5) requires the orange conductor with a distinct stripe to be connected to the receptacle terminal normally intended for the grounded (neutral) conductor. **See Figure 14-23.**

The most common practice is to install isolated power system EGCs in the same raceway as the circuit conductors. However, because the system is ungrounded, Section 517.19(F) permits these EGCs to be installed outside of the conduit or enclosure containing the circuit conductors. This is a modification to the general requirements in Sections 300.3(B) and 250.134(B), where the EGC is required to be installed in the same raceway or cable or otherwise run with the ungrounded circuit conductors.

Grounding and Bonding Requirements for Panelboards

Grounding Requirements

Section 517.13 applies only to branch circuits serving patient care areas. With an understanding of the term *branch circuit*, it is clear that the requirement for two equipment grounding paths is not applicable to feeders. The general requirements for feeder EGCs are found in Section 215.6.

FIGURE 14-21 Isolated power systems are an optional method of protection installed in health care facilities, unless required for wet location patient care areas.

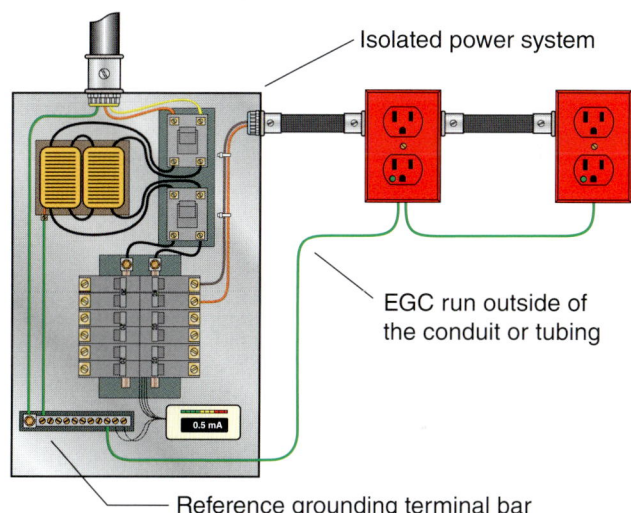

FIGURE 14-22 EGCs installed with circuits of isolated power systems must be connected to the reference grounding bus within the isolated power system enclosure.

FIGURE 14-23 EGC connections at a receptacle are connected to an isolated power system in an operating room of a hospital. Note there is no grounded conductor connected to the receptacle.

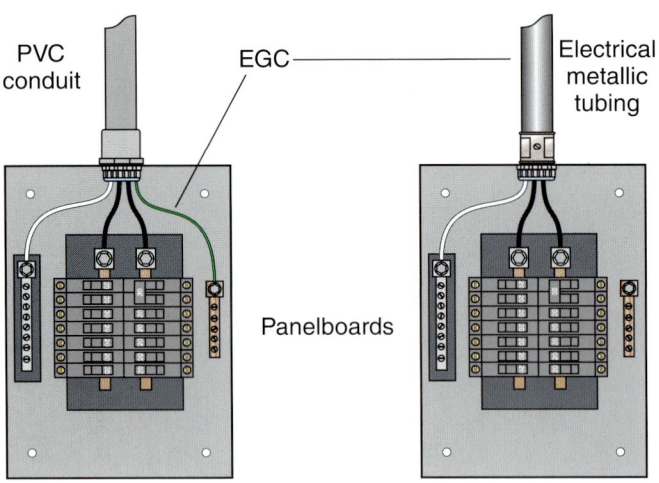

FIGURE 14-24 Feeders must include an equipment grounding means, as required in Section 215.6.

This section requires feeders to provide an EGC where the feeder supplies branch circuits in which EGCs are required. **See Figure 14-24.** This requirement applies to all feeders regardless of occupancy. Each feeder supplying branch circuits serving patient care areas must include an EGC in accordance with the provisions of Section 250.134 for connection of branch circuit EGCs served by the feeder.

If a feeder in the critical branch is supplied from a grounded distribution system in a health care facility and is installed using a metallic wiring method, the grounding of the switchboard or panelboard is required to be assured by one of the methods in Sections 517.19(D)(1), (2), or (3). This enhanced bonding requirement applies to all junction points (connection points at enclosures, junction and pull boxes, and so on) of the metal raceway or metallic Type MC or mineral-insulated–(Type MI) cable armor. Enhancing the bonding requirements for feeder equipment grounding means helps ensure effective operation of branch circuit and feeder overcurrent devices in ground-fault conditions. Because only one EGC is required to meet the minimum *NEC* requirements for feeders in health care facilities, enhanced and more restrictive bonding is essential. This bonding at termination points can be accomplished by a grounding (bonding) bushing and a properly sized bonding jumper, threaded bosses or hubs, or other approved devices, such as bonding-type locknuts or bushings.

Bonding Requirements

In addition to the feeder EGC rules in Sections 215.6 and 517.19(D), there are bonding requirements for panelboards that provide branch circuits serving patient care areas. Again, this is an effort to minimize potential differences among conductive equipment and parts serving the same patient care vicinity. The *NEC* requires the equipment grounding terminal bars of the normal and essential branch circuit panelboards serving the same individual patient vicinity to be bonded together. Section 517.14 requires they be connected (bonded) together with an insulated copper conductor no smaller than 10 AWG.

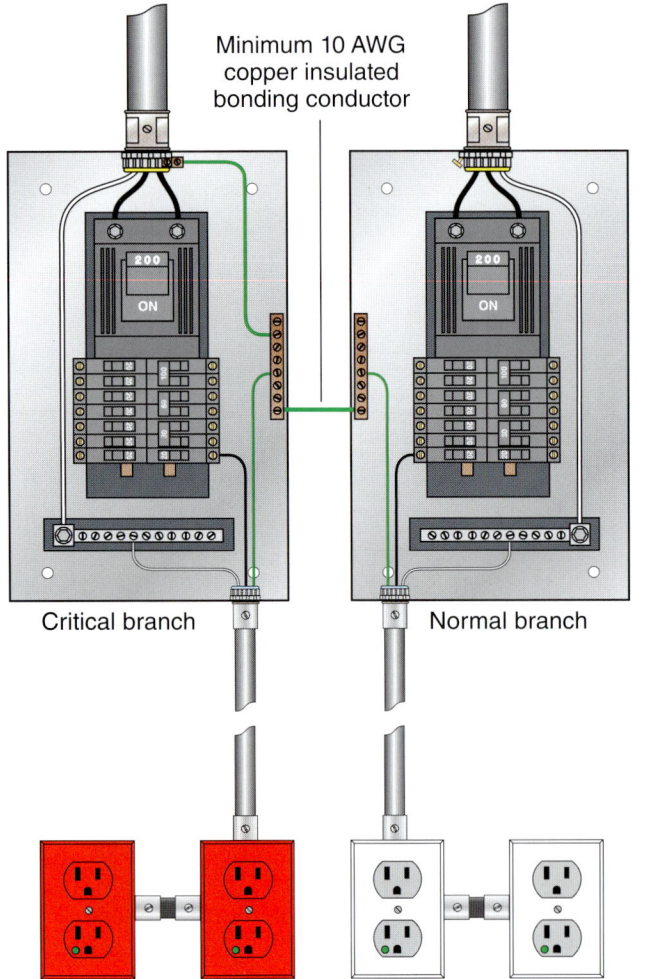

FIGURE 14-25 Panelboards serving the same patient care location have to be bonded together with, at minimum, 10 AWG insulated copper conductor.

See Figure 14-25. The bonding conductor installed between the panelboard equipment grounding terminal bars must be continuous, except it may be broken where it connects to the terminal bar in the panelboard. This applies to all situations in which two or more panelboards supply the same patient vicinity and are served through separate transfer equipment on the emergency system. The reason for this additional bonding is that the normal branch circuits and the critical branch circuits serving the same individual patient vicinity are generally supplied from different separately derived systems. This situation creates the possibility of potential differences between the branch circuit EGCs of the normal and emergency systems.

Bonding all equipment grounding terminal bars of all panelboards serving the same patient vicinity helps minimize the possible potential differences. Note that the *Code* requires this bonding conductor to be at least a 10 AWG copper conductor and insulated. It may be larger depending on the design. For example, if the panelboards serving the same patient care vicinity were located a considerable distance apart, voltage drop concerns could be a problem.

Special Rules for Agricultural Installations

Article 547 provides special requirements for agricultural buildings or any part of a building or structure with the conditions associated with these types of buildings or structures. **See Figure 14-26.** The common adverse conditions encountered in an agricultural building are excessive dust and dust with water accumulations, along with corrosive atmospheres from animal excretion. Often, corrosion is accelerated in these areas because of moisture and wet conditions related to periodic washing operations and sanitizing by the use of water and cleansing agents. There are three primary concerns, in these types of occupancies, for electrical wiring systems. The first concern is related to the excessive amount of corrosion and maintaining the integrity of the grounding path for electrical wiring in these buildings or areas. The other two concerns are about voltage gradients and maintaining the integrity of the grounding paths with the feeders and branch circuits installed in these locations. **See Figure 14-27.**

FIGURE 14-26 Agricultural buildings and property have special grounding and bonding requirements.

FIGURE 14-27 Special bonding requirements apply to animal containment areas in agricultural facilities.

Voltage Gradients

In agricultural facilities, an important concern is about voltage gradients that can be present in the Earth. Neutral-to-Earth stray current can result in significant behavior problems, loss of production, and even death of farm animals. These stray currents are often referred to as *tingle voltages* and can appear between various conductive components: metal piping systems, conductive flooring, metal rails, feeding troughs, and so forth. Some common causes of stray currents are current present in primary distributions systems, faulty wiring on or in the farm buildings, and inappropriate neutral-to-ground connections on the load side of the service grounding point or grounding point of a separately derived system. Ground-fault events can also introduce current into the Earth. Where electrical wiring systems and equipment are effectively grounded and bonded, a phase-to-ground fault condition should operate an overcurrent protective device and clear the abnormal condition quickly.

> Neutral-to-Earth stray current can result in significant behavior problems, loss of production, and even death of farm animals.

Site-Isolating Device

> **Site-Isolating Device.** A disconnecting means installed at the distribution point for the purposes of isolation, system maintenance, emergency disconnection, or connection of optional standby systems.[2]

The *Code* requires that any portion of an underground EGC that is installed to or in the building or structure has to be covered or insulated. **See Figure 14-28.** The *Code* is more restrictive in Section 547.5(F) and requires copper EGCs rather than providing a choice between copper and aluminum.

An important requirement for electrical systems in agricultural buildings is to isolate the neutrals, EGCs, and other non–current-carrying parts of equipment for milking parlors, cattle corral and feeding areas in buildings, and so forth. The *NEC* generally requires isolation of grounded (neutral) conductors at separate buildings or structures in accordance with Section 250.32(B). These general requirements are supplemented by the provisions in Section 547.9(B)(3). Each agricultural building or structure has to meet the provisions in Section 250.32 and has to meet the sizing and connection rules in Sections 547.9(B)(1) and (2). Here, the minimum size required for the EGC must be no smaller than the largest supply conductor for the building if the same conductor material is used. **See Figure 14-29.**

If the EGC is of a different material, it must be adjusted in size in accordance with the equivalent sizes in the appropriate columns of Table 250.122. Another important requirement deals with the connections between the grounded (neutral) conductor and the EGC and the non–current-carrying metal parts of equipment. This connection must only be made at the site-isolation device at the distribution point of the premises.

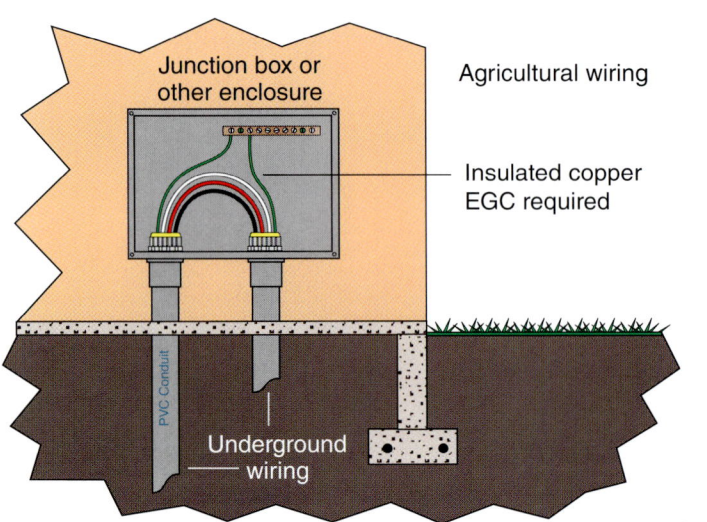

FIGURE 14-28 EGCs of underground wiring for agricultural installations must be insulated copper.

The site-isolating device on an agricultural property is installed to provide a means to disconnect and isolate the agricultural premises' wiring system from the serving utility typically under emergency conditions or for maintenance of the load-side wiring system. It also provides an accessible point for the connection of an alternate power source during power outages. In accordance with Section 547.9(A)(2), the site-isolating device must be pole mounted, must be an isolating switch, and is not considered to be the service disconnecting means for the agricultural premises. A grounding electrode system or electrode is required to be connected to the grounded conductor at a site-isolating device. **See Figure 14-30.**

When feeders are routed, either overhead or underground from the site-isolating device to buildings or structures on the premises, the feeder must provide an insulated grounded conductor and a separate EGC and meet all other applicable requirements in Section 250.32.

Equipotential Bonding Planes

A common method of protection for farm animals is to establish an equipotential bonding plane in agricultural buildings or structures. The term *equipotential plane* is defined in Section 547.2.

> **Equipotential Plane.** An area where wire mesh or other conductive elements are embedded in or placed under concrete, bonded to all metal structures and fixed nonelectrical equipment that may become energized, and connected to the electrical grounding system to prevent a difference in voltage from developing within the plane.[3]

The *Code* requires an equipotential plane to be installed in livestock areas as prescribed in Section 547.10(A).

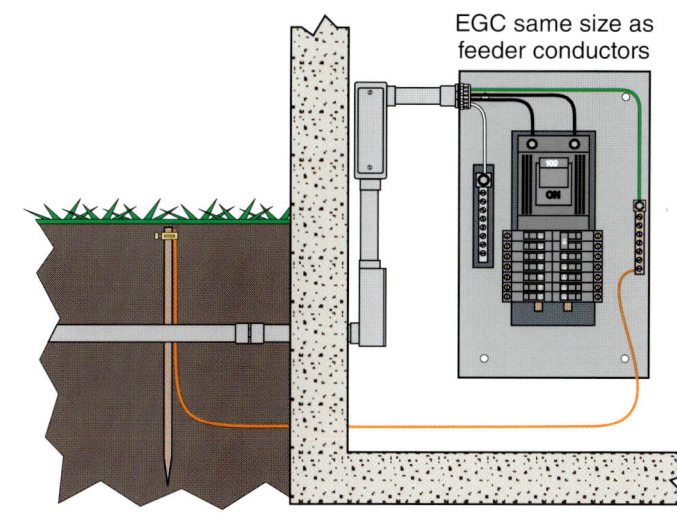

FIGURE 14-29 The EGC(s) of feeders supplying separate buildings are required to be sized the same as the ungrounded conductors of the circuit.

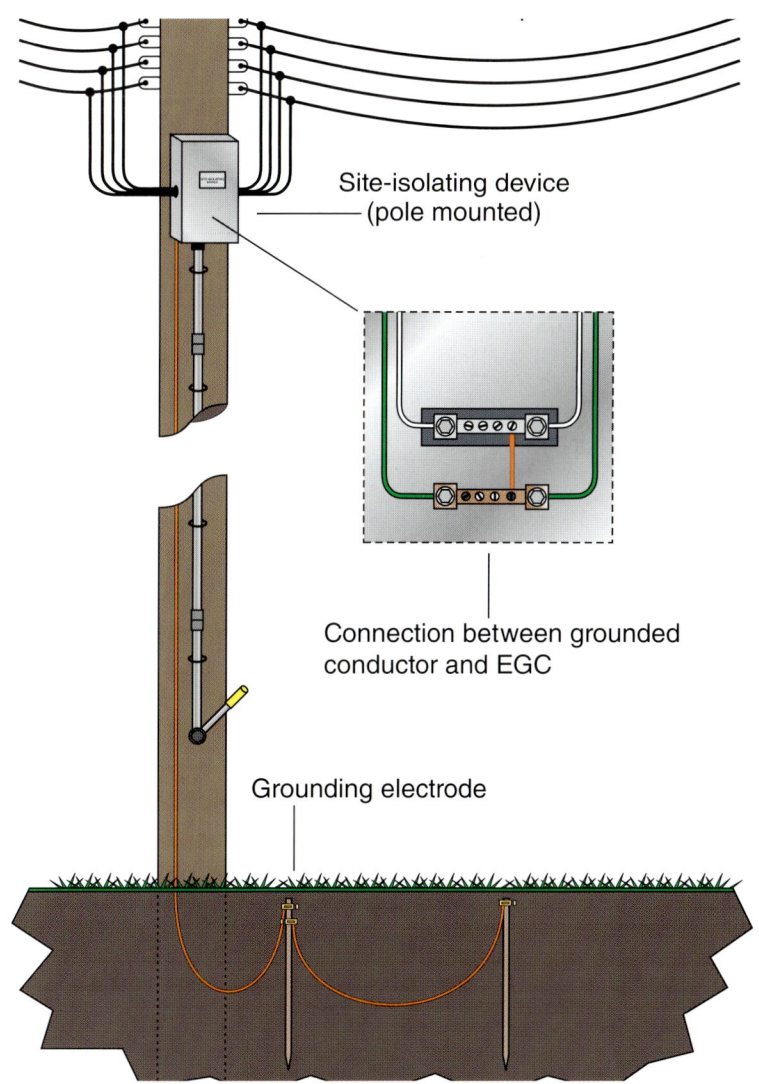

FIGURE 14-30 A site-isolating device has to be provided with a grounding electrode.

The livestock covered by this rule does not include poultry but does include cows, bulls, horses, and other four-legged animals. This type of bonding system is usually installed before the concrete is poured. It can be installed afterward but is a far more extensive project and usually less effective. The idea is to let the concrete cure and bond simultaneously with the conductive elements contained in it. Areas required to be equipped with an equipotential bonding plane are indoors in animal confinement areas with conductive floors where metallic equipment that could become energized is located accessible to the livestock area. It is important to remember that concrete floors are considered conductive. An equipotential plane is also required for outdoor animal confinement areas with concrete slabs and metallic equipment that could become energized and is accessible to livestock. The bonding plane has to encompass the entire area where livestock would normally be standing and are subject to contact with metallic equipment that could become energized. Animals with four legs present unique problems because of the multiple connection points to the Earth, which provide multiple paths for current to traverse the animal's body. This is why establishing an equipotential bonding plane is so important. The equipotential bonding plane can be fabricated using wire mesh, reinforcing bars, and other steel rods and conductive elements. **See Figure 14-31.** The grid is created when all of these conductive elements are bonded together by, at minimum, an 8 AWG solid copper conductor. This conductor must connect all concrete-encased steel and must extend up to metal stanchions and the building grounding electrode system.

Listed copper wire mesh products are manufactured and available specifically for this purpose, which can assist installers in establishing a *Code*-compliant equipotential bonding plane for agricultural buildings and confinement areas where required by Article 547. See Annex C of this textbook for additional information about listed copper wire mesh products that is addressed in the *UL Guide Information for Electrical Equipment category (KDER)*. Creating an equipotential bonding plane requires a connection to the building electrical grounding system, which includes the grounding electrodes and the EGCs. The bonding conductor has to be a solid copper conductor no smaller than 8 AWG. All connections between the bonding conductor and the conductive elements or wire mesh have to be made using pressure connectors or clamps made of brass, copper alloy, or other equivalent and substantial connection means.

An important resource for information about establishing equipotential bonding planes is provided in the American Society of Agricultural and Biological Engineers (ASABE) EP473.2-2001, *Equipotential Planes in Animal Containment Areas*. Electrically heated livestock watering troughs can present an electric shock hazard for livestock and personnel. Methods for safe installation of livestock watering equipment are described in American Society of Agricultural and Biological Engineers (ASABE) EP342.2-1995, *Safety for Electrically Heated Livestock Waterers*.[4] See the informational notes following Section 547.10.

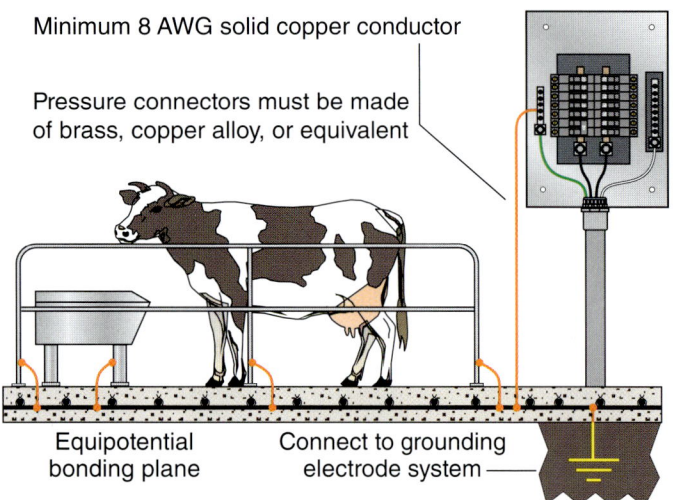

FIGURE 14-31 An equipotential bonding plane must be created in concrete slabs of animal confinement areas.

Mobile and Manufactured Home Grounding and Bonding Rules

Mobile Homes

Specific grounding and bonding rules apply to mobile home installations. The main concerns are that the grounded (usually the neutral) conductors are isolated from grounded equipment and the EGCs. This means that an insulated grounded conductor and a separate EGC must be used to supply a mobile home. Section 550.32(A) has requirements for installing a service disconnecting means adjacent to, within sight of, and no more than 30 feet from the mobile home. At this location, the grounded conductor is connected to the service disconnecting means enclosure and to the grounding electrode. This is typical of all grounded services, and this is the service equipment for the mobile home. This equipment could be a power outlet that is supplied by a feeder from a service in another location. The 30-foot distance requirement for a disconnecting means still applies in this situation.

Where the service equipment is located elsewhere on the property, the required disconnecting means located no more than 30 feet from the mobile home must be suitable for use as service equipment. This disconnecting means must also be rated for no less than what is required for a service disconnecting means for the mobile home. If the disconnecting means supplying the mobile home is supplied by a feeder and the service equipment is elsewhere on the property, the grounding and bonding at the disconnecting means enclosure must be according to the requirements in Section 250.32. The feeder run from the service disconnecting means or the feeder disconnecting means located no more than 30 feet from the mobile home must contain an insulated grounded conductor and a separate EGC. **See Figure 14-32.** The feeder conductors are required to be either a listed cord assembly that is factory-installed on the mobile home or a permanently installed feeder consisting of four insulated, color-coded conductors. Identification of the conductors shall be done by the factory or field marked to comply with Section 310.110. Where the feeder is installed in the field, the requirements in Section 250.32(B) apply. This means incorporates two ungrounded phase conductors: one insulated grounded (neutral) conductor and a separate EGC to supply the panelboard in a mobile home. An exception recognizes grounding alternatives only for existing wiring systems and strictly in compliance with the exception to Section 250.32(B).

Manufactured Homes

The service equipment for a manufactured home is permitted to be installed in or mounted on the structure. Multiple conditions must be met for service equipment installed either in or on the manufactured home. First, instructions provided with the home must provide information about securing the home by an anchoring system or to a permanent foundation. The requirements for services in Parts I through VII of Article 230 must be met.

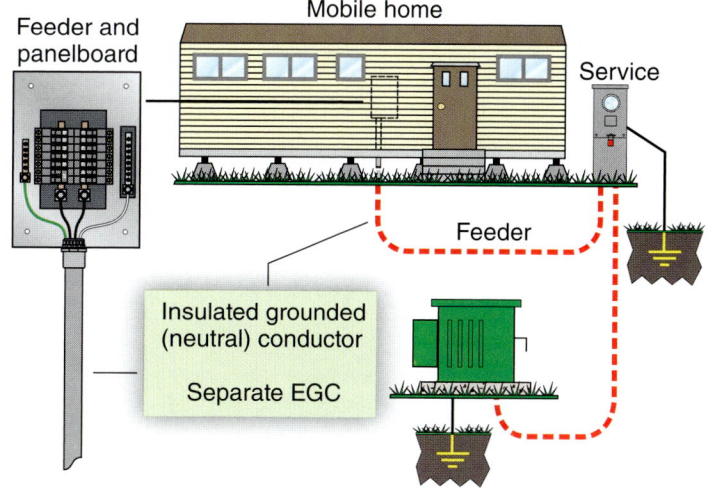

FIGURE 14-32 Feeders to mobile homes must contain a separate EGC and an insulated grounded conductor.

The service disconnecting means must have provisions for connection of a grounding electrode conductor, and it must be routed outside the structure. All grounding and bonding for the service has to meet the requirements in Parts I through V of Article 250. The manufacturer's instructions must indicate one method of grounding the service equipment but that the other methods of grounding in Article 250 are permitted, and reference must be made to Article 250 for those alternate methods to achieve the same compliance. Another unique requirement for manufactured homes is that the following warning label be mounted on the structure:

> WARNING DO NOT PROVIDE ELECTRICAL POWER UNTIL THE GROUNDING ELECTRODE(S) IS INSTALLED AND CONNECTED (SEE INSTALLATION INSTRUCTIONS)[5]

If the service equipment for the manufactured home is not installed on or within the structure, the other provisions of Section 550.32 must be followed. The neutral conductor(s) in mobile home and manufactured home installations must be insulated from ground, grounded metal parts, and EGC(s).

Summary

Electrical installations in hazardous locations require a more comprehensive design than usual, including robust bonding and grounding systems. Health care facilities have many special electrical requirements that require grounding and bonding methods that exceed the basics for general wiring. Specific grounding requirements for agricultural buildings, mobile homes, and manufactured homes were covered. These subjects were explored in depth, and the *Code* requirements were described in concise, understandable language.

Agricultural facilities often include hazardous (classified) locations because of grain dust.

Equipment such as motors and electrical enclosures installed in dust environments of agricultural facilities must be suitable for the location.

References

1. NFPA 70 National Electrical Code 2011, Article 100 (National Fire Protection Association, Quincy, MA 2010), p. 70–27.
2. NFPA 70 National Electrical Code 2011, Section 547.2 (National Fire Protection Association, Quincy, MA 2010), p. 70–478.
3. NFPA 70 National Electrical Code 2011, Section 547.2 (National Fire Protection Association, Quincy, MA 2010), p. 70–478.
4. NFPA 70 National Electrical Code 2011, Section 547.10 (National Fire Protection Association, Quincy, MA 2010), p. 70–480.
5. NFPA 70 National Electrical Code 2011, Section 550.32(B) (National Fire Protection Association, Quincy, MA 2010), p. 70–489.

Review Questions

1. Chapter 5 of the *NEC* provides the requirements for which of the following?
   a. Special conditions
   b. Special equipment
   c. Special occupancies
   d. Communications systems

2. The rules in Chapters 5 through 7 of the *NEC* are in addition to and do not amend or modify the general requirements in Chapters 1 through 4.
   a. True b. False

3. Regardless of the voltage, the electrical continuity and conductivity of non–current-carrying metal parts of equipment, raceways, and other enclosures in any hazardous (classified) location, as defined in Article 500, are required to be ensured by any of the methods specified for _____ in Section 250.92(B)(2) through (4) that are approved for the wiring method used.
   a. Services
   b. Separately derived systems
   c. Load-side equipment
   d. Special equipment

4. Specific methods of bonding of metal raceways installed in a hazardous (classified) location are required even if there is an EGC or bonding conductor installed inside the raceway.
   a. True b. False

5. The more restrictive bonding requirements of Section 250.100 for installations in hazardous (classified) locations are necessary to reduce the likelihood that a line-to-ground fault will cause arcing and sparking at connection points of metallic raceways and boxes or other enclosures.
   a. True b. False

6. Which of the following methods of bonding a raceway to a metal enclosure is not acceptable in a hazardous (classified) location?
   a. Connections using threaded couplings or threaded bosses on enclosures where they are made up wrenchtight
   b. Double locknuts, one on the inside and one on the outside of the enclosure
   c. Threadless couplings and connectors made up tight for metal raceways and metal-clad cables
   d. Other listed devices, such as bonding-type locknuts, bushings, or bushings with bonding jumpers

7. Where flexible metal conduit or liquidtight flexible metal conduit is installed for flexibility as permitted in hazardous (classified) locations, a wire-type _____ must be installed in accordance with Section 250.102.
   a. Equipment bonding jumper
   b. System bonding jumper
   c. Grounding electrode conductor
   d. Grounded (neutral) conductor

8. Branch circuits installed in patient care locations of health care facilities are required to be installed in a metal raceway or using another metallic wiring method that qualifies as an EGC in accordance with Section 250.118, in addition to a(n) _____.
   a. Bare copper conductor
   b. Insulated aluminum or copper-clad aluminum conductor
   c. Insulated copper conductor
   d. Bare aluminum conductor

9. In a patient care location, the branch circuit wiring can be installed in PVC conduit as long as two copper insulated EGCs are included with the other circuit conductors.
   a. True b. False

10. The term *patient vicinity* refers to a distance of _____ feet from the patient bed location in a health care facility.
    a. 2
    b. 3
    c. 5
    d. 6

11. A luminaire installed no less than 7.5 feet above the floor in a patient care area does not have to be provided with two EGC paths as required by *NEC* in Section 517.13.
    a. True b. False

12. Isolated power systems installed in health care facilities are required to be installed and operated as _____.
    a. Solidly grounded systems
    b. Impedance grounded neutral systems
    c. Ungrounded systems
    d. Any of the above

13. Where an isolated power system is used, the patient equipment grounding point is typically connected to the reference grounding bar in the _____.
    a. Operating room
    b. Service equipment enclosure
    c. Transformer enclosure
    d. Isolated power panel

14. Where isolated circuit conductors supply 125 volt, single-phase, 15- and 20-ampere receptacles are supplied from an isolated power panel, the _____ conductor is required be connected to the terminals on the receptacles that are identified in accordance with Section 200.10(B) for connection to the grounded circuit conductor.
    a. White
    b. Gray
    c. Green
    d. Striped orange

15. The redundant EGC requirements of Section 517.13 apply only to branch circuits serving patient care areas, not to the feeders.
    a. True
    b. False

16. When a critical branch feeder is supplied from a grounded distribution system in a health care facility and is installed using a metallic wiring method that qualifies as an EGC according to Section 250.118, the grounding of the switchboard or panelboard is required to be ensured by any of the following methods except _____.
    a. A grounding bushing and a continuous copper bonding jumper, sized in accordance with Table 250.122, with the bonding jumper connected to the junction enclosure or the ground bus of the panel
    b. A connection of feeder raceways or Type MC or MI cable to threaded hubs or bosses on terminating enclosures
    c. Other approved devices, such as bonding-type locknuts or bushings
    d. Two locknuts, one on the inside and one on the outside of the enclosure

17. The *NEC* requires the equipment grounding terminal bars of the normal and essential branch circuit panelboards serving the same individual patient vicinity to be bonded together with an insulated copper conductor no smaller than _____ AWG.
    a. 4
    b. 6
    c. 8
    d. 10

18. _____ is an area where wire mesh or other conductive elements are embedded in or placed under concrete, bonded to all metal structures and fixed nonelectrical equipment that may become energized, and connected to the electrical grounding system to prevent a difference in voltage from developing within the plane.
    a. An equipotential bonding plane
    b. An equipotential bonding grid
    c. A signal reference grid
    d. A common grounding electrode

19. Which of the following areas of an agricultural facility must have an equipotential bonding plane installed?
    a. Indoor animal confinement areas with conductive floors where metallic equipment that could become energized is located accessible to the livestock area
    b. Outdoor animal confinement areas with concrete slabs where metallic equipment that could become energized is accessible to livestock
    c. Outdoor areas on farms that are not intended to confine animals
    d. Both a and b

20. The equipotential bonding plane can be built using wire mesh, reinforcing bars, and other steel rods and conductive elements. The grid is created when all of these conductive elements are bonded together by, at minimum, a(n) _____ copper conductor.
    a. 10 AWG stranded
    b. 8 AWG solid
    c. 14 AWG stranded
    d. 12 AWG solid

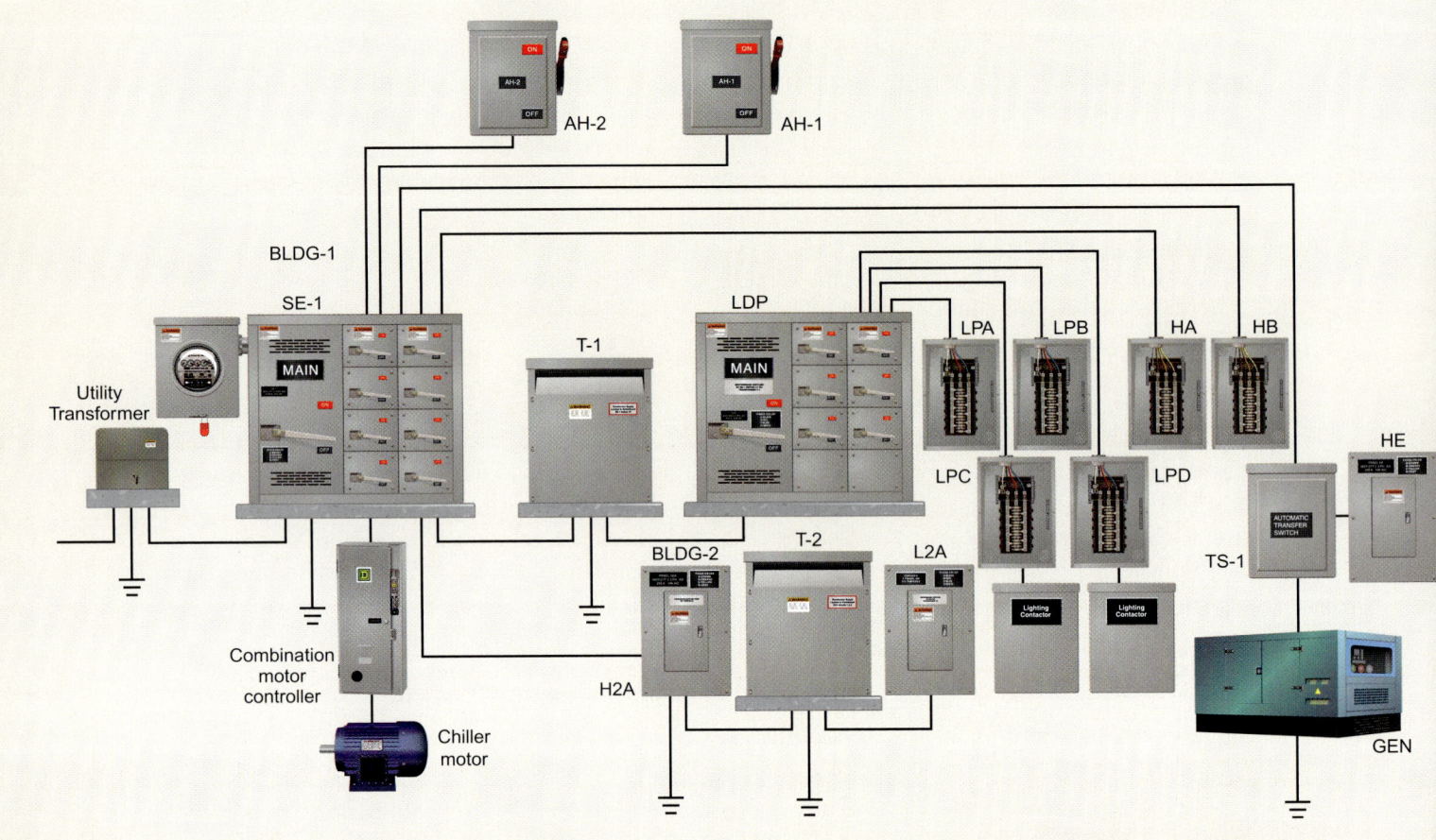

CHAPTER 15
Grounding for Special Equipment

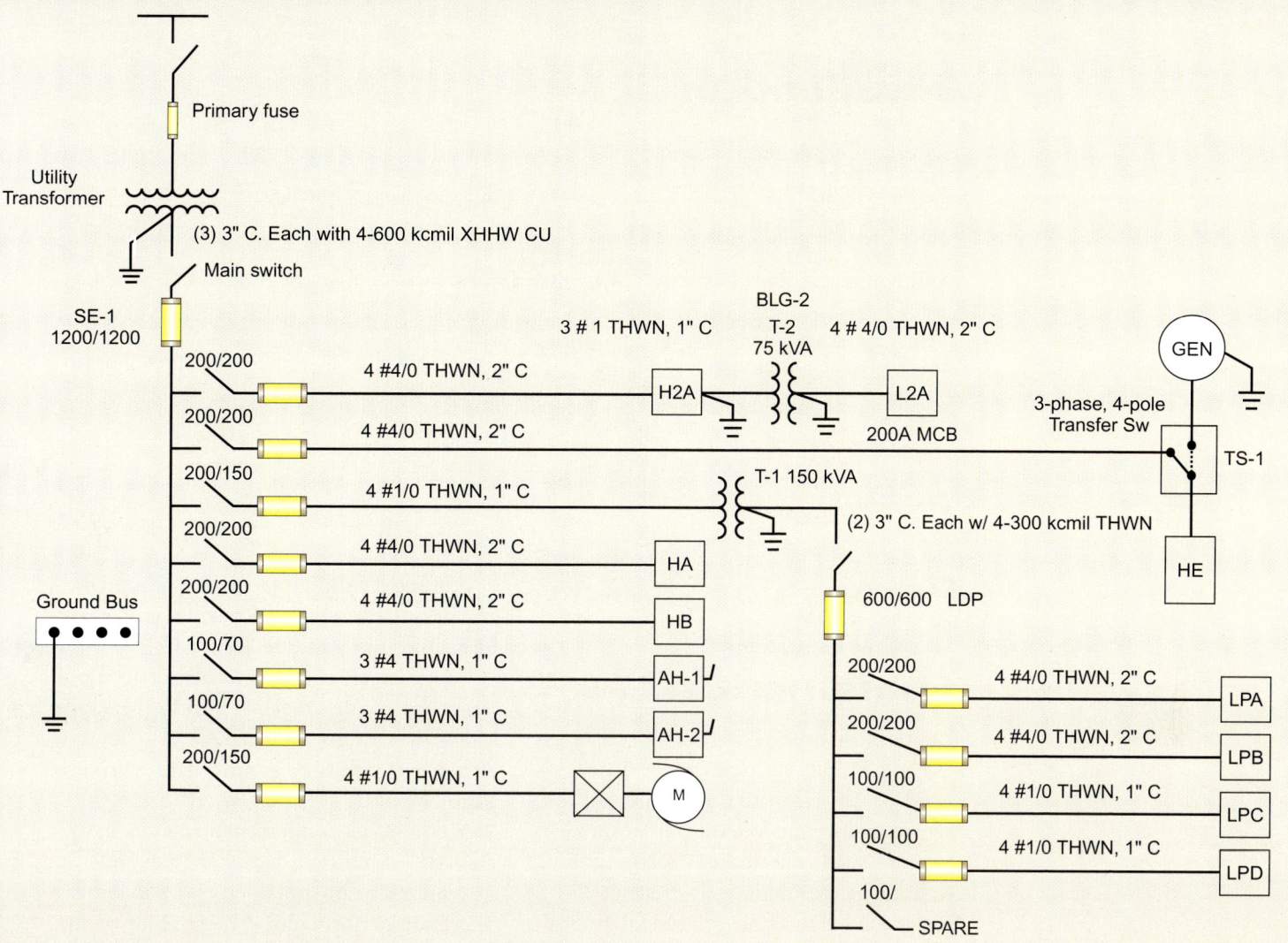

Objectives

- Understand general grounding and bonding requirements for special equipment and determine the grounding and bonding rules for special equipment as provided in Chapter 6 of the *NEC*
- Determine the more restrictive electrical grounding and bonding requirements for electric signs and outline lighting systems, including light-emitting diode systems
- Understand the specific grounding requirements that apply to sensitive electronic equipment
- Understand the purpose of enhanced grounding and bonding requirements for swimming pools and similar equipment in aquatic environments
- Understand special grounding and bonding rules for photovoltaic systems

Outline

Purpose of Grounding Equipment

Electric Signs and Outline Lighting Systems

Electric Cranes and Elevators

Information Technology Equipment and Sensitive Electronic Equipment

Grounding and Bonding Requirements for Swimming Pools and Similar Installations

Grounding Requirements for Solar PV Systems

Introduction

Equipment grounding is an important aspect of electrical safety for people and property. Chapter 6 of the *Code* provides the requirements for special equipment, such as electric signs, cranes, elevators, audio systems, sensitive electronic equipment, information technology systems, swimming pools, and solar photovoltaic (PV) systems. Enhanced grounding and bonding rules often apply due to increased shock hazards and the unique construction and operation of special equipment. Grounding special equipment generally requires a connection to an equipment grounding conductor as specified in *NEC*® Section 250.118.

Purpose of Grounding Equipment

The purpose of the *NEC* is to protect people and property from electrical hazards arising from electricity use. The purpose of grounding and bonding is covered in Section 250.4. The reasons for grounding equipment are outlined in Section 250.4. For grounded systems, equipment grounding and equipment bonding performance requirements are set forth in Sections 250.4(A)(2) and (3). These rules focus specifically on grounding and bonding equipment, not on system grounding, which is covered in Section 250.4(A)(1). Equipment is grounded to limit potential to ground on the normally non–current-carrying parts of such equipment. **See Figure 15-1.**

The conductive materials of equipment are bonded together and to the supply source in an arrangement that establishes an effective ground-fault current path. Bonding is an electrical connection that establishes continuity and conductivity between conductive parts. The equipment is required to be grounded and bonded to meet the performance criteria required in Section 250.4. In some cases, additional bonding is necessary because of unique hazards encountered in some special equipment installations, such as pools, hot tubs, electric signs, and outline lighting systems.

Electric Signs and Outline Lighting Systems

Electric signs and outline lighting systems are common. Just about every building or structure, other than residential occupancies, has a sign or outline lighting system installed on it. The signs and special lighting systems provide direction to the public and are an essential part of businesses. **See Figure 15-2.** There are a variety of electrical signs and outline lighting systems installed today using technology such as incandescent, fluorescent, neon, cold cathode, fiber optics, and light-emitting diode (LED) systems.

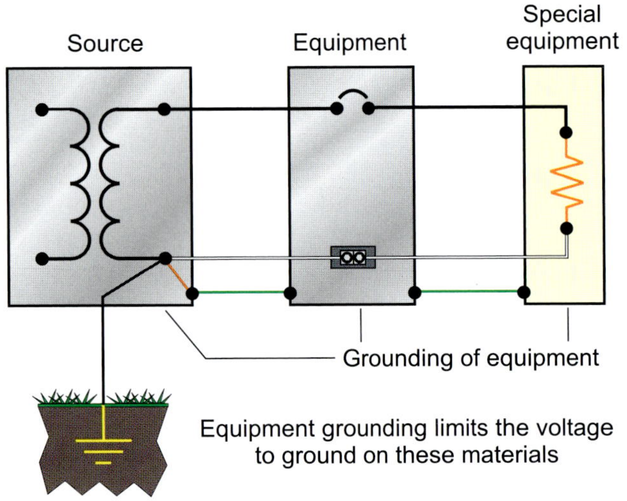

FIGURE 15-1 Grounding places equipment at the same potential as ground (the Earth) thereby limiting aboveground potentials on such equipment.

It's easy to see the many technologies used for signage both in this country and around the world. Just visit any major city or attraction; the amount of electric signs and outline lighting used becomes evident. Article 600 of the *NEC* provides the special requirements for electric signs and outline lighting systems regardless of the voltage. Specific grounding and bonding requirements for signs and outline lighting are provided in Section 600.7. This section is subdivided into two parts. Subdivision (A) includes the grounding requirements, and subdivision (B) provides the special bonding requirements.

Grounding Signs and Outline Lighting Equipment

Article 250 includes equipment grounding rules in Section 250.112, which requires the normally non–current-carrying parts of equipment to be connected to an equipment grounding conductor (EGC), regardless of the voltage. **See Figure 15-3.** This is the general grounding rule for signs and outline lighting. Section 250.112(G) recognizes that more specific grounding and bonding has to be applied for this type of equipment, and a reference is made to Section 600.7. Electric signs and metal equipment of outline lighting systems have to be grounded by connecting them to the EGC of the supplying branch circuit or feeder.

Courtesy of Young Electric Sign Company (YESCO)

FIGURE 15-2 Electric signs and outline lighting systems are installed on many buildings or structures.

FIGURE 15-3 The electric signs and outline lighting systems installed on buildings or structures convey a variety of messages and provide illumination and draw the attention of the public.

See Figure 15-4. Any of the types of EGCs provided in Section 250.118 can be used to accomplish the grounding, and if a wire-type EGC is used, it has to be sized in accordance with Section 250.122 based on the rating of the fuse or circuit breaker protecting the circuit.

All EGC connections have to be made using a method specified in Section 250.8. It is important to recall that sheet metal screws are not acceptable for grounding and bonding connections. The *Code* relaxes the sign grounding requirement by exception, but only for portable cord-connected signs that are manufactured using a system of double insulation or an equivalent and are so marked. Sometimes electric signs are pole mounted, as is the case for billboards and pole signs in parking lots. It is often desirable to establish a grounding electrode at these pole locations. See Figure 15-5. The grounding objectives in this case are to connect the electric sign to the EGC of the supply circuit and connect it to an auxiliary grounding electrode, usually a ground rod, as covered by Section 250.54. See Figure 15-6. This creates a local ground reference for the pole-mounted sign while it is still connected to the EGC of the supply circuit, which is the required effective ground-fault current path.

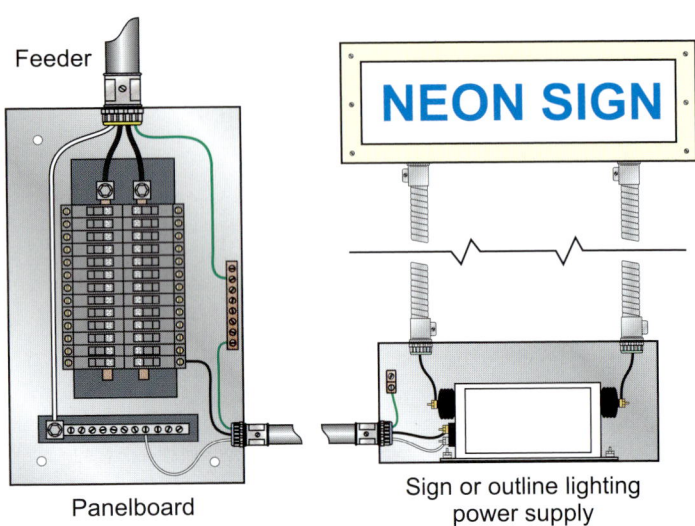

FIGURE 15-4 Electric signs and metal parts of outline lighting systems are required to be grounded.

I-Stock Photo Courtesy of NECA

FIGURE 15-5 Pole-mounted electric signs are required to be grounded by connection to an EGC. They are often also grounded by use of an auxiliary grounding electrode installed at the base of the pole.

The Earth alone cannot be used as an effective ground-fault current path. Metal parts of a building or structure are not permitted as an EGC or as a high-voltage secondary return circuit for a neon lighting system.

Bonding Metal Parts of Signs and Outline Lighting Systems

The requirements for bonding metal parts and equipment for electric signs and outline lighting systems are provided in Section 600.7(B). The objectives are to connect all metal parts together and to the EGC supplying the transformer or power supply. **See Figure 15-7.** Many electric signs and outline lighting systems have high-voltage transformers that are used to ignite gases in glass tubing, such as that used in neon and cold cathode systems. The branch circuit supplies the primary of the transformer, and the voltage is then raised by a transformer to levels as high as 15 000 volts (7500 volts line-to-ground). The secondary (high voltage) conductors are routed in suitable wiring methods to the electrodes molded into the ends of the glass tubing.

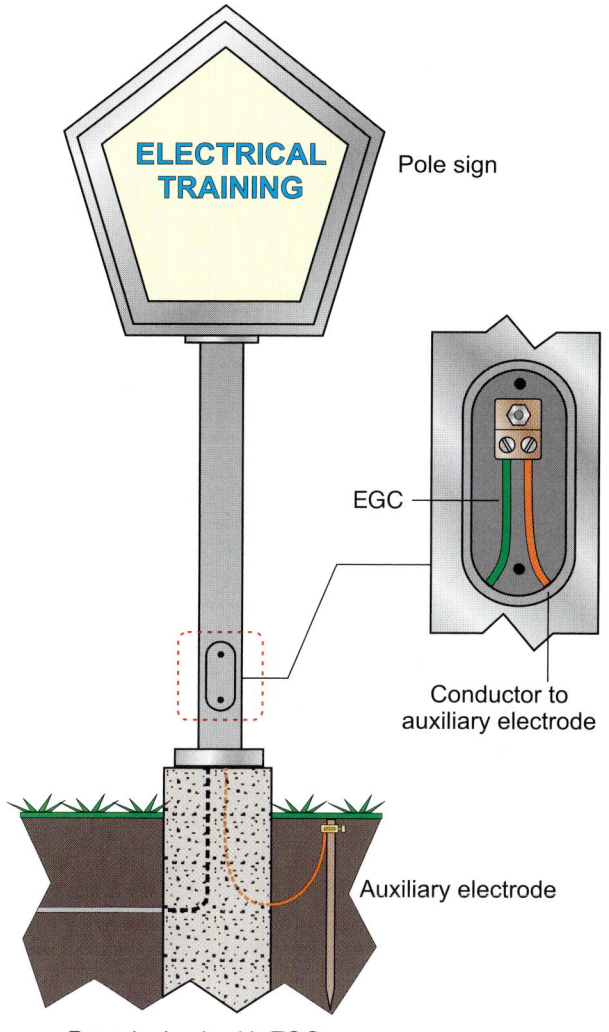

FIGURE 15-6 Pole signs grounded by connection to an EGC are often connected to an auxiliary grounding electrode that establishes a direct (local) reference to ground at the pole location.

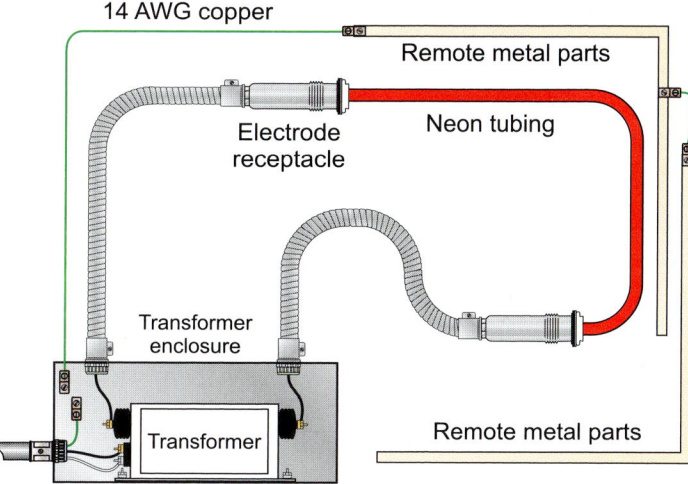

FIGURE 15-7 Bonding requirements apply to neon signs and outline lighting systems with remote metal parts.

Bonding all metal parts associated with such systems is an important safety requirement. If the metal is not bonded, the high-voltage secondary circuits can actually induce voltages in the "dead metal," raising potentials and creating shock and fire hazards. **See Figure 15-8.**

The *Code* does relax this bonding requirement for remote metal parts of section signs or outline lighting systems that are supplied by a Class 2 power supply. Here, the secondary circuit is at the Class 2 voltage levels and does not present the same hazards to people and property as the high-voltage secondary systems. All bonding connections to metal parts and equipment associated with signs and outline lighting must be made using a method provided in Section 250.8, and metal parts of buildings or structures cannot be used as the required bonding for signs and outline lighting systems. Bonding conductors for signs and outline lighting systems are required to be no smaller than 14 AWG copper, and where installed external of a sign enclosure or raceway system, they have to be protected from physical damage. This external bonding jumper does not have to be sized according to Section 250.102, nor is it limited to 6 feet in length. The reason is that the secondary circuits of these transformers and power supplies are operating at high voltage and current is in the milliamp range. A ground fault on the transformer or power supply secondary causes operation of the secondary circuit ground-fault protection (SCGFP) device that is integral to the transformer or power supply.

Where high-voltage secondary circuits are installed from transformers or power supplies to glass tubing of neon lighting or cold cathode systems, the length of high-voltage conductors is limited. Basically, these runs are limited to 20 feet in length for metal conduit from each output hub on the transformer to the first tubing electrode and a 50-foot maximum length for nonmetallic conduit by Section 600.32(I). The secondary high-voltage cable has to be installed using a suitable wiring method or listed sleeve material. When listed flexible metal conduit (FMC) or listed liquidtight flexible metal conduit (LFMC) is used as the raceway to enclose the high-voltage gas tube oil ignition (GTO) cables, the metal parts of a sign or outline lighting system are permitted to be bonded through the flexible metal raceway wiring method, provided that the total accumulative length of the conduit does not exceed 100 feet. Any small metal parts that do not measure more than 2 inches in any dimension are not required to be bonded.

FIGURE 15-8 Flexible metal conduit (FMC) enclosing a high-voltage neon secondary circuit is required to be bonded. This photo shows a violation in that a few sections of FMC are not bonded or grounded.

If listed nonmetallic conduit is used to enclose high-voltage secondary circuits to neon tubing systems, bonding of metal parts associated with the sign or outline lighting is required.

A common example of this is a reverse pan channel letter sign where the individual letters of the sign are all isolated from one another and isolated from a remote transformer or power supply. **See Figure 15-9.** In this case, a bonding conductor is required to be installed separate and remote from the nonmetallic conduit containing the high-voltage secondary circuit. This bonding conductor has to be a minimum of 14 AWG copper. Many neon signs and outline lighting systems are supplied by electronic power supplies operating above 60 Hz. Section 600.7(B)(6) specifies a distance of at least 1½ inches from a conduit containing a high-voltage circuit of 100 Hz or less and no less than 1¾ inches from a nonmetallic conduit containing a circuit operating above 100 Hz. **See Figure 15-10.** This is to minimize the effects of unbalanced stress and allow the magnetic lines of flux to remain symmetrical in the high-voltage secondary circuit during normal operation. If bonding conductors are installed closer to the conduit than these prescribed distances, the magnetic lines of force will tend to become asymmetrical, which creates additional stress and can lead to premature conductor failure.

If the bonding conductor or any grounded metal is close to the nonmetallic conductor, the lines of flux can become asymmetrical and cause stress and even premature failure of the contained high-voltage GTO cable and the transformer or power supply. Section 600.32 also requires spacing from grounded or bonded metal parts for nonmetallic conduits containing high-voltage secondary circuits, for the same reasons. The minimum spacing requirements are the same as those provided in Section 600.7(B)(6), which are no less than 1½ inches from circuits of 100 Hz or less and no less than 1¾ inches from circuits operating above 100 Hz.

FIGURE 15-9 Reverse pan channel letter signs with metal parts remote from one another are required to be bonded.

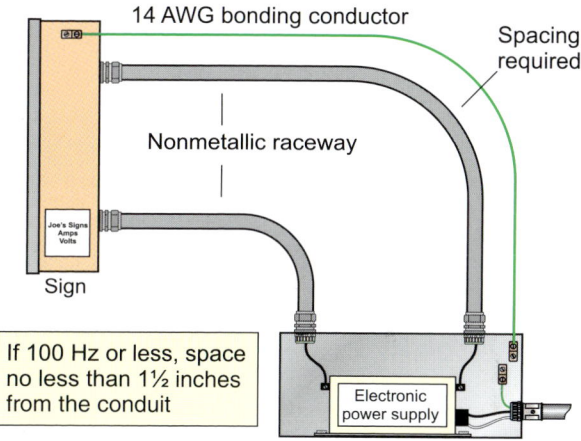

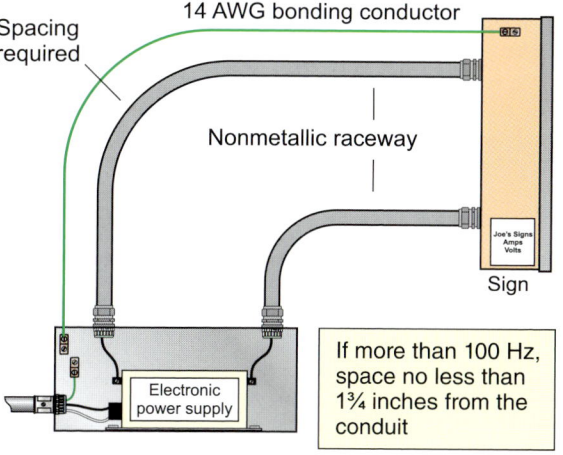

FIGURE 15-10 Spacing is required for bonding conductors installed with neon secondary circuits.

It is important to remember that in higher-frequency circuits the skin effect is amplified and greater care is needed to ensure that spacing is maintained to reduce failures in these types of systems while maintaining effective bonding and grounding of all associated metal parts and equipment for such installations.

Electric Cranes and Elevators

Electric cranes and hoists are covered by Article 610 of the *NEC*. Electric cranes are typically supplied by branch circuits that include an EGC. **See Figure 15-11.** The EGC can be any of the types listed in Section 250.118. If it is a wire type, it must be sized according to Section 250.122. Section 610.61 requires grounding and bonding of all exposed non–current-carrying metal parts of cranes, monorail hoists, and any associated pendant controls.

The bonding can be accomplished either by mechanical connections of the conductive parts or by suitable bonding jumpers. The objective is to create a common conductive path for ground-fault current that includes all metal parts associated with the crane or hoist. Moving parts that have metal-to-metal bearing contact are considered to be bonded together through such contact. A trolley frame and bridge are not considered grounded through the bridge and trolley wheels and tracks; a separate bonding jumper is required around the wheel-to-track contact.

As a reminder, systems supplying crane circuits that operate over Class III locations are not permitted to be grounded due to the restriction in Section 250.22(1). The reason relates to reducing the probabilities of arcing and sparks that could result from a ground fault in the equipment, which could send hot particles into accumulations of fibers and flyings below the crane creating a fire hazard. **See Figure 15-12.** The metal parts associated with such equipment have to be grounded and bonded, but the system supplying the crane is not permitted to be grounded. Ground detection is required on such ungrounded systems to alert qualified people of a first phase-to-ground fault condition. See Section 503.155 for more requirements.

Elevators and associated equipment are covered by Article 620 of the *NEC*. Equipment grounding is required for elevators as specified by Section 250.112(E), and specific rules are provided in Part IX of Article 620. **See Figure 15-13.**

FIGURE 15-11 Equipment grounding and bonding are required for electric cranes and hoists.

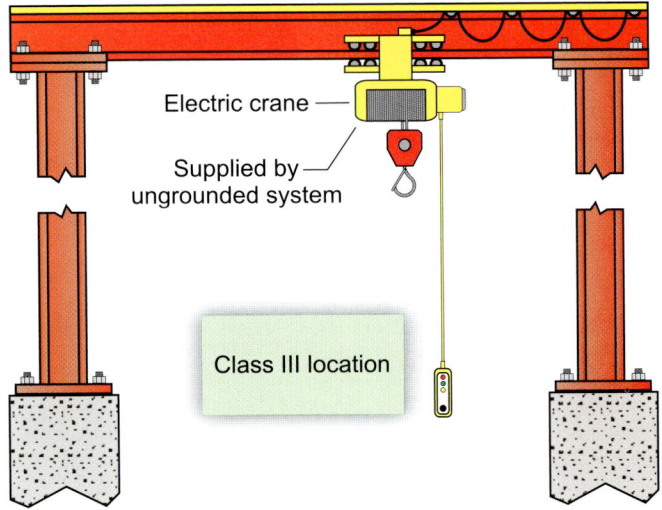

FIGURE 15-12 Cranes installed and operating over Class III locations (fire hazards due to accumulations of fibers and flyings) are required to be supplied by ungrounded systems.

Metal raceways, mineral-insulated cables, metal tape–type cables (Type MC), or armored cables (Type AC) attached to elevator cars must include an EGC with the supply circuit for elevator equipment.

Elevators and hoists are typically supplied by a circuit(s) sized according to the applicable requirements in Part II of Article 620. The rating of the overcurrent device supplying this equipment determines the minimum size required for wire-type EGCs. Often, more than a single circuit supplies elevator equipment. Usually, a larger circuit supplies the pump equipment for a hydraulic elevator or the motors for cable elevator equipment and smaller individual circuits supply car lights and receptacle power. In addition, circuits for servicing the elevator car and pit include EGCs. The frames and metal equipment of electric elevators including motors, machines, and controllers, and all metal equipment on the elevator car must be grounded and bonded. Non-electrical elevators that have any electrical circuits in contact with the car must have the frame of the car grounded and bonded in accordance with Parts V and VII of Article 250. Escalators and moving walks are also covered by the rules in Article 620. Any metal parts associated with these types of equipment have to be grounded and bonded.

sulated equipment grounding circuits in accordance with Section 250.146(D). Information technology rooms are usually equipped with power distribution units (PDUs) that supply all equipment associated with the information technology system. **See Figure 15-14.**

FIGURE 15-13 Elevators and associated equipment are required to be grounded by connection to an EGC of the supply circuit.

Information Technology Equipment and Sensitive Electronic Equipment

Grounding rules for information technology equipment are provided in Section 645.15. This rule requires all non–current-carrying metal parts of an information technology system to be grounded and bonded to an EGC or to be double insulated. Often, information technology equipment is grounded by isolated or in-

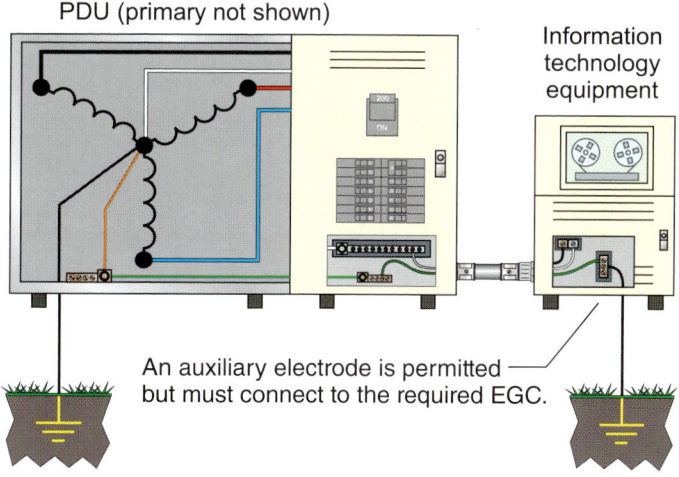

FIGURE 15-14 PDUs are commonly installed and operated in information technology rooms that are wired according to Article 645 of the *NEC*.

While listed PDUs typically include transformers, they are usually not treated as separately derived systems for grounding purposes in accordance with Section 250.30(A). They must be grounded according to specific instructions provided by the manufacturer. Any signal reference grids installed in information technology rooms must be connected to the EGC provided with the circuits supplying the information technology equipment.

Sensitive Electronic Equipment

Article 647 provides specific rules applicable to sensitive electronic equipment installations. Some electronic equipment is sensitive to common voltages such as 120 or 240 volts. The supply systems for sensitive electronic equipment are three-wire, single-phase systems. The ungrounded conductors have a voltage of 120 volts between them, and from each ungrounded conductor to the neutral of such systems, the voltage is 60 V. **See Figure 15-15.** These systems must be grounded in accordance with the requirements in Section 250.30(A) for separately derived systems and must be connected to a grounding electrode as specified in Section 250.30(A)(4).

The purpose of such special low-voltage systems is to reduce objectionable noise in sensitive electronic equipment systems and locations. This type of system is permitted only in commercial and industrial establishments, provided that the system is restricted to areas under supervision of qualified persons. All wiring rules in Sections 647.4 through 647.8 must be applied to such installations. Sensitive electronic systems are required to be installed using listed equipment for this specific purpose or using standard rated equipment and specific field-applied markings to indicate the voltage and use of the equipment. All junction boxes, equipment, and conductors have to be clearly identified to show they are associated with such systems. Branch circuits and feeders have to be provided with a disconnecting means that simultaneously opens all ungrounded conductors. The voltage drop on these systems cannot exceed 1.5% on the branch circuits and 2.5% for the combined voltage drop of the feeder and branch circuit supplied by the system. The voltage is already low on these systems, and the equipment supplied is vulnerable to damage resulting from lower supply voltages. The *NEC* generally does not address voltage drop as a requirement in Chapters 1 through 4; however, it is addressed as a rule for these systems.

Equipment and receptacles supplied by these systems must be grounded by an EGC run with the supply circuit conductors and terminated on the equipment grounding terminal bus of the panelboard that is marked "Technical Equipment Ground." This equipment grounding bus is required to be connected to the grounded conductor of the system on the line side of the separately derived system disconnecting means. **See Figure 15-16.** The EGC must be sized according to Table 250.122 and installed with the feeder conductors. The technical equipment grounding bus does not have to be bonded to the equipment enclosure. Alternative grounding methods recognized in the *NEC* are permitted for these systems if the impedance of the effective ground-fault return path does not exceed the impedance of the EGC sized according to the requirements in Article 647. These special grounding provisions limit the impedance of a ground-fault path because only 60 volts would be present in a phase-to-ground fault condition, rather than the usual 120 volts.

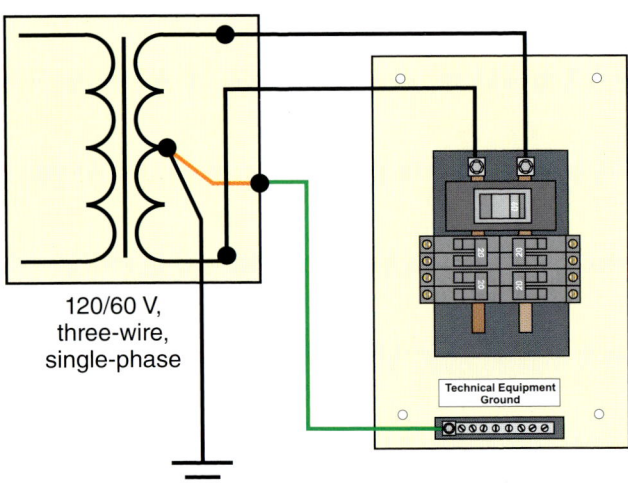

FIGURE 15-15 Special grounding requirements apply to systems supplying sensitive electronic equipment in accordance with Article 647.

Section 647.7 includes special rules for receptacles used with this type of system. All 15- and 20-ampere receptacles must be ground-fault circuit interrupter (GFCI) protected. All receptacle outlet strips, adaptors, covers, and faceplates have to be marked as provided in the following example:

> WARNING: TECHNICAL POWER
> Do not connect lighting equipment.
> For electronic equipment use only.
> 60/120 V, Single-phase AC system
> GFCI protected

All receptacles used with systems covered in Article 647 must have unique configurations and be identified for use with this type of system. See the specific rules for receptacles installed with sensitive electronic equipment installations as provided in Section 647.7. If isolated grounding circuits and receptacles are used with these systems, the installation has to be in accordance with Section 250.146(D), except that the EGC of these circuits can be terminated according to Section 647.6(B). There are also specific rules for lighting equipment installed for the purpose of reducing electrical noise originating in lighting equipment. These installations have to meet the requirements in Section 647.8. See Article 647 for all rules pertaining to sensitive electronic equipment.

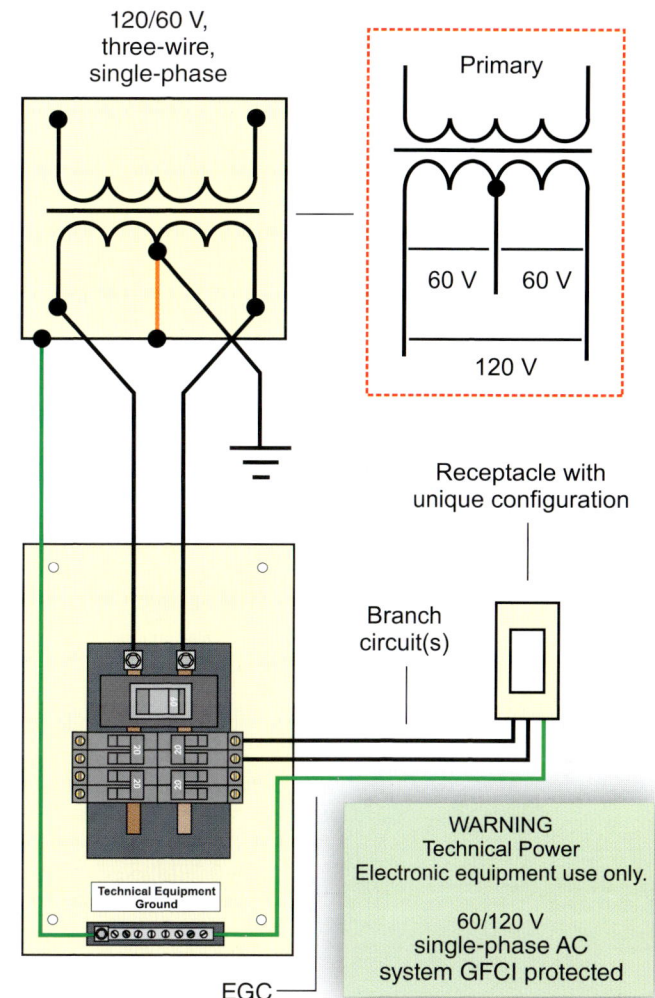

FIGURE 15-16 Branch circuits for sensitive electronic equipment must include an EGC.

Grounding and Bonding Requirements for Swimming Pools and Similar Installations

In wet locations electric shock hazards are present. Swimming pools and similar installations are aquatic environments and involve applications using special equipment, and more restrictive requirements apply to these installations.

Swimming pools are constructed in a variety of configurations and typically require special grounding and bonding methods and protection in accordance with *NEC* Article 680.

See Figure 15-17. Special emphasis is placed on electrical grounding and bonding rules for these types of equipment and installations.

Article 680 of the *NEC* provides rules related to receptacle locations, overhead conductor clearances, GFCI protection, and special grounding and bonding. It is important to recognize the arrangement and application of Article 680. Part I applies generally to all other parts of the article. Part II includes special requirements that apply specifically to permanently installed pools. Part III applies to storable pools. Note that Section 680.30 indicates that both Parts I and III apply to storable pools. This is typical of how the various parts of Article 680 have to be applied to swimming pools and similar equipment such as hot tubs. **See Figure 15-18.**

Equipment Grounding

Section 680.6 provides a general requirement for grounding equipment associated with pools or similar installations. The equipment grounding requirement applies to all equipment wired using the methods covered in Chapter 3 of the *NEC*. Equipment grounding is required for through-wall lighting assemblies and underwater luminaires. Note that low-voltage lighting equipment listed for use without an EGC is permitted. Any electrical equipment located within 5 feet of the inside walls of a pool or specified body of water must be grounded. The pool pump motor or equipment associated with the water recirculation system must be grounded. Any junction boxes, transformer enclosures, and GFCIs have to be grounded by connection to an EGC. Any panelboards, in addition to service panelboards that supply electrical equipment associated with the pool or other body of water covered by Article 680, must be grounded by connection to an EGC. So, it is safe to summarize that electrical equipment associated with pools and all similar equipment are generally required to be grounded.

The equipment grounding is accomplished by connection to an EGC of the supply circuit. Equipment associated with pools and similar installations are typically required to be connected to an EGC that is insulated copper conductor. **See Figure 15-19.** Insulation on the EGC is required as an extra level of protection against corrosive conditions associated with chemically treated water.

Another benefit of installing insulated EGCs is the reduced fault paths during a ground fault condition in the circuits supplying associated equipment. The main purpose for insulation is related to protection and integrity of the EGC of the circuit.

FIGURE 15-17 Swimming pools require special grounding and bonding.

FIGURE 15-18 Hot tubs require special grounding and bonding.

Insulated copper EGCs are generally required for circuits installed in rigid metal conduit (RMC), intermediate metal conduit (IMC), electrical metallic tubing (EMT), and LFMC. When nonmetallic wiring methods are installed for pools or other equipment covered by Article 680, an EGC is required to be installed. It is typically required to be an insulated copper conductor and sized in accordance with Section 250.122 but no smaller than 12 AWG.

Pool Pump Motors

The branch circuit for a pool pump motor must be installed using the methods specified in Section 680.21(A)(1), which includes RMC, IMC, rigid polyvinyl chloride (PVC) conduit, reinforced thermosetting resin conduit (Type RTRC), or Type MC cable listed for the location. These wiring methods must include an insulated copper EGC sized according to Section 250.122 but no smaller than 12 AWG. **See Figure 15-20.**

Note that aluminum EGCs are not permitted for these circuits. Where the wiring is installed on or within a building or structure, EMT is permitted as the wiring method. If flexible wiring methods are installed, LFMC or liquidtight flexible nonmetallic conduit is permitted. Insulated copper EGCs must be installed with the circuits in these flexible wiring methods. **See Figure 15-21.**

Where wiring for pools or other equipment covered in Article 680 is installed within a dwelling unit or any accessory building associated with a dwelling unit, any wiring method in Chapter 3 of the *NEC* is permitted. If a cable assembly is installed in these locations, the EGC of the circuit can be bare (uninsulated) but has to be covered by the sheath of the assembly. Nonmetallic sheathed cable or service-entrance (Type SE) cable assemblies are examples of wiring that falls into this category. There are two types of service-entrance cable assemblies, Type SEU and Type SER. The *R* in *Type SER* stands for *round* and includes an insulated neutral conductor, and the *U* in *Type SEU* indicates that the neutral of the cable is an uninsulated conductor.

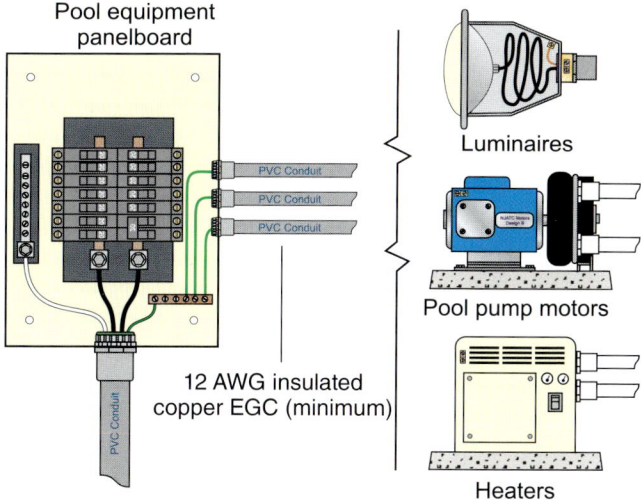

FIGURE 15-19 Equipment is generally required to be grounded by an insulated copper EGC no smaller than 12 AWG.

FIGURE 15-20 Pool pump motors are required to be grounded by connection to an EGC of the supply circuit.

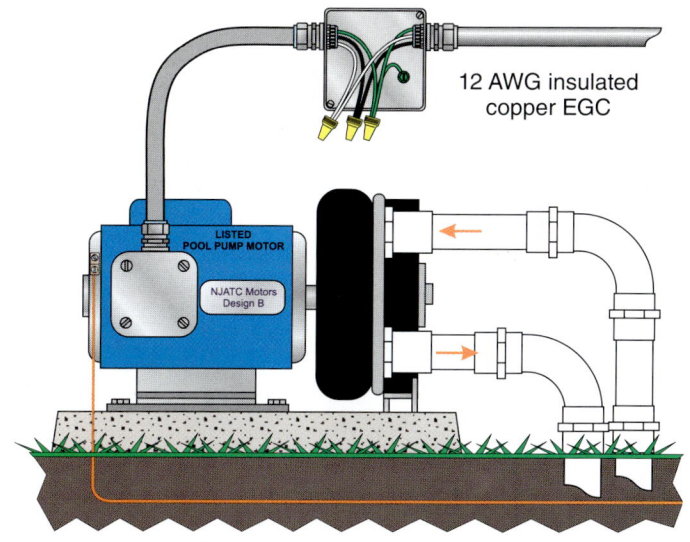

FIGURE 15-21 Flexible conduit wiring methods for pool pump motors must include an insulated copper EGC no smaller than 12 AWG.

Pool pump motors are permitted to be cord connected if the cord is limited to 3 feet in length. The cord must include a copper EGC sized according to Section 250.122, and it must be connected to a grounding-type attachment plug.

Grounding and Bonding Underwater Luminaires

Underwater luminaires installed for swimming pools and similar equipment are required to be grounded by connection to the supply circuit EGC. **See Figure 15-22.** Through-wall lighting assemblies and wet-, dry-, or no-niche luminaires are required to be connected to an insulated copper EGC installed with the circuit conductors.

The EGC generally must be installed without joint or splice except as permitted in Sections 680.23(F)(2)(a) and (b). The EGC must be sized in accordance with Table 250.122 but no smaller than 12 AWG. Where rigid nonmetallic conduit or liquidtight flexible nonmetallic conduit (LFNC) is installed between a forming shell for a wet-niche luminaire and a junction box or other enclosure, an 8 AWG insulated copper bonding jumper is required to be installed in the conduit to provide electrical continuity between the forming shell and the junction box or other enclosure. **See Figure 15-23.** The conduit must be sized large enough to enclose both the 8 AWG insulated copper bonding jumper and the flexible cord that supplies the wet-niche luminaire. The flexible cord assembly provides the EGC connection to the metal luminaire parts. The 8 AWG copper conductor provides the bonding connection for the forming shell. The connection must be encapsulated in a listed potting compound to protect it from chemically treated water and related corrosion.

Junction Boxes and Other Enclosures

Junction boxes, transformer enclosures, and enclosures containing GFCIs connected to conduit systems that extend to forming shells or mounting brackets for no-niche luminaires must be provided with a quantity of grounding terminals that exceed the number of conduit entries by at least one. The purpose is to include provisions for any bonding conductor that may also be required for the enclosure.

Boxes for underwater luminaires are required to be listed and equipped with threaded hubs or entries or nonmetallic hub entries. They must be made of copper, brass, plastic, or other corrosion-resistant material. The junction boxes must provide continuity between all metallic conduit entries and grounding terminals inside the box. **See Figure 15-24.** The EGC terminals of junction boxes for transformers and electrical enclosures for GFCIs must be connected to the equipment grounding terminal of the panelboard supplying any connected dry- or wet-niche luminaire.

FIGURE 15-22 Underwater luminaires are required to be grounded by connection to the EGC of the supply circuit.

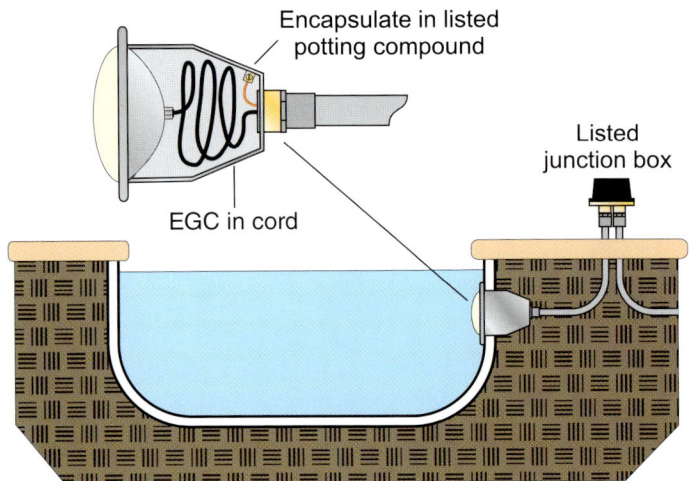

FIGURE 15-23 Grounding and bonding for underwater luminaires includes an EGC for the luminaire assembly and an 8 AWG copper bonding conductor that connects the forming shell to the equipotential bonding grid for the pool.

Feeders Supplying Pool Equipment

Feeder conductors for panelboards supplying pool equipment must meet the requirements in Section 680.25. The wiring methods must be installed using RMC, IMC, rigid PVC conduit, or Type RTRC. EMT is permitted where installed on or within a building, and electrical nonmetallic tubing is permitted for use inside buildings only. Type MC cable is permitted within a building if not subject to a corrosive environment. Aluminum conduit is not permitted in pool areas that are subject to corrosion. An EGC is required with the feeder conductors supplying panelboards. The EGC generally has to be insulated.

Existing feeders between service equipment and a panelboard are permitted in FMC or an approved cable assembly that includes an EGC within its sheath. An insulated EGC is not required for existing feeders covered in the exception to Section 680.25(A) or as covered in Section 680.25(B)(2) for separate buildings.

The EGC size must be according to Section 250.122 and no smaller than 12 AWG. If the panelboard is supplied by a separately derived system, the supply-side bonding jumper is sized according to Section 250.30(A)(2), and it cannot be smaller than 8 AWG copper. This EGC must provide an effective ground-fault current path as specified in Section 250.4(A)(5).

Separate Buildings

Section 680.25(B)(2) includes requirements for pool wiring supplied from separate buildings or structures that are supplied by a feeder or branch circuit in accordance with Section 250.32. Where a feeder supplies a separate building or structure, it is also permitted to supply pool equipment if all grounding requirements of Section 250.32(B) have been satisfied. The EGC installed with the feeder to a separate building or structure supplying pool equipment must be an insulated conductor when supplying pool associated loads, unless the feeder is existing and meets the provisions of the exception to Section 680.25(A).

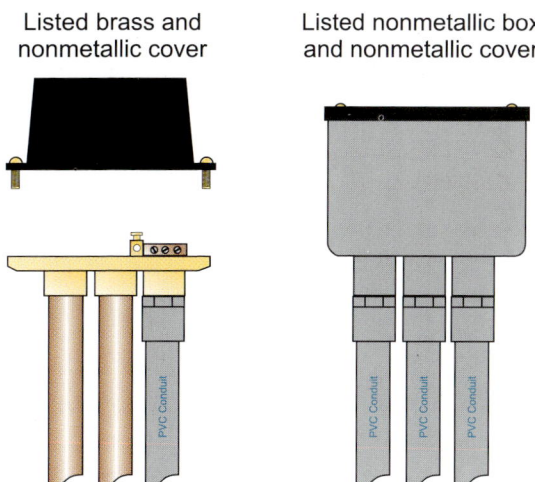

FIGURE 15-24 Junction boxes for pool equipment are required to be listed for this use.

Equipotential Bonding Requirements

Section 680.26 provides detailed requirements related to equipotential bonding for pools and similar installations. This equipotential bonding is required to reduce voltage gradients in the pool area. Bonding metallic parts together places them at the same potential electrically, reducing the ability for current to be present in any path between them. **See Figure 15-25.**

FIGURE 15-25 Equipotential bonding is used to reduce voltage gradients in pool areas.

334 APPLIED GROUNDING AND BONDING

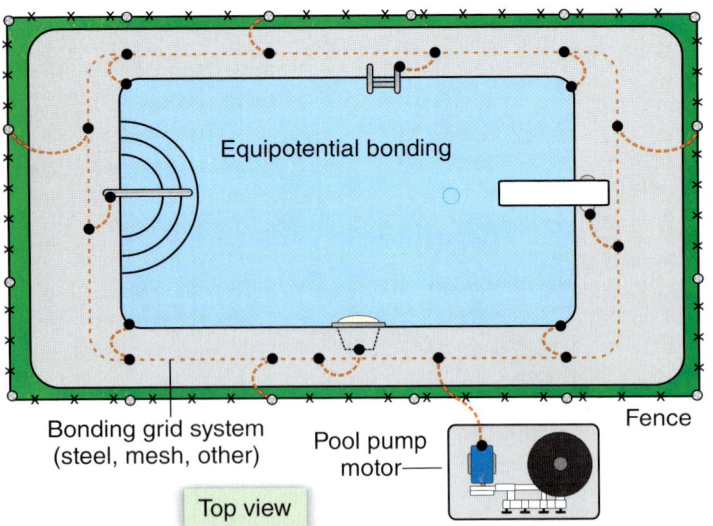

FIGURE 15-26 An 8 AWG solid copper conductor is required for the equipotential bonding of metal parts and equipment.

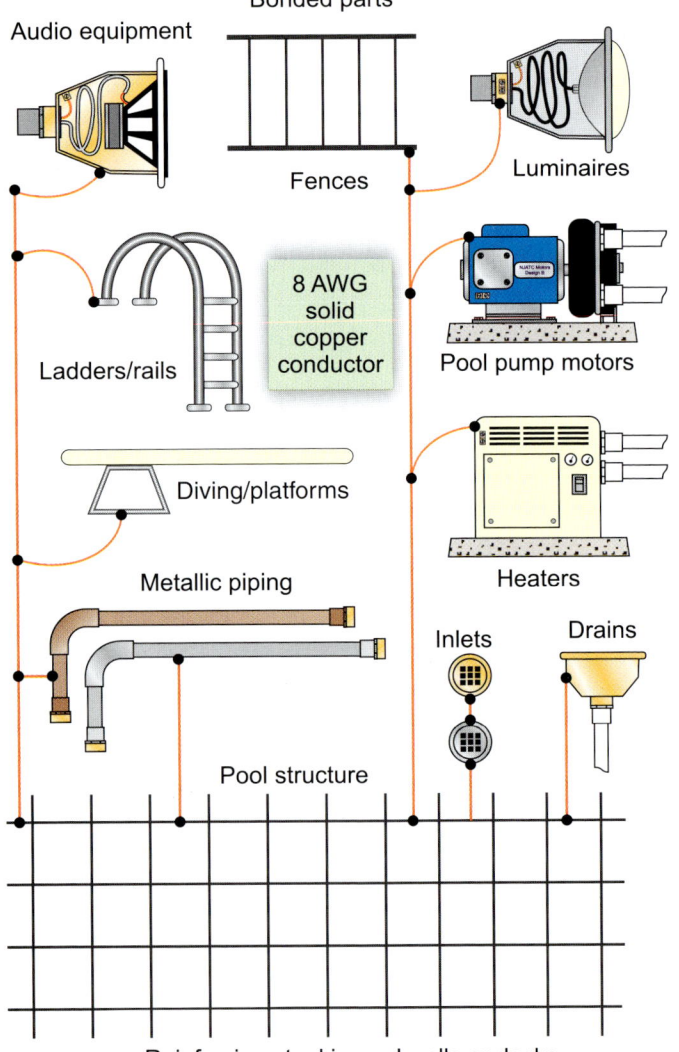

FIGURE 15-27 Equipotential bonding requirements for pools apply to multiple conductive surfaces and equipment related to the installation.

This is the philosophy behind establishing the equipotential bonding requirements for swimming pools and similar installations.

Section 680.26 first looks at what conductive parts and equipment are required to be connected together to form the equipotential bonding structure. Section 680.26(B) indicates that the conductive components provided in Sections 680.26(B)(1) through (7) have to be bonded together using solid copper conductors that are 8 AWG or larger. **See Figure 15-26.**

The reason for solid copper wire is that stranded conductors are more vulnerable to corrosive influences associated with pool water chemicals. Solid conductors are more resistant to corrosion effects as compared to the stranding of an 8 AWG stranded copper conductor. The bonding conductor can be insulated, covered or bare. The grid can be connected by use of RMC made of brass or another corrosion-resistant material. All connections between the 8 AWG solid conductor and the parts required to be bonded must be made using connection methods that comply with Section 250.8. This means they must be suitable for the materials and the location. Installers should understand that no requirement exists for the equipotential bonding conductor to be run to the panelboard or service equipment, nor is it required to be connected to any grounding electrode. These electrically conductive components and materials have to be bonded together to establish equipotential bonding in the pool area. **See Figure 15-27.**

Pool Shells (Conductive)

Conductive pool shells are typically made of poured concrete, pneumatically applied concrete, and concrete block with painted or plastered coatings. These surfaces all have varying degrees of water permeability and porosity. Pools with vinyl liners remove the contact between the chemical treated water and the pool shell. Structural reinforcing steel that is not encapsulated must be bonded together by steel tie wire or an equivalent method.

This rebar cage is typically the largest conductive component of an in-the-ground swimming pool and is a key component of the equipotential bonding system. Obviously, if the rebar is coated with an encapsulating compound, the rebar is ineffective for use in the equipotential bonding system. Section 680.26(B)(1)(b) provides the requirement for conductive pool shells where the rebar is coated. In this case, a copper grid is required to be constructed for the conductive pool shell. The copper conductor grid must be composed of, at minimum, 8 AWG solid copper wire that is bonded together to form a mesh that conforms to the contour of the pool and pool decking surface. The 8 AWG copper conductors have to be arranged in a 12 × 12-inch network of conductors uniformly spaced in a perpendicular grid pattern with a 4-inch tolerance. The connections at intersecting portions of the 8 AWG conductors must be made using a suitable connection means provided in Section 250.8. The copper grid must be embedded in the concrete no more than 6 inches from the outer contour of the pool shell. In simple terms, a copper wire mesh basket is created to be used as the grid embedded in the concrete shell of the pool structure. Listed copper wire mesh products are manufactured and available specifically for this purpose, which can assist installers in establishing a *Code*-compliant equipotential bonding system for swimming pools and similar installations. See Annex C of this textbook for additional information about listed copper wire mesh products in the UL Guide Information for Electrical Equipment category KDER.

Perimeter Surfaces

The perimeter surfaces are also required to be part of the equipotential bonding system. These surfaces are contacted by persons entering and leaving the body of water. The potential between the perimeter surface and the rest of the conductive components of the grid should be the same or as close as possible.

Perimeter surfaces around a pool can include paved, unpaved, and poured concrete surfaces that surround it and provide a deck for walking and so forth. The bonding requirement applies to paved and unpaved surfaces extending a minimum of 3 feet horizontally from the inside walls of the pool. **See Figure 15-28.** The 3-foot distance is about the maximum step distance as a person exits the pool water. Bonding of perimeter surfaces is required to be accomplished by the usual reinforcing steel embedded in the concrete perimeter surface or by installing a copper wire or wires that are arranged and installed to form a grid. Where perimeter surfaces are separated by a permanent wall or building, the perimeter surface bonding only has to be applied to the pool side of the wall or building. The conductive bonding for perimeter surfaces has to be constructed in accordance with Section 680.26(B)(2)(a) or (b) using a minimum of one 8 AWG bare solid copper conductor. The perimeter surface bonding has to follow the perimeter surface contour of the pool, and any splices must be made using listed connectors. This perimeter surface grid conductor has to be bonded to the conductive pool shell structure at least four times at uniformly spaced intervals around the contour of the pool.

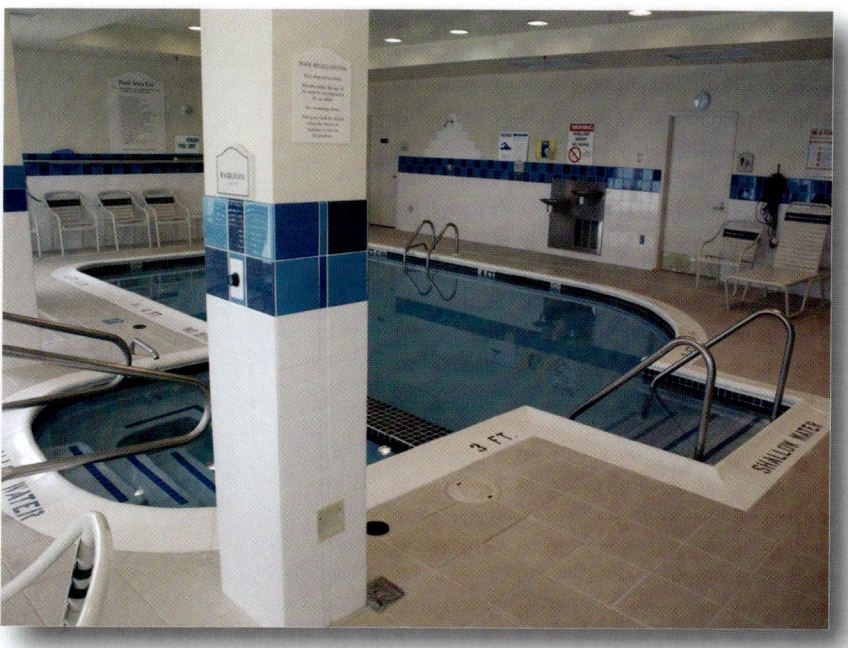

FIGURE 15-28 Equipotential bonding has to include the perimeter surfaces around a pool.

The bonding grid for perimeter surfaces must be either installed within the poured deck surface or installed under the deck surface at a depth no more than 4 to 6 inches from the underside of the deck material. To simplify the requirements for bonding a perimeter surface, it can be done by embedded reinforcing steel, which could be wire mesh (steel or copper), or constructed using a minimum 8 AWG solid copper conductor that forms a grid (mesh), and is arranged to meet the requirements in 680.26(B)(1)(b)(3). The specific criteria for constructing a bonding grid for pool perimeter surfaces is provided in Section 680.26(B)(2)(a) or (b).

Other Conductive Components

All metal parts of the pool structure, including reinforcing rods that are not encapsulated, must be bonded into the equipotential bonding system. Any metal forming shells for underwater luminaires and brackets of no-niche luminaires must be bonded, along with all metal fittings within or attached to the pool structure, with the exception of metal parts no larger than 4 inches in any dimension and not penetrating the pool structure. All metal equipment associated with the pool circulating system, including pump motors and heaters, and metal parts of pool covers and motors for pool covers must be bonded. An exception relaxes the bonding requirement for listed equipment with a system of double insulation.

Double-Insulated Pool Pump Motors and Water Heaters

Double-insulated pool pump motors are addressed in Section 680.26(B)(6)(a) and generally do not have to be bonded to the equipotential bonding system of the pool. However, an 8 AWG solid copper conductor has to be connected to the grid and extended to the pool pump motor vicinity to serve as a means of connecting any replacement pump motors that are not double insulated. **See Figure 15-29.** If there is no electrical connection between the equipotential bonding grid and the equipment grounding system for the premises, the 8 AWG conductor installed in the vicinity of the double-insulated pool pump motor must be connected to the EGC of the branch circuit supplying the motor. If pool water heaters rated more than 50 amperes have specific instructions about grounding and bonding requirements, only those parts designated must be bonded and grounded.

Fixed Metal Parts

Any fixed metal parts have to be bonded to the equipotential bonding system. This includes metal items such as fences, metal sheathing of cables, raceways, piping, awnings, gutters, and door and window frames. Bonding is not required for such metal parts when they are separated from the pool by a permanent barrier or a distance of no less than 5 feet. If metal parts are located more than 12 feet vertically from the maximum water level or from the top of observation stands, towers, platforms, or diving structures, they are not required to be bonded to the equipotential bonding grid system.

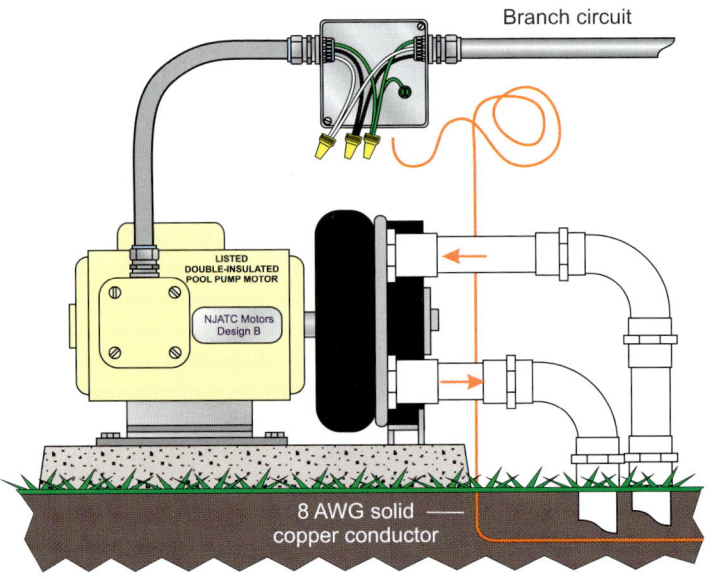

FIGURE 15-29 The 8 AWG solid copper bonding conductor from the pool bonding grid system must be routed to the double-insulated pool pump motor location and connected to any replacement motor that is not a double-insulated type.

Pool Water

Section 680.26(C) requires the pool water to be bonded. The chemically treated pool water must be bonded to the equipotential bonding system by contact between the water and the other bonded metal parts.

A minimum of 9 square inches of contact is considered sufficient for establishing a bonding connection between the water and the bonding system of conductive parts. This is not an electrical connection but a connection between water and electrically conductive metal parts that are bonded to the equipotential grid of the pool structure. An example of pool water bonding is a metal handrail or metal ladder in contact with the water and bonded to the grid. **See Figure 15-30.** If there is no metal in contact with the water, bonding the pool water to the equipotential bonding system for the pool is still a requirement. Some manufacturers of pre-formed nonmetallic or fiberglass pool shells make provisions for establishing a bonding connection for the water. This can be accomplished by embedding a small brass or other corrosion-resistant plate in the wall of a nonmetallic pool structure. This should be done by the pool manufacturer.

Specialized Pool Equipment

Section 680.27 includes specific requirements for special underwater equipment such as speaker systems. Such special equipment installed in pools must be identified for the purpose. Underwater speakers must be mounted in a metal forming shell that is connected to a listed junction box as specified in Section 680.24.

The forming shell and screen have to be of brass or another corrosion-resistant material. The wiring methods that may be used are RMC of brass or another identified corrosion-resistant metal, LFNC-B, rigid PVC conduit, or Type RTRC. Where any of the preceding nonmetallic wiring methods are used, an 8 AWG insulated solid or stranded bonding conductor has to be installed in the conduit to bond the forming shell to the equipotential bonding grid system. **See Figure 15-31.** The bonding conductor must terminate to the shell and has to be coated or encapsulated with a listed potting compound to protect the connection from corrosion.

Spas and Hot Tubs

The spa or hot tub installations have to meet the requirements in Parts I and IV of Article 680. This incorporates all general equipment grounding requirements specified in Section 680.6.

FIGURE 15-30 Pool water is required to be bonded, which can be accomplished by contact with metal parts of handrails, ladders, and so forth.

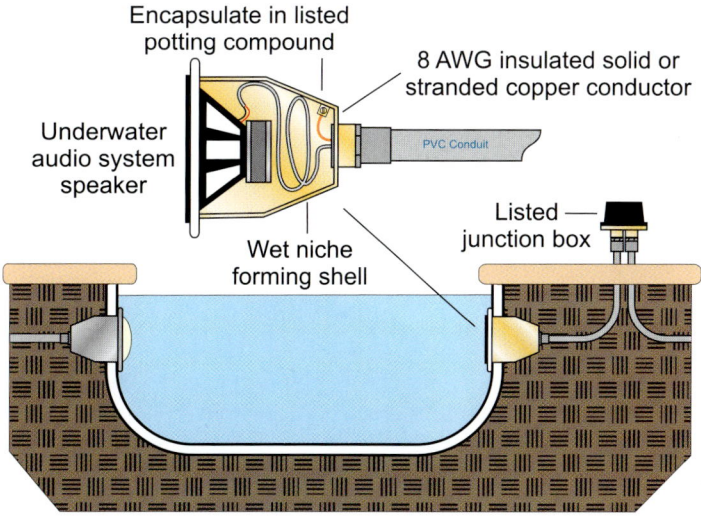

FIGURE 15-31 Bonding is required for forming shells associated with special equipment such as underwater speakers.

Sections 680.42(B) and (C) provide specific bonding and grounding requirements for indoor and outdoor installations. **See Figure 15-32.** The metal frames of spas or hot tubs, specifically, the package units, are permitted to accomplish bonding by metal-to-metal contact on a common conductive frame. Any metal strapping that secures wood to the structure does not have to be bonded to the equipotential bonding system. If the spa or hot tub is located outdoors and the wiring is run through a dwelling unit or other associated dwelling unit structure, any of the wiring methods in Chapter 3 of the *NEC* can be installed. The wiring must include a copper EGC no smaller than 12 AWG that is enclosed within the sheath of the cable assembly. This wiring method is permitted for packaged hot tub or spa assemblies. Wiring to any underwater luminaire must be in accordance with Section 680.23 or 680.33. For a spa or hot tub located indoors, the requirements in Parts I and II of Article 680 apply except as modified by Section 680.43. They must be wired using wiring methods in Chapter 3 of the *NEC,* with the exception of listed cord- and plug-connected units rated 20 A or less. The bonding requirements for indoor spas and hot tub installations are provided in Section 680.43(D). All metal fittings within or attached to the spa or hot tub structure must be bonded. Metal parts of electrical equipment for the spa or hot tub—including the pump motor or motors, the metal raceway and metal piping within 5 feet of the inside walls of the spa or hot tub if not separated by a permanent barrier, and all metal surfaces within 5 feet of the inside walls of the spa or hot tub—must be bonded. Any electrical controls or devices within 5 feet of the inside walls of the spa or hot tub, and not associated with the spa or hot tub installation, must be bonded. By exception, small conductive surfaces unlikely to become energized and metal parts of listed package assemblies are not required to be bonded. Bonding has to be accomplished by any of the following methods:

1. Interconnected fittings or threaded metal piping
2. Metal-to-metal contact on common framing
3. Solid copper conductor sized 8 AWG or larger[1]

Spa and hot tub equipment has to be grounded. All electrical equipment within 5 feet of the inside walls of the spa or hot tub must be grounded, including the equipment associated with the circulating system. The grounding is accomplished by connection to the EGC of the supply circuit wiring.

Fountains

The specific requirements for fountains are located in Part V of Article 680. Section 680.50 indicates that Parts I and V apply to all permanently installed fountains. **See Figure 15-33.** Fountains that have water common to a pool must also comply with the requirements in Part II of Article 680.

Sections 680.53 and 680.54 provide specific grounding and bonding rules for fountains covered by Article 680. Essentially, the requirements are similar to those for spas and hot tub installations.

FIGURE 15-32 Special bonding rules apply to both indoor and outdoor spas and hot tubs.

Equipment grounding is required for all equipment associated with the fountain, and metal parts such as piping systems associated with the fountain must be bonded to the EGC of the fountain supply circuit or circuits. See these sections for specific requirements. Portable fountains must meet the requirements in Article 422.

- Any metal parts of electrical equipment associated with the tub water circulating system, including pump motors
- Any metal-sheathed cables and raceways and metal piping that are within 1.5 m (5 feet) of the inside walls of the tub and not separated from the tub by a permanent barrier

Therapeutic Pools and Tubs for Health Care Use

Part VI of Article 680 includes requirements for pools and tubs intended for therapeutic use. The requirements in Parts I and VI of Article 680 apply to these installations in health care facilities, gymnasiums, training centers, and similar areas. **See Figure 15-34.** Grounding and bonding requirements for these installations are located in Sections 680.62(B) through (D). Any portable therapeutic appliances must meet the requirements in Article 422.

Bonding

Special bonding requirements for therapeutic tubs (hydrotherapeutic tanks) are provided in Section 680.62(B) of the *Code*. When reviewing bonding in its simplest form, it is the process of establishing continuity and conductivity between conductive parts. The objective is to have them become electrically common to one another by either bonding jumpers or mechanical connections. Hydrotherapeutic tubs and tank assemblies often include a variety of conductive objects as part of the entire assembly or installation. The manufacturer of this type of equipment often provides the necessary bonding connections for all conductive parts of the assembly to meet the requirements of applicable product standards. Where any of the conductive parts associated with the equipment are remote, bonding must be provided in accordance with the requirements in Section 680.62(B). The following metal parts of therapeutic tubs (hydrotherapeutic tanks) must be bonded together:

- Any metal fittings within or attached to the tub structure

Courtesy of Jim Dollard, IBEW Local 98

FIGURE 15-33 Rules for fountains are provided in Parts I and V of Article 680.

FIGURE 15-34 Special bonding requirements apply to therapeutic pools and tubs for health care use.

- Any metal surfaces that are within 1.5 m (5 feet) of the inside walls of the tub and not separated from the tub area by a permanent barrier
- Any electrical devices and controls that are not associated with the therapeutic tubs and located within 1.5 m (5 feet) from such units[2]

Section 680.62(C) includes the various methods permitted for accomplishing the bonding requirements for therapeutic tub installations and associated equipment. Remember the main objectives of bonding as provided in the general definition: Bonding is provided to establish continuity and conductivity between conductive parts. All metal or conductive parts identified in this section as required to be bonded must be bonded by one of the following methods:

- Interconnection of threaded metal piping and fittings
- Metal-to-metal mounting on a common frame or base
- Connections by suitable metal clamps
- Provision of a solid copper bonding jumper—insulated, covered, or bare—no smaller than 8 AWG[3]

Equipment Grounding

The specific equipment grounding requirements for therapeutic tubs (hydrotherapeutic tanks) are provided in Section 680.62(D) of the *Code*. These grounding requirements apply to this type of equipment whether it is stationary or fixed. Any equipment located within 5 feet of the inside walls of the tub or tank must be grounded. This is typically accomplished by connection to the supply circuit EGC. Because Section 680.60 indicates that Parts I and VI apply to this type of equipment, the EGC size rule is provided in Section 680.7(B). The EGC within the supply cord for this type of equipment must not be smaller than 12 AWG copper and must meet the minimum sizing requirements in Section 250.122.

Therapeutic tubs or tanks typically include a water circulation system and can include air pumps and blowers. Any of this type of equipment associated with the therapeutic tub must also be grounded by connection to an EGC. This EGC is typically included within the supply cord of listed package units or assemblies. Once the equipment is plugged into a grounding-type receptacle, the equipment grounding requirements of the *NEC* are satisfied, according to Sections 680.62(D)(1)(a) and (b).

Portable Therapeutic Appliances and Equipment

Any portable therapeutic appliances used in health care facilities are required to meet the grounding requirements in Section 250.114. This section provides general equipment grounding requirements for non–current-carrying metal parts of any cord- and plug-connected equipment that is likely to become energized. Where any of this type of equipment is protected by a system of double insulation (double insulated), the *Code* relaxes the grounding requirements, according to the exception to Section 250.114.

Hydromassage Bathtubs

The specific rules for hydromassage bathtubs are provided in Part VII of Article 680. This equipment requires at least one individual branch circuit and may require more depending on the load served. Some of these units are equipped with heaters requiring more than one branch circuit. **See Figure 15-35.** GFCI protection is required for this equipment, and the GFCI has to be located so as to be readily accessible, as does other equipment for the hydromassage bathtub. The GFCI protective device cannot be located under the unit. Equipment grounding is accomplished by connecting the EGC of the branch circuit to the units. Bonding requirements for hydromassage bathtubs are found in Section 680.74.

All metal piping systems and all grounded metal parts in contact with the circulating water must be bonded with, at minimum, an 8 AWG solid copper conductor. This conductor can be insulated, covered, or bare. The bonding jumper must terminate on the circulating pump motor.

Grounding for Special Equipment

A terminal lug is typically installed there for this purpose. A bonding jumper connection to a double-insulated motor is not required. The purpose of this bonding requirement is to establish equipotential bonding in the hydromassage bathtub area. As indicated in Section 680.26(A) for pools, the 8 AWG or larger copper bonding conductor does not have to be run to a panelboard or to the service equipment of the building or structure, nor is it required to be connected to a grounding electrode. This conductor is for equipotential bonding purposes only, not grounding, although interconnection with the EGC is established by connecting the EGC of the branch circuit wiring to the equipment.

Grounding Requirements for Solar PV Systems

Solar PV systems and installations are becoming more popular as efforts increase to reduce foreign oil dependency and expand the installation and use of alternative renewable energy sources. PV systems and equipment have specific grounding and bonding requirements provided in Part V of Article 690. **See Figure 15-36.**

System Grounding

Grounded PV systems must be installed according to the provisions of Part V of Article 690. Ungrounded systems must meet the rules in Section 690.35.

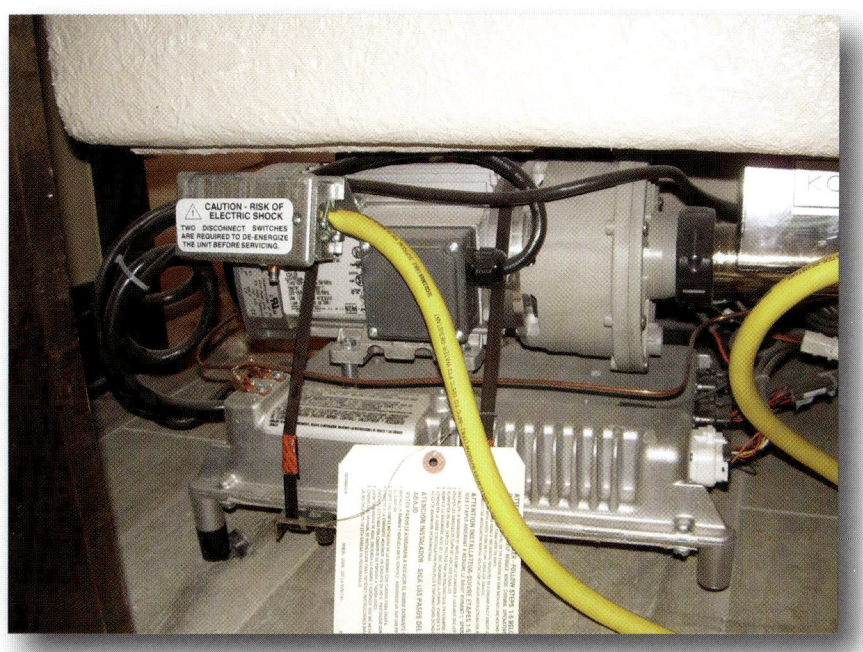

FIGURE 15-35 Special bonding and equipment grounding requirements apply to hydromassage bathtubs.

I-Stock Photo Courtesy of NECA

FIGURE 15-36 PV systems have specific grounding and bonding requirements in Article 690.

342 APPLIED GROUNDING AND BONDING

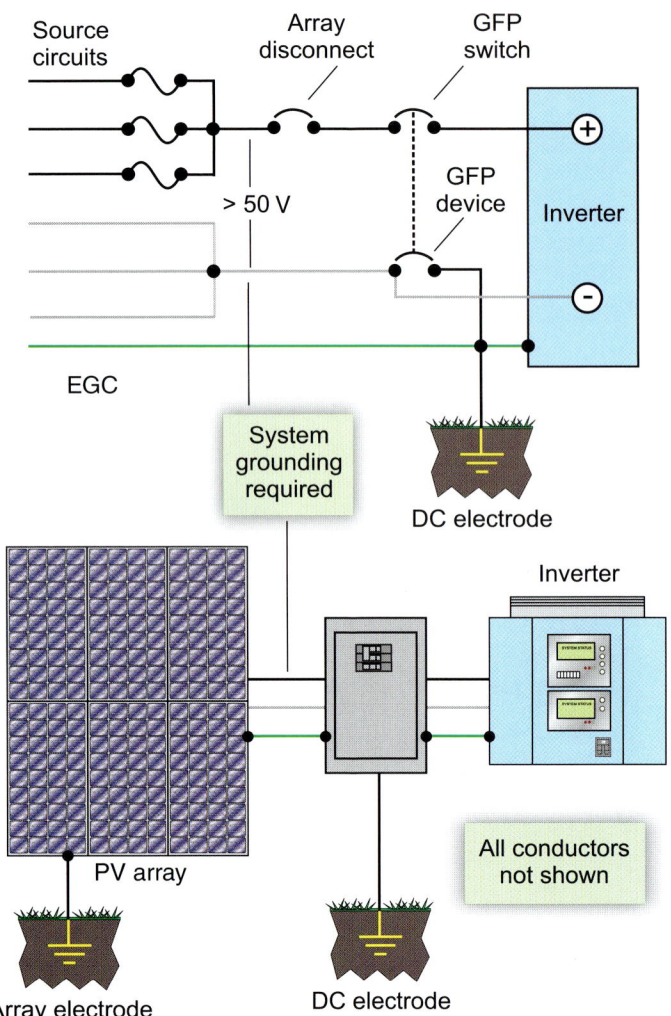

FIGURE 15-37 Grounding PV systems is required for systems with an output of more than 50 V.

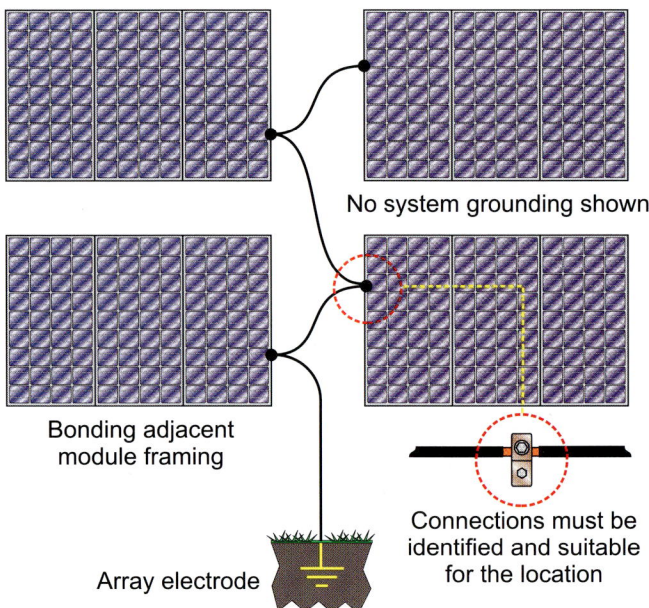

FIGURE 15-38 Exposed metal module framing is required to be bonded to adjacent exposed metal module frames.

General requirements for system grounding are provided in Section 250.20. The grounding and bonding rules in Article 250 apply to PV systems and installations except as modified or amended by Article 690. A PV system producing a voltage of more than 50 volts is required to be solidly grounded. This includes two-wire systems with more than 50 volts output and bipolar systems. **See Figure 15-37.**

The grounding of PV systems is accomplished by connection to a grounding electrode or grounding electrode system. The system grounding connection for the direct current (DC) circuit has to be made at a single point on the output circuit and preferably close to the output connection of the array for optimum protection against line surges and lightning events. By exception, systems with GFP devices are allowed to establish the grounding connection through the GFP device.

Equipment Grounding

Regardless of the system voltage, all equipment, metal frames, module frames, and conductor enclosures must be grounded in accordance with Sections 250.134 and 250.136(A). One or more EGCs are required between the PV array and other equipment, as specified in Section 250.110. The exposed metal frames of PV modules are permitted to be grounded and bonded with devices listed and identified for such use. Identified devices are also permitted to bond the exposed metal module framing to adjacent exposed metal module frames. **See Figure 15-38.** The bonding jumpers and the connections have to be made tight and effective, and they must be suitable for the location. **See Figure 15-39.**

EGCs for the PV array must be installed using the same wiring method with the associated circuit conductors. The size of the EGC for PV source and output circuits must be in accordance with Section 690.45(A) or (B). The general rule is to size the EGC with PV output or source circuits using Table 250.122.

If no overcurrent device is present to accomplish the sizing requirement, the assumed overcurrent device rating and the marked rated short circuit current must be used to size the EGC. The EGC for PV output or source circuits can be no smaller than 14 AWG. A PV system installed on a structure other than a dwelling unit is not required to be equipped with GFP. The EGC or EGCs must have an ampacity of at least two times the temperature and conduit fill corrected circuit conductor ampacity. Note that the amount of short circuit current from PV systems is very low compared to that from utility-supplied systems. Typically, the amount of short circuit current is slightly higher than the normal output rating. This means overcurrent devices will take much longer to react to ground-fault or short circuit events. The EGCs must be sized to withstand the amount of PV output short circuit current for the length of time it takes an overcurrent device to react to the abnormal condition. EGCs smaller than 6 AWG must be protected according to the provisions in Section 250.120(C).

Grounding Electrode Systems

Section 690.47 provides the requirements for grounding electrodes installed in PV systems. Alternating current (AC) systems must be grounded using a grounding electrode system according to Sections 250.50 through 250.60. DC systems that are grounded have to be connected to a grounding electrode system in accordance with Section 250.166, and for ungrounded DC systems, the grounding electrode requirements of Section 250.169 apply. The grounding electrode conductor for both AC and DC systems must meet the installation rules in Section 250.64. For AC systems, the grounding electrode system is usually established for the structure or building. The same electrode system has to be used for the PV equipment and system grounding. In systems with AC modules only and no field-installed or accessible DC circuits, the grounding electrode system for the AC power service to the building must be used. **See Figure 15-40.**

For DC-only systems, such as small, stand-alone units, the grounding electrode system has to meet the provisions in Part VIII of Article 250. The DC system grounding electrode conductor has to be as large as the largest output conductor supplied by the PV system or at least 8 AWG, whichever is larger.

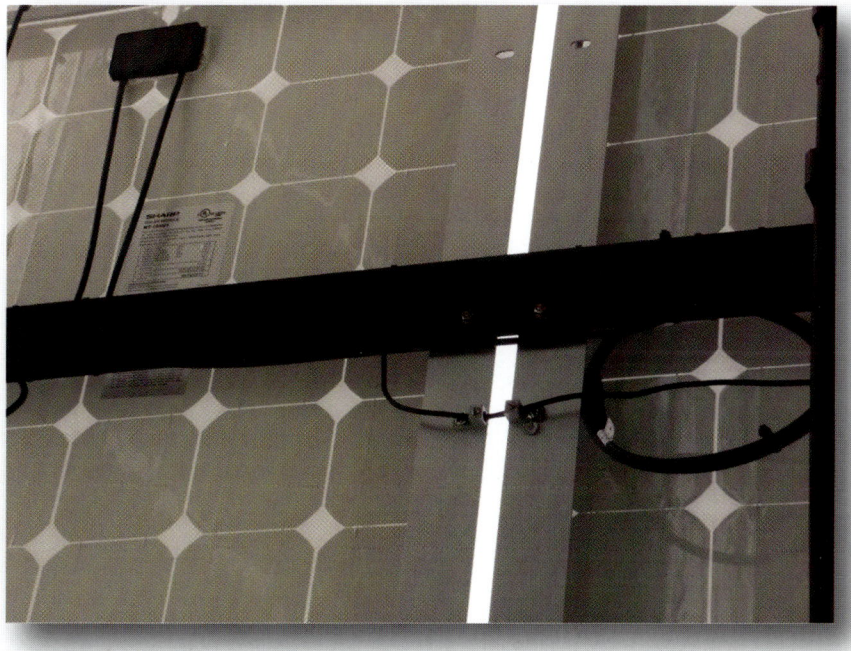

Courtesy of IBEW Local 26 Training Center

FIGURE 15-39 Bonding jumpers between equipment are required to be effective and suitable for the location in which they are installed.

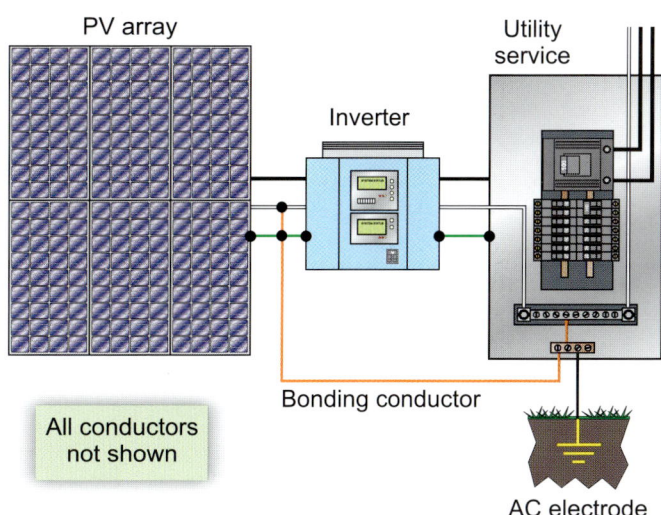

FIGURE 15-40 Grounding for PV system can be accomplished using the same electrode that is used for the building electrical service supplied by the utility.

If the grounding electrode is a sole connection to a rod, pipe, or plate electrode, the largest grounding electrode conductor required is 6 AWG copper.

Three methods of grounding for PV systems include both AC and DC voltages. The requirements for these grounding methods are located in Section 690.47(C). The first method is to use two separate grounding electrode systems, one for the DC side and one for the AC side, and bond the two electrode systems together. **See Figure 15-41.** The size of the bonding conductor can be no smaller than the larger of the two grounding electrode conductors required for either the AC or the DC system.

The second method of grounding involves using the AC system grounding electrode by bonding the DC system grounding electrode conductor to it. **See Figure 15-42.** This is a fairly common method of grounding, especially at buildings or structures that are already supplied by a utility service and for which a grounding electrode system exists.

This method is common for utility interactive systems. The grounding electrode conductor size must meet the requirements for both AC and DC system grounding electrode conductor sizing.

The third method uses a combined DC grounding electrode conductor and AC EGC. This conductor must be unsliced, or irreversibly spliced, and connect the marked DC grounding electrode connection point (along with the AC circuit conductors) to the grounding bus in the AC equipment. It must be the larger of that specified by Section 250.122 or Section 250.166 and be installed according to Section 250.64(E).

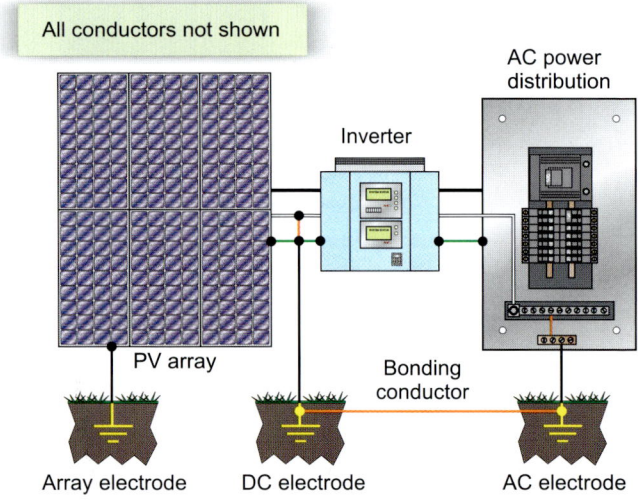

FIGURE 15-41 Grounding for the DC system and the AC service can be accomplished using two separate grounding electrode systems. Note that if separate electrode systems are installed, they must be bonded together.

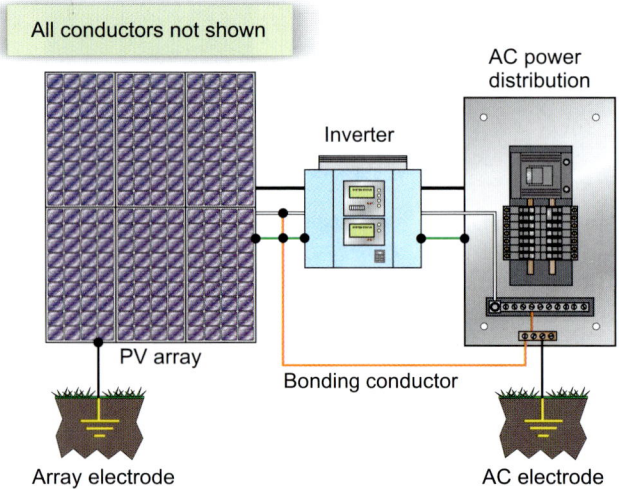

FIGURE 15-42 Grounding for the DC system can be accomplished using the same electrode for the AC service or supply system.

Equipment Grounding System Continuity

Sections 690.48 through 690.50 include requirements for EGC continuity for PV systems and equipment. The requirements in these sections anticipate removal of equipment that disconnects bonding connections. The *Code* requires a bonding jumper to be installed between the grounding electrode conductor and the exposed conductive surfaces of the PV source or output circuit equipment. The bonding jumper must be installed and sized according to the requirements in Section 250.120(C), based on the largest ungrounded circuit conductor for either the DC or the AC system. This requirement is a safety provision for service personnel who could remove equipment for repair or replacement. Grounding connections for PV arrays and all associated equipment are essential for property and personnel safety.

Note that PV systems will be energized when exposed to the sun, and special electrical safety work practices must be used when working with such systems and equipment. Always conform to OSHA safety regulations and the requirements provided in NFPA 70E, *Standard for Electrical Safety in the Workplace.*

Summary

Chapter 6 of the *NEC* includes some special grounding and bonding rules that are more restrictive than the general rules in Chapters 1 through 4 of the *NEC*. The special equipment described, in many cases, has unique operating characteristics and can present additional electrical hazards to people and property if specific grounding and bonding methods are not applied. Some special equipment bonding and grounding rules in Chapter 6 address shock protection in addition to equipment protection and facilitating overcurrent device operation.

References

1. NFPA 70 National Electrical Code 2011, Section 680.43(E) (National Fire Protection Association, Quincy, MA 2010), p. 70–587.
2. NFPA 70 National Electrical Code 2011, Section 680.62(B) (National Fire Protection Association, Quincy, MA 2010), p. 70–590.
3. NFPA 70 National Electrical Code 2011, Section 680.62(C) (National Fire Protection Association, Quincy, MA, p. 70–590.

Review Questions

1. Signs and metal equipment of outline lighting systems shall be grounded by connection to the _____ of the supply branch circuit or circuits or the feeder using the types of equipment grounding conductors specified in Section 250.118.
    a. Grounding electrode conductor
    b. Equipment bonding jumper
    c. Grounded conductor
    d. Equipment grounding conductor

2. A wire-type equipment grounding conductor for a 30-ampere branch circuit supplying an outline lighting system must be sized a no less than _____.
    a. 12 AWG copper
    b. 10 AWG aluminum
    c. 10 AWG copper
    d. 8 AWG copper

3. All equipment grounding conductor connections for signs and outline lighting systems have to be made using a method specified in Section 250.8, which includes all but which of the flowing?
    a. Sheet metal screws with two full threads engaged
    b. Listed pressure connectors
    c. Listed lugs
    d. Pressure connectors listed as grounding and bonding equipment

4. Metal parts of a building or structure are not permitted as an equipment grounding conductor or as a high-voltage secondary return circuit for a neon lighting system.
    a. True
    b. False

5. Copper bonding conductors for metal parts associated with high-voltage neon secondary circuits of outline lighting systems are not permitted to be smaller than _____.
    a. 12
    b. 14
    c. 8
    d. 6

6. When listed flexible metal conduit or listed liquidtight flexible metal conduit is used as the raceway to enclose the high-voltage GTO cables, the metal parts of a sign or outline lighting system are permitted to be bonded through the flexible metal raceway wiring method, provided that the total accumulative length of the conduit does not exceed _____ feet.
    a. 25
    b. 50
    c. 6
    d. 100

7. Where an external 14 AWG copper bonding conductor for metal parts of outline lighting systems is installed, a distance of at least 1½ inches must be maintained from a nonmetallic conduit containing a high-voltage circuit operating at _____ Hz or less, and no less than 1¾ inches must be maintained from a nonmetallic conduit containing a circuit operating at greater than _____ Hz.
    a. 50, 50
    b. 60, 60
    c. 100, 100
    d. 400, 400

8. The bonding for electric cranes and hoists can be accomplished either by mechanical connections of the conductive parts or by connection of suitable bonding jumpers.
    a. True
    b. False

9. The rating of the overcurrent device supplying cranes and hoists determines the minimum size required for wire-type equipment grounding conductors. The minimum size of the aluminum equipment grounding conductor for an electric crane supplied by a 125-ampere branch circuit is _____ AWG.
    a. 10
    b. 8
    c. 6
    d. 4

10. The frames and metal equipment of electric elevators are required to be grounded, and this includes which of the following?
    a. Motors and controllers
    b. Machines
    c. All metal equipment on the elevator car
    d. All of the above

11. Any signal reference grids installed in information technology rooms have to be connected to the _____ provided with the circuits supplying the information technology equipment.
    a. Grounded conductor
    b. Grounding electrode conductor
    c. Ungrounded conductor
    d. Equipment grounding conductor

12. The voltage drop on sensitive electronic equipment covered in Article 647 cannot exceed _____% on the branch circuits and _____% for the combined voltage drop of the feeder and branch circuit supplied by the system.
    a. 3, 5
    b. 2, 4
    c. 1.5, 2.5
    d. 5, 10

13. Equipment and receptacles supplied by sensitive electronic equipment covered in Article 647 have to be grounded by an equipment grounding conductor run with the supply circuit conductors and terminating on the equipment grounding terminal bus of the panel that is marked _____.
    a. Technical equipment ground
    b. Reference grounding bus
    c. Isolated grounding terminal bus
    d. Equipment grounding point

14. Any electrical equipment located within _____ feet of the inside walls of a pool or specified body of water must be grounded.
    a. 6
    b. 10
    c. 5
    d. 15

15. Pool pump motors have to be connected using wiring methods that include an insulated copper equipment grounding conductor sized according to Section 250.122 but no smaller than _____ AWG.
    a. 14
    b. 12
    c. 10
    d. 8

16. What is the minimum size of the copper conductor that is installed for an equipotential bonding grid for a swimming pool?
    a. 10 solid
    b. 8 stranded
    c. 12 solid
    d. 8 solid

17. Aluminum equipment grounding conductors are permitted for pool pump motor circuits.
    a. True
    b. False

18. Where rigid nonmetallic conduit or liquidtight flexible nonmetallic conduit is installed between a forming shell for a wet-niche luminaire and a junction box or other enclosure, a(n) _____ AWG insulated copper bonding jumper is required to be installed in the conduit to provide electrical continuity between the forming shell and the junction box or other enclosure.
    a. 14
    b. 12
    c. 10
    d. 8

19. The pool water is required to be bonded to the equipotential bonding grid.
    a. True
    b. False

20. Junction boxes for pool equipment are required to be listed and equipped with threaded hubs or entries or nonmetallic hub entries and must be made of any of the following materials *except* _____.
    a. Copper or brass
    b. Plastic
    c. Aluminum
    d. Other corrosion-resistant material

21. Equipotential bonding is required for pools and similar installations in an effort to eliminate voltage gradients in the pool area.
    a. True
    b. False

22. Double-insulated pool pump motors are addressed in Section 680.26(B)(6)(b) and generally do not have to be bonded to the equipotential bonding system of the pool. However, a(n) _____ conductor has to be connected to the grid and extended to the pool pump motor vicinity to serve as a means of connecting any replacement pump motors that are not double insulated, and the replacement motor requires bonding to the grid.
    a. 10 AWG solid copper
    b. 8 AWG solid copper
    c. 6 AWG stranded copper
    d. 10 AWG stranded copper

23. All electrical equipment within _____ feet of the inside walls of the spa or hot tub must be grounded, including the equipment associated with the circulating system.
    a. 3
    b. 5
    c. 6
    d. 10

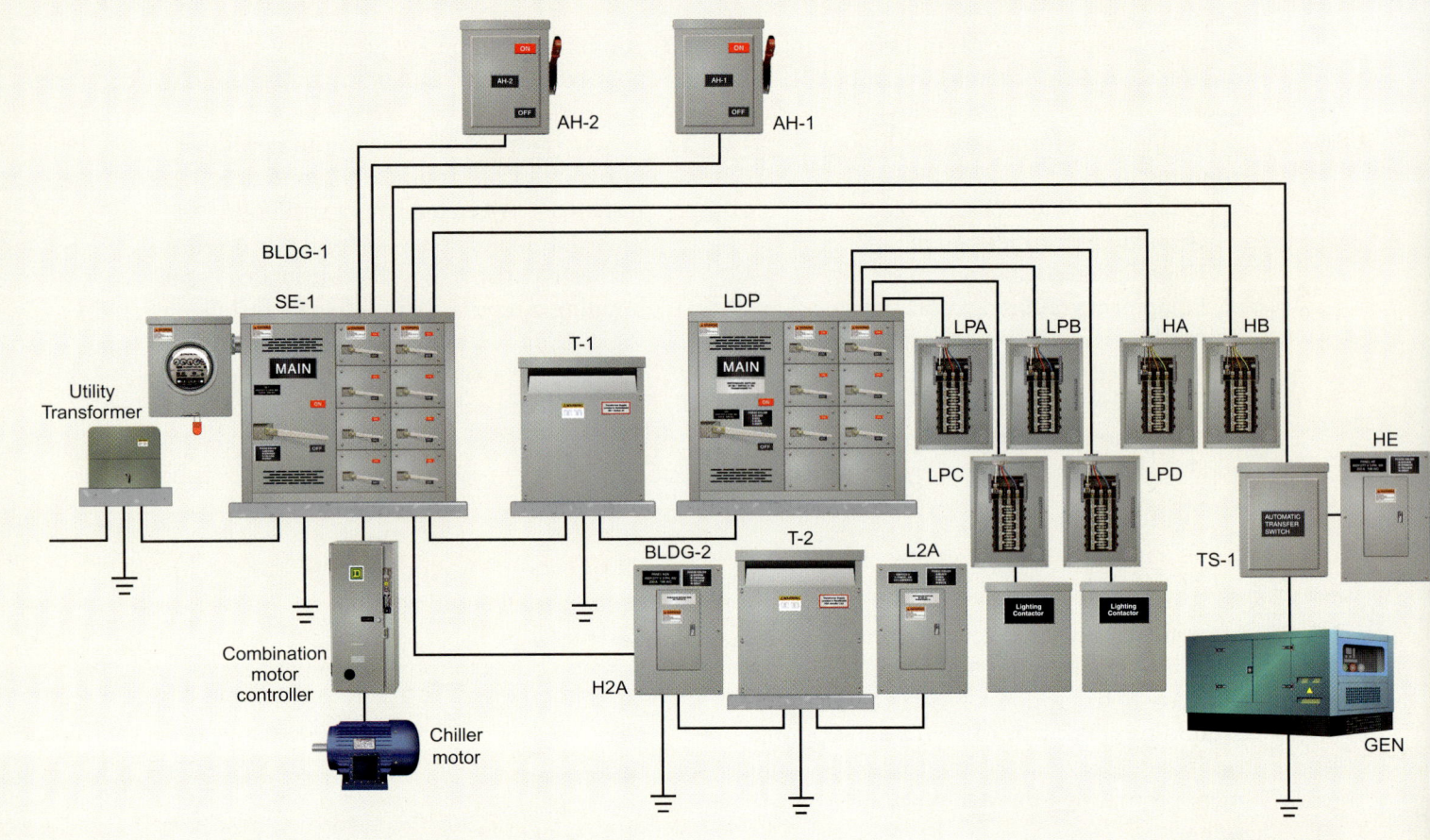

CHAPTER 16

Grounding and Bonding for Limited-Energy Systems

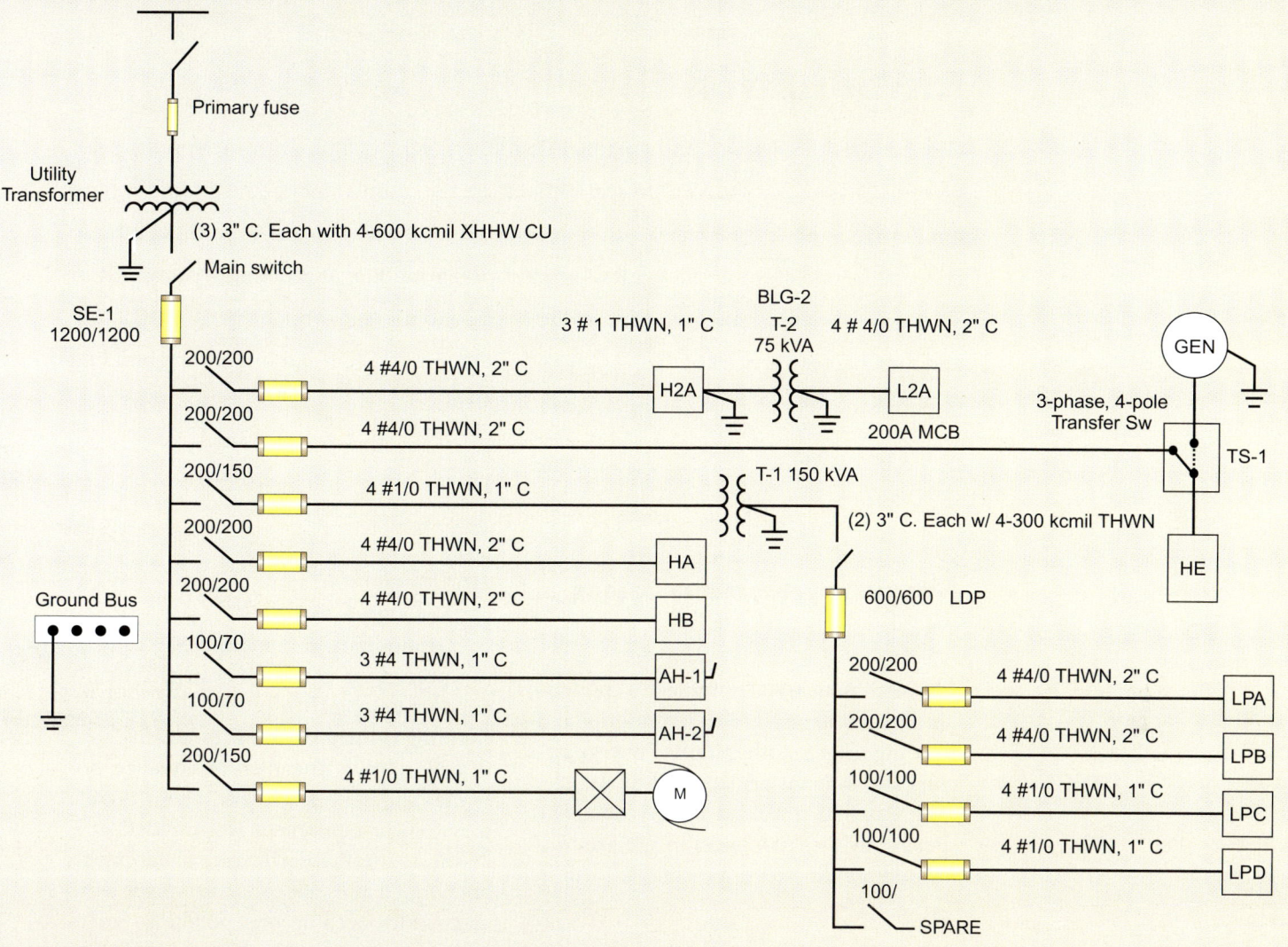

Objectives

- Determine the purpose of grounding and bonding requirements for limited-energy communications systems and equipment
- Understand how the grounding and bonding requirements in Article 770 and Chapter 8 of the *NEC* apply to the communications systems
- Determine the requirements for grounding electrodes to be used for intersystem equipment grounding
- Determine minimum sizes required for grounding electrode conductors and bonding conductors installed for limited-energy communications systems and equipment
- Understand the means of connection requirements for grounding electrode conductors installed for limited-energy communications systems and equipment

Outline

Performance and Concepts

Definitions

Grounding and Bonding Performance

Connecting to a Grounding Electrode

Grounding Electrode Conductor Installation

Intersystem Grounding and Bonding

Common Grounding and Bonding Rules for Communications Systems

Grounding and Bonding at Mobile Homes

Radio and Television Equipment and Antennas

Overvoltages and Lightning Events

Introduction

Communications systems and circuits in buildings must comply with the specific rules given in Chapter 8 of the *NEC®*. It seems that there is a complacency about grounding and bonding requirements for communications equipment and systems. Even though these systems operate at lower energy levels, improper grounding and bonding can result in severe consequences for equipment and property and present shock hazards. Article 770 and the Chapter 8 articles of the *NEC* provide unique and specific grounding and bonding requirements for communications systems installations.

Performance and Concepts

Although Article 800 is limited to communications systems, similar grounding and bonding rules for other limited energy circuits and systems are covered in Articles 770, 810, 820, 830, and 840. Article 840, specifically 840.100, refers to the grounding and bonding requirements set forth in 770.100, 800.100, and 820.100 as applicable. **See Figure 16-1.**

Grounding, in the simplest form, is the process of connecting an electrically conductive object to ground (the Earth). Bonding is the process of connecting conductive objects together to equalize potential differences between them. When a system or equipment is grounded, it is connected to the Earth, and when objects are bonded, they are connected together to electrically become one potential—or as close to the same potential as possible. **See Figure 16-2.** These two processes work in unison to provide safety for communications systems, equipment, and property. Grounding and bonding for limited-energy circuits and systems provide operational grounding and protective grounding functions.

Definitions

The definitions in Article 100 of the *NEC* provide a foundation on which grounding and bonding requirements are built. The meanings of defined terms used in Articles 770, 800, 810, 820, 830, and 840 are given here.

FIGURE 16-1 There are specific grounding and bonding requirements that apply to communications systems and equipment covered in Article 770 and Chapter 8 of the *NEC*.

Ground. The earth.[1]

Bonded (Bonded). Connected to establish electrical continuity and conductivity.[2]

Grounded (Grounding). Connected to *ground* or to a conductive body that extends the ground connection.[3]

Bonding Conductor or Jumper. A reliable conductor to ensure the required electrical conductivity between metal parts required to be electrically connected.[4]

Grounding Electrode. A conducting object through which a direct connection to earth is established.[5]

Grounding Electrode Conductor. A conductor used to connect the system grounded conductor or the equipment to a grounding electrode or to a point on the grounding electrode system.[6]

Intersystem Bonding Termination. A device that provides a means for connecting bonding conductors for communications systems to the grounding electrode system.[7]

Grounding and Bonding Performance

Grounding is the process of connecting a system or equipment to ground or to a conductive body that extends the ground connection. Grounding requires a connection to the Earth through a grounding electrode. Bonding is the process of connecting objects or entities together. Bonding electrically means that conductive objects are connected to establish continuity and conductivity between them.

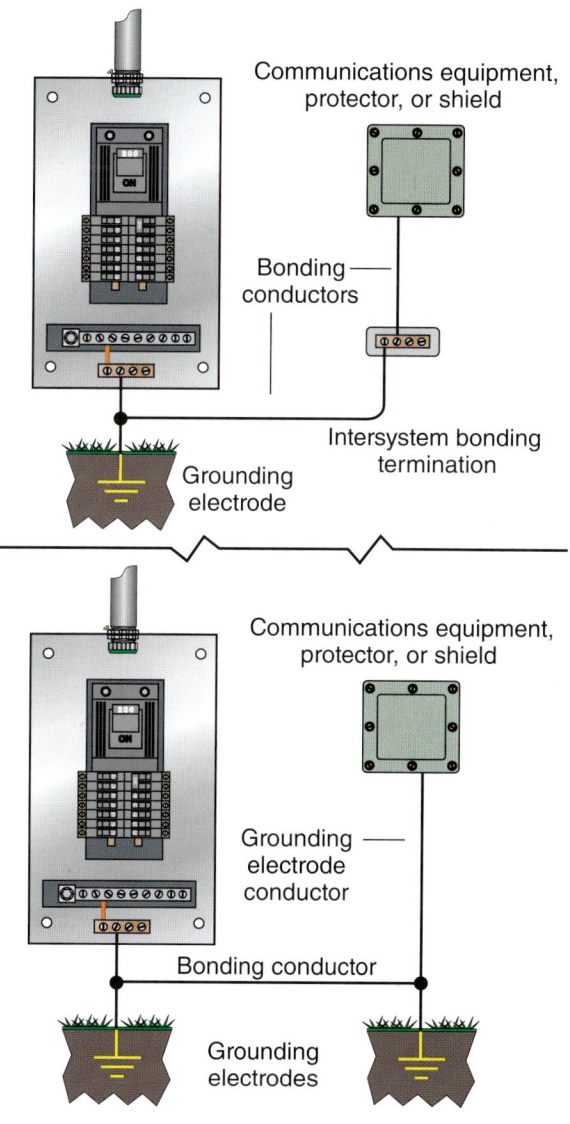

FIGURE 16-2 All communication systems equipment must be effectively bonded together and connected to ground (the Earth).

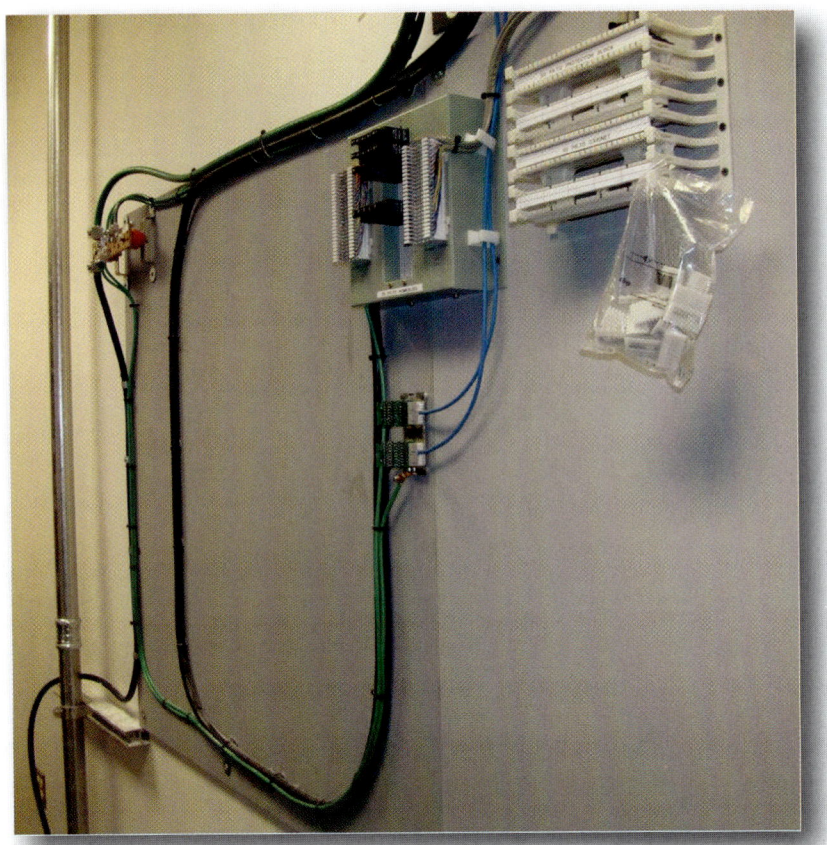

FIGURE 16-3 Grounding and bonding requirements for communications systems provide a level of protection by providing a path to ground for lightning and surge events.

Both grounding and bonding are functions necessary for safety when installing communications or other limited-energy systems. The purpose of grounding and bonding for limited-energy systems and equipment is to provide a level of shock protection and limit damage from voltage surges created by lightning, line surges, or unintentional contact with higher-voltage lines. **See Figure 16-3.**

Grounding protects the equipment and provides a path to the Earth for lightning events. The grounding and bonding requirements of Chapter 8 in the *NEC* should not be confused with the requirements for lightning protection systems as provided in NFPA 780, *Standard for Installation of Lightning Protection Systems.* Remember Section 90.1 indicates that the purpose of the *NEC* is to protect persons and property from hazards that arise from the use of electricity. Lightning is an unpredictable force that is not used by persons. Electrical grounding and bonding requirements in the *NEC* provide varying degrees of protection from lightning events; it is typically not the primary purpose but one of the functional benefits.

> The *NEC* provides the minimum requirements for safe installations of communications systems grounding and bonding.

Connecting to a Grounding Electrode

System or equipment grounding is accomplished by establishing a connection to the Earth. This connection is made through a grounding electrode. **See Figure 16-4.** Section 800.100(B) requires the grounding electrode conductor for communications systems to be connected to a grounding electrode—specifically, the same grounding electrode that the building electrical system is connected to.

This requirement ensures that both systems and connected equipment are at the same ground potential. Attempts to install separate grounding electrodes and not bond them to the power system grounding electrode is a recipe for disaster. This is not permitted by the *NEC* and creates unsafe conditions for persons and property.

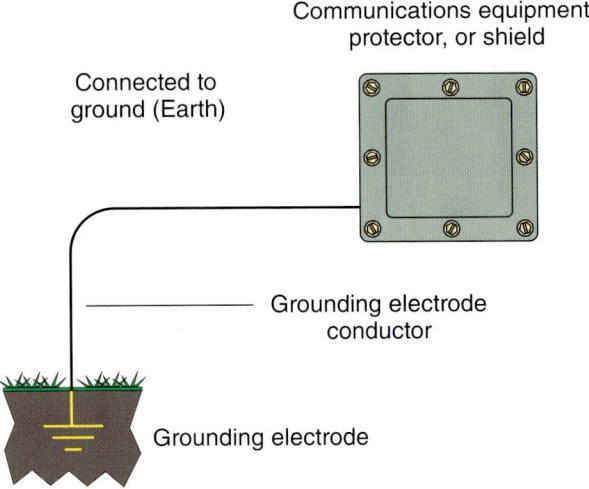

FIGURE 16-4 Grounding communications systems involve a connection to the Earth through a grounding electrode.

A revision in the 2008 *NEC* requires an *intersystem bonding termination (IBT)* to be installed at the service location, or at the disconnecting means for other buildings, for connecting these other systems. **See Figure 16-5.** It is intended specifically for connecting communications system grounding and bonding conductors. In many designs for buildings or structures other than dwelling units, there is often a telephone mounting board designated for all communications equipment and service-point connections.

The designation on blueprints is typically "TMGB" and stands for telecommunications main grounding busbar. **See Figure 16-6.** The *NEC* provides the minimum requirements for safe installations of communications systems grounding and bonding. Engineering designs for limited-energy systems may meet or exceed these minimums. The engineering designs take precedent in such situations.

Grounding Electrode Conductor Installation

Preceding the 2011 *NEC*, the term *grounding conductor* was used in Articles 770 through 830. The term was changed to *grounding electrode conductor* to reflect how it functions in the grounding system. (There was no need to use two different defined terms that apply to the same component of the grounding system.) Communications system grounding electrode conductors must be 14 AWG or larger and be made of copper or another corrosion-resistant material. They can be solid or stranded and must be insulated. Grounding electrode conductors for communications systems should be kept short, and specifically for one- and two-family dwelling installations, they must not exceed 20 feet. An exception permits a separate grounding electrode to be installed where the grounding conductor length of 20 feet is exceeded.

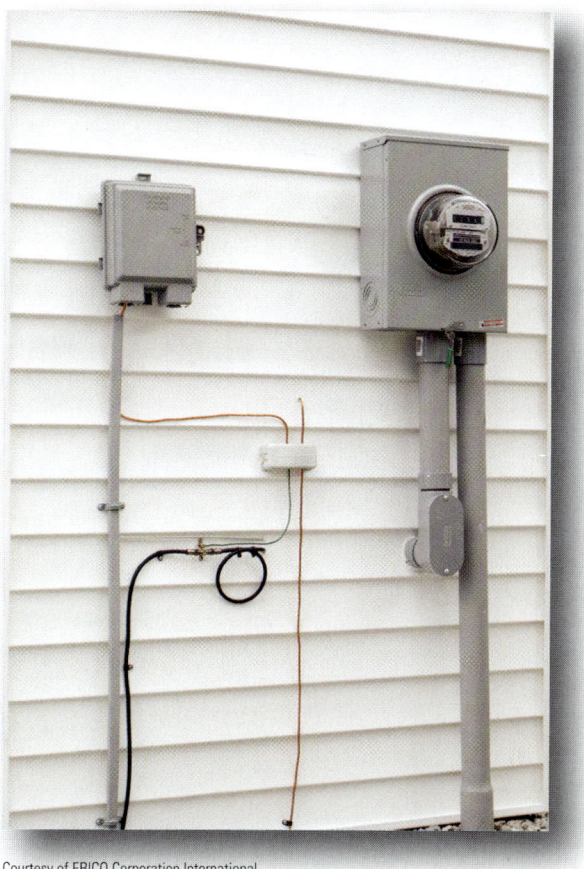

Courtesy of ERICO Corporation International

FIGURE 16-5 Intersystem bonding terminations are required to provide a connection point for not less than three bonding conductors for communications systems.

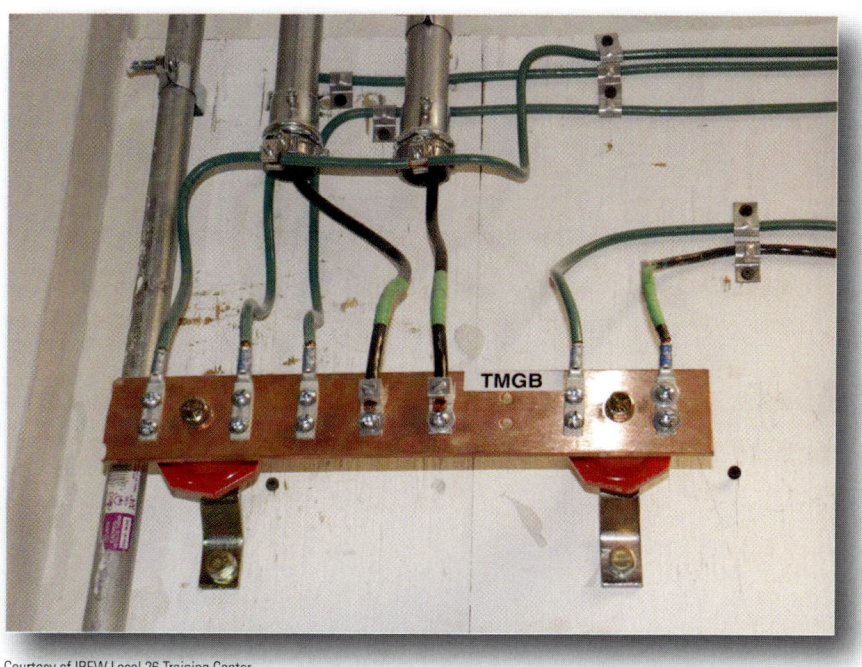

Courtesy of IBEW Local 26 Training Center

FIGURE 16-6 A telecommunications main grounding busbar (TMGB) for communications grounding is often supplied at the telephone mounting board.

354 APPLIED GROUNDING AND BONDING

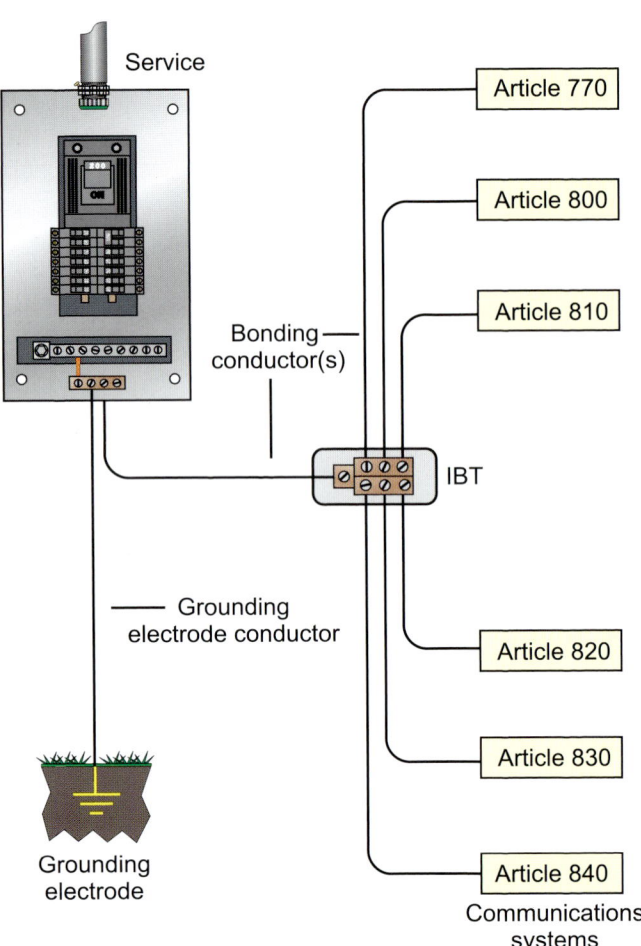

FIGURE 16-7 An IBT is required for connecting all communications systems.

Courtesy of Thomas and Betts

FIGURE 16-8 No less than three terminals mounted directly to the electrical power meter can serve as the IBT provision.

In this case, any separate electrode must be bonded to the power system grounding electrode for the building with a copper or equivalent conductor sized 6 AWG at minimum. Connections to grounding electrodes for communications circuits must meet the requirements in Section 250.70.

Intersystem Grounding and Bonding

General requirements for intersystem grounding and bonding are located in Section 250.94. This section contains a general requirement that an IBT be installed at the service equipment at a building or structure served. **See Figure 16-7.** An IBT is also required at each separate building or structure supplied by one or more feeders or branch circuits. The IBT must be installed in a way that leaves it accessible for connection and inspection.

Note that not all limited-energy systems (typically communications systems) are installed when the electrical service is installed; these systems, such as cable TV and other antenna systems, usually are installed later. The idea of the IBT is that it is in place when a limited-energy system is installed so that the grounding or bonding connection can be made to it. The IBT is connected to ground (the Earth) when it is first installed, and it is connected to the building electrode for the power service of the building. A variety of IBT products are manufactured specifically for this use. The installation of an IBT must not interfere with the opening of a meter enclosure or any electrical equipment enclosure. This termination device must provide the means for connection of no less than three intersystem bonding conductors and is permitted to be any of the following:

1. The IBT can be a set of terminals listed as grounding and bonding equipment securely mounted to the meter or service equipment enclosure. **See Figure 16-8.**

2. The IBT can be a bonding bar near the service equipment enclosure, meter enclosure, or service raceway. The bonding bar has to be connected to the equipment grounding conductor(s) (EGC) in the service equipment or meter enclosure. The connection must be made with, at minimum, a 6 AWG copper conductor. **See Figure 16-9.**
3. The IBT can also be a bonding bar located near the grounding electrode conductor for the service. The bonding bar has to be connected to the grounding electrode conductor with, at minimum, a 6 AWG copper conductor.[8]

Courtesy of ERICO Corporation International

FIGURE 16-9 An intersystem bonding termination (IBT) in the form of a copper bonding bar is often mounted near the service equipment supplying the building or structure.

The installation of an IBT is not optional. It has been a requirement since the 2008 edition of the *NEC*. The *Code* does still recognize that existing building or structures may not have an IBT. There are two choices to achieve effective grounding for communications systems in existing buildings: (1) install an IBT in conformance with Section 250.94, or (2) achieve intersystem grounding and bonding by connection to an accessible grounding electrode conductor connection point that is external to the service equipment enclosure or at the disconnecting means at a separate building or structure supplied by one or more feeders or branch circuits. The connection can be made to a nonflexible metal raceway (location 1), at an exposed grounding electrode conductor (location 2), or at another approved means of external connection to a grounded raceway or equipment (location 3). **See Figure 16-10.**

The conductor used to make this connection has to be made of copper or another corrosion-resistant material. The terminals must be listed as grounding and bonding equipment. The UL Guide Information for Electrical Equipment category KDSH provides information about grounding clamps for communications system grounding connections. See Annex C of this textbook for additional information.

Common Grounding and Bonding Rules for Communications Systems

Articles 770, 800, 810, 820, 830, and 840 of the *NEC* all provide specific grounding and bonding requirements for communications and antenna systems covered by the respective article. All of these articles, with the exception of Article 810, provide similar grounding and bonding requirements. Article 810 covers radio and television equipment and provides specific grounding and bonding rules in Sections 810.20 and 810.21. These rules are covered separately. Articles 770, 800, 820, 830, and 840 incorporate a parallel numbering sequence to enhance usability of the *Code*. This training material uses the requirements in Article 800 as the basis for covering all common grounding and bonding requirements in these articles and includes some of the differences in tabular form following this section. Part III of Articles 770, 800, 820, 830, and 840 includes information about protection. Protection is provided by the installation of primary protectors that are typically required to be installed at the communications system point of entrance to the building or structure served. The primary protector can be inherent to communications equipment or landing blocks, or it can be an externally installed separate device.

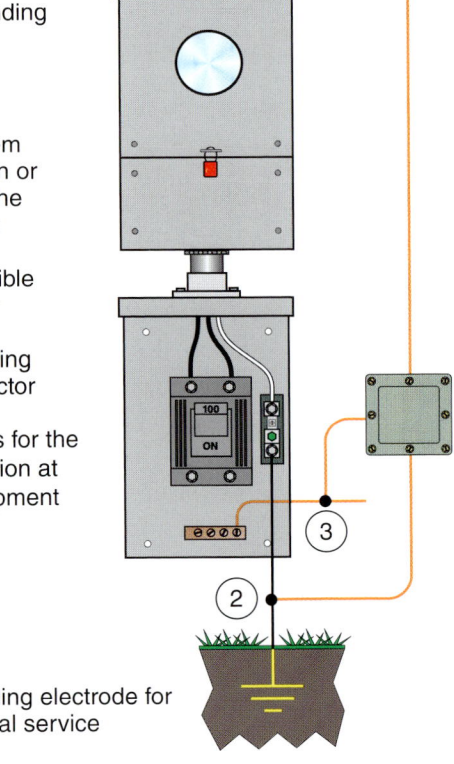

FIGURE 16-10 By exception, the grounding electrode conductors of communications systems can be connected to any of three locations.

The purpose of primary protectors and discharge units is to provide a level of surge protection and for ground discharge of transient overvoltages and unintentional contact with higher-voltage lines. Protectors can be either fused or nonfused types. For specific information about primary protectors, refer to the UL Guide Information for Electrical Equipment category QVGV in Annex C of this textbook.

Part IV of Articles 770, 800, 820, 830, and 840 provides the grounding and bonding methods for communications systems covered by each article. The requirements for installation are similar except that the minimum size of the grounding electrode conductors may be different in each article. Each of these articles presents the same grounding and bonding requirements and methods for these communications systems.

Part IV of each article is titled "Grounding Methods" (excluding Article 810). The first section in Part IV of each article (except Article 810) is .100, and each is arranged into four subdivisions, labeled (A) through (D). These subdivisions provide the minimum requirements for grounding and bonding communications systems. The first component of the grounding scheme is the grounding electrode conductor. This is the conductor installed to connect the primary protector, other metal parts of equipment, and any metallic cable shields to the grounding electrode system. There are six installation requirements for this conductor:

1. The grounding electrode conductor could be insulated, covered, or bare, and it must be listed.
2. The conductor must be solid or stranded copper or another corrosion-resistant material.
3. The conductor must be, at minimum, 14 AWG, and it shall have a current-carrying capacity of no less than the grounded metallic sheath member and protected conductor of the communications cable. This conductor is not required to exceed 6 AWG in size.
4. The length must be kept as short as practicable, and in dwelling units, it generally cannot exceed 20 feet in length. By exception, the length can exceed 20 feet if an electrode meeting the criteria in Section 800.100(B)(3)(2) is installed and bonded to the grounding electrode for the power system supplying the building or structure.
5. This conductor must be run as straight as practicable.
6. Where subject to physical damage, it must be protected. If it is installed in a ferrous metal raceway, both ends of the raceway must be bonded to the contained grounding electrode conductor. **See Figure 16-11.**

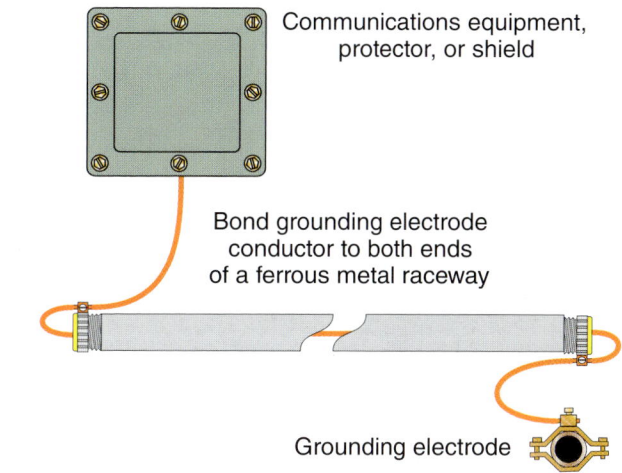

FIGURE 16-11 Grounding electrode conductor installation requirements for communications systems are provided in Section 800.100(A).

The next component of the grounding scheme is the grounding electrode. By definition, the grounding electrode is a conducting object through which a direct connection to the Earth is established. Section 800.100(B) provides three options for grounding electrodes that must be used. These are provided in somewhat of a hierarchy. The driving language indicates that an electrode, according to Sections 800.100(B)(1) through (3), be used to connect the grounding electrode conductor.

The first option indicates that if an IBT is present for use, the communications system–bonding conductor must connect to it. If an IBT is not present for use, the second option indicates that the grounding electrode connection methods provided in Section 800.100(B)(2) can be applied.

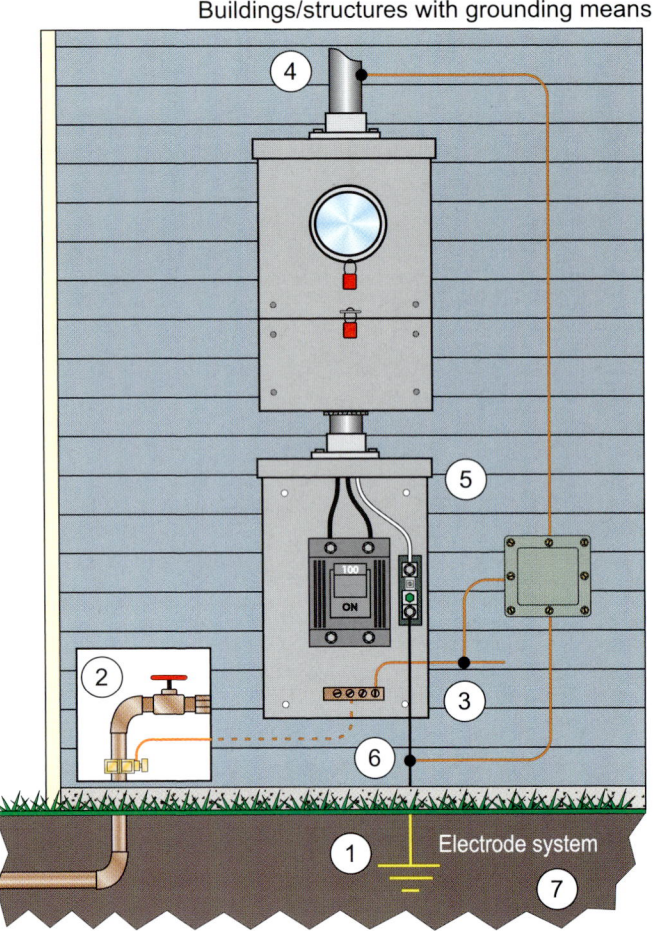

FIGURE 16-12 A grounding electrode conductor can be connected to a grounding electrode or one of the locations identified in the Exception to Section 250.94 if an IBT is not present (existing installations only).

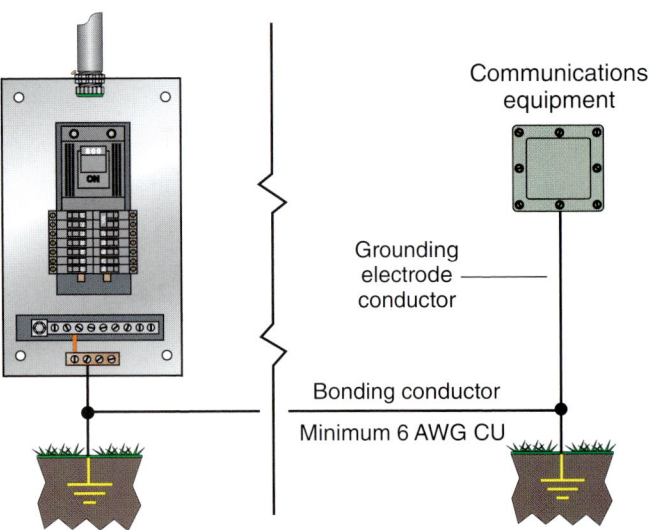

FIGURE 16-13 If the option of installing a grounding electrode for communications systems is used, this grounding electrode has to be bonded to the grounding electrode system for the power service supplying the building or structure.

This grounding electrode conductor connection must be made to the nearest accessible location on any of the following seven points (the numbers correlate with those in **Figure 16-12**):

1. The grounding electrode system for the building or structure electrical service
2. The grounded interior metal water piping system within 5 feet of the point where it enters the building
3. The accessible means external to the service equipment, as provided in Section 250.94
4. The nonflexible metallic power service raceway
5. The service equipment enclosure
6. The grounding electrode conductor or metal enclosure for a grounding electrode conductor
7. The grounding electrode conductor or grounding electrode for a separate building or structure, as required in Section 250.32

The third electrode option addresses buildings or structures without an IBT or any grounding means. In this case, a grounding electrode that is present must be used; otherwise, one has to be installed. Any grounding electrode described in Sections 250.52(A)(1) through (4) can be used as the grounding electrode. If no electrodes previously mentioned are present for use, an electrode meeting the requirements in Section 250.52(A)(7) and (8) can be installed and used. See **Figure 16-13**. Section 800.100(B)(3)(2) also allows a ground rod no less than 5 feet long and no smaller than ½ inch in diameter to be driven where practicable. The *NEC* indicates that this separate short electrode must be spaced a minimum of 6 feet from any electrode of another system. It is required to bond this electrode to other electrodes that are present.

The third component in the communications system grounding scheme is covered in Section 800.100(C) and is the grounding electrode conductor connection to the grounding electrode. See **Figure 16-14**.

A connection method specified in Section 250.70 must be used to make this connection. This requirement specifies that grounding electrode conductors be connected to the grounding electrode by exothermic welding, listed lugs, listed pressure connectors, listed clamps, or other listed means. **See Figure 16-15.**

The ground clamps must be listed for the materials of the grounding electrode and the grounding electrode conductor. Also, where a grounding electrode conductor is connected to a pipe, rod, or other buried electrodes, the connection means must be listed for direct soil burial or concrete encasement.

The fourth and final requirement, in Section 800.100(D), addresses bonding of grounding electrodes. This rule indicates that if grounding electrodes for communications system installations are separate from the electrode system for the power supply to the building, the two grounding electrode systems must be bonded together to become one electrically and a complete system. The minimum size for the bonding jumper must not be smaller than 6 AWG copper. **See Figure 16-16.**

The grounding and bonding installation requirements in Articles 770 through 840 are similar to those already expressed.

FIGURE 16-14 The grounding electrode conductor connection means is required to be listed and the rod must be driven flush with the Earth to ensure a minimum 8 feet of contact.

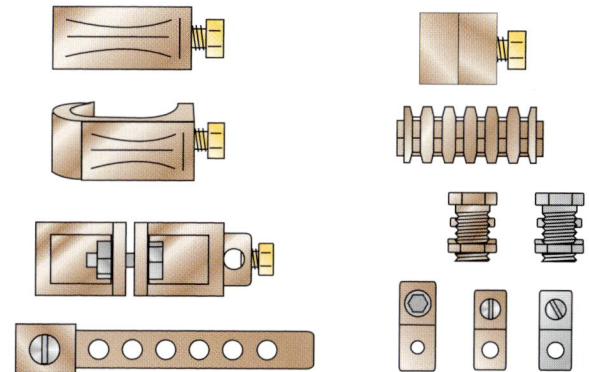

Exothermic welding permitted in addition to listed means

FIGURE 16-15 Connections to grounding electrodes are generally required to be listed.

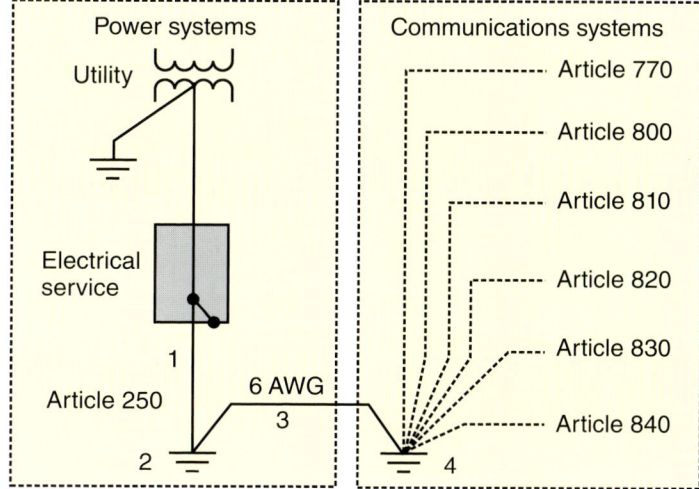

FIGURE 16-16 A separate grounding electrode for a communications system must be bonded to the grounding electrode system for the power supply to the building or structure served.

Article of *NEC*	Minimum Size of Grounding Electrode Conductor	Grounding Electrode Conductor Material	Section of Article	Electrode System Bonding Jumper
770	14	Copper or equivalent	.100(A)(3)	6 AWG copper
800	14	Copper or equivalent	.100(A)(3)	6 AWG copper
810	10	Copper	.21(H)[a]	6 AWG copper
	8	Aluminum		
	17	Bronze		
820	14	Copper or equivalent	.100(A)(3)	6 AWG copper
830	14	Copper or equivalent	.100(A)(3)	6 AWG copper
840	14	Copper or equivalent	.100(A)(3)	6 AWG copper

[a] Section 810.58 requires, at minimum, 10 AWG copper, bronze, or copper-clad steel for a protective grounding electrode conductor and, at minimum, 14 AWG copper or equivalent for the operating grounding electrode conductor.

FIGURE 16-17 Grounding electrode conductor and bonding jumper sizes and material are provided within Article 770 and each Chapter 8 article.

> GEC connections must be made by exothermic welding, listed lugs, listed pressure connectors, listed clamps, or other listed means.

The differences between these specific requirements in each respective article relate to grounding electrode conductor sizes and conductor materials. **Figure 16-17** provides the minimum sizes required for grounding electrode conductors for Articles 770, 800, 810, 820, 830, and 840. This table also provides the permitted conductor material for each conductor, the section of each article in which the sizing requirement is located, and the minimum size required for a bonding jumper between separate grounding electrode systems.

Grounding and Bonding at Mobile Homes

Articles 770, 800, 810, 820, 830, and 840 all include rules for communications system grounding and bonding at mobile homes. These requirements are located in the same section (.106) of each article, following a parallel numbering sequence. It is advisable to refer to each *NEC* article for additional information specific to that particular communications system.

Grounding Requirements

The grounding requirements for communications systems installed in mobile homes are similar to the requirements for permanent structures. Where the service equipment for a mobile home is located within sight of and no more than 30 feet from the mobile home, the grounding electrode conductor from the primary protector must be connected to the grounding electrode for the service equipment. If the service equipment is located elsewhere on the premises and a feeder disconnect is installed within sight of and no more than 30 feet from the mobile home, the grounding electrode conductor from the primary protector must connect to that grounding electrode. Where there is no mobile home service within 30 feet of the mobile home, the primary protector's grounding electrode conductor must be connected to a grounding electrode in accordance with the provisions of the .100(B)(2) section of the relevant article.

If there is no mobile home disconnecting means that is grounded according to Section 250.32 and located within sight of and no more than 30 feet from the mobile home, the primary protector's grounding electrode conductor must be connected to a grounding electrode in accordance with the .100(B)(3) section of the relevant article. These conditions specifically address the need to establish a connection to a grounding electrode from the primary protection at the point of entrance of the communications system to the mobile home structure.

Bonding Requirements

The primary protector's grounding terminal or grounding electrode must be bonded to the mobile home frame or an available grounding terminal on the mobile home frame. The minimum size required is a 12 AWG copper conductor. This bonding conductor connection must be made if there is no mobile home service equipment or disconnecting means or if the mobile home is supplied by a cord-and-plug connection.

Radio and Television Equipment and Antennas

Radio and television equipment is covered by Article 810. Section 810.21 provides the specific grounding and bonding rules for radio and television receiving stations. **See Figure 16-18.** Grounding electrode conductor installations must be in accordance with Sections 810.21(A) through (K). The conductor material must be copper, aluminum, copper-clad steel, or another corrosion-resistant material. Aluminum or copper-clad aluminum conductors must not be terminated within 18 inches of the Earth due to the corrosion influences. The grounding electrode conductor does not have to be an insulated conductor; it can be bare. It must be secured on the surface on which it is run, but insulated support brackets are not required. If proper support cannot be achieved, the size of the grounding electrode conductor must be increased proportionally to allow for less supports by brackets.

FIGURE 16-18 Television equipment antennas must comply with the grounding and bonding rules in Article 810.

If this conductor is subject to physical damage, it has to be protected. Where run in a ferrous metal raceway, the contained grounding electrode conductor must be bonded to both ends of the raceway at the points of conductor entrance and emergence. The grounding electrode conductor must be run in a straight line, as much as practicable, from the mast or discharge unit to the grounding electrode connection. **See Figure 16-19.** The grounding electrode conductor for these systems must connect to an IBT device located as specified in Section 250.94.[9] If there is no IBT device, the conductor must be connected to the nearest of the following:

1. The grounding electrode system for the building or structure electrical service, per Section 250.50
2. The grounded interior metal water piping system within 5 feet of the point where it enters the building
3. The accessible means external to the service equipment, as provided in Section 250.94
4. The nonflexible metallic power service raceway
5. The service equipment enclosure
6. The grounding electrode conductor or metal enclosure for a grounding electrode conductor

The other electrode option for radio and television system receiving stations addresses buildings or structures without an IBT or any grounding electrode system. In this case, a grounding electrode that is present must be used; otherwise, one has to be installed. Any grounding electrode described in Section 250.52 can be installed and used. If the building or structure has no grounding provision as described in Section 810.21(F)(1) or (2), a connection to an effectively grounded metal structure is permitted. The grounding electrode conductor for these systems can be run either inside or outside a building. The minimum size of the grounding electrode conductor is based on the type of conductor material: 10 AWG copper, 8 AWG aluminum, or 17 AWG copper-clad steel or bronze. The same grounding electrode conductor is permitted to serve as the operating ground and provide protective functions. It is always the best choice to connect the grounding electrode conductor of these systems to the power system grounding electrode for the building.

If separate electrodes are used, they must be bonded together with, at minimum, a 6 AWG copper conductor. **See Figure 16-20.** All grounding electrode conductor connections must meet the requirements in Section 250.70 as previously discussed for the other limited-energy systems.

The grounding requirements for amateur and citizens band transmitting and receiving stations are provided in Section 810.58. This section basically indicates that all the requirements in Sections 810.21(A) through (K) are applicable except the minimum size for the protective and operating grounding electrode conductors.

FIGURE 16-19 A grounding electrode conductor for antenna equipment is generally required to be connected to the IBT.

The minimum size required for a protective grounding electrode conductor is 10 AWG copper, bronze, or copper-clad steel. The minimum size required for the normal operating grounding electrode conductor is 14 AWG copper or equivalent.[10] Refer to Section 810.20 for specific requirements pertaining to discharge units that are required for receiving stations.

Overvoltages and Lightning Events

The reasons communications systems must be connected to the building power system grounding electrode are quite simple, yet such connections are not always made correctly. Using the same grounding electrode as the building electrical service keeps the conductive parts of communications and equipment at or close to the same ground (the Earth) potential in normal operation. In abnormal events, such as surges related to lightning strikes to the building or close to the building, the objective is to keep conductive parts of electrical power systems and limited-energy communications systems at the same potential as the potentials rise and fall. This minimizes the possibilities of destructive flashover events within electronic equipment and between electrically conductive parts and equipment within buildings or structures. If the grounding conductors of a communications system are connected to an electrode separate from the building power service grounding electrode, a lightning event on the building or close to the building can cause conductive parts of equipment in the power system and the communications system to rise at different potentials, creating possible flashovers that can damage equipment or even cause a fire.

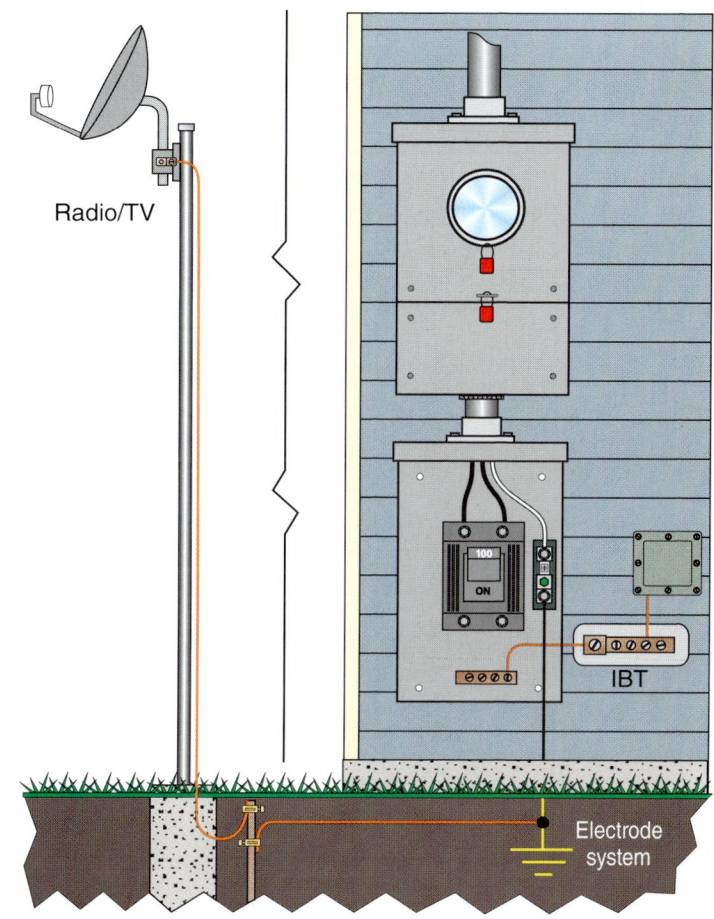

FIGURE 16-20 If a separate grounding electrode is installed, it must be bonded to the electrode for the power system supplying the building or structure.

Summary

There are important grounding and bonding rules for communications systems. The rules are intended to protect persons and property from electrical hazards in normal operation and minimize differences of potential during abnormal events, such as line surges or lightning. Lightning is an unpredictable force, so meeting the *Code* requirements is the minimum plan against damage from these natural and unpredictable events. A lightning protection system in accordance with NFPA 780 provides another degree of protection above the minimum grounding protection required by the *NEC*. Communications system grounding electrode conductors must be electrically common to the grounding electrode used for the electrical power system for system and equipment safety and for personnel safety. Articles 770, 800, 810, 820, 830, and 840 provide rules specific to the grounding and bonding schemes for communications systems. These articles provide the specific minimum sizing requirements for grounding electrode conductors and bonding conductors installed for these systems. The minimum size conductors were covered, along with specific rules that address bonding all grounding electrodes together to become one electrically.

References

1. NFPA 70 National Electrical Code 2011, Article 100 (National Fire Protection Association, Quincy, MA 2010), p. 70–29.
2. NFPA 70 National Electrical Code 2011, Article 100 (National Fire Protection Association, Quincy, MA 2010), p. 70–26.
3. NFPA 70 National Electrical Code 2011, Article 100 (National Fire Protection Association, Quincy, MA 2010), p. 70–29.
4. NFPA 70 National Electrical Code 2011, Article 100 (National Fire Protection Association, Quincy, MA 2010), p. 70–26.
5. NFPA 70 National Electrical Code 2011, Article 100 (National Fire Protection Association, Quincy, MA 2010), p. 70–29.
6. NFPA 70 National Electrical Code 2011, Article 100 (National Fire Protection Association, Quincy, MA 2010), p. 70–29.
7. NFPA 70 National Electrical Code 2011, Article 100 (National Fire Protection Association, Quincy, MA 2010), p. 70–29.
8. NFPA 70 National Electrical Code 2011, Section 250.94 (National Fire Protection Association, Quincy, MA 2010), p. 70–117.
9. NFPA 70 National Electrical Code 2011, Section 810.21 (National Fire Protection Association, Quincy, MA 2010), p. 70–684.
10. NFPA 70 National Electrical Code 2011, Section 810.58 (National Fire Protection Association, Quincy, MA 2010), p. 70–685.

Review Questions

1. Communications systems and circuits in buildings must comply with the applicable rules in Chapter 8 of the *NEC*, in addition to the requirements in *NEC* Chapters 1 through 4 only where referenced from the Chapter 8 articles.
    a. True
    b. False
2. The _____ is a device that provides a means for connecting bonding conductors for communications systems to the grounding electrode system.
    a. Grounding electrode system
    b. Intersystem bonding termination
    c. Grounding electrode conductor
    d. Grounding electrode
3. The purpose of grounding and bonding for limited-energy communications systems and equipment is to provide a level of shock protection and damage from voltage surges created by lightning, line surges, or unintentional contact with higher-voltage lines.
    a. True
    b. False
4. Communications system and equipment grounding is accomplished by all but which of the following?
    a. Connection to ground through an electrode
    b. Connection to a conductive body that extends the ground connection
    c. Connection to an intersystem bonding termination
    d. Connecting to a conductive body that serves in place of the Earth
5. Communications system grounding electrode conductors must be _____ AWG or larger and be made of copper or another corrosion-resistant material. They can be solid or stranded but must be insulated.
    a. 8
    b. 6
    c. 14
    d. 12
6. An intersystem bonding termination device has to provide the means for connection of no less than _____ intersystem bonding conductors.
    a. Two
    b. Three
    c. Four
    d. Five

7. If the intersystem bonding termination is provided through a set of terminals securely mounted to the meter or service equipment enclosure, the terminals have to be _____.
   a. Listed as grounding and bonding equipment
   b. Identified as grounding and bonding equipment
   c. Approved as grounding and bonding equipment
   d. Marked

8. By definition, a(n) _____ is a conducting object through which a direct connection to the Earth is established.
   a. Grounding electrode conductor
   b. Equipment grounding conductor
   c. Intersystem bonding termination
   d. Grounding electrode

9. The *NEC* specifies that grounding electrode conductors for communications systems are required to be connected to the grounding electrode by all but which of the following means?
   a. Sheet metal screws
   b. Exothermic welding
   c. Listed lugs or listed pressure connectors
   d. Listed clamps or other listed means

10. In any installation other than a dwelling, if a separate grounding electrode is installed for a communications system, this grounding electrode has to be bonded to the grounding electrode system for the electrical power supplying the building with, at minimum, a _____ AWG copper conductor.
    a. 10
    b. 8
    c. 6
    d. 4

11. The copper conductor required for a protective grounding electrode conductor installed for a television antenna system and equipment must not be smaller than _____ AWG.
    a. 10
    b. 8
    c. 17
    d. 14

12. Where the service equipment for a mobile home is located within sight of and no more than 30 feet from the mobile home, the grounding electrode conductor from the primary protector is required to be connected to the grounding electrode for the service equipment.
    a. True
    b. False

13. The purpose of primary protectors and discharge units is to provide a level of surge protection and ground discharge of transient overvoltages and unintentional contact with higher-voltage lines.
    a. True
    b. False

14. The length of a communications system grounding electrode conductor has to be kept as short as practicable in dwelling units, and it generally cannot exceed _____ feet in length. By exception, the length can exceed _____ feet if an electrode meeting the criteria in Section 800.100(B)(3)(2) is installed and bonded to the grounding electrode for the power system supplying the building or structure.
    a. 25, 25
    b. 20, 20
    c. 10, 10
    d. 100, 100

15. The grounding electrode conductor for a communications system has to be insulated, covered, or bare and is required to be _____.
    a. Identified
    b. Approved
    c. Listed
    d. Provided with green insulation or marked with green marking tape

16. If a communications system grounding electrode conductor is subject to physical damage, it has to be protected. If it is installed in a ferrous metal raceway, both ends of the raceway have to be bonded to the contained grounding electrode conductor.
    a. True
    b. False

17. Where a bonding jumper or conductor is installed for bonding a separate communications system grounding electrode to the power system grounding electrode, the minimum size required is _____ AWG copper.
    a. 10
    b. 8
    c. 6
    d. 4

18. The primary protector's grounding terminal or grounding electrode has to be bonded to a mobile home frame or an available grounding terminal on the mobile home frame, and the minimum size required for the bonding conductor is _____ AWG copper.
    a. 14
    b. 12
    c. 10
    d. 6

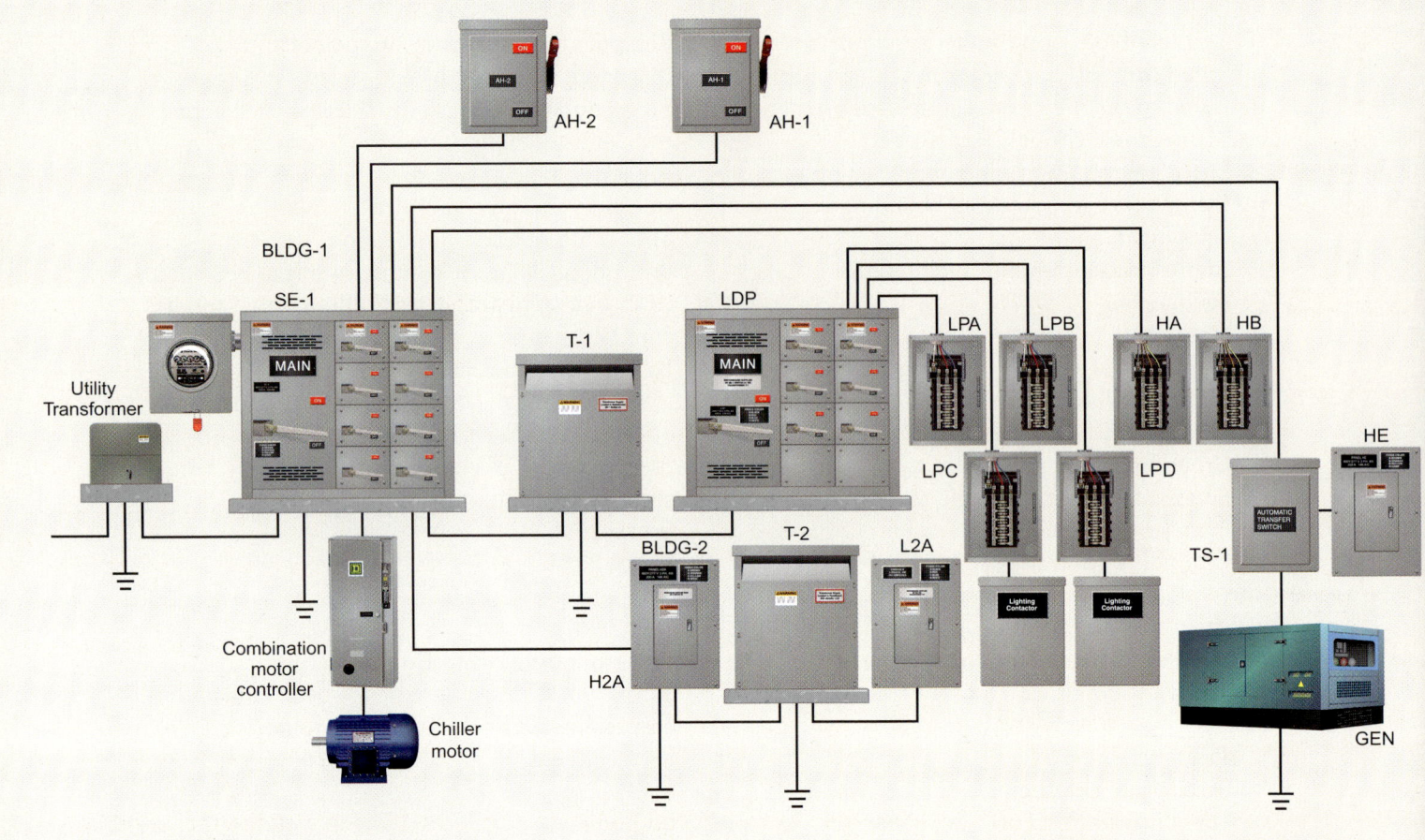

CHAPTER 17

Ground-Fault Circuit Interrupters and Equipment Ground-Fault Protection

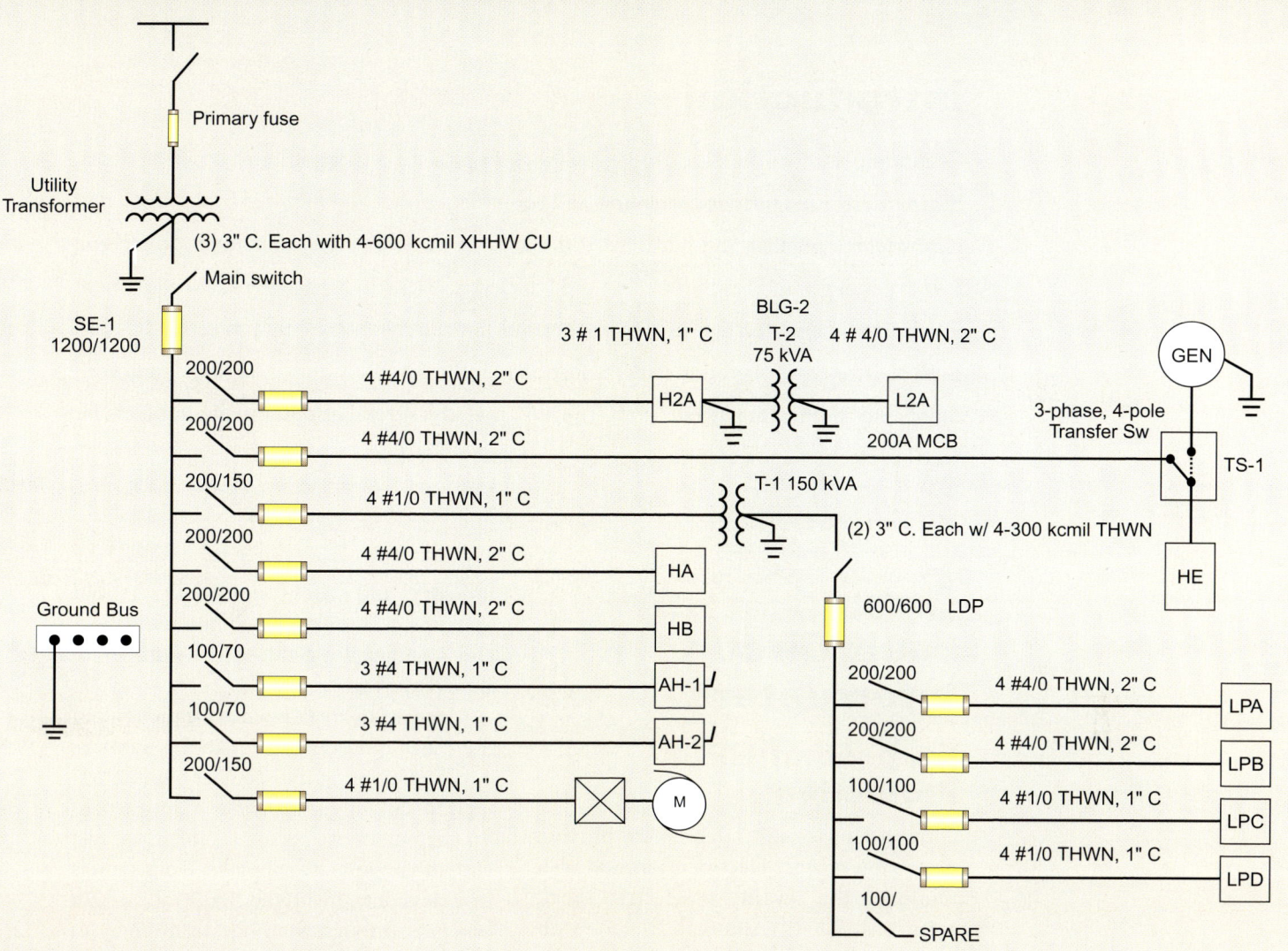

Objectives

- Understand the purpose of ground-fault circuit interrupter (GFCI) protection
- Understand the factors related to the severity of electric shock in humans
- Understand the operating principles of GFCI protection and equipment ground-fault protection (EGFP)
- Determine *NEC* requirements for GFCIs and GFPE

Outline

Ground-Fault Circuit Interrupters

Equipment Ground-Fault Protection

Introduction

Grounding and bonding are proven methods to provide protection in electrical systems, but there are other forms of personnel and equipment protection required by the *NEC®* that have become popular and have raised the level of personnel and equipment protection. Two types of ground fault protection used in electrical distribution systems are equipment grounding fault protection (EGFP) and ground-fault circuit interrupter protection (GFCI). Both of these types of ground fault protection function similarly during a ground fault condition, but at different current levels. The *NEC* includes several important requirements for GFCI protection and equipment grounding fault protection.

Ground-Fault Circuit Interrupters

Purpose of GFCI Protection

Ground-fault circuit interrupters provide protection against electrocution and also minimize the severity of electric shock a person receives. Although there are no actual statistics available, data collected by the Consumer Product Safety Commission (CPSC) indicates that electrocutions in the United States have been decreasing since the introduction of GFCI protective devices, specifically Class A GFCIs. Most industry and safety experts agree that GFCI protection has saved numerous lives and significantly reduced the number of shock injuries.

Insulation breakdown or failure is often the cause of electrical contact by humans. A person coming into contact with an energized circuit can become a series path for current, or the person can be in parallel with other fault current paths. If a person is in series contact with the circuit, all current can pass through the body from entry to exit points. Persons in parallel contact with the circuit will become a current path that is in addition to other current paths over which all the fault current will divide. The severity of shock a person receives is related to the length of time the current is present through the body, the amount of current, the path through the body, the frequency of the current, and the size of the person involved. **See Figure 17-1.** There are many requirements for GFCI protection provided throughout the *NEC*. These devices operate at a low level (4 to 6 mA). The phrase *ground-fault circuit interrupter* is defined in Article 100 of the *NEC*.

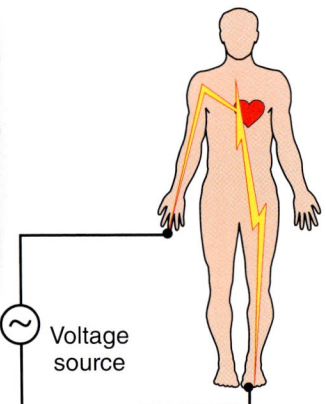

Common shock severity factors

(1) Length of time
(2) Amount of current
(3) Path through the body
(4) Frequency of the circuit
(5) Size of the person

Voltage source

FIGURE 17-1 There are a number of factors that contribute to the severity of electric shock.

Ground-Fault Circuit Interrupter (GFCI). A device intended for the protection of personnel that functions to de-energize a circuit or portion thereof within an established period of time when a current to ground exceeds the values established for a Class A device.[1]

Function of GFCIs

The informational note following this definition indicates that Class A GFCIs trip when the current to ground has a value in the range of 4 to 6 mA. Additional information about GFCIs is located in UL 943, *Standard for Ground-Fault Circuit Interrupters.*[2] See Annex C of this textbook for GFCI information from the UL Guide Information for Electrical Equipment Category KCXS. GFCI protection is available in a few forms. **See Figure 17-2.** There are GFCI circuit breakers, outlet devices, portable cord sets, portable plug-in devices, and so forth. **See Figure 17-3.** All GFCI protection devices generally function in the same way.

The circuit breaker types of GFCI open the entire circuit if a ground fault occurs, while outlet types only open a portion of the circuit. The principles of operation for a GFCI device are simple. The circuit conductors are monitored by the GFCI-sensing device for equal circuit current in both directions (supply and return). If the return current (on the neutral conductor) of the circuit becomes lower that the current present in the ungrounded conductor, there is a fault current path. When this imbalance reaches the level of a 4- to 6-mA difference, the device opens. **See Figure 17-4.**

GFCI devices do not directly limit the level of current through the body during a contact event; they reduce the amount of time the current is present. Extensive research about preventing electrocution of humans resulted in established current and time levels for Class A GFCI devices. GFCI protection monitors current in the complete electrical circuit for an imbalance; therefore, it can be used in grounded systems and ungrounded circuits. An equipment grounding conductor (EGC) is not necessary for a GFCI to operate and provide protection, but the system must be grounded.

Code Requirements for GFCI Protection

There are multiple requirements for GFCI protection throughout the *Code*. The first requirement for GFCI protection in the *NEC* was for swimming pool installations where a Class B GFCI device was required for the circuit supplying underwater luminaires.

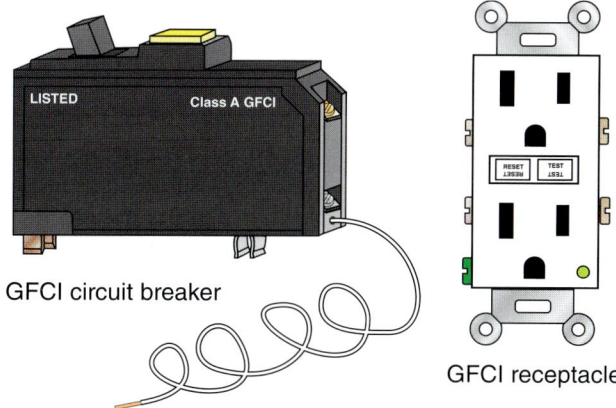

FIGURE 17-2 Two types of hardwired GFCI devices (receptacle and breaker) are available for use in locations where GFCI protection is required by the *NEC*.

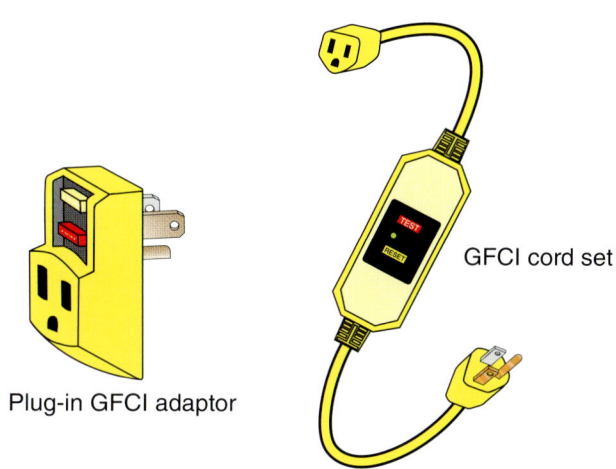

FIGURE 17-3 Portable GFCI devices such as adaptors and cord sets are available for workers to provide protection at the point of use.

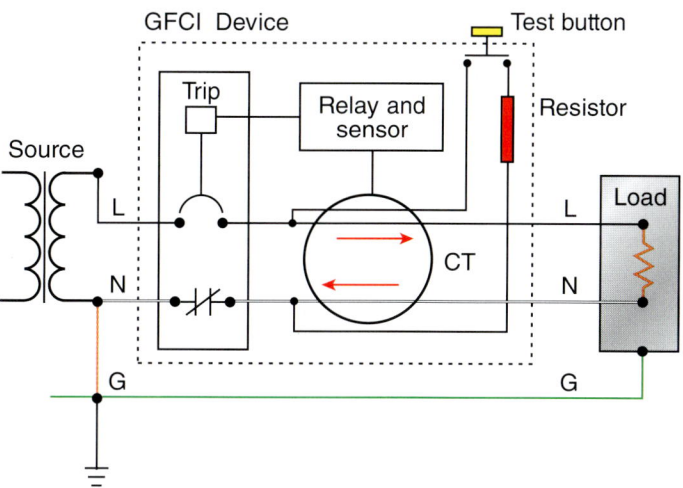

FIGURE 17-4 GFCIs operate by measuring the current in the supply and return current-carrying conductors of the circuit and by monitoring for an imbalance that exceeds 4 to 6 mA.

The requirements for GFCI have increased significantly over the years and resulted in reduced shock and electrocutions in the locations in which they are installed. In some of the areas in various occupancies, GFCI protection is a requirement. Section 210.8(A) requires GFCI protection for 125 volt, 15 or 20 ampere, single-phase receptacles in various locations throughout dwelling units. The GFCI protection must be in a readily accessible location.

Section 210.8(A) lists eight locations where GFCI protection is a requirement for dwelling units. This list does not provide the GFCI protection requirements for special equipment such as pools or hot tubs. The protection can be provided by either a circuit breaker or an outlet-style GFCI device, but each of these types is required to be readily accessible. GFCI protection is required for dwellings at the locations specified in **Figure 17-5**.

Section 210.8(B) provides a list of locations for structures other than dwelling units where GFCI protection is required for all 125-volt, 15- or 20-ampere, single-phase receptacles. **See Figure 17-6.** GFCI protection installed in locations other than dwelling units must also be in a readily accessible location.

There are other requirements for GFCI protection throughout the *Code*, as provided in **Figure 17-7**.

Wet Procedure Locations in Health Care Facilities

Section 517.20 requires special protection from electric shock in wet procedure, patient care areas in the form of two choices:

1. A power distribution system that inherently limits the possible ground-fault current due to a first fault to a low value, without interrupting the power supply
2. A power distribution system in which the power supply is interrupted if the ground-fault current exceeds a value of 6 mA

All dwelling unit bathrooms	Unfinished basements[3]
Garages and accessory buildings[1]	Kitchens (receptacles for countertops)
Outdoor receptacles[2]	Sinks (other for kitchen countertops)
All crawl spaces at or below grade	Boathouses

[1] GFCI protection requirements apply to all garages and to those accessory buildings that have a floor located at or below grade level not intended as habitable rooms and limited to storage areas, work areas, and areas of similar use.
[2] Receptacles that are not readily accessible and are supplied by a dedicated branch circuit for electric snow-melting or deicing equipment and installed according to Section 426.28, meaning the circuit is equipped with EGFP.
[3] A receptacle supplying a permanently installed burglar alarm or fire alarm system is not permitted to have GFCI protection. Sections 760.41 and 760.121 prohibit GFCI and arc-fault circuit-interrupter (AFCI) protection for the power circuits supplying this equipment.

FIGURE 17-5 GFCI protection is required for 15- and 20-ampere 125 volt, single-phase receptacle outlets in multiple locations of dwelling occupancies.

All bathrooms	At (within 6 feet of) sinks[3,4,5]
All kitchens	Indoor wet locations
Rooftops[1]	Locker rooms with associated shower facilities
Outdoor locations[2]	Garages, service bays, similar locations (diagnostic tools, hand tools, portable lighting)

[1] Receptacles installed on rooftops are required to have GFCI protection.
[2] Receptacles that are not readily accessible and are supplied from a dedicated branch circuit for electric snow-melting or deicing equipment are permitted without GFCI protection, by exception.
[3] In industrial establishments only, where the conditions of maintenance and supervision ensure that only qualified personnel are involved, an assured EGC program as specified in Section 590.6(B)(2) is permitted for only those receptacle outlets that are used to supply equipment that would create a greater hazard if power is interrupted or that have a design that is not compatible with GFCI protection, by exception.
[4] In industrial laboratories, receptacles used to supply equipment where removal of power would introduce a greater hazard are permitted to be installed without GFCI protection, by exception.
[5] For receptacles in patient bed locations of critical care areas of health care facilities other than those covered under Section 210.8(B)(1), GFCI is not required, by exception.

FIGURE 17-6 GFCI protection is required for 15- and 20-ampere, 125 volt, single-phase receptacle outlets in multiple locations within occupancies other than dwelling units.

Section 517.21 allows GFCI protection for receptacles to be omitted in critical care areas where the toilet and basin are in the patient room. In many critical care areas, there are continuity of power concerns for life support equipment and other electrical medical equipment necessary. **See Figure 17-8.** In such locations, the *NEC* relaxes the GFCI protection requirement. It is logical that patients in such locations are generally incapacitated to the point where they are restricted to beds. Such locations include intensive care units and operation recovery units of hospitals.

The general requirements, in Section 210.8(B), that all 125 volt, 15 and 20 ampere, single-phase receptacles have GFCI protection such as for bathrooms, kitchens, sinks, and indoor wet locations still are applicable to health care facilities.

GFCI Protection for Receptacle Replacements

Receptacles installed on 15- and 20-ampere circuits are generally required to be of the grounding type provided in Section 406.4(A). Section 406.4(D)(3) requires that replacement receptacles have GFCI protection if such protection is a requirement by the *NEC*. Installers must be aware of the locations where GFCI protection is required when servicing and replacing receptacles. Section 406.4(D) provides the alternatives that can be used for receptacle replacements.

Location	Section
Aircraft hangars	513.12
Audio system equipment	640.10(A)
Boathouses	555.19(B)(1)
Carnivals, circuses, fairs, and similar events	525.23
Commercial garages	511.12
Electric vehicle charging systems	625.22
Electronic equipment, sensitive	647.7(A)
Elevators, escalators, and moving walkways	620.85
Feeders	215.9
Fountains	680.51(A)
Health care facilities	517.20(A), 517.21
High-pressure spray washers	422.49
Hydromassage bathtubs	680.71
Marinas	555.19(B)(1)
Mobile and manufactured homes	550.13(B) and (E), 550.32(E)
Natural and artificially made bodies of water	682.15
Park trailers	552.41(C)
Pools, permanently installed	680.22(A)(1), 680.22(A)(4), 680.22(B)(4), 680.23(A)(3)
Pools, storable	680.32
Sensitive electronic equipment	647.7(A)
Signs with fountains	680.57(B)
Signs, mobile or portable	600.10(C)(2)
Recreational vehicles	551.40(C), 551.41(C)
Recreational vehicle parks	551.71
Replacement receptacles	406.4(D)(3)
Temporary installations	590.6
Vending machines	422.51

FIGURE 17-7 The *NEC* includes multiple requirements for GFCI protection. Reproduction of Commentary Table 210.1 (portions) from the 2011 *National Electrical Code Handbook* by permission from NFPA.[4]

FIGURE 17-8 GFCI protection is not required for receptacles in bathrooms in critical care areas. Sometimes the bathroom is part of the critical care area.

372 APPLIED GROUNDING AND BONDING

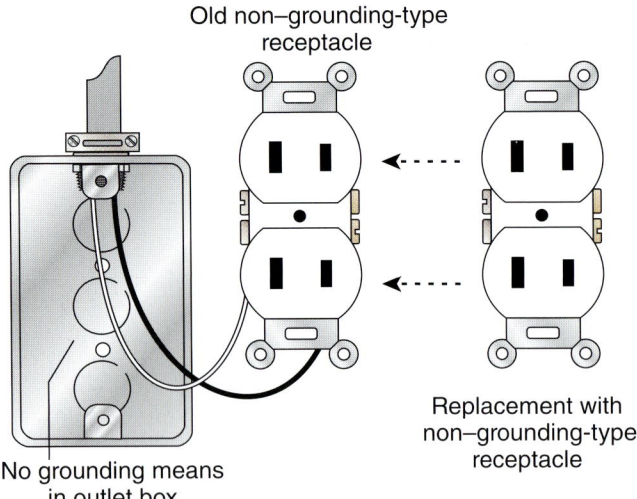

FIGURE 17-9 A non–grounding-type receptacle is permitted to be replaced with another non–grounding-type receptacle.

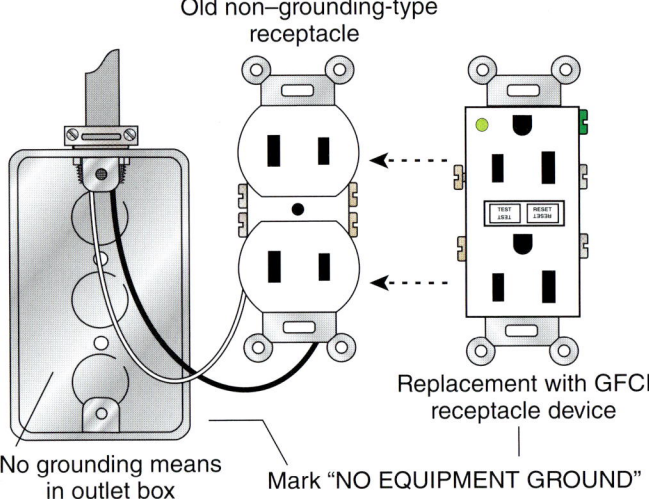

FIGURE 17-10 A non–grounding-type receptacle is permitted to be replaced with a GFCI receptacle (GFCI receptacle equipped with pilot light).

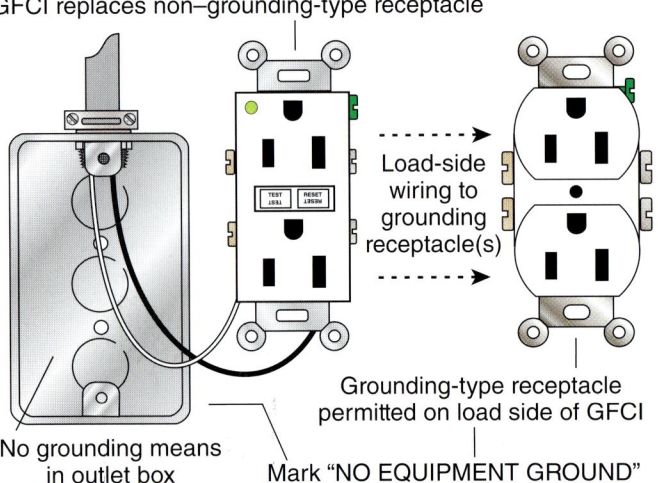

FIGURE 17-11 A non–grounding-type receptacle is permitted to be replaced with a GFCI device that supplies grounding-type receptacles downstream where marked "GFCI Protected and NO EQUIPMENT GROUND" (GFCI receptacle equipped with pilot light).

If a receptacle is being replaced and a grounding means exists in the outlet box, then a grounding-type receptacle must be installed as the replacement. The grounding terminal of the receptacle replacement must be connected to the EGC of the branch circuit. In many older installations, two-wire branch circuits were installed using knob-and-tube wiring, older nonmetallic-sheathed (NM) cable, or legacy alternating current cable wiring systems that did not provide an EGC in accordance with Section 250.118. In these installations without EGCs at the outlets, there are three choices for receptacle replacements as follows:

1. A non–grounding-type receptacle can be replaced with another non–grounding-type receptacle or receptacles. **See Figure 17-9.**
2. A non–grounding-type receptacle can be replaced with a GFCI receptacle device that is marked "No Equipment Ground" and if no EGC is run to any receptacles on the load side of the GFCI receptacle device. **See Figure 17-10.**
3. A non–grounding-type receptacle can be replaced with a grounding-type receptacle only where it is supplied through a GFCI device. Where installed using this alternative, the receptacles supplied through the GFCI device must be marked "GFCI Protected" and "No Equipment Ground." An EGC is not permitted to be installed between the GFCI device and the grounding-type receptacles on the load side of the GFCI device. **See Figure 17-11.**

Temporary Wiring Receptacles

Temporary wiring installations must have GFCI protection if used by personnel during construction, maintenance, demolition, and similar activities, according to Section 590.6. GFP is provided by the use of GFCIs or an assured EGC program, under the specific details provided.

Permanent 125-volt; 15-, 20-, and 30-ampere; single-phase receptacle outlets, existing or new, must have GFCI protection when used with temporary wiring for the activities previously mentioned. GFCI devices listed and identified for portable use, such as cord sets, are acceptable for providing this protection. **See Figure 17-12.**

An exception relaxes this requirement for industrial establishments where conditions of maintenance and supervision ensure that qualified persons service the installation and interruption by a GFCI would present a greater hazard for persons and if an assured EGC program is implemented according to Section 590.6(B)(2).

Temporary 125-volt; 15-, 20-, and 30-ampere; single-phase receptacle outlets must have GFCI protection. **See Figure 17-13.** Other receptacles (different voltages and ampere ratings) must have GFCI protection or the assured EGC program must be used.

All 125- and 125/250-volt; 15-, 20-, and 30-ampere; single-phase receptacles that are part of portable generators, 15 kW or smaller, must have GFCI protection. All 125- and 250-volt, 15- and 20-ampere receptacles, with or without a generator, used in damp or wet locations must have GFCI protection unless the assured equipment grounding program can be used. GFCI devices listed and identified for portable use, such as cord sets, are acceptable for providing this protection, but only for generators manufactured before January 1, 2011.

Assured EGC Program

The *NEC* allows the assured EGC program as an alternative to the requirement for providing GFCI protection, but only for limited situations. The requirements of the assured EGC program are restrictive. These programs must be in writing and continuously enforced at the site by one or more designated persons. Their responsibilities are to ensure that all cord sets and receptacles not part of the permanent building wiring, and equipment connected by cord and plug, are installed and maintained according to the provisions of Sections 250.114, 250.138, 406.4(C), and 590.4(D). The testing requirements for cord sets and receptacles are as follows:

1. Test all EGC continuity of cord sets and cord and plug connected equipment.
2. All receptacles have to be tested to verify EGC continuity and that the EGC is connected to the appropriate terminal of the device.
3. Testing has to be performed before the first use on site, when there is evidence of damage, before returning equipment to service after a repair and at intervals not exceeding 3 months.

The testing has to be documented and available to the applicable authority having jurisdiction.

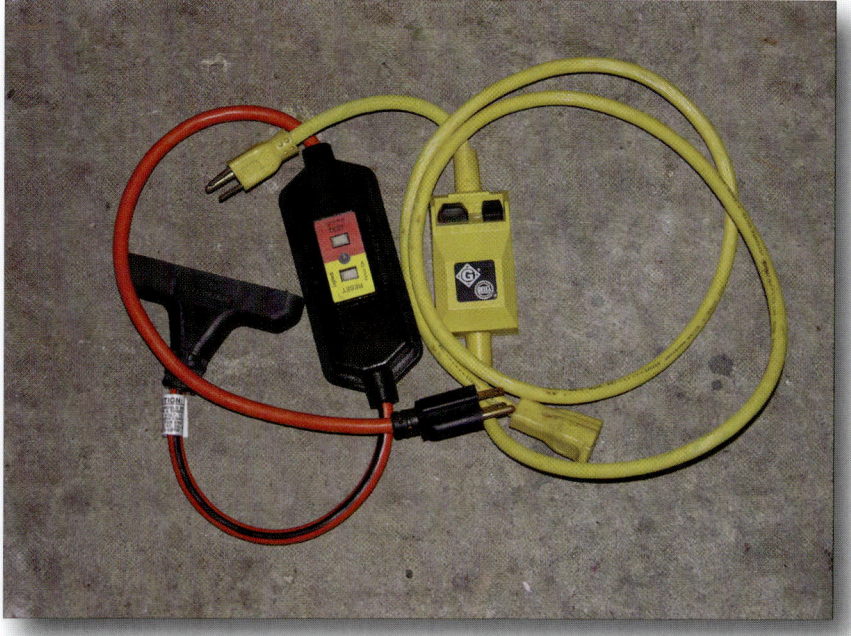

Courtesy of Jim Dollard, IBEW Local 98

FIGURE 17-12 Portable GFCI protection is a common tool that can be carried by persons and used as an added level of shock protection at the point of use.

Courtesy of Bill McGovern, City of Plano, TX

FIGURE 17-13 GFCI protection is required for temporary receptacles installed and used on construction sites in accordance with *NEC* Section 590.6.

374 APPLIED GROUNDING AND BONDING

Courtesy of Bill McGovern, City of Plano, TX

FIGURE 17-14 Equipment can be destroyed or severely damaged as a result of an arcing fault.

> During an arcing ground-fault event, the air becomes ionized, creating a conductive plasma effect.

Courtesy of Bill McGovern, City of Plano, TX

FIGURE 17-15 An arcing ground fault event can quickly develop into a phase-to-phase short circuit and cause rapid destruction of electrical equipment that is not equipped with ground fault protection.

Equipment Ground-Fault Protection

Purpose of GFPE

Ground-fault protection of equipment (GFPE) protects large equipment from devastating arcing events and destructive burn-downs. **See Figure 17-14.** GFPE requirements are provided in Sections 215.10, 230.95, 240.13, and 517.17. **See Figure 17-15.** The terms *ground fault protection of equipment* (GFPE) and *equipment ground fault protection* (EGFP) are used interchangeably in this text, but the *NEC* defines the term *Ground-fault Protection of Equipment*.

Electric arcs generate significant amounts of heat. At 277 volts to ground, an arcing fault is readily sustained. A ground fault is typically not a solid or "bolted fault" condition so dynamic arcing impedance is introduced in the circuit reducing the fault current seen by a standard overcurrent device and increasing the time the fault can exist. This allows arcing faults to manifest into destructive events. During an arc event, ionized gas is dispersed creating a conductive gas or plasma in the atmosphere surrounding the busbars within the equipment. This condition often rapidly escalates from a phase-to-ground fault event to a phase-to-phase short circuit condition. This is why the *NEC* requires ground fault protection of equipment. GFPE is generally required for solidly grounded wye services and feeders of more than 150 volts to ground but not exceeding 600 volt phase-to-phase for each disconnect rated at or above 1000 amperes. GFPE is required for nominal 480Y/277 volt, three-phase, four-wire, wye-connected systems. The maximum settings are 1200 amperes and not longer than 1 second for fault currents 3000 amperes or more. **See Figure 17-16.** GFPE is not permitted for fire pumps or in systems where a non-orderly shutdown or interruption would introduce additional hazards.

Types of GFP Equipment

Two types of GFP equipment are *ground-strap type* GFP equipment, and *zero-sequence* (residual) GFP equipment. The main bonding jumper is implied by the term *ground strap* or *neutral ground strap*. Both types provide protection from load-side ground faults. Equipment ground fault protection installed in a service does not provide protection on the line side of the GFP equipment. A line side ground fault is not detected by the GFP sensors and equipment can be severely damaged or destroyed by the ground fault event.

Ground-Fault Protection of Equipment. A system intended to provide protection of equipment from damaging line-to-ground fault currents by operating to cause a disconnecting means to open all ungrounded conductors of the faulted circuit. This protection is provided at current levels less than those required to protect conductors from damage through the operation of a supply circuit overcurrent device.[5]

Neutral Ground-Strap System

Equipment ground fault protection systems typically include a current transformer (sensor), a control relay, and a shunt-trip breaker of fused switch. In this type of system the main bonding jumper passes through a sensing window (CT). **See Figure 17-17.**

This type of protection is permitted for use in service equipment and typically

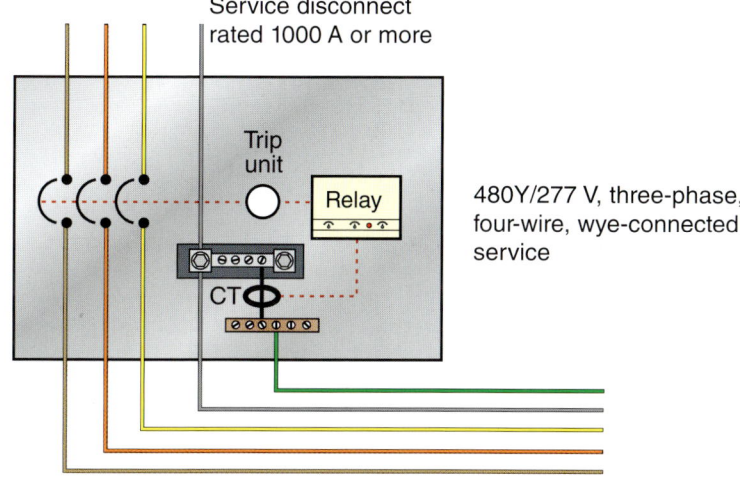

FIGURE 17-16 EGFP is required for service disconnects rated 1000 amperes or more that have an applied nominal voltage of more than 150 volts to ground not exceeding 600 volts phase-to-phase.

FIGURE 17-17 EGFP systems can be in the form of a circuit breaker that includes internal EGFP sensing and relaying devices.

would not be installed on the load side of the service due to the restrictions of load-side neutral-to-ground connections as provided in Section 250.24(A)(5). The ground-strap type system functions by sensing a high amount of fault current passing over the main bonding jumper during a ground fault. At predetermined values of current and time, the ground fault protection system sends a signal through a relay that activates the shunt trip mechanism to rapidly open the disconnect or breaker. The pickup ampere setting cannot exceed 1200 amperes and the maximum time permitted is not to exceed 1 second. This really is a significant amount of time (60 cycles). Reviewing the diagram of this type of GFP, one can easily see the complete circuit from the utility transformer to the service equipment. Follow the neutral from the utility transformer to the service disconnect and locate the main bonding jumper. **See Figure 17-18.** This is the effective ground-fault current path from the service to the utility source. Any ground fault current will be present in the neutral conductor for the duration of time it takes the GFP equipment to open the circuit. All of the fault current will pass

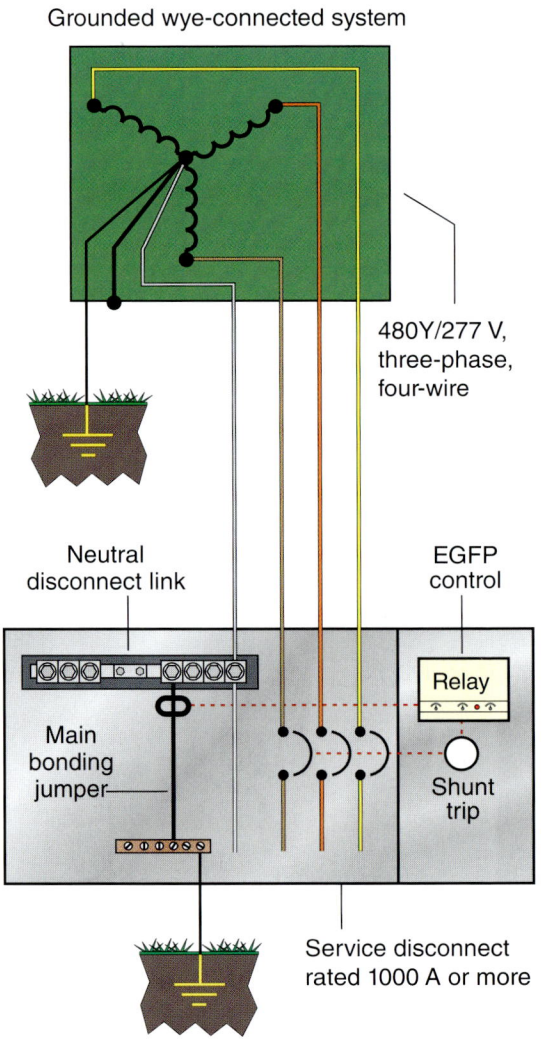

FIGURE 17-18 A neutral ground-strap EGFP system includes a current transformer (CT) around the main bonding jumper.

through the current transformer around the main bonding jumper in the service equipment. As required, grounding electrodes are installed and connected at the utility transformer and the service equipment. Even though the Earth is in this circuit, only a minimum amount of current will be present in this path because of the high impedance in the Earth. During a ground fault event, current will divide over any and all paths that are present and common to the power source. The amount of current in each path is limited by the impedance in that particular path. The inductive reactance and high-impedance of those other fault current paths results in most of the fault current being carried through the main bonding jumper at amounts far greater than the current on other paths that are separate from the service-entrance conductors.

Zero-Sequence Transformer-Type System

A common type of equipment ground fault protection system is the zero-sequence type, sometimes referred to as a residual GFP system. The term *residual* is used because this type of equipment sums up the all current in the phases and grounded conductor and any excess current is residual or left over.

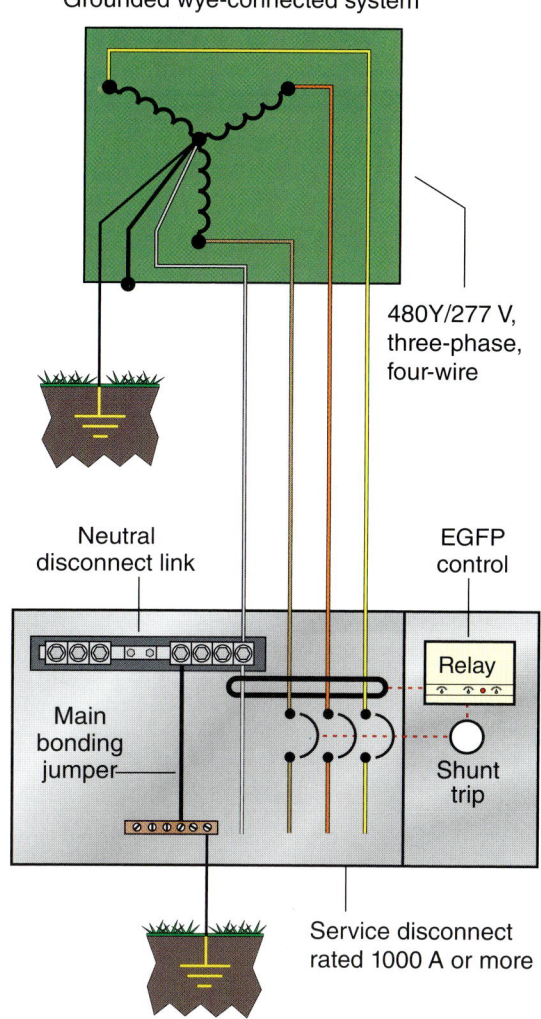

FIGURE 17-19 EGFP can be provided using a basic zero-sequence system.

FIGURE 17-20 All ungrounded phase conductors and the neutral conductor pass through the zero-sequence-sensing current transformer.

378 APPLIED GROUNDING AND BONDING

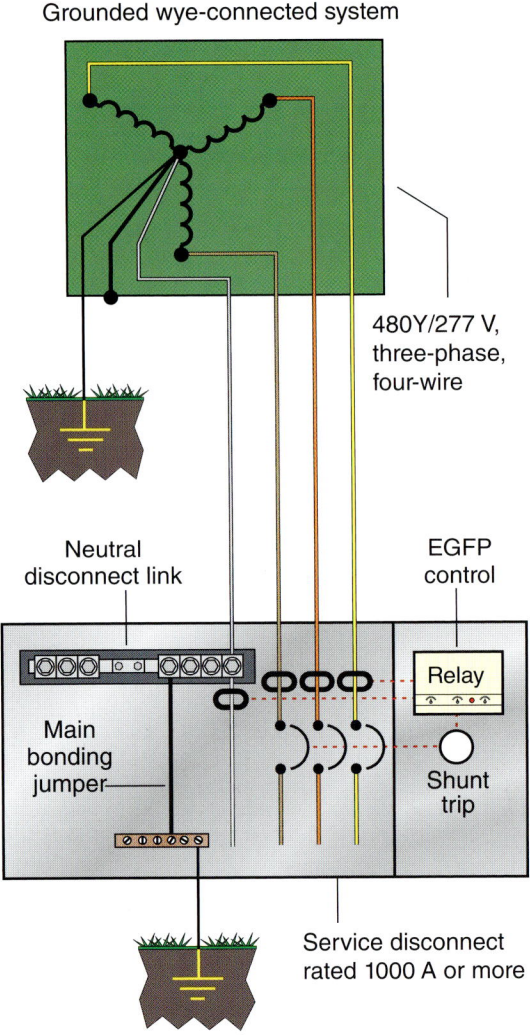

FIGURE 17-21 EGFP can be provided using a basic zero-sequence system with optional neutral sensing window.

This type of system is equipped with a control relay, a sensing current transformer(s), and a shunt-trip breaker or disconnecting means. **See Figure 17-19.** In this type, all ungrounded phase conductors and the neutral conductor are routed through a single current transformer. **See Figure 17-20.** The main bonding jumper and equipment grounding conductor do not pass through the CT. **See Figure 17-21.** Zero sequence types are also available with a separately-installed CT around the neutral bus. The difference between the two types is that in the latter of the two mentioned, the neutral CT is optional. **See Figure 17-22.** This type of GFP equipment can be used on three-phase systems or three-phase, four-wire, wye-connected systems. Many zero-sequence GFP systems have the CTs built into the breaker and the external CT is installed in the equipment around the neutral bus. **See Figure 17-23.** In the zero-sequence system, the vector sum of normal current in the system ungrounded conductors and neutral conductor is at around zero due to the cancelling effect between the conductors of a three-phase, 4-wire, wye-connected system. When a ground fault event occurs, the fault current path is through the main bonding jumper and other fault current paths outside of the CT. This imbalance activates the GFP relay and shunt-trip mechanism to open

Courtesy of IBEW Local 26 Training Center

FIGURE 17-22 EGFP is commonly provided using a zero-sequence GFPE (breaker type).

the breaker or switch. The output of the current transformer is proportional to the level of fault current passing through and the ground-fault relay is field-adjustable to a maximum of 1200 amperes. The time-delay settings of GFP equipment are usually at levels less than 1 second (typically from 2 to 35 cycles). Keep in mind the maximum ampere setting permitted in the *NEC* is 1200 amperes and the maximum time is 1 second.

Selective Coordination

Coordinating the trip sequence of overcurrent and GFP devices in power distributions systems is necessary where selective coordination is required by the *NEC*. Selective coordination can be accomplished by various combinations of overcurrent and GFP devices that are carefully applied and engineered into the power distribution system. Selective coordination can be accomplished using circuit breakers, fuses, or combinations of fuses and circuit breakers. The last of these three methods mentioned is probably the most common in today's electrical systems.

Where GFP systems are installed, they should also be designed such that they are selective to the point where the offending ground-fault event opens only the closest upstream device from the fault event. A definition of the term *selective coordination* is included in Article 100 of the *NEC* and is applicable for selective coordination of overcurrent devices for emergency systems specified in Section 700.27 and legally required standby systems in Section 701.27.

> **Coordination (Selective).** Localization of an overcurrent condition to restrict outages to the circuit or equipment affected, accomplished by the choice of overcurrent protective devices and their ratings or settings.[6]

Where overcurrent and GFP devices are selectively coordinated, they provide the benefits of restricting outages to the circuit or equipment closest to the ground-fault or short circuit event by operating the local overcurrent or GFP device, rather than causing the entire system to suffer a failure. **See Figure 17-24.**

Courtesy of IBEW Local 26 Training Center

FIGURE 17-23 EGFP can include a zero-sequence current transformer for circuit breaker-type GFP.

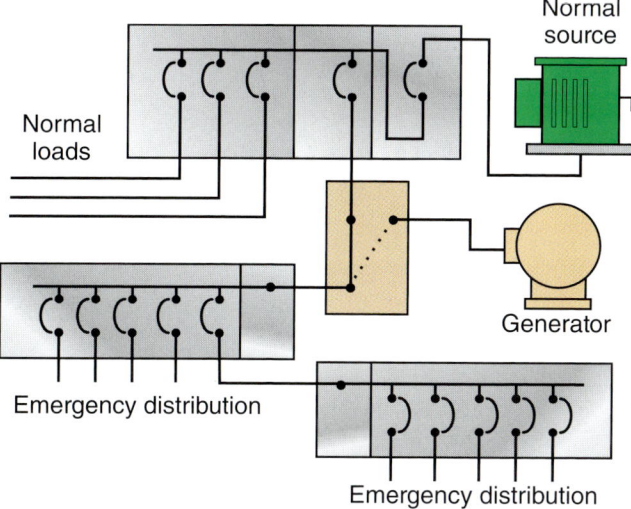

FIGURE 17-24 Selective coordination is required for overcurrent devices used in emergency systems in accordance with *NEC* Section 700.27.

The requirements for selective coordination in Section 700.27 apply to all emergency systems. Article 517 includes specific rules for emergency systems in health care facilities. It is important to remember that the emergency system in a hospital includes the life safety branch and the critical branch. Both of these branches are required to be selectively coordinated to minimize power failures to only those local areas directly affected by the faulted condition. Section 517.26 establishes the essential *Code* correlation between Article 517 and Article 700 that requires application of the selective coordination rules in Section 700.27 to both the life safety branch and the critical branch of health care facilities. Section 517.17 requires that all feeders with GFP be equipped with a second level of GFP installed in "subfeeder" disconnecting means; it also requires this additional protection be selectively coordinated. This is logical and consistent with the objectives of NFPA 99, *Standard for Health Care Facilities,* that are concerned with continuity of electrical service in health care facilities where critical care and life support are essential for patients.

GFP System Coordination

System coordination is often desired and necessary to avoid power outages and interruption of service. Coordination of power in distribution systems can be easily accomplished using selectively coordinated standard overcurrent devices or equipment ground fault protection. The point of a ground fault in any system is never known, so system coordination is all about anticipating a ground fault event and strategically locating GFP equipment at each level desired in the system. Selective coordination is defined in the *NEC* and applying GFP equipment in multiple levels of feeders on the load side of the service GFP device can effectively localize ground faults and simultaneously provide power continuity. A coordinated system of ground fault protection in cascading feeder levels affords the ability to isolate an offending fault to one location or feeder level. The coordination often incorporates zone selective interlocks or differential relay settings. Zone selective interlocking involves installing signaling circuit wiring between each level of GFP equipment. **See Figure 17-25.** GFP equipment operates well below the normal ratings of the overcurrent protection feature in the same device. GFP systems are affected only by ground-fault currents; they do not interfere with normal operation of the overcurrent device. Where necessary or desired, the tripping function of the service GFP device can be delayed while the feeder GFP closest to the offending fault can be activated. A magnetic current relay initiates the required time delay relay providing effective system coordination continuity of service. GFP equipment coordination involves analyzing the system characteristics and applying appropriately selected GFP equipment in a way that effectively provides a deliberate separation of the tripping bands from the lowest rated device to the highest rated GFP system device. This type of system provides excellent protection of property for large low voltage electrical distribution equipment.

Applicability in Health Care Facilities

Special rules for GFP of electrical systems apply to health care facilities. Section 517.17(B) indicates when GFP is provided at the service or feeder disconnecting means, as specified by Section 230.95 or 215.10, an additional level of GFP must be installed in the next level of feeder disconnecting means downstream toward the load.

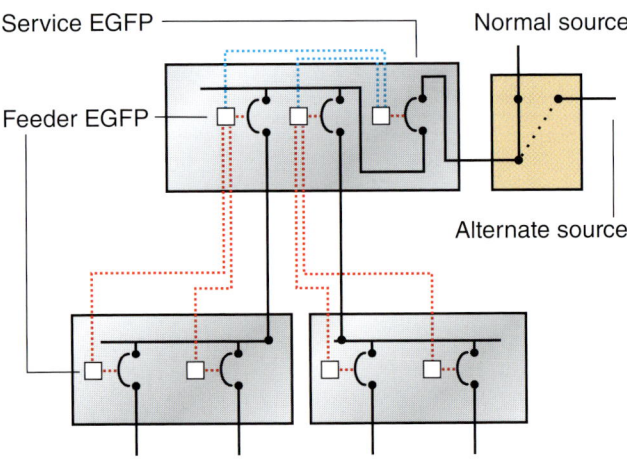

FIGURE 17-25 System coordination can be accomplished using zone-interlocking of the EGFP devices at multiple levels.

See Figure 17-26. The primary purpose is so that a ground-fault event in the electrical system does not open the service EGFP disconnecting means but opens the device closest to the fault, thus isolating the offending circuit while maintaining continuity of power to the rest of the facility.

Section 517.17 clarifies that, regardless of whether or not the health care facility is supplied by a service or feeder, where EGFP is provided at the service or feeder disconnecting means, a second level is required downstream.

According to Section 517.17, additional levels of GFP are not permitted to be installed on the load side of an essential electrical system transfer switch.

This feeder protection is required to be 100% selective so that if a ground fault occurs downstream from the feeder overcurrent device, only the closest overcurrent device will open and the upstream devices will remain closed. This requirement is in place to prevent the blackout of a facility caused when a main opens for a fault that should be isolated to a single feeder, for example.

Each level of the GFP system must be performance tested when first installed. It is critical to GFP system functionality that the grounded (neutral) conductor be completely isolated from any grounding connections downstream from the service. This is a general requirement of the *NEC* in Section 250.24(A)(5) and is a specific requirement of the manufacturer's installation and testing instructions for the GFP equipment.

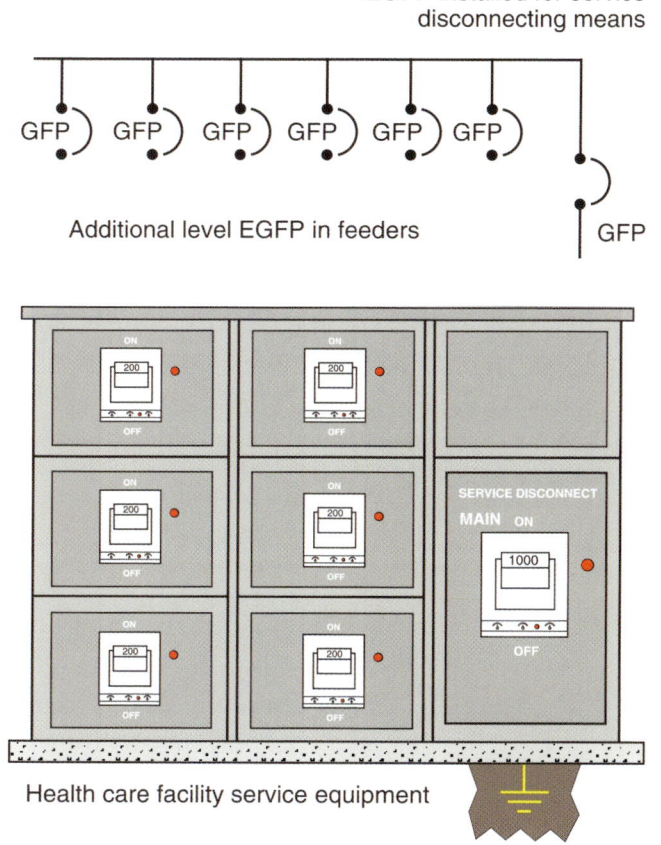

FIGURE 17-26 An additional level of EGFP is required in health care facilities according to Section 517.17.

This is especially important where transfer switches and alternate power sources are installed in the premises wiring system. In most cases, four-pole transfer switches are installed on three-phase, four-wire systems to allow the system to function properly without grounding connections to the neutral or grounded conductor downstream from the service disconnecting means that includes GFP equipment for that particular separately derived system.

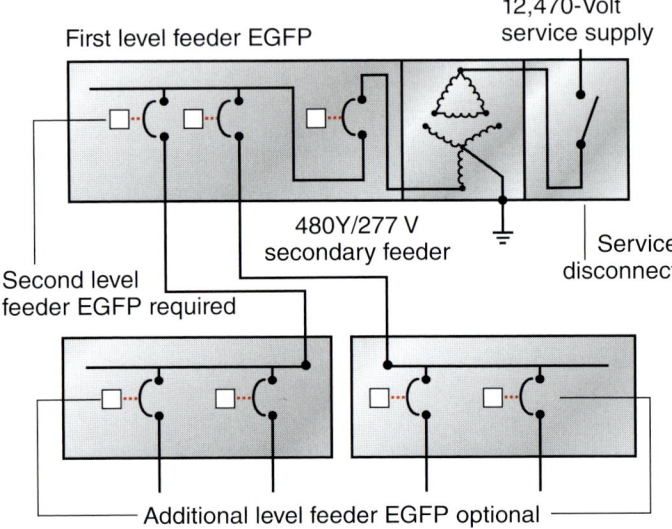

FIGURE 17-27 EGFP is required for feeders in accordance with Section 215.10.

Courtesy of IBEW Local 26 Training Center

FIGURE 17-28 A neutral disconnect means (link) is required by Section 230.75 and is typically used for testing associated with EGFP.

Feeder GFP

Section 215.10 requires equipment ground fault protection for feeders that are of the same voltage and current ratings that qualify services for GFP equipment. An example is where the service to a property is provided at 12,470 volts. **See Figure 17-27.** In this case the electrical system is delivered at more than 600 volts and has the service disconnecting means at that voltage level. A transformer steps the voltage down to 480Y/277 creating the need for equipment ground-fault protection in the feeder position. The exceptions that are provided in Section 230.95 for services are also provided in Section 215.10 for feeders. Each separately derived system with feeders at voltages and configurations requiring GFPE must be protected if it qualifies because of the ampacity level of the feeder and equipment supplied. Remember that equipment ground fault protection is generally required for each disconnecting means rated at 1000 amperes or greater installed on 480Y/277-volt systems.

Testing of GFPE Systems

Equipment ground fault protection systems must be performance tested when first installed to ensure proper operation. Testing GFP equipment verifies the system will interrupt a ground fault event at selected current pick-up and time settings. The performance testing must be in accordance with the manufacturer's instructions. Section 230.75 provides the requirement for a means to disconnect the neutral within the service equipment. This is for testing purposes. Although the term *neutral disconnect link* is not used in the *NEC*, this term is used by manufacturers of equipment that is suitable for use as service equipment. Once the link is removed, a test can verify that the neural is isolated from grounding connections on the load side of the service disconnect. **See Figure 17-28.** The test records must be made available to the authority having jurisdiction.

Summary

Two types of ground fault protection addressed in the *NEC* are ground-fault circuit interrupter protection and equipment ground fault protection. Common traits of these two types of GFP are that they require a ground fault to operate, they interrupt current when they operate, and they open the faulted circuit within a prescribed time frame. GFCIs protect people from serious shock and electrocution, while EGFP protects electrical equipment from arcing burn-downs and destruction. The principles of operation of GFCI and EGFP are similar except for a difference in current and time pickup levels. There are several *NEC* rules that require GFCI protection for persons and requirements for installing equipment grounding fault protection (EGFP). Performance testing requirements are required for EGFP that is first installed on site.

References

1. NFPA 70 National Electrical Code 2011, Article 100 (National Fire Protection Association, Quincy, MA 2010), p. 70–29.
2. NFPA 70 National Electrical Code 2011, Article 100 (National Fire Protection Association, Quincy, MA 2010), p. 70–29.
3. NFPA 70 National Electrical Code 2011, Section 210.8(B)(5) Exceptions (National Fire Protection Association, Quincy, MA 2010), p. 70–51.
4. NFPA 70 National Electrical Code HB 2011, Table 210.1 (National Fire Protection Association, Quincy, MA 2010), p. 87.
5. NFPA 70 National Electrical Code 2011, Article 100 (National Fire Protection Association, Quincy, MA 2010), p. 70–27.
6. NFPA 70 National Electrical Code 2011, Article 100 (National Fire Protection Association, Quincy, MA 2010), p. 70–27.

Review Questions

1. GFCIs and EGFP provide GFP. What are the common operating characteristics for both of these types of GFP?
   a. They require a ground fault to operate.
   b. They interrupt current when they operate.
   c. They open the faulted circuit within a prescribed time frame.
   d. All of the above.

2. A GFCI operates within which of the following current ranges?
   a. 1 to 5 A
   b. 15 to 20 A
   c. 4 to 6 mA
   d. 1000 to 1200 A

3. Ground-fault circuit-interrupters for dwelling units and structures other than dwelling units are required to be readily accessible.
   a. True
   b. False

4. GFCI provides protection for people from electrocution and minimizes shock hazards.
   a. True
   b. False

5. The _____ is a device intended for the protection of personnel that functions to de-energize a circuit or portion thereof within an established period when a current to ground exceeds the values established for a Class A device.
   a. Equipment ground-fault protection
   b. Arc-fault circuit interrupter protective device
   c. Overload protective device
   d. GFCI

6. GFCI devices do not limit the magnitude of current; they limit the time the current is present.
   a. True
   b. False

7. Which of the following locations in dwelling units does not require GFCI protection for 125-volt 15- and 20-ampere receptacles?
   a. Bathrooms
   b. Kitchen countertops
   c. Living areas
   d. Garages

8. Which of the following locations in structures other than dwelling units does not require GFCI protection for 125-volt, 15- and 20-ampere receptacles?
   a. Outdoor locations
   b. Indoor locations in general use areas
   c. Bathrooms
   d. Within 6 feet of a sink

9. In industrial establishments only, where the conditions of maintenance and supervision ensure that only qualified personnel are involved, an assured EGC program as specified in Section 590.6(B)(2) is not permitted for receptacle outlets used to supply equipment that would create a greater hazard if power is interrupted or that have a design that is not compatible with GFCI protection, by exception.
   a. True
   b. False

10. In wet procedure locations of health care facilities where interruption by a GFCI cannot be tolerated, GFCI protection is not permitted to be installed. In these instances, which type of protective equipment is required?
    a. Equipment ground-fault protection (EGFP)
    b. Ground detectors only
    c. Overload protection
    d. Isolated power systems in accordance with Section 517.160

11. In older installations without EGCs at the outlets, which of the following is acceptable?
    a. A non–grounding-type receptacle can be replaced with another non–grounding-type receptacle or receptacles.
    b. A non–grounding-type receptacle can be replaced with a GFCI receptacle device as long as it is marked "No Equipment Ground" and no EGC is run to any receptacles on the load side of the GFCI receptacle device.
    c. A non–grounding-type receptacle can be replaced with a grounding-type receptacle only where it is supplied through a GFCI device. Where installed using this alternative, the receptacles supplied through the GFCI device have to be marked "GFCI Protected" and "No Equipment Ground." An EGC is not permitted to be installed between the GFCI device and the grounding-type receptacles on the load side of the GFCI device.
    d. Any of the above is acceptable.

12. On construction sites, GFCI protection is required for 125 volt, single-phase receptacles that are not part of the permanent wiring of a building or structure. This GFCI protection is not applicable to _____ A receptacle outlets.
    a. 15
    b. 20
    c. 30
    d. 50

13. EGFP is required to protect large equipment from arcing burn-downs and destruction caused by phase-to-ground faults.
    a. True
    b. False
14. Equipment ground-fault protection is generally required for solidly grounded wye electrical services and feeders of more than 150 volts to ground but not exceeding 600 volts phase-to-phase for each service disconnect rated _____ amperes or more.
    a. 400
    b. 800
    c. 1000
    d. 1200
15. Equipment ground-fault protection is required for fire pumps or continuous industrial processes where a nonorderly shutdown would introduce additional or increased hazards, as indicated in the exceptions in the *NEC* to Section 230.95.
    a. True
    b. False
16. The maximum current setting for service disconnect equipment ground-fault protection is _____ amperes, and the maximum time delay for operation is _____ second(s) for ground-fault currents equal to or greater than 3000 amperes.
    a. 1000, 5
    b. 3000, 1
    c. 600, 2
    d. 1200, 1
17. _____ is a system intended to provide protection of equipment from damaging line-to-ground fault currents by operating to cause a disconnecting means to open all ungrounded conductors of the faulted circuit. This protection is provided at current levels less than those required to protect conductors from damage through the operation of a supply circuit overcurrent device.
    a. Equipment ground-fault protection
    b. GFCI protection
    c. Arc-fault circuit interrupter protection
    d. Short circuit protection
18. The neutral ground-strap type of EGFP consists of a current transformer, control relay, and usually, a shunt trip circuit breaker. In this type of EGFP system, the _____ passes through the current transformer for sensing ground-fault current.
    a. Grounding electrode conductor
    b. Main bonding jumper or system bonding jumper
    c. Grounded conductor
    d. EGC
19. A zero-sequence EGFP system typically consists of a control relay, a shunt trip circuit breaker, and a current transformer that is placed around all of the circuit conductors, including the grounded (neutral) conductor.
    a. True
    b. False
20. _____ is the localization of an overcurrent condition to restrict outages to the circuit or equipment affected, accomplished by the choice of overcurrent protective devices and their ratings or settings.
    a. Short circuit protection
    b. Overload protection
    c. Selective coordination
    d. GFP
21. Which of the following systems does not require selective coordination of the overcurrent devices?
    a. Emergency systems
    b. Legally required standby systems
    c. Life safety systems in hospitals
    d. Optional standby systems
22. Section 517.17(B) indicates when EGFP is provided at the service or feeder disconnecting means, as specified by Section 230.95 or 215.10, an additional level of GFP is not required to be installed in the next level of feeder disconnecting means downstream toward the load.
    a. True
    b. False
23. The additional level of equipment ground-fault protection required in Section 517.17(B) is not permitted to be installed in which of the following locations or applications?
    a. On the load side of an essential electrical system transfer switch
    b. Between the on-site generating unit or units described in Section 517.35(B) and the essential system transfer switch or switches
    c. On electrical systems that are not solidly grounded wye systems with greater than 150 volts to ground but not exceeding 600 volts phase-to-phase
    d. All of the above
24. Sections 230.95(C) and 517.17(D) of the *NEC* require EGFP systems to be performance-tested when first installed on site to ensure proper operation.
    a. True
    b. False

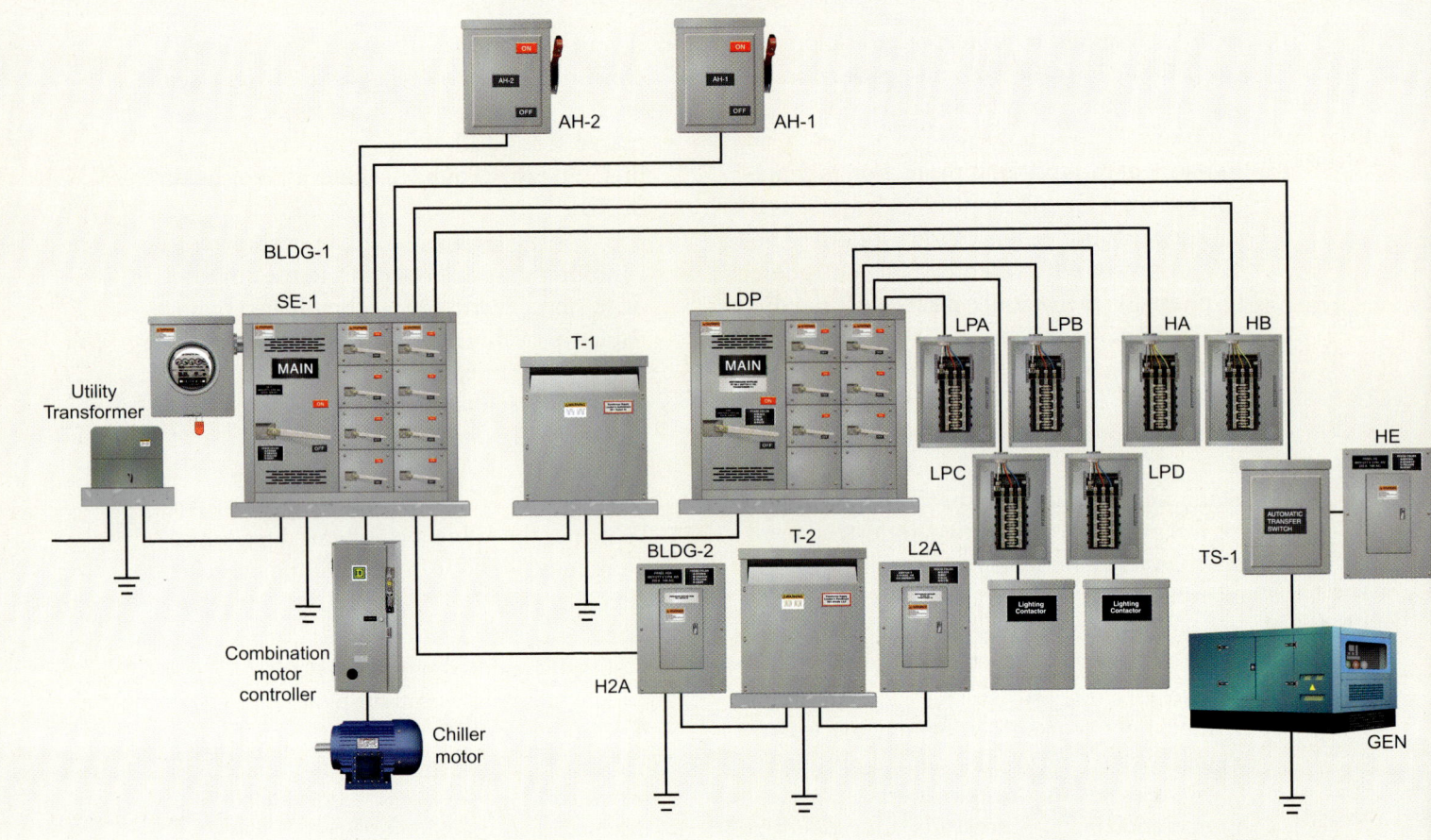

CHAPTER 18

Grounding Rules for Medium- and High-Voltage Systems

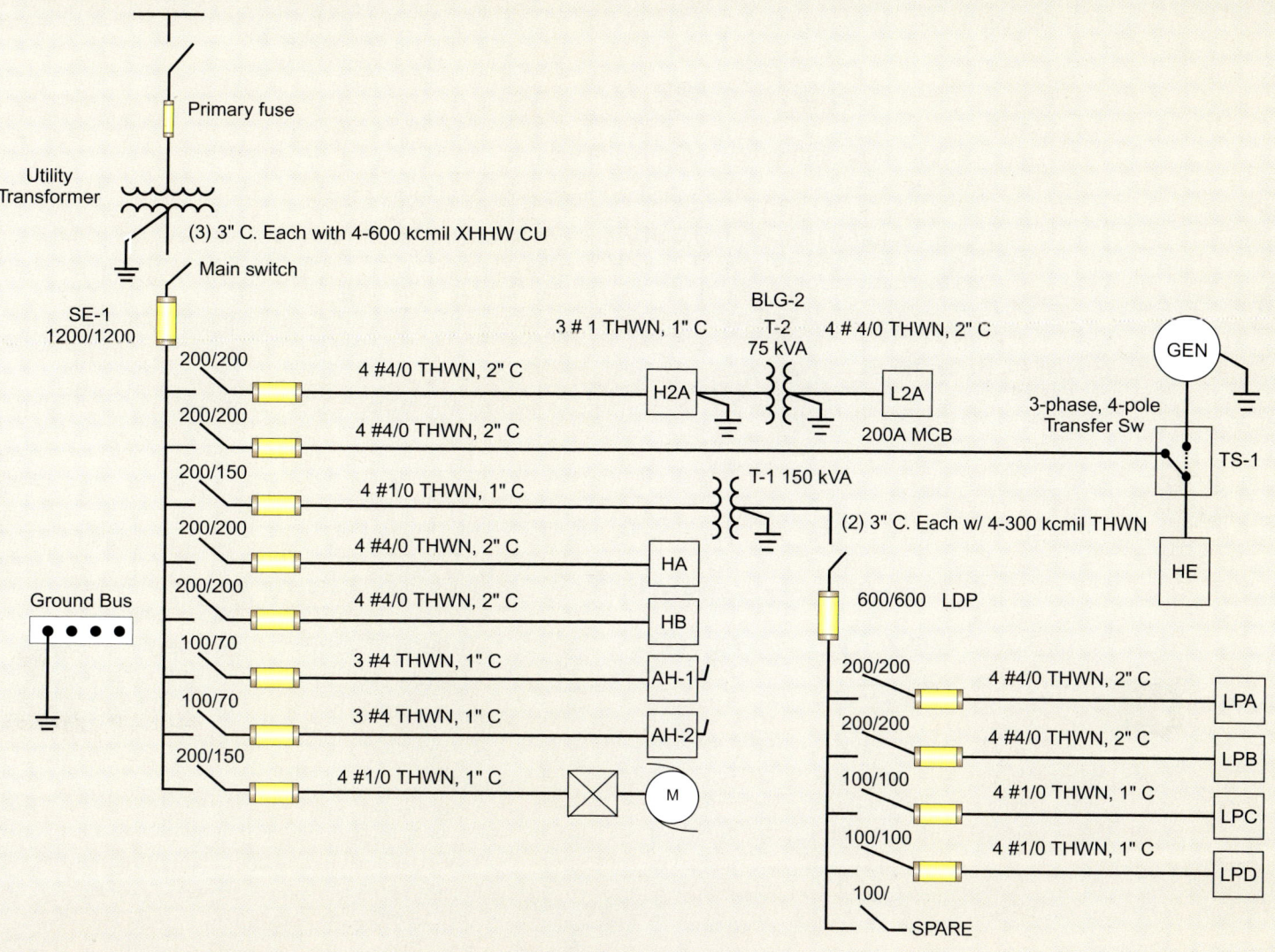

Objectives

- Understand the grounding methods used for medium- and high-voltage systems
- Determine the requirements for alternating current substation grounding electrode systems
- Identify the grounding rules in the *NEC* that apply to medium- and high-voltage systems
- Understand that the purpose of grounding cable shielding is to confine the voltage stresses
- Understand the purpose of grounding through surge arresters

Outline

Requirements for Grounding Systems

Grounding Methods for Systems Over 1 kV

Solidly Grounded Systems

Impedance Grounding

Portable or Mobile Equipment Grounding

Grounding Equipment

Substation Grounding Requirements

Conductor Shielding and Stress Reduction

Grounding through Surge Arresters

Introduction

System grounding requirements and methods were discussed in Chapter 12, but the information was generally limited to systems of 1000 volts and less. The grounding requirements for systems of greater than 1000 volts are provided in Part X of *NEC* Article 250. There are various methods to accomplish the grounding required for medium- and high-voltage systems. Equipment grounding and specific rules for grounding cable shielding are necessary for medium- and high-voltage installations and systems. Important rules for grounding medium-voltage cable shields are included on part X of Article 250.

Requirements for Grounding Systems

Grounding methods and requirements for systems operating at more than 1000 volts, such as 5 and 15 kV systems, differ slightly from those for systems of 600 volts or less. Systems in these voltage ranges are commonly referred to as medium-voltage systems. A typical 5 kV system is one with a phase-to-phase voltage of 4160 volts. The phase-to-ground voltage of this system is roughly 2400 volts. An example of a 15 kV system is one with a phase-to-phase voltage of 12 470 volts. The phase-to-ground voltage in this system is roughly 7200 volts. These systems are typically referred to as either medium- or high-voltage systems. The *NEC* provides several rules related to the grounding of these systems and associated equipment. Part X of Article 250 provides the rules for grounding and bonding systems of more than 1000 volts. The reasons for grounding systems of more than 1000 volts are the same as the reasons for applying grounding on systems of 1000 volts or less. These systems are grounded to limit voltages imposed by lightning events, line surges, or unintentional contact with higher-voltage lines and to provide voltage stabilization during normal operation of the system. Where systems of more than 1000 volts are grounded, the requirements in Sections 250.182 through 250.191 must apply accordingly, depending on the type of grounding employed for the system. It is important to realize that grounding and bonding provisions in Parts I through IX are only modified or supplemented by Part X of Article 250. It is important to remember that 90.3 indicates Chapters 5, 6, and 7 of the *NEC* could modify or amend any of the requirements within Article 250.

Grounding Methods for Systems Over 1 kV

A few grounding methods are permitted to create a ground reference for such systems. Electrical systems greater than 1000 volts are usually grounded by one of three ways. They can be solidly grounded, grounded through an impedance device, or grounded through a set of grounding transformers that create a reference to ground. **See Figure 18-1.** Part X of Article 250 provides specific rules for systems grounded at a single point and those systems that are grounded at multiple locations. The following are requirements for single-point grounded neutral systems.

Solidly Grounded Systems

A solidly grounded electrical system is one that has a direct electrical connection to ground with no intentional impedance installed between the Earth connection and the system. A common solidly grounded system operating beyond 1000 volts is a 4160 volt, three-phase, four-wire, wye-connected system. **See Figure 18-2.** In this system, there is a derived neutral that is the grounded conductor.

The requirements for grounding such systems are found in Section 250.184(A). There, the neutral of such systems generally has to be an insulated conductor with 600 volts–rated insulation. Bare neutral conductors of such systems are only permitted if the following specific conditions can be satisfied:

1. The bare neutral is installed with service entrance conductors.
2. The bare neutral is installed with a service lateral.
3. The bare neutral is installed with the direct-buried portion of a feeder.

The neutral conductor of solidly grounded neutral systems can also be bare when installed as overhead conductors. Only the portion installed overhead is permitted to be bare in this case.

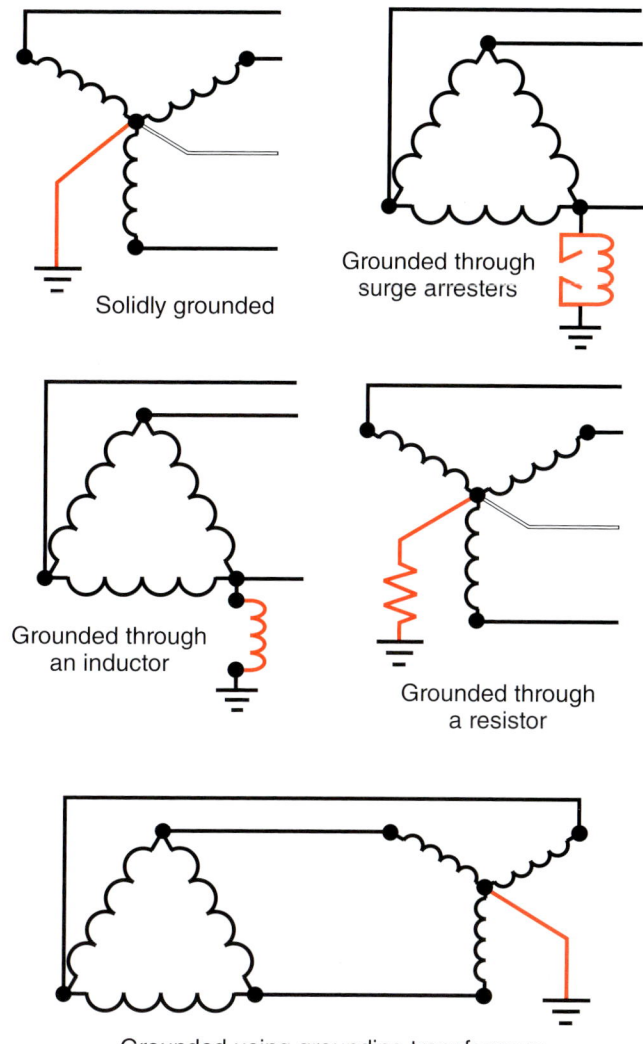

FIGURE 18-1 There are a few different grounding methods for systems of more than 1000 V.

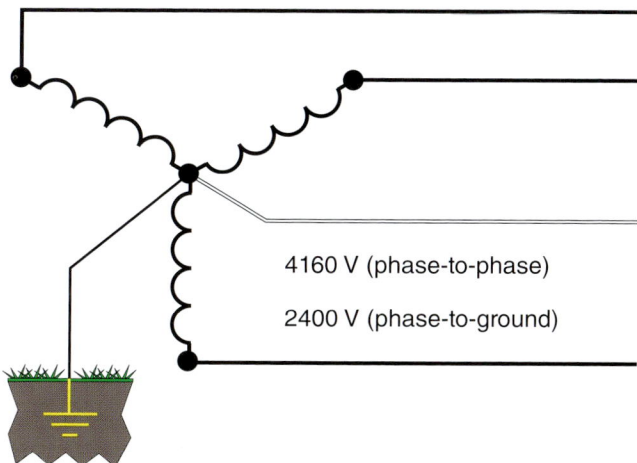

FIGURE 18-2 Solid grounding is often applied for a 5000 (4160/2400) volt system.

Exception No. 3 to Section 250.184 also permits a bare neutral conductor for solidly grounded neutral systems if the neutral is isolated from the phase conductors and protected from physical damage. See Exceptions 1 through 3 to Section 250.184.

The neutral conductor of a solidly grounded system has to be of sufficient current-carrying capacity for the load served and generally must not be smaller than one third of the ampacity of the ungrounded phase conductors supplied by the system. The *Code* does permit the neutral for these systems to be sized no smaller than 20% of the ungrounded phase conductor ampacity only in commercial and industrial establishments where conditions of engineering supervision are in place. **See Figure 18-3.**

Single-Point Grounding

Single-point grounding of a system means the system is grounded at only one point and no neutral-to-ground connections can be made downstream of that initial grounding location. A common aspect of each of the grounded systems at lower voltages is that they are all grounded at one point, unless grounding the neutral downstream is permitted by exception. In a single-point grounded neutral system, the neutral is grounded typically at the source. Then, an EGC is run from the single point of grounding of the system, with allowable grounding connections to the Earth from that EGC. **See Figure 18-4.**

Section 250.184(B) indicates that single-point grounded neutral systems can be supplied from a separately derived system or a source of a multigrounded neutral system. The multiple grounding connections are from the EGC run with the supply conductors derived from a single-point grounded neutral system. The connection to the Earth for a single-point grounded neutral system is made through a grounding electrode meeting the applicable requirements in Part III of Article 250. A grounding electrode conductor is required from the neutral conductor of such systems to the grounding electrode. The EGC for single-point grounding neutral systems is installed to connect the neutral of the system to the grounding electrode, just as it does for other separately derived systems. This bonding jumper is essentially a system bonding jumper, but note the difference in terminology in Part X of Article 250. Functionally, these components of the grounding and bonding system perform in the same way. If feeders are routed to each load, such as equipment or separate buildings or structures, an EGC must be installed.

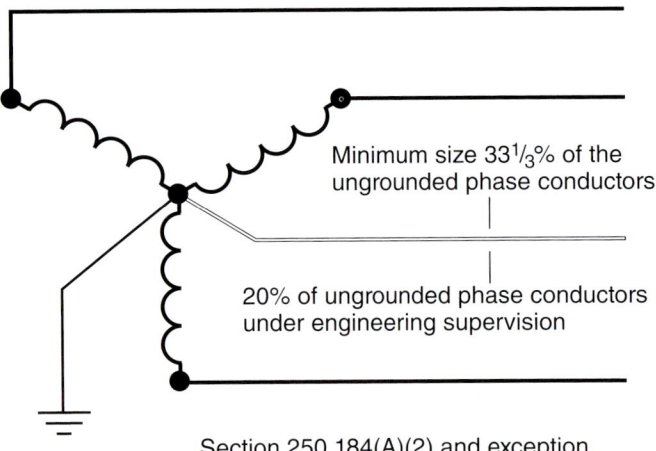

FIGURE 18-3 The neutral of solidly grounded systems is generally not permitted to be smaller than 33⅓% of the ungrounded conductor supplied by the system.

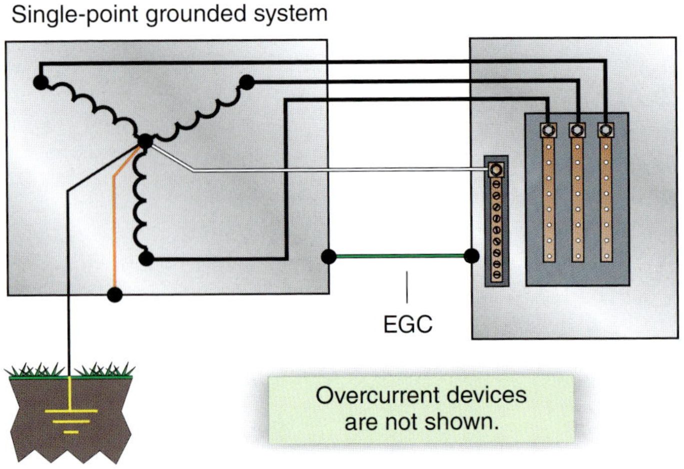

FIGURE 18-4 Single-point grounded systems are connected to a grounding electrode, and an EGC must be installed with the circuit supplying equipment.

The EGC must be routed with the ungrounded phase conductors of the system and cannot carry any continuous load current. This EGC can be insulated or bare and has to have sufficient current-carrying capacity for the maximum fault likely to be imposed on it. A word of caution here is that ribbon shielding or metal tape shielding on medium- and high-voltage cables is usually not of sufficient size to serve as an EGC. The shielding serves a different purpose. A neutral conductor is not required to be run with the feeder conductors unless there is a load requiring it. Section 250.184(B) requires the neutral of single-point grounding neutral systems to be isolated from ground except at the grounding point, the source location in most cases.

Multipoint Grounding

The *NEC* also recognizes multigrounded neutral systems. The rules for multigrounded neutral systems are provided in Section 250.184(C). As the term implies, there are multiple grounding points to the neutral of such systems. In multigrounded neutral systems, the system neutral is typically derived and grounded at the source and then distributed for long distances. Grounding is required from the neutral at multiple points along its route. **See Figure 18-5.**

Three common applications for multigrounded neutral systems are installations where the system supplies buildings or structures, such as in a campus distribution system. Multigrounded neutral systems are also permitted for use in underground systems where the neutral conductor is exposed and run as an overhead circuit conductor—between poles, for example. The grounding of such a system has to be accomplished at each transformer and at additional locations by connection to the Earth through a grounding electrode. Connection to an EGC of the circuit is not permitted. The neutral conductor of a multigrounded neutral system has to be connected to ground (the Earth) at intervals not exceeding 1,300 feet. This maximum distance between grounding points on these systems is often referred to as the *four grounds per mile* method of grounding. Where a multigrounded neutral system employs shielded cables, the cable shielding has to be grounded at each cable connection point where the shields are exposed and subject to contact by persons.

Impedance Grounding

Another method of grounding systems of more than 1000 volts is through an impedance device. Impedance grounding means that there is intentional opposition to current inserted between the grounded (neutral) conductor of the system and the grounding electrode conductor. These systems are referred to in the *NEC* as *impedance grounded neutral systems*. The impedance intentionally limits the amount of current that will return to the source in ground-fault conditions, thus increasing continuity of service. The impedance device is typically a resistor or impedance coil inserted in the AC circuit. Impedance grounded neutral systems are installed in an effort to limit current to the source in ground-fault conditions, thereby providing a measure of protection for equipment. By limiting the current through an intentional impedance device, any arcing condition is kept to a lower magnitude.

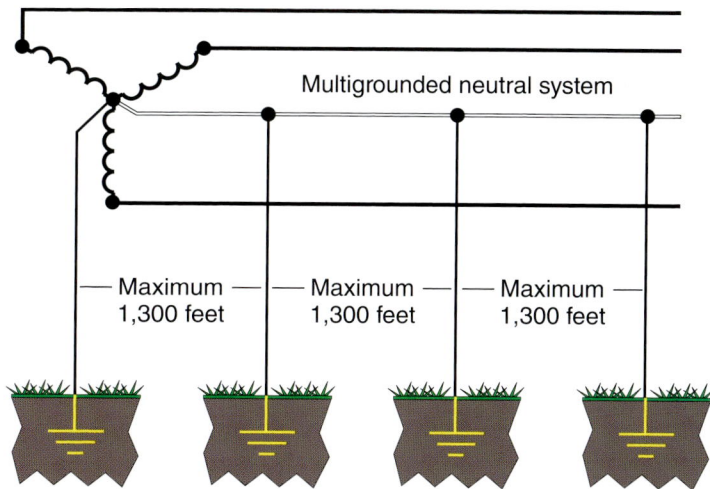

FIGURE 18-5 Multigrounded neutral systems of more than 1000 volts include a system neutral that is grounded in more than one location.

Special equipment is manufactured for use on these types of installations. Impedance grounded neutral systems are only permitted under controlled conditions. First, conditions of maintenance and supervision must be in place, ensuring that only qualified persons service these installations. This is an important condition, because in a ground-fault condition, qualified persons must understand the proper response and course of action to remove the faulted condition. Second, ground-fault detection systems are installed to notify qualified persons of a first phase-to-ground fault event. The third condition that must be met for use of this type of system is that no line-to-neutral loads are served. See Sections 250.186(1) through (3).

The grounding impedance device for an impedance grounded neutral system has to be installed in series with the grounding electrode conductor and the neutral point of the system source, which could be located at a transformer or a generator. The system neutral grounding connection is only permitted to be made through the impedance device. **See Figure 18-6.** The neutral conductor has to be fully insulated, with an insulation equivalent to that of the ungrounded phase conductors supplied by the system. The grounded conductor (neutral) of these types of systems is also required to be identified. The means of identification is typically white or gray or in accordance with the applicable provisions in Article 200. The neutral conductor of these systems extends from the center point of the wye connection to the line side of the impedance device, and a grounding electrode conductor extends from the load side of the impedance device to the grounding electrode. The neutral conductor of an impedance grounded neutral system is not a circuit conductor. It does not supply a load. It is the conductor that connects from the impedance device to the system neutral point. It is the point of grounding for this type of system. Just as with other supply circuits from grounded systems, an EGC is required to be installed. The EGC connection at the source has to be made on the load side of the impedance device.

The EGC can be a bare conductor, or it can be insulated. It has to connect to the grounding electrode conductor and the equipment grounding terminal bus of the source equipment enclosure. The EGC is installed for grounding equipment supplied by the system and serves three important functions: it grounds equipment, it performs bonding, and in ground-fault conditions, it serves as an effective ground-fault current path. **See Figure 18-7.**

Portable or Mobile Equipment Grounding

An important factor related to grounding a system rated more than 1000 volts is whether the system supplies equipment that is portable or mobile. Section 250.20(C) indicates that if portable or mobile equipment is supplied by a medium- or high-voltage system (more than 1 kV), the system has to be grounded in accordance with Section 250.188. Sections 250.188(A) through (F) provide the requirements for grounding systems that supply portable or mobile equipment. Portable or mobile equipment must be supplied by a system with a neutral grounded through an impedance device unless the system is delta-connected, in which case a grounded neutral must be derived.

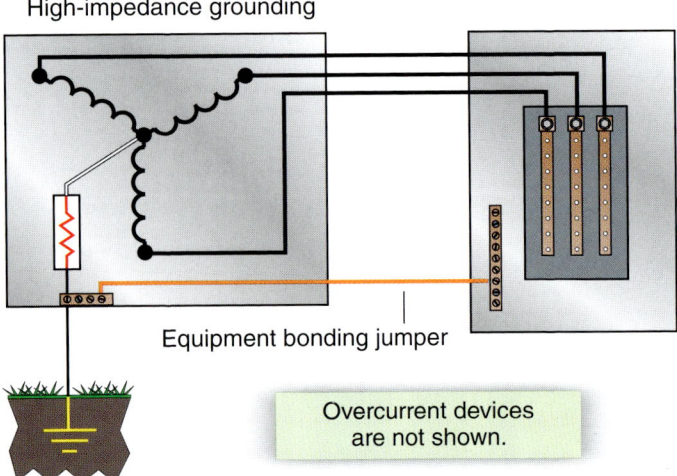

FIGURE 18-6 Impedance grounded neutral systems of more than 1000 volts are connected to ground (the Earth) through an impedance device.

This section of the *Code* requires both system and equipment grounding. Exposed non–current-carrying metal parts of portable or mobile equipment have to be grounded by connection to an EGC that is connected to the point at which the system neutral impedance is grounded.

Ground-fault detection and relaying has to be provided to automatically de-energize any high-voltage system component that has developed a ground-fault condition. The continuity of the EGC must be continuously monitored and has to cause automatic de-energizing of the supply system upon loss of continuity of the EGC. This rule emphasizes the importance placed on the EGCs of such equipment by having to monitor its continuity. The EGC is a safety circuit and must be effective at all times while the system is live. If the continuity of the EGC is not established, the system supplying the mobile or portable equipment cannot be energized. The impedance device has to be connected to a grounding electrode to establish the ground connection. The grounding electrode for systems supplying portable or mobile equipment has to be isolated, separated at least 6.0 m (20 feet) from any other system or equipment grounding electrode, and there can be no direct connection between the grounding electrodes, such as buried pipe and fence. High-voltage trailing cable and couplers for interconnection of portable or mobile equipment are required to meet the specific requirements of Part III of Article 400 for cables. The cable couplers must also meet the requirements in Section 490.55.

Grounding Equipment

Grounding of equipment associated with medium- and high-voltage systems is required for fences, enclosures, housings, support structures, and so forth, and is required for all non–current-carrying metal parts of fixed, portable, or mobile equipment. Note that equipment that is isolated from ground, and that cannot be contacted by persons in contact with the ground, is not required to be grounded. This exception applies to pole-mounted equipment such as transformer and capacitor cases that are elevated. Grounding is accomplished through a grounding electrode conductor. Section 250.190(B) provides installation requirements for grounding electrode conductors for systems of more than 1000 volts. The sizing requirements for grounding electrode conductors is based on the use of Table 250.66, using the largest ungrounded service, feeder, or branch circuit conductor supplying the equipment. **See Figure 18-8.**

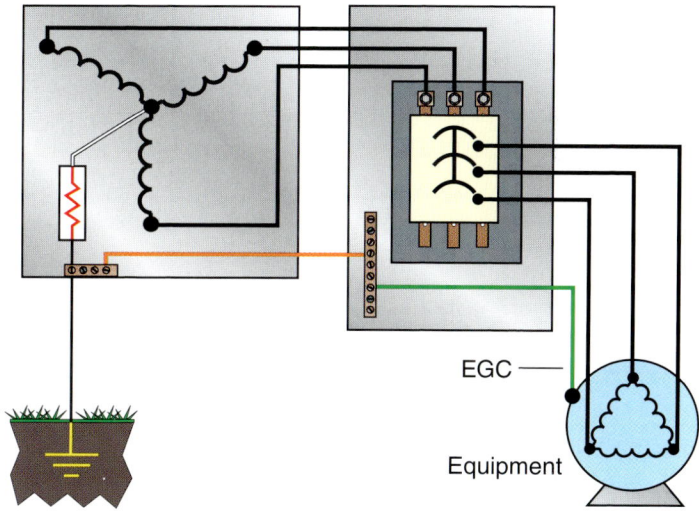

FIGURE 18-7 EGCs are required to be installed for impedance grounded neutral systems.

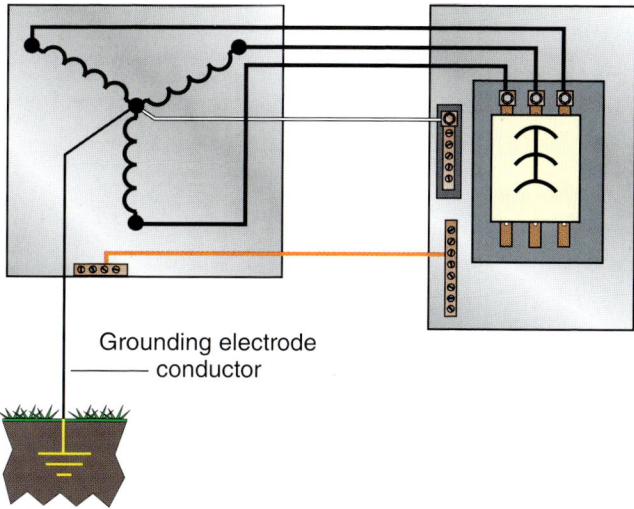

FIGURE 18-8 The grounding electrode conductor for a system over 1 kV must generally be sized using Table 250.66.

The minimum size required for the grounding electrode conductor is 6 AWG copper or 4 AWG aluminum. Feeders and branch circuits of more than 1000 volts often include EGCs.

The EGCs have to be of sufficient capacity. As previously reviewed, EGCs installed with circuits of more than 1000 volts cannot be smaller than 6 AWG copper or 4 AWG aluminum unless they are an integral part of a cable assembly. If a cable assembly shield is a concentric neutral type and suitable for ground-fault current performance, the shield can serve as the required EGC. For solidly grounded systems, a cable ribbon shield or tape shield of the cable assembly is not permitted as an EGC because of its inadequate size. This shielding material of cable assemblies typically is under the minimum capacity necessary to perform during ground-fault conditions. EGCs must provide an effective ground-fault current path to facilitate overcurrent device operation. The EGC is sized using Table 250.122 based on the rating of the overcurrent device protecting the feeder circuit. **See Figure 18-9.**

As an example, if a pad-mounted transformer is single-point grounded and includes overcurrent protection at the bushing of the output side of the transformer, the rating of the overcurrent protection integral with the busing establishes the minimum size of the EGC. A 150-ampere bushing results in a 4 AWG copper EGC for this feeder circuit. The overcurrent rating for a circuit breaker in these types of systems is typically the combination of the current transformer and current pickup setting of a protective relay system in the breaker assembly. Remember that the minimum size of the EGC for systems of 1000 volts and higher is 6 AWG copper or 4 AWG aluminum if an EGC is not an integral part of a cable assembly.

Article 490 of the *NEC* provides grounding requirements for equipment rated more than 600 volts. The frames of switchgear and control assembly enclosures are generally required to be grounded. This type of equipment is usually equipped with an equipment grounding terminal bar for landing all grounding electrode conductors and EGCs supplied for the equipment. Equipment rated more than 1000 volts often includes surge protection as an integral part of the assembly; it could also be provided as an accessory feature. **See Figure 18-10.** Medium- and high-voltage systems are vulnerable to line surges and events that are high magnitude and can be destructive. Surge arresters are often installed on these systems and provide a level of protection against such events.

Substation Grounding Requirements

The *NEC* falls short of providing many specific details and requirements for substation grounding. Section 250.191 requires a grounding electrode system in accordance with the applicable provisions of Part III of Article 250.[1]

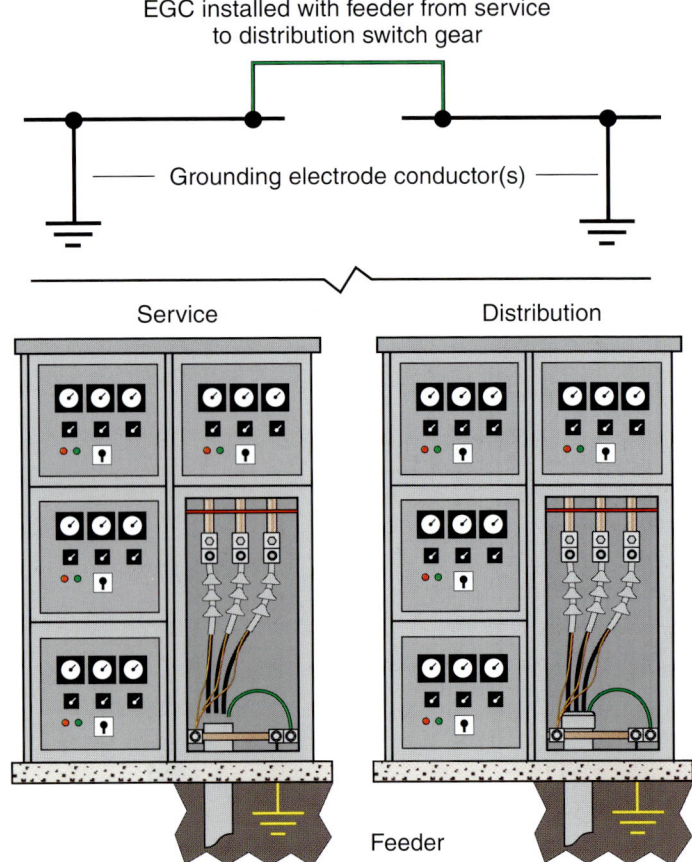

FIGURE 18-9 EGCs with feeders of more than 1000 volts must generally be sized using Table 250.122.

Common grounding electrodes installed for alternating current (AC) outdoor substation grounding are concrete-encased electrodes, ground rings, and ground rods. IEEE 80, *Guide for Safety in AC Substation Grounding,* provides specific information about outdoor AC substation grounding. **See Figure 18-11.**

A few key requirements for substation grounding are provided. When dealing with substation grounding, both system and equipment grounding are accomplished. System grounding requirements were covered previously in this chapter. Outdoor AC substations typically include a variety of conductive parts and equipment that must be grounded. **See Figure 18-12.** One of the primary objectives in the grounding of metal parts at an outdoor substation is to establish an equipotential grid to which all conductive parts can be connected. This grid is installed underground outside of a fenced enclosure, typically about 3 feet from the fence, and completely encircles the enclosure. There are usually several grounding electrodes (usually ground rods) driven and connected to this ring system. **See Figure 18-13.** Any underground metal structures, such as piping and framing, should be bonded to this grid. The minimum size of the grid conductor should be no less than 4/0 copper and no less than 25% of the capacity of the system.

Courtesy of Jim Dollard, IBEW Local 98

FIGURE 18-10 Surge arresters are often included in metal-clad switchgear for medium-voltage systems.

I-Stock Photo Courtesy of NECA

FIGURE 18-11 Grounding is required for substation installations.

Courtesy of Donny Cook, Shelby County, AL

FIGURE 18-12 Grounding for substation installations is typically designed using IEEE 80 *Guide for Safety in AC Substation Grounding.*

396 APPLIED GROUNDING AND BONDING

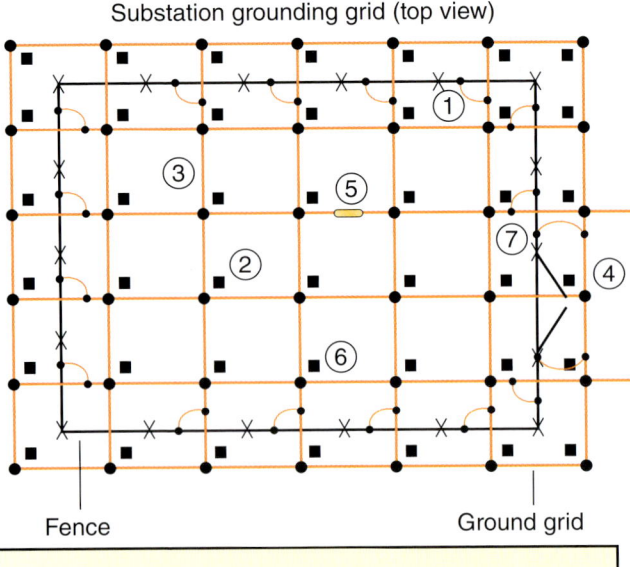

FIGURE 18-13 Substation fences and other conductive parts must be grounded and bonded to the substation grounding grid system.

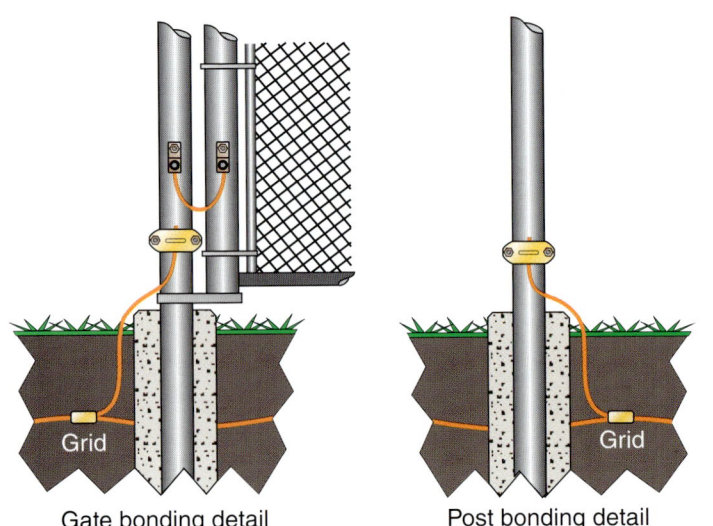

FIGURE 18-14 Substation grounding of fences and other conductive parts is accomplished using suitable bonding connection devices and bonding jumpers.

Connections from metal enclosures and structures within the fenced area and from the fence should be made using 4/0 AWG copper conductors at a minimum and no less than 25% of the output conductors of the system. **See Figure 18-14.**

There are different philosophies on fence grounding at outdoor substations. Most designs require the fence to be commonly connected to the grounding grid. Another approach is to isolate the fence from the grid system in the event of a phase-to-ground fault that elevates the potential of all connected metal parts. In this case, the potential of the fence is raised, which can present a hazard for persons coming in contact with the fence during the abnormal event. Most designs require the fence to be grounded common to the grid grounding system. Sometimes the grid system for a substation is created by laying copper conductors in a checkerboard arrangement at 3-foot intersections to create a mat that is buried in the ground on which the substation is constructed. This establishes a convenient point of connection for any conductive metal in the substation. The length of bonding jumper connections is reduced when this method of grid construction is used. The connections for substation grounding grid systems have to be effective and strong. The common methods used for these connections are exothermic welding processes and irreversible compression connectors. It is important to use compression connectors that are listed as grounding and bonding equipment. These connectors are evaluated to endure stresses of rise and fall of potentials that are caused by various events occurring on these systems in normal operation and abnormal events. One form of protection provided by the grid system for outdoor substations is the ability to dissipate lightning strikes effectively to minimize possibilities of damage.

Conductor Shielding and Stress Reduction

Cables installed for medium- and high-voltage systems are generally required to be of a shielded type. The cable shields can be in a concentric strand or conductive tape arrangement. **See Figure 18-15.** Cable shielding provides a method to evenly distribute voltage stress and drain it off to ground at termination points of the cable. **See Figure 18-16.** Section 310.10(E) provides *NEC* requirements for cable shields. The primary purposes of shielding are to confine the voltage stresses to the insulation, dissipate insulation leakage current, drain off the capacitive charging current, and carry the ground-fault current to facilitate operation of ground-fault protective devices in the event of an electrical cable fault.

Solid dielectric insulated conductors operated above 2000 volts in permanent installations are required to have ozone-resistant insulation and must be of a shielded type. All metallic shields (ribbon tape shields or concentric stranding) shall be connected to a grounding electrode conductor, grounding busbar, EGC, or grounding electrode. **See Figure 18-17.** In enclosed equipment, the cable shield connections are typically made to the equipment grounding bus in the enclosure. At a pole installation, the cable shields are typically connected to a grounding electrode conductor or directly to a grounding electrode. There are stress reduction kits that include the provisions for connecting cable shields to ground. Load-break elbow assemblies also include such provisions to bleed off stress at termination points.

Nonshielded cables are permitted for circuits up to 2400 volts but only under the following restrictive conditions:

1. They must be listed by a qualified electrical testing laboratory.

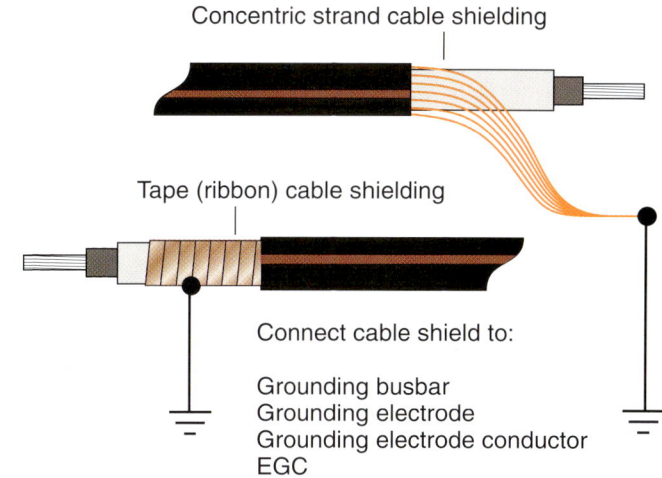

FIGURE 18-15 Cable shielding can be concentric stranding or ribbon (tape).

Courtesy of Cogburn Brothers Electric, Inc.

FIGURE 18-16 Cable shields are required to be connected to ground through a grounding electrode conductor, grounding busbar, EGC, or grounding electrode.

Courtesy of Cogburn Brothers, Inc.

FIGURE 18-17 Cable shields are required to be connected to ground using suitable conductors and connectors.

2. The insulation has to be resistant to electric discharge and surface tracking or covered with a material that is ozone resistant and resistant to surface tracking and electric discharging.
3. Where installed in wet locations, the conductor must have an overall nonmetallic jacket or metallic sheath.
4. The conductor insulation thickness has to be no less than the values in Table 310.104(D).

Nonshielded cables are also permitted for circuits up to 5000 volts in replacement applications only and under the following restrictive conditions:

1. They must be listed by a qualified electrical testing laboratory.
2. Conditions of maintenance and supervision ensure that only qualified persons service the system.
3. The insulation has to be resistant to electric discharge and surface tracking or covered with a material that is ozone resistant and resistant to surface tracking and electric discharging.
4. Where installed in wet locations, the conductor has to have an overall nonmetallic jacket or metallic sheath.
5. The conductor insulation thickness has to be no less than the values in Table 310.104(D).

This allowance of nonshielded cables can be applied in situations where existing nonshielded cables have to be replaced and connected to existing equipment. This allowance is not applicable to situations in which existing equipment is relocated or reinstalled on the same property and new wiring is installed to the old equipment.

Directly buried cables are required to be of a type suitable for the location and are generally required to be shielded if rated above 2000 volts. Cables rate 2001 to 2400 volts are permitted to be nonshielded types if that cable has an overall metallic sheath or armor. The metallic sheath or armor has to be connected to a grounding electrode conductor or grounding terminal bar in the equipment, EGC, or grounding electrode. There is an exception that relaxes the cable shielding requirement for airfield lighting equipment and cable installations. **See Figure 18-18.** In these types of installations, series circuits rated up to 5000 volts that are controlled by regulators are permitted to be installed using nonshielded cable as long as they meet the criteria set forth in the Federal Aviation Administration Advisory Circulars that include guidelines and practices for installing airport lighting systems. **See Figures 18-19.**

FIGURE 18-18 Regulators supplying airport runway lighting with 5000 volt cable that is unshielded.

FIGURE 18-19 Regulators for airport runway lighting can be wired using 5000 volt cable that is not a shielded type.

Grounding through Surge Arresters

Surge arresters are devices that protect electrical equipment and installations from transients (voltage spikes). This is accomplished by either limiting or shunting to ground all voltage events above a set threshold.

> **Surge Arrester.** A protective device for limiting surge voltages by discharging or bypassing surge current; it also prevents continued flow of follow current while remaining capable of repeating these functions.[2]

The term *clamping voltage* is often used and associated with surge arresters. The clamping voltage is that voltage at which the surge arrester initiates its transient protection operation. Grounding of medium- and high-voltage systems can be accomplished by using surge arresters. Surge arresters are a method of providing an effective means for protection against overvoltages and events such as lightning strikes. Surge arresters are often installed for overhead lines because of their vulnerability to lightning strikes. **See Figure 18-20.** A common application of surge arresters is found on the tops of power poles where medium- and high-voltage lines are installed. **See Figure 18-21.** The surge arresters are connected directly to the medium- or high-voltage line conductors. They are connected to ground by a grounding electrode conductor.

They operate during overvoltage events, such as a lightning strike or significant line surge. The principles of operation are such that this device is a spark gap configuration that allows the contacts of the device to close and drains excessive overvoltages to the Earth during the duration of such line surges or even unintentional contact with higher-voltage lines. Surge arresters typically protect the primary of higher-voltage systems. Surge arrester grounding conductors have to be connected to any of the following:

1. A grounding electrode
2. A grounding electrode conductor
3. The grounded conductor at the service
4. The equipment grounding terminal bus within medium- or high-voltage equipment enclosures

Surge arresters are not required by the *NEC*. Article 280 provides requirements for surge arresters installed on electrical systems covered by the *NEC*.

Courtesy of Bill McGovern, City of Plano, TX

FIGURE 18-20 Surge arresters are usually installed for overhead lines for systems operating at more than 1000 volts.

Courtesy of Bill McGovern, City of Plano, TX

FIGURE 18-21 Surge arresters installed for overhead lines of more than 1000 volts are connected to ground through a grounding electrode conductor.

Where they are installed on a premises system, a surge arrester has to be installed in each of the ungrounded supply conductors. Surge arresters are not permitted where the phase-to-ground power frequency voltage exceeds the rating of the surge arrester. The surge arresters applied in an electrical system must have a rating equal to or greater than the maximum continuous operating voltage available at the point it is connect to the system. Grounding through surge arresters is often employed for outdoor systems distributed on pole installations. Surge arrester installation is covered in Part II of Article 280. Surge arresters can be installed inside or outside of buildings or structures, but they must be inaccessible to unqualified persons.

The conductor for connecting surge arresters must be as short as practicable and must not have any unnecessary bends. Where surge arresters are connected to line conductors or equipment, the connecting conductor cannot be smaller than 6 AWG. Copper or aluminum is permitted for this connection, but the rules for aluminum conductors must be followed. Where a surge arrester is installed in an ungrounded primary system, the spark gap or listed device must have a 60 Hz breakdown voltage of at least twice the primary circuit voltage but not necessarily more than 10 kV. In addition, there must be at least one other grounding connection on the grounded conductor of the secondary that is located no less than 20 feet from the surge arrester grounding electrode. Where surge arresters are connected to multigrounded neutral primary systems, the spark gap or listed device must have a 60 Hz breakdown of no more than 3 kV, and there must be at least one other grounding connection on the grounded conductor of the secondary that is no less than 20 feet from the surge arrester grounding electrode. Where the grounding conductor for a surge arrester is installed in a ferrous metallic raceway or enclosure, the grounding conductor must be bonded to the enclosure at points of entrance and emergence from the raceway or enclosure. **See Figure 18-22.** This type of installation of physical protection for surge arrester grounding conductors has to meet the requirements in Section 250.64(E). This section requires grounding electrode conductors installed in ferrous metal raceways to be bonded at both ends of the raceway where the grounding electrode conductor enters or emerges from the raceway. This minimizes the choke effect on the grounding electrode conductor.

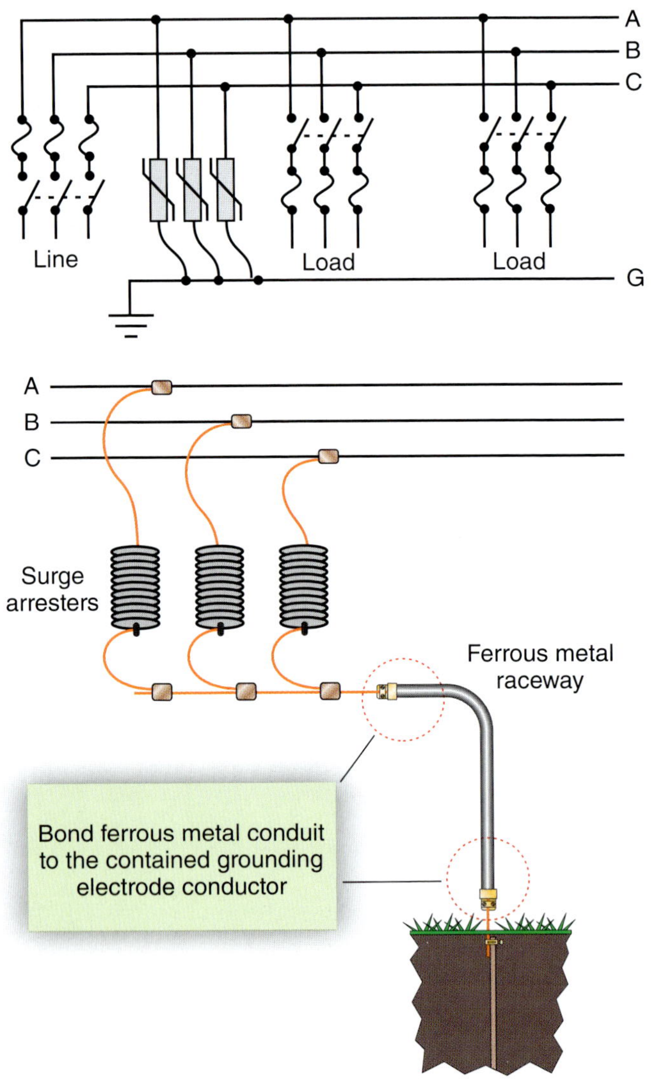

FIGURE 18-22 Ferrous metal sleeves for surge protection grounding conductors must be bonded to the conductor it contains at both ends of the sleeve.

Summary

Medium- and high-voltage systems are required to be grounded under certain conditions—more specifically, where they supply portable or mobile equipment. There are a few different grounding methods and requirements for systems operating at more than 1000 volts, such as 5 and 15 kV systems. A typical 5 kV system is one with a phase-to-phase voltage of 4160 volts. The phase-to-ground voltage of this system is roughly 2400 volts. An example of a 15 kV system is one with a phase-to-phase voltage of 12 470 volts. The phase-to-ground voltage in this system is roughly 7200 volts. These systems are typically referred to as either medium- or high-voltage systems. The *NEC* provides some methods and several rules related to grounding of these systems and associated equipment. Three methods of grounding for medium- and high-voltage systems are solid grounding, impedance grounding, and grounding accomplished through grounding transformers. Sizing requirements for grounding electrode conductors and equipment grounding conductors used with systems over 1 kV are similar to those rules for systems operating at 1000 volts or less.

References

1. NFPA 70 National Electrical Code 2011, Section 250.191 (National Fire Protection Association, Quincy, MA 2010), p. 70–131.

2. NFPA 70 National Electrical Code 2011, Article 100 (National Fire Protection Association, Quincy, MA 2010), p. 70–32.

Review Questions

1. Which part(s) of Article 250 provides requirements for grounding systems of more than 1000 volts?
   a. Part X and all other parts of Article 250 as modified by Part X
   b. Part X only
   c. Parts I and X only
   d. Parts I, III, and X only

2. Ungrounded systems greater than 1000 volts are permitted to be _____.
   a. Solidly grounded
   b. Grounded through an impedance device
   c. Grounded through a set of grounding transformers
   d. Grounded by any of the above methods

3. A solidly grounded electrical system of more than 1000 volts is one that has a direct electrical connection to ground with a grounding impedance device installed between the Earth connection and the system.
   a. True   b. False

4. The neutral conductor of a solidly grounded system of more than 1000 volts has to be of sufficient current-carrying capacity for the load served and generally must not be smaller than _____ of the ampacity of the ungrounded phase conductors supplied by the system.
   a. 10%
   b. 25%
   c. 33⅓%
   d. 50%

5. Single-point grounding of a system of more than 1000 volts means the system is grounded at only one point and no neutral-to-ground connections can be made downstream of that initial grounding location.
   a. True   b. False

6. Where a multigrounded neutral system of more than 1000 volts is installed, the connections to ground from the neutral conductor must be made at intervals not exceeding _____ feet.
   a. 1,300
   b. 5,280
   c. 2,500
   d. 100

7. The connection to the Earth for a single-point grounded neutral system is made through a grounding electrode meeting the applicable requirements in Part _____ of Article 250.
   a. I
   b. II
   c. III
   d. X

8. The grounding impedance device for an impedance grounded neutral system has to be installed in series with the grounding electrode conductor and the neutral point of the system source, which could be at a transformer or a generator. Any equipment grounding conductor connections at the source have to be made on the _____ of the impedance device.
   a. Line side
   b. Load side
   c. Either side
   d. Any of the above

9. For systems of more than 1000 volts, the sizing requirements for grounding electrode conductors are based on _____, using the largest ungrounded service, feeder, or branch circuit conductor supplying the equipment.
   a. Table 250.66
   b. Table 250.122
   c. Table 250.4
   d. Table 8, Chapter 9

10. Equipment grounding conductors installed with circuits of more than 1000 volts cannot be smaller than 6 AWG copper or 4 AWG aluminum, unless they are an integral part of a cable assembly.
    a. True   b. False

11. Where a separate equipment grounding conductor is installed with feeder conductors supplied from a 4160 volts system, the minimum size shall not be smaller than the sizes provided in _____.
    a. Table 250.66
    b. Table 250.122
    c. Table 310.16
    d. Table 1, Chapter 9

12. If a 4160-volt feeder is protected by a 100-ampere fuse, the minimum size wire-type copper equipment grounding conductor for the circuit is _____ AWG.
    a. 8
    b. 6
    c. 4
    d. 2
13. Section 250.191 requires a grounding electrode system in accordance with the applicable provisions of Part III of Article 250.
    a. True
    b. False
14. A common size for grounding grids installed at substations is no less than _____ copper.
    a. 1/0
    b. 2/0
    c. 3/0
    d. 4/0
15. The primary purposes of shielding are to confine the voltage stresses to the insulation, dissipate insulation leakage current, drain off the capacitive charging current, and carry the ground-fault current to facilitate operation of ground-fault protective devices in the event of an electrical cable fault.
    a. True
    b. False
16. Nonshielded cables are permitted for circuits up to 2400 volts but only under which of the following conditions?
    a. They must be listed by a qualified electrical testing laboratory, and the conductor insulation thickness has to be no less than the values in Table 310.104(D).
    b. The insulation has to be resistant to electric discharge and surface tracking or covered with a material that is ozone resistant and resistant to surface tracking and electric discharging.
    c. Where installed in wet locations, the conductor has to have an overall nonmetallic jacket or metallic sheath.
    d. All of the above.
17. The _____ is a protective device for limiting surge voltages by discharging or bypassing the surge current; it also prevents continued flow of follow current while remaining capable of repeating these functions.
    a. Transient voltage surge suppressor
    b. Surge protective device
    c. Surge arrester
    d. Primary protector

ANNEX A

Investigation and Testing of Footing-Type Grounding Electrodes for Electrical Installations

H. G. Ufer, Associate Member IEEE

Summary: Footing-type grounding electrodes installed in the concrete foundations of residential and small commercial buildings designed to meet a maximum ground resistance value of 5 ohms are described. Bare solid copper electrode wires of various lengths, steel reinforcing rods, and 10-foot lengths of hot galvanized rigid steel conduit were used to determine the resistance values.

City Electrical Inspector members reported to the Southwestern Section of the International Association of Electrical Inspectors that a continuous metallic water piping system, as recommended by the National Electrical Code, is not always available. A committee was appointed to investigate this condition and report its findings at subsequent meetings.

The National Electrical Code, which is prepared by a committee of the National Fire Protection Association, is the recognized American standard for the safe installation of electrical equipment in the United States. It is based on the combined experience of all groups in the electrical industry and all factual information that is available at the time of each edition's preparation. One of the oldest basic requirements of the code is the protection of electric installations by grounding; it recommends that a metallic underground water piping system, either local or supplying a community, should always be used as the grounding electrodes when such a piping system is available.

Paper **63-105**, recommended by the IEEE Safety Committee and approved by the IEEE Technical Operations Committee for presentation at the IEEE Western Appliance Technical Conference, Los Angeles, Calif., November 4, 1963. Manuscript submitted August 1, 1963; made available for printing May 5, 1964.

H. G. Ufer is with Underwriters' Laboratories, Inc., Santa Clara Calif.

There is increasing concern over the fact that a continuous metallic water piping system is not presently available in some areas for electrical grounding purposes. This is partially a result of the use of nonmetallic water pipe mains and laterals to bring water into a building, in addition, it has become standard practice in the installation of cast-iron water pipe mains and laterals to use neoprene gaskets to join the pipe sections and discontinue the poured-lead joint formerly used.

Inspection authorities have also reported that the ground is interrupted by the installation of insulating joints on the plumbing system. When copper pipe is employed for the hot water side and galvanized steel pipe for the cold water side, such joints are used to prevent decomposition of the copper pipe resulting from electrolysis. These insulating joints are also used for water-softening tanks. Bonding jumpers are not always used, and often they are removed.

The driven grounds required by the code in lieu of connection to a water system are not always dependable. In the paper "Grounding of Electric Services to Water Piping Systems" presented to the American Water Works Association on October 26, 1960, A. G. Clark of Los Angeles reported that the use of driven grounds had not proven satisfactory. The minimum resistance obtained on representative ground rods was 10 ohms, and most of the driven representative rods tested were several times the minimum value of 10 ohms.

In view of these conditions and failures, it seemed that an adequate means could be provided for grounding, one which is not likely to be disturbed, which requires very little maintenance, and which does not require connection to the water pipe systems to provide an adequate low resistance ground.

The purpose of this paper is to report the development and testing of such a grounding method and to record the results of tests on installations in residential and small commercial buildings since 1961.

Background

A number of military installations of ammunition and pyrotechnic storage facilities were built during World War II and provided with lightning protection systems which required a permanent ground connection. The author conducted the field inspection of the lightning protection for Underwriters' Laboratories, Inc., and had first-hand knowledge of the construction details. Installations in Arizona have been selected for this paper because the climate is normally hot and dry during most of the year.

Tucson Installation

The installation at the Davis-Monthan Air Force Base, Tucson, Ariz., consists of six bomb-storage vaults and four pyrotechnic storage sheds. The bomb storage vaults are steel-reinforced concrete, while the storage sheds are steel angle frame construction covered with corrugated galvanized sheet steel.

The soil at this location is sand and gravel, and the average rainfall is 10.91 inches. Low resistance grounding had to be provided for each of these buildings in order to discharge to ground any static charge of electricity caused by wind and sandstorms or by exposure to lightning during a rain storm or thunder shower.

In the areas where these bomb storage vaults and storage sheds were located, an underground water piping system is usually not available. Driven-ground rods or a bare copper counterpoise could be used, but at the time these installations were being rushed to completion, strategic materials were being conserved. It was decided to provide grounding by the use of the steel angle iron or the reinforcing rods of the building structure.

For the bomb storage vaults (igloos as designated by the Armed Forces) with the dimensions 10 by 30 or 40 feet, the two parallel exterior walls were provided with footings. The front and rear walls of these vaults were not provided with footings.

The footings were dug to a depth of 2 feet. In these footings, ½-inch vertical reinforcing rods approximately 30 inches long were spaced about 12 inches apart. They were pushed a few inches into the earth to maintain their position and spacing while the concrete was poured. Across the upper ends of these vertical rods, ½-inch steel reinforcing bars were laid horizontally and welded to every alternate vertical rod embedded in the concrete of the footings.

From this base, an umbrella or igloo-shaped structure was formed of ½-inch steel reinforcing rods welded to the rod at the top of the footing. The lightning rod terminals were erected on the top of each igloo by welding them to the reinforcing rod assembly. The umbrella was then covered with concrete.

For the 10- by 4-foot storage sheds, grounding was obtained through the angle iron framework, which had the ends of the vertical angle iron at each corner of the building embedded in concrete to a depth of 1 foot.

Complete foundations were not provided for these small storage sheds. As a bond between the various metal parts of the frame and metal covering for these buildings, welding or bolts were used; the adjacent edges of the corrugated steel sheets were overlapped and secured by rivets or sheet metal screws.

Flagstaff Installation

The installation at the Navajo Ordnance Depot is located adjacent to U.S. Highway 66, approximately 15 miles west of Flagstaff, Ariz. This ordnance depot has an area of about 56 square miles. The soil is clay, shale, gumbo, and loam, with small-area stratas of soft limestone. The average rainfall is 20.27 inches.

The original contract in 1942 called for the construction of 800 bomb storage vaults. These were of the same construction and dimensions as those erected at the Davis-Monthan Air Force Base at Tucson.

TABLE I. Beverly Hills Tests

Type of Ground Electrode	Installation Date	Testing Date	Resistance (Ohms)
Electrode 1	11/1/60	11/1/60	5.1
	11/1/60	11/2/60	9
	11/2/60	11/10/60	9
	11/2/60	11/30/60	9
Electrode 2	11/1/60	11/1/60	90
	11/1/60	11/2/60	
		11/10/60	100
		11/30/60	
Bare pipe and pipe in concrete	11/1/60	11/1/60	
		11/2/60	
		11/10/60	14
		11/30/60	
Electrode 2 in concrete	1/9/61	1/9/61	8.7
	1/9/61	1/11/61	9
Electrodes 1 and 2 in concrete (series)	1/9/61	1/9/61	5
		1/11/61	
Cast-iron sewer and gas pipes-bonded		1/11/61	30
Sewer, gas, and water pipes-bonded		1/11/61	1
Sewer, gas, and water pipes connected to electrode 1		1/11/61	2

A water piping system either local or serving a community was not available for grounding at this location; accordingly, the method used at Tucson was utilized.

Ground Resistance Tests

Each bomb storage vault and storage shed erected in Tucson was inspected, and ground resistance tests were made on each of these structures. The 800 bomb storage vaults erected at the Navajo Ordnance Depot were inspected during the construction, and then individual ground resistance tests were performed.

Ground resistance readings were taken between each lightning protection terminal and ground on all igloos and storage sheds. Tests were also made between all exposed conductive material and ground on each installation described in this paper.

Readings were taken by using a heavy-duty megger ground tester with a scale from 0 to 50 ohms graduated in increments of 1 ohm for each scale division. Each vault and shed was required to have a ground resistance reading of not more than 5 ohms.

The method of measuring the ground resistance values was a 3-point method.[1] The sequence of ground resistance tests for the two lengths of conduit recorded in Table I were as follows: Two lengths of rigid-steel ¾-inch galvanized conduit were laid 5 inches apart in a trench 36 inches deep, 11 feet long, and 15 inches wide. One length of conduit was enclosed in 2 inches of concrete (electrode 1), while the second length was not enclosed in concrete (electrode 2); the second length was then enclosed in 2 inches of concrete and placed back in the trench beside the other length of conduit, and the dirt was replaced in the trench. The resistance measurements were repeated with the results recorded in Table I.

Check Tests

In order to determine the adequacy of the ground resistance values which were made in 1942, it was decided to conduct check tests. Permission was obtained from the Office of Chief of Operations in Washington, D.C., to check the grounds at Flagstaff Army Base and Tucson Air Force Base.

Check tests were made at the Navaho Ordnance Depot on July 12, 1960, and at the Davis-Monthan Air Force Base on August 23, 1960. Additional check tests were made at the Navajo Ordnance Depot on October 20, 1960.

At both bases, the readings taken measured from 2 to 5 ohms in all instances. All readings taken in 1942 when the original installations were made did not exceed 5 ohms to ground.

The Surveillance Officer at the Navaho Ordnance Depot must regularly check and record the ground resistance for each igloo, shell loading, and storage building. These tests are made at approximately 30–60-day intervals. During the 20 years since these ground systems were installed and accepted by the author, the maintenance department at the Navajo Ordnance Depot has not been requested to replace or repair any grounding components on the 800 bomb storage vaults or igloos.

As far as could be determined, there was no report of repairs or maintenance on grounding systems which had been installed on igloos and sheds at the Davis-Monthan Air Force Base.

Preliminary Investigation

Beverly Hills Tests

Test A

In view of the previous findings, an installation was made using conduit electrodes laid horizontally in the ground and enclosed in 2 inches of concrete at a depth of 3 feet.

Two 10-foot lengths of ¾-inch galvanized steel rigid conduit were laid side by side horizontally in a single trench 15 inches wide, 11 feet long, and 36 inches deep. The installation was made in Beverly Hills, Calif., on property which has not been cultivated during the past 30 years.

The resistivity of the soil in the area was 3,830 ohms per cm^3 (cubic centimeter). A ground rod driven to a depth of 20 feet was calculated to have a resistance of 5 ohms based on this resistivity test. (Table II shows the soil resistivity of the various sites mentioned in the paper.)

In the first series of the conduit electrode tests, one length of conduit (electrode 1) was enclosed in 2 inches of concrete; the other (electrode 2) was not enclosed in concrete. These two lengths of conduit were then covered with the soil removed to prepare the test trench.

In the second series of tests, the soil in the trench was removed and the bare conduit electrode 2 was enclosed in 2 inches of concrete.

The ground resistance of the water, gas, and sewer drainage piping systems bonded together was 1 ohm. The resistance was also 2 ohms when connected to electrode 1. This low resistance value is apparently due to the fact that the local water pipe system in galvanized steel pipe for the mains and the lateral. The lateral extends underground 40 feet from the water meter to the supply connection in the building.

The overhead electric service is a 3-wire a-c 110–240-volt system. The neutral of this service is grounded to the water pipe in the building. The utility ground for this service is provided at the utility pole which is 25 feet from the water meter.

The results of the ground resistance tests on the two conduit electrodes are recorded in Table I.

Test B

On January 16, 1960, one 10-foot length of 3½-inch galvanized rigid steel conduct was installed vertically to a depth of 5 feet and enclosed in 2 inches of concrete.

The ground resistance was 10 ohms; this value has remained constant during the 2-year period of these tests, and the readings are apparently stable.

In 1961, during a discussion of these supplemental tests, it was suggested that a test should be made on the property where the supplemental tests were conducted to be sure that the ground did not contain any foreign items which would favor these tests. In September 1961, the same Biddle ground megger, electrodes, and wire leads were used for this test as were used to measure the ground resistance of electrodes 1 and 2 (Table I).

TABLE II. Resistivity of Soils

Installation	Soil	Resistivity (Ohms/Meter^2)
Beverly Hills	Loam and sand	38.30
Bishop	Gravel and sand	120
Riverside	Red adobe	50
Livermore	Loam and Sand	35
Hayward	Large rocks and gravel	150
Portland, Ore.	Loam and sand	40
Long Beach	Heavy loam and sand	25* / 20†
Twenty-nine Palms	Decomposed granite and sand	70
Burbank	Sandy loam	40

* Metal sign on 12-inch steel I-beams.
† Concrete floor slab in garage.

The megger was placed adjacent to the ground in which pipe electrodes 1 and 2 were placed for the supplemental tests, and it was connected to these electrodes for a ground resistance test. The two 25-foot *No. 14* American wire gage (AWG) stranded wire leads were connected to the proper terminals on the megger. The probes were 36-inch ⅜-inch copper weld rods, and two were now placed in the ground 50 feet apart. On a 30-foot arc, ground resistance readings were taken at 3-foot intervals until probe 1 was adjacent to probe 2.

The probes were returned to their original positions. The readings were repeated with the wire leads reversed on the terminals of the megger. Six readings measured 11 ohms, two measured 12 ohms, and two measured 13 ohms; the average reading was 11.1 ohms. The measurements were repeated with the wire leads reversed on the megger, with no change recorded in these measurements.

The megger was then connected to electrodes 1 and 2 with the probes separated 50 feet and attached to the proper terminals on the ground megger. The ground resistance of electrodes 1 and 2 measured 5 ohms, as has been reported in Table I.

Building Installations

The military installation tests demonstrate that a low-resistance ground electrode can be obtained in an area where conditions are unfavorable and that such a ground can be continuously effective over many years without maintenance. Encasement of electrodes in concrete and compaction provided by building weight may be the controlling factors in these installations. The supplementary tests on a limited-size electrode indicates that a concrete encasement contributes to the reduction of ground resistance.

A number of tests on footing-type electrode installations in building constructions were made during 1961, 1962, and 1963. Arrangements were made to have this type of electrode installed in conventional residential and commercial buildings. Locations selected were widely separated and these buildings were provided with a footing-type concrete building foundation. *No. 4* AWG solid bare copper wires were embedded in the center of the concrete footing approximately 2 inches above the base of the footing.

The concrete building foundations are T-shaped. The base is about 12 inches wide, while the top of the footing where the foundation still is attached is 6 inches wide. The depth of these footings varies from 2 to 4 feet below the grade level of the building plot.

Two separate lengths of wire were installed in each foundation: one wire approximately 30 feet and the other 100 feet in length. The ends of each wire extended out of the foundation walls to provide ready access for testing and for connection to the neutral terminal in the service switch if and when authorized by the local electrical department.

The area of these buildings on which these installations were made varied from 500 to 8,000 square feet. They are 1-story residential- and commercial-type buildings.

An installation was made with five 20-feet lengths of *No. 4* AWG bare solid copper wire embedded in a concrete building foundation. The ends of each 20-foot length of wire extended from the walls of the concrete footing to be available for measurements. This special installation was made to determine the length of the grounding wire required to provide 5-ohm maximum resistance values.

Bishop Installation

An experienced electrical contractor was contacted in Bishop, Calif., and requested to assist in this program of installing footing-type grounding electrodes in a residential or small commercial building. He had previously received a contract to wire a new residence at Bishop, and the owner agreed to have the footing-type grounding electrode installed.

The California Electric Power Company is the serving public utility for this area. They were notified that this installation was to be made, and their engineer conducted the ground resistance tests when the installation was completed.

The electrical contractor installed two *No. 4* AWG bare copper wires in the concrete forms for the building foundation, one 76 feet long, the other 82 feet. The ends of both wires were brought out of the side of the concrete foundation below the location for the installation of the main switch. The grounding wires were run in opposite directions, and the ends were brought out of the side of the foundation at a location diametrically opposite the wall on which the main switch was installed.

The contractor was requested to install one long wire in the foundation about 100 feet long and one short wire about 25 feet long. This was suggested so that test results would show how much wire was required to provide a ground resistance of not more than 5 ohms.

This installation was made during September 1961. The building is a one-story masonry residence with an area of approximately 1,500 square feet. The concrete footing is T-shaped. The base is 14 inches wide, while the top of the footing where the wood sills are installed is 8 inches wide. The footing is approximately 12 inches below grade level. The *No. 4* AWG bare copper grounding conductor was installed in the center of the concrete foundation about 2 inches above the base of the footing.

On November 1, 1961, 60 days after the concrete had been poured for the foundation, the test engineer for the California Electric Power Company measured the ground resistance with a Siemens & Halske ground megger and recorded a ground resistance of 4.6 and 4.0 ohms (see Table III).

TABLE III

Location	Installation Date	Testing Date	Ground-Wire Resistance (Ohms)	Ground-Wire Length (Feet)	Soil Condition	Average Rainfall (Inches)
Bishop	9/61	11/1/61	4.0, 4.6	76, 82	Gravel, sand	11
		12/2/62	1.2	76 + 82		
Riverside	10/61	11/30/61	2.8, 2.8	100, 100	Red adobe	8
		11/16/62	0.3, 0.3			
Livermore	3/62	4/62	1.3, 1.3	25, 75	Loam, sand	14
		9/62	1.55	25 + 75		
Hayward	9/62	9/62	4.2, 2.8	22, 73	Gravel, large rocks	14
		10/62	3.2, 1.8	22, 22 + 73		
Hayward	3/26/62	4/3/62	1.29, 1.25	25, 75	Adobe, topsoil	14
Portland	2/62	7/62	1.1, 0.8	78, 90	Loam, sand	42.67
		12/62	1.1, 0.8			
Long Beach*	12/61	12/62	0.6		Heavy loam, sand	13.32
		3/25/63†	0.8			
Long Beach‡	8/61	8/62	4.5	30	Heavy loam, sand	13.32
		3/25/63†	5.4			
Twenty-nine Palms	2/8/63	3/22/63	4.3, 2.2§	50, 100	Decomposed granite, sand	0.5
			6.5, 4.4‖			
Palm Springs, Calif.¶	9/20/63	11/21/63**	1	100	Loam, sand	6.74

* Billboard of two 12-inch steel I-beams 8 feet in ground.
† Four days after rain.
‡ No. 6 wire embedded in garage floor.
§ Biddle megger.
‖ Siemens and Halske megger.
¶ More recent data.
** Heavy rain 10/20/63.

The serving public utility is the authority responsible for the inspection and test of all electrical installations in the area. Their engineer authorized the electrical contractor to use this footing-type grounding electrode for grounding the electrical installation of this building and permitted the contractor to connect the free ends of both grounding conductors together to provide a grounding conductor with a total length of approximately 158 feet in the concrete footing. The ground resistance for the 158 feet measured 4.6 ohms.

During a regular trip to Bishop in December 1962, the engineer for the California Electric Power Company made the second test on the ground resistance of the footing-type grounding electrode and recorded a reading of 1.2 ohms.

Riverside Installation

The Chief Electrical Inspector of Riverside, Calif., arranged for the installation of footing-type grounding electrodes in the concrete footings of a residence erected in Riverside. Two *No. 4* AWG bare copper wire electrodes, each 100 feet in length, were installed in the concrete footings of this residence. In all locations, the electrode wire was placed in the concrete forms to be spaced about 2 inches above the base of the footing. The author was at this location during construction to see that the electrode wires were properly placed above the base of the foundation footing before the concrete was poured for the foundation.

The Chief Electrical Inspector of Riverside supervised these installation. The first ground resistance readings were taken November 30, 1961, about 60 days after the concrete was poured, and the second readings were taken 1 year later; all readings are recorded as 2.8 ohms, for the years 1961 and 1962.

Livermore Installation

The City Engineer for Alameda County, Calif., selected the locations for the installation of footing-type grounding electrodes. Two residential buildings were provided with these grounding electrodes. One building was in Livermore, and two electrode wires, 25 and 71 feet long, were installed in the concrete foundation.

Ground resistance measurements were made by the Testing Engineer for the Pacific Gas and Electric Company. The first measurements were made April 3, 1962, 1 week after the concrete was poured and 12 days after rain. The second measurements were made September 20, 1962, 6 months after the concrete was poured. The engineer reported that the adobe and topsoil was very dry and cracked, since the last rain in the area consisted of 0.22 inch in April.

All readings were recorded as less then 2 ohms with very slight difference between the long and short wires in the first readings. For the second measurements, the resistance for the short and long wires were measured separately and when connected together were recorded as 1.5 ohms. Fig 1(A) shows the position of the wires, while Fig. 1(B) gives the results of the measurements.

Hayward Installations

The City engineer of Alameda County selected these locations in Hayward for a footing-type grounding electrode, since the soil was mostly rock and not dependable for a good ground. The last rain in this area was in April 1962.

Two *No. 4* AWG bare copper wires, 22 and 73 feet long, were placed in the concrete foundation for this residence, as shown in Fig. 2(A). The foundation was poured on September 1, 1962.

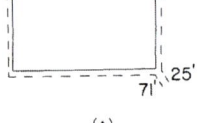

(A)

	APRIL 3, 1962			SEPT. 20, 1962		
LOCATION	25'	71'	BOTH	25'	71'	BOTH
#1	1.2 Ω	1.15 Ω	1.15 Ω	1.55 Ω	1.55 Ω	1.55 Ω
#2	1.27 Ω	1.25 Ω	1.22 Ω	1.55 Ω	1.55 Ω	1.55 Ω
#3	1.29 Ω	1.26 Ω	1.25 Ω	1.55 Ω	1.55 Ω	1.55 Ω

(B)

FIGURE 1 Livermore installation

Ground resistance readings were measured on September 20, 1962, by a test engineer of the Pacific Gas and Electric Company, the serving public utility in this area. As shown in Fig. 2(B), the 22-foot wire measured 4.2 ohms; the 73-foot wire, 3.3 ohms; both wires (95 feet), 2.75 ohms. Measurements taken October 1962 were 3.2 ohms for the 22-foot wire, 1.8 for the 73-foot wire, and 1.8 ohms for both wires.

At a second location, where the earth was a mixture of adobe and topsoil, two wires were placed in the foundation wall as shown in Fig. 3. Readings were made by an engineer from the Pacific Gas and Electric Company on April 3, 1962, 1 week after the footing was poured; the weather was dry, since the last rain had been on March 22, 1962. See Fig. 3 and Table III for the results. A 25-foot ground wire is apparently adequate in this location as the resistance measured less than 2 ohms for both the long and short wires.

Portland Installation

The Chief Electrical Inspector of Portland, Ore., selected the location for a footing-type grounding electrode. Two *No. 4* AWG bare copper wires, approximately 78 and 100 feet long, were installed in the concrete foundation of a 1-story commercial warehouse with dimensions of 80 by 100 feet. The building foundation, the floor slab and the 6-inch precast walls were reinforced concrete, and a wood truss roof on wood columns was provided.

The ground resistance tests recorded 0.8 ohm. The power company then tested the water pipe ground and recorded 1.1 ohms. They decided that the probable cause of the 0.8-ohm reading was that the footing ground wire might possibly have been touching the water pipe. The author visited this installation to check the test results but found that the ground wires had been cut off where they extended beyond the faces of the concrete foundation.

The *No. 4* AWG wires installed in the concrete foundation contacted the steel reinforcing rods, but apparently this did not affect the values of the ground resistance readings.

One of the engineers of the serving public utility cut away the concrete where the wire had been cut and was able to attach the testing conductor to this wire; he recorded a ground resistance of about 1 ohm. Because of the location of this building and the availability of an adequate water pipe system for grounding the electrical installation, the water pipe is being used for the ground since the ends of the *No. 4* grounding wires are not accessible.

Long Beach Installations

The first installation consists of a steel frame sign mounted on two 12-inch steel I-beams spaced 25 feet apart. The ends of the I-beams stand 8 feet in the ground. The ground resistance measured 0.6 ohms.

The second Long Beach installation consists of 30 feet of *No. 6* AWG bare copper wire embedded in the floor slab of a 2-car garage. The wire is in the shape of a circular loop with one end extending beyond the face of the floor slab for connection to the neutral terminal of the service switch for this installation.

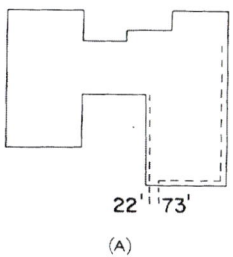

FIGURE 2 Hayward installation

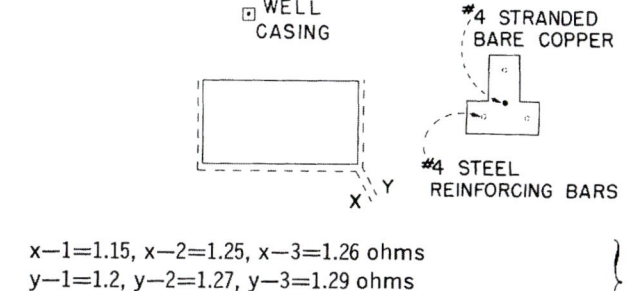

FIGURE 3 Second Hayward installation

Ground resistance readings were made by the Chief Electrical Inspector of Long Beach with a *Model 263A* Vibroground. The resistance measured 4.5 ohms.

The second resistance tests for both Long Beach installations were made 4 days after rain.

Twenty-Nine Palms Installation

The Chief Electrical Inspector for San Bernardino County arranged to have the footing-type ground electrode installed in the concrete foundation of the Methodist Church of Twenty-nine Palms. This is desert location, and ground resistance tests recorded for this area by other engineers indicated that an adequate ground is not available.

The church is a 1-story building approximately 35 by 70 feet with a wood-frame stucco construction on a poured concrete foundation. The concrete footing is T-shaped 12 inches below grade; the base is 14 inches wide, while the upper end is 6 inches wide with ½-inch bolts embedded in the concrete for attaching the foundation sills.

Two *No. 4* AWG bare copper wire electrodes were used, one 50 and the other 100 feet long. One end of the 50-foot wire extended beyond the face of the foundation wall on the north elevation of the building (point C), while the other end of this 50-foot conductor extended beyond the face of the foundation on the west elevation (point A). At point A, one end of the 100-foot electrode wire extends from the face of the foundation wall, while the other end of this 100-foot wire extends itself from the face of the east elevation (point B) of the concrete foundation wall of this building (refer to Fig. 4).

Ground resistance measurements were taken by the Test Engineer for the California Electric Power Company using a Siemens & Halske ground megger *No. 2585269* with a range of 0–25 ohms and 0–2500 ohms. These readings were checked with a Biddle ground megger No. *167273* having three ranges, 0–3, 0–30, and 0–300 ohms.

As bare wire leads were used to record the readings taken with the Siemens & Halske megger, it was decided to use bare and insulated wire for connections from the megger to each of two reference grounds and to connect the third lead to the ground being tested. The results are recorded in Table IV.

An engineer for the California Electric Power Company reported a grounding installation in a residence in San Bernardino County where a 4-inch concrete floor slab was poured over a membrane. The hot and cold water pipes enclosed in concrete were the grounding conductors. An insulating joint was installed in the water supply pipe lateral. The engineer for the California Electric Power Company measured the ground resistance of the copper water pipes enclosed in the concrete floor slab as over 10 ohms. He then measured the ground resistance of the underground water pipe lateral a short distance from the insulating joint as 1.3 ohms.

Length of Electrodes

Installations were made at Alameda and San Bernardino Counties where the soil conditions were not favorable for an adequate ground. At Hayward and Livermore, the soil was poor and filled with large rocks. In these locations, electrode wires 22 and 25 feet long were installed (Table III).

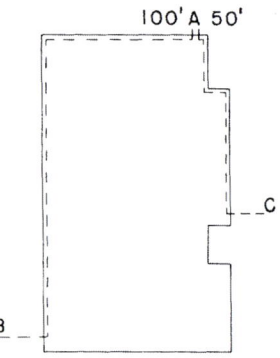

FIGURE 4 Twenty-nine Palms installation

An installation was also made in the desert at Twenty-nine Palms where the soil conditions are not favorable for an adequate ground. The ground resistance readings for these installations, including the 50-foot electrode at Twenty-nine Palms, record measurements of 5 ohms or less using a Biddle megger and a recognized test method.[2]

Burbank Installation

During 1952, a steel floodlight pole was installed in the athletic field of John Burroughs High School, Burbank, Calif. The pole is 14 inches in diameter at eh base and tapers to a diameter of 8 inches at the top; it is 100 feet long over-all and is set in a centering sleeve from the base to a height of 16 feet. This assembly is set in a reinforced concrete footing 8 by 8 by 18 feet. (See Fig. 5.)

Since installation, the resistance to ground has been checked at least once each year by the Burbank city engineers. On July 2, 1963, the ground resistance was checked with their ground megger and measured 1.8 ohms. The soil is sandy loam, the soil resistivity in this area is 4,000 ohms per cm^3, (Table II), and the average rainfall is about 10 inches.

Conclusion

The 20-year history of the military installation in Arizona, together with the shorter records of installations in conventional residential and small commercial buildings, apparently establishes the adequacy and advisability of footing-type grounding electrodes where a continuous underground water system is not available or dependable; in addition, it indicates that the grounding conductor ought to be placed about 2 inches above the base of the concrete foundation footing.

References

1. Guide for Measuring Ground Resistance and Potential Gradients in the Earth. *AIEE Report No. 81,* July 1960.
2. Master Test Code for Resistance Measurement. *AIEE Standard No. 550,* May 1949.

TABLE IV

Location of Ground Lead	Type of Lead	Length of Grounding Conductor (Feet)	Resistance (Ohms)	
			Siemens and Halske Megger	Biddle Megger
Point C	Insulated wire	50	6.0	4.3
		150*	4.4	2.2
		50	6.5	4.3
Point A	Bare wire	50	3.1	5.5
		100	2.0	3.2
		150*	2.0	3.2

*Short and long wires.

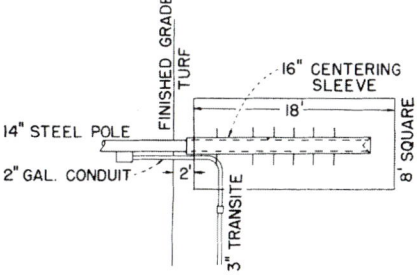

FIGURE 5 Burbank installation

ANNEX B
Steel Conduit and EMT for Equipment Grounding

In addition to *system* grounding, electrical systems require *equipment* grounding and bonding to safeguard personnel and protect equipment. Properly sized and bonded equipment grounding conductors (EGCs) ensure that all metal parts of electrical equipment are at the same electrical potential as the earth to prevent electric shock and provide a low impedance path to facilitate the operation of the circuit protective devices.

Section 250.118 of the National Electrical Code® (*NEC*) recognizes several types of conductors that are permitted to be used as EGCs. In other words, they qualify as EGCs because they provide an effective path for ground-fault current. Rigid metal conduit (RMC), intermediate metal conduit (IMC), and electrical metallic tubing (EMT) are included in 250.118 as list items (2), (3), and (4) respectively, followed by other types of metal raceways, metal cable trays, and metal cables that qualify as EGCs.

Steel conduit and EMT are widely used in secondary power distribution systems both indoors and outdoors. Electrical systems on the load side of the service point are required to be designed in such a way that the steel conduit or EMT does not carry any appreciable electric current under normal operating conditions but performs grounding and bonding functions. During ground-fault conditions, the metal conduit or EMT, acting as an EGC, will carry most of the return fault current, or, in some cases, it will be the only return path of the fault current to the source. This is due to the conductive path of the conduit or tubing, which is usually larger in surface area compared to any wire-type EGC contained within the conduit or tubing run. The ground-fault current divides over all paths available as it returns to the source. The metal conduit around a contained EGC provides impedance in circuit, which will limit or choke the conductor as it is subjected to the heavier levels of ground-fault current. As a result, the majority of the fault current will be present in the metallic raceway containing the wire-type EGC.

In reality, the conduit is only one of the fault current return paths. **See Figure 1.**

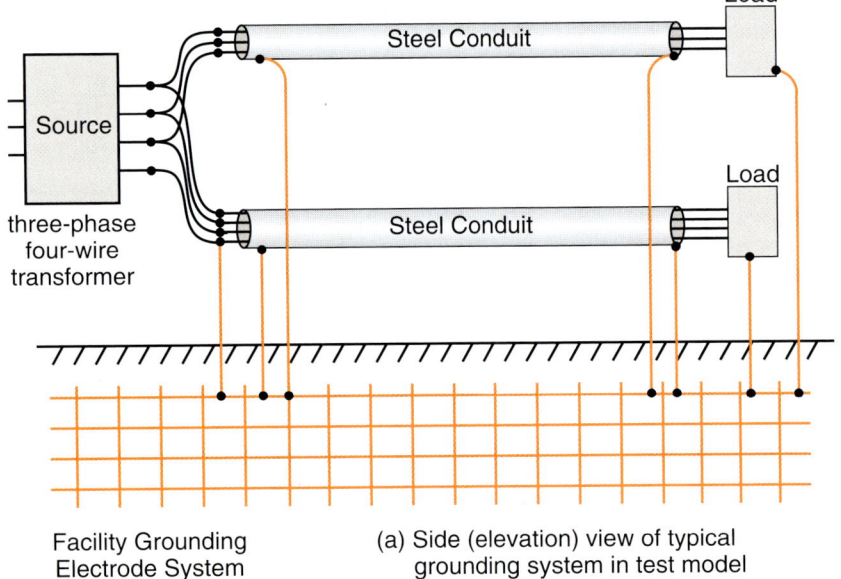

(a) Side (elevation) view of typical grounding system in test model

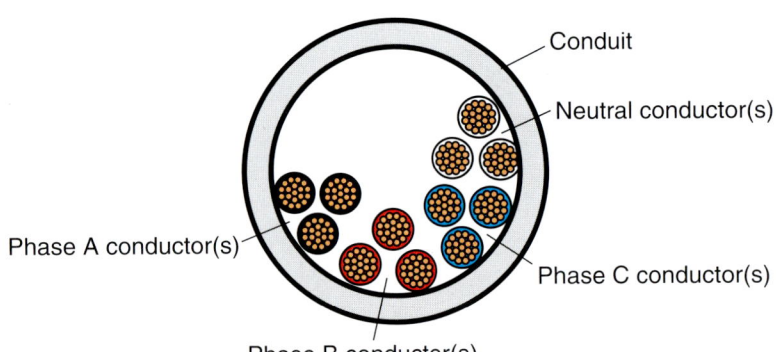

(b) Cross section of a typical steel conduit enclosed distribution circuit

Reproduction of Figure 1. Illustration of steel conduit enclosed secondary distribution system

FIGURE 1 This illustration shows the testing model of steel conduit enclosed secondary distribution system.

Specifically, in a practical system, the fault current will split among several parallel paths when returning to the source. For example, for a phase- to neutral-to-ground fault in the bottom conduit of Figure 1, the fault current will return to the source through the conduit (path BA), neutral (path DC), and facility ground (path FE).

For a phase-to-conduit fault, the fault current will return to the source through the conduit and facility ground. It is possible that for a phase-to-conduit fault, the only return path for the fault current is the conduit. For a single phase-to-neutral conductor fault, the fault-current return path is through the neutral conductor; the neutral conductor may have higher impedance than the previously noted fault-current paths.

Therefore, the type of fault becomes the limiting factor. This report is focused on the performance of steel conduit as the EGC, so the situation where the only fault path is the steel conduit was used as the worst-case condition. It is important to discuss other fault conditions that represent worst conditions and determine the design procedure of electrical installations. **See Figure 2.** For a single phase-to-neutral conductor fault, the fault-current return path is through the neutral conductor only. The steel conduit does not participate in the fault circuit. For neutral conductor sizes, as recommended in standards, the impedance of the fault-current path for these cases is higher than the impedance for a fault to the steel conduit. Consequently, for these cases, the maximum allowable length of the circuit is dictated by the phase-to-neutral conductor fault.

NEC Requirements

Article 250 of the *NEC* contains the general requirements for grounding and bonding of electrical installations, as well as other specific requirements. Sections 250.4(A) and (B) titled "General Requirements for Grounding and Bonding" set forth in detail what must be accomplished by the grounding and bonding of non–current-carrying parts of the electrical system. The conductive parts must form an effective low-impedance path to the source in order to safely conduct any fault current and facilitate the operation of overcurrent devices protecting the enclosed circuit conductors and any equipment in the circuit.

Part VI of Article 250 specifically covers equipment grounding. This part of the article includes the list of acceptable EGCs in Section 250.118.

In order for metallic conduit and EMT to perform effectively as EGCs, it is crucial that they are installed properly with tight joints. Workmanship in electrical construction is extremely important for the integrity of the safety system. If a fault occurs, this helps ensure a continuous, low-impedance path back to the overcurrent protective device, which will open the circuit.

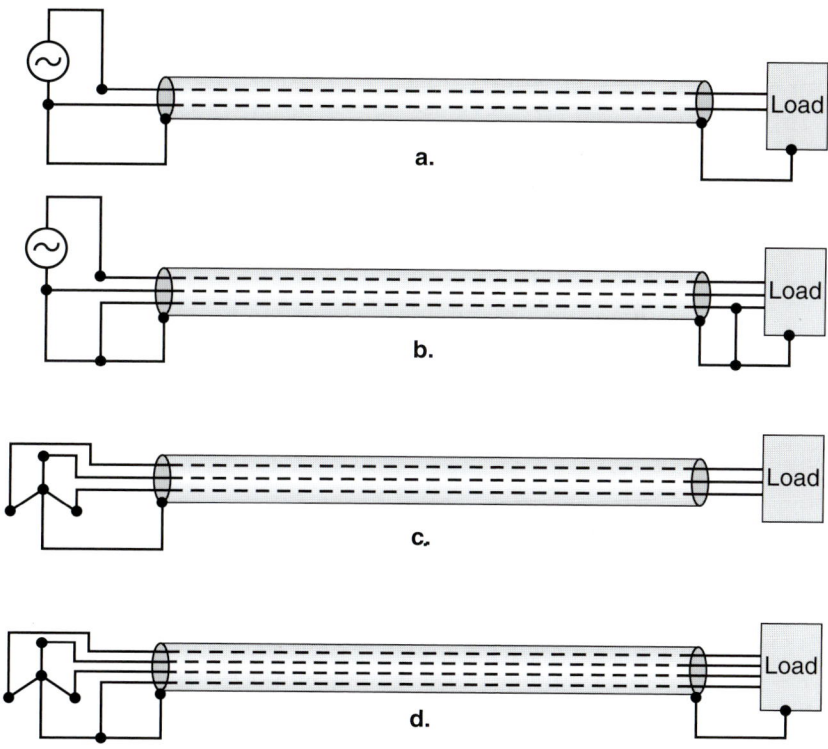

Reproduction of Figure 2.1 a. through d. Illustration of simplified installations

a. Single-phase circuit without equipment grounding conductor
b. Single-phase circuit with equipment grounding conductor
c. Three-phase circuit without neutral conductor
d. Three-phase circuit with neutral conductor

FIGURE 2 Current, be it normal current or ground-fault current, will take any and all paths to return to the power source. The amount of current in a particular path is related to the amount of impedance in that path.

If the joints are not made up tight and there is a break in the circuit under fault conditions, there is a possibility of electric shock for anyone who comes in contact with the conduit. *NEC* Sections 300.10 (Electrical Continuity) and 300.12 (Mechanical Continuity) state that metal raceways, cable armor, and other metal enclosures for conductors shall be metallically joined together into a continuous electrical conductor and shall be connected to all boxes, fittings, and cabinets so as to provide effective electrical continuity. Section 250.120 of the *NEC* requires that all connections, joints, and fittings *"shall be made tight using suitable tools."*

Section 250.122 covers the sizing of wire-type EGCs and includes Table 250.122 titled, *Minimum Size Equipment Grounding Conductors for Grounding Raceway and Equipment*. An important, mandatory note at the bottom of Table 250.122 makes it clear that the conductor sizes given in Table 250.122 may not be adequate to comply with 250.4(A)(5), *Effective Ground-Fault Current Path* (Grounded Systems) and 250.4(B)(4), *Path for Fault Current* (Ungrounded Systems). These installations may have to be evaluated to ensure that they can provide the effective ground-fault current path.

Conduit as Effective Ground-Fault Current Path

The *NEC* does not dictate any particular size of conduit or tubing to serve as the EGC for an upstream overcurrent device. The metallic raceway that is sized properly for the conductor fill will provide an adequate equipment ground-fault return path. In 1966, Eustace C. Soares, a renowned expert in the area of grounding electrical systems, published the first edition of his book, *Grounding Electrical Distribution Systems for Safety*, which included Tables showing acceptable lengths of steel conduit and EMT (metal raceways?) for equipment grounding, based on his calculations. The work of Soares did not include any field testing of conduit systems to measure performance as effective paths for fault current.

In the early 1990s, U.S. steel conduit producers decided to undertake a research project in order to confirm the work expressed by Soares to provide scientific proof that steel conduit and EMT do provide an adequate equipment ground-fault return path and to develop software that would assist engineers and others in determining the appropriate maximum lengths of steel conduit and EMT when they are serving as EGCs.

Georgia Institute of Technology Grounding Research

One of the top experts in the field of grounding, Dr. Sakis Meliopoulos, professor of Electrical and Computer Engineering at Georgia Tech, facilitated the EGC research project, which was completed in 1994. This research represented the first substantial update on the impedance and permeability of steel conduit in over forty years.

The first phase of the grounding research at Georgia Tech consisted of resistance and permeability testing of various steel raceways that were purchased from local distributors. Based on this information, Dr. Meliopoulos and his team developed a computer model and validated the results through actual field testing. The next step was to develop a computer software program that allowed the user to calculate the appropriate length of steel conduit or EMT runs necessary to meet *NEC* requirements.

It is important to discuss other fault conditions that represent worst conditions and determine the design procedure of electrical installations. For this reason consider the simplified installations of Figure 2.1. For a single phase-to-neutral conductor fault in the systems of Figures 2.1a, 2.1b, and 2.1d, the fault-current return path is through the neutral conductor only.

The steel conduit does not participate in the fault circuit. For neutral conductor sizes, as recommended in standards, the impedance of the fault-current path for these cases is higher than the impedance for a fault to the steel conduit. Consequently, for these cases, the maximum allowable length of the circuit is dictated by the phase-to-neutral conductor fault.

In recent years, fault-current levels have increased. For this reason, it is important to examine existing parameters for EGCs. A program was initiated to evaluate the performance of steel EMT, IMC, or RIGID conduit during faults in secondary distribution systems. A relevant issue is that of grounding of steel conduit. For many technical and safety reasons, electric power installations must be grounded. Two main considerations are: (a) protection in the event of faults and (b) avoidance of electric shocks.

As the capacity of the secondary distribution systems increases, so do the short circuit capacity (available fault current levels) and associated protection and safety concerns. Performance of EGCs can be best determined by exact modeling and testing of the system under various excitation and fault conditions. This research project did exactly that.

The above issues have been investigated by performing the following tasks: (1) modeling of steel conduit enclosed multi-conductor systems, (2) laboratory testing of several representative steel conduit types under low currents for the purpose of characterizing the magnetic material, (3) computer simulation of steel conduit enclosed secondary distribution systems, (4) full-scale testing of steel conduit enclosed power systems under high-fault current, (5) full-scale measurement of arc voltage for various fault current levels, and (6) analysis of full-scale test results and conclusions. Details of these tasks are included in the Georgia Institute report.

A summary of the results of the investigation is as follows:

- A model of steel conduit enclosed power distribution systems has been developed and validated with laboratory and full-scale measurements 256 feet in length.
- Comparably sized steel EMT, IMC, and RIGID conduit will allow the flow of higher fault current than an EGC as listed in *NEC* Table 250.122.
- Steel EMT, IMC, and RIGID conduit, not exceeding the maximum allowable length, meets the performance requirements of Section 250.4 of the *NEC*. As a matter of fact, the performance of steel conduit sized in accordance with Chapter 9, Table 1 of the *NEC*, compared to the minimum size EGCs in Table 250.122, allows the flow of higher fault current. This is due to the lower impedance of the steel conduit.
- Steel EMT, IMC, and RIGID conduit are of sufficiently low impedance to limit the voltage to ground and facilitate the operation of the circuit protective devices in runs not exceeding the maximum allowable lengths detailed in this report. In most cases, the maximum allowable lengths exceed those permitted by the IAEI Soares *Book on Grounding* using the same arc voltage and ground-fault current.
- The arc voltage of 50 V and the ground-fault current of 500% of the overcurrent device rating, as stated in the IAEI Soares *Book on Grounding*, 1993 edition [1], is overly conservative as a design guide. Testing for this project confirmed the actual voltage across a fault arc to be up to 30 V for an arc length of 75 mils and fault current ranging from 400 to 2500 A.
- A validated computer model has been developed that computes the maximum allowable lengths for specific conductor size enclosed by specific size steel EMT, IMC, or RIGID conduit under fault conditions.
- Where lengths do not exceed the maximum allowable computed by the method, supplemental grounding conductors in secondary power systems enclosed in steel EMT, IMC, or RIGID conduit are not necessary. The supplemental conductor is sometimes required by the *NEC* in critical installations such as healthcare areas where dual protection for patients is considered prudent.

- The recommended maximum allowable lengths listed in the IAEI Soares *Book on Grounding* were compared to those computed with the validated model for the same arc voltage and ground-fault current. In general, it was found that the IAEI Soares *Book on Grounding* numbers are in reasonable agreement except in some cases where errors were found. The detailed comparison and explanation is given in Section 5.
- The maximum allowable length for a specific system depends on conductor size, steel conduit size, and fault type. In many cases, the maximum allowable length for a phase-to-neutral fault is shorter than the maximum allowable length for a phase-to-steel conduit fault. Thus in most cases, the steel conduit is *not* the limiting factor. In these cases, use of a supplemental grounding conductor will not increase the maximum allowable length. The use of the validated computer model to compute the maximum allowable length in specific systems is recommended.

A few years later, Georgia Tech conducted research on steel conduit to show how steel conduit reduces electromagnetic fields. This research was added to the grounding research and ultimately rolled into the GEMI (**G**rounding and **E**lectro **M**agnetic **I**nterference) software analysis program now available for free downloading at www.steelconduit.org.

Findings in Georgia Tech Grounding Research

The GEMI research project and resulting software analysis program proved that listed steel conduit and EMT clearly exceed the minimum equipment grounding requirements of the *NEC*. In addition, the GEMI research on grounding verified the following:

- Comparably sized steel rigid metal conduit, IMC, and EMT allow the flow of higher fault current than an aluminum or copper EGC as provided in *NEC* 250.118.
- Rigid metal conduit, IMC, and EMT provide a low impedance path to ground and facilitate the operation of the overcurrent devices in runs not exceeding the maximum allowable lengths detailed in the research report.
- Where lengths do not exceed the maximum allowable computed by the GEMI software, supplemental grounding conductors in secondary power systems enclosed in steel EMT, IMC, or rigid conduit do not add to safety in a phase-to-neutral fault. The use of supplementary EGCs, when participating in the fault circuit, reduce the overall impedance and may or may not increase the allowable length of the run, depending on the size and system design.
- The maximum allowable length for a specific system depends on conductor size, steel conduit size, and fault type. In many cases, the maximum allowable length for a phase-to-neutral fault is shorter than the maximum allowable length for a phase-to-steel conduit fault. Thus, in most cases, the steel conduit is not the limiting factor in a conductor-to-neutral fault.

Summary

A computer model has been developed that computes the impedance of steel conduit with enclosed power conductors. The computer model has been validated with full-scale tests. The model is capable of predicting the effect of temperature on the total impedance. The measured impedances during the full-scale tests are within the values predicted with the model in the temperature range 25°C to 55°C. It is important to note that the full-scale tests consisted of repeated short circuit tests on samples of steel conduit systems, and therefore the pre-fault temperature was different for each test but generally in the above stated range of temperatures.

Tests of arc voltage were performed. These tests consisted of generating an arc between two electrodes. The current through the arc was controlled by current-limiting impedance. The separation distance between the two electrodes was measured before and after the test. This separation distance was always longer than the thickness of power conductor insulation and, therefore, represents worst-case (or conservative) results.

Full-scale tests were performed with and without supplementary grounding conductors. **See Figure 3.** Supplementary grounding conductors, when participating in the fault circuit, reduce the overall impedance. However, it is important to note that the limiting factor in the capability of the system to interrupt a fault is the size of the phase conductor and neutral. Specifically, for systems designed with present standards, the fault circuit for a phase conductor to neutral fault (steel conduit or supplementary ground conductor is not involved in the fault) presents the maximum impedance and, therefore, will draw the minimum fault current as compared to other faults at the same location. Use of supplementary EGCs does not add to the safety of the systems in these instances. The maximum allowable length of a steel conduit run can be increased to some degree by the addition of a supplemental grounding conductor. This varies by size and system designs. An additional project is undertaken to expand the model to compute the maximum allowable length in these cases.

The power source utilized for the tests had the following short circuit characteristics:

- Available fault current at the 120 V tap: 85k A
- Available fault current at the 277 V tap: 83k A

The waveforms of the electric current and voltage were recorded by means of a strip chart recorder, as well as by means of a digitizing scope and subsequent transfer of the data to a personal computer for further analysis. Georgia Tech provided the digitizing oscilloscope and personal computer.

In this report, the Georgia Institute also examined the practice as provided by the original Soares *Book on Grounding*. The findings suggested to use 500% of the protective device rating as the ground-fault current for interruption of the fault to be overly conservative. Present industry experience with protective devices indicates that a fault current of 300% of protective device rating will result in a reliable operation of the protective device. Current practice of protective device testing is to test the protective device for an electric current up to 300% of its rating. Based on these observations, it could be concluded that even the value of 400% of the protective device rating for ground-fault current would be conservative.

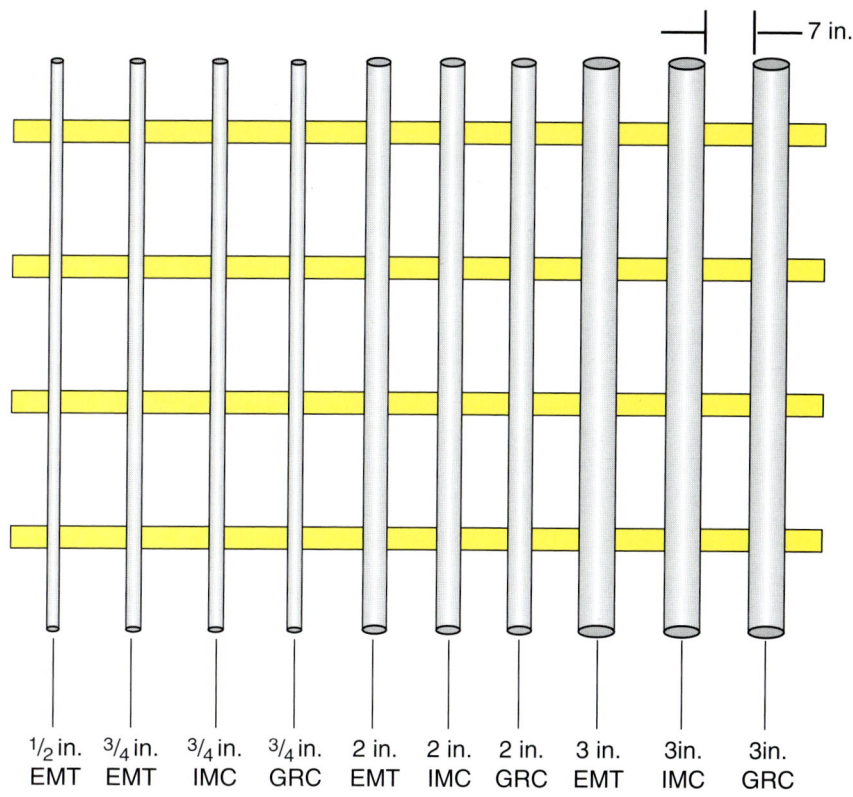

Reproduction of the illustration showing the layout of ten runs of conduit assembled for the full scale testing (the conductors are not shown). The total length of each conduit was 256 feet and wood blocks were used to space the conduit systems.

FIGURE 3 These conduit runs, installed for testing purposes, are 256 feet in length and spaced 7 inches apart.

Workers pulling wire and preparing the conduit models for ground fault testing.

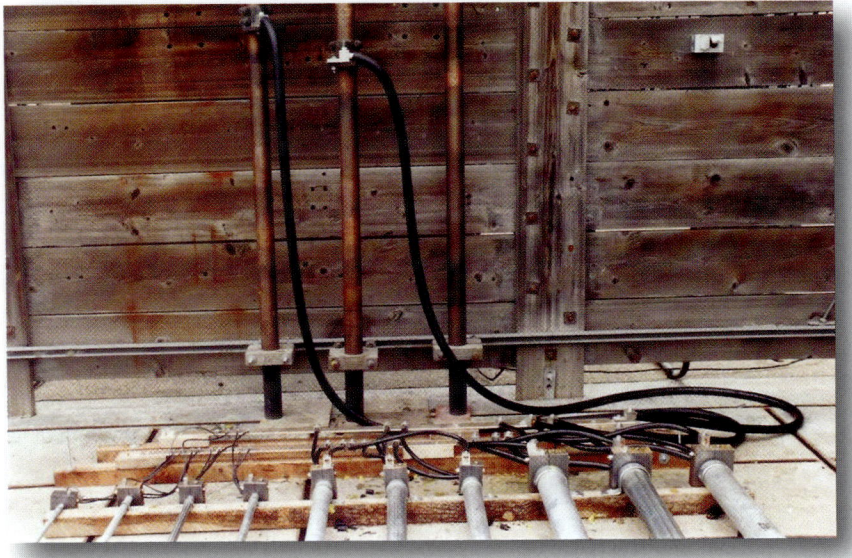
The conduit testing model was prepared to simulate a variety of ground-fault current injection tests.

All computations of maximum run and conclusions are based on the impedance of the steel conduit enclosed power distribution system. For this reason, it is expedient to base the comparison of the experimental results and the modeling results on the impedance of the steel conduit enclosed distribution system. This comparison is summarized in Tables 5.1 and 5.2 for systems with 120 and 277 volt nominal voltage, respectively. In addition, Tables 5.3 and 5.4 summarize the measured data for all cases. The Tables represent the summary of all tests done at Kearney Laboratories in McCook, IL (the May 1993 and August 1993 tests). For the complete NEMA Report of the Georgia Institute of Technology research project on Modeling and Testing of Steel EMT, IMC and RIGID (GRC) Conduit visit www.steelconduit.org.

Tables Reflecting Maximum Length of Metallic Conduit as Effective Ground-Fault Current Path

TABLE 5.1. Summary of Test Results and Model Comparison for 120 Volt Systems. Steel Conduit is the Only Return Path. Impedance Values are Given in mOhms per 100 Feet.

Steel Conduit Type and Conductor Size	Electric Current Level (Amperes)	Measured Impedance R, X	Computed Impedance R, X
EMT - 1/2Z CU-#10	247	187.9, 30.55	188.71, 24.72
EMT - 1/2S CU-#10	228	199.3, 32.15	189.05, 26.65
EMT - 3/4 CU-#8	345	130.1, 25.33	121.11, 21.59
IMC - 3/4 CU-#8	437	102.0, 29.63	105.29, 27.55
GRC - 3/4 CU-#8	417	106.8, 30.13	105.31, 31.65
EMT - 2 CU-3/0	1822	21.56, 10.64	22.37, 8.96
IMC - 2 CU-3/0	1849	19.78, 12.89	18.83, 12.56
GRC - 2 CU-3/0	1758	21.22, 13.04	18.82, 13.41
EMT - 3 CU-350	2578	13.15, 9.86	13.02, 10.44
IMC - 3 CU-350	2609	12.69, 10.21	11.56, 10.43
GRC - 3 CU-350	2433	13.98, 10.86	11.81, 10.41

Z zinc diecast coupling
S Steel coupling
CU Copper Conductor

TABLE 5.2. Summary of Test Results and Model Comparison for 277 Volt Systems. Steel Conduit is the Only Return Path. Impedance Values are Given in mOhms per 100 Feet.

Steel Conduit Type and Conductor Size	Electric Current Level (Amperes)	Measured Impedance R, X	Computed Impedance R, X
EMT - 1/2Z CU-#10	537	193.1, 17.76	187.49, 14.84
EMT - 1/2S CU-#10	546	193.7, 18.6	187.47, 14.66
EMT - 3/4 CU-#8	851	123.9, 15.5	119.64, 12.20
IMC - 3/4 CU-#8	1066	95.1, 18.98	100.79, 15.57
GRC - 3/4 CU-#8	997	95.9, 20.25	97.63, 18.36
EMT - 2 CU-3/0	4628	20.39, 7.28	21.95, 6.17
IMC - 2 CU-3/0	5422	15.78, 8.40	16.64, 7.53
GRC - 2 CU-3/0	5271	15.71, 9.59	15.12, 8.15
EMT - 3 CU-350	7092	11.33, 6.94	11.72, 6.84
IMC - 3 CU-350	6775	9.71, 8.35	8.69, 7.52
GRC - 3 CU-350	6920	10.18, 8.68	8.40, 7.87

Z zinc diecast coupling
S Steel coupling
CU Copper Conductor

TABLE 5.3. Summary of Test Results and Model Comparison for 120 Volt Systems. Effect of Other Parallel Fault Current Return Paths. Impedance Values are Given in mOhms per 100 Feet.

Steel Conduit Type and Conductor Size	Electric Current Level (Amperes)	Measured Impedance R, X	Measured Impedance with a Supplementary Conductor	Measured Impedance with a Supplementary Conductor and Earth Return
EMT - 1/2Z CU-#10	247	187.9, 30.55	157.2, 17.53	137.77, 62.59
EMT - 1/2S CU-#10	228	199.3, 32.15	163.8, 17.53	137.8, 62.6
EMT - 3/4 CU-#8	345	130.1, 25.33	105.6, 15.56	99.2, 60.05
IMC - 3/4 CU-#8	437	102.0, 29.63	89.84, 16.53	90.43, 16.33
GRC - 3/4 CU-#8	417	106.8, 30.13	91.19, 15.87	91.69, 16.08
EMT - 2 CU-3/0	1822	21.56, 10.64	11.34, 10.34	11.38, 10.35
IMC - 2 CU-3/0	1849	19.78, 12.89	11.29, 10.57	11.30, 10.58
GRC - 2 CU-3/0	1758	21.22, 13.04	11.42, 10.46	11.44, 10.43
EMT - 3 CU-350	2578	13.15, 9.86	6.89, 10.15	6.93, 10.20
IMC - 3 CU-350	2609	12.69, 10.21	6.95, 9.96	7.03, 10.00
GRC - 3 CU-350	2433	13.98, 10.86	6.89, 10.15	6.94, 10.13

Z zinc diecast coupling
S Steel coupling
CU Copper Conductor

TABLE 5.4. Summary of Test Results and Model Comparison for 277 Volt Systems. Effect of Other Parallel Fault Current Return Paths. Impedance Values are Given in mOhms per 100 Feet.

Steel Conduit Type and Conductor Size	Electric Current Level (Amperes)	Measured Impedance R, X	Measured Impedance with a Supplementary Conductor	Measured Impedance with a Supplementary Conductor and Earth Return
EMT - 1/2Z CU-#10	537	193.1, 17.76	161.92, 13.77	135.88, 48.66
EMT - 1/2S CU-#10	546	193.7, 18.6	161.4, 13.86	134.6, 48.62
EMT - 3/4 CU-#8	851	123.9, 15.5	102.44, 12.51	89.70, 47.00
IMC - 3/4 CU-#8	1066	95.1, 18.98	88.48, 15.14	88.54, 15.55
GRC - 3/4 CU-#8	997	95.9, 20.25	88.16, 15.91	90.06, 15.23
EMT - 2 CU-3/0	4628	20.39, 7.28	11.87, 8.87	11.90, 8.87
IMC - 2 CU-3/0	5422	15.78, 8.40	11.23, 8.97	11.18, 8.95
GRC - 2 CU-3/0	5271	15.71, 9.59	11.23, 9.23	11.28, 9.24
EMT - 3 CU-350	7092	11.33, 6.94	7.15, 8.63	7.11, 8.59
IMC - 3 CU-350	6775	9.71, 8.35	6.74, 8.32	6.81, 8.36
GRC - 3 CU-350	6920	10.18, 8.68	7.14, 8.70	6.91, 8.69

Z zinc diecast coupling
S Steel coupling
CU Copper Conductor

Complete Report and Other Information

The GEMI software is available for free downloading at www.steelconduit.org. Click on "resources/downloads/GEMI analysis research reports" where you will also find copies of the two complete Georgia Institute of Technology research reports on grounding and on shielding against electromagnetic fields. This Annex provided simplified information related to testing the effectiveness of steel conduit systems as a ground-fault current path. Full information is provided in the GEMI analysis reports developed by the Georgia Institute of Technology.

All information contained in Annex B is courtesy of Steel Tube Institute.

ANNEX C

UL Guide Information for Electrical Equipment (2009 White Book)

Grounding and Bonding Equipment KDER

USE

This category covers bonding devices, ground clamps, grounding and bonding bushings and locknuts, ground rods, armored grounding wire, protector grounding wire, grounding wedges, ground clips for securing the ground wire to an outlet box, water meter shunts, and similar equipment.

Some devices are to be assembled to wire using a special tool specified by the manufacturer. Such special tooling is identified by appropriate marking on or within the device shipping carton.

Armored Grounding Wire — Armored grounding wire consisting of a single corrosion-resistant copper, aluminum or copper-clad aluminum conductor within helically-formed steel armor is marked with the size of the conductor "Bare Armored Grounding Wire."

Ground Rods — Ground rods and pipe electrodes are suitable for use as grounding electrodes in accordance with ANSI/NFPA 70, "National Electrical Code" (NEC), and are also suitable for use in installation of lightning protection equipment.

Ground rods are solid copper, solid stainless steel, copper-jacketed steel, stainless-steel jacketed, galvanized steel, and chemically charged. They are not less than 1/2 in. diameter and not less than 8 ft long and capable of being driven to a depth of 8 ft. If other than circular, they have a periphery not less than 1.6 in. and a minimum thickness of not less than 3/8 in.

Ground rods are marked with the rod length, and manufacturer's name and catalog number within 12 in. of the top of the rod.

The ground rods of a sectional ground rod kit consisting of two four-foot sections of ground rods, a driving sleeve, and a ground rod coupling are marked with the manufacturer's name, catalog number, rod size and length, and "Sectional Ground Rod" within 12 in. of the top of each rod.

Ground rod couplings are intended for connection of two ground rods and are suitable for direct burial.

Ground Clamps — Strap-type ground clamps are not suitable for attachment of the grounding conductor of an interior wiring system to a grounding electrode.

Ground clamps and other connectors suitable for use where buried in earth or embedded in concrete are marked for such use. The marking may be abbreviated "DB" (for "Direct Burial").

Ground clamps are also suitable for telecommunication applications, such as telephone, radio, CATV and the like, in accordance with Articles 800, 810, 820 and Section 250.94 of the NEC, in addition to those covered under Grounding and Bonding Equipment, Communication (KDSH).

Ground clamps are intended for use with ground rods and/or pipe electrodes in accordance with the NEC and are marked with the size of electrode and electrode grounding conductor with which the clamp is intended to be used. Clamps suitable for use on copper water tubing are marked "Copper Water Tubing," or the equivalent, preceded or followed by the size of tubing. Ground rods, pipe electrodes and water tubing trade sizes are stated in fractions, such as 1/2, 5/8, etc.

Ground clamps intended for use with re-bar are marked with the size of re-bar with which the clamp is intended. Re-bar sizes may be specified in fractions, such as 1/2, 5/8, etc., or a number, such as 3, 4, 5, etc., where the number represents the numerator of the fraction when stated in eighth-inch increments, e.g., 4 = 4/8.

Grounding and Bonding Bushings — Bonding bushings for use with conduit fittings, tubing (EMT) fittings, threaded rigid metal and intermediate metal conduit, or unthreaded rigid metal and intermediate metal conduit are provided with means (usually one or more set screws) for reliably bonding the bushing (and the conduit on which it is attached) to the metal equipment enclosure or box. They provide the electrical continuity required by the NEC at service equipment and for circuits rated over 250 V. Means for connecting a grounding or bonding conductor are not provided and if there is need for such a conductor a grounding bushing should be used.

Grounding bushings for use with conduit fittings, tubing (EMT) fittings, threaded rigid metal and intermediate metal conduit, or unthreaded rigid metal and intermediate metal conduit have provision for the connection of a bonding or grounding wire or have means for mounting a wire connector available from the manufacturer. Such a bushing may also have means (usually one or more set screws) for reliably bonding the bushing to the metal equipment enclosure or box in the same manner that this is accomplished by a bonding bushing. Grounding bushings provide the electrical continuity required by the NEC at service equipment and for circuits rated over 250 V. They may be used with or without a bonding or grounding conductor as determined by the bonding or grounding function that is intended to be accomplished.

Insulating throat liners in grounding or bonding bushings are suitable for temperatures of 150°C if they are black or brown in color. Unless otherwise marked, insulating throat liners of any other color are suitable for temperatures of 90°C.

Grounding and Bonding Locknuts — Grounding and bonding locknuts serve in a manner similar to grounding and bonding bushings except they do not provide abrasion protection for the conductor at the end of the conduit.

Grounding and Bonding Hubs — Grounding and bonding hubs are Listed hubs (see DWTT) provided with a Listed grounding or bonding locknut. They serve in a manner similar to grounding and bonding bushings except they are only for use with threaded rigid metal and intermediate metal conduit. Grounding hubs provide the electrical continuity required by NEC 250.92 at service equipment and the electrical continuity required by NEC 250.97 for circuits rated over 250 V.

Ground Clips — Ground clips are intended to be pressed on the flat surface of a square, rectangular, or octagonal box to hold a grounding conductor against the sidewall of the box. Ground clips are not intended for use with round boxes. Ground clips are typically used for connecting the grounding conductor of various wiring methods to outlet boxes or for connecting the bonding jumper from a receptacle, switch or other device to an outlet box.

Ground Mesh — The ground mesh consists of a copper wire mesh that is intended to be installed in ground or embedded in concrete and bonded to the grounding electrode system for the purpose of improving ground planes, such as an equipotential plane as described in Sections 547.2, 547.10 and 680.26 of the NEC. Ground mesh is not intended to serve as a required grounding electrode as described in Article 250 of the NEC.

Fittings — A fitting such as a hub, bushing or locknut intended to provide a raintight or liquidtight connection is marked "Raintight," "Type 3R," "Type 4" or "Wet Locations."

Protector Grounding Wires — Protector grounding wires are intended for use in accordance with Article 800 of the NEC. They are marked with the manufacturer's name, size, and "Protector Grounding Wire."

Water Meter Shunts — Consists of a 4 AWG or larger solid copper wire connected between two ground clamps that comply with requirements for such ground clamps.

Miscellaneous Devices — Grounding and bonding equipment not specifically mentioned above, such as bonding locknuts, gaskets, grounding wedge lugs, adapters, grounding grids and the like, are investigated under the intent of the requirements in the standard.

PRODUCT MARKINGS

Some of the markings referred to above may be on a tag attached to the product.

Grounding and bonding devices are intended for use only with copper conductors unless they are marked "AL" or "AL-CU."

RELATED PRODUCTS

Hospital grounding jacks and grounding cord assemblies are covered under Hospital Ground Jacks and Grounding Cord Assemblies (KEVX).

Equipment for grounding and bonding for telecommunication applications is covered under Grounding and Bonding Equipment, Communication (KDSH).

Grounding and bonding hubs may additionally be covered as a hub under Conduit Fittings (DWTT).

ADDITIONAL INFORMATION

For additional information, see Electrical Equipment for Use in Ordinary Locations (AALZ).

REQUIREMENTS

The basic standard used to investigate products in this category is ANSI/UL 467, "Grounding and Bonding Equipment."

UL MARK

The Listing Mark of Underwriters Laboratories Inc. on the product, on a tag securely attached to the product or container, or on the smallest unit container in which the product is packaged is the only method provided by UL to identify products manufactured under its Listing and Follow-Up Service. The Listing Mark for these products includes the UL symbol (as illustrated in the Introduction of this Directory) together with the word "LISTED," a control number, and one of the following product names: "Grounding Equipment," "Bonding Equipment," "Bonding Jumper," "Ground Clamp," or other appropriate product name as shown in the individual Listings.

Grounding and Bonding Equipment, Communication KDSH

USE

This category covers grounding devices intended for use in telecommunication applications, such as telephone, radio, CATV and the like, in accordance with Articles 800, 810, 820 and Section 250.94 of ANSI/NFPA 70, "National Electrical Code" (NEC).

Strap-type ground clamps constructed of perforated or expanded metal are suitable for grounding conductor connections to electrodes for indoor telecommunications purposes only. Where permitted by the NEC, they are also suitable in both indoor and outdoor applications when used for bonding purposes only.

Strap-type ground clamps are intended for use with pipe electrodes in accordance with the NEC and are marked with the size of electrode and electrode grounding conductor with which the clamp is intended to be used. Clamps suitable for use on copper water tubing are marked "Copper Water Tubing" or the equivalent, preceded or followed by the size of tubing. Pipe electrodes and water tubing trade sizes are stated in fractions, such as 1/2, 5/8, etc.

PRODUCT MARKINGS

Some of the required markings may be on a tag attached to the product.

RELATED PRODUCTS

Ground clamps covered under Grounding and Bonding Equipment (KDER) are also suitable for use in applications as specified in this category.

ADDITIONAL INFORMATION

For additional information, see Electrical Equipment for Use in Ordinary Locations (AALZ).

REQUIREMENTS

The basic standard used to investigate products in this category is ANSI/UL 467, "Grounding and Bonding Equipment."

UL MARK

The Listing Mark of Underwriters Laboratories Inc. on the product, on the smallest unit container in which the product is packaged, or on a tag securely attached to the product or container, is the only method provided by UL to identify these products manufactured under its Listing and Follow-Up Service. The Listing Mark for these products includes the UL symbol (as illustrated in the Introduction of this Directory) together with the word "LISTED," a control number, and the product name "Ground Clamp – Communication."

Grounding Equipment, Neutral Grounding Devices, Over 600 Volts KDZC

GENERAL

This category covers neutral grounding devices intended for use on systems having ac voltage ratings from 601 V to 38 kV. Neutral grounding devices are used for the purpose of controlling the ground current or the potentials to ground of an alternating-current system.

These devices are grounding transformers, ground-fault neutralizers, resistors, reactors, capacitors, or a combination of these. In addition, these devices may include current sensors, relays, audible and visual signaling and similar accessories.

PRODUCT MARKINGS

Devices suitable for outdoor use are marked "Outdoor."

Enclosures are marked to indicate the exposure category (A, B or C) for which they are intended. Enclosures marked "Category A" are intended to be installed in areas accessible to the unsupervised general public; enclosures marked "Category B" are intended to be installed in areas accessible to authorized personnel only; enclosures marked "Category C" are intended for use in areas accessible to qualified personnel only.

Devices covered under this category are marked with the following information: Name of manufacturer, serial number, name of device, type designation, impedance (except resistors), number of phases as applicable, rated current, rated frequency, rated time, rated voltage, BIL of line, indoor or outdoor service, weight, volume of oil (as applicable), instruction book number or equivalent.

ADDITIONAL INFORMATION

For additional information, see Electrical Equipment for Use in Ordinary Locations (AALZ).

REQUIREMENTS

The basic standards used to investigate products in this category are ANSI/IEEE 32-1972, "IEEE Standard Requirements, Terminology, and Test Procedure for Neutral Grounding Devices," and ANSI/IEEE C37.20.3- 2001, "Metal-Enclosed Interrupter Switchgear."

Health Care Facilities Equipment KEVQ

GENERAL

This category covers appliances, utilization equipment and construction materials which have been judged to be particularly applicable to a health care facility as defined by Article 517 of ANSI/NFPA 70, "National Electrical Code."

The general information under the specific categories indicate the areas in which the individual Listings are intended to apply in health care facility installations.

This equipment, unless otherwise indicated, is for installation in unclassified (ordinary) areas of health care facilities.

Hospital Ground Jacks and Grounding Cord Assemblies KEVX

USE

This category covers hospital ground jacks and mating grounding cord assemblies intended for use in hospital rooms or other in health care facilities to connect equipment to a patient grounding point or other appropriate reference grounding point.

The visible face of a grounding jack is green.

PRODUCT MARKINGS

The cover of a hospital grounding jack having a twist-to-lock configuration is marked "Locked – for Grounding" or "Twist to Lock – for Grounding."

RELATED PRODUCTS

General equipment for grounding and bonding is covered under Grounding and Bonding Equipment (KDER).

Equipment for grounding and bonding for telecommunication applications is covered under Grounding and Bonding Equipment, Communication (KDSH).

ADDITIONAL INFORMATION

For additional information, see Electrical Equipment for Use in Ordinary Locations (AALZ).

REQUIREMENTS

The basic standard used to investigate products in this category is ANSI/UL 467, "Grounding and Bonding Equipment."

UL MARK

The Listing Mark of Underwriters Laboratories Inc. on the product or on the smallest unit container in which the product is packaged is the only method provided by UL to identify products manufactured under its Listing and Follow-Up Service. The Listing Mark for these products includes the UL symbol (as illustrated in the Introduction of this Directory) together with the word "LISTED," a control number, and the product name "Grounding Jack" or "Grounding Cord Assembly," or other appropriate product name as shown in the individual Listings.

Isolated Power Systems Equipment KEWV

These listings include isolated power centers which incorporate complete assemblies of isolation transformers and one or more isolated secondary circuits terminated in integrally mounted grounding type load receptacles in an overall enclosure which are intended for use in health care facilities where it is considered desirable to minimize available leakage and short-circuit currents.

Line isolation monitors may be included in the assembly to indicate the "condition" of the isolated circuit and its connected components with respect to electrical ground.

Other distribution panels listed as isolated power panelboards incorporate the same features as described above except that they may be supplied with power from a separate isolation transformer. They are connected by an approved wiring method to remote receptacles located in operating rooms or other anesthetizing location areas of health care facilities.

Accessory equipment, such as terminal assemblies located in patient care areas, are also included in these listings.

For additional information, see Electrical Equipment for Use in Ordinary Locations (AALZ).

The basic standard used to investigate products in this category is UL 1047, "Isolated Power Systems Equipment."

The Listing Mark of Underwriters Laboratories Inc. on the product is the only method provided by UL to identify products manufactured under its Listing and Follow-Up Service. The Listing Mark for these products includes the UL symbol (as illustrated in the Introduction of this Directory) together with the word "LISTED," a control number, and the product name "Isolated Power Systems Equipment".

Ground-fault Circuit Interrupters KCXS

GENERAL

This category covers ground-fault circuit interrupters (GFCI) for use in accordance with ANSI/NFPA 70, "National Electrical Code" (NEC).

A GFCI is a device whose function is to interrupt the electric circuit to the load when a fault current to ground exceeds some predetermined value that is less than that required to operate the overcurrent protective device of the circuit.

GFCIs are intended to be used only in circuits where one of the conductors is solidly grounded.

Class A GFCIs trip when the current to ground has a value in the range of 4 through 6 mA. Class A GFCIs are suitable for use in branch and feeder circuits, including swimming pool circuits. However, swimming pool circuits installed before local adoption of the 1965 NEC may include sufficient leakage current to cause a Class A GFCI to trip.

Class B GFCIs trip when the current to ground exceeds 20 mA. These devices are suitable for use with underwater swimming pool luminaires installed before the adoption of the 1965 NEC.

GFCIs of the enclosed type that have not been found suitable for use where they will be exposed to rain are so marked.

The "TEST" and "RESET" buttons on the GFCIs are only intended to check for the proper functioning of the GFCI. They are not intended to be used as "ON/OFF" controls of motors or other loads unless the buttons are specifically marked "ON" and "OFF." Products with "ON" and "OFF" markings have been additionally Listed under Motor Controllers, Mechanically-operated and Solid-state (NMFT).

Ground-fault Circuit Interrupters KCXS Receptacle GFCIs

Some GFCIs include flush receptacles and are intended to be installed in an outlet box for fixed installation on a branch circuit similar to a conventional receptacle.

Receptacle-type GFCIs for use in wet and damp locations in accordance with Articles 406 of the NEC are identified by the words "Weather Resistant" or the letters "WR" where they will be visible after installation with the cover plate secured as intended.

Weather-resistant receptacle-type GFCIs installed in wet locations are intended to be installed with an enclosure that is weatherproof, whether or not the attachment plug cap is inserted.

Receptacle-type GFCIs for use in dwelling units in accordance with Section 210.52 of the NEC, or pediatric patient care areas in accordance with Article 517 of the NEC, are identified by the words "Tamper Resistant" or the letters "TR" where they will be visible after installation with the cover plate removed.

Receptacle-type GFCIs that have additionally been found to meet appropriate receptacle requirements are marked "Hospital Grade" and/or "CO/ALR."

Receptacle-type GFCIs with receptacles rated 15 or 20 A that are provided with more than one set of terminals for the connection of line and neutral conductors are suitable for through wiring on 20 A branch circuits.

The standard horsepower ratings for specific general-use receptacle configurations are also applicable to the receptacle portion of a GFCI employing the same receptacle configuration.

See Receptacles for Plugs and Attachment Plugs (RTRT) for further information.

Portable GFCIs

This category also covers portable GFCIs. These are plug-in type ground-fault circuit interrupters provided with male blades or an integral power-supply cord for connection to a receptacle outlet . Portable GFCIs are also provided with one or more receptacle outlets located on the GFCI or on a cord-connector body at the end of a length of flexible cord .

REBUILT PRODUCTS

This category also covers rebuilt or refurbished portable GFCIs that are rebuilt or refurbished by the original manufacturer or another party having the necessary facilities, technical knowledge and manufacturing skills. Rebuilt or refurbished portable GFCIs are rebuilt or refurbished to the extent necessary by disassembly and reassembly using new or reconditioned parts. Rebuilt or refurbished portable GFCIs are subject to the same requirements as new portable GFCIs.

ADDITIONAL INFORMATION

For additional information, see Electrical Equipment for Use in Ordinary Locations (AALZ).

REQUIREMENTS

The basic standard used to investigate products in this category is ANSI/UL 943, "Ground-Fault Circuit-Interrupters."

UL MARK

The Listing Mark of Underwriters Laboratories Inc. on the product is the only method provided by UL to identify products manufactured under its Listing and Follow-Up Service. The Listing Mark for these products includes the UL symbol (as illustrated in the Introduction of this Directory) together with the word "LISTED," a control number, and the product name "Ground-fault Circuit Interrupter."

For portable GFCIs, the word "Portable" precedes the product name, and the Listing Mark may be of the flag type applied to the cord.

For rebuilt products the word "Rebuilt" or "Refurbished" precedes the product name.

Special-purpose Ground-fault Circuit Interrupters KCYC

USE

This category covers ground-fault circuit interrupters for use in applications where equipment grounding is provided or is required by ANSI/NFPA 70, "National Electrical Code," or where the voltage to ground is greater than 150 V.

PRODUCT CHARACTERISTICS

These ground-fault circuit interrupters trip when the current to ground has a value in the range of 15 through 20 mA. Let-go protection is not provided by the ground-fault circuit interrupter; however, a person touching the protected equipment and earth would have a low-impedance equipment grounding path in parallel with the person's body.

These ground-fault circuit interrupters rely upon equipment grounding for let-go protection. The reliability of the grounding circuit may be demonstrated by a system that monitors the grounding path to the service and to the load, such that an unacceptable increase in the resistance of the grounding path will cause the circuit to be opened, or by some other method that demonstrates, by investigation, that the grounding circuit is reliable or that faults are unlikely because of the level of insulation that is provided (double insulation).

CLASSES

These ground-fault circuit interrupters are divided into classes based upon voltage rating and the quality of the grounding circuit. Some may be used in circuits where grounding is not provided to the load but double insulation is provided.

A Class C ground-fault circuit interrupter (GFCI) is intended to be used in circuits with voltage not exceeding 300 V AC to ground on any conductor. The Class C GFCI is intended to be used in circuits where reliable equipment grounding or double insulation is provided or is required by the National Electrical Code.

A Class D GFCI is intended to be used in circuits with one or more conductors over 300 V to ground, where specially sized reliable equipment grounding, to provide a low impedance path so that the voltage across the body during a fault does not exceed 150 V, is provided for the protected equipment in the system.

A Class E GFCI is intended to be used in circuits with one or more conductors over 300 V to ground but with conventional equipment grounding or double insulation provided for the protected equipment in the system. These GFCIs respond rapidly to open the circuit before the magnitude and duration for the current flowing through a person's body exceeds the limits for ventricular fibrillation.

RELATED PRODUCTS

For additional information, see Ground-fault Circuit Interrupters (KCXS) and Electrical Equipment for Use in Ordinary Locations (AALZ).

REQUIREMENTS

The basic standard used to investigate products in this category is UL 943, "Ground-Fault Circuit Interrupters," as modified by UL Subject 943C, "Outline of Investigation for Special Purpose Ground-Fault Circuit Interrupters."

UL MARK

The Listing Mark of Underwriters Laboratories Inc. on the product is the only method provided by UL to identify products manufactured under its Listing and Follow-Up Service. The Listing Mark for these products includes the UL symbol (as illustrated in the Introduction of this Directory) together with the word "LISTED," a control number, and the product name "Class ___ Ground-Fault Circuit Interrupter, Special Purpose."

Ground-fault Circuit Interrupters for Use in Hazardous Locations KCYN

GENERAL

This category covers ground-fault circuit interrupters (GFCI) intended for use in accordance with ANSI/NFPA 70, "National Electrical Code." These devices are mounted in explosion-proof and/or dust-ignition-proof enclosures.

GFCIs interrupt the electric circuit to the load when a fault current to ground exceeds some predetermined value that is less than that required to operate the overcurrent protective device of the circuit.

GFCIs are intended to be used only in circuits where one of the conductors is solidly grounded.

Class A GFCIs trip when the current to ground has a value in the range of 4 through 6 mA. Class A GFCIs are suitable for use in branch and feeder circuits.

The "TEST" and "RESET" buttons on GFCIs are only intended to check for the proper functioning of the GFCI. They are not intended to be used as "ON" and "OFF" controls of motors or other loads unless the buttons are specifically marked "ON" and "OFF."

ADDITIONAL INFORMATION

For additional information, see Equipment for Use in and Relating to Class I, II and III, Division 1 and 2 Hazardous Locations (AAIZ).

REQUIREMENTS

The basic unclassified (ordinary) locations standard used to investigate products in this category is ANSI/UL 943, "Ground-Fault Circuit Interrupters."

The basic hazardous (classified) locations standards used to investigate products in this category are referenced in Equipment for Use in and Relating to Class I, II and III, Division 1 and 2 Hazardous Locations (AAIZ).

UL MARK

The Listing Mark of Underwriters Laboratories Inc. on the product is the only method provided by UL to identify products manufactured under its Listing and Follow-Up Service. The Listing Mark for these products includes the UL symbol (as illustrated in the Introduction of this Directory) together with the word "LISTED," a control number, and the product name "Ground Fault Circuit Interrupter for Use in Hazardous Locations" or "Ground-Fault Interrupter for Use in Hazardous Locations," or other appropriate product name as shown in the individual Listings.

Lightning Conductors, Air Terminals and Fittings OVTZ

GENERAL

Lightning protection components are intended to be installed to provide a lightning protection system complying with UL 96A, "Installation Requirements for Lightning Protection Systems."

ADDITIONAL INFORMATION

For additional information, see Electrical Equipment for Use in Ordinary Locations (AALZ).

REQUIREMENTS

The basic standard used to investigate products in this category is UL 96, "Lightning Protection Components."

UL MARK

The Listing Mark of Underwriters Laboratories Inc. on the smallest unit container in which the product is packaged or on the product, when size or shape permits, is the only method provided by UL to identify products manufactured under its Listing and Follow-Up Service. The Listing Mark for these products includes the UL symbol (as illustrated in the Introduction of this Directory) together with the word "LISTED," a control number, and one of the following product names, as appropriate: "Lightning Conductor," "Air Terminal," "Fitting," or other appropriate product name.

Lightning Protection System Installations OWAY

GENERAL

This category covers the installation of lightning protection systems on structures (as limited by UL 96A) to protect them from damage by lightning. The issuance of a Certificate is evidence that the installation of the lightning protection system (1) has been made by an installer that Subscribes to UL's Follow-Up Service, (2) employs materials subject to factory inspection service and bears the UL Mark, and (3) is subject to a field inspection program covering proper installation of the system. The components of the system are described in UL 96A, "Installation Requirements for Lightning Protection Systems" and UL 96, "Lightning Protection Components."

RELATED PRODUCTS

For manufacturers of Listed ground rods suitable for use in installations of lightning protection equipment, see Grounding and Bonding Equipment (KDER).

ADDITIONAL INFORMATION

For additional information, see Electrical Equipment for Use in Ordinary Locations (AALZ).

REQUIREMENTS

The basic standard used to investigate products in this category is UL 96A, "Installation Requirements for Lightning Protection Systems."

LOOK FOR THE CERTIFICATE FOR NEW INSTALLATIONS

The Certificate of Underwriters Laboratories Inc. is the only method provided by UL to identify lightning protection systems covered under its Listing and Follow-Up Service. The Certificate is limited to the number of years for which it has been issued and must be renewed to remain in effect.

Underwriters Laboratories Inc. maintains a factory inspection service for counterchecking conductors, air terminals and fittings, and also a field inspection service for counterchecking installations.

Surge Arresters OWHX

USE

This category covers surge arresters intended to afford protection against surge-related damage to secondary distribution wiring systems and/or to equipment connected thereto. These devices are for use on alternating current power circuits and are intended to be installed in accordance with Article 280 of ANSI/NFPA 70, "National Electrical Code."

RELATED PRODUCTS

Transient voltage surge suppressors are intended for use only on the load side of the main service disconnect and are covered under Transient Voltage Surge Suppressors (XUHT).

ADDITIONAL INFORMATION

For additional information, see Electrical Equipment for Use in Ordinary Locations (AALZ).

REQUIREMENTS

The basic standard used to investigate products in this category is ANSI/IEEE C62.11, "Standard for Metal-Oxide Surge Arresters for AC Power Circuits." All other types of surge arresters are investigated to IEEE C62.1-1989, "Standard for Gapped Silicon-Carbide Surge Arresters for AC Power Circuits."

UL MARK

The Listing Mark of Underwriters Laboratories Inc. on the product is the only method provided by UL to identify products manufactured under its Listing and Follow-Up Service. The Listing Mark for these products includes the UL symbol (as illustrated in the Introduction of this Directory) together with the word "LISTED," a control number, and one of the following product names, as appropriate: "Surge Arrester," "Secondary Surge Arrester," "Secondary MOV Surge Arrester," "Secondary Metal-Oxide Surge Arrester," "Secondary Valve Type Surge Arrester" or "Distribution Light Duty Surge Arrester."

Metallic Outlet Boxes QCIT

GENERAL

This category covers metallic flush device boxes, conduit bodies, conduit boxes, floor boxes, outlet boxes, special-purpose boxes, extension rings, covers, and cover plates for flush-mounted wiring devices, intended for installation in accordance with Article 314 of ANSI/NFPA 70, "National Electrical Code." These products are also intended for installation and use in accordance with the following information.

EXTENSION RINGS

Extension rings are suitable for extending properly secured flush- or surface-mounted boxes. One or more extensions may be used. An extension ring is intended to increase the box depth, volume, or both.

USE IN FIRE-RATED ASSEMBLIES

Listed single- and double-gang metallic outlet and switch boxes with metallic or nonmetallic cover plates may be used in bearing and nonbearing wood stud and steel stud walls with ratings not exceeding 2 h. These walls have gypsum wallboard facings similar to those shown in Design Nos. U301, U411 and U425, as covered under Fire Resistance Ratings – ANSI/UL 263 (BXUV). The boxes are intended to be fastened to the studs with the openings in the wallboard facing cut so that the clearance between the boxes and the wallboard does not exceed 1/8 in. The boxes are intended to be installed so that the surface area of individual boxes does not exceed 16 sq in, and the aggregate surface area of the boxes does not exceed 100 sq in per 100 sq ft of wall surface.

Boxes located on opposite sides of walls or partitions are intended to be separated by a minimum horizontal distance of 24 in. This minimum separation distance between the boxes may be reduced when Wall Opening Protective Materials (QCSN) are installed according to the requirements of their Classification.

The boxes are not intended to be installed on opposite sides of walls or partitions of staggered stud construction unless Wall Opening Protective Materials (QCSN) are installed with the boxes in accordance with Classification requirements for the protective materials.

Listed metallic outlet and switch boxes with metallic or nonmetallic cover plates may be used in floor-ceiling and roof-ceiling assemblies with ratings not exceeding 2 h when these assemblies have gypsum wallboard membranes. The boxes are intended to be fastened to the joists with the openings in the wallboard facing cut so that the clearance between the boxes and the gypsum wallboard does not exceed 1/8 in. The boxes are intended to be installed so that the surface area of individual boxes does not exceed 16 sq in, and the aggregate surface area of the boxes does not exceed 100 sq in per 100 sq ft of ceiling surface.

CONDUIT BODIES

Conduit bodies that are provided with a volume marking can enclose splices, taps or devices. Conduit bodies that are not provided with a volume marking are covered under Conduit Fittings (DWTT). Conduit bodies Classified for use with specific conduit body covers and conduit body covers Classified for use with specific conduit bodies are covered under Conduit Bodies and Covers Classified for Use with Specified Equipment (QCKW).

CONCENTRIC OR ECCENTRIC KNOCKOUTS

All boxes with concentric or eccentric knockouts have been investigated for bonding and are suitable for bonding without any additional bonding means around concentric (or eccentric) knockouts where used in circuits above or below 250 V, and may be marked as such.

CLAMPS

Boxes may or may not be provided with clamps. When clamps are provided, the carton is marked to indicate the type of wiring system or combination of systems for which they have been tested. The clamps are marked with the following letters or combinations thereof to indicate that they are suitable for use with armored cable ("A"): flexible metal conduit – "F," nonmetallic-sheathed cable – "N," or flexible tubing (loom) – "T." Clamps suitable for Type MC metal-clad cable are marked "MCI" for metal-clad interlocking armored cable, "MCI-A" for metal-clad interlocking armor ground cable, "MCS" for metal-clad continuous smooth-sheath cable, and "MCC" for metal-clad continuous corrugated-sheath cable.

If suitable for all seven types, the clamp is marked "ALL." Clamps suitable for non-metallic-sheathed cable are also suitable for multiconductor underground feeder and branch circuit cable where used in dry locations.

Clamps have been tested for securing only one cable per clamp, except multiple section clamps are considered suitable for securing one cable under each section of the clamp, each cable entering a separate knockout.

GROUNDING

Clamps for armored cable, flexible metal conduit, metal-clad interlocking armor ground cable, metal-clad continuous smooth-sheath cable, or metal-clad continuous corrugated-sheath cable are considered suitable for grounding where installed in accordance with ANSI/NFPA 70, "National Electrical Code" (NEC).

FIXTURE SUPPORT

A box, with or without a bracket or bar hanger, intended for support of a fixture weighing 50 lbs or less is marked "FOR FIXTURE SUPPORT" on the carton to indicate that the box is intended for fixture support. A box, with or without a bracket or bar hanger, intended for support of a fixture weighing more than 50 lbs is marked with the weight of the fixture to be supported. Metallic device boxes and device plaster rings have not been investigated for support of a ceiling fixture unless marked for use in ceilings, walls, and with the weight of the product to be supported. Metallic device boxes or metallic device boxes intended to be installed in an existing structure have been investigated for the support of fixtures, smoke detectors and carbon monoxide detectors weighing not more than 6 lbs.

INTEGRAL CONNECTORS

Boxes with integral connectors for electrical metallic tubing or for unthreaded rigid metallic conduit are provided with a marking on the carton to indicate the specific type or types of wiring system for which the boxes have been tested.

CEILING-SUSPENDED FAN SUPPORT

A box, or a box with a bracket or bar hanger intended for support of a ceiling-suspended (paddle) fan weighing 35 lbs or less is marked "ACCEPTABLE FOR FAN SUPPORT" on the product. A box, or a box with a bracket or bar hanger intended for support of a ceiling-suspended (paddle) fan weighing more than 35 lbs but not more than 70 lbs is marked "ACCEPTABLE FOR FAN SUPPORT OF 70 LBS OR LESS" on the product. A box, or a box with a bracket or bar hanger intended for support of a ceiling-suspended (paddle) fan is acceptable for use with a fixture when provided with the above fixture-support markings.

CONCRETE TIGHT

All metal boxes, except aluminum alloy boxes, are provided with corrosion protection suitable for installation in concrete. Aluminum alloy boxes covered under this category are not considered acceptable for installation in concrete or cinder fill unless protected by asphalt paint or the equivalent. Boxes designated as "concrete tight" may have no means of support other than the concrete and often accommodate covers at top and bottom.

FLOOR BOXES

Floor boxes designed for floor installation as covered in the NEC are provided with covers and gaskets to exclude surface water and sweeping compounds that might be present in floor cleaning operations. Covers with gaskets may be shipped separately from the boxes. Both products are provided with installation instructions. Those boxes intended for installation in concrete floors are frequently provided with leveling screws, threaded hubs or both, and are provided with a marking on the carton to identify boxes of this type such as "Floor Box Cover," "Floor Box" or "Floor Box, Concrete Tight" as appropriate.

WET AND DAMP LOCATIONS

Boxes and covers intended for use in wet locations as defined by the NEC are marked "Wet Location." Damp location boxes and covers are intended to be so located or equipped as to prevent water from entering or accumulating in the box and are marked "Damp Location." Boxes with threaded conduit hubs will normally prevent water from entering except for condensation within the box or connected conduit.

Box and device cover combinations, and flush device covers that provide protection from the weather only when the cover is closed, are marked "Wet Location Only When Cover Closed" and may be marked "Damp Location."

ADDITIONAL INFORMATION

For additional information, see Electrical Equipment for Use in Ordinary Locations (AALZ).

REQUIREMENTS

The basic standards used to investigate products in this category are ANSI/UL 514A, "Metallic Outlet Boxes," and ANSI/UL 514D, "Cover Plates for Flush-Mounted Wiring Devices."

UL MARK

The Listing Mark on the product or the UL symbol on the product and the Listing Mark of Underwriters Laboratories Inc. on the smallest unit container in which the product is packaged is the only method provided by UL to identify products manufactured under its Listing and Follow-Up Service. The Listing Mark for these products includes the UL symbol (as illustrated in the Introduction of this Directory) together with the word "LISTED," a control number, and one of the following product names: "Outlet Box," "Outlet Box and Cover," "Extension Ring," "Flush Device Box," or other appropriate product name as shown in the individual Listings.

Outlet Bushings and Fittings QCRV

GENERAL

This category covers supports for outlet and flush device boxes; bushings for use in metal studs; fittings for use in or on outlet and flush device boxes, such as knockout reducers, seals and insulating inserts, and cord grip attachments; insulating gaskets used behind cover plates for flush-mounted wiring devices to stop drafts; pulling grips, strain-relief grips and support grips; locknuts for conduit; sealing gaskets (washers), service-entrance heads for rigid conduit or electrical metallic tubing; cable riser supports; and bushings for use on the ends of rigid or flexible conduit, or electrical metallic tubing, where a change to open wiring is made.

Service-entrance heads or hoods are intended to be used on rigid conduit or electrical metallic tubing that is mounted with the conductor openings facing toward the ground.

Armored Cable Bushings — These bushings are used on armored cable between the conductors and the outer armor. They are a readily distinguishable bright color such as red, orange or yellow.

Bushings — These bushings are suitable for temperatures of 150°C if they are black or brown in color, 90°C if they are any other color unless specifically marked for a higher temperature. Other bushings are covered under Insulating Bushings (NZMT) and Conduit Fittings (DWTT). Service entrance heads for use with service entrance cable are covered under Service Entrance Cable Fittings (TYZX). Temporary wiring, such as round flexible cables or cords may be secured by the use of a connector suitable for use with flexible cord.

Floor Outlet Fittings — Floor outlet fittings are for use in concrete floors for coupling short lengths of exposed conduit to concealed systems when so installed that floor couplings do not come below surface of floor in which they are embedded and subject to the following restrictions: Elbow to be used only where conduit wires pass through fitting without splice, joint, or tap within fitting, and only where no more than one elbow is used in any conduit run.

Tees to be used only where conductors are not drawn in until after main conduit installation is complete. If splices, joints, or taps are used in tees, conductors are intended to be looped that upon removing exposed conduit at floor coupling, splices, joints, or taps can readily be disconnected without interfering with other wiring within fitting.

Sealing Gaskets (Washers) — Sealing gaskets are intended for use with threaded rigid metal conduit and intermediate metal conduit with one sealing gasket on the outside and an ordinary locknut or sealing locknut on the opposite side of the enclosure for wet locations or liquid-tight applications. Sealing gaskets may also be used with Listed wet location or liquid-tight fittings where so marked on the fitting carton.

GROUNDING

Metal reducing washers are considered suitable for grounding for use in circuits over and under 250 V and where installed in accordance with ANSI/NFPA 70, "National Electrical Code." Reducing washers are intended for use with metal enclosures having a minimum thickness of 0.053 in. for non-service conductors only. Reducing washers may be installed in enclosures provided with concentric or eccentric knockouts, only after all of the concentric and eccentric rings have been removed. However, those enclosures containing concentric and eccentric knockouts that have been Listed for bonding purposes may be used with reducing washers without all knockouts being removed.

CARTON MARKINGS

Fittings for use with flexible cords and marked "Liquid-Tight" on the carton indicates suitability for the use where directly exposed to oil spray or to rain.

ADDITIONAL INFORMATION

For additional information, see Electrical Equipment for Use in Ordinary Locations (AALZ).

REQUIREMENTS

The basic standards used to investigate products in this category are ANSI/UL 514A, "Metallic Outlet Boxes," ANSI/UL 514B, "Conduit, Tubing, and Cable Fittings," ANSI/UL 514D, "Cover Plates for Flush-Mounted Wiring Devices," and ANSI/UL 651, "Schedule 40 and 80 Rigid PVC Conduit and Fittings."

UL MARK

The UL symbol on the product and the Listing Mark of Underwriters Laboratories Inc. on the smallest unit container in which the product is packaged is the only method provided by UL to identify products manufactured under its Listing and Follow-Up Service. The Listing Mark for these products includes the UL symbol (as illustrated in the Introduction of this Directory) together with the word "LISTED," a control number, and one of the following product names: "Outlet Bushing," "Outlet Fitting," "Offset Adapter," "Bar Hanger," or other appropriate product name as shown in the individual Listings.

Primary Protectors for Communications Circuits QVGV

GENERAL

This category covers protectors intended for use on communication circuits as defined in Article 800 of ANSI/NFPA 70, "National Electrical Code" (NEC).

These protectors are intended to suppress abnormal voltage conditions that may exist on the circuit due to accidental contact with electric light or power conductors operating at or over 300 V to ground as defined in the NEC. These devices may also be used to protect against electrical transients from an electromagnetic disturbance or higher than normal voltages induced on the communication circuits due to close proximity of the protected circuit to electric light or power conductors.

This category includes both fuse and fuseless protectors. Requirements for the location and installation of primary protectors are contained in the NEC. The individual Listings provide the following information: Protector block number, catalog numbers of arresters that may be employed in a Listed block, types of arresters, design features, maximum fusing wire that is used in series with the block, and indoor or outdoor use.

The maximum size fusing wire is indicated by the following alphabetical designations:

- A – 24 AWG, copper wire with thermoplastic insulation
- B – 22 AWG, copper wire with thermoplastic insulation
- C – 20 AWG, 40 percent copper-clad wire
- D – 26 AWG, copper wire with thermoplastic insulation

Protector blocks suitable for outdoor use are also suitable for use indoors. Blocks marked for indoor use are suitable for installation only indoors.

This category also covers network interface devices, which are two-compartment enclosures that serve to provide a demarcation between the equipment of the private residence and the outside plant. The first compartment, located on the incoming side of the telephone line, may employ a Listed compatible telephone protector, where the compatibility is determined by UL. The second compartment employs terminals and standard telephone jacks for use by the resident. Indoor and outdoor Listing is subject to the same requirements used in the investigation of telephone protectors.

RELATED PRODUCTS

Separate network interface devices intended for use without a protector are covered under Communication Circuit Accessories (DUXR).

Protectors intended for use with municipal fire alarm circuits are covered under Miscellaneous Devices (UXKV).

Secondary protectors intended for telephone, telegraph, fire alarm and similar signaling circuits are covered under Secondary Protectors for Communications Circuits (QVRG).

ADDITIONAL INFORMATION

For additional information, see Protectors (QVGK) and Electrical Equipment for Use in Ordinary Locations (AALZ).

REQUIREMENTS

The basic standard used to investigate products in this category is ANSI/UL 497, "Protectors for Paired-Conductor Communications Circuits."

Protectors that have been subjected to an 8/20, 10 kA surge have additionally been investigated to ANSI/NFPA 780, "Standard for the Installation of Lightning Protection Systems" (2004).

UL MARK

The Listing Mark of Underwriters Laboratories Inc. on the product or on the smallest unit container in which the product is packaged is the only method provided by UL to identify products manufactured under its Listing and Follow-Up Service. The Listing Mark for these products includes the UL symbol (as illustrated in the Introduction of this Directory) together with the word "LISTED," a control number, and one of the following product names as appropriate: "Signal Circuit Protector," "Telephone Protector," "Network Interface Device" or "Signal Circuit Protector Enclosure."

The product name for protectors that comply with the 8/20, 10 kA surge test as required by ANSI/NFPA 780 includes "10 kA."

Primary Protectors for Coaxial Communications Circuits QVKC

Primary Protectors for Coaxial Communications Circuits QVKC

GENERAL

This category covers primary coaxial protectors intended for use on coaxial communication circuits and network-powered broadband communications systems as defined in Article 830 of ANSI/NFPA 70, "National Electrical Code" (NEC). The protectors are typically installed by the public utility company that provides the service and are installed at the point of entry where the coaxial circuit enters the subscriber premises.

The primary coaxial protectors are intended to suppress abnormal voltage conditions that may exist on the circuit due to accidental contact with electric light or power conductors operating at over 300 V to ground as defined in Articles 800 and 830 of the NEC. These protectors may also be used to protect against electrical transients produced from electromagnetic disturbance on the communication circuits.

The primary coaxial protectors may also be used in low- and medium-network-powered sources as defined in the Limitations for Network-Powered Broadband Communications Systems Table of Article 830 of the NEC. The protectors are Listed for use with a current-limiting or extinguishing device, or current-limiting or extinguishing component specified in the individual Listings and installation instructions. The current-limiting or extinguishing device, or current-limiting or extinguishing component may be employed within the protector or may be a separate device or component coordinated externally with the protector.

Coaxial protectors may be used indoors or outdoors. Coaxial protectors marked for outdoor use are also suitable for use indoors. Protectors marked for indoor use are intended for indoor installation only. The coaxial protectors may be installed within a Listed enclosure or network interface device or may be installed as a stand-alone device.

ADDITIONAL INFORMATION

For additional information, see Electrical Equipment for Use in Ordinary Locations (AALZ).

REQUIREMENTS

The basic standard used to investigate products in this category is ANSI/UL 497C, "Protectors for Coaxial Communications Circuits."

Protectors that have been subjected to an 8/20, 10 kA surge have additionally been investigated to ANSI/NFPA 780, "Standard for the Installation of Lightning Protection Systems" (2004).

UL MARK

The Listing Mark of Underwriters Laboratories Inc. on the product or on the smallest unit container in which the product is packaged is the only method provided by UL to identify products manufactured under its Listing and Follow-Up Service. The Listing Mark for these products includes the UL symbol (as illustrated in the Introduction of this Directory) together with the word "LISTED," a control number, and the product name "Primary Coaxial Protector."

The product name for protectors that comply with the 8/20, 10 kA surge test as required by ANSI/NFPA 780 includes "10 kA."

Receptacles RTDV

GENERAL

This category covers the following attachment plug products: 1) receptacles for plugs and attachment plugs, 2) stage-type receptacles, 3) combination receptacles with switches, and 4) utility service receptacles.

The above products include the following:

Appliance, Equipment or Fixture Outlet — A female contact device for mounting on utilization equipment.

Receptacle — A female contact device intended to be installed on a wiring system to supply current to utilization equipment.

This category may also cover the following types of products of a nonstandard configuration blade or slot configuration type, which are part of a manufacturer's line of wiring devices, including receptacles. Other similar devices are covered under Attachment Plugs, Fuseless (AXUT), Attachment Plugs with Switches (AYIR) and Attachment Plugs with Overload Protection (AYVZ).

Attachment Plug — A male contact device for the temporary connection of a flexible cord or cable to a receptacle, cord connector, or other female outlet device.

Cord Connector — A female contact device intended to be wired on flexible cord for use as an extension from an outlet to make a detachable electrical connection to an attachment plug or, as an appliance coupler, to a male inlet.

Male Inlet (Equipment Inlet, Motor Attachment Plug) — A male contact device intended to be mounted on utilization equipment to provide a detachable electrical connection to an appliance coupler or cord connector.

This category does not cover devices intended to be molded on flexible cord or wire, or unassembled devices intended to be factory assembled on flexible cord or wire. Such devices are complete only after installation of the flexible cord or wire and are investigated as part of a complete assembly.

RATINGS

These devices are rated 600 V or less, ac or dc; and 200 A or less. They may also be rated in horsepower as noted in the individual product categories.

Devices rated 250 V are tested on circuits involving a nominal potential to ground of 125 V. Devices having other voltage ratings are tested on circuits involving full rated potential to ground, except for multiphase rated devices, which are tested on circuits consistent with their voltage ratings (e.g., a 120/208 V, 3-phase device is tested on a circuit involving 120 V to ground).

Devices marked "Not for Current Interruption" are not intended to be disconnected while under load. They are intended to be installed in series with switches or other appropriate disconnecting means.

TERMINALS

The terminations of devices intended to be wired to flexible cord are based on the use of flexible cord or cable having copper conductors, in accordance with Article 400 of ANSI/NFPA 70, "National Electrical Code" (NEC). The ampacity of the flexible cord and cable is based on Section 400.5, Tables 400.5(A) and 400.5(B). The conductors are sized as specified on the product or in the manufacturer's instructions provided with the device. The terminations are based on the use of 60°C flexible cord or cable.

Unless stated otherwise in the individual product categories, the termination provisions of all other devices are based upon the use of 60°C insulated conductors in circuits rated 100 A or less, and the use of 75°C insulated conductors in circuits rated more than 100 A, as specified in Table 310.16 of the NEC.

GROUNDING

Devices having a terminal identified by a green-colored finish, the words "Green" or "Ground" (or the letters "G" or "GR"), or the symbol with or without the circle are grounding types. The blade, pin or contact member connected to this terminal is for equipment grounding only.

ENCLOSURES

In general, devices having integral enclosures or installed as intended have been investigated for use indoors, in dry locations. All such Listed products provide a degree of protection against ordinary corrosion, accidental contact with live parts, and a limited amount of falling dirt.

Some devices have been investigated for use in other operating environments when unmated and when mated with other devices in the same manufacturer's line of products. They are marked with one of the type designations 2 through 6, 12 and 13 indicated in Electrical Equipment for Use in Ordinary Locations (AALZ). All outdoor types provide a degree of protection against rain, snow and sleet. Outdoor types are also suitable for use indoors if they meet the environmental conditions present. A device that complies with the requirements for more than one type of enclosure may be marked with multiple designations. Complete use and mating information is provided in the installation instructions provided with each device.

WET AND DAMP LOCATIONS

Receptacles provided with integral outlet box covers or cover plates for flush-mounted wiring devices may be identified for use in damp or wet locations as defined in the Nec. If the cover provides protection only when it is closed, the combination is marked "Wet Location When Cover Closed" and may be marked "Damp Location."

RELATED PRODUCTS

This category does not cover pin-and-sleeve-type devices; see Attachment Plugs, Pin-and-Sleeve Type (QLHN) and Receptacles, Pin-and-Sleeve Type (QLIW).

NOTICE: Applicable to all categories included in Annex C:

UL, in performing its functions in accordance with its objectives, does not assume or undertake to discharge any responsibility of the manufacturer or any other party. UL shall not incur any obligation or liability for any loss, expense or damages, including incidental or consequential damages, arising out of or in connection with the use, interpretation of, or reliance upon this Guide Information. Reprinted from the 2009 UL Guide Information for Electrical Equipment (White Book) by permission from: Underwriters Laboratories Inc. 333 Pfingston Rd. Northbrook, IL 60062-2096

ANNEX D
Short-Circuit Current Calculations

Introduction

Several sections of the National Electrical Code® relate to proper overcurrent protection. Safe and reliable application of overcurrent protective devices based on these sections mandate that a short circuit study and a selective coordination study be conducted. These sections include, among others:

- 110.9 Interrupting Rating
- 110.10 Component Protection
- 110.24 Available Fault Current
- 240.4 Conductor Protection
- 250.122 Equipment Grounding Conductor Protection
- Marked Short-Circuit Current Rating;
  - 230.82 (3) Meter Disconnect
  - 409.22 Short-Circuit Current Rating
  - 430.8 Motor Controllers
  - 440.4(B) Air Conditioning & Refrigeration Equipment
  - 670.5 Short-Circuit Current Rating
- Selective Coordination
  - 517.17 Health Care Facilities - Selective Coordination
  - 517.26 Essential Electrical Systems In Healthcare Systems
  - 620.62 Selective Coordination for Elevator Circuits
  - 700.27 Emergency Systems
  - 701.27 Legally Required Standby Systems
  - 708.54 Critical Operations Power Systems

Compliance with these code sections can best be accomplished by conducting a short circuit study as a start to the analysis. The protection for an electrical system should not only be safe under all service conditions but, to insure continuity of service, it should be selectively coordinated as well. A coordinated system is one where only the faulted circuit is isolated without disturbing any other part of the system. Once the short circuit levels are determined, the engineer can specify proper interrupting rating requirements, selectively coordinate the system and provide component protection. See the various sections of this book for further information on each topic.

Low voltage fuses have their interrupting rating expressed in terms of the symmetrical component of short-circuit current. They are given an RMS symmetrical interrupting rating at a specific power factor. This means that the fuse can interrupt the asymmetrical current associated with this rating. Thus only the symmetrical component of short-circuit current need be considered to determine the necessary interrupting rating of a low voltage fuse. For listed low voltage fuses, interrupting rating equals its interrupting capacity.

Low voltage molded case circuit breakers also have their interrupting rating expressed in terms of RMS symmetrical amps at a specific power factor. However, it is necessary to determine a molded case circuit breaker's interrupting capacity in order to safely apply it. See the section Interrupting Rating vs. Interrupting Capacity in this book.

110.16 now requires arc-flash hazard warning labeling on certain equipment. A flash hazard analysis is required before a worker approaches electrical parts that have not been put into a safe work condition. To determine the incident energy and flash protection boundary for a flash hazard analysis the short-circuit current is typically the first step.

General Comments on Short Circuit Calculations

Sources of short-circuit current that are normally taken under consideration include:

- Utility Generation
- Local Generation
- Synchronous Motors
- Induction Motors
- Alternate Power Sources

Short circuit calculations should be done at all critical points in the system. These would include:

- Service Entrance
- Transfer Switches
- Panel Boards
- Load Centers
- Motor Control Centers
- Disconnects
- Motor Starters
- Motor Starters

Normally, short circuit studies involve calculating a bolted 3-phase fault condition. This can be characterized as all 3-phases "bolted" together to create a zero impedance connection. This establishes a "worst case" (highest current) condition that results in maximum three phase thermal and mechanical stress in the system. From this calculation, other types of fault conditions can be approximated. This "worst case" condition should be used for interrupting rating, component protection and selective coordination. However, in doing an arc-flash hazard analysis it is recommended to do the arc-flash hazard analysis at the highest bolted 3 phase short circuit condition and at the "minimum" bolted three-phase short circuit condition. There are several variables in a distribution system that affect calculated bolted 3-phase short-circuit currents. It is important to select the variable values applicable for the specific application analysis. In the Point-to-Point method presented in this section there are several adjustment factors given in Notes and footnotes that can be applied that will affect the outcomes. The variables are utility source short circuit capabilities, motor contribution, transformer percent impedance tolerance, and voltage variance.

In most situations, the utility source(s) or on-site energy sources, such as on-site generation, are the major short-circuit current contributors. In the Point-to-Point method presented in the next few pages, the steps and example assume an infinite available short-circuit current from the utility source. Generally this is a good assumption for highest worst case conditions and since the property owner has no control over the utility system and future utility changes. And in many cases a large increase in the utility available does not increase the short-circuit currents a great deal for a building system on the secondary of the service transformer. However, there are cases where the actual utility medium voltage available provides a more accurate short circuit assessment (minimum bolted short-circuit current conditions) that may be desired to assess the arc-flash hazard.

When there are motors in the system, motor short circuit contribution is also a very important factor that must be included in any short-circuit current analysis. When a short circuit occurs, motor contribution adds to the magnitude of the short-circuit current; running motors contribute 4 to 6 times their normal full load current. In addition, series rated combinations can not be used in specific situations due to motor short circuit contributions (see the section on Series Ratings in this book).

For capacitor discharge currents, which are of short time duration, certain IEEE (Institute of Electrical and Electronic Engineers) publications detail how to calculate these currents if they are substantial.

Procedures and Methods

To determine the fault current at any point in the system, first draw a one-line diagram showing all of the sources of short-circuit current feeding into the fault, as well as the impedances of the circuit components.

To begin the study, the system components, including those of the utility system, are represented as impedances in the diagram.

The impedance tables include three-phase and single-phase transformers, cable, and busway. These tables can be used if information from the manufacturers is not readily available.

It must be understood that short circuit calculations are performed without current-limiting devices in the system. Calculations are done as though these devices are replaced with copper bars, to determine the maximum "available" short-circuit current. This is necessary to project how the system and the current-limiting devices will perform.

Also, multiple current-limiting devices do not operate in series to produce a "compounding" current-limiting effect. The downstream, or load side, fuse will operate alone under a short circuit condition if properly coordinated.

The application of the point-to-point method permits the determination of available short-circuit currents with a reasonable degree of accuracy at various points for either 3Ø or 1Ø electrical distribution systems. This method can assume unlimited primary short-circuit current (infinite bus) or it can be used with limited primary available current.

Basic Point-to-Point Calculation Procedure

Step 1. Determine the transformer full load amps (F.L.A.) from either the nameplate, the following formulas or Table 1:

3Ø Transformer $\quad I_{F.L.A.} = \dfrac{kVA \times 1000}{E_{L-L} \times 1.732}$

1Ø Transformer $\quad I_{F.L.A.} = \dfrac{kVA \times 1000}{E_{L-L}}$

Step 2. Find the transformer multiplier. See Notes 1 and 2

$\quad$ Multiplier $= \dfrac{100}{*\%Z_{transformer}}$

* **Note 1.** Get %Z from nameplate or Table 1. Transformer impedance (Z) helps to determine what the short circuit current will be at the transformer secondary. Transformer impedance is determined as follows: The transformer secondary is short circuited. Voltage is increased on the primary until full load current flows in the secondary. This applied voltage divided by the rated primary voltage (times 100) is the impedance of the transformer.

Example: For a 480 Volt rated primary, if 9.6 volts causes secondary full load current to flow through the shorted secondary, the transformer impedance is 9.6/480 = .02 = 2%Z.

* **Note 2.** In addition, UL (Std. 1561) listed transformers 25kVA and larger have a ± 10% impedance tolerance. Short circuit amps can be affected by this tolerance. Therefore, for high end worst case, multiply %Z by .9. For low end of worst case, multiply %Z by 1.1. Transformers constructed to ANSI standards have a ±7.5% impedance tolerance (two-winding construction).

Step 3. Determine by formula or Table 1 the transformer let-through short-circuit current. See Notes 3 and 4.

$\quad I_{S_C}$ = Transformer$_{F.L.A.} \times$ Multiplier

Note 3. Utility voltages may vary ±10% for power and ±5.8% for 120 Volt lighting services. Therefore, for highest short circuit conditions, multiply values as calculated in step 3 by 1.1 or 1.058 respectively. To find the lower end worst case, multiply results in step 3 by .9 or .942 respectively.

Note 4. Motor short circuit contribution, if significant, may be added at all fault locations throughout the system. A practical estimate of motor short circuit contribution is to multiply the total motor current in amps by 4. Values of 4 to 6 are commonly accepted.

Step 4. Calculate the "f" factor.

3Ø Faults
$$f = \frac{1.732 \times L \times I_{3\emptyset}}{C \times n \times E_{L\text{-}L}}$$

1Ø Line-to-Line (L-L) Faults
See Note 5 & Table 3
$$f = \frac{2 \times L \times I_{L\text{-}L}}{C \times n \times E_{L\text{-}L}}$$

1Ø Line-to-Neutral (L-N) Faults
See Note 5 & Table 3
$$f = \frac{2 \times L \times I_{L\text{-}N}^{\dagger}}{C \times n \times E_{L\text{-}N}}$$

Where:
- **L** = length (feet) of conductor to the fault.
- **C** = constant from Table 4 of "C" values for conductors and Table 5 of "C" values for busway.
- **n** = Number of conductors per phase (adjusts C value for parallel runs)
- **I** = Available short-circuit current in amperes at beginning of circuit.
- **E** = Voltage of circuit.

† **Note 5.** The L-N fault current is higher than the L-L fault current at the secondary terminals of a single-phase center-tapped transformer. The short-circuit current available (I) for this case in Step 4 should be adjusted at the transformer terminals as follows: At L-N center tapped transformer terminals, $I_{L\text{-}N} = 1.5 \times I_{L\text{-}L}$ at Transformer Terminals.

At some distance from the terminals, depending upon wire size, the L-N fault current is lower than the L-L fault current. The 1.5 multiplier is an approximation and will theoretically vary from 1.33 to 1.67. These figures are based on change in turns ratio between primary and secondary, infinite source available, zero feet from terminals of transformer, and $1.2 \times \%X$ and $1.5 \times \%R$ for L-N vs. L-L resistance and reactance values. Begin L-N calculations at transformer secondary terminals, then proceed point-to-point.

Step 5. Calculate "M" (multiplier) or take from Table 2.

$$M = \frac{1}{1 + f}$$

Step 6. Calculate the available short circuit symmetrical RMS current at the point of fault. Add motor contribution, if applicable.

$$I_{S.C.\ sym.\ RMS} = I_{S.C.} \times M$$

Step 6A. Motor short circuit contribution, if significant, may be added at all fault locations throughout the system. A practical estimate of motor short circuit contribution is to multiply the total motor current in amps by 4. Values of 4 to 6 are commonly accepted.

Calculation of Short-Circuit Currents at Second Transformer in System

Use the following procedure to calculate the level of fault current at the secondary of a second, downstream transformer in a system when the level of fault current at the transformer primary is known.

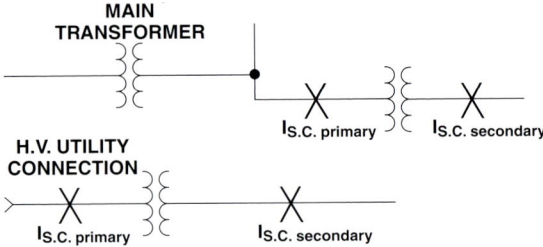

Procedure for Second Transformer in System

Step A. Calculate the "f" factor ($I_{S.C.\ primary}$ known)

3Ø Transformer
($I_{S.C.\ primary}$ and $I_{S.C.\ secondary}$ are 3Ø fault values)

$$f = \frac{I_{S.C.\ primary} \times V_{primary} \times 1.73\ (\%Z)}{100{,}000 \times kVA_{transformer}}$$

1Ø Transformer
($I_{S.C.\ primary}$ and $I_{S.C.\ secondary}$ are 1Ø fault values:
$I_{S.C.\ secondary}$ is L-L)

$$f = \frac{I_{S.C.\ primary} \times V_{primary} \times (\%Z)}{100{,}000 \times kVA_{transformer}}$$

Short Circuit Current Calculations

Step B. Calculate "M" (multiplier).

$$M = \frac{1}{1+f}$$

Step C. Calculate the short-circuit current at the secondary of the transformer. (See Note under Step 3 of "Basic Point-to-Point Calculation Procedure".)

$$I_{S.C.\ secondary} = \frac{V_{primary}}{V_{secondary}} \times M \times I_{S.C.\ primary}$$

System A

Available Utility Infinite Assumption

1500 KVA Transformer, 480V, 3Ø, 3.5 % Z, 3.4 5% X, 0.56%R

$I_{f.l.} = 1804A$

25' - 500 kcmil
6 Per Phase
Service Entrance
Conductors in Steel Conduit

2000A Switch

KRP-C-2000SP Fuse

Fault X_1

400A Switch

LPS-RK-400SP Fuse

50' - 500 kcmil
Feeder Cable
in Steel Conduit

Fault X_2
Motor Contribution

One-Line Diagram

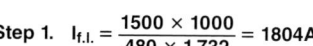

Fault X_1

Step 1. $I_{f.l.} = \frac{1500 \times 1000}{480 \times 1.732} = 1804A$

Step 2. Multiplier $= \frac{100}{3.5} = 28.57$

Step 3. $I_{S.C.} = 1804 \times 28.57 = 51,540A$

$I_{S.C.\ motor\ contrib} = 4 \times 1804^* = 7216A$

$I_{total\ S.C.\ sym\ RMS} = 51,504 + 7216 = 58,720A$

Step 4. $f = \frac{1.732 \times 25 \times 51,540}{22,185 \times 6 \times 480} = 0.0349$

Step 5. $M = \frac{1}{1 + 0.0349} = 0.9663$

Step 6. $I_{S.C.sym\ RMS} = 51,540 \times 0.9663 = 49,803A$

$I_{S.C.motor\ contrib} = 4 \times 1804^* = 7\,216A$

$I_{total\ S.C.\ sym\ RMS} = 49,803 + 7\,216 = 57,019A$
(fault X_1)

Fault X_2

Step 4. Use $I_{S.C.sym\ RMS}$ @ Fault X_1 to calculate "f"

$$f = \frac{1.732 \times 50 \times 49,803}{22,185 \times 480} = 0.4050$$

Step 5. $M = \frac{1}{1 + 0.4050} = 0.7117$

Step 6. $I_{S.C.sym\ RMS} = 49,803 \times 0.7117 = 35,445A$

$I_{sym\ motor\ contrib} = 4 \times 1804^* = 7216A$

$I_{total\ S.C.\ sym\ RMS} = 35,445 + 7216 = 42,661A$
(fault X_2)

*Assumes 100% motor load. If 50% of this load was from motors,
$I_{S.C.\ motor\ contrib.} = 4 \times 1804 \times 0.5 = 3608A$

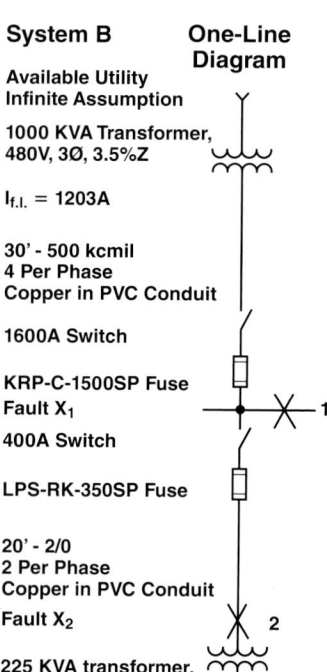

System B

Available Utility Infinite Assumption

1000 KVA Transformer, 480V, 3Ø, 3.5%Z

$I_{f.l.} = 1203A$

30' - 500 kcmil
4 Per Phase
Copper in PVC Conduit

1600A Switch

KRP-C-1500SP Fuse
Fault X_1

400A Switch

LPS-RK-350SP Fuse

20' - 2/0
2 Per Phase
Copper in PVC Conduit
Fault X_2

225 KVA transformer, 208V, 3Ø 1.2%Z
Fault X_3

Fault X1

Step 1. $I_{f.l.} = \frac{1000 \times 1000}{480 \times 1.732} = 1203A$

Step 2. Multiplier $= \frac{100}{3.5} = 28.57$

Step 3. $I_{S.C.} = 1203 \times 28.57 = 34,370A$

Step 4. $f = \frac{1.732 \times 30 \times 34,370}{26,706 \times 4 \times 480} = 0.0348$

Step 5. $M = \frac{1}{1 + 0.0348} = 0.9664$

Step 6. $I_{S.C.sym\ RMS} = 34,370 \times 0.9664 = 33,215A$

Fault X_2

Step 4. $f = \frac{1.732 \times 20 \times 33,215}{2 \times 11,424 \times 480} = 0.1049$

Step 5. $M = \frac{1}{1 + 0.1049} = 0.905$

Step 6. $I_{S.C.sym\ RMS} = 33,215 \times 0.905 = 30,059A$

Fault X_3

Step A. $f = \frac{30,059 \times 480 \times 1.732 \times 1.2}{100,000 \times 225} = 1.333$

Step B. $M = \frac{1}{1 + 1.333} = 0.4286$

Step C. $I_{S.C.sym\ RMS} = \frac{480 \times 0.4286 \times 30,059}{208} = 29,731A$

Single-Phase Short Circuits

Short circuit calculations on a single-phase center tapped transformer system require a slightly different procedure than 3Ø faults on 3Ø systems.

1. It is necessary that the proper impedance be used to represent the primary system. For 3Ø fault calculations, a single primary conductor impedance is only considered from the source to the transformer connection. This is compensated for in the 3Ø short circuit formula by multiplying the single conductor or single-phase impedance by 1.73.

   However, for single-phase faults, a primary conductor impedance is considered from the source to the transformer and back to the source. This is compensated in the calculations by multiplying the 3Ø primary source impedance by two.

2. The impedance of the center-tapped transformer must be adjusted for the half-winding (generally line-to-neutral) fault condition.

   The diagram at the right illustrates that during line-to-neutral faults, the full primary winding is involved but, only the half-winding on the secondary is involved. Therefore, the actual transformer reactance and resistance of the half-winding condition is different than the actual transformer reactance and resistance of the full winding condition. Thus, adjustment to the %X and %R must be made when considering line-to-neutral faults. The adjustment multipliers generally used for this condition are as follows:

   - 1.5 times full winding %R on full winding basis.
   - 1.2 times full winding %X on full winding basis.

   Note: %R and %X multipliers given in "Impedance Data for Single Phase Transformers" Table may be used, however, calculations must be adjusted to indicate transformer kVA/2.

3. The impedance of the cable and two-pole switches on the system must be considered "both-ways" since the current flows to the fault and then returns to the source. For instance, if a line-to-line fault occurs 50 feet from a transformer, then 100 feet of cable impedance must be included in the calculation.

   The calculations on the following pages illustrate 1Ø fault calculations on a single-phase transformer system. Both line-to-line and line-to-neutral faults are considered.

Note in these examples:
a. The multiplier of 2 for some electrical components to account for the single-phase fault current flow,
b. The half-winding transformer %X and %R multipliers for the line-to-neutral fault situation, and
c. The kVA and voltage bases used in the per-unit calculations.

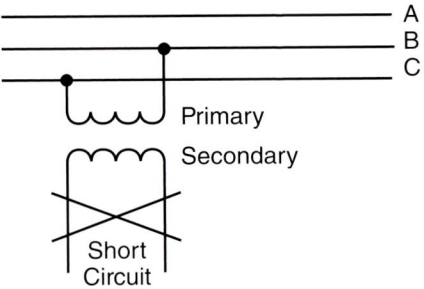

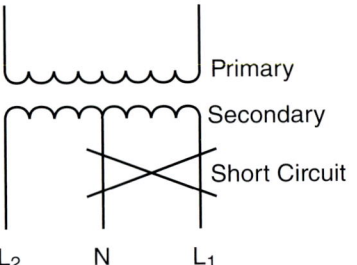

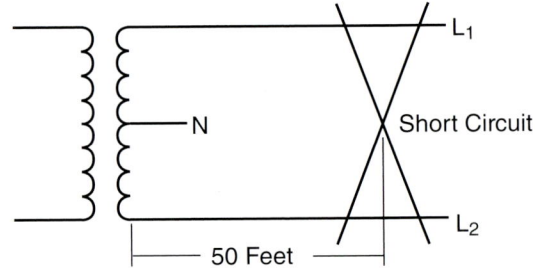

Single-Phase Short Circuits

Line-to-Line Fault @ 240V — Fault X_1

One-Line Diagram

Available Utility
Infinite Assumption

75KVA, 1∅ Transformer,
1.22%X, 0.68%R
1.40%Z
120/240V

Negligible Distance

400A Switch

LPN-RK-400SP Fuse

25' - 500kcmil

Magnetic Conduit

Fault X_1

Step 1. $I_{f.l.} = \dfrac{75 \times 1000}{240} = 312.5A$

Step 2. Multiplier $= \dfrac{100}{1.40} = 71.43$

Step 3. $I_{S.C.} = 312.5 \times 71.43 = 22{,}322A$

Step 4. $f = \dfrac{2 \times 25 \times 22{,}322}{22{,}185 \times 240} = 0.2096$

Step 5. $M = \dfrac{1}{1 + 0.2096} = 0.8267$

Step 6. $I_{S.C.\ L\text{-}L\ (X_1)} = 22{,}322 \times 0.8267 = 18{,}453A$

Line-to-Neutral Fault @ 120V — Fault X_1

One-Line Diagram

Available Utility
Infinite Assumption

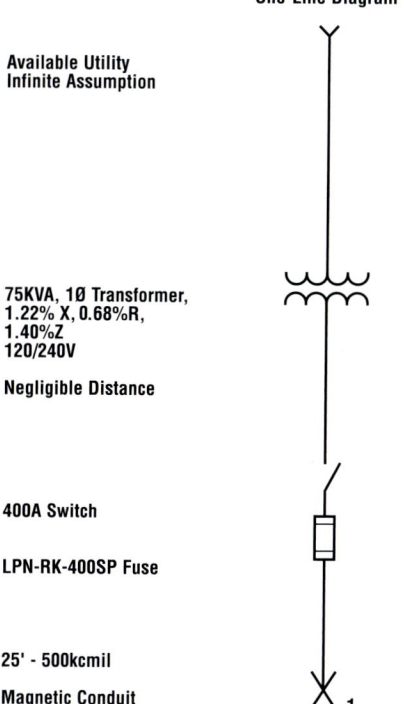

75KVA, 1∅ Transformer,
1.22% X, 0.68%R,
1.40%Z
120/240V

Negligible Distance

400A Switch

LPN-RK-400SP Fuse

25' - 500kcmil

Magnetic Conduit

Fault X_1

Step 1. $I_{f.l.} = \dfrac{75 \times 1000}{240} = 312.5A$

Step 2. Multiplier $= \dfrac{100}{1.40} = 71.43$

Step 3. $I_{S.C.\ (L\text{-}L)} = 312.5 \times 71.43 = 22{,}322A$

$I_{S.C.\ (L\text{-}N)} = 22{,}322 \times 1.5 = 33{,}483A$

Step 4. $f = \dfrac{2^* \times 25 \times 22{,}322 \times 1.5}{22{,}185 \times 120} = 0.6288$

Step 5. $M = \dfrac{1}{1 + 0.6288} = 0.6139$

Step 6. $I_{S.C.\ L\text{-}N\ (X_1)} = 33{,}483 \times 0.6139 = 20{,}555A$

*Assumes the neutral conductor and the line conductor are the same size.

Impedance & Reactance Data

Transformers

Voltage and Phase	kVA	Full Load Amps	% Impedance†† (Nameplate)	Short Circuit Amps†
120/240 1 ph.*	25	104	1.5	12175
	37.5	156	1.5	18018
	50	208	1.5	23706
	75	313	1.5	34639
	100	417	1.6	42472
	167	696	1.6	66644
120/208 3 ph.**	45	125	1.0	13879
	75	208	1.0	23132
	112.5	312	1.11	31259
	150	416	1.07	43237
	225	625	1.12	61960
	300	833	1.11	83357
	500	1388	1.24	124364
	750	2082	3.50	66091
	1000	2776	3.50	88121
	1500	4164	3.50	132181
	2000	5552	4.00	154211
	2500	6940	4.00	192764
277/480 3ph.**	75	90	1.00	10035
	112.5	135	1.00	15053
	150	181	1.20	16726
	225	271	1.20	25088
	300	361	1.20	33451
	500	602	1.30	51463
	750	903	3.50	28672
	1000	1204	3.50	38230
	1500	1806	3.50	57345
	2000	2408	4.00	66902
	2500	3011	4.00	83628

*Single-phase values are L-N values at transformer terminals. These figures are based on change in turns ratio between primary and secondary, 100,000 KVA primary, zero feet from terminals of transformer, 1.2 (%X) and 1.5 (%R) multipliers for L-N vs. L-L reactance and resistance values and transformer X/Rratio = 3.

**Three-phase short-circuit currents based on "infinite" primary.

††UL listed transformers 25 KVA or greater have a ±10% impedance tolerance. Short-circuit amps shown in Table 1 reflect −10% condition. Transformers constructed to ANSI standards have a ±7.5% impedance tolerance (two-winding construction).

†Fluctuations in system voltage will affect the available short-circuit current. For example, a 10% increase in system voltage will result in a 10% greater available short-circuit currents than as shown in Table 1.

TABLE 1. Short-Circuit Currents Available from Various Size Transformers
(Based upon actual field nameplate data or from utility transformer worst case impedance)

f	M	f	M	f	M
0.01	0.99	0.50	0.67	7.00	0.13
0.02	0.98	0.60	0.63	8.00	0.11
0.03	0.97	0.70	0.59	9.00	0.10
0.04	0.96	0.80	0.55	10.00	0.09
0.05	0.95	0.90	0.53	15.00	0.06
0.06	0.94	1.00	0.50	20.00	0.05
0.07	0.93	1.20	0.45	30.00	0.03
0.08	0.93	1.50	0.40	40.00	0.02
0.09	0.92	1.75	0.36	50.00	0.02
0.10	0.91	2.00	0.33	60.00	0.02
0.15	0.87	2.50	0.29	70.00	0.01
0.20	0.83	3.00	0.25	80.00	0.01
0.25	0.80	3.50	0.22	90.00	0.01
0.30	0.77	4.00	0.20	100.00	0.01
0.35	0.74	5.00	0.17		
0.40	0.71	6.00	0.14		

TABLE 2. "M" (Multiplier)

kVA 1Ø	Suggested X/R Ratio for Calculation	Normal Range of Percent Impedance (%Z)*	Impedance Multipliers** For Line-to-Neutral Faults for %X	for %R
25.0	1.1	1.2–6.0	0.6	0.75
37.5	1.4	1.2–6.5	0.6	0.75
50.0	1.6	1.2–6.4	0.6	0.75
75.0	1.8	1.2–6.6	0.6	0.75
100.0	2.0	1.3–5.7	0.6	0.75
167.0	2.5	1.4–6.1	1.0	0.75
250.0	3.6	1.9–6.8	1.0	0.75
333.0	4.7	2.4–6.0	1.0	0.75
500.0	5.5	2.2–5.4	1.0	0.75

*National standards do not specify %Z for single-phase transformers. Consult manufacturer for values to use in calculation.

**Based on rated current of the winding (one-half nameplate kVA divided by secondary line-to-neutral voltage).

Note: UL Listed transformers 25 kVA and greater have a ± 10% tolerance on their impedance nameplate.

This table has been reprinted from IEEE Std 242-1986 (R1991), IEEE Recommended Practice for Protection and Coordination of Industrial and Commercial Power Systems, Copyright© 1986 by the Institute of Electrical and Electronics Engineers, Inc. with the permission of the IEEE Standards Department.

Impedance Data for Single-Phase Transformers

kVA		%Z	Suggested X/R Ratio for Calculation
1Ø	3Ø		
10	—	1.2	1.1
15	—	1.3	1.1
	75	1.11	1.5
	150	1.07	1.5
	225	1.12	1.5
	300	1.11	1.5
333	—	1.9	4.7
	500	1.24	1.5
500	—	2.1	5.5

†These represent actual transformer nameplate ratings taken from field installations.
Note: UL Listed transformers 25kVA and greater have a ±10% tolerance on their impedance nameplate.

Impedance Data for Single-Phase and Three-Phase Transformers-Supplement†

Three Phase Bolted Fault	100%
Line-to-Line Bolted Fault	87%
Line-to-Ground Bolted Fault	25-125%* (Use 100% near transformer, 50% otherwise)
Line-to-Neutral Bolted Fault	25-125% (Use 100% near transformer, 50% otherwise)
Three Phase Arcing Fault	89% (maximum)
Line-to-Line Arcing Fault	74% (maximum)
Line-to-Ground Arcing Fault (minimum)	38% (minimum)

*Typically much lower but can actually exceed the Three Phase Bolted Fault if it is near the transformer terminals. Will normally be between 25% to 125% of three phase bolted fault value.

TABLE 3. Various Types of Short-Circuit Currents as a Percent of Three Phase Bolted Faults (Typical).

Conductors & Busways "C" Values

AWG or kcmil	Three Single Conductors Conduit						Three-Conductor Cable Conduit					
	Steel			Nonmagnetic			Steel			Nonmagnetic		
	600V	5kV	15kV	600V	5kV	15kV	600V	5kV	15kV	600V	5kV	15kV
Copper												
14	389	—	—	389	—	—	389	—	—	389	—	—
12	617	—	—	617	—	—	617	—	—	617	—	—
10	981	—	—	982	—	—	982	—	—	982	—	—
8	1557	1551	—	1559	1555	—	1559	1557	—	1560	1558	—
6	2425	2406	2389	2430	2418	2407	2431	2425	2415	2433	2428	2421
4	3806	3751	3696	3826	3789	3753	3830	3812	3779	3838	3823	3798
3	4774	4674	4577	4811	4745	4679	4820	4785	4726	4833	4803	4762
2	5907	5736	5574	6044	5926	5809	5989	5930	5828	6087	6023	5958
1	7293	7029	6759	7493	7307	7109	7454	7365	7189	7579	7507	7364
1/0	8925	8544	7973	9317	9034	8590	9210	9086	8708	9473	9373	9053
2/0	10755	10062	9390	11424	10878	10319	11245	11045	10500	11703	11529	11053
3/0	12844	11804	11022	13923	13048	12360	13656	13333	12613	14410	14119	13462
4/0	15082	13606	12543	16673	15351	14347	16392	15890	14813	17483	17020	16013
250	16483	14925	13644	18594	17121	15866	18311	17851	16466	19779	19352	18001
300	18177	16293	14769	20868	18975	17409	20617	20052	18319	22525	21938	20163
350	19704	17385	15678	22737	20526	18672	22646	21914	19821	24904	24126	21982
400	20566	18235	16366	24297	21786	19731	24253	23372	21042	26916	26044	23518
500	22185	19172	17492	26706	23277	21330	26980	25449	23126	30096	28712	25916
600	22965	20567	17962	28033	25204	22097	28752	27975	24897	32154	31258	27766
750	24137	21387	18889	29735	26453	23408	31051	30024	26933	34605	33315	29735
1,000	25278	22539	19923	31491	28083	24887	33864	32689	29320	37197	35749	31959
Aluminum												
14	237	—	—	237	—	—	237	—	—	237	—	—
12	376	—	—	376	—	—	376	—	—	376	—	—
10	599	—	—	599	—	—	599	—	—	599	—	—
8	951	950	—	952	951	—	952	951	—	952	952	—
6	1481	1476	1472	1482	1479	1476	1482	1480	1478	1482	1481	1479
4	2346	2333	2319	2350	2342	2333	2351	2347	2339	2353	2350	2344
3	2952	2928	2904	2961	2945	2929	2963	2955	2941	2966	2959	2949
2	3713	3670	3626	3730	3702	3673	3734	3719	3693	3740	3725	3709
1	4645	4575	4498	4678	4632	4580	4686	4664	4618	4699	4682	4646
1/0	5777	5670	5493	5838	5766	5646	5852	5820	5717	5876	5852	5771
2/0	7187	6968	6733	7301	7153	6986	7327	7271	7109	7373	7329	7202
3/0	8826	8467	8163	9110	8851	8627	9077	8981	8751	9243	9164	8977
4/0	10741	10167	9700	11174	10749	10387	11185	11022	10642	11409	11277	10969
250	12122	11460	10849	12862	12343	11847	12797	12636	13115	13236	13106	12661
300	13910	13009	12193	14923	14183	13492	14917	14698	13973	15495	15300	14659
350	15484	14280	13288	16813	15858	14955	16795	16490	15541	17635	17352	16501
400	16671	15355	14188	18506	17321	16234	18462	18064	16921	19588	19244	18154
500	18756	16828	15657	21391	19503	18315	21395	20607	19314	23018	22381	20978
600	20093	18428	16484	23451	21718	19635	23633	23196	21349	25708	25244	23295
750	21766	19685	17686	25976	23702	21437	26432	25790	23750	29036	28262	25976
1,000	23478	21235	19006	28779	26109	23482	29865	29049	26608	32938	31920	29135

Note: These values are equal to one over the impedance per foot and based upon resistance and reactance values found in IEEE Std 241-1990 (Gray Book), IEEE Recommended Practice for Electric Power Systems in Commercial Buildings & IEEE Std 242-1986 (Buff Book), IEEE Recommended Practice for Protection and Coordination of Industrial and Commercial Power Systems. Where resistance and reactance values differ or are not available, the Buff Book values have been used. The values for reactance in determining the C Value at 5 KV & 15 KV are from the Gray Book only (Values for 14-10 AWG at 5 kV and 14-8 AWG at 15 kV are not available and values for 3 AWG have been approximated).

TABLE 4. "C" Values for Conductors

Ampacity	Busway				
	Plug-In		Feeder		High Impedance
	Copper	Aluminum	Copper	Aluminum	Copper
225	28700	23000	18700	12000	—
400	38900	34700	23900	21300	—
600	41000	38300	36500	31300	—
800	46100	57500	49300	44100	—
1000	69400	89300	62900	56200	15600
1200	94300	97100	76900	69900	16100
1350	119000	104200	90100	84000	17500
1600	129900	120500	101000	90900	19200
2000	142900	135100	134200	125000	20400
2500	143800	156300	180500	166700	21700
3000	144900	175400	204100	188700	23800
4000	—	—	277800	256400	—

Note: These values are equal to one over the impedance per foot for impedance in a survey of industry.

TABLE 5. "C" Values for Busway (Reprinted with permission from: Cooper Bussman P.O. Box 14460 St. Louis, MO 63178-4460)

ANNEXE

NEC 2011 Table 8, Chapter 9 reprinted by permission from National Fire Protection Association, One Batterymarch Park, Quincy, MA, 02169.

Table 8 Conductor Properties

Size (AWG or kcmil)	Area mm²	Area Circular mils	Stranding Quantity	Stranding Diameter mm	Stranding Diameter in.	Overall Diameter mm	Overall Diameter in.	Overall Area mm²	Overall Area in.²	Copper Uncoated ohm/km	Copper Uncoated ohm/kFT	Copper Coated ohm/km	Copper Coated ohm/kFT	Aluminum ohm/km	Aluminum ohm/kFT
18	0.823	1620	1	—	—	1.02	0.040	0.823	0.001	25.5	7.77	26.5	8.08	42.0	12.8
18	0.823	1620	7	0.39	0.015	1.16	0.046	1.06	0.002	26.1	7.95	27.7	8.45	42.8	13.1
16	1.31	2580	1	—	—	1.29	0.051	1.31	0.002	16.0	4.89	16.7	5.08	26.4	8.05
16	1.31	2580	7	0.49	0.019	1.46	0.058	1.68	0.003	16.4	4.99	17.3	5.29	26.9	8.21
14	2.08	4110	1	—	—	1.63	0.064	2.08	0.003	10.1	3.07	10.4	3.19	16.6	5.06
14	2.08	4110	7	0.62	0.024	1.85	0.073	2.68	0.004	10.3	3.14	10.7	3.26	16.9	5.17
12	3.31	6530	1	—	—	2.05	0.081	3.31	0.005	6.34	1.93	6.57	2.01	10.45	3.18
12	3.31	6530	7	0.78	0.030	2.32	0.092	4.25	0.006	6.50	1.98	6.73	2.05	10.69	3.25
10	5.261	10380	1	—	—	2.588	0.102	5.26	0.008	3.984	1.21	4.148	1.26	6.561	2.00
10	5.261	10380	7	0.98	0.038	2.95	0.116	6.76	0.011	4.070	1.24	4.226	1.29	6.679	2.04
8	8.367	16510	1	—	—	3.264	0.128	8.37	0.013	2.506	0.764	2.579	0.786	4.125	1.26
8	8.367	16510	7	1.23	0.049	3.71	0.146	10.76	0.017	2.551	0.778	2.653	0.809	4.204	1.28
6	13.30	26240	7	1.56	0.061	4.67	0.184	17.09	0.027	1.608	0.491	1.671	0.510	2.652	0.808
4	21.15	41740	7	1.96	0.077	5.89	0.232	27.19	0.042	1.010	0.308	1.053	0.321	1.666	0.508
3	26.67	52620	7	2.20	0.087	6.60	0.260	34.28	0.053	0.802	0.245	0.833	0.254	1.320	0.403
2	33.62	66360	7	2.47	0.097	7.42	0.292	43.23	0.067	0.634	0.194	0.661	0.201	1.045	0.319
1	42.41	83690	19	1.69	0.066	8.43	0.332	55.80	0.087	0.505	0.154	0.524	0.160	0.829	0.253
1/0	53.49	105600	19	1.89	0.074	9.45	0.372	70.41	0.109	0.399	0.122	0.415	0.127	0.660	0.201
2/0	67.43	133100	19	2.13	0.084	10.62	0.418	88.74	0.137	0.3170	0.0967	0.329	0.101	0.523	0.159
3/0	85.01	167800	19	2.39	0.094	11.94	0.470	111.9	0.173	0.2512	0.0766	0.2610	0.0797	0.413	0.126
4/0	107.2	211600	19	2.68	0.106	13.41	0.528	141.1	0.219	0.1996	0.0608	0.2050	0.0626	0.328	0.100
250	127	—	37	2.09	0.082	14.61	0.575	168	0.260	0.1687	0.0515	0.1753	0.0535	0.2778	0.0847
300	152	—	37	2.29	0.090	16.00	0.630	201	0.312	0.1409	0.0429	0.1463	0.0446	0.2318	0.0707
350	177	—	37	2.47	0.097	17.30	0.681	235	0.364	0.1205	0.0367	0.1252	0.0382	0.1984	0.0605
400	203	—	37	2.64	0.104	18.49	0.728	268	0.416	0.1053	0.0321	0.1084	0.0331	0.1737	0.0529
500	253	—	37	2.95	0.116	20.65	0.813	336	0.519	0.0845	0.0258	0.0869	0.0265	0.1391	0.0424
600	304	—	61	2.52	0.099	22.68	0.893	404	0.626	0.0704	0.0214	0.0732	0.0223	0.1159	0.0353
700	355	—	61	2.72	0.107	24.49	0.964	471	0.730	0.0603	0.0184	0.0622	0.0189	0.0994	0.0303
750	380	—	61	2.82	0.111	25.35	0.998	505	0.782	0.0563	0.0171	0.0579	0.0176	0.0927	0.0282
800	405	—	61	2.91	0.114	26.16	1.030	538	0.834	0.0528	0.0161	0.0544	0.0166	0.0868	0.0265
900	456	—	61	3.09	0.122	27.79	1.094	606	0.940	0.0470	0.0143	0.0481	0.0147	0.0770	0.0235
1000	507	—	61	3.25	0.128	29.26	1.152	673	1.042	0.0423	0.0129	0.0434	0.0132	0.0695	0.0212
1250	633	—	91	2.98	0.117	32.74	1.289	842	1.305	0.0338	0.0103	0.0347	0.0106	0.0554	0.0169
1500	760	—	91	3.26	0.128	35.86	1.412	1011	1.566	0.02814	0.00858	0.02814	0.00883	0.0464	0.0141
1750	887	—	127	2.98	0.117	38.76	1.526	1180	1.829	0.02410	0.00735	0.02410	0.00756	0.0397	0.0121
2000	1013	—	127	3.19	0.126	41.45	1.632	1349	2.092	0.02109	0.00643	0.02109	0.00662	0.0348	0.0106

Notes:
1. These resistance values are valid **only** for the parameters as given. Using conductors having coated strands, different stranding type, and, especially, other temperatures changes the resistance.
2. Equation for temperature change: $R_2 = R_1 [1 + \alpha (T_2 - 75)]$ where $\alpha_{cu} = 0.00323$, $\alpha_{AL} = 0.00330$ at 75°C.
3. Conductors with compact and compressed stranding have about 9 percent and 3 percent, respectively, smaller bare conductor diameters than those shown. See Table 5A for actual compact cable dimensions.
4. The IACS conductivities used: bare copper = 100%, aluminum = 61%.
5. Class B stranding is listed as well as solid for some sizes. Its overall diameter and area is that of its circumscribing circle.

Reprinted with permission from NFPA 70-2011, National Electrical Code®, Copyright 2010, National Fire Protection Association, Quincy, MA 02169. This printed material is not the complete and official position of the NFPA on the referenced subject, which is represented only by the standard in its entirety.

Index

Page numbers followed by *f* denote figures.

A

AC. *See* Alternating current (AC) circuit; Alternating current (AC) system
Accessibility
  of bonding jumper for water piping system, 159
  exception to rule, 49
  of grounding electrode conductor, 97, 97*f*
  of grounding electrode connections, 131, 131*f*
  NEC requirement (Section 250.68), 49
Agricultural installations
  adverse conditions in agricultural buildings, 309
  special rules for, 309–312, 309*f*
    EGCs of underground wiring, 310, 310*f*
    equipotential bonding planes, 311–312, 312*f*
    site-isolating device, 310–311, 311*f*
    voltage gradients, 310
AHJ (authority having jurisdiction), 48, 50, 50*f*
Air-conditioning
  liquid-tight flexible metal conduit to minimize vibration effects, 173, 173*f*
Alternating current (AC) circuit
  conductor installation, 27
  impedance, 3
  Ohm's law, 24
  opposition to current in, 26–27
Alternating current (AC) system
  grounding electrode conductor sizing, 122–123, 123*f*
  solar PV systems, 343–344, 343*f*, 344*f*
  ungrounded, 15, 15*f*
Aluminum
  busbar, 127, 127*f*
  conductors and busways "C" values, 453–454
  equipment grounding conductor (EGC), 180–181, 331
  grounding electrode conductor, 120–121, 120*f*, 122, 123*f*, 124
  prohibition against use as grounding electrode, 82–83
  signal reference grid, 225
American Society of Agricultural and Biological Engineers, 312
Ampacity tables, 39
Amperes, described, 24, 24*f*
Appliances, grounding, 205–206, 205*f*
Arc-flash hazard analysis, 444, 445
Arc-flash hazard warning label, 444
Arcing fault
  effect of equipment, 374, 374*f*
  in hazardous locations, 296, 298
Armored-clad (Type AC) cable
  as EGC, 174, 174*f*, 175*f*, 220, 303–304, 304*f*
  in patient care areas, 303–304, 304*f*
Armored cable bushings, 437
Armored grounding wire, UL Guide information on, 425
Asphalt sealants or coatings, 69
Assured EGC program, 373
Attachment plug, 203, 442
Authority having jurisdiction (AHJ), 48, 50, 50*f*
Autotransformers, 270, 270*f*
Auxiliary electrode
  lightning protection, 287
  wind turbine tower grounding, 286, 287

Auxiliary grounding electrode, 206, 206*f*
  installation, 82
  for pole-mounted electric signs, 322, 322*f*, 323*f*
  use with isolated/insulated circuits, 222, 222*f*
Available fault current, 29

B

Bathtubs, hydromassage, 340–341, 341*f*
Batteries
  as DC system, 287, 287*f*
  as separately derived system, 269
Bolted 3-phase fault condition, 444
Bonded (bonding), definition, 4, 12, 52, 144, 268, 351
Bonding
  basic concepts, 4–5, 4*f*
  connections
    cleaning coated surfaces, 149
    methods, 148*f*, 149
    wire-type conductors, 147–149
  continuity, 147, 147*f*
  definition, 11–12, 12*f*
  described, 9*f*
  effectiveness of, 12–13
  of electrical equipment, 11–12
  of electrically conductive materials, 12–13, 12*f*
  electric cranes and elevators, 326–327, 326*f*
  equipotential for swimming pools, 333–337, 333*f*–336*f*
  equipotential plane in agricultural buildings, 311–312, 312*f*
  ferrous metal raceways and grounding electrode conductors, 136–138, 137*f*, 138*f*
  fittings, 146, 146*f*
  function of equipment grounding conductors, 169, 169*f*
  functions of, 9
  grounding electrode system, 71–72, 71*f*
  in hazardous locations, 153, 297–300, 297*f*–299*f*
  in health care facilities and hazardous locations, 153
  lightning protection systems, 163
  to service electrode system, 84, 84*f*
  for limited-energy system
    common rules for communication systems, 356–360
    description, 351–352, 351*f*, 352*f*
    intersystem bonding, 354–356, 355*f*, 356*f*
    intersystem bonding termination (IBT), 353, 353*f*–355*f*, 354–356
  link to grounding function, 145
  load side
    equipment bonding jumpers, 156–157, 156*f*, 157*f*
    general rules, 155–156, 155*f*
    manufactured homes, 313–314
  metal building frames, 278–279, 279*f*
  metal parts of signs and outline lighting systems, 323–326, 323*f*–325*f*
  metal piping systems, 158–161, 278–279, 279*f*
    multiple buildings/structures supplied by feeder or branch circuit, 160, 160*f*
    in multiple occupancy buildings, 159–160, 160*f*

    non-water piping, 160–161
    sizing and accessibility, 159, 159*f*
  mobile homes, 313–314, 361
  NEC requirements, 12
  panelboards in patient care location, 308–309, 308*f*
  performance criteria, 145–146
  for separate buildings or structures
    of grounding electrodes, 235*f*
    metal water pipe, 240, 240*f*
    purpose, 233–234
  for separately derived systems, 278–279, 279*f*
  service bonding rules (line side), 151–153, 152*f*
  shock hazard reduction, 4
  signal reference grid, 226
  swimming pools, 329–337
  terms, 144–145
  therapeutic pools and tubs for health care use, 339–340, 339*f*
  underwater luminaires, 332, 332*f*
Bonding bar, for intersystem bonding termination, 355, 355*f*
Bonding bushings, 138, 297–298, 297*f*, 298*f*
  bonding continuity and, 147, 147*f*
  on conduits entering a switchboard enclosure, 150, 150*f*
  service disconnecting means and, 152*f*
  UL Guide information on, 426
Bonding conductor. *See also* Bonding jumper
  definition, 52, 52*f*
  length, 149–150
  for signs and outline lighting systems, 324–325, 325*f*
  sizing requirements for the supply side and load side, 150–151
Bonding jumper
  bonding continuity and, 147, 147*f*
  in communications systems, 359–360, 360*f*
  connection to equipment grounding terminal bar, 150, 150*f*
  definition, 52, 52*f*, 144, 351
  equipment, 52–53, 53*f*, 144
    definition, 52–53, 53*f*, 144
    expansion fittings and, 156, 156*f*, 157–158, 158*f*
    function and purpose, 150, 150*f*
    in hazardous locations, 299, 299*f*
    of high-impedance grounded neutral systems, 262
    load side
      installation, 156, 156*f*, 157*f*
      sizing, 151, 151*f*, 157, 157*f*
    materials, 156
    in patient care areas, 306
    receptacle grounding connection and, 198–201, 198*f*, 199*f*, 200*f*
    wire-type, 299, 299*f*
  to form grounding electrode system, 71–72, 71*f*
  function and purpose, 150, 150*f*
  in hazardous locations, 298–299, 299*f*
  installation, 126–127, 126*f*
  integrity of grounding connections and, 132, 132*f*
  length, 149–150

Bonding jumper (*continued*)
main
busbar, 99*f*, 100
color, 100
definition, 53, 53*f*, 99, 144
neutral ground-strap EGFP system, 375–376, 376*f*
requirement, 98
screw, 99, 99*f*, 100
in service equipment enclosure, 97, 97*f*, 98, 98*f*, 99–100, 99*f*
size, 100
suitable for use as service equipment, 108–109, 108*f*
wire, 100, 100*f*
for metal building frames, 162, 162*f*
for metal piping systems, 159–161, 159*f*, 160*f*
for metal water pipe bonding at separate building or structure, 240
sizing, 72, 126, 126*f*
for load side, 150–151, 151*f*
for supply side, 150–151, 151*f*, 154–155, 154*f*, 155*f*
wire type, 154–155, 154*f*, 155*f*
for solar PV systems, 342, 343*f*, 344
supply-side, 154–155, 154*f*
definition, 57, 144, 268
for separately derived system, 273–274, 274*f*
sizing, 150–151, 151*f*, 154–155, 154*f*, 155*f*
system
DC system, 288–289, 289*f*
definition, 56–57, 145, 268
neutral ground-strap EGFP system, 375–376
for separately derived system, 271–273, 271*f*, 272*f*, 273*f*
for water pipe grounding electrode installation, 82
wire-type supply-side
"daisy-chain" arrangement, 154*f*
sizing, 154–155, 154*f*, 155*f*
Bonding locknuts, 297–298, 297*f*, 298*f*, 426
Bonding wedge, 152*f*, 153
Boxes. *See also* Outlet boxes
with concentric or eccentric knockouts, 154, 154*f*
for underwater luminaires, 332, 333*f*
Bracketed information, *NEC*, 50
Branch circuit
bonding, 155, 156
bonding jumper for, 160
definition, 198, 232, 305
EGCs, 281, 300–302
isolated/insulated, 218–222
feeders and, 196–197, 196*f*
in health care facility, 300–309, 300*f*–302*f*, 306*f*
multiwire, 232, 235, 235*f*
in patient care locations, 222
for pool pump motor, 331
for sensitive electronic equipment, 328, 329*f*
for separate building or structure, 236–237, 236*f*
remote disconnecting means, 240–241
Broadband grounding systems, 225
Buck-boost transformer, 270
Building, defined, 232
Bushings
armored cable, 437
grounding and bonding, 426
outlet (category QCRV), 437–438
Busway, "C" values for, 454

C

Cable assemblies, EGC size in, 184
Cable shielding, 397–398, 397*f*, 398*f*
Capacitance, 26
Capacitive reactance, 26
Capacitor
described, 26
grounding requirements for pole-mounted, 192, 192*f*
Ceiling-suspended fan support, 436

Choke effect, 136–137
Circuit
capacitance, 26
components of basic, 22, 22*f*
current in, 24
current return, 5
EGC for multiple in a single raceway or cable tray, 182, 182*f*
electrical safety, 22
through human body
parallel circuit, 7
series circuit, 6, 7*f*
impedance of, 26–27, 28
interrupted, 27
resistance in, 23, 23*f*
safety, 7
voltage of, 23, 23*f*
Circuit breaker. *See also* Overcurrent protection device
instantaneous trip range, 32
interrupting rating, 443
NEC code requirements, 31
time current curves, 32, 34*f*, 35*f*
Circuit conductor
equipment grounding conductor installation and, 176–177, 176*f*
in parallel arrangements, 180, 180*f*
Circulating currents
effect on electronic equipment, 214–215
ground loops, 214
Clamping voltage, 399
Clamps
listed for grounding electrode conductor connections, 134–136, 134*f*, 135*f*, 136*f*
in metallic outlet boxes, 435–436
strap-type ground, 427
Classified locations. *See* Hazardous locations
Clean ground, 219–220, 221
Cleaning coated surfaces, 149, 149*f*
Coatings, removal for bonding continuity, 149, 149*f*
Coaxial protectors, 440
Code-Making Panels, *NEC*, 51
Common grounding electrode conductor, 129–130, 129*f*–131*f*
Communications systems
definitions, 350–351
grounding and bonding performance, 351–352
grounding and bonding rules, 356–360, 357*f*, 358*f*, 359*f*, 360*f*
grounding electrode conductor for, 118, 353–354, 356*f*, 357*f*
grounding electrode conductor connection, 352, 352*f*
intersystem bonding termination (IBT), 353, 353*f*, 354–356, 354*f*, 355*f*
intersystem grounding and bonding, 354–356, 355*f*, 356*f*
mobile homes, 360–361
bonding requirements, 361
grounding requirements, 360–361
overvoltages and lightning events, 363
radio and television equipment and antennas, 361–363, 361*f*–363*f*
telecommunications main grounding busbar (TMGB), 353, 353*f*
UL Guide information
category KDSH grounding and bonding equipment, 427–428
requirements, 428
UL Mark, 428
use, 427
category QVGV primary protectors for communications circuits, 438–439
general, 438–439
related products, 439
requirements, 439
UL Mark, 439
category QVKC primary protectors for coaxial communications circuits, 440

Compressed air systems, bonding of piping for, 161, 161*f*
Computer equipment. *See* Information technology (IT) centers
Concrete-encased electrodes
described, 74–75, 74*f*
grounding electrode conductor sizing, 122, 122*f*
Conductive materials, bonding of, 12–13, 12*f*
Conductor. *See also* Equipment grounding conductor (EGC); Grounded conductor; Grounding electrode conductor; Neutral conductor
annealing of, 38–39
"C" values for, 452
Earth as, 70
ICEA charts, 39
installed in same raceway, cable, or trench, 27, 27*f*
in lightning protection systems, 85*f*, 86, 86*f*, 87*f*
properties, table of, 455
in service equipment enclosure, 97–98, 97*f*
ungrounded, 14
Conductor enclosure, grounding requirements for, 194–195
Conductor insulation, protecting, 38–39
Conductor shielding, 397–398
Conduit. *See also specific conduit types*
as effective ground-fault current path, 416, 421–424
Conduit bodies, for metallic outlet boxes, 435
Connection point for grounding electrode conductors, 127–128
Consumer Product Safety Commission (CPSC), 7
Converter windings, as separately derived system, 269
Copper
busbar, 127, 127*f*
conductors and busways "C" values, 452–453
equipment grounding conductor (EGC), 180–181, 331
equipotential bonding plane, 312, 312*f*
grid in swimming pool shell, 335
grounding electrode conductor, 120, 122, 123*f*, 124
pool water chemicals and corrosion, 334
signal reference structures, 224, 225, 225*f*
Copper-clad aluminum
grounding electrode conductor, 120–121, 123*f*
Copper- or zinc-coated steel electrodes, 76
"Copper Water Tubing" marking, 425, 427
Cord connector, 442
Core and coil ballast, 270
Corner-grounded conductor, 55, 55*f*
Corner-grounded delta system
grounded conductor size requirement, 103, 103*f*
grounded phase conductor, 101
phase-to-ground voltage, 257, 257*f*
phase-to-phase voltage, 257, 257*f*
Corner-grounded system, grounding, 256
Corrosion
in agricultural installations, 309
of grounding electrodes, 82–83
Corrugated stainless steel tubing (CSST) gas piping, bonding requirements for, 161
CPSC (Consumer Product Safety Commission), 7
Cranes
grounding and bonding, 326, 326*f*
grounding requirements for, 192, 193*f*
system grounding prohibitions, 261
Current
in AC circuit, 26
ampere as measure of, 24
available fault, 29
determining with Watt's Wheel, 25–26
in equipment grounding conductors, 185
intensity of, 24
in Ohm's law, 24, 25*f*

INDEX

opposition to in circuits, 26–27
path in a circuit, 28, 28f
path through the human body, 5–7, 6f, 7f
   parallel circuit, 7
   series circuit, 6, 7f
power (watts) in, 25
presence in a circuit, 27, 27f
return to source, 27
time current curves, 32, 33f–37f
Current-limiting overcurrent protective devices, 32
Current transformer, 375–377, 376f, 378

D

DC. *See* Direct current (DC) circuit; Direct current (DC) system
Delta system
   closed, 255
   corner-grounded
      grounded conductor size requirement, 103, 103f
      grounded phase conductor, 101
      phase-to-ground voltage, 257, 257f
      phase-to-phase voltage, 257, 257f
      grounding, 254–255, 255f, 256–257, 257f
   open, 255
   ungrounded, 262–263
Derived phase conductors, 272–273, 273f, 275
Direct current (DC) circuit
   equipment grounding conductor (EGC), 196
   Ohm's law, 24
   resistance, 3
Direct current (DC) switchboard, grounding requirements, 192
Direct current (DC) system
   battery systems, 287, 287f
   grounding, 287–289, 288f, 289f
      grounding electrode conductor, 288, 289f
      point of grounding connection, 288
      system bonding jumper, 288–289, 289f
   grounding electrode conductors, 122, 123–124, 124f, 288, 289f
   solar PV systems, 343–344, 344f
   unidirectional current in, 287
   ungrounded, 289
Dirty ground, 219–220
   as EGC, 220, 220f, 221f
Disconnecting means
   DC systems, 289, 289f
   for a feeder or branch circuit supplying the separate building or structure, 239–240, 239f, 240f
   manufactured home, 314
   mobile home, 313
   remote from building or structure, 240–241, 241f
   separately derived systems, 272–273, 272f, 273f
   site-isolating device, 310–311
Distributed leakage capacitance, 14, 15f, 263, 263f
Dryer, grounding, 205–206, 205f
Dual-fed service equipment, 100, 101f

E

Earth
   conductivity of, 3, 3f, 54, 70, 120
   in the electrical circuit, 2–3
   as ground, 2, 3, 3f, 53–54, 53f
   impedance of, 119, 120
   opposition to current, 14
   soil resistivity, 78–80
Effective ground-fault current path
   definition, 14, 51, 51f, 57–58, 169
   described, 13–14, 13f
   Earth, nonpermissibility of, 14
   equipment grounding conductors as, 169–170, 170f
   function of, 13
   grounded conductor as, 102, 105
   in hazardous locations, 297
   importance of, 5
   metallic conduit as, 416, 421–424
   overcurrent device operation, 31, 31f
EGC. *See* Equipment grounding conductor
EGFP. *See* Equipment ground-fault protection (EGFP)
Electric shock
   electrocution, 6–7, 7f
   hazard for livestock, 312
   severity factors, 368, 368f
   in wet procedure, patient care areas, 370
Electric signs
   bonding, 323–326, 323f–325f
   described, 320–321, 321f
   grounding, 321–323, 322f, 323f
   grounding requirements, 193, 193f
   high-voltage secondary circuits, 324–325, 324f
   reverse pan channel letter sign, 325, 325f
Electrical equipment, grounding of, 9, 9f, 11, 11f
Electrical metallic tubing (EMT)
   bonding, 146, 146f
   as EGC, 175, 175f, 220
   equipment bonding jumper between sections of, 150, 150f
   for equipment grounding, 414–424
   as equipment grounding conductor, 171
   in feeders, 300
   for feeders supplying pool equipment, 333
   fittings, 175, 175f
   in patient care areas, 300, 301
   for pool pump motor wiring, 331
Electrical noise in grounding circuits, 214–215
Electrically conductive materials, bonding of, 16
Electrolytic cells, system grounding prohibitions, 261
Electromagnetic interference (EMI), 214, 216, 222, 306
Electromotive force, 23. *See also* Voltage
Electronic equipment, sensitivity to noise in grounding circuits, 214–216
Elevators
   grounding and bonding, 326–327, 327f
   grounding requirements for, 192, 193f
EMI (electromagnetic interference), 214, 216, 222, 306
EMT. *See* Electrical metallic tubing (EMT)
Equipment and antennas, 361–363, 361f–363f
Equipment bonding jumper
   definition, 52–53, 53f, 144
   expansion fittings and, 156, 156f, 157–158, 158f
   function and purpose, 150, 150f
   in hazardous locations, 299, 299f
   of high-impedance grounded neutral systems, 262
   load side
      installation, 156, 156f, 157f
      sizing, 151, 151f, 157, 157f
   materials, 156
   in patient care areas, 306
   receptacle grounding connection and, 198–201, 198f, 199f, 200f
   wire-type, 299, 299f
Equipment ground-fault protection (EGFP), 368–383
   applicability in health care facilities, 380–382, 381f
   definition, 375
   feeder GFP, 382, 382f
   performance testing, 110
   purpose of, 374, 375f
   selective coordination, 379–380, 379f
   system coordination, 380, 380f
   testing of systems, 382–383, 382f
   types of, 375–379
      neutral ground-strap system, 375–377, 375f, 376f
      zero-sequence transformer-type system, 377–379, 377f–379f
Equipment grounding, 318–347
   appliances, 205–206, 205f
   auxiliary ground electrode requirements, 206, 206f
   conductor enclosure and raceways, 194–195, 195f
   electric cranes and elevators, 326–327, 326f, 327f
   electric signs and outdoor lighting systems, 320–326
   fountains, 338–339, 339f
   general rules for, 190–194, 191f–194f
   IT equipment and sensitive electronic equipment, 327–329, 327f–329f
   medium- and high-voltage systems, 393–394, 393f–395f
   methods, 195–198
      feeders and branch circuits, 196–197, 196f
      isolating grounded conductor from ground, 197–198, 197f
      panelboard equipment grounding terminal bars, 196f, 197, 197f
   nonelectrical equipment, 206–207
   portable or mobile equipment, 392–393
   purpose of, 190, 190f, 320, 320f
   receptacles, 198–205
   by secure metal supports, 207, 207f
   solar PV systems, 341–345
      equipment grounding, 342–343, 342f, 343f
      equipment grounding system continuity, 344–345
      grounding electrode systems, 343–344, 343f, 344f
      system grounding, 341–342, 342f
   spas and hot tubs, 330f, 337–338, 337f, 338f
   steel conduit and EMT for, 414–424
      effective ground-fault current path, 416, 421–424
      Georgia Institute of Technology research, 416–420
      NEC requirements, 415–416
   swimming pools, 329–337
      equipment grounding, 330–331, 331f
      equipotential bonding, 333–337, 333f–336f
      feeders, 333
      junction boxes and enclosures, 332, 333f
      overview, 329–330, 330f
      pool pump motors, 331–332, 331f
      underwater luminaires, 332, 332f
      wiring from separate building or structures, 333
   therapeutic pools and tubs for health care use, 339–341
      bonding, 339–340, 339f
      grounding, 340
      hydromassage bathtubs, 340–341, 341f
      portable appliances and equipment, 340
   wind electrical system, 286
Equipment grounding bus, 328, 328f
Equipment grounding conductor (EGC)
   for agricultural installations, 310, 310f, 311f
   appliance grounding, 205–206, 205f
   auxiliary grounding electrodes and, 82, 206, 206f
   bonding function of, 169, 169f
   branch circuits and, 155, 156
   capacity, 40–41
   connections, 177, 177f, 198
      receptacle, 198–203, 198f, 200f
      at services, 203, 203f
      wiring device terminal, 202–203
   continuity, 200–201, 200f, 201f, 203
   current in, 185
   of DC circuits, 196
   definition, 55–56, 168
   as effective ground-fault current path, 169–170, 170f, 217, 237
   electrical metallic tubing, 168f
   for electric signs and outline lighting equipment, 321–323, 322f, 323f

Equipment grounding conductor (EGC) (*continued*)
  electrocution and, 6, 7
  for elevators, 327, 327*f*
  feeders and, 155, 155*f*, 195, 196, 196*f*
  functions of, 56, 56*f*, 169–170
  grounding equipment by connection to, 190, 191, 191*f*
  grounding function of, 169, 169*f*
  in health care facility, 300–309, 301*f*–307*f*
  identification, 178–179, 178*f*
  for impedance grounded neutral systems, 392, 393*f*
  installations, 175–177, 175*f*, 176*f*
    with circuit conductors, 176–177, 176*f*
    protection from physical protection, 176
    in same raceway, cable, or trench, 27, 27*f*
    tightness of joints/fittings, 176, 176*f*, 177
  isolated/insulated, 216–217, 216*f*, 217*f*, 218–219
  material, 170
  mobile home, 313, 313*f*
  objectionable currents through, 217
  parallel, 182–183, 182*f*, 183*f*
  protecting wire-type, 40, 40*f*
  for sensitive electronic equipment, 328–329, 329*f*
  for separately derived system, 271–273
  service equipment enclosure, 97, 97*f*
  short circuit protection, 39
  for single-point grounding neutral systems, 390–391, 390*f*
  sizing, 61–62, 61*f*, 179–185
    cable assemblies in parallel, 184
    criteria, 179
    examples, 184
    feeder taps and, 185, 185*f*
    increases in size proportionate to increase of associated ungrounded conductors, 181, 181*f*
    for motor circuits, 184, 185*f*
    multiple circuits in single raceway or cable tray, 182, 182*f*
    overcurrent device size and, 180–181, 180*f*
    in parallel installation, 180, 182–184, 183*f*
    for wire-type EGCs, 179–180, 179*f*, 180*f*, 183, 184, 191*f*, 322
  for solar PV systems, 342–343, 344
    continuity, 344
  splicing, 200–201
  for swimming pools, 330–333, 331*f*, 332*f*
  for therapeutic pools and tubs, 340
  types, 170–174, 414
    armored-clad (AC) cable, 174, 174*f*
    electrical metal tubing (EMT), 171, 171*f*
    flexible metal conduit, 171–173, 171*f*–173*f*
    liquid-tight flexible metal conduit, 172–173, 172*f*, 173*f*
    metal-clad (MC) cable, 173–174, 173*f*, 174*f*
    NEC Section 20.118 list of acceptable, 170, 170*f*
  ungrounded systems and, 281
  for wind electrical system, 286–287
  wire-type
    connection, 168, 168*f*
    connection requirements, 205
    identification of, 218*f*, 219
    panelboard equipment grounding terminal bar and, 197, 197*f*
    in patient care areas, 304–305, 305*f*
    protecting, 40, 40*f*
    sizing, 179–180, 179*f*, 180*f*, 183, 184, 191*f*, 322
Equipment grounding terminal bar
  bonding connection, 148*f*, 149
  panelboard, 196*f*, 197, 197*f*
Equipotential bonding for pools and similar installations, 333–337, 333*f*–336*f*
Equipotential bonding plane, in agricultural buildings, 311–312, 312*f*
Exceptions to *NEC* rules, 49, 49*f*
Exothermic welding

bonding connection, 148*f*, 149
in communications systems, 359, 359*f*
connections, 80, 80*f*, 81*f*
for grounding electrode conductor splicing, 125–126
use for grounding electrode conductor connections, 134, 134*f*
Expansion fittings, equipment bonding jumper installation around, 156, 156*f*, 157–158, 158*f*
Extension rings, 435

F

"Fall-of-potential" grounding resistance testing method, 79
Fan, ceiling-suspended support, 436
Fault current
  available, 29
  bonding continuity and, 147
  path for in ungrounded systems, 16
Feeders
  bonding, 155, 155*f*
  bonding jumper for, 160, 160*f*
  branch circuits and, 196–197
  definition, 232
  EGC and, 155, 155*f*, 195, 196, 196*f*, 236, 236*f*, 238, 238*f*, 281, 300, 307–308, 394, 394*f*
  equipment ground-fault protection (EGFP), 382, 382*f*
  in health care facilities, 307–308, 308*f*
  to mobile home, 313, 313*f*
  for parallel runs, 182–183
  for pool equipment, 333
  for separate building or structure, 236–237, 236*f*, 237*f*
  remote disconnecting means, 241*f*, 245–241
Feeder taps, equipment grounding conductors with, 185, 185*f*
Fences, substation, 395–396, 396*f*
Ferrous metal raceways, grounding electrode conductors and, 136–138, 137*f*, 138*f*
Fine print notes, *NEC*, 8, 50
Fire alarm circuits, grounding requirements for, 194
Fire-rated assemblies, metallic outlet boxes in, 435
Fittings
  bonding continuity and, 146, 146*f*
  category QCRV, 437–438
  for electrical metallic tubing, 175, 175*f*
  floor outlet, 437–438
  in hazardous locations, requirements, 296–297, 297*f*
  MC cable, 304
  nonmetallic raceway, 221, 221*f*
  tight, 176, 176*f*
  Type AC, 304
  UL Guide information on, 426
Flash hazard analysis, 444, 445
Flexible metal conduit (FMC)
  as equipment grounding conductor, 171–173, 171*f*–173*f*
  in hazardous locations, 299, 299*f*
  for high-voltage electric sign cables, 324, 324*f*
  liquid-tight, 172–173, 172*f*, 173*f*
  in patient care areas, 302, 302*f*
Floor boxes, metallic outlet, 437
Floor outlet fittings, 437–438
FMC. *See* Flexible metal conduit (FMC)
Footings
  building, 3, 3*f*
  grounding electrodes, 68–69, 75
Foundations
  building, 3, 3*f*
  connection to Earth, 68–69
Fountains, 338–339, 339*f*
Fuses. *See also* Overcurrent protection device
  NEC code requirements, 31
  operating characteristics, 32
  time current curves, 32, 33*f*, 36*f*, 37*f*

G

Gas piping
  bonding of, 161, 161*f*
  prohibition against use as grounding electrode, 82*f*, 83*f*
Gas tube oil ignition (GTO) cables, 324, 325
Gas water heater, bonding of, 161, 161*f*
GEMI software, 424
General Information of Electrical Equipment Directory (UL White Book), 304
Generators, 243–244, 243*f*–245*f*
  DC system, 287, 287*f*
  in outdoor locations, 279
  as separately derived system, 269, 269*f*, 281–285, 281*f*–285*f*
Georgia Institute of Technology Grounding Research, 416–420
GFCI. *See* Ground-fault circuit interrupter (GFCI)
Green marking tape, used at terminations to identify EGCs, 178, 178*f*
Ground, definition, 2, 3*f*, 53–54, 53*f*, 250, 350
Ground clamp, UL Guide information on, 425
Ground clips, 426
Ground detection, 326
  requirements for, 258–259, 258*f*
Ground detection system, 17, 17*f*
  ungrounded systems and, 110–111
Ground detectors, 242
Ground fault
  causes, 5, 5*f*
  definition, 11, 29, 54, 250
  high-impedance grounded neutral systems and, 261
  opening of overcurrent devices, 30*f*, 31
  short circuit condition compared, 29*f*
  on ungrounded systems, 16–17, 17*f*
Ground loops, 214
Ground mesh, 426
Ground resistance, measuring, 79–80, 80*f*
Ground ring electrode
  described, 75–76, 75*f*
  grounding electrode conductor sizing, 122, 122*f*
Ground rod, 69, 96
  grounding electrode sizing, 121, 121*f*
  UL Guide information on, 425
Grounded (grounding)
  definition, 2, 3*f*, 54, 54*f*, 168, 250, 268, 351
  solidly, 54–55, 55*f*, 250
Grounded conductor
  agricultural facilities, 310
  appliance grounding, 205–206, 205*f*
  color, 106, 106*f*
  definition, 55, 55*f*, 94, 250
  described, 102
  disconnect requirement for services, 109–110, 109*f*
  identification, 58–59, 59*f*, 106–107, 106*f*, 107*f*
  isolation from ground in load-side equipment, 197, 197*f*
  load-side use, 105–106, 105*f*
  NEC Article 200 information on use and identification of, 58–59, 59*f*
  neutral, 55, 102, 102*f*
  non-neutral, 261
  phase conductor, 55
  routing and connections, 98–99, 98*f*, 99*f*
  routing with service-entrance conductors, 101, 101*f*
  for separately derived system, sizing, 274–275, 275*f*
  in service equipment enclosure, 97–98, 97*f*
  sizing for parallel installations, 104, 104*f*
  sizing requirements, 103–104, 103*f*
  system grounding, 251–252, 251*f*–252*f*
    identification of conductors, 259, 259*f*
    rules for, 259–261, 259*f*, 260*f*
    selection of conductor to be grounded, 259, 259*f*
    use for grounding, 207–208

Grounded electrical system, described, 9–14
Grounded service conductor
    associated ungrounded service conductors, 101, 101f
    functions of, 101–102, 102f
    sizing requirements, 103–104, 103f
Ground-fault circuit interrupter (GFCI), 368–373
    Class A, 430, 432
    Class B, 430
    Class C, 432
    Class D, 432
    Class E, 432
    Code requirements for GFCI protection, 369–373, 370f, 371f
        assured EGC program, 373
        receptacle replacements, 371–372, 372f
        temporary wiring receptacles, 372–373, 373f
        wet procedure locations in health care facilities, 370–371, 371f
    definition, 368
    function of, 369, 369f
    for hydromassage bathtubs, 340
    portable, 369, 369f, 372, 373f, 431
    purpose of, 368
    rebuilt or refurbished, 431
    receptacle, 430–431
        marked as "No Equipment Ground," 204, 204f
        replacing non-grounding-type receptacles with, 204, 204f
    for sensitive electronic equipment, 329
    "TEST" and "RESET" buttons, 430, 432
    types, 369, 369f
    UL Guide Information for Electrical Equipment
        ground-fault circuit interrupters (category KCXS), 430–431
            general, 430
            receptacle GFCIs, 430–431
        ground-fault circuit interrupters for use in hazardous locations (category KCYN), 432–433
        special-purpose ground-fault circuit interrupters (category KCYC), 431–432
            classes, 432
            product characteristics, 431
            related products, 432
            requirements, 432
            UL Mark, 432
            use, 431
Ground-fault current, return of, 5, 5f
Ground-fault current path, defined, 16, 58
Ground-fault event
    arcing, 374, 374f
    current amount present in grounding electrode conductor, 119, 120, 120f
Ground-fault protection. *See also* Equipment ground-fault protection (EGFP); Ground-fault circuit interrupter (GFCI)
    feeder, 382, 382f
    in health care facilities, 380–382, 381f
    selective coordination, 379–380, 379f
    system coordination, 380, 380f
    testing of systems, 382–383, 382f
Grounding
    basic concepts, 3–4
    equipment grounding, 9, 9f, 11, 11f, 318–347
        for special equipment, 318–347
    to limit potentials (voltages imposed by lightning, line surges, or unintentional contact with lines of higher-voltage lines), 8, 8f
    shock hazard reduction, 4–5
    system grounding, 9, 9f, 10, 10f, 248–265
Grounding bushing, UL Guide information on, 426
Grounding clamps, 134–136, 134f, 135f, 136f

Grounding conductor. *See* Grounding electrode conductor
Grounding Electrical Distribution Systems for Safety (Soares), 40, 416
Grounding electrode
    for communications systems, 357–360, 357f–359f
    connection locations, 132–133, 132f, 133f
    corrosion of, 82–83
    definition, 56, 57f, 68–69, 232, 268, 351
    description, 3–4, 4f
    footing-type investigation and testing (Ufer publication), 404–413
        Beverly Hills tests, 406, 407–408
        Bishop installation, 408–410
        Burbank installation, 413, 413f
        electrode length and, 412–413
        Flagstaff installation, 405–406
        ground resistance test, 406–413
        Hayward installation, 410–411, 411f
        Livermore installation, 410, 410f
        Long Beach installation, 411–412
        Riverside installation, 410
        Tucson installation, 405
        Twenty-nine Palms installation, 412–413, 412f
    grounding symbol as representation of, 56
    installation requirements, 77–84
        auxiliary electrodes, 82
        bonding of lightning protection systems to electrical service electrode systems, 84, 84f
        common grounding of utility services, 83, 83f
        electrode spacing, 80–81, 81f
        metals prohibited from use, 82–83
        plate electrode, 80, 80f
        rod and pipe electrode, 77–78, 77f, 78f
        soil resistivity and, 78–80, 78f, 79f
        water pipe electrodes, 81–82
    for limited energy system, 352, 352f
    mandatory, 72
    multiple, 70, 70f
    performance, 69–70
    purpose of, 69, 69f
    for separately derived system
        criteria for, 275–276, 275f
        for individual system, 276–277, 276f, 277f
        for multiple systems, 277–278, 277f, 278f
    for site-isolating device, 311
    spacing requirements, 80–81, 81f
    types of, 72–76
        concrete-encased electrode, 74–75, 74f
        ground ring electrode, 75–76, 75f
        metal building frame electrode, 73–74, 73f
        metal underground water pipe electrode, 72–73, 73f
        plate electrode, 76, 76f
        rod and pipe electrode, 76
Grounding electrode conductor
    accessibility of connections to, 131, 131f
    accessible location requirement, 97, 97f
    common, 277–278, 277f, 278f
    for communications systems, 353–354, 356f, 357f–360f
    connection methods, 134–136, 134f–136f
    connection point, 127–128, 127f, 128f
    current in, 120
        ground-fault event, 119, 119f, 120
        normal system operation, 119, 119f, 120
    for DC system, 122, 123–124, 124f, 288, 289f
    definition, 56, 57f, 118, 232, 268, 351
    of high-impedance grounded neutral systems, 262
    installation, 124–131
    lightning protection, 287
    for limited-energy systems, 118
    magnetic field concerns, 136–138
    for manufactured home, 314
    materials, 120–121, 120f

    methods of installation, 126–127, 126f, 127f
    protection from physical damage, 125, 125f
    purpose of, 119, 119f
    for radio equipment and antennas, 361–363, 362f
    securing and protecting from physical damage, 125–126, 125f
    for separate buildings or structures supplied by feeders or branch circuits, 235–236, 235f
    for separately derived system
        multiple systems, 277–278, 277f, 278f
        single system, 276–277, 277f
    at service disconnection means, 128
        multiple disconnection means in separate enclosures, 129–131
        single disconnection means, 128–129
    for services supplied by ungrounded system, 110
    sizing, 60–61, 61f, 110, 121–123, 121f–123f
        for DC systems, 122, 124, 124f
        Table 250.66, 122–123, 123f
    splicing, 125–126
    for system over 1 kV, 393–394, 393f
    taps, 277–278, 277f, 278f
    in wet locations, 121
    for wind electrical system, 286–287
Grounding electrode conductor tap, 129–130, 129f–131f
Grounding electrode system
    concrete-encased electrode, 46
    definition, 70, 70f
    establishing, 71–72, 71f
    as foundation of electrical system, 46, 47f
    grounding symbol as representation of, 56
    interconnecting grounding electrode, 126–127, 126f, 127f
    for lowering soil resistivity, 79
    requirements, 71, 71f
    at separate building or structure supplied by feeders or branch circuits, 234–235, 234f, 235f
    for solar PC systems, 343–344, 343f, 344f
    ungrounded supply systems and, 243, 243f
Grounding equipment. *See* Equipment grounding
Grounding path, effectiveness (integrity) of, 131–132, 132f
Grounding symbol, 62, 62f, 202f
Grounding terminal bar, EGC connection to, 177f
Grounding through surge arresters, 252, 253f
GTO (gas tube oil ignition) cables, 324, 325
Guide for Safety in AC Substation Grounding, 395, 395f

H

Hazardous locations, special rules for, 296–300, 296f–299f
Health care facilities
    equipment (category KEVQ), 428–429
    GFCI protection in wet procedure locations, 370–371, 371f
    ground-fault protection, 380–382, 381f
    grounding of receptacles and fixed equipment, 301–307
        armored cable (Type AC) use, 303–304, 304f
        isolated ground receptacle installation, 306, 306f
        isolated power system grounding, 307, 307f
        patient equipment grounding point, 306–307
        wore-type EGC requirement, 304–305
    grounding receptacles and fixed equipment
        electrical metallic tubing (EMT) use, 300, 301
        flexible metal conduit (FMC) use, 302, 302f
        liquidtight flexible metal conduit (LFMC) use, 302, 302f
        MC cable use, 302–303, 303f

INDEX

alth care facilities (*continued*)
  hospital ground jacks and grounding cord assemblies (category KEVX), 429
  isolated grounding circuits in, 221–222, 222*f*
  panelboards, 307–309
    bonding, 308–309, 308*f*
    grounding, 307–308, 308*f*
  special rules for, 300–309
High-frequency effects in grounding circuits, 215–216
High-impedance grounded neutral systems, 55, 261–262, 262*f*
High-leg delta system, grounding of, 252, 252*f*
High-resistance grounding, 252, 253*f*
Hospital ground jacks and grounding cord assemblies (category KEVX), 429
Hot tubs, 330*f*, 337–338, 337*f*, 338*f*, 426
Human body, path for current through, 5–7, 6*f*, 7*f*
Hydromassage bathtubs, 340–341, 341*f*

I

ICEA. *See* Insulated Cable Engineers Association (ICEA)
IMC. *See* Intermediate metal conduit (IMC)
Impedance, 26–27
  in AC circuit, 3
  of Earth, 119, 120
  effect on current, 28
  NEC rules concerning, 26–27
  resonance effect on, 216
  in service grounded conductors, minimizing, 100–101
  short-circuit calculations and, 446, 449, 451–452
Impedance grounded neutral systems, 391–392, 392*f*, 393*f*
Impedance grounding, of systems of more than 1000 volts, 391–392, 392*f*, 393*f*
Incident energy, 32
Inductive reactance, 26, 27, 136
Inductor, grounding through, 252, 253*f*, 389*f*
Information technology (IT) centers
  auxiliary grounding electrode use, 222, 222*f*
  electrical noise in grounding circuits
    circulating currents, 214–215
    high-frequency effects (resonance), 215–216
    practical solutions, 215, 215*f*
  grounding and bonding in, 223–224
  grounding rules, 327–328
  isolated/insulated grounding circuit use, 216, 219, 221
  power distribution units, 214, 214*f*, 215, 215*f*, 223–224, 223*f*, 327*f*, 328
  signal reference structures (grids), 215*f*, 224–226, 224*f*, 225*f*, 328
  surge protection, 226
Insulated Cable Engineers Association (ICEA), 39, 179
Insulated copper EGC
  in patient care areas, 302–306, 302*f*–306*f*
  for swimming pools, 330–331, 331*f*
Insulated grounded conductor, mobile home, 313, 313*f*
Insulation
  failures, causes of, 38
  protecting conductor, 38–39
Insulation tests, 38, 38*f*
Intermediate metal conduit (IMC)
  as EGC, 220, 414
  for feeders, 236
  for feeders supplying pool equipment, 333
Interrupting rating
  circuit breaker, 443
  fuse, 443
  overcurrent protective devices, 28–29
Intersystem bonding termination (IBT), 351, 353, 353*f*–355*f*, 354–356, 362, 362*f*
Intersystem grounding and bonding, 354–356, 355*f*, 356*f*
Inverse time operation, 30
Inverter windings, 271*f*

Investigation and Testing of Footing-Type Grounding Electrodes for Electrical Installations (Ufer), 404–413
Irreversible compression, 148*f*
Irreversible compression connectors, 125–126
Isolated grounding terminal bars, 220
Isolated ground receptacle, 199–200, 199*f*, 200*f*
  in patient care locations, 306, 306*f*
  wiring rules, 219–220, 219*f*, 220*f*
Isolated/insulated grounding circuits
  auxiliary grounding electrodes, 222, 222*f*
  for equipment, 221, 221*f*
  in health care facilities, 221–222, 222*f*
  isolated grounding-type receptacle wiring rules, 219–220, 219*f*, 220*f*
  panelboards and, 220, 221*f*
  power distribution unit and, 214, 224
  purpose of, 216–217
Isolated power systems
  equipment (category KEWV), 429–430
  in health care facilities, 307, 307*f*
IT centers. *See* Information technology (IT) centers

J

Joints, in hazardous locations, 296–297, 297*f*
Junction boxes, for underwater luminaires, 332, 333*f*

K

Kaufmann, R. K., 27
Knockouts
  bonding requirements and, 153, 154, 154*f*
  concentric, 153, 154, 154*f*, 435
  eccentric, 153, 154, 154*f*, 435

L

Leakage capacitance, 14, 15
LFMC. *See* Liquidtight flexible metal conduit (LFMC)
LFNC (liquidtight flexible nonmetal conduit), for underwater luminaires, 332
Lighting pole bases, grounding electrodes installed at, 82
Lightning events, communication systems and, 363
Lightning Protection Institute (LPI) Certified System program, 87, 88
Lightning protection system
  bonding, 163
  bonding to electrical service grounding electrode system, 84, 84*f*
  for communications systems, 352
  components of, 85–86, 85*f*
  definition, 85
  installation methods and criteria, 86–87
  purpose of, 84–85, 84*f*
  quality control, 87–88
  surge protective devices, 87, 87*f*
  UL Guide Information for Electrical Equipment
    category OVTZ, 433
    category OWAY, 433–434
  for wind electrical system, 287
Limited-energy systems, 348–365
  definitions, 350–351
  grounding and bonding performance, 351–352
  grounding and bonding rules for communications systems, 356–360, 357*f*, 358*f*, 359*f*, 360*f*
  grounding electrode conductor for, 118, 353–354, 356*f*, 357*f*
  grounding electrode connection, 352, 352*f*
  intersystem bonding termination (IBT), 353, 353*f*, 354–356, 354*f*, 355*f*
  intersystem grounding and bonding, 354–356, 355*f*, 356*f*
  mobile homes, 360–361
    bonding requirements, 361
    grounding requirements, 360–361

overvoltages and lightning events, 363
radio and television equipment and antennas, 361–363, 361*f*–363*f*
telecommunications main grounding busbar (TMGB), 353, 353*f*
Line isolation monitors, 429
Liquidtight flexible metal conduit (LFMC)
  as EGC, 172–173, 172*f*, 173*f*
  in hazardous locations, 299, 299*f*
  for high-voltage gas tube oil ignition cables, 324
  in patient care areas, 302, 302*f*
Liquidtight flexible nonmetal conduit (LFNC), for underwater luminaires, 332
Livestock areas, equipotential plane installation in, 311–312, 312*f*
Livestock watering equipment, 312
Load-side grounding, 105, 105*f*
Local metal underground systems or structures, as grounding electrodes, 76
Locknuts
  bonding, 153
  grounding and bonding, 426
Low-voltage systems
  grounding electrode conductor for, 118
  system grounding prohibitions, 261
Lugs
  listed for grounding electrode conductor connections, 134, 134*f*
  torque valves for, 177
Luminaires (lighting fixtures), grounding requirements for, 194, 194*f*

M

Machine screw fastener, bonding connection, 148*f*, 149
Magnetic field
  in AC circuit, 26
  effect on grounding electrode conductors, 136–138
Main bonding jumper
  busbar, 99*f*, 100
  color, 100
  definition, 53, 53*f*, 99, 144
  neutral ground-strap EGFP system, 375–376, 376*f*
  requirement, 98
  screw, 99, 99*f*, 100
  in service equipment enclosure, 97, 97*f*, 98, 98*f*, 99–100, 99*f*
  size, 100
  suitable for use as service equipment, 108–109, 108*f*
  wire, 100, 100*f*
Male inlet, 442
Mandatory exceptions to *NEC* rules, 49
Manufactured homes, grounding and bonding rules, 313
Marking tape, for grounded conductor identification, 59
MC cable. *See* Metal-clad cable (Type MC)
Medium- and high-voltage systems, 386–403
  conductor shielding and stress reduction, 397–398, 397*f*, 398*f*
  grounding equipment, 393–394, 393*f*–395*f*
  grounding methods for systems over 1 kV, 388, 388*f*
  grounding requirements, 388
  impedance grounding, 391–392, 392*f*, 393*f*
  portable or mobile equipment grounding, 392–393
  solidly grounded systems, 389–391, 389*f*
    bare neutral conductors, 389–390
    multipoint grounding, 391, 391*f*
    single-point grounding, 390–391, 390*f*
  substation grounding requirements, 394–396, 395*f*, 396*f*
  surge arresters, 399–400, 399*f*, 400*f*
Megohmmeters, 38
Meliopoulos, Dr. Sakis, 416

Metal building frames
 bonding, 162, 162f
 separately derived system, 278–279, 279f
 grounding electrode conductor connection to, 133, 133f
Metal elbows, grounding of, 112, 112f, 195, 195f
Metal frame, equipment grounded by, 207, 207f
Metal frame electrode, described, 73–74, 73f
Metal piping system
 bonding, 158–161
  multiple buildings/structures supplied by feeder or branch circuit, 160, 160f
  in multiple occupancy buildings, 159–160, 160f
  non-water piping, 160–161
  at separate building or structure, 240, 240f
  separately derived system, 278–279, 279f
  sizing and accessibility, 159, 159f
 grounding electrode conductor connection to, 132–133, 132f, 133f
Metal raceway
 as EGC, 236
 grounding requirements, 194–195, 195f
Metal-clad cable (Type MC)
 as EGC, 173–174, 174f, 184, 220, 302–303, 303f
 for feeders supplying pool equipment, 333
 fittings, 173, 173f
 in patient care areas, 302–303, 303f
Meter enclosures
 bonding, 152, 152f
 grounding, 197
Meter socket enclosures, accessibility of, 97
Micro-inverters, 271f
Mobile equipment, grounding, 392–393
Mobile homes
 bonding requirements, 361
 grounding and bonding rules, 313, 313f
 grounding requirements, 360–361
Motion picture projection equipment, grounding requirements, 194
Motor circuit, equipment grounding conductors for, 184, 185f
Motor frames, grounding requirements, 192, 192f, 193f
Multi-conductor cables
 identification of EGCs in, 178
 size of EGCs in, 180
Multigrounded neutral systems, 391, 391f
Multiple occupancy building, bonding of metal water piping in, 159–160, 160f
Multisection switchboards, 98f, 99
Multiwire branch circuit, 235, 235f

N

National Electrical Code (NEC), 44–65
 arrangement and application, 46–48
 Article 100, 51–57
 Article 200 (identification and use of grounded conductors), 58–59, 59f
 Article 250 (arrangement and use), 59
  description of parts of, 59
  grounding figure, 62, 62f
  order of, 60, 60f
  Part I, 59–60
  Part X, 62
  Table 250.66 (Grounding Electrode Conductor for Alternating-Current Systems), 60–61, 61f
  Table 250.122 (Minimum Size Equipment Grounding Conductors for Grounding Raceway and Equipment), 61–62, 61f
 Chapter 8, 63
 Code-Making Panels, 51–52
 defined terms
  for grounding and bonding in Article 100, 52–57
  in Section 250.2, 57–58
  slang terms, avoidance of, 52
  use of, 50–51, 51f
 enforcement and approvals, 48
 exceptions to rules, 49, 49f
 explanatory information, 50, 50f
 modifications, 62–63
  Section 501.30, 47–48, 48f
  Section 517.13, 62, 63f
  Section 600.7, 62–63, 63f
  Section 770.100, 63
 notes, 8
 performance language, 8–9
 permissive rules, 49–50, 50f
 purpose of, 2, 46, 46f, 47
 requirements, 48–49
 Section 250.4 requirements, 145, 145f
National Electrical Contractors Association (NECA), 145
National Electrical Installation Standards (NEIS), 145
National Fire Protection Association (NFPA), 70
National Fuel Gas Code, 161
Natural gas piping systems, prohibition against use as grounding electrode, 82f, 83f
NEC. *See National Electrical Code (NEC)*
NECA (National Electrical Contractors Association), 145
NEC Style Manual, 51
NEIS (National Electrical Installation Standards), 145
Neon lighting, 324–325, 325f. *See also* Electric signs
Neutral conductor
 agricultural facilities, 310
 current in, 261
 definition, 102, 260
 described, 102
 disconnect requirement for services, 109–110, 109f
 grounded conductors of systems, 259–261, 260f
 system grounding, 251–252, 251f–252f
  identification of conductors, 259, 259f
  rules for, 259–261, 259f, 260f
  selection of conductor to be grounded, 259, 259f
Neutral current, parallel paths for, 105
Neutral disconnect means (link), 109–110, 109f
 for testing of EGFP systems, 382, 382f
Neutral grounding devices (category KDZC), 428
Neutral ground-strap system, 375–377, 375f, 376f
Neutral point, 102, 102f, 103
 definition, 260
 high-impedance grounded neutral systems and, 262
Neutral-to-Earth stray current, in agricultural facilities, 310
NFPA (National Fire Protection Association), 70
"No Equipment Ground" marking, 372, 372f
Nonelectrical equipment, grounding, 206–207
Non-grounding-type receptacle, replacement of, 372, 372f
Notes, NEC, 50, 50f

O

Objectionable current, 214, 217–218, 217f
Ohm
 described, 23, 23f
 symbol for, 23
Ohm, George Simon, 22
Ohm's law, 22–23, 23f
 applying, 24, 25f
 Earth resistance testing and, 79
 forms of, 24
Outdoor sources, of separately derived systems, 279–280, 279f
Outlet
 definition, 198
 EGC attachment, 198
 equipment bonding jumper connection to grounding-type receptacle, 150, 150f
Outlet boxes
 bonding of boxes, 153, 154, 154f
 concentric or eccentric knockouts, 153, 154, 154f, 436
 equipment grounding conductor
  connections, 198–203, 198f
  continuity, 200–201, 200f, 201f
 UL Guide Information for Electrical Equipment (category QCIT), 434–437
  ceiling-suspended fan support, 436
  clamps, 435–436
  concentric or eccentric knockouts, 435
  concrete tight, 436
  conduit bodies, 435
  extension rings, 435
  fixture support, 436
  floor boxes, 436
  general, 434
  grounding, 436
  integral connectors, 436
  requirements, 437
  UL Mark, 437
  use in fire-rated assemblies, 435
  wet and damp locations, 437
Outlet bushings and fittings (category QCRV), 437–438
 carton markings, 438
 general, 437–438
 grounding, 438
 requirements, 438
 UL Mark, 438
Outline lighting systems
 bonding, 323–326, 323f–325f
 described, 320–321, 321f
 grounding, 321–323, 322f, 323f
 grounding requirements, 193, 193f
Overcurrent protection
 NEC requirements, 28
 in separately derived systems, 238, 238f, 242, 242f
 short-circuit current calculations, 443–453
Overcurrent protection device
 current-limiting, 32
 fusion or tripping point, 30
 in generators, 244, 244f
 in hazardous locations, 296
 interrupting ratings, 28–29
 inverse time operation, 30
 NEC size requirement, 30
 response to current, 30–31, 30f
 selection of, 28, 28f
 selective coordination for, 379, 379f
 sizing requirements, 31
 speed of function, 31
 time current curves, 32, 33f–37f
Overvoltage protection. *See* Surge arrester
Overvoltages, in communication systems, 363

P

Panelboard
 bonding, 152
 equipment grounding terminal bars, 196f, 197, 197f
 in health care facilities
  bonding, 308–309, 308f
  grounding, 307–308, 308f
 isolated grounding circuits and, 211f, 220
 isolation of grounded conductor from ground, 197–198, 197f
 marking of ungrounded system equipment, 111, 111f
 in power distribution units, 223f, 224
 "Technical Equipment Ground" terminal bus, 328, 328f
Parallel circuit through the human body, 7
Parallel conductor, sizing requirements for, 104, 104f
Parallel paths for neutral current, 105
Parallel runs, equipment grounding conductors for, 182–184, 182f, 183f
Path to ground, 118

INDEX

[i]ent care vicinity, defined, 305
[pa]tient equipment grounding point, 306–307
PDUs. See Power distribution units (PDUs)
Permissive rules, NEC, 49–50, 50f
Phase conductor
   grounded conductor of a service, 101
   mobile home, 313
Phase-to-conduit fault, 415
Photovoltaic systems
   DC system, 287, 287f
   in outdoor locations, 279
Pipe or conduit used as grounding electrodes, 76
Plate electrodes
   described, 76, 76f
   grounding electrode conductor sizing, 121
   installation, 80, 80f
Pole-mounted electric signs, grounding of, 322, 322f, 323f
Polyvinyl chloride (PVC)
   EGC inside of conduit, 221
   fittings, 221
   rigid conduit for feeders supplying pool equipment, 333
Pool pump motors, 331–332, 331f, 336, 336f
Pool shells, conductive, 334–335
Pool water bonding, 337, 337f
Portable (mobile) equipment grounding, 392–393
Power distribution system, 38
Power distribution units (PDUs), 214, 214f, 215, 215f, 223–224, 223f
   in information technology rooms, 327f, 328
   transformers, 328
Power-limited circuits, grounding requirements for, 194
Power meter
   intersystem bonding termination, 354, 354f
Power quality system grounding analysis, 218
Power supply
   secondary circuit ground-fault protection, 324
Power systems, in health care facilities, 261, 307, 307f
Pressure connectors
   bonding connection, 148f, 149
   listed for grounding electrode conductor connections, 134f
Primary protector
   for communications circuits (category QVGV), 438–439
   general, 438–439
   related products, 439
   requirements, 439
   UL Mark, 439
   for communications systems, 356–357, 360–361
   for coaxial communications circuits (category QVKC), 440
Protection from physical damage, for equipment grounding conductors, 176
Protector grounding wires, 426
PVC. See Polyvinyl chloride (PVC)

Q

Quiet ground, 199

R

Raceway
   equipment bonding jumper installation, 156, 156f
   ferrous metal raceways, grounding electrode conductors and, 136–138, 137f, 138f
   grounding of, 111–112
   grounding requirements, 194–195
   service bonding rules, 151–153, 152f
Radio equipment and antennas, 361–363, 361f–363f
Range, grounding, 205–206, 205f
Reactance grounding, 252, 253f
Rebar, equipotential bonding in swimming pools and, 335

Receptacle
   "CO/ALR," 431
   equipment bonding jumper connection from outlet, 150, 150f
   ground-fault circuit interrupter (GFCI), 430–431
   grounding connection, 198–203, 198f
      attachment plugs, 203
      EGC continuity, 200–201, 200f, 201f, 203
      equipment bonding jumper requirements, 198–201, 198f, 199f, 200f
      isolated grounding receptacles, 199–200, 199f, 200f
      listed floor boxes, 199, 199f
      self-grounding receptacles, 199, 199f
      surface mounted boxes, 198f, 199, 199f
   "Hospital Grade," 430
   identification of wiring device terminals, 202–203, 202f
   isolated grounding
      for electrical noise reduction, 216, 216f
      wiring rules, 219–220, 219f, 220f
   "No Equipment Ground" marking, 372, 372f
   in patient care areas, 305, 305f, 306, 306f
   replacements, 203–205
      GFCI protection for, 371–372, 372f
      grounding-type, 204
      non-grounding-type, 204–205, 204f
   for sensitive electronic equipment, 328–329
   "Tamper Resistant," 430
   temporary wiring, 372–373, 373f
   UL Guide category RTDV, 440–442
      enclosures, 441
      general, 440–441
      grounding, 441
      ratings, 441
      related products, 442
      terminals, 441
      wet and damp locations, 442
   "Weather Resistant," 430
Reducing washers, bonding and, 153–154, 153f
Refrigeration equipment
   liquidtight flexible metal conduit to minimize vibration effects, 173, 173f
Reinforced thermosetting resin conduit (Type RTRC), for feeders supplying pool equipment, 333
Remote-control circuits, grounding requirements for, 194
Research Institute and the Illinois Institute of Technology (IITRI), 7
Resistance
   DC circuit, 3
   between the Earth and a grounding electrode, 70
   ohm as unit of measure, 23
   in Ohm's law, 24, 25f
   soil resistivity, 78–80, 78f–80f
Resistance grounding, 252, 253f
Resistor, grounding through, 252, 253f, 389f
Resonance, 215–216
Reverse pan channel letter signs, 325, 325f
Rigid metal conduit (RMC)
   bonding, 146
   as EGC, 196, 196f, 220, 236, 414
   for feeders, 236, 300
   for feeders supplying pool equipment, 333
   in patient care areas, 300, 301
RMS symmetrical interrupting rating, fuse, 443
Rod and pipe electrodes
   described, 76
   installation, 77–78, 77f, 78f

S

Safety
   by design, 32, 38
   grounding and bonding for, 2
Safety circuit
   "weakest link in the chain" concept, 147
Safety grounds, 225
Secondary circuit ground-fault protection (SCGFP) device, 324

Selective coordination, 379–380, 379f
Sensitive electronics
   grounding requirements for, 328–329, 328f, 329f
   in patient care areas, 306
Separate buildings or structures
   grounding and bonding
      disconnecting means remote from structure, 240–241, 241f
      disconnecting means requirements, 239–240, 239f, 240f
      feeder and branch circuit requirements, 236–237, 236f, 237f
      generators, 243–244, 243f, 244f, 245f
      grounding electrode conductor, 235–236, 235f
      grounding electrode requirement, 234–235, 234f, 235f
      metal water pipe bonding, 240, 240f
      purpose of, 233–234
      separately derived systems with overcurrent protection, 238, 238f, 242, 242f
      separately derived systems without overcurrent protection, 238–239, 239f, 242, 242f
      ungrounded systems supplying services, 237–238, 237f, 242–243, 243f
   pool wiring supplied from, 333
   power supply to, 233
   supply by generators, 243–244, 243f–245f
   supply by separately derived systems, 238–239, 241–242
      with overcurrent protection, 238, 238f, 242, 242f
      without overcurrent protection, 238–239, 239f, 242, 242f
   supply by ungrounded system, 237–238, 242–243
Separately derived systems, 266–293
   bonding water piping and building steel, 278–279, 279f
   DC systems, 287–289, 287f
      grounding, 287–289, 288f, 289f
      ungrounded, 289, 289f
   definitions, 268
   determining, 269–270, 269f
   examples of, 269–270, 269f, 270f
   generators and transfer equipment, 281–285, 281f–285f
   grounded systems, components of, 270–275, 270f
      grounded conductor sizing, 274–275, 275f
      supply-side bonding jumper, 273–274, 274f
      system bonding jumper, 271–273, 271f, 272f, 273f
   grounding, 257
   requirements, 270
   grounding electrodes, 275–278
      criteria for, 275–276, 275f
      for individual system, 276–277, 276f, 277f
      for multiple systems, 277–278, 277f, 278f
   outdoor sources, 279–280, 279f
   ungrounded systems, 280–281, 280f, 281f
   wind electrical systems, 286–287, 286f
      equipment grounding, 286
      grounding connections, 286–287
      system grounding, 286
      tower grounding, 286
Series circuit through the human body, 6, 7f
Service
   definition, 94
   equipment grounding conductor connections at, 203, 203f
Service conductor, definition, 94
Service disconnecting means. See also
   Disconnecting means
   methods of connection, 128
   multiple, 129–131, 129f–131f
   single, 128–129

Service enclosures
  bonding rules, 151–153, 152*f*
    knockouts and, 153, 154, 154*f*
    reducing washer use, 153–154, 153*f*
  grounding of, 111–112
Service equipment
  connecting conductors together, 97–98, 97*f*
  described, 94, 94*f*
  dual-fed, 100, 101*f*
  grounded conductor routing and connections, 98–99, 98*f*, 99*f*
  grounding electrode sizing and, 121, 121*f*
  identification as
    suitable for use as, 107–109, 107*f*, 108*f*
    suitable for use only as, 108, 108*f*, 109
  main bonding jumpers in, 97, 97*f*–99*f*, 98, 99–100
  multisection switchboards, 98*f*, 99
  neutral disconnect link, 109–110, 109*f*
  service disconnection means, 128–131
    multiple, 129–131, 129*f*–131*f*
    single, 128–129
Service equipment enclosure, 97–98
  EGC connection, 203, 203*f*
  routing grounding electrode conductors from, 127–128, 127*f*
Service grounded conductor, minimizing impedance in, 100–101
Service point, definition, 145
Service raceways, grounding of, 111–112
Service-entrance cable
  for pool pump motor wiring, 331
  Type SER, 331
  Type SEU, 331
Shock
  hazard for livestock, 312
  minimizing hazard, 4–5, 11
  severity, 6, 368, 368*f*
  in wet procedure, patient care areas, 370
Short circuit
  described, 29
  ground fault condition compared, 29*f*
  opening of overcurrent devices, 30–31, 30*f*
  protection for EGCs, 39
Short-circuit current calculations, 443–453
  basic point-to-point calculation procedure, 445–446
  bolted 3-phase fault condition, 445
  conductors and busways "C" values, table of, 452–453
  locations for, 444
  overview, 444
  at second transformer in system, 446–447
  single-phase short circuits, 448–449
  transformer impedance and reactance data, 450–451
Short circuit current rating, 28–29
Signal reference structures (grids), 215*f*, 224–226, 224*f*, 225*f*, 328
Single-phase short circuits, 448–449
Single-point grounding, 390–391, 390*f*
Site-isolating device, in agricultural installations, 310–311, 311*f*
Slang terms, 52
Snap switches, grounding, 201–202, 202*f*
Soares, Eustace C., 40, 416
Soil conditions, effect on grounding electrodes, 70
Soil resistivity, 78–80
  measuring, 79–80, 80*f*
  methods for lowering, 78–79, 79*f*
  variability in, 78, 78*f*
Solar photovoltaic systems, 341–345
  equipment grounding, 342–343, 342*f*, 343*f*
  equipment grounding system continuity, 344–345
  grounding electrode systems, 343–344, 343*f*, 344*f*
  as separately derived system, 269
  system grounding, 341–342, 342*f*

Solid grounding, 252, 253*f*
  definition, 250
  medium- and high-voltage systems, 389–391, 389*f*–391*f*
Source of the system, 251
Sparking, in hazardous location, 296, 298
Spas, 330*f*, 337–338, 337*f*, 338*f*
Speakers, underwater, 337, 337*f*
Splicing, equipment grounding conductors, 200–201
Stainless steel grounding electrodes, 76
Static electricity, in hazardous locations, 299–300
Steel building frames. *See* Metal building frames
Steel conduit, for equipment grounding, 414–424, 414*f*
Stray current, in agricultural facilities, 310
Stress reduction, 397–398
Strike termination device, 85, 85*f*, 86, 86*f*
Structure, definition, 232
Submersible pump, grounding requirements for, 194, 194*f*
Substation grounding, 394–396, 395*f*, 396*f*
Supplemental electrode, for water pipe grounding electrodes, 81–82, 81*f*
Supply-side bonding, 151, 151*f*
Supply-side bonding jumper, 154–155, 154*f*
  definition, 57, 144, 268
  for separately derived system, 273–274, 274*f*
Surge arrester
  definition, 399
  ferrous metal sleeves, 400, 400*f*
  grounding through, 252, 253*f*, 389*f*, 399–400, 399*f*, 400*f*
  in medium- and high-voltage systems, 394, 395*f*, 399–400, 399*f*, 400*f*
  in metal-clad switchgear for medium-voltage systems, 395*f*
  UL Guide Information for Electrical Equipment (category OWHX), 434
Surge protection
  devices for, 87, 87*f*
  for IT equipment, 226, 226*f*
Surges, vulnerability of ungrounded systems to, 10
Swimming pools, 329–337. *See also* Hot tubs; Spas; Therapeutic pools and tubs for health care use
  equipment grounding, 330–331, 331*f*
  equipotential bonding, 333–337, 333*f*–336*f*
    conductive pool shells, 334–335
    double-insulated pool pump motors and water heaters, 336, 336*f*
    fixed metal parts, 336
    overview, 333–334, 333*f*, 334*f*
    perimeter surfaces, 335–336, 335*f*
    pool water, 337, 337*f*
    specialized pool equipment, 337, 337*f*
  feeders, 333
  junction boxes and enclosures, 332, 333*f*
  overview, 329–330, 330*f*
  pool pump motors, 331–332, 331*f*
  underwater luminaires, 332, 332*f*
  wiring from separate building or structures, 333
Switch
  grounding snap, 201–202, 202*f*
  identification of wiring device terminals, 202–203, 202*f*
  transfer, 281–282, 281*f*
Switchboard
  bonding, 152
  marking equipment for ungrounded systems, 111, 111*f*
System, defined, 71
System bonding jumper
  DC system, 288–289, 289*f*
  definition, 56–57, 145, 268
  neutral ground-strap EGFP system, 375–376
  for separately derived system, 271–273, 271*f*, 272*f*, 273*f*

System conductors, 251–252, 251*f*–252*f*, 259–261, 259*f*, 260*f*
System coordination, 380, 380*f*
System equipment, marking of ungrounded, 111, 111*f*
System grounding, 248–265
  definitions, 250
  grounded conductors
    identification of, 259, 259*f*
    rules for, 259–261, 260*f*
    which conductor to ground, 259, 259*f*
  high-impedance grounded neutral systems, 261–262, 262*f*
  mandatory, 253–257, 254*f*, 255*f*
    grounded system voltages, 255–256, 256*f*
    using transformers, 256–257, 257*f*
  methods of, 252, 253*f*
  optional, 257–261
    ground detection requirements, 258–259, 258*f*
    grounded conductors, 259–261
    separately derived systems, 257–258
  overview of, 251–252, 251*f*, 252*f*
  prohibited systems, 261
  requirements, 253
  for solar PV systems, 341–342, 342*f*
  ungrounded systems, 262–263, 263*f*
  wind electrical system, 286

T

T-connected transformers, 256
"Technical Equipment Ground," 328, 328*f*
Telecommunications main grounding busbar (TMGB), 353, 353*f*
Terminal lugs, torque valves for, 177
Therapeutic pools and tubs for health care use, 339–341
  bonding, 339–340, 339*f*
  grounding, 340
  hydromassage bathtubs, 340–341, 341*f*
  portable appliances and equipment, 340
Threaded hubs, bonding use of, 152*f*
Threaded joints
  requirements in hazardous locations, 296–297, 297*f*
  sparking at, 296
Three-terminal grounding resistance testing method, 79
Throat liners, insulating, 426
Time current curves, 32, 33*f*–37*f*
Tingle voltages, 310
Torque valves, for terminal lugs and equipment, 177
Tower, grounding wind turbine, 286, 286*f*
Transfer equipment, generator grounding and, 281–284, 281*f*–285*f*
Transformer
  autotransformers, 270, 270*f*
  current, 375–377, 376*f*, 378
  for electric signs and outline lighting systems, 323, 323*f*, 324
  ground-fault protection
    neutral ground-strap EGFP system, 375–377, 375*f*, 376*f*
    zero-sequence transformer type EGFP system, 377–379, 377*f*–379*f*
  grounding, 96, 96*f*, 97*f*, 252, 254, 256–257, 257*f*
  components, 270–275, 270*f*
  methods for systems of more than 1000 V, 389
  requirements for pole-mounted, 192, 192*f*
  impedance and reactance data, 451–452
  isolation, 261
  liquidtight flexible metal conduit to minimize vibration effects, 173, 173*f*
  in outdoor locations, 279
  in power distribution units, 223*f*, 224, 328
  secondary circuit ground-fault protection, 324
  as separately derived system, 269, 269*f*

Transformer (*continued*)
   short-circuit calculations, 445–452
   single-phase center tapped, 449–450
   T-connected, 256
   zigzag, 256, 257*f*

U

Ufer, Herbert G., 75, 404
UFER electrode, 75
UL Guide Information for Electrical Equipment (2009 White Book), 303, 312, 425–443
   ground-fault circuit interrupters for use in hazardous locations KCYN, 432–433
   ground-fault circuit interrupters KCXS, 430
      general, 430
      receptacle GFCIs, 430–431
   grounding and bonding equipment, communication KDSH, 427–428
      requirements, 428
      UL Mark, 428
      use, 427
   grounding and bonding equipment KDER, 425–427
      product marking, 427
      requirements, 427
      UL Mark, 427
      use, 425–426
   grounding equipment, neutral grounding devices, over 600 volts KDZC, 428
   health care facilities equipment KEVQ, 428–429
   hospital ground jacks and grounding cord assemblies KEVX, 429
   isolated power systems equipment KEWV, 429–430
   lightening conductors, air terminals, and fittings OVTZ, 433
   lightening protection system installations OWAY, 433–434
   metallic outlet boxes QCIT, 434–437
      ceiling-suspended fan support, 436
      clamps, 435
      concentric or eccentric knockouts, 435
      concrete tight, 436
      conduit bodies, 435
      extension rings, 435
      fixture support, 436
      floor boxes, 436
      general, 434
      grounding, 436
      integral connectors, 436
      requirements, 437
      UL Mark, 437
      use in fire-rated assemblies, 435
      wet and damp locations, 437
   outlet bushings and fittings QCRV, 437–438
      carton markings, 438
      general, 437–438
      grounding, 438
      requirements, 438
      UL Mark, 438
   primary protectors for communications circuits QVGV, 438–439
      general, 438–439
      related products, 439
      requirements, 439
      UL Mark, 439
   primary protectors for coaxial communications circuits QVKC, 440
   receptacles RTDV, 440–442
      enclosures, 441–442
      general, 440–441
      grounding, 441
      ratings, 441
      related products, 442
      terminals, 441
      wet and damp locations, 442
   special-purpose ground-fault circuit interrupters KCYC, 431–432
      classes, 432
      product characteristics, 431
      related products, 432
      requirements, 432
      UL Mark, 432
      use, 431
   surge arresters OWHX, 434
Underwater luminaires, grounding and bonding, 332, 332*f*
Underwriters Laboratories (UL)
   fittings for bonding continuity, 146
   Master Label program, 87, 87*f*
Ungrounded, definition, 250
Ungrounded systems
   bonding of electrical equipment, 15–16
   bonding of electrically conductive materials, 16
   DC systems, 289, 289*f*
   described, 14–17, 14*f*
   disadvantages of, 263, 263*f*
   ground detection requirements, 258–259, 258*f*
   grounding electrical equipment, 14–15, 15*f*
   marking equipment for, 111, 111*f*
   marking of, 258–259
   path for fault current, 16
   reasons for choosing, 262
   requirements for services supplied by, 110–111, 110*f*
   separately derived systems, 280–281, 280*f*, 281*f*
   supplying separate building or structure, 237–238, 237*f*, 242–243, 243*f*
Utility services
   common grounding of, 83, 83*f*
   electrical service requirements of, 95
   as grounded system, 95, 95*f*
   grounding scheme for, 96–98
   requirements for services supplied by ungrounded systems, 110–111, 110*f*
   transformers, 96, 96*f*, 97*f*

V

Vapor barrier, 69
Vibration, minimizing with liquid-tight flexible metal conduit, 173, 173*f*
Voltage
   in AC circuit, 26
   described, 23, 23*f*
   in Ohm's law, 24, 25*f*
Voltage gradients, in agricultural buildings, 310
Voltage to ground, 250, 263

W

Washers, reducing, 153–154, 153*f*
Water meter shunts, 426
Water pipe electrodes
   bonding jumper use, 82
   described, 72–73, 73*f*
   installation, 81–82
   supplemental electrode, 81–82, 81*f*
Water piping system
   bonding, 158–160
      at separate building or structure, 240, 240*f*
      separately derived system, 278–279, 279*f*
   grounding electrode conductor connection to, 132–133, 132*f*, 133*f*
Water pumps, grounding requirements for, 194, 194*f*
Watering troughs, electric shock and, 312
Wattage, symbol for, 25
Watt's Wheel, 25–26, 25*f*
"Weakest link in the chain" concept, 147
Wet and damp locations
   metallic outlet boxes, 437
   receptacles, 442
Wind electrical systems, 286–287, 286*f*
   equipment grounding, 286
   grounding connections, 286–287
   lightning protection, 287
   system grounding, 286
   tower grounding, 286
Wire nuts, for splicing EGCs in outlet boxes, 200–201
Wire pressure connectors, 148*f*, 149
Wiring device terminals, identification of, 202–203, 202*f*
Workmanship, importance of good, 8, 8*f*, 13
Wye-connected systems, grounding of, 255–256, 256*f*

X

XO terminal, 256, 256*f*

Z

Zero-sequence transformer-type system, 377–379, 377*f*–379*f*
Zigzag transformers, 256, 257*f*